Fuel Cell Technology for Vehicles 2002-2004

PT-96

Edited by
Richard Stobart

Published by
Society of Automotive Engineers, Inc.
400 Commonwealth Drive
Warrendale, PA 15096-0001
U.S.A.
Phone: (724) 776-4841
Fax: (724) 776-5760
www.sae.org
January, 2004

For permission and licensing requests contact:

SAE Permissions
400 Commonwealth Drive
Warrendale, PA 15096-0001-USA
Email: permissions@sae.org
Fax: 724-772-4891
Tel: 724-772-4028

Global Mobility Database®

All SAE papers, standards, and selected books are abstracted and indexed in the Global Mobility Database.

For multiple print copies contact:

SAE Customer Service
Tel: 877-606-7323 (inside USA and Canada)
Tel: 724-776-4970 (outside USA)
Fax: 724-776-1615
Email: CustomerService@sae.org

ISBN 0-7680-1502-2
Library of Congress Catalog Card Number: 2001012345
SAE/PT-96
Copyright © 2004 SAE International

Positions and opinions advanced in this publication are those of the author and not necessarily those of SAE. The authors are solely responsible for the content of the book.

SAE Order No. PT-96

Printed in USA

IV. Component Development

Air Supply

Electrical Systems

Fuel Cell Stacks

Water and Thermal Management

Fuel Processors

V. Development, Testing and Life Cycle Issues

Life Cycle

Test Methods

VI. Modelling, Control and Diagnosis

I. POLICY, FLEET TRIALS, PUBLIC REACTIONS

I. Policy, Fleet Trials, Public Reactions

Public policy remains one of the strongest forces to bring about change in propulsion technology. Taxation policy has been shown to produce major changes in when tax is applied preferentially to a particular type of fuel. Vehicle tax incentives can also create a significant shift in customers' buying decision. One of the most significant issues facing policy makers is how to create an environment in which consumers make major changes, such as from a liquid fuel to hydrogen. The adoption of fuel cell vehicles will depend on policy makers, members of the public, suppliers of fuel and the vehicle manufacturers.

Paper 2002-01-1973 is based on the SAE research report entitled "Fuel Cell Vehicles: Technology, Market and Policy Issues". The author sets the technical progress of the fuel cell in the context of the business proposition represented by the wide scale introduction of fuel cell powered passenger cars.

2002-01-3057 is a description of a specific project in which the feasibility of the introduction of a fuel cell powered bus fleet is investigated. This paper highlights a process that is underway in a number of cities in which fuel cell technology is seen as an important factor in the formulation of transport policy.

A Preliminary Assessment of the Possible Acceptance of Fuel Cell Bus Technology by Current Fleet Vehicle Operators

Timothy Simmons, Daniel Betts and Vernon Roan
University of Florida

Paul Erickson
University of California, Davis

ABSTRACT

Fuel cell engines are expected to deliver greater efficiency and lower emissions than conventional transit bus powertrains in the near future. Although experimental vehicles have demonstrated the emission and efficiency benefits of fuel cell power, the next step toward implementation is widespread fleet demonstrations to prove the technology in the field. In order to aid in the start of new demonstrations and speed fuel cell technology towards the fleet vehicle marketplace, an assessment of the needs, risks, and advantages of using fuel cell power must be obtained from a consumer perspective. It has been assumed that the increased fuel efficiency that is inherent to fuel cell systems will lower operating costs as compared with conventional diesel powertrains. A comparison of two fuel cell buses and a diesel bus was completed in order to quantify the operational cost benefits and identify potential cost deterrents to fuel cell bus implementation. A limited survey of Florida mass transit operators was also conducted to gauge the desire for some of the intangible benefits of fuel cell buses. Some of these issues included reduced emissions, decreased noise, lack of oil and minimized vibration. This information can be used to help analyze the status of fuel cell power for transit applications and begin to form economic and technological goals for future design of fuel cell transit buses.

INTRODUCTION

Diesel fueled buses currently supply nearly 90% of the mass transit needs in the United States. Although diesel engines are efficient sources of power, they release substantial quantities of pollutant gases and particulate matter during operation. The use of fuel cell power in buses has proven to dramatically lower vehicle emissions and increase efficiency when compared to diesel engines. [1, 2]

A fuel cell bus is an electric bus that derives its electrical power from fuel cells. Batteries may be used in conjunction with the fuel cell in a hybrid arrangement to allow use of a smaller fuel cell engine, recapture braking energy, and to mitigate fuel cell system transients. Fuel cell buses have operating characteristics that differ from those of traditional diesel buses. They tend to be quieter, with less vibration than diesel buses. [3] Additionally, the electric drives tend to have very good low speed torque with diminishing torque at higher speeds. Fuel cell buses do not require engine oil, as do traditional diesel engines. Finally, it is expected that the start-up and shutdown periods for fuel cell engines will require minutes and perhaps tens of minutes. Although there are several operational differences between fuel cell buses and traditional diesel buses, it is foreseen that the driver controls, exterior and interior size and overall weight will be similar to current bus technology.

BACKGROUND

A fuel cell is an electrochemical power source that works much like a battery using an anode, a cathode and an electrolyte. However in fuel cells, the reactants are fed continuously to generate the electrical power. Low temperature fuel cells (Polymer Electrolyte Membrane Fuel Cell – PEMFC, Alkaline – AFC, and Phosphoric Acid Fuel Cells – PAFC) provide electrical power from electrochemical potential generated through the oxidation of hydrogen, where air is used as the oxygen source. Other primary products of this reaction include heat and water vapor. High temperature fuel cells operate in a similar manner but are viewed as less compatible with transportation applications due to transient limitations. Previous fuel cell bus demonstrations have used either the PAFC, or the PEMFC due primarily to the sensitivity of the AFC to ambient concentrations of CO_2.

Hydrogen is the optimal fuel for fuel cell vehicles. When hydrogen is used in the fuel cell, its byproducts are water vapor, heat and electricity. Unfortunately, because there are no abundant natural sources of pure hydrogen on earth, this fuel must be obtained from other

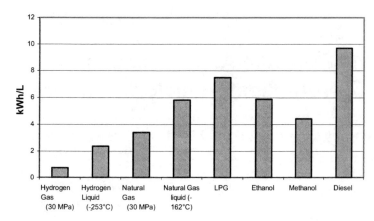

Figure 1: Fuel Energy Density Comparison

compounds. In order to support fuel cell buses, hydrogen generation and storage are critical issues. Hydrogen generation can take place on the vehicle, at a local filling station, or at a remote fuel processing plant. Because hydrogen lacks the volumetric energy density of commonly available transportation fuels (fig. 1), on-board hydrogen production from a more energy dense fuel increases vehicle range with the same tank volume. [4]

Also, the introduction of a new infrastructure, hydrogen filling stations for example, will be complicated expensive as compared with current diesel, or gasoline refueling in order to meet safety codes. [2, 5] This cost may be prohibitive to the introduction and use of direct hydrogen fuel cell vehicles.

Hydrogen produces a colorless flame with a very wide flammability range (4 to 75% in air) and the small molecule diffuses easily, increasing the possibility of leakage. Thus careful attention to storage and to garage design is required if hydrogen is stored on a vehicle. Beyond filling stations, retrofitting existing garages for hydrogen storage can also prove to be extremely cost intensive and must not be overlooked for fuel cell bus introduction if hydrogen is to be used as the fuel. Although transit authorities have long been adept at handling various fuels for vehicles, the perception of storing and using large amounts of hydrogen still suffers from catastrophes associated with its use in past applications. Thus education of the safety and benefits of hydrogen may have to be supplied for hydrogen-fueled buses.

The application of typical transportation fuels for on-board hydrogen generation may reduce the cost of the introduction of fuel cell vehicles into the market place by limiting required changes to the current fuel supply infrastructure. Of the few past demonstrations of fuel cell bus technology, methanol has been used in several prototypes. [1, 3] Methanol may be converted into hydrogen at low (~300°C) temperatures using existing reformer technology. Unfortunately, some of the safety codes for methanol differ from that of diesel fuel. Methanol is also more corrosive than diesel fuel and thus cannot simply be placed in existing diesel rated hoses, or tanks. Thus, there would be an additional cost

to retrofit existing diesel fuel lines, or purchase new fueling stations for methanol service. Although expensive (in 1998, $300,000 for garage and $400,000 for filling station for 200 buses), the cost is much less than for propane, hydrogen, or natural gas. [6, 7] Methanol is also toxic if ingested, or comes in skin contact for an extended period of time. Finally, methanol produces a nearly colorless flame and thus it is difficult to detect flame propagation. Although the on-board reformation available fuels involves less infrastructure changes, this process adds CO_2 to the list of primary products of fuel cell vehicle systems. It is important to note that since nearly all hydrogen is currently produced from natural gas, there is a corresponding portion of production plant CO_2 released for the use of direct hydrogen fuel cell vehicles as well. Whether plant, or on-board reforming is used, the CO_2 production rate for a given power output is typically far less than in conventional diesel engines due to the increase in efficiency using fuel cell engines. [1] Further reductions in fuel cell system CO_2 can come through the use of alternative methods of fuel production such as solar, or other renewables to hydrogen, or through the use of cellulose for methanol.

For on-board fuel conversion to hydrogen, a fuel processor must be included as part of the fuel cell system. The fuel processor, which is commonly called a reformer, uses either a thermal, or thermo-catalytic process by which hydrogen is stripped from the hydrocarbon fuel. The hydrogen rich reformer product gas is termed reformate. Due to the temperature of the reaction, reformate typically contains CO, which has been shown to reduce fuel cell power output. [8] In order to reduce this effect, additional post reformer clean-up steps are typically used for a PEMFC to oxidize CO to CO_2, which acts as a diluent in the fuel stream. PAFCs have a higher tolerance for CO and thus may not require additional clean-up steps. The addition of the on-board fuel processor and CO clean-up devices decreases the overall fuel cell system efficiency, adds carbon monoxide and carbon dioxide to the system products, increases the size and complexity of the system and also hinders transient performance of the fuel cell engine.

The requirements for the implementation of fuel cells are well matched by the current operating practices and infrastructure common to mass transit systems. Buses provide a large engine compartment and are suitable for carrying larger and heavier powertrains that may result from the implementation of experimental fuel cell engines. Buses are also items for which governmental subsidies are common, thus the potentially higher initial cost of fuel cell buses may be offset through such subsidies. Because of the new technology, special maintenance education is required. The staff for such maintenance already exists in a central location for most transit companies. Additionally, the single refueling point that fleet operators commonly use is ideal for a specialty fuel such as methanol, or hydrogen since it would not require widespread changes in fuel supply infrastructure as it would for automotive consumer vehicles.

There are still several engineering and economic aspects of fuel cell buses that are unknown. Some of these issues include the initial cost, type of fuel to be used, as well as the durability of the powertrain in actual service. Another issue not yet understood is the incentive for consumers to purchase a fuel cell powered vehicle. Currently fuel cell engines are projected to be much more expensive than the diesel, or natural gas engines they would replace. In order to validate the greater cost of a fuel cell bus, the value of its benefits and potential benefits must be quantified.

Although the initial cost of a fuel cell bus and its maintenance costs are currently somewhat unclear due to the small number of test buses produced and the rapid advancement of the technology, fuel cells are more efficient, quieter and produce less emissions than traditional diesel powertrains. The characteristic of decreased emissions is extremely beneficial to attaining the increasingly stringent transit bus emission limits imposed by the government. [9] Since there are relatively few financial benefits for nearly emission free and quiet transit, the efficiency of fuel cell buses must be examined to determine the present benefit. A comparison of methanol-fueled and hydrogen-fueled fuel cell buses and a common diesel fueled bus follows in order to investigate the financial benefit of increased efficiency.

RESULTS AND DISCUSSION

In order to perform the comparison, fuel cell bus data was obtained from demonstrations of pre-prototype buses. These included a methanol fueled 30-foot bus powered by a 55kW PAFC and a hydrogen fueled 40-foot bus powered by a 100kW PEMFC. [10, 11] The PAFC powered bus included an on-board methanol steam reformer as well as a 216V Ni-Cad battery pack that was used to help support transient power loads and recapture braking energy. No such battery pack was used in the PEMFC bus because the bus was capable of following the transient loads of typical driving. These two buses were compared to a typical 40-foot bus powered by an 185kW (250 hp) compression ignition engine (CIE) bus using diesel fuel. The gross weight vehicle rating of the PEMFC bus was 41,450 pounds while the CIE and PAFC buses were 38,000 and 29,500 pounds, respectively.

Although fuel cell power is more efficient than conventional diesel power, at steady state, the PEMFC bus should be more efficient than the PAFC bus primarily due to the methanol reforming system on-board the PAFC bus. The reforming system allows for a range of approximately 200 miles with a fuel tank similar in size to the diesel bus. The PEMFC bus has a 250 mile range, but because of the energy density of hydrogen, a large portion of its roof is used for

pressurized hydrogen fuel tanks. Finally, due to the high reformer temperatures required to operate the PAFC, start-up takes tens of minutes as opposed to the PEMFC bus, which may be started almost immediately.

Using these three bus types, the data provided in table 1 was gathered. Calculations were made assuming a 12-year bus life and a driving distance consistent with 2000 averages (30,860 miles per year). [10]

Table 1: The efficiency effect on fuel cost

Bus Type	Fuel Type	Miles/ gallon (diesel equiv.)	Fuel Cost/ gallon (diesel equiv.) ($)	Yearly Fuel Cost ($)	Change in Lifetime Fuel Costs ($)
CIE	Diesel	3.7*	0.86[a]	6,755	0
PAFC	Methanol	4.57*	1.34[b]	9,040	19,700
PEMFC	Hydrogen	6.9**	3.50[c]	15,640	89,500

*[11], **[12], [a] [13], [b][14], [c][15]

Note: fuel costs are before tax.

While the CIE and PAFC buses were placed on a Central Business District (CBD) test cycle in order to calculate fuel economy, this value for the PEMFC bus was found in actual service. There is also a substantial weight advantage for the PAFC bus over both the diesel and hydrogen buses, which benefits its fuel economy. The fuel costs for hydrogen and methanol shown in Table 1 are current as of July 2002 while the diesel cost is from April of 2002. Each fuel price includes distribution costs. It is important to note that the fuel economy values will change for different driving cycles and that fuel costs can be highly variable. The total fuel cost was calculated by placing a 2% rate of inflation on future fuel costs over a 12-year period. Although these costs are only approximates, these values show trends and current hurdles to commercialization of fuel cell buses.

TANGIBLE BENEFITS

As table 1 shows, although the PEMFC bus has nearly twice the fuel economy of the diesel bus, one would have to pay $89,500 less for the bus initially in order to recover increased fuel costs over the lifetime of the bus. In order for hydrogen to compete with the diesel fueled CIE shown, the price of diesel must escalate to $1.88 per gallon (without tax). In order to breakeven, the PEMFC bus would have to reach 15 miles per gallon if diesel prices remain unchanged. Since it is unlikely that the efficiency of hydrogen-fueled fuel cell engines will double in the near future, hydrogen fuel prices must decrease for fuel cost competition.

It is important to note that currently hydrogen is produced as a specialty gas, thus the future cost of hydrogen may decrease considerably if production and transportation methods are changed in order to supply hydrogen in larger quantities and as a fuel. In addition, hydrogen cost could be lowered with increase utilization of the gas.

Several projections of hydrogen costs have been made, which reduce the cost of delivered hydrogen from various feedstocks. Many of these projected costs are below the current value of diesel costs with over 20,000kg/day production, making the future of direct hydrogen fuel cells quite promising. [16,17,18]

Fueling stations and garages for hydrogen vehicles will also most likely be quite costly due to the low energy density and stringent regulations on hydrogen transferal and storage as compared with diesel. Another possible option is on-site reforming of a fuel such as natural gas for hydrogen. This has been found to be a financially feasible method of supplying hydrogen above a production of about 2.5 metric tons a day, or refueling over 222 PEMFC buses. [2] As stated previously, using on-board reforming of a fossil fuel changes in the infrastructure may be reduced. Nonetheless some changes will be necessary due to the toxicity and corrosive nature of methanol, if that fuel is to be used.

The current hydrogen fuel costs should be partially compensated for by the methanol-fueled PAFC bus, which offers a smaller fuel cost. This cost advantage will materialize provided that the initial cost of including the reformer on the bus does not exceed that which may be lost in fuel costs by using a hydrogen-fueled bus system. It is important to note that although the PAFC bus has a greater fuel economy than the CIE, its gross weight vehicle rating (GWVR) is very different. Although this is the case, when the CBD test weight is considered, the PAFC retains a 3% edge in fuel economy over the diesel bus. [11] Thus an equivalently weighted PAFC bus would achieve approximately 3.8 miles per gallon. If the PAFC bus is to compete with the CIE bus in fuel costs, diesel prices must rise to approximately $1.30 per gallon (without tax). If diesel prices remain unchanged, the fuel economy of the PAFC bus must increase to 5.75 miles per gallon to compete with the cost of fueling a diesel CIE.

While the current costs of methanol and hydrogen remove them from current competition with diesel, there are several methods of increasing fuel cell system efficiency during the drive cycle. The PAFC bus was built as a proof of concept bus. It uses a system that was based on a stationary power plant and thus does not perform as efficiently as desired during cycle testing. One of the primary causes of poor performance is thermal control of the reformer during transients. There are many methods, which may be imposed to enhance reformer thermal control that are currently being studied such as internally heated reformers (partial oxidation and autothermal reformers), enhanced heat transfer within the catalyst bed and more dedicated logic

controls. Many of these concepts could also reduce weight from the system, which allows for decreased power demands during the driving cycle. [19] The PAFC bus increases its efficiency by using batteries for load leveling and regenerative braking. This concept may also be employed in the PEMFC bus for potentially higher efficiency during the drive cycle depending on the added weight of the batteries. This advantage would be more expensive to integrate into a non-electrically driven bus.

Although there is great potential for decreasing fuel costs with production, the initial costs of fuel cell buses are foreseen to be more expensive than conventional buses. Ballard, a fuel cell company in Canada, has calculated a final production price of between $500,000 and $550,000 per bus. [20] This cost increase over a conventional diesel bus is more than what can be saved over a 12-year life of the bus by using the most optimistic hydrogen fuel price predictions. Thus either the price of the buses must be reduced, or the cost of diesel fuel must increase for fuel cell buses to become financially beneficial assuming similar maintenance costs of the buses are similar. A final production cost was not found for the methanol fueled bus for comparison although reformer integrated fuel cell buses are expected to be more expensive than direct hydrogen buses due to the additional hardware and design necessary to incorporate a reformer into the system. Thus, the financial benefit of a methanol over a hydrogen-fueled bus arises from the reduced manufacture, storage, and transport changes in the current fueling infrastructure.

In order to justify the initial cost increase, a brief survey of Florida mass transit companies was completed, which analyzed the desire for the benefits of fuel cells. Respondents were asked to rank their interest on a scale from 1 to 10 with 10 indicating most interested (fig. 2).

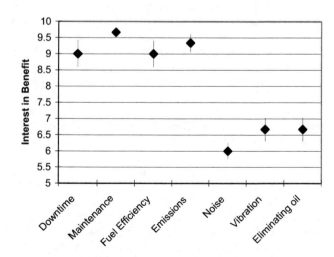

Figure 2: Interest in the potential benefits of fuel cell power

8

Since overall cost is the primary factor for transit companies, fuel cell advantages for which there is little financial benefit such as reduction in noise and vibration garnered relatively low scores. Of greatest importance was the potential for decreased maintenance, decreased emissions and increased fuel efficiency. Since the purchase and disposal of used engine oil amounted to less than 2% of the cost of bus transit operation per year, the possible elimination of oil was not of great interest for any of the companies.

Fuel cells are an inherently clean technology. Thus the emissions do not rely upon operation of emissions control equipment and therefore should provide environmentally friendly operation over the lifetime of the unit. The interest in this potential benefit was quite surprising in that there are currently few financial advantages to lowering emissions for transit operators. This finding was most likely due to the increased pressure the Environmental Protection Agency (EPA) places on cities to lower harmful emissions. Although this is not a direct financial benefit, it may have some influence in the interest in lower emission vehicles.

While the emissions of fuel cell buses have proven to range between ZEV (for hydrogen-fueled bus) and ULEV (for methanol-fueled bus) standards, other technologies exist, which compete with fuel cell power. Diesel electric and other hybrids have emerged to not only decrease emissions, but also decrease maintenance and increase fuel efficiency. Although these technologies cannot be neglected due to their ease of integration into the current fueling infrastructure and proven reliability, they may not be able to meet future emission standards due primarily to NO_x and particulate emissions. Particulate matter (PM) has been known to aggravate existing respiratory health problems. NO_x, on the other hand is one of the primary precursors to smog. Several correlations have been formulated between pollutant emission and healthcare as well as environmental damage costs. While the uncertainty of such calculations remains large, the reduction of these emissions is currently an overriding concern for many urban areas. [21] In response to these concerns, the fiscal year 2000 found nearly 10% of all US buses in operation were alternatively fueled which emphasizes the dedication of transit companies to low emissions. [10]

In order for fuel cell buses to compete with diesel buses in fleet vehicle operations a cost advantage is vital to the consumer. Unfortunately for fuel cell buses, there is not enough current information to explicitly determine a financial benefit.

As has been previously noted, fuel cell buses will initially be more expensive than conventional diesel buses. It is generally assumed that as production increases, fuel cell technology has the potential to be less costly to manufacture than conventional diesel bus engines due to the possibility of fewer machined parts. It is also foreseen that the fuel cell system will have a longer service life than a diesel engine because there are fewer moving parts to wear out. Thus the life cycle cost of the power plant could potentially be reduced. Also, the use of regenerative braking for a fuel cell hybrid system and the lack of a transmission may reduce maintenance and fuel costs for fuel cell buses. The salvage value of a fuel cell bus is unknown, but should be very high because the precious metals typically used for catalyst can be recycled for another application.

INTANGIBLE BENEFITS

The efficiency and emission benefits, which have been previously stated, do not improve the current financial benefit of fuel cells when fuel cost is considered. Thus the fuel cell must use intangible cost benefits to provide additional incentive. Unfortunately, the intangible cost benefits of fuel cell power have yet to be determined and indeed may be the most difficult variable to quantify.

There are many intangible benefits of fuel cells. Aspects such as the lack of oil, and increased passenger comfort (less noise and vibration) are defining attributes of fuel cells. Since there has been no previously acknowledged financial allocation for these benefits and little opportunity for fleet managers to experience them, they did not rank high on the survey.

New subsidies may also develop when a monetary value is placed on pollution prevention. The subsidy should incorporate the increased initial and fuel cost as well as the maintenance support necessary to service the vehicles. It is important to note that the importance placed on pollution prevention varies from region to region. Thus the total subsidy possible will inevitably vary as well.

CONCLUSION

Diesel engines have been designed and optimized for over a hundred years for high efficiency and low cost. Thus competitive price comparisons at this stage of development are somewhat premature. If fuel cell technology is to reach mass transit, continued and expanded fleet demonstrations will be necessary in order to improve the technology and to educate end users of the benefits of fuel cell power. It must also be noted that there exists a general desire for decreasing dependence on imported oil which may also lead to governmental measures that promote nationally produced fuels.

Fuel cell developers must use the opportunity the technology provides to tackle the current primary concerns of high initial cost, fuel cost and proving reliability while providing and possibly exceeding the fuel efficiency, emissions and performance expectations of transit bus managers. It is understood that if field trials are to occur, heavy purchasing subsides and continued maintenance support will probably be required in order to balance the current financial burden for early deployment of the technology.

ACKNOWLEDGEMENTS

The authors would like to thank Graduate Assistants in Areas of National Need (GAANN) Program for their support of this endeavor.

REFERENCES

1. Simmons, Timothy, et. al., "The Effects of Start-up and Shutdown of a Fuel Cell Transit Bus" SAE 2002 International Congress Paper No. 2002-01-0101, 2002.
2. Raman, Venki, "Chicago Develops Commercial Hydrogen Bus Fleet" Oil & Gas Journal, 1999.
3. Matheny, Michael S., et. al., "Interior and exterior noise emitted by a fuel cell transit bus" Journal of Sound and Vibration, 251(5) 2002.
4. Erickson, Paul A., "Enhancing the steam-reforming process with acoustics: an investigation for fuel cell vehicle applications" Dissertation (Ph.D. in press) University of Florida, 2002.
5. Bokow, Jacquelyn C. "Chicago Chooses Nonpolluting Fuel Cell Buses for City's Public Transit" National Hydrogen Association Newsletter, Autumn 1997.
6. Report 38 Guidebook for Evaluating, Selecting, and Implementing Fuel Choices for Transit Bus Operations. Transit Cooperative Research Program National Academy Press, Washington D.C. 1998.
7. Chandler, Kevin and Norman Malcosky, "Alternative Fuel Transit Bus Evaluation Program Results" SAE International Spring Fuels and Lubricants Meeting, 1996.
8. Betts, Daniel, "Modeling and analysis of fuel cell engines for transportation" Thesis (M.S.) University of Florida, 2000.
9. DieselNet web site http://www.dieselnet.com/ standards/us/hd.html July 25, 2002.
10. American Public Transportation Association website, http://www.apta.com/stats/modesumm/bussum.htm, July 15, 2002.
11. Wimmer, Robert et al. "Emissions Testing of a Hybrid Fuel Cell Bus" SAE International Congress Technical Paper No. 980680, 1998.
12. Krom, Larry S. "Renewable Hydrogen for Transportation Study" Final Report, Wisconsin Energy Bureau contract 87052, 1998.
13. Energy Information Service web site http://www.eia.doe.gov/pub/oil_gas/petroleum/data_ publications/petroleum_marketing_monthly/current/t xt/tables16.txt July 24, 2002.
14. Methanex web site http://www.methanex.com/ methanol/currentprice.htm July 18, 2002.
15. Praxair, Personal Communication on July 18, 2002.
16. Mauro, Robert L. "NETL Hosts Workshop of Hydrogen Production from Fossil Fuels" National Hydrogen Association, Spring, 2000.
17. Mann, Margaret K. "Sensitivity Study of the Economics of Photoelectrochemical Hydrogen Production" National Renewable Energy Laboratory, 1999.
18. "Hydrogen fueling station's time has come, researchers conclude" Oak Ridge National Laboratory, 1996
19. Betts, Daniel et al. "Identification of the Dynamically Limiting Systems in an Indirect Methanol Fuel Cell Bus Powertrain" SAE Powertrain & Fluid Systems Conference & Exhibition Paper No. 2002-01-2855, 2002.
20. "Mass Transit: Use of Alternative Fuels in Transit Buses" Report to Congressional Committees, United States General Accounting Office, 1999.
21. Spadaro, Joseph V. and Ari Rabl. "Damage costs due to automotive air pollution and the influence of street canyons" Atmospheric Environment, Vol. 35, Is. 28 2001.

2002-01-1973

Fuel Cell Vehicles: Technology, Market, and Policy Issues

John M. DeCicco
Environmental Defense

ABSTRACT

This paper provides a synopsis of a SAE Research Report of the same title that takes a comprehensive look at the status of fuel cell technology for highway vehicles. The report gives an overview of the technology along with a reasoned view of the challenges to be confronted before a business case for fuel cell vehicles can be made. In addition to assesing the status of technical progress and the barriers remaining, a unique aspect of the study is its examination of fuel cells within the larger context of automotive market trends, policy drivers, and competing technologies as they stand at the beginning of the twenty-first century.

SYNOPSIS

Over the past decade, the fuel cell has risen to prominence as a future option for many energy using systems. For motor vehicles, polymer electrolyte membrane (PEM) fuel cells promise to address concerns about air pollution, petroleum dependence, and global warming while providing ample on-board electricity to efficiently meet growing customer needs. Rapid R&D progress has yielded hydrogen-fueled stack designs suitable for mass production. Automakers have pledged initial vehicle applications, including buses, by 2003–2005.

In assessing fuel cell technology, it is necessary to look at both the progress made over the past decade and the challenges still to be confronted before commercialization can occur. Barriers of varying degrees remain for all fuel cell vehicle systems, including both cost and infrastructure hurdles. Prospects for success depend not only on the time and effort needed to overcome these barriers, but also on the market trends and policy requirements that shape automotive design as well as the capabilities of competing technologies for meeting similar needs.

Hydrogen fuel cell stack development has progressed rapidly. Nevertheless, time is still needed to prove out low-cost, easy-to-manufacture designs and ancillary systems. It will take several production cycles at low-to-medium volumes to gain experience, establish reliability, and achieve maturity. Manufacturers may be loath to commit to extensive production, since early designs could become obsolete before costs are fully recovered.

Complex issues surround the choice of fuel for fuel cell vehicles. Direct hydrogen systems offer the best environmental performance, but fundamental technical barriers must be overcome to demonstrate fully onboard storage systems; even then it will take many years to establish a refueling infrastructure. Liquid fuels (gasoline, methanol, or possibly ethanol) are easy to store, but workable onboard fuel processing systems do not yet exist and would add to cost and detract from environmental performance. Liquid fuels face infrastructure barriers ranging from nil or mild (for gasoline) to substantial (for alcohols).

Sufficient time is also needed for cost reductions in motor/controller systems and high-power batteries. The costs of these common components of any electrodrive vehicle compound the costs of fuel cell systems even when counting the savings from a conventional powertrain. Optimistic learning rates suggest a 10–15 year time frame before electrodrive component costs fall to acceptable levels. Thus, fuel cell powertrains may meet widespread affordability targets over a time frame similar to that required for changing the fuel infrastructure.

Commercialization strategies must reckon with competing technologies as well as market and policy drivers. Advances in gasoline vehicle emissions controls make air quality alone an insufficient reason to abandon internal combustion engines over the next two decades. Urban bus applications hold more promise, since difficult-to-clean-up diesel engines, high population exposures, and central fueling abilities give fuel cells an advantage. For cars and light trucks, market forces do not provide a strong enough call for higher efficiency to motivate even modest technology improvements, let alone radical technology change. As pressure grows to address global warming and petroleum security, improved conventional and hybrid-electric vehicles will continue to offer lower-cost, lower-risk ways to stretch fuel economy while meeting growing needs for onboard electricity.

Public policy regarding transportation energy use presents another set of issues. Fuel-neutral measures have not initiated significant market transformation to date. Fuel taxation policies sustain some niche applications and a limited subsidized market for gasohol, but also fail to motivate real change. Aggressive measures, such as California's zero-emissions vehicle mandate, have stimulated R&D. But they may not trigger investments beyond low-volume production because the broader business case for ZEV technologies remains weak. Government-industry research programs have also advanced R&D. The lack of compelling policies for higher fuel economy inhibits extensive production investments in any technology oriented to improving vehicle efficiency, including fuel cells.

In short, fuel cells do hold enormous long-run potential for highway vehicles and enthusiasm has been generated by automakers' promises to "put fuel cell vehicles on the road" by 2003–2005. Nevertheless, the current status of the technology and related issues suggest a "deployability gap" of at least 15 years before a business case seems feasible for mass-market automotive applications. Closing this deployability gap will entail accelerating the convergence of three key pathways, each of which involves inherent time lags: engineering development and maturation of stacks and ancillaries; cost reduction for common components of electrodrive systems; and resolution of technical and infrastructure barriers related to the fuel.

It will take either a coordinated commitment by the auto and oil industries or major changes in public policy (or both) to break out of the box of weak market drivers and feeble transportation energy policies that may otherwise keep fuel cell technology in a largely R&D status for years to come. During this time, demonstrations and limited applications (such as buses) may occur, but neither profitability nor significant impacts on oil consumption and greenhouse gas emissions are likely. Other technologies will be needed to address transportation energy problems over the next two decades. Nevertheless, while the R&D portfolio should be hedged with work on low-carbon fuel options for other powertrains, automotive fuel cells clearly warrant steady efforts in light of their long-term promise.

ACKNOWLEDGMENTS

The report summarized here was supported by grants from the U.S. Environmental Protection Agency, Office of Transportation and Air Quality, and the U.S. Department of Energy, Office of Transportation Technologies. The U.S. Government retains a non-exclusive, royalty-free license to reproduce this work for government purposes. The author has sole responsibility for the content and conclusions.

CONTACT

John M. DeCicco, Ph.D., completed this work as an independent consultant. He is now a Senior Fellow with Environmental Defense, a national, non-profit organization representing more than 300,000 members that links science, economics and law to create innovative, equitable and cost-effective solutions to society's most urgent environmental problems; see www.EnvironmentalDefense.org. The author can be reached at JDeCicco@EnvironmentalDefense.org or 1875 Connecticut Avenue NW, Washington, DC 20009.

ADDITIONAL SOURCES

See the full report, *Fuel Cell Vehicles: Technology, Market, and Policy Issues,* which provides a comprehensive and up-to-date look at what promises to be the most exiting automotive technology for the next generation. With 35 tables, 44 figures and photos, a glossary, and an extensive 230-entry bibliography, this report is a must-have reference for researchers, policy makers, investors, and anyone seeking in-depth knowledge of future transportation energy technology. *Fuel Cell Vehicles: Technology, Market, and Policy Issues* (SAE Research Report RR-010) is available from SAE Publications, 1-877-606-7323, or via the web at www.sae.org.

II. FUEL ISSUES

II. Fuel Issues

The preferred type of fuel for fuel cell vehicles is the subject of a wide ranging debate that includes the technical merits of the potential fuels, the overall energy performance (often referred to as "well to wheels") and the social aspects that include perceptions of the fuel and how it should be handled. Because of its relation to carbon dioxide (CO_2) emissions, probably the most significant issue with fuel is the well to wheels efficiency and its constituent well-to-tank and tank-to-wheels efficiencies. Predictions of efficiency in each aspect of the life cycle of fuels depend heavily on the assumptions made by the technical team responsible for the particular study.

Hydrogen is widely regarded as having strong attributes as a source of energy for transportation . Hydrogen is not a fuel, but as a means of converting and transmitting energy is sometimes termed an "energy vector". Hydrogen can be used to fuel both internal combustion (IC) engines and fuel cell stacks. Hydrogen must be created and stored and can be transported by pipeline. While technical demonstrations of the use of hydrogen have been made, the acceptance of new fuels by the buying public remains an uncertain factor in the evaluation of its potential.

The papers in this section cover the technical aspects of fuel supply, and an economic evaluation of the life cycle cost of a particular fuel strategy. While the respective production and usage efficiencies are important the full picture includes the manufacturing, operation and disposal of a fleet of vehicles.

Fuel Technology

Paper 2002-01-0098 describes a new storage method in which hydrogen. Hydrogen is stored in the form of an aqueous solution of sodium borohydride from which the gas is extracted before the residual solution is recycled. The paper illustrates the kind of novel development that is being pursued to identify practical and safe methods for the storage of hydrogen.

Paper 2003-01-0414 considers the performance of a range of fuels in a reforming function. This work points to the need to develop new systems hand-in-hand with new fuels. The current range of fuels has been heavily customized to the needs of IC engines and does not offer the features that lead to efficient reformer operation.

The paper of Mercuri, Bauen and Hart is taken from the 2001 Grove Fuel Cell symposium and describes the results of a EU funded study into the types of re-fuelling methods that would suit conditions in Italy. The work illustrates that even if there is consensus in the type of fuel, local solutions are still going to be needed.

The paper 2003-01-0413 considers the whole impact of a fuel choice in the context of a number of Canadian cities. Fuel supply scenarios as well as vehicle disposal makes a complex picture that needs to be considered in its entirety before decisions can be made.

Paper 2004-01-0788 is a report from a discussion panel held during the SAE 2003 World Congress on the subject of hydrogen. It covers a wide range of opinions and possibilities and reviews the thermodynamics of hydrogen as a source of energy for transportation.

Fuel/Vehicle Analysis

In paper 2003-01-0412 the authors analyse the results of a number of studies into the well to wheels efficiency of a range of advanced technology vehicles. They consider the variation in the results and the underlying reasons. Once the source of variations is taken into account, the authors conclude that the differences between the studies are reduced, although further work is needed to gain a detailed understanding of the potential of new propulsion technologies.

The authors of paper 2004-01-1011 describe a detailed analysis of the efficiency of conventional and advanced powerplant. They make projections of the potential efficiency gains and conclude that the efficiency differences between conventional engines and fuel cell systems are small once the energy cost of the fuel supply is taken into account. The analysis continues with projections of the fleet fuel consumption for the United States.

II. FUEL ISSUES

Fuel Technology

2002-01-0098

Performance Bench Testing of Automotive-Scale Hydrogen on Demand™ Hydrogen Generation Technology

Richard M. Mohring, Ian A. Eason and Keith A. Fennimore
Millennium Cell, Inc.

ABSTRACT

Millennium Cell has developed a novel catalytic process (called Hydrogen on Demand™) that generates high purity hydrogen gas from air-stable, non-flammable, water-based solutions of sodium borohydride, $NaBH_4$. This paper discusses initial performance bench testing of an automotive-scale hydrogen generation system based on our proprietary technology. Our system was coupled to a hydrogen flow controller system designed to simulate the hydrogen draw of a large (>50 kW net) fuel cell engine. The controller was programmed to emulate various driving cycles, and behavior of the Hydrogen on Demand™ system under realistic load transient conditions was measured. The testing indicates that the Hydrogen on Demand™ system successfully provides hydrogen under realistic load conditions, and that the data are qualitatively indistinguishable from the runs performed with compressed hydrogen gas.

INTRODUCTION

A general description of the Millennium Cell Hydrogen on Demand™ technology can be found in Reference [1]. In summary, the "hydrogen" is stored on-board at ambient conditions in a liquid fuel – a stabilized aqueous solution of sodium borohydride, $NaBH_4$. The solution is non-flammable, and can be handled similarly to common household chemicals.

The high purity, humidified hydrogen produced by this system can be used for numerous applications, addressing a wide range of power requirements. In particular, these systems can be implemented to supply hydrogen for fuel cells and internal combustion engines in automobiles.

The Hydrogen on Demand™ system releases the hydrogen stored in sodium borohydride solutions by passing those liquid fuels through a chamber containing a proprietary catalyst bed. The hydrogen is liberated in the reaction:

$$NaBH_4 + 2\ H_2O \xrightarrow{\text{catalyst}} NaBO_2 + 4\ H_2 + \text{heat}$$

A noteworthy point is that there is no heat input to the reaction. In fact, there is enough heat generated to vaporize a portion of the excess water in the fuel solution, resulting in a naturally humidified hydrogen stream. The product, sodium metaborate, can be recycled as the starting material for the generation of sodium borohydride.

In its solid form, $NaBH_4$ can store greater than 10% hydrogen by weight. In our current designs, $NaBH_4$ is dissolved in water to form a fuel solution. As an example, a 30 wt% $NaBH_4$ solution contains approximately 6.7% hydrogen by weight.

SYSTEM SCHEMATIC AND OPERATION

A general schematic for an automotive-scale hydrogen generation system is shown in Figure 1. A fuel pump directs fuel from a tank of sodium borohydride solution into a catalyst chamber. Upon contacting the catalyst bed, the fuel solution generates hydrogen gas and sodium metaborate (in solution). The hydrogen and metaborate solution separate in a second chamber, which also acts as a small storage ballast for hydrogen gas. The humidified hydrogen is processed through a heat exchanger to achieve a specified dewpoint, and is then sent through a regulator to the fuel cell or internal combustion engine, or for this work, to the fuel cell emulator (FCE) device.

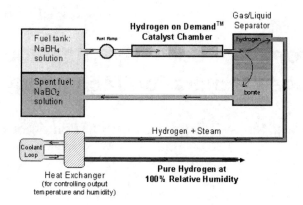

Figure 1. General schematic for a typical Hydrogen on Demand™ system

In operation, the rate at which hydrogen gas is generated is directly proportional to the rate at which the borohydride solution is pumped into the catalyst chamber. This operational simplicity translates into relatively straightforward control strategies.

For these tests, the system was run in a simple "on/off" mode using the system pressure as a process variable to control the state of the fuel pump (analogous to the way an air compressor operates). The delivery pressure setpoint for our system was 150 psig (~10 bar). This low operation pressure for Hydrogen on Demand™ is in contrast to the high pressures of 2200-5000 psig (~150-350 bar) that are present in typical compressed hydrogen tanks. In a compressed hydrogen system, the pressure is dropped through a regulator to deliver hydrogen to the fuel injectors at a much lower pressure of 75-150 psig (~5-10 bar). Operating the entire Millennium Cell system at the proper pressure for the fuel injectors eliminates the need to have highly pressurized hydrogen on board.

When the FCE demand for hydrogen gas increases, hydrogen is drawn off from the ballast tank, and the system pressure drops below the setpoint. This triggers the fuel pump to start up. As fuel reaches the catalyst chamber, hydrogen gas is generated causing the pressure to increase above the setpoint, and the fuel pump is turned off (see Reference [1]).

More sophisticated strategies will be implemented at the integration stage with the vehicle to achieve higher degrees of control. For example, in a fuel cell vehicle, a signal proportional to the electric current load demanded from the fuel cell will be used to directly control the rate at which hydrogen is produced by the generator. For an IC engine vehicle, a signal related to the already-existing instantaneous injector pulse widths can be used.

SYSTEM TESTING

The system under test was designed to power a Millennium Cell internal combustion vehicle, a Ford Crown Victoria, which originally ran on compressed natural gas (see Figure 2).

Figure 2. Crown Victoria natural gas internal conbustion engine, modified to run on hydrogen

Millennium Cell made minor modifications to the car so that it would run on hydrogen. Initially, we installed two cylinders of compressed hydrogen in the trunk area and took measurements of the gas flow rates under driving conditions. During these tests, it was noted that there was a significant occurrence of autoignition (pinging) while running the system. A simple water-injector was added to the fuel manifold, which essentially eliminated the autoignition problem.

The hydrogen generator system is shown mounted in the rear of the vehicle in Figures 3 and 4. The catalyst chamber is the near-horizontal cylinder mounted in the foreground of the photo and is easily visible in Figure 4. Sodium borohydride fuel is stored in a standard ABS plastic tank mounted underneath the vehicle in front of the trunk area (not shown) and is metered into the catalyst chamber from its right side through a flexible line by a piston pump. The hydrogen and borate solutions separate via gravity in the vessel directly behind the catalyst chamber. The borate solution is periodically pulsed (using the hydrogen pressure as a driver) from the separation vessel into an ambient pressure tank located below the trunk area.

Figure 3. Hydrogen on Demand™ system mounted in trunk of vehicle

The hydrogen is taken from the top of the separation vessel, passed through a heat exchanger, and sent into the engine (or in the bench testing, into the Fuel Cell Emulator). It should be noted that the hydrogen generation system is mounted such that the passenger compartment was not compromised, and the majority of the trunk area has been retained.

Prior to beginning the installation into the vehicle, the prototype Hydrogen on Demand™ system was run on the bench and connected to the hydrogen inlet line of the engine. The engine ran with no difficulty, in fact, an interesting consequence of the humidified hydrogen stream was that the water injection system could be removed completely (as the humidity provided plenty of moisture to defeat the autoignition effect).

The fuel cell emulator (FCE) is a device commissioned from ATC, Inc. (Indianapolis, IN) and is shown in Figure 5. The system is essentially a mass flow controller coupled via a PID control loop to a precision hydrogen mass flowmeter. A hydrogen source is connected to the inlet, and the hydrogen flow rate is controlled to meet a flow schedule that is programmed into the computer system. With proper modeling of the vehicle systems, one can convert a speed profile (for example) into realistic fuel cell or engine power demands, and further into a realistic profile for hydrogen gas demand. At that stage, various hydrogen sources can be compared identically as to their performance under these conditions.

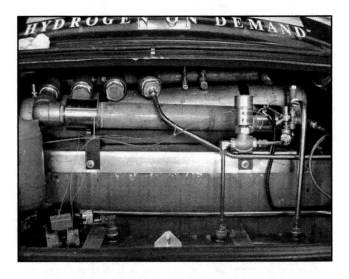

Figure 4. Close-up of Hydrogen on Demand™ system, showing catalyst chamber and separation vessel

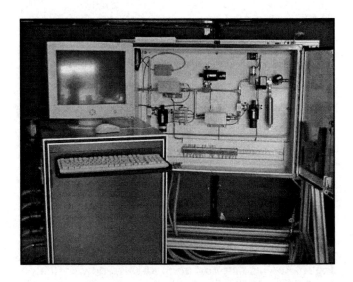

Figure 5. Fuel Cell Emulator (FCE) device with data acquisition system

This paper compares a specific FCE flow profile for both compressed hydrogen gas and the Hydrogen on Demand™ prototype system. The flow profile was manually generated to be very aggressive in order to demonstrate system response under transient and heavy load conditions, although it is not modeled for an specific vehicle. The magnitude and frequency of the transients was chosen to be representative of conditions similar to highway driving with accelerations to >60 mph (> ~95 kph) as well as a short period of aggressive stop-and-go driving. The range of flow rates was chosen to be consistent with our measured flow rates for the Crown Victoria running on compressed hydrogen.

In order to establish a baseline, we connected a manifold of six "standard" 2200 psig (~150 bar) compressed hydrogen cylinders to the FCE and ran the selected profile. Figure 6 shows our results for compressed hydrogen. The profile setpoint for hydrogen flow rate (in standard liters per minute, SLM) is shown as the solid line, while the measured flow rate is the line with hollow triangles.

Of note in Figure 6 are the over- and under-shooting of the measured flow relative to the setpoints as well as a slight delay in the response. This is simply an effect of the PID control loop of the FCE being calibrated in a slightly "underdamped" state, and bears no reflection on the ability of the hydrogen cylinders to follow the transients. As one can see, the compressed hydrogen cylinders are quite able to provide a satisfactory amount of hydrogen to meet the demand of the profile.

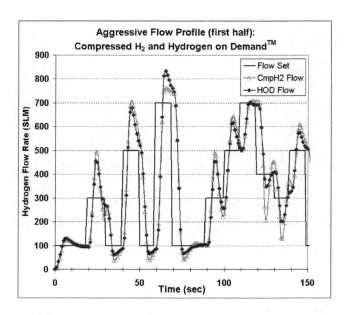

Figure 7. First half of aggressive flow profile shown with data from the Hydrogen on Demand™ (HOD) system overlaid atop the flow profile and baseline run performed with compressed hydrogen gas at 2200 psig (~150 bar)

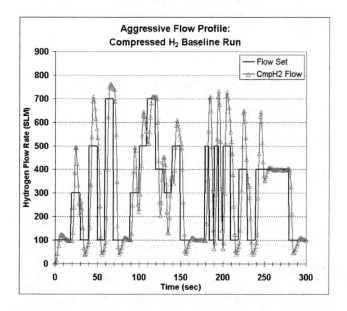

Figure 6. Baseline run of aggressive flow profile on the FCE using the manifold of compressed hydrogen cylinders at 2200 psig (~150 bar)

After establishing the baseline with compressed hydrogen gas, we removed the manifold and connected the outlet of our prototype Hydrogen on Demand™ system to the FCE. The results are shown in Figures 7 and 8, broken into two graphs (first half and second half) so as to show better detail. Our system is shown as the line with solid circles, while the profile itself and compressed hydrogen baseline are depicted as before.

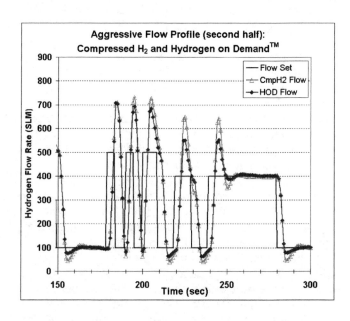

Figure 8. Second half of aggressive flow profile shown with data from the Hydrogen on Demand™ system overlaid atop the flow profile and baseline run performed with compressed hydrogen gas at 2200 psig (~150 bar)

These data are qualitatively indistinguishable from the runs made using our prototype system. As with the compressed hydrogen baseline run, the PID control on the FCE was such that there are delays and over- and under-shooting of the setpoints our the prototype runs. However, the data from the Hydrogen on Demand™ system are qualitatively indistinguishable from the baseline compressed hydrogen data.

As of this writing, we are currently completing the integration of Hydrogen on Demand™ with our Crown Victoria, and will begin in-vehicle evaluation. In addition to verifying our dynamic load performance, we expect to investigate the benefits of the humidified hydrogen stream on autoignition as well as a potential reduction of NO_x emissions.

CONCLUSIONS

In this paper, we have demonstrated on the bench that Hydrogen on Demand™ technology under a realistic power profile provides excellent delivery of hydrogen to meet demand requirements. There was no difference detected between compressed hydrogen at 2200 psi (~150 bar) and the HOD system at 150 psi (~10 bar). Hydrogen flow rates of up to ~800 SLM were demonstrated, without requiring operation at high pressure. With proper system integration with a vehicle, we expect our technology to provide sufficient response to supply the power demand of the engine or fuel cell. From a packaging perspective, Hydrogen on Demand™ systems have the potential for relatively high energy storage density (estimated to be greater than 4.5 wt% for a system as described herein) and do not suffer from the (real or perceived) safety issues associated with many technologies for storing hydrogen. The proof of concept design in our Crown Victoria potentially represents a commercially viable strategy for retrofitting existing cars to give low emissions performance. In-vehicle testing will be presented at a later time.

ACKNOWLEDGMENTS

The authors would like to acknowledge the dedicated efforts of the Millennium Cell team. Special thanks to Rex Luzader and Ying Wu for assistance and insightful commentary on the draft.

REFERENCES

[1]   Mohring, R. M., Luzader, R. E., "A Sodium Borohydride On-Board Hydrogen Generator for Powering Fuel Cell and Internal Combustion Vehicles," SAE Future Transportation Technologies Conference, Society of Automotive Engineers, August 2001, Paper Number 2001-01-2529.

CONTACT

Dr. Mohring is Program Director of the Hydrogen on Demand™ efforts at Millennium Cell, and can be reached by email at *mohring@millenniumcell.com*, by phone at 732-544-5706, or by mail at 1 Industrial Way West, Eatontown, NJ, 07724. Further information can be found at *www.millenniumcell.com*.

2003-01-0414

Fuel-Cell Vehicle Fuels: Evaluating the Reforming Performance of Gasoline Components

Osamu Sadakane, Kenichiro Saitoh and Koji Oyama
Nippon Oil Corporation

Noboru Yamauchi and Hiroshi Komatsu
Nissan Motor Co., Ltd

ABSTRACT

Fuel cell vehicles (FCVs) are an emerging transportation technology, with a potential to provide very low vehicle emissions and significant improvements in fuel efficiency. The choice of a fuel for FCVs must consider several critical issues, including the availability of a distribution and storage infrastructure, manufacturing cost and capital requirements,energy efficiency, and performance.

Gasoline, one of the candidate fuels, is noteworthy not only for its existing infrastructure, but because it has the possibility of usage for both FCVs and conventional internal combustion engines. Gasoline consists of different types of hydrocarbons, including paraffins, naphthens, olefins and aromatics - - with carbon numbers distributed over a wide range. Gasoline also contains a variety of sulfur compounds and selected additives in small amounts., In some cases gasoline also contains oxygenates such as MTBE and ethanol. To develop the optimal fuels and associated catalysts for FCVs it is necessary to understand the effects of these constituents on FCV fuel processing.

Reforming performance has been evaluated with a variety of gasoline components, as well as conventional gasoline provided by refineries. Tests of initial reformer catalyst activity and durability at auto-thermal reforming (ATR) conditions have been conducted.

Results showed similar reforming performance among most hydrocarbon components except aromatics. The addition of MTBE improved reforming reactivity. Both aromatics and sulfur compounds caused rapid deterioration of reforming activity, while the magnitude of the reactivity loss varied between different aromatic compounds.

INTRODUCTION

FCVs are an emerging transportation technology, with a potential to provide very low vehicle emissions and significant improvements in fuel efficiency which can result in reduction of greenhouse gases (GHGs). The choice of a fuel for FCVs must consider several critical issues, including the availability of a distribution and storage infrastructure, manufacturing cost and capital requirements, energy efficiency, and performance.

There are three strong candidates for FCV fuels: hydrogen, methanol, and gasoline. Much discussion has been carried out about these candidates concerning their GHGs emissions on a well-to-wheels basis, costs (vehicle and fuel manufacturing and fuel infrastructure), and technical barriers[1-11]. So far, it has been generally accepted that hydrogen will be the ultimate FCV fuel because of its potential for GHGs reduction and that gasoline is an interim FCV fuel until the hydrogen infrastructure is consolidated. Gasoline is noteworthy not only for its existing infrastructure but because it has the possibility of usage for both FCVs and conventional internal combustion engines. However, both hydrogen and gasoline have significant technical barriers for use as FCV fuels. Hydrogen needs breakthroughs in storage technology and manufacturing cost, and gasoline needs the development of proper processing technology.

Gasoline consists of different types of hydrocarbons—including paraffins, naphthenes, olefins, and aromatics—with carbon numbers distributed over a wide range. Gasoline also contains a variety of sulfur compounds, selected additives, and, in some cases, oxygenates. Each of these constituents has a different effect on processing. Therefore the fuel and fuel processor must be developed as a package of technologies. It might be possible to mitigate the technical barriers for processing by means of fuel optimization. To develop the optimal fuels and associated catalysts, it is necessary to understand the effects of these constituents on FCV fuel reforming.

EXPERIMENTAL

TEST FUELS

We evaluated a variety of gasoline components and gasoline products, including both conventional types and prototypes. The properties of the fuels tested are shown in Table1A and 1B. The components and blending stocks were purchased from a refinery to represent their real-world properties. The test fuels matrix was designed so that we could estimate the effect of the fuel properties, specifically the hydrocarbon type and carbon number. The fuel matrix is shown in Figure1. The test fuels were characterized as follows.

Naphtha: Distillates from a crude distillation unit in the range of 20 to 200 deg C (whole range naphtha). Whole range naphtha is subdivided into light and heavy. Generally, naphthas are hydrotreated for sulfur removal. According to the composition, naphthas are classified into the "paraffin type" and "naphthene type." We prepared the paraffin type based on naphtha refined from Arabian light crude, which is a typical paraffinic crude, and the naphthene type based on naphtha from Sumatran light crude, which is a typical naphthenic crude. The sulfur content was usually less than 1 ppm after hydrotreating.

Isomerate: The product of an isomerization process that reforms straight-chain hydrocarbons to branched-chain hydrocarbons. The main components are C5 and C6 isoparaffins. The sulfur content is usually less than 1 ppm.

Alkylate: The product of an alkylation unit that polymerizes butylenes and isobutane. The main components are C8 isoparaffins. It sometimes contains low molecular weight sulfur compounds at a content of less than 5 ppm.

Reformate: A reformed heavy naphtha that is upgraded in octane number by means of catalytic reforming of C4 to C10 range compounds. It contains over 50% aromatics. The sulfur content is usually less than 1 ppm.

Catalytic Cracked Gasoline (CCG): Gasoline distillation range products from a catalytic cracking process. CCG contains 30 to 50% olefins. As a refinery component, it contains 10 to 200 ppm sulfur. For the purpose of the experiments, the test samples were desulfurized by an adsorption process.

CARB Phase 3: A prototype gasoline that is suitable for the California Air Resources Board (CARB) Phase 3 (P3) specifications with ethanol. P3 gasoline will be the next generation of California's cleaner burning gasoline and will be required at the beginning of 2004. Another prototype blend that met the Phase 3 requirements, but without an oxygenate compound, i.e., without ethanol, was also prepared. For the purpose of the experiments, these samples were desulfurized with hydrotreating.

Dual Purpose Gasoline (DPG): Prototype gasoline with low aromatics, low olefins, and low sulfur but a sufficient octane number for internal combustion engines (ICEs). These samples are intended as dual purpose for FCVs and ICE vehicles. Three types of DPG were prepared. High RON gasoline (HR) with MTBE had 98.9 Research Octane Number (RON), High RON gasoline without MTBE had 96.7 RON, and Low RON gasoline (LR) had 92.1 RON.

In the sulfur series samples, a sulfur compound mixture was blended into whole range naphtha. The sulfur compound mixture was a 1:3:1 blend of C1 thiophene : benzothiophene : C8 sulfide.

Table 1A. Test Fuels

Sample No.		1	2	3	4	5	6	7	8	9	10	11	12
Name		Naphtha - Paraffinic			Naphtha - Naphthenic			Isomerate	Alkylate	Reformate			CCG
Character													
Main composition		n-Paraffin			Naphthene			Iso-Paraffin		Aromatics			Olefin
Range		Light	Heavy	Whole-range	Light	Heavy	Whole-range	Light	Heavy	Light	Heavy	Whole-range	Light
Distillation													
10%	deg C	33.5	82.5	62.5	36.5	86.5	66.5	40.0	64.0	54.5	142.5	71.5	33.5
50%	deg C	38.5	100.5	91.5	42.0	102.5	94.5	45.5	106.0	67.0	147.0	122.5	45.5
90%	deg C	54.0	127.0	124.5	57.5	126.5	124.0	57.5	122.0	107.0	162.0	154.0	67.5
Composition													
n-Paraffin	vol%	52.9	34.5	38.2	46.7	28.3	31.8	16.2	10.3	18.9	0.0	11.1	5.3
Isoparaffin	vol%	43.5	40.0	40.7	38.6	30.8	32.3	82.1	89.0	35.3	0.0	20.5	41.2
Naphthene	vol%	2.9	16.6	13.8	14.1	32.7	29.1	1.6	0.1	1.2	0.0	0.7	3.0
Olefin	vol%	0.2	0.2	0.2	0.1	0.1	0.1	0.1	0.1	2.2	0.0	1.2	1.1
Aromatics	vol%	0.6	8.8	7.1	0.5	8.1	6.6	0.0	0.5	42.4	100.0	66.5	49.4
H/C Ratio		2.4	2.1	2.2	2.3	2.1	2.1	2.4	2.3	1.7	1.3	1.5	2.1
Sulfur	mass ppm	<1.0	<1.0	<1.0	<1.0	<1.0	<1.0	<1.0	<1.0	<1.0	<1.0	<1.0	<1.0

Table 1B. Test Fuels

Sample No.			13	14	15	16	17	18	19	20	21	22	23	24	25	26
Name			Oxygenates						Dual Purpose Gasoline			CARB Phase 3 (Hydrotreated)		Sulfur Effect		
Character			MTBE	MTBE + Light Naphtha		Ethanol	Iso-Propanol	Iso-Butanol	High RON with MTBE	High RON without MTBE	Low RON	With Ethanol	Without Ethanol	Naphtha - Paraffinic (Whole-range)		
														S=10 ppm	S=30 ppm	S=80 ppm
Distillation																
10%	deg C		-	35.5	36.0	-	-	-	51.5	52.0	44.5	59.0	67.0	62.5	62.5	62.5
50%	deg C		bp=56	40.0	40.5	bp=78	bp=82	bp=108	92.0	95.0	82.5	100.0	101.0	91.5	91.5	91.5
90%	deg C		-	51.5	52.0	-	-	-	112.5	111.5	110.5	141.0	150.0	124.5	124.5	124.5
RON			-	-	-	-	-	-	98.9	96.7	92.1	88.6	87.5	-	-	-
Composition																
n-Paraffin	vol%		-	49.2	44.8	-	-	-	8.2	9.2	19.9	10.3	9.3	38.2	38.2	38.2
Isoparaffin	vol%		-	40.9	37.2	-	-	-	69.4	75.0	63.6	48.9	55.9	40.7	40.7	40.7
Naphthene	vol%		-	2.7	2.5	-	-	-	0.5	0.5	1.1	9.5	8.4	13.8	13.8	13.8
Olefin	vol%		-	0.1	0.1	-	-	-	5.0	4.9	4.8	0.3	0.7	0.2	0.2	0.2
Aromatics	vol%		-	0.5	0.5	-	-	-	10.5	10.4	10.6	26.4	25.7	7.1	7.1	7.1
H/C Ratio			2.4	2.4	2.4	3.0	2.7	2.5	2.1	2.1	2.1	2.0	1.9	2.2	2.2	2.2
Sulfur	mass ppm		<1.0	<1.0	<1.0	<1.0	<1.0	<1.0	2.0	2.3	2.3	<1.0	<1.0	10.0	30.0	80.0
MTBE	vol%		100.0	6.5	14.9	-	-	-	6.4	0.0	-	-	-	-	-	-
Ethanol	vol%		-	-	-	100.0	-	-	-	-	-	4.8	0.0	-	-	-

where: bp = boiling point (deg C) , RON = Research Octane Number

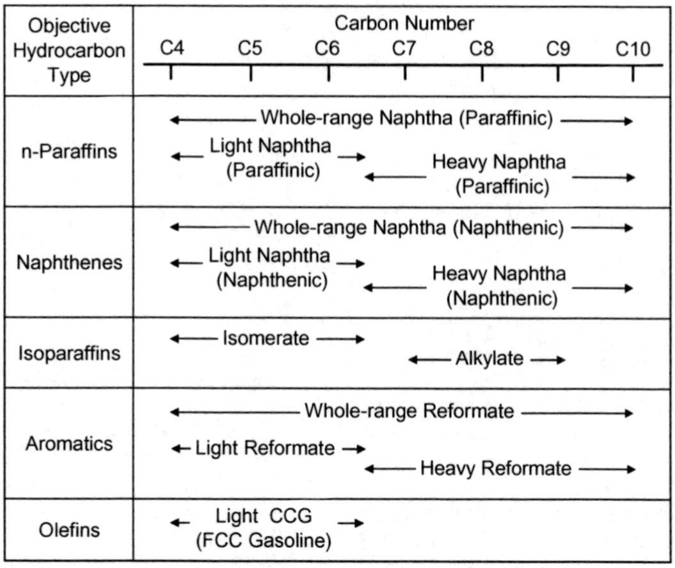

Figure 1. Test Fuel Matrix (Gasoline Component)

TEST APPARATUS

The evaluation was done using an auto-thermal reforming (ATR) test apparatus. Because of the start-up requirements of FCVs, ATR is the most favorable method of reforming. Figure 2 shows a schematic of the evaluation system. A 19 mm in diameter, stainless reactor was filled with a ruthenium (Ru) pellet-type catalyst. The content of catalyst filled in the reactor was 10ml (7.3g). The reactor was temperature-controlled with an electric heater; thermocouples were located at the top, middle, and bottom of the catalyst layer. Gas sampling lines were located at the top (feed gas) and the end (product gas) of the catalyst. Gas samples were analyzed by gas chromatography with TCD. The conversion of each sample was calculated according to equation (1).

$$\text{Conversion} = 1 - \left[\frac{[\text{Fuel}]_{out} / [N_2]_{out}}{[\text{Fuel}]_{in} / [N_2]_{in}} \right] \times 100 \quad ----- (1)$$

where:

$[\text{Fuel}]_{in}$ = Concentration of fuel in feed gas

$[\text{Fuel}]_{out}$ = Concentration of fuel in product gas

$[N_2]_{in}$ = Concentration of N_2 in feed gas

$[N_2]_{out}$ = Concentration of N_2 in product gas

The ATR conditions were $H_2O/C = 1 - 3$, $O_2/C = 0.2$, LHSV = 3 h^{-1} (WHSV = 2.6 – 3.6 h^{-1}), temperature = 723 – 873 deg K (450 – 600 deg C).

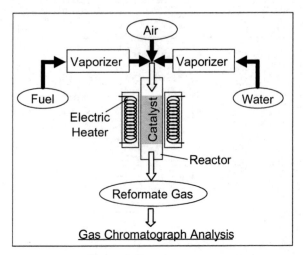

Figure 2. Test Apparatus

RESULTS

The evaluation was divided into two steps: initial reactivity and durability.

INITIAL REACTIVITY

Naphtha (Paraffins, Naphthenes)

Both paraffin and naphthene are saturated compounds. However, they have different chemical structures: one is straight-chain and the other is cyclic. It was expected that this difference would have an effect on the reactivity. The relationships between reactor temperature and reforming conversion were evaluated. Figures 3 and 4 show the results for the paraffin-type naphtha and naphthene-type naphtha, respectively. Contrary to expectations, there was no significant difference between the paraffin type and naphthene type. It seems that the difference in the naphthene content between paraffin type and naphthene type was not enough to affect the reactivity. In other words, all kinds of naphthas from the various available crude oils may have almost the same reactivity.

As for comparisons of the distillation or carbon number, conversions of "light" naphtha at low temperatures were higher than of "whole-range" or "heavy." However, the 97% conversion temperatures, that mean almost 100% conversion temperature because there were a few percent error span at the repeating accuracy of these experiments, were almost the same: 778 to 798 deg K. It seems that the influence of the distillation range is negligible at practical conversion temperatures.

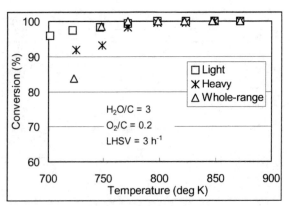

Figure 3. Temperature Dependence of Reforming Conversion: Paraffinic Naphtha

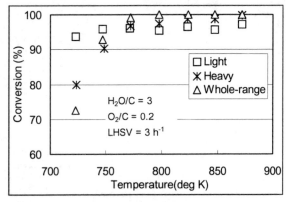

Figure 4. Temperature Dependence of Reforming Conversion: Naphthenic Naphtha

Isomerate, Alkylate (Isoparaffins)

Figure 5 shows the results for isomerate and alkylate, which are regarded as typical isoparaffin components. At low temperatures, isomerate (= light isoparaffin) shows a higher conversion than alkylate (= heavy isoparaffin). Isomerate and alkylate showed almost the same temperature dependence of reactivity as did the naphthas.

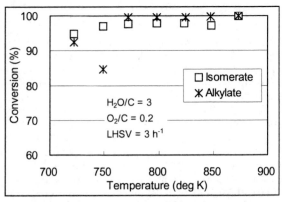

Figure 5. Temperature Dependence of Reforming Conversion: Isomerate, Alkylate (Isoparaffin-rich)

Reformate (Aromatics)

Aromatics have completely different chemical structures from saturates. The results shown in Figure 6 revealed their influence on the reactivity. A substantial deterioration in conversion efficiency was observed at low temperatures. The deterioration with "heavy" aromatics was greater than with "light" aromatics, apparently because of the influence of the higher carbon number aromatic compounds. However, the 97% conversion temperature was 773 to 798 deg K, which was only 30 deg higher than for the naphthas.

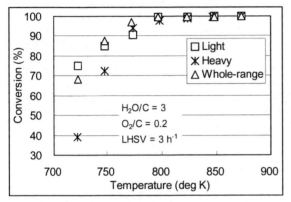

Figure 6. Temperature Dependence of Reforming Conversion: Reformate (Aromatic-rich)

CCG (Olefin)

Figure 7 shows the results for olefin-rich catalytic cracked gasoline (CCG). As mentioned above, this sample was desulfurized for the experiment by means of adsorption (without a hydrocarbon composition change). The conversion behavior was almost the same as for the naphthas. CCG is difficult to use as a component of FCV gasoline because of its high sulfur content. However, these results revealed that olefins have almost the same reactivity as saturates.

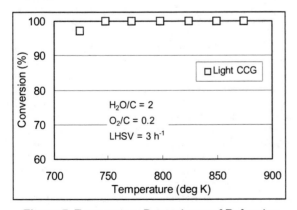

Figure 7. Temperature Dependence of Reforming Conversion: Light CCG (Olefin-rich)

Oxygenates

Methanol is known to be easy to reform at low temperature, so other oxygenates were expected to have good reactivity as well. Figure 8 shows the results for MTBE and light naphtha with MTBE. Good reactivity of reforming was observed; even at 700 deg K, MTBE showed 100% conversion. As for the MTBE mixing effect, the trend of conversion for naphtha with 6.5% MTBE was almost the same as for naphtha alone. However, an improvement in the conversion was observed with the naphtha that contained 15% MTBE. Figure 9 shows the results for the other oxygenates: ethanol, isopropanol, and isobutanol. All of them had good reactivity. However, some coke formation on the reactor wall was observed in the ethanol experiment. It was inferred that ethanol has a tendency for coking in the presence of high-temperature air. Unlike methanol, these oxygenates can be used as gasoline components.

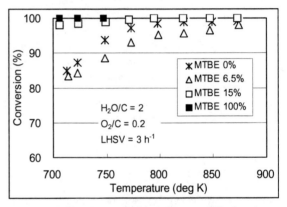

Figure 8. Temperature Dependence of Reforming Conversion: Naphtha with MTBE

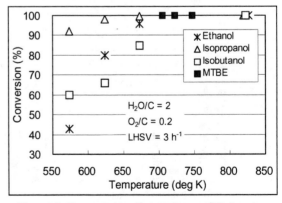

Figure 9. Temperature Dependence of Reforming Conversion: Oxygenate

Prototype Gasolines

Figure 10 shows the results for the prototype gasolines: CARB Phase 3 and DPG. At low temperatures, CARB Phase 3 had substantially lower reactivity than DPG. This might be the result of the higher aromatic content and the inclusion of higher boiling components in the CARB Phase 3 gasoline. At temperatures over 800 deg K, their reactivities were almost equal to each other's and to the naphthas'. This result was expected from the results described above for the individual components.

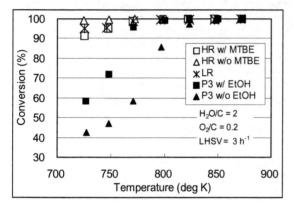

Figure 10. Temperature Dependence of Reforming Conversion: DPG,Phase3 (Prototype Gasoline)

Reforming Gas Composition

Figure 11 shows reforming gas compositions for heavy fractions. Reforming gas compositions at high conversion were almost the same as the results of equilibrium calculation. Reformates produced less hydrogen gas than the other gasoline components because of the high aromatic content of reformates.

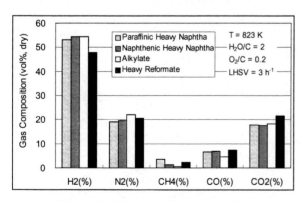

Figure 11. Reforming Gas Composition

DURABILITY

It is expected that some fuel properties would have an influence on the reforming durability, so the sulfur and aromatic contents were evaluated. The effect of olefin, which is also expected to have some influence on the durability, will be reported at a later time.

Sulfur

Sulfur is a big concern for FCV systems because of its poisoning effect on the reforming catalysts and fuel cell stacks. Figure 12 shows the sulfur effect of the reforming catalyst that was chosen for this research. As expected, sulfur caused a rapid and significant deterioration in conversion efficiency. Even at a sulfur content of 10 ppm, which is the lowest level for conventional gasoline, the conversion declined quickly. Though the degree of sulfur poisoning may change depending on the catalyst type, it is probably necessary to decrease the sulfur content from the level of conventional gasoline before introducing gasoline into the reforming catalyst.

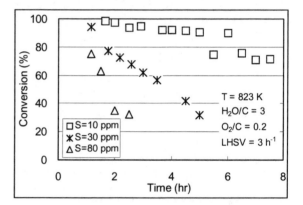

Figure 12. Time Dependence of Reforming Conversion: Naphtha with Sulfur Compounds

Aromatic content

Aromatic compounds have a tendency to deteriorate the catalyst performance due to coke formation. Figure 13 shows the reactivity trend in 8-17hours of operation with three different aromatic content samples. Deterioration was observed with 100% (heavy reformate) and 40% (light reformate) aromatic components. CARB Phase 3, which has an aromatic content of 26%, did not show deterioration in 17 hours. Despite this limited operating period, it seems that gasoline with an aromatic content of about 26% (CARB Phase 3) may be used for FCV fuel. To determine the upper limit for FCV gasoline's aromatic content, further evaluations, including longer-duration tests and experiments with various catalysts—are necessary.

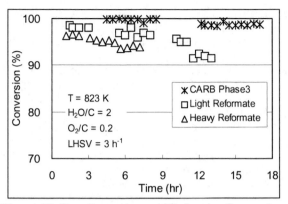

Figure 13. Time Dependence of Reforming Conversion: CARB Phase3, Reformate

DISCUSSION

Among the gasoline components, the naphthas, isomerate, and alkylate showed almost the same reactivity. All of them reached 97% or higher conversion at 800 deg K, while heavy fractions showed a bigger decline in conversion than did the light ones as the temperature decreased. These components contain various types of hydrocarbons, including paraffins, isoparaffins naphthenes, and aromatics. Due to their different chemical structures, these hydrocarbon groups may have different reactivities during reforming. However, their concentrations in gasoline are limited. It seems that the difference in the chemical components of gasoline components, except aromatics, is likely to have a negligible effect on reforming reactivity. As for distillation range, light fractions are favorable but heavy fractions are also usable because of they have the same conversion efficiency at temperatures where greater-than-97% conversion occurs.

Reformates showed a larger decline in conversion efficiency than the other hydrocarbon gasoline components. With regard to the reforming gas composition, reformates produced less hydrogen gas because of the high aromatic content of reformates. The durability tests also showed a strong influence of aromatics on catalyst deterioration. However, reformates are also usable for FCV gasoline due to their almost identical conversion efficiency as other gasoline components at high reforming temperatures. However, it seems necessary to control the aromatic content to approximately 25% in order to maintain the durability. The exact figure for the upper limit of the aromatic content must be evaluated in further research with other types of catalysts.

Desulfurized CCGs showed almost the same reactivity as the saturated-type gasoline components, suggesting the possibility of using CCG for FCV gasoline. However, CCG is the least likely component for FCV gasoline because of its high sulfur content. Moreover, the effect of olefin on durability is yet to be evaluated.

Oxygenates are promising candidates for FCV gasoline components. They have not only good reactivity but also a possibility that they improve the reactivity at low temperatures by blending into the hydrocarbon components. However, it must be noted that ethanol may have some tendency for coking.

Sulfur must be reduced for gasoline to be useful as a reformer feed. From the point of view of current gasoline components, it seems to be possible to produce <1 ppm sulfur FCV fuel while excluding CCGs. The level of sulfur needed will have to be determined through further research.

DPG may be technically feasible if the proper components are chosen. The candidates include isomerate, alkylate, and oxygenates that have low sulfur and low aromatic contents but high octane numbers. To enhance the octane number, reformates can be used at a limited blending ratio. However, the manufacturing capacity will be extremely limited without CCGs which have been eliminated from candidates. CCGs are the principal components of current gasoline, but on the other hand, alkylate is the by-product of CCGs and isomerate is produced only at highly integrated refineries. (Figure 14 shows Japan's total capacity of gasoline component production. The estimation is based on OGJ's[12]data.) Because of the special requirements for use in FCVs it is possible that an exclusive purpose gasoline for FCVs with sulfur-free distribution will be the practical solution.

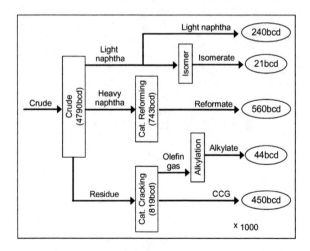

Figure 14. Japan total gasoline component production capacities (Estimation based on OGJ's[12]data)

CONCLUSION

For the purpose of FCV fuel usage, various gasoline components have been evaluated with ATR.

All components can be reformed at conversions greater than 97% at 800 deg K.

Components with higher molecular weight compounds and aromatic-rich components showed lower reactivity than the light ones did at low temperature.

Oxygenates had a possibility of improving the reactivity at low temperature by blending into hydrocarbon components.

Sulfur and aromatics affected the catalyst durability. Further evaluation and discussion are needed to determine the acceptable levels of each.

ACKNOWLEDGMENTS

We would like to thank Steve Welstand and Bill Cannella of ChevronTexaco for providing test fuels and for helpful discussions.

REFERENCES

1. Stodolsky, F., et al., "Total Fuel Cycle Impacts of Advanced Vehicles" SAE1999-01-0322 (1999)
2. Casten, S., et al., "Fuels for Fuel Cell-Powered Vehicles" SAE2000-01-0001 (2000)
3. Berlowitz, P. J., et al., "Fuel Choices For Fuel Cell Powered Vehicles" SAE2000-01-0003 (2000)
4. Contadini, J. F., et al., "Life-Cycle Emissions of Alternative Fuels for Transportation : Dealing with Uncertanties" SAE2000-01-0597 (2000)
5. He, D., et al., "Contribution Feedstock and Fuel Transportation to Total Fuel-Cycle Energy Use and Emissions" SAE2000-01-2976 (2000)
6. Louis, J.J.J., "Well-to Wheel Energy Use and Greenhouse Gas Emissions for Various Vehicle Technologies" SAE2001-01-1343 (2001)
7. Ellinger, R., et al., "Comparison of CO2 Emission Levels for Internal Combustion Engine and Fuel Cell Automotive Propulsion Systems" SAE2001-01-3751(2001)
8. Massachusetts Institute of Technology, "ON THE ROAD IN 2020, A life-cycle analysis of new automobile technologies" Energy Laboratory Report # MIT EL 00-003 (2000)
9. General Motors Corporation, Argonne National Laboratory, BP Amoco, ExxonMobil and Shell, "Well-to-Wheel Energy Use and Greenhouse Gas Emissions of Advanced Fuel/Vehicle Systems – North American Analysis-" (2001)
10. Bevilacqua-Knight, Inc., "Bringing Fuel Cell Vehicles to Market : Scenarios and Challenges with Fuel Alternatives" Consultant study report prepared for California Fuel Cell Partnership (2001)
11. Arthur D. Little, Inc., "Guidance for Transportation Technologies : Fuel Choice for Fuel Cell Vehicles, Phase2 Final Deliverable to DOE"(2002)
12. Nakamura, D., "Worldwide refining capacity declines slightly in 2001" Oil & Gas Journal / Dec. 24, 2001

DEFINITIONS AND ACRONYMS

FCV: Fuel Cell Vehicle

ATR: Auto Thermal Reforming

CCG: Catalytic Cracked Gasoline

P3:   CARB Phase3 Gasoline

DPG: Dual Purpose Gasoline

HR: High RON Gasoline

LR: Low RON Gasoline

RON: Research Octane Number

ELSEVIER

Journal of Power Sources 106 (2002) 353–363

JOURNAL OF
POWER
SOURCES

www.elsevier.com/locate/jpowsour

Options for refuelling hydrogen fuel cell vehicles in Italy

R. Mercuri[a], A. Bauen[b,*], D. Hart[b]

[a]EniTecnologie, Strategic Planning and Technology Innovation, 00016 Monterotondo, Rome, Italy
[b]Imperial College, Centre for Energy Policy and Technology, RSM Building, Prince Consort Road, London SW7 2BP, UK

Abstract

Hydrogen fuel cell vehicle (H_2 FCV) trials are taking place in a number of cities around the world. In Italy, Milan and Turin are the first to have demonstration projects involving hydrogen-fuelled vehicles, in part to satisfy increasing consumer demand for improved environmental performance. The Italian transport plan specifically highlights the potential for FCVs to enter into the marketplace from around 2005.

A scenario for FCV penetration into Italy, developed using projected costs for FCV and hydrogen fuel, suggests that by 2015, 2 million Italian cars could be powered by fuel cells. By 2030, 60% of the parc could be FCVs. To develop an infrastructure to supply these vehicles, a variety of options is considered. Large-scale steam reforming, on-site reforming and electrolysis options are analysed, with hydrogen delivered both in liquid and gaseous form. Assuming mature technologies, with over 10,000 units produced, on-site steam reforming provides the most economic hydrogen supply to the consumer, at US$ 2.6/kg. However, in the early stages of the infrastructure development there is a clear opportunity for on-site electrolysis and for production of hydrogen at centralised facilities, with delivery in the form of liquid hydrogen. This enables additional flexibility, as the hydrogen may also be used for fuel refining or for local power generation. In the current Italian context, energy companies could have a significant role to play in developing a hydrogen infrastructure.

The use of hydrogen FCVs can substantially reduce emissions of regulated pollutants and greenhouse gases. Using externality costs for regulated pollutants, it is estimated that the use of hydrogen fuel cell buses in place of 5% of diesel buses in Milan could avoid US$ 2 million per year in health costs. The addition of even very low externality costs to fuel prices makes the use of untaxed hydrogen in buses and cars, which is slightly more expensive for the motorist than untaxed gasoline or diesel, competitive on a social cost basis. © 2002 Elsevier Science B.V. All rights reserved.

Keywords: Hydrogen infrastructure; Fuel cell vehicle; Hydrogen energy; External cost; Cost of hydrogen; FCV; Emissions reduction

1. Introduction

Consumer demand for local and global environmental quality, in the form of reduced noise and air pollution, including greenhouse gas emissions, may help drive the introduction of hydrogen fuel cell vehicles (H_2 FCVs). Furthermore, the demand for environmental quality is likely to rise as demand for road transport increases worldwide, together with demand for improved vehicle performance and comfort and affordable mobility. The success of H_2 FCVs will depend on their ability to satisfy different consumer requirements.

This paper aims to understand possible drivers for H_2 FCV demand in Italy, develop a scenario for H_2 FCV introduction, analyse different supply routes for H_2 fuel and assess the potential environmental benefits of H_2 FCVs.

2. Road transport and environmental quality

About 70% of the European population lives in urban centres and is exposed to air and noise pollution as a result of road transport. A survey conducted by the Italian Ministry of Environment indicates that 80% of families in Milan, 75% in Bologna and 65% in Florence and Turin consider air pollution in the area in which they live to be "very or quite high" [1]. Road transport accounts for 24% of the total CO_2 emissions in Italy [2] and 17% worldwide [3]. It is the principal contributor to noise in urban areas, with 50% of the families in Milan, Florence and Turin considering noise pollution to be "very or quite high" and 86% attributing it to road traffic. The World Health Organisation estimates that about 97% of Europe's population is exposed to noise levels above 55 dB (A), of which about 72% to levels above 65 dB (A) and 27% to levels above 75 dB (A). This is of concern since it can, amongst other things, affect people's verbal communication and sleep and result in nervous disorders [4]. Notwithstanding growing concern over the environmental implications of road transport, private transport appears to be expanding gradually at the expense of public transport [5].

---

* Corresponding author. Tel.: +44-20-7594-9332;
fax: +44-20-7594-9334.
E-mail address: a.bauen@ic.ac.uk (A. Bauen).

0378-7753/02/$ – see front matter © 2002 Elsevier Science B.V. All rights reserved.
PII: S 0378-7753(01)01060-6

2.1. The role of fuel cells

Automotive and energy companies have been making considerable efforts to produce cleaner vehicles and fuels based on the internal combustion engine (ICE) and using improved petroleum fuels. However, more innovative solutions are likely to be required for sustained mobility while meeting different consumer demands. Battery electric vehicles (BEVs), FCVs and some hybrid ICE or FC/battery designs hold promise.

The Italian General Transport Plan (Piano Generale dei Trasporti, PGT) highlights these potential technology solutions, while indicating that no significant market penetration is likely to occur before 2005 and that recent rapid developments in FCVs may lead them to prevail over other technologies beyond that date [6]. Both BEVs and H_2 FCVs offer the advantage that they are truly zero emission vehicles (ZEVs) at the point of use, and that their operation results in very low noise levels. However, BEV market penetration has been hampered by low range and long refuelling time, and by the high cost of the vehicle and of electricity, in Italy in particular. FCVs may provide a solution to the issues of range and refuelling time and FC engines have been projected to achieve greater cost reductions through mass manufacturing compared to batteries, and possibly ICEs [7].

FCV development worldwide is progressing rapidly, but a number of issues remain to be addressed, with regard to the fuelling of cars in particular. The fuelling of buses is less of an issue because space constraints are less important and their refuelling generally takes place at depots. While direct hydrogen fuelling of cars is likely to result in the simplest design, intensive R&D is directed to the development of on-board reformers to allow fuelling with fuels, such as methanol, gasoline and, possibly, natural gas. On-board reforming, if technically viable, could provide a transitional solution to issues associated with the development of a hydrogen infrastructure and its storage on-board cars.

The question of fuel choice for FCVs remains open, but the viability and benefits of on-board reforming appear strongly questionable partly because of cost, complexity, start-up time, transient response and power density issues, in particular if regarded as a transitional solution. H_2 FC buses have been extensively demonstrated and a number of H_2 FC car prototypes are being tested on the road (e.g. H_2 Ford Focus) [8].

H_2 FCVs may be an important element in moving towards sustainable transport. They appear as the most likely option capable of simultaneously ensuring long-term sustained mobility while ensuring high levels of environmental quality, vehicle performance and comfort. However, different factors will affect the success of H_2 FCVs and the rate of their uptake.

- Technology readiness, availability and costs.
- Associated refuelling infrastructure availability and costs.
- Market structure, norms and regulations.
- Public acceptance.
- Government policy.

A number of perceived barriers may affect H_2 FCV market penetration.

- H_2 health and safety issues.
- Public perception.
- Technology readiness.
- H_2 fuel and FCV cost.
- H_2 fuel infrastructure.
- Norms and standards (for vehicles and refuelling infrastructure).

Hydrogen fuel is not thought to represent additional risks compared to conventional or other alternative fuels and most health and safety issues could be resolved by setting adequate norms and regulations, training qualified personnel and educating the public [9]. The public perception of hydrogen remains of concern, though surveys of potential FCV users and passengers seem to disprove this myth [10]. Technology readiness is still an issue, in particular with regard to FC engine system reliability and durability and advanced on-board hydrogen storage. Hydrogen is produced at large scale for captive use or for export as a commodity chemical, transported in gaseous or liquid form by truck or pipeline. On-site hydrogen production and refuelling systems should not present particular technical challenges. FCV cost appears as a barrier, with FC buses currently about 10 times more expensive than a conventional diesel engine bus, but several studies indicate that the mass manufacture of FCs could make them competitive with ICE vehicles on a life cycle cost basis. The establishment of a H_2 infrastructure and associated costs are commonly perceived as strong barriers. However, other studies have shown that a H_2 infrastructure need not be created overnight and that different options, as illustrated in this paper, exist to adapt the fuelling infrastructure to a potential phased-in demand for FCVs [11]. Also, as discussed later, hydrogen could be produced at a cost that would make FCVs competitive with other vehicles under certain circumstances. It is imperative that norms and regulations related to H_2 FCVs and refuelling be developed early on as these may delay and discourage uptake. Finally, the commercialisation of H_2 FCVs is not likely to happen without adequate government policies aimed at realising their potential social benefits.

2.2. Alternative fuels and vehicles and policy developments

Most Italian cars are gasoline fuelled, although other fuels, diesel in particular, are becoming more common. Liquid petroleum gas (LPG) and compressed natural gas (CNG) use is growing, albeit slowly. Most buses are diesel-fuelled, followed by gasoline and LPG buses, although other fuels, such as white diesel[1], biodiesel in certain regions and

---

[1] Consists of a mixture of diesel, water (10%) and special additives (1.7%). It has been commercialised under the name GECAM by CAM Tecnologie, part of the Pirelli Group.

Table 1
Split of Italian vehicle population according to fuel (1999)

	Gasoline	Diesel	LPG	CNG	Others	Total
Cars	26,386,617	4,132,262	1,253,774	256,739	8,899	32,038,291
Buses	1,203	84,052	67	92	348	85,762

CNG are all experiencing growth. Table 1 provides a split of Italian cars and buses according to their fuel [12].

The number of CNG cars is expected to grow from about 300,000 to 600,000 and LPG cars from about 1,170,000 to 1,500,000 by 2005 [13]. However, the numbers of LPG cars may not grow as predicted, both because of a lack of sufficient communication and competition from CNG and diesel cars [14]. Some growth is predicted in CNG buses, and a growing number of buses are fuelled by "white diesel", which already fuels around 3,600 buses mainly in the Lombardy region, but with prospects of expanding to other regions [15].

A number of economic incentives, mainly in the form of tax breaks, are aimed at alternative fuels, as shown in Table 2. While taxes represent about 70% of the price of gasoline and 60% of the price of diesel, their level is of about 43% for LPG, 27% for "white diesel" and it is negligible for CNG [16]. Biodiesel also benefits from tax breaks, with a production of up to 300,000 ton per year, to be used pure or as a blend, totally tax exempt [17]. The principal objective of the tax breaks is to encourage switching to fuels that produce lower emissions than petrol and diesel.

Economic incentives are also directed to vehicle purchases or conversions. Legislation has been passed to provide incentives aimed at renewing the bus fleet (Law no. 194 dated 18 June 1998), about 60% of which is over 10 years old. About US$ 7 million (LIT 15 billion) have been allocated annually between 2001 and 2003 to promote purchase of electric cars and motorbikes (BEV, FCV, hybrid) and conversion of cars to CNG or LPG fuel (Ministerial Decree 5 April 2001).

2.3. FCV activities in Italy

Several H_2 fuel and FCV activities are planned for the short-term in Italy.

FIAT has revealed a prototype H_2 FC car (spring 2001), the 600 H_2 FC Elettra, and has announced the development of a FC Punto with multi-fuel on-board reforming [18].

Table 2
Fuel price and taxation

Fuels	Price	Tax (%)
Gasoline (no lead) (US$/l)	1.0	70
Diesel (US$/l)	0.8	60
LPG (US$/l)	0.5	43
CNG (US$/m³)	0.4	1
White diesel (US$/l)	0.8	27

ENEA in collaboration with De Nora, Centro Ricerche Fiat, CNR-TAE, and the Universities of Milan, Brescia, Genova and Rome has plans to develop a FCV with a natural gas reformer on-board [19].

The city of Turin "City-Class Fuel Cell" project aims to demonstrate a fleet of Irisbus buses fuelled with hydrogen produced from a small-scale electrolysis plant [20]. The project, co-funded by the Italian Ministry of Environment, involves the city of Turin, ATM (municipal transport company of Turin) IRISBUS (joint-venture between IVECO and Renault), International Fuel Cells (IFC), SAPIO Group (Industrial gas company in partnership with Air Products), Ansaldo Ricerche, TÜV (German safety organisation) and Compagnia Valdostana delle Acque (hydroelectricity generator-CVA).

The "Milano Bicocca" project plans to demonstrate a stationary FC for power generation, and part of the hydrogen generated on-site will be used to fuel a fleet of buses and cars. The main partners are the city of Milan, Zincar (subsidiary of the municipal energy company of Milan AEM), ENEA (Italian agency for new technologies, energy and the environment) and other fuel cell, automotive and industrial gas companies [21]. The city of Florence, through a consortium led by i²t³, also has plans to demonstrate an FC bus [22].

2.4. FCV penetration scenario for Italy

A H_2 FCV penetration scenario has been developed for Italy, based on a FCV penetration model described in [23], to provide an indication of the potential H_2 FCV growth and associated H_2 demand in Italy over the period extending to 2030 (Fig. 1).

The following main assumptions are made in modelling H_2 FCV penetration.

- Fuel price: H_2 price per distance driven is initially half that of gasoline (taxed). The price of gasoline is assumed to increase by 1.8% per annum. The cost of hydrogen is assumed to increase by 2% per annum.
- Vehicle price: The cost of a FCV is initially 20% higher than that of a conventional vehicle. A learning factor of 10% is applied to the initial real (unsubsidised) cost of the vehicle. Conventional vehicle prices are assumed to grow at a rate of 1% per annum.
- Willingness to pay: 2% of vehicle owners are willing to pay a premium for ZEVs (this means that 2% of new vehicles are FCV whatever the cost).
- Refuelling stations: The number of H_2 refuelling stations is assumed to grow by 1% per annum.

The assumptions used provide a relatively optimistic scenario for fuel cell penetration (shown in Fig. 1) which indicates that H_2 FCV could represent about 65% of the vehicle parc by 2030. This corresponds to an annual H_2 fuel demand of about 1.4 million ton by the year 2030. However, even under optimistic conditions uptake will be gradual,

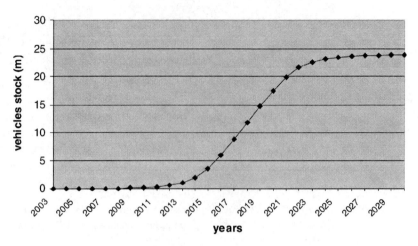

Fig. 1. A penetration scenario for FCVs in Italy.

with about 6% of the vehicle parc (about 2 million cars) converted to H_2 FCV by about 2015, corresponding to an annual H_2 fuel demand of about 150,000 ton.

2.5. FCVs and the environment

The demand for environmental quality is a main driver behind the development of FCVs. Use of FCVs should result in very low well-to-wheels emissions of regulated pollutants, the H_2 FCV being the only true fuel cell ZEV with regard to on-board emissions, and will generally offer significant benefits over conventional and other alternative vehicles. The benefits of FCVs over other vehicles in terms of energy efficiency and GHG emissions depend largely on the fuel from which hydrogen is derived and, in the case of electrolysis, the source of electricity used. Tables 3 and 4 provide a well-to-wheels comparison of emissions and energy use for different cars and buses based on [24]. These values are indicative and based on UK data. Emissions for an Italian situation would differ somewhat due to the different network associated with fossil fuel extraction, processing and transportation, while emissions associated with derived fuel (e.g. methanol and hydrogen) production and, if necessary, distribution would be very similar. On a well-to-wheels basis the emissions and energy use for the cases considered are likely to differ little between the geographic regions.

FCVs, like BEVs, are characterised by much lower noise emissions than IC engines; experimental vehicles emit about 65 dB. The difference in noise levels is more significant for the larger vehicles, such as trucks and buses.

The reduced emissions and noise levels of FCVs compared to conventional and other alternative ICE vehicles can result in significant social benefits. The ExternE study provides the greatest effort to date to attribute a monetary value to the impacts on the environment (externalities) of energy and transport activities, air pollution in particular

Table 3
Well-to-wheels emissions and energy use comparison for different cars

Application		NO_x (g/km)	SO_x (g/km)	CO (g/km)	NMHC (g/km)	CO_2 (g/km)	CH_4 (g/km)	PM (g/km)	Energy (MJ/km)
Gasoline ICE car	Absolute values	0.26	0.2	2.3	0.77	209	0.042	0.01	3.16
Diesel ICE car	Absolute values	0.57	0.13	0.65	0.25	154	0.03	0.05	2.36
	Relative to gasoline (%)	219	64	28	33	74	72	489	75
CNG ICE car	Absolute values	0.10	0.01	0.05	0.05	158	0.12	<0.0001	2.74
	Relative to gasoline (%)	39	5	2	6	76	277	<0.5	87
Hydrogen ICE car[a]	Absolute values	0.11	0.03	0.04	0.05	220	0.15	0.0001	4.44
	Relative to gasoline (%)	43	17	2	7	105	364	1	141
MeOH fuel cell car	Absolute values	0.04	0.006	0.014	0.047	130	0.072	0.0015	2.63
	Relative to gasoline (%)	15	3	0.6	6.1	62	169	14	83
Gasoline fuel cell car	Absolute values	0.08	0.13	0.01	0.41	147	0.03	0.0002	2.24
	Relative to gasoline (%)	30	68	0.4	53	70	71	2	71
Hydrogen fuel cell car[a]	Absolute values	0.04	0.01	0.02	0.02	87.6	0.06	<0.0001	1.77
	Relative to gasoline (%)	16	7	1	3	42	145	<0.5	56
Battery car	Absolute values	0.17	0.06	0.08	0.02	88.1	0.06	0.0001	1.71
CCGT electricity	Relative to gasoline (%)	67	32	4	3	42	150	1	54

[a] Assumes steam–methane reforming at the refuelling station.

Table 4
Well-to-wheels emissions and energy use comparison for different buses

Application		NO_x (g/km)	SO_x (g/km)	CO (g/km)	NMHC (g/km)	CO_2 (g/km)	CH_4 (g/km)	PM (g/km)	Energy (MJ/km)
Diesel bus	Absolute values	5.8	0.78	2.2	3.2	962	0.19	0.11	14.6
SPFC bus	Absolute values	0.43	0.11	0.17	0.18	588	0.33	0.0031	11.7
Central reformer	Relative to diesel (%)	7	14	8	6	61	175	3	80
SPFC bus	Absolute values	0.27	0.08	0.11	0.13	560	0.39	0.0001	11.3
Depot reformer	Relative to diesel (%)	5	10.8	5	4.2	58	206	<0.5	78
CNG bus	Absolute values	0.56	0.05	0.57	0.20	826	0.56	0.01	15.4
	Relative to diesel (%)	10	7	25	6	86	296	12	105
Battery bus	Absolute values	1.20	0.44	0.58	0.15	608	0.43	0.0009	11.8
CCGT electricity	Relative to diesel (%)	21	56	26	4	63	231	1	81

Table 5
Range of external costs for emissions in European urban centres (US$/ton)

PM	158,875–1,063,636
SO_2	5,540–28,636
NO_x	337–8,573
NMHC	237–1,561
CO	1.2–18
CO_2	14–38

[28]. Table 5 provides a range of externalities estimates expressed per unit of pollutant based on the ExternE methodology [29]. The range represents estimates of externalities for the cities of Brussels, Helsinki, Paris, Stuttgart, Athens, Amsterdam and London. The lowest values apply to cities with low population, such as Helsinki and the highest values to cities with high population, such as Paris.

Tables 6 and 7 provides estimates of the externalities for different types of vehicles operating in urban centres.

The city of Milan, representative of a high-population urban centre, has about 1,500,000 cars and 1,500 buses. The potential avoided emissions and social benefit of substituting 0.5% of gasoline cars (7,500 vehicles based on penetration rate in Fig. 1) and 5% of diesel buses (75 buses) by H_2 FCVs by 2010 are shown in Table 7. The calculations assume average annual urban travel distances of 20 and 200 km per day for cars and buses, respectively.

Table 6
Range of external costs for different vehicles in European urban centres (US$/km)

	External cost (US$/km)	
	Including CO_2 equivalent	Excluding CO_2 equivalent
Gasoline car	0.0059–0.027	0.0030–0.020
Diesel bus	0.038–0.34	0.024–0.19
H_2 FC car[a]	0.0013–0.0050	0.00009–0.00077
H_2 FC bus[a]	0.0084–0.026	0.00058–0.0049

[a] Assumes steam–methane reformer at the refuelling station.

Table 7
Avoided emissions and external costs from partial introduction of H_2 FCVs in Milan (H_2 fuel from steam–methane reforming at refuelling station)

Conventional fuel replaced	Cars (gasoline)	Buses (diesel)
Avoided emissions (kg per year)		
PM	542	602
SO_2	10,403	3,833
No_x	12,045	30,277
NMHC	41,063	16,808
CO	124,830	11,443
CO_2	6625,955	21,77,955
Social benefit estimate (inclusive CO_2) (US$ per year)	412,535	29,13,998
Social benefit estimate (exclusive CO_2) (US$ per year)	288,713	2,079,757

H_2 FCVs could contribute significantly to meeting a demand for increased local and global environmental quality.

3. H_2 supply economics

The price of a FCV will need to be similar to that of a conventional vehicle of similar performance for it to be accepted by consumers. The cost of H_2 fuel will affect the lifetime cost of a FCV and could be a determining factor in influencing consumer choice, especially for fleet vehicles. The cost of the fuel is currently perceived as one of the major barriers to its introduction. Hence, this section provides an analysis of H_2 supply costs in the context of Italy. Three scenarios for H_2 supply have been considered, distinguished mainly on the basis of centralised or on-site hydrogen production:

- centralised steam reforming of natural gas for hydrogen production and liquid hydrogen transport to refuelling station;
- on-site production and storage of gaseous hydrogen from steam reforming of natural gas;

- on-site production and storage of gaseous hydrogen from electrolysis of water.

The following main assumptions have been made.

- the refuelling station has a capacity to refuel about 50 buses or 300 cars per day, corresponding to a H_2 requirement of about 900 kg per day. The choice of refuelling station capacity is based on requirements compatible with dedicated bus depot refuelling and retail refuelling stations for cars. The case of a smaller refuelling station (180 kg per day) is also studied, in order to evaluate transitional aspects related to the build-up of H_2 FCVs and related refuelling infrastructure;
- hydrogen is dispensed in compressed gaseous form, a solution presently adopted for the refuelling of existing FCV bus demonstrations and which appears as a promising option through advanced compressed gas storage or storage in metal hydride or possibly carbon structures on-board cars. Various prototype H_2-fuelled cars currently store hydrogen on-board as a compressed gas (e.g. FIAT 600 H_2 Elettra);
- in the case of centralised H_2 production from steam reforming of natural gas, liquid H_2 transport, practised routinely, has been considered because of the substantially lower cost per unit of energy transported compared with compressed H_2. The latter may be viable for the transport of smaller quantities of hydrogen, because of the significant impact on cost of liquefaction facilities, while dedicated pipelines for the transport of compressed H_2 may become viable, either for captive users situated at short distances from the reforming plant or once a large stable H_2 demand becomes established.
- Natural gas has been considered as the feedstock of choice because of its widespread availability in Italy. Other feedstock could be considered, in particular for centralised production, such as refinery residues.

Investment costs and operation and maintenance costs for the different hydrogen fuel supply options are based on Berry [25], updated using an inflation rate value and adapted for the Italian context, in particular with regard to energy prices [26]. The initial costs and costs projections have been compared with other cost projections in the literature for small-scale electrolysis, reforming equipment and refilling station equipment [27]. Small differences were found, but these do not greatly affect the cost of hydrogen to the consumer, especially because of the high influence of the variable costs of electricity, gas and labour.

A progress ratio (learning factor) of 15% has been used to determine the cost reduction of new technologies, which corresponds to a 15% reduction in the cost of a product for every doubling in cumulative production capacity. The progress ratio is assumed to account for labour productivity gains, product and process optimisation and management

efficiency gains. The progress ratio has been applied to the current cost of new hydrogen production technologies (i.e. small-scale electrolysis, small-scale steam reforming and H_2 refuelling station equipment) to project the cost of single units up to a cumulative global production of 10,000 units. A discount rate of 20% has been used, reflecting the rate of capital recovery.

Typical natural gas and electricity prices for industrial and large commercial users have been used. Table 8 summarises the main economic parameters used in the analysis.

Tables 9–11 show a breakdown of capital, maintenance and operation costs for the different options.

Fig. 2 illustrates the H_2 supply options considered and the contributions of the different production and handling stages to the final cost of H_2.

Table 8
Main economic parameters used in H_2 production analysis

Scaling factor[a]	0.8
Utilisation factor (%)	95
Discount rate (%)	20
Inflation USA 1996–2000 (% per year)	2.5
Learning curve factor (%)	15
Exchange rate (LIT per US$)	2,100
Diesel price (US$/l)	0.8
Electricity price (US$/kWh)	
Large scale liquefaction plant	0.12
Small plant (electrolyser and filling station)	0.09
Natural gas price (US$/kWh)	
Large scale centralised plant (SMR)	0.017
Medium scale on-site SMR plant	0.021
Small scale on-site SMR plant	0.026

[a] Used for determining economies of scale for liquefaction plant.

Table 9
Centralised SMR option

Investment cost centralised SMR facility (237 H_2 ton per day) (US$)	220,000,000
Natural gas fuel cost 5 US$/GJ (US$ per day)	197,000
Labour & other O&M (US$ per day)	55,000
Investment cost liquefaction plant (237 H_2 ton per day) (US$)	153,000,000
Electricity cost 0.09 US$/kWh (US$ per day)	260,000
Labour & other O&M (US$ per day)	20,000
Transport investment cost (4 ton LH_2 per truck) (US$ per truck)	500,000
Fuel cost (US$/l)	0.8
Fuel consumption (km/l)	2.2
Labour (US$/h)	28
Refuelling station (900 H_2 kg per day)	
Initial capital cost (storage tanks, pump and vaporiser, dispenser) (US$)	575,000
Final capital cost (10,000 units) (US$)	133,200
Electricity cost 0.12 US$/kWh (US$ per year)	11,000
Labour (US$ per year)	265,000
Other O&M (US$ per year)	9,000

Table 10
On-site SMR option (900 H$_2$ kg per day)

Initial capital cost:	
Reformer (US$)	2,420,000
Refuelling station (compressor, vessels, dispenser) (US$)	1,254,000
Final capital cost (up to 10,000 units)	
Reformer (US$)	555,000
Refuelling station (compressor, vessels, dispenser) (US$)	290,000
Natural gas fuel cost (US$/GJ)	5.9
(US$ per year)	260,000
Electricity cost (US$/kWh)	0.12
(US$ per year)	110,000
Labour (US$ per year)	265,000
Other O&M (US$ per year)	15,000

Table 11
On-site electrolysis option (1170 H2 kg per day)

Initial capital cost	
Electrolyser plant (US$)	5,030,000
Filling station (compressor, vessels, dispenser) (US$)	1,254,000
Final capital cost (up to 10,000 units)	
Electrolyser plant (US$)	1,430,000
Refuelling station (compressor, vessels, dispenser) (US$)	290,000
Electricity cost 0.12 US$/kWh (US$ per year)	2,000,000
Labour (US$ per year)	265,000
Other O&M (US$ per year)	15,000

In the case of on-site electrolysis the investment cost contributes only about 13% of the price of the hydrogen at the refuelling station, assuming mass production effects on small-scale polymer membrane electrolysers and on refuelling station equipment. The bulk of the cost is attributable to the cost of the electricity (74%) needed by the electrolysis process and for H$_2$ compression. Because of the major contribution of the cost of electricity, a sensitivity analysis has been performed to determine the effect of its variation on the cost of H$_2$ production (Fig. 3). Variations in the cost of electricity could depend on a variety of factors, such as market liberalisation effects, specific contractual agreements and purchase of electricity at particular times (e.g. off-peak).

In the case of on-site natural gas steam reforming the investment cost contributes about 21% of the cost of H$_2$ to the consumer, while about 32% is attributable to the cost of the natural gas feedstock. The natural gas price considered is that typical of medium industrial users, and a sensitivity analysis has been performed to investigate the effect of its variation on the cost of H$_2$ production (Fig. 4). Refuelling station equipment for H$_2$ compression, storage and dispensing is assumed to be the same as that for the on-site electrolysis plant.

The total capital cost of supplying hydrogen from on-site SMR (US$ 0.5/kg H$_2$) appears to be significantly lower to that of electrolysis (US$ 0.8/kg H$_2$), furthermore the variable

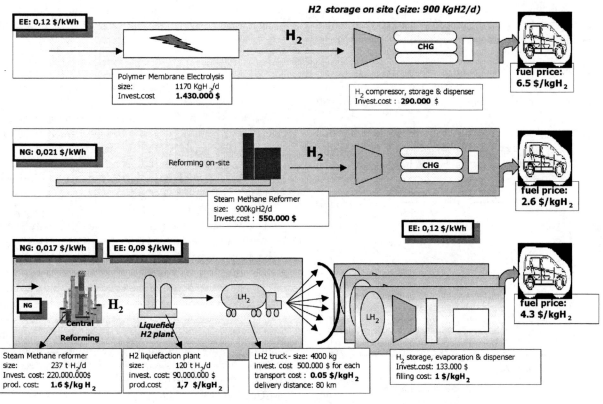

Fig. 2. Different possible hydrogen supply options and cost contributions.

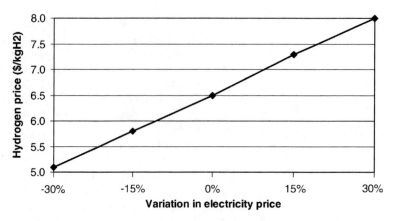

Fig. 3. Sensitivity of on-site electrolytic H_2 production to the cost of electricity (capacity 900 kg H_2 per day).

costs associated with natural gas inputs are also significantly lower than those associated with electricity inputs for the production of hydrogen.

In the case of centralised liquid H_2 production from natural gas steam reforming and H_2 liquefaction, the technologies are assumed to be mature and a progress ratio has been applied only to the refuelling station as per the previous cases. A range of liquefaction plant sizes (from 6 to 237 ton per day H_2) has been considered, to accommodate the fact that only part of the hydrogen may not be exported from the plant for the transport market (there may be some complementary use for electricity production or refining of fuels) and to allow for flexibility in meeting the requirements of a potentially growing H_2 fuel market. For example, a 6 ton per day H_2 liquefaction plant would be enough to supply about six refuelling stations of the size considered. A delivery distance of 80 km has been considered as representative of a regional H_2 fuel market, with hydrogen assumed to be stored as a liquid at the refuelling site and then vaporised and compressed to refuel vehicles.

The costs of natural gas and electricity at the production site are typical of large industrial users and lower than those for the on-site production facilities. H_2 liquefaction has an

important impact on the cost of H_2 and its energy consumption is estimated at about 35% of the energy content (LHV) of the H_2 produced. The liquefaction stage is characterised by high investment costs and variable costs in the form of electricity costs. Fig. 5 illustrates the cost of H_2, as a function of the liquefaction plant size (for a fixed steam reformer unit of 237 ton H_2 per day) and for a range of natural gas and electricity prices for a refuelling station of 900 kg H_2 per day.

It can be seen that there are significant reductions in liquid H_2 production costs associated with economies of scale in the liquefaction plant. The cost of H_2 fuel is also highly sensitive to the cost of the natural gas feedstock and to the cost of the electricity used for liquefaction. This is because of their large contribution to the cost of H_2 as illustrated in Fig. 3.

3.1. Assessing the options and addressing transitional aspects of hydrogen infrastructure development

The costs of hydrogen from different options could change significantly as a H_2 fuel infrastructure develops (Fig. 6). The analysis shows that costs for on-site H_2 production options can be reduced significantly as the market for H_2 fuel becomes established, with on-site steam reforming potentially the lowest cost H_2 fuel production option. This is likely to be the case, especially if significant price differentials between gas and electricity persist. However, hydrogen from large-scale SMR plants appears to be a cost effective option in the very short-term. The greatest dilemma remains the development of a hydrogen refuelling infrastructure in the short- to medium-term. For early H_2 FCV demonstration programmes and fleets of a few vehicles, electrolysis may represent the most easily implemented on-site solution offering suitable operating flexibility. However, liquid hydrogen imported from large-scale production plants may represent the most economic solution, possibly to be challenged soon by the on-site SMR option. Also, although electrolysis appears to be the most expensive option, it may be a very interesting one

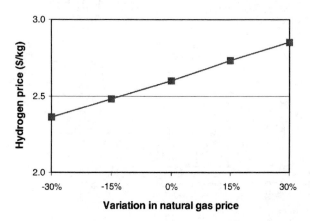

Fig. 4. Sensitivity of on-site SMR H_2 production to the cost of natural gas.

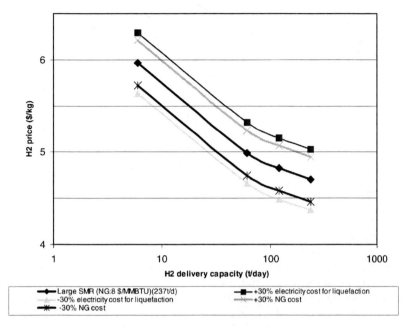

Fig. 5. Delivered H_2 cost at a 900 kg per day H_2 refuelling station for the centralised production option, as a function of liquefaction plant scale and natural gas and electricity prices.

long-term, in particular in relation to producing H_2 from renewable electricity (possibly even directly on-board the vehicle).

One plausible evolution of a hydrogen infrastructure is described below.

3.1.1. Short-term

Electrolysis at small scale offers an interesting option, more readily available and flexible in operation than small-scale steam reforming. Furthermore, electricity generators

interested in the potential market for H_2 fuel production may become involved early on offering electricity at special rates. In addition, bulk H_2 from large-scale SMR plants operated by industrial gas companies and from a variety of industrial processes, such as refineries and chlor alkali plants, may be competitive.

3.1.2. Medium-term

On-site SMR establishes itself as the most viable option where gas is available. H_2 supply from large-scale SMR

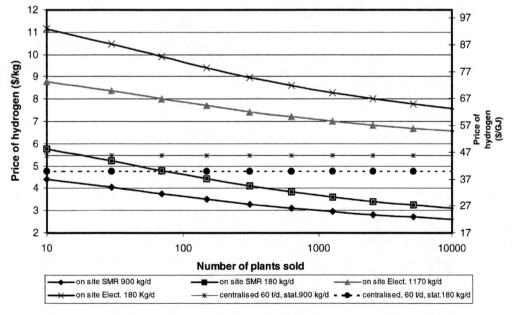

Fig. 6. Comparison of delivered hydrogen cost for different production options and as a function of learning curves.

Table 12
Fuel consumption and fuel consumption costs for ICE and FC vehicles

	Fuel consumption (MJ/km)	Fuel related cost (US$/km)
Diesel bus	13	0.117
Gasoline car	2.6	0.0247
H₂ FC bus	6.8	0.147
H₂ FC car	1.2	0.0258

where on-site SMR not viable because of demand (low) or location (no gas grid connection).

3.1.3. Long-term

A mix of steam reforming and electrolysis options at different scales are used, with electrolysis possibly an increasingly interesting option if hydrogen can be produced at relatively low cost from renewable electricity (perhaps including incentives, such as carbon taxes, arising from environmental concerns) or other competitive clean electricity sources.

3.2. The cost of H_2 to the consumer

To provide a useful indication of the cost of H_2 to the consumer, it has been expressed in US$/km driven by a FC bus and car, and compared to equivalent costs for diesel buses and gasoline cars (based both on taxed and untaxed diesel and gasoline costs). Table 12, Figs. 7 and 8 provide the energy consumption assumptions for FC, diesel and gasoline buses and cars assumed to have similar performance requirements and a fuel cost comparison. The cost of H_2 (US$ 2.6/ kg H_2) is based on the on-site SMR option (900 kg H_2 per day) for a cumulative production of over 10,000 units.

The private cost comparisons above show H_2 FCV fuel costs could be competitive with current ICE fuel costs, based on untaxed costs of diesel and gasoline, on a per distance travelled basis.

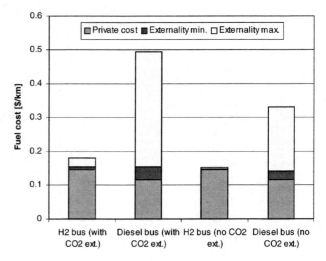

Fig. 8. Fuel cost comparison (buses).

3.3. Accounting for externalities

The taxation levels of diesel and gasoline fuels are generally not meant to reflect the external costs associated with their use, although environmental concerns have been part of the rationale for increasing fuel taxes in some countries. Possible reductions in externalities associated with cleaner fuel consumption for transport could be a powerful argument for the introduction of H_2 fuel. The potential economic benefits associated with the substitution of conventional fuels with hydrogen could effectively be translated into price signals favouring H_2.

Figs. 7 and 8 provide an indication of the fuel cost, including a range of externality costs (see Table 6), for diesel, gasoline and H_2 in the case of buses and cars. The inclusion of externalities could significantly improve the economic viability of H_2 fuel. In the case of gasoline cars and diesel buses, the external costs appear to be larger than the private cost per unit distance travelled differential between these fuels and hydrogen. Thus, on a social cost basis H_2 would appear to be a competitive fuel.

4. Conclusion

There seems to be some potential for H_2 FCVs in Italy. FCVs could help meet a number of consumer demands, in particular an increasing demand for environmental quality. In fact, reducing urban pollution from transport could be one of the main initial drivers behind the uptake of FCVs. Due to issues of logistics and economics, fuel cells are likely to be introduced first in buses and other fleet vehicles, which allow for centralised refuelling and possibly more favourable economics than private cars over the lifetime of the vehicle. In Italy, a number of FCV demonstration projects are planned and, under favourable conditions, FC cars could represent over 60% of cars on Italian roads in 2030. However, even under optimistic conditions, uptake will be gradual and

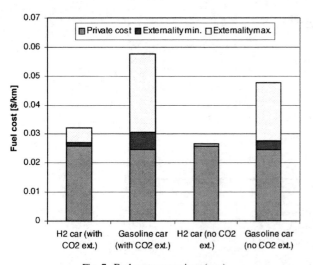

Fig. 7. Fuel cost comparison (cars).

at the most a 6% FC car penetration (2 million cars) could be expected for around the year 2015. H_2 fuel costs need not be a barrier in the long-term as it has been shown that on a distance travelled basis, H_2 fuel costs, based on mature on-site SMR H_2 production technologies, would only be slightly higher than those of untaxed gasoline or diesel for ICE vehicles. Also, a preliminary analysis of externalities associated with gasoline, diesel and hydrogen fuels, appears to indicate that the social benefit of introducing H_2 as a fuel could largely outweigh the higher private costs. The benefits that could be derived from the use of H_2 fuel, and its potential for long-term economic competitiveness with gasoline and diesel could justify incentives aimed at facilitating its introduction, together with FCVs, in the early stages where cost reductions need to be achieved.

All H_2 supply options appear to have future market potential. In the case of Italy in particular, large-scale SMR could supply a significant part of the short-term H_2 fuel demand and capture part of the long-term market. However, based on Italian energy market conditions and widespread gas infrastructure, on-site SMR appears to be the most interesting option to supply a FCV mass market. H_2 supply from electrolysis plants is likely to remain marginal in the medium-term, but could become an increasingly interesting long-term option if it can be achieved at relatively low cost from renewable electricity.

Acknowledgements

We would like to thank Vittorio Meda of EniTecnologie for helpful discussions on some of the technical and economic aspects of this paper.

References

[1] Ministero dell'Ambiente Relazione sullo Stato dell'Ambiente, 2001, p. 298. www.miniambiente.it/sito/pubblicazioni/pubblicazioni.asp

[2] Ministero dell'Ambiente, Relazione sullo Stato dell'Ambiente, 2001, p. 56.

[3] IEA, CO_2 emissions from fuel combustion 1971–1998, International Energy Agency, Paris, 2000.

[4] ENEA, Rapporto Energia e Ambiente, 2000, p. 93.

[5] Ministero dell'Ambiente, Relazione sullo Stato dell'Ambiente, 2001, p. 53.

[6] ENEA, Rapporto Energia e Ambiente, 2000, p. 99.

[7] L.D. Franklin Jr., B.D. James, G.N. Baum, C.E. (Sandy) Thomas, Detailed Manufacturing Cost Estimates for Polymer Electrolyte Membrane (PEM) Fuel Cells for Light Duty Vehicles, Directed Technologies, Inc., Arlington, 1997.

[8] www.h2forum.org

[9] K. Pehr, Aspects of safety and acceptance of LH_2 tank system in passenger cars, Int. J. Hydrogen Energy 21 (5) (1996) 387–395.

[10] Ford Motor Company, Direct hydrogen fuelled PEM Fuel cell system for trasportation applications: hydrogen vehicle safety report, DOE/CE/50389-502, US Department of Energy.

[11] J.M. Ogden, Developing an infrastructure for hydrogen vehicle: a Southern California case study, Int. J. Hydrogen Energy 24 (8) (1999) 709–730.

[12] Automobile Club Italia, Autoritratto 1999, www.aci.it.

[13] Staffetta Quotidiana 7/04/2001.

[14] Staffetta Quotidiana 28/07/2001.

[15] Associazione Transporti ASSTRA (Ed.), Trasporti pubblici, Aprile 2001, www.asstra.it/stampa/rivista_tp.asp, www.gasoliobianco.com

[16] Staffetta quotidiana 5/07/2001, 18/07/2001.

[17] EnergiaBlu, n. 2, March 2001.

[18] Staffetta quotidiana 3/03/2001.

[19] M. Ronchetti and A. Iacobazzi, Celle a combustibile, ENEA, 2000.

[20] www.comune.torino.it.

[21] www.aem.it, www.ecotrasporti.it/bmw/idrogeno.htm, Conference Tecnologia e unificazione delle prestazioni e dei requisiti di sicurezza per le motorizzazione a FC, Milan, 1/6/2001.

[22] www.firenze.net.

[23] D. Hart, A. Bauen, R. Fouquet, M. Leach, P. Pearson, D. Anderson, Hydrogen supply for SPFC vehicles ETSU F/02/00176/REP, 2000.

[24] A. Bauen, D. Hart, Assessment of the environmental benefit of transport and stationary fuel cell, J. Power Sources, 86 (1/2) (2000) 482–494.

[25] G.D. Berry, Hydrogen as a transportation fuel: Costs and benefits, UCRL-ID-123465, Lawrence Livermore National Lab. CA (United States), 1996.

[26] Autorità per l'Energia, Report Stato dei servizi: settore gas ed elettrico (tav. 3.5/2.16) (2001).

[27] C.E. Thomas, B.D. James, F.D. Lomax Jr., I.F. Kuhn Jr., Fuel options for the fuel cell vehicle: hydrogen, methanol or gasoline?, International Journal of Hydrogen Energy, 25 (6) (2000) 551–567.

[28] Commission of the European Communities (CEC), DG, XII Edition, 1998a, Final Reports of ExternE Phase III (preliminary versions available at http://ExternE.jrc.es), in press.

[29] R. Friedrich, P. Bickel (Eds.), Environmental External Costs of Transport, Springer, Berlin, 2001.

2003-01-0413

Life-Cycle Value Assessment (LCVA) of Fuel Supply Options for Fuel Cell Vehicles

Jesse Row and Marlo Raynolds
Pembina Institute

Renato Legati
Ballard Power Systems

Ron Monk
BC Hydro

Grant Arnold
Suncor Energy

ABSTRACT

The fuel cell vehicle (FCV) has the potential to revolutionize the world's transportation systems. As choices are made on sources of fuel for FCVs it is important to consider the life-cycle implications of each option or system. This paper summarizes the methodology and results of a joint initiative to evaluate the life-cycle performance of 72 vehicle and fuel scenarios in 3 Canadian cities, comparing Proton Exchange Membrane (PEM) fuel cell vehicles and fuelling infrastructure with conventional and alternative fuel vehicles. The analysis is based on actual performance data of commercial and near-commercial technologies. The specific fuels investigated were gasoline, diesel, natural gas, methanol, hydrogen and electricity. The Pembina Institute's Life-Cycle Value Assessment (LCVA) methodology was used to compare the environmental, economic and social performance of each system. The stressor categories quantified include emissions of greenhouse gases and criteria air contaminants, resource consumption, fuel cost, and vehicle cost. A wide number of other stressor categories were identified as important to vehicle / fuel supply performance and should be considered when comparing systems. This paper details only the air emissions results from the full LCVA report[1]. The conclusions reached within the LCVA highlight that there are considerable differences in the life-cycle performance of vehicle systems and that fuel supply choices play a critical role.

DISCLAIMER

This Life-Cycle Value Assessment is based on data available to the project team at the time of data collection. Consequently, the project team was required to apply judgment in comparing technologies at different maturity levels and vehicles of different body type and performance. Significant effort has been made to maintain consistency between the systems compared, and, where this was not possible, the inconsistencies have been identified in the report.

INTRODUCTION

The fuel cell vehicle (FCV) could revolutionize the world's transportation systems. With the potential for zero tailpipe emissions, greater vehicle performance, increased fuel efficiency and lack of dependence on crude oil compared to conventional internal combustion engines (ICEs), FCVs could effect a dramatic improvement in urban air quality, climate change, overall energy consumption and energy security. However, as choices are made about sources of fuel for FCVs, it is important to consider the life-cycle performance of each option or fuel supply system, since for many fuel choices the majority of emissions occur before the fuel reaches the FCV.

The objective of this Life-Cycle Value Assessment (LCVA) is to quantify and evaluate the life-cycle environmental and economic factors of a wide range of technically advanced (i.e., poised for commercialization with actual performance data available) options for operating light-duty vehicles and buses in Canada. In addition, this LCVA qualitatively identifies technical design challenges and improvement opportunities, along with social impacts. The relative performance of the technologies studied and the resulting conclusions are expected to change in the future as the technologies continue to mature. The performance of not-yet-commercialized technologies are believed to have the

45

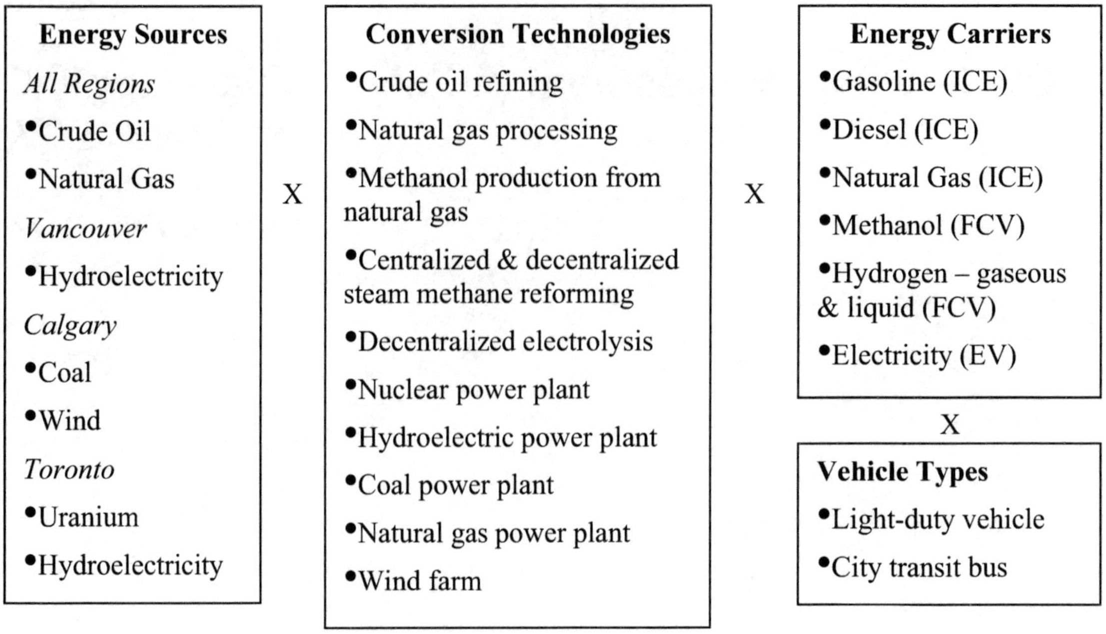

Energy Sources		Conversion Technologies		Energy Carriers
Energy Sources		**Conversion Technologies**		**Energy Carriers**
All Regions		•Crude oil refining		•Gasoline (ICE)
•Crude Oil		•Natural gas processing		•Diesel (ICE)
•Natural Gas	X	•Methanol production from natural gas	X	•Natural Gas (ICE)
Vancouver		•Centralized & decentralized steam methane reforming		•Methanol (FCV)
•Hydroelectricity				•Hydrogen – gaseous & liquid (FCV)
Calgary		•Decentralized electrolysis		•Electricity (EV)
•Coal		•Nuclear power plant		
•Wind		•Hydroelectric power plant		X
Toronto		•Coal power plant		**Vehicle Types**
•Uranium		•Natural gas power plant		•Light-duty vehicle
•Hydroelectricity		•Wind farm		•City transit bus

Figure 1 Energy and Technology Types Selected for Assessment

greatest opportunity to improve, given their relatively low level of maturity.

This LCVA looks specifically at three Canadian cities – Toronto, Vancouver and Calgary – and at both light-duty vehicles and buses. Fig. 1 illustrates the system components. A number of other potential technologies were initially considered, but were eliminated from this study owing to limited available data. In total, 72 different scenarios (unique city, vehicle and fuelling pathway combinations) were evaluated in this study.

SYSTEM BOUNDARIES

Fig. 1 also demonstrates the system boundaries that were established for this LCVA. The objective of selecting system boundaries is to minimize the number of unit processes that require intensive data collection while ensuring that a fair and thorough comparison is made. In general, three types of unit processes were placed outside of the system boundaries: product construction or production, maintenance, and decommissioning or disposal. The focus of the LCVA is therefore placed on the fuel production, distribution and consumption.

By not evaluating the effects of infrastructure and vehicle manufacturing, maintenance and disposal, the results of the LCVA will not reflect a complete life-cycle assessment of the systems investigated. However, the relative environmental impacts of these unit processes are expected to be small for the stressor categories investigated when considered over their lifetime as compared with the impacts of the daily operation of the system. Given that approximately 15% of the life-cycle air emissions of a vehicle result from manufacturing and 85% from the use of the vehicle[2], and the relative difference between the vehicles being evaluated is very

small (e.g., the bodies, frame, wheels, interior and glass are all relatively equal in amounts between the vehicles), the overall environmental implications of the manufacturing of the vehicles plays a relatively small role in the life-cycle emissions when compared with the fuel production and use. This LCVA is therefore limited in the areas where manufacturing and disposal of the selected vehicles are significantly different, such as in the use of unique materials (e.g., batteries for EVs and HEVs). Ultimately, the results will show the potential environmental, economic and social impacts of the daily operation of light-duty vehicles and buses in three Canadian cities.

STRESSOR CATEGORIES

In any evaluation of competing transportation systems, it is essential to consider the environmental, economic and social aspects of each option. This LCVA brings together both qualitative and quantitative information in its evaluation.

QUANTITATIVE

The environmental stressors quantified by this LCVA are limited to the following stressor categories:

- greenhouse gases – GHGs
- acid deposition precursors – ADPs
- criteria air contaminants
  - ozone and ground-level ozone precursors – GLOPs
  - particulate matter – PM;
  - secondary particulate matter precursors – SPMPs
- sulphur dioxide – SO_2
- nitrogen oxides – NO_X

- carbon monoxide – CO
- resource consumption

Sources of environmental performance data include industry experts and technology developers for information on emerging technologies, and public resources and industry experts for conventional technologies. Whenever possible, actual performance data, either in testing environments or from actual field observations of commercial technologies, demonstration equipment, or working prototypes, has been used.

The economic performance of each option has been quantified through estimates of vehicle and fuel costs. These costs are based on both historical prices and, when historical prices do not exist, publicly available literature attempting to project future costs is employed. The economic analysis, therefore, provides a comparison of potential future vehicle and fuel costs for each system assuming mass production quantities are achieved.

QUALITATIVE

There are many other environmental, social, economic and technical aspects that have also been identified as important to consider when comparing the performance of vehicle / fuel supply systems. For example, looking solely at the quantified results, battery electric vehicles using power from nuclear power plants appear very attractive – near zero air emissions, and relatively low fuel costs, although the vehicle costs are quite high. However, when a wider range of stressor categories is investigated, it becomes clear that this vehicle / fuel supply systems has many negative aspects that may outweigh its initially obvious advantages – low vehicle range, relatively short battery lifetime and long recharge times, lack of refuelling infrastructure, environmental, health and safety risks of radioactive waste, and the political sensitivities to using nuclear power. A list of the stressor categories identified for qualitative assessment is presented below.

Environmental Considerations

- additional air pollutants from combustion (eg. benzene, polyaromatic hydrocarbons, lead, mercury, etc.)
- oil & gas well venting (eg. hydrogen sulphide)
- waterbody, groundwater and sewer pollutants through the release of the following:
  o gasoline
  o diesel
  o methanol
  o motor oil
  o automotive coolant
  o refinery chemicals and byproducts
  o water effluents from power plants
  o crude oil
  o drilling rig chemical

- wildlife and plantlife disturbance, habitat fragmentation, biodiversity impacts particularly through the following activities:
  o hydroelectricity reservoirs including upstream and downstream areas
  o oil and gas exploration
  o coal, oilsands and uranium mining
  o pipelines
  o electricity transmission lines
  o roads and railways, along with their associated traffic
- solid waste with particular attention to:
  o industrial facilities
  o fuel distribution infrastructure
  o resource extraction
  o electric and hybrid electric vehicle batteries
- radioactive waste
- land use with particular interest in:
  o hydroelectricity reservoirs
  o coal, oilsands and uranium mines
  o oil and gas wells, and production facilities
  o electricity transmission lines
  o pipelines
  o centralized hydrogen production plants
  o electricity generating plants
  o wind farms
  o oil refineries
  o increased fuelling station size for storage of a variety of fuels, gaseous fuels, and distributed hydrogen generation
  o waste disposal sites
  o roads for access to remote sites
  o electric power cables over existing city roadways for trolley bus operation
- land, water body and underground formation disturbance.

Social Considerations

Consumer expectations and needs

- vehicle performance
- vehicle refuelling including safety and convenience

Community expectations

- fuel and infrastructure safety
- employment
- aesthetics (appearance, noise and odour);
- energy security
- resource consumption
- land use

Economic considerations

- personal and societal value generated through vehicle operation
- air pollutant emissions and impacts on human, environmental and infrastructure health

- greenhouse gas emissions
- employment and tax revenue benefits
- safety risks and insurance costs
- subsidies and tax incentives.

<u>Technical considerations</u>

- safety
- refuelling
- fuel quality
- integration with existing infrastructure
- fuel storage
- on-site hydrogen production
- lifetime
- start-up time
- emerging technologies
- fuel flexibility.

DATA INVENTORY

The following conditions apply to the data used in this LCVA, and provide context for the results that are presented:

- The vehicles assessed are similar in body type and overall efficiency but have not been normalized to include only fuel-related differences. Each vehicle is considered typical for their fuel and class.
- Real-world performance data of both specific operations and entire industrial sectors have been used whenever possible.
- Historical performance data of well-established processes may or may not reflect the performance of these processes currently, or in the future. The most current data available have been used.
- Pre-commercial technologies use information based on the actual performance of prototypes available at the time of data collection as opposed to theoretical performance predictions. The results, therefore, reflect the state of technology development that companies are willing to release at this time.
- Likely performance improvements to technologies, such as legislated emission reductions in vehicles or continuing optimization of emerging products, have not been taken into account. The results represent a "snapshot" in time.
- Light-duty vehicle systems reflect the performance of new vehicles operated over standard dynamometer tests. Bus systems reflect the performance of in-use buses operated both in vehicle fleets and over standard dynamometer tests. This does not take into account performance degradation over time.
- The influence of weather and adverse road conditions has not been accounted for in vehicle performance. Each vehicle type will operate differently under variations in temperature, humidity, precipitation and roadway characteristics.

- The costs of hydrogen, methanol, time-of-use electricity and fuel cell vehicles have been estimated using assumptions based on current technology in a developed market.
- The potential availability of natural resources or equipment has not been evaluated and may limit the possibility of systems being introduced on a large scale. This is a larger concern for systems that have not yet been established in the transportation market.

DATA SUMMARY

For the purpose of comparing actual performance data of vehicle / fuel supply systems, it was necessary to select specific vehicle and fuel production technologies to be used in the assessment. The light-duty FCVs for which data were available for analysis were DaimlerChrysler's NECAR 4 and 5. The remaining vehicles selected for the study are intended to represent a typical vehicle within a class comparable to the NECARs. The specific diesel, hybrid electric, natural gas and electric vehicles were selected because they were the only vehicles available within North America that had data available to the project team, and that provide adequately comparable performance to the NECARs. A Honda Civic HEV has recently become available to the consumer (2002); however, specific emission data were not available at the time the data inventory was being completed. The gasoline vehicle, on the other hand, was selected as a mid-range performer from a group of compact cars that are considered comparable to other vehicles in this study, as shown in Fig. 2.

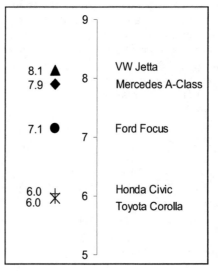

Notes:

All vehicle models are from 2001.

All values are based on standard driving cycles and have not been adjusted to reflect real-world driving.

Mercedes A-Class data based on the New European Driving Cycle, other vehicles based on EPA combined city / highway driving cycle.

Figure 2   Fuel Consumption of Gasoline Light-Duty Vehicles (litres / 100 km)

Table 1 lists the average fuel consumption used for each of the vehicles whereas Table 2 describes the hydrogen production technologies used in the study. The majority of the remaining technologies are well-established and therefore either typical performance figures or industry averages are used.

Table 1 Summary of Vehicle Fuel Consumption

Vehicle	Fuel Consumption	MPGGE
Ford Focus (gasoline)	7.15 litres gasoline / 100 km*	33
VW Jetta TDI (diesel)	5.24 litres diesel / 100 km*	40
Honda Civic GX (natural gas)	3.9 kg natural gas / 100 km*	39
Toyota Prius (HEV)	4.05 litres gasoline / 100 km*	58
NECAR 5 (methanol)	Confidential	
NECAR 4 (hydrogen)	1.1 kg hydrogen / 100 km	59
GM EV1 (electricity)	20 kWh electricity / 100 km	105

*EPA combined city / highway driving cycle
MPGGE: miles per gallon gasoline equivalent

Table 2 Description of Hydrogen Production Technologies

Technology Description	Natural Gas Consumption	Electricity Consumption
Centralized steam methane reformer (Caloric)	3.2 kg natural gas / kg hydrogen	0.32 kWh / kg hydrogen
Decentralized steam methane reformer (H$_2$Gen Innovations)	3.2 kg natural gas / kg hydrogen	1.4 kWh / kg hydrogen
Decentralized Electrolyzer	N/A	50 kWh / kg hydrogen

DISCUSSION OF RESULTS

For the sake of brevity, only the air emissions results from the Toronto light-duty vehicle scenarios will be discussed in detail. Specific references will be made, however, to both the Vancouver and Calgary scenarios where they contribute a unique perspective.

A summary of the life-cycle air emissions for the Toronto scenario is presented in Fig. 3 at the end of the paper. The chart has been normalized to show the relative performance of each scenario compared to the incumbent technology (gasoline ICE LDVs).

REGIONAL CONSIDERATIONS

Most of the air pollutants investigated are of a regional nature; their impacts can occur several hundred kilometers from the emission source. Therefore, the following discussion differentiates emissions based on location. For the Toronto scenario, any emissions occurring between Windsor, Ontario and Quebec City, Quebec, including emissions in Toronto itself, are considered to be within the same region, namely the Windsor-Quebec City region. The pollutants that will not be discussed on a regional basis are GHG emissions (these have global impacts).

BASE CASE – GASOLINE ICEV (FORD FOCUS)

The base case, or basis for comparison, throughout the LDV scenarios is a Ford Focus gasoline ICEV.

In reviewing the life-cycle air emissions from the base case scenario, it can be seen that there are four primary emission sources, as shown in Fig. 4. The portion of the life-cycle emissions that remains undefined can be attributed to the unit processes not shown here and is typically less than 10% of the total. For the Toronto scenario, both vehicle operation and crude oil refining occur in the Windsor-Quebec City region.

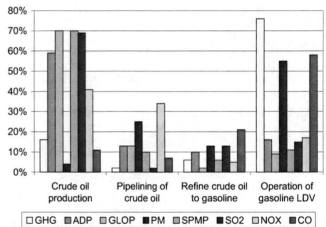

Percentage indicates the fraction of total life-cycle emissions allocated to each unit process.

Figure 4   Major Emission Sources for the Gasoline ICE Light-Duty Vehicle Base Case - Toronto Scenario

For GHG emissions, the majority (76%) are produced during vehicle operation. The remainder of emissions are a result of upstream fuel production. In populated centres, where air quality is a significant concern, gasoline light-duty vehicles, along with gasoline production and distribution, have been identified as important sources of many of the criteria air pollutants. The significant emissions of certain air stressors at the source of crude oil production (ADP, GLOP, SPMP, SO$_2$

and NO$_X$) are also cause for concern. Air quality issues will likely worsen in these areas as oil production continues to expand.

DIESEL ICE LDV

In the comparison of the diesel Volkswagen Jetta to the gasoline Ford Focus, the increased fuel efficiency of the compression ignition diesel engine is reflected in the 20% reduction in life-cycle greenhouse gases, as shown in Fig. 3, but the potential impact on air quality is not as well defined. Fig. 5 shows the increase in local air pollutant emissions by the diesel ICE system in the Windsor-Quebec City region. ADP, GLOP, SPMP and NO$_X$ emissions increase primarily because NO$_X$ emissions are 8.5 times higher in the diesel Jetta than in the gasoline Focus. PM emissions in the cities are similarly affected because it is the nature of diesel combustion to produce particulates, unlike gasoline combustion. There are currently regulations in place in Canada to reduce diesel vehicle emissions to standards set for gasoline vehicles by 2008; however, gasoline vehicles are likely to remain lower emitters given their current advantage over diesel vehicles.

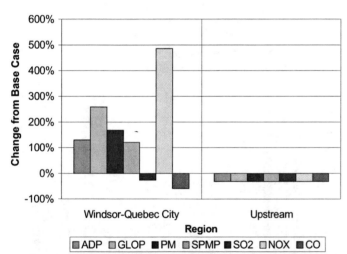

Figure 5   Change in Regional Emissions for the Diesel Jetta Compared with the Gasoline Focus – Toronto Scenario

The most noticeable improvement in criteria air pollutants is in carbon monoxide emissions, which are 78% less from the tailpipe for the diesel Jetta than for the gasoline Focus. The relative levels of sulphur dioxide emissions are dependent primarily on emissions from oil production, with some influence also from refinery and vehicle emissions. The overall result is a 26% decrease in life-cycle SO$_2$ emissions when compared with the base case.

Overall, selection of a diesel ICE LDV in any of the cities is likely to significantly increase ADP, GLOP, PM, SPMP and NO$_X$ emissions within the city. A doubling of air pollutants from light-duty vehicles would have a significant impact on air quality. At the same time, selection of a diesel ICE LDV instead of a gasoline ICE LDV will significantly benefit regions of upstream fuel

production, through reduced crude oil demand, and the global environment, through the reduction of life-cycle GHG emissions.

It should be kept in mind that a gasoline Jetta is approximately 14% less fuel-efficient than the baseline vehicle, a comparable gasoline Focus.

NATURAL GAS ICE LIGHT-DUTY VEHICLE

The natural gas–powered Honda Civic GX offers significant life-cycle emissions reduction in all of the stressor categories evaluated compared with the gasoline-powered Ford Focus. The greenhouse gas emission reductions amounted to 34% in Toronto, while the reductions in Vancouver and Calgary were 39% and 35% respectively. The two primary reasons for the difference in emission reductions for the three cities are the different electricity sources used to meet the compression requirements of NGVs, and the varying distances of natural gas transport. Fig. 6 shows the change in regional emissions for all three scenarios when compared with the base case. In the major urban regions a greater than 80% decrease in emissions occurs in almost every stressor category with the exception of CO (greater than 50% reduction) and PM (brake and tire wear is assumed to be identical for the two vehicles). Fig. 3 shows the reductions in life-cycle emissions for the natural gas LDV scenario in Toronto, with the largest decreases (greater than 60%) occurring in GLOP, SPMP and CO.

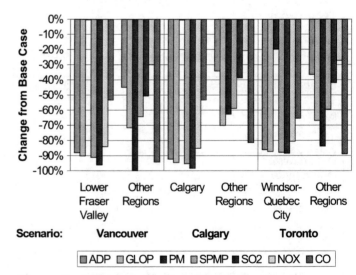

Figure 6   Change in Regional Emissions for the Natural Gas Civic Compared with the Gasoline Focus

It should be kept in mind that a gasoline Civic is approximately 16% more fuel-efficient than the baseline vehicle, a comparable gasoline Focus.

GASOLINE HYBRID ELECTRIC VEHICLE (HEV)

The Toyota Prius HEV consumes 43% less gasoline than the Ford Focus, allowing it to correspondingly reduce its life-cycle greenhouse gas emissions and its upstream non-greenhouse gas emissions. However, the non-greenhouse gas emissions from the Prius itself are, in some cases, higher than those of the Focus. The result is an increase in NO_X tailpipe emissions within the regions of vehicle operation as shown in Fig. 7, while others decrease substantially; PM, and SO_2 have the largest decrease. Overall, this drives the life-cycle air emission reductions to a minimum of 24% for NO_X and greater than 37% for GHG, ADP, PM, SPMP and SO_2 emissions.

It should be kept in mind that a gasoline Toyota Corolla, with similar body size to the Prius, is approximately 16% more fuel-efficient than the baseline vehicle, a comparable gasoline Focus.

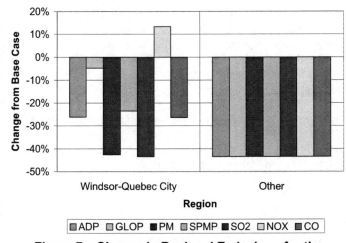

Figure 7   Change in Regional Emissions for the Prius HEV Compared with the Gasoline Focus – Toronto Scenario

METHANOL FCV

Fuel cell vehicles with on-board methanol processing (MFCVs), also known as Indirect Methanol FCVs, have nearly zero tailpipe emission of criteria air contaminants, but on a life-cycle basis, current technologies result in an increase or little change in some stressor categories, while the majority decrease significantly when compared with the gasoline Focus base case. Because the NECAR 5 MFCV is currently in the testing and development stage, the methanol fuel consumption has not yet been fully optimized. The authors roughly estimate a potential reduction of 17% from the results presented through a vehicle weight reduction program for the NECAR 5. Future technology advances may reduce the fuel consumption of MFCVs even further.

Nonetheless, the NECAR 5 does have the potential to significantly reduce vehicle tailpipe emissions in the city of operation, as shown in Fig. 8. Exceptions are in PM emissions, where brake and tire wear are equal for both vehicles, and VOC emissions, which contribute to ozone and secondary PM production. VOC emissions from the evaporation of methanol were assumed to be 60% of gasoline on a volume basis, but, since the energy density of gasoline is higher than methanol, a larger volume of methanol is required per kilometre travelled, thereby producing more VOC emissions during fuel storage and transmission. A more varied change in air emissions occurs in the fuel production and distribution unit processes. The majority of these emissions occur in the categories marked "Other Regions," as shown in Fig. 8. The largest spatial change in emissions will undoubtedly occur from the regions of oil refining to the regions of methanol production. The 66% increase in CO emissions in the upstream regions of the Toronto scenario are due to the fact that gasoline system's refinery emissions are included within the city region for this scenario while the methanol plant emissions does not. There are also noticeable changes in both the oil- and gas-producing regions, due to the change in production from crude oil to natural gas, and along transportation routes where rail is used instead of pipelines to transport methanol long distances.

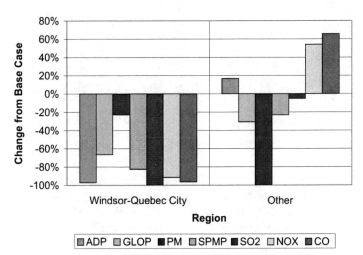

Figure 8   Change in Regional Emissions for the Methanol NECAR 5 Compared with the Gasoline Focus – Toronto Scenario

On a life-cycle basis, as shown in Fig. 3, GHG emissions decrease approximately 8%, while ADP decreases 14%. NO_X emissions are of particular concern since they increase 20%, mostly due to methanol and natural gas production. The remainder of the stressor categories present noticeable decreases, with CO seeing by far the largest drop (63% to 70%), due primarily to the high CO emissions from gasoline vehicles.

In short, there are significant reductions in both vehicle emissions and some life-cycle emissions resulting from the operation of the NECAR 5, however, there is also little change in life-cycle GHG and ADP emissions and a significant increase in upstream NO_X emissions.

It should be kept in mind that a gasoline Mercedes A-Class, which the NECAR 5 is based, is approximately 10% less fuel-efficient than the baseline vehicle, a comparable gasoline Focus.

DECENTRALIZED SMR-BASED FCV

The life-cycle emissions of the decentralized steam methane reformer (SMR)–based NECAR 4a scenarios are dependent on the electricity source used to power both the reformer and the hydrogen compressor (approximately 4.4 kWh per kg hydrogen).

The Vancouver scenario demonstrates a system that uses zero-emission electricity production to supply 90% of the electricity requirements. As shown in Fig. 9, this results in substantial decreases in emissions in all stressor categories ranging from 34% to 97% when compared with the gasoline Focus base case.

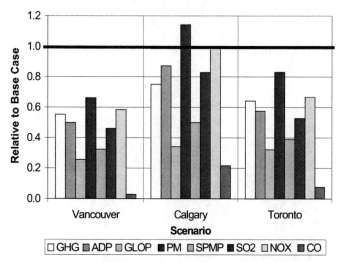

Figure 9 Life-Cycle Emissions for the Hydrogen NECAR 4a Using Decentralized SMR Relative to the Gasoline Focus

Alternatively, the Calgary scenario demonstrates the effect that electricity generation from almost solely fossil fuels has on the decentralized SMR system. The life-cycle performance of the Calgary scenario results in a range of emission levels when compared with the gasoline Focus base case, from a 14% increase in PM to a 78% decrease in CO, also shown in Fig. 9. As seen in Fig. 10, some of the air emission reductions in the Calgary region result in increases of air emissions in other regions, mostly those of electricity generation and resource extraction within Alberta.

In the Toronto scenario, the combination of low emission sources of electricity and fossil fuel–based electricity results in life-cycle emissions between the Vancouver and Calgary results. However, the emission reduction in the Windsor-Quebec region is not nearly as substantial as those in the Lower Fraser Valley and in Calgary. This is due to the fact that a significant amount of electricity comes from fossil fuel power plants within the region of vehicle operation.

Given the relatively low level of maturity of fuel cell vehicles and small-scale steam methane reformers, they are expected to have efficiency improvements that will decrease the life-cycle emissions of this system. It

should be kept in mind, however, that the small-scale SMR technology used for this system is still a prototype unit. It is claimed to have lower natural gas consumption but higher electricity consumption than many commercial large-scale SMR units.

In comparison with other vehicle and fuel supply systems, the decentralized SMR-based system using the NECAR 4a has life-cycle emissions ranging from slightly lower than the natural gas Civic system in the Vancouver scenario to noticeably higher in the Calgary scenario.

Ultimately, the environmental performance of the decentralized SMR FCV system will depend on the overall system design. If low-emission electricity sources are used to power the reformer and compressor, the system provides major emission reductions in all stressor categories. If a mix of low emission and fossil fuel–based electricity sources are used, then the environmental burden can either be decreased or increased and, depending on where the electricity generation takes place, a large portion of the emissions can be shifted into other regions.

It should be kept in mind that a gasoline Mercedes A-Class, which the NECAR 4a is based, is approximately 10% less fuel-efficient than the baseline vehicle, a comparable gasoline Focus.

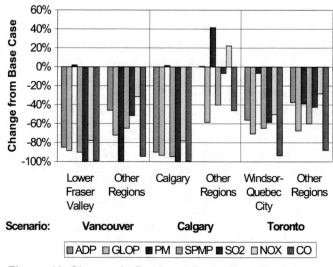

Figure 10 Change in Regional Emissions for the Hydrogen NECAR 4a Using Decentralized SMR Compared with the Gasoline Focus

CENTRALIZED SMR-BASED FCV

The centralized production of hydrogen using steam methane reformers presents life-cycle results that are very similar to the decentralized system, except that there is approximately 47% less electricity required for the reformer and the hydrogen compressors, and 8% more natural gas required due to the SMR fuel conversion efficiencies. Once again, the Vancouver and Toronto scenarios show lower life-cycle emissions than the Calgary scenario, as shown in Fig. 11, due to the source of electricity used in each province. Also similar to the decentralized SMR system is the fact that the centralized SMR system compares closely with the natural gas Civic system. Accordingly, fuel cell vehicles are expected to have efficiency improvements as the technology matures, thus decreasing the life-cycle emissions of this system. likely to have similar life-cycle results are photovoltaics, solar thermal power, low-impact hydroelectricity, geothermal power, tidal power and wave power (all of these technologies will have different life-cycle emissions from their raw material extraction, manufacturing and maintenance, which are outside the system boundary in this LCVA).

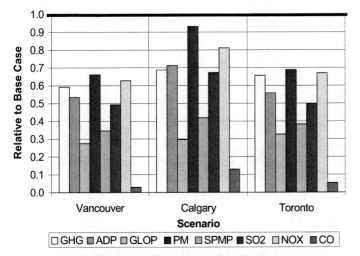

Figure 11 Life-Cycle Emissions for the Hydrogen NECAR 4a Using Centralized SMR Relative to the Gasoline Focus

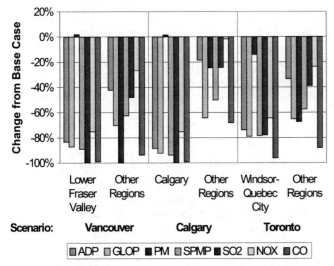

Figure 12 Change in Regional Emissions for the Hydrogen NECAR 4a Using Centralized SMR Compared with the Gasoline Focus

Currently, the regional breakdown of results, as shown in Fig. 12, shows a reduction in emissions for every stressor category in all regions with a few exceptions:

PM emissions in Vancouver and Calgary and NO_X emissions in Calgary's upstream regions show little change owing to the vehicle brake and tire wear, and the coal power plant emissions, respectively.

ELECTRICITY-BASED VEHICLE OPERATION (FCV & EV)

The main advantage of using electrolysis to produce hydrogen for fuel cell cars or using battery electric cars is that they produce zero emissions at the vehicle, with the exception of particulate matter from brake and tire wear. These systems shift most of the environmental air emissions of vehicle operation away from the vehicle and to the production of electricity. It should be kept in mind that the following discussion is limited to the specific stressor categories chosen for quantification in this study and will not reflect their entire life-cycle environmental, social and economic impacts. For example, wind-powered vehicle systems for FCVs and EVs will have nearly identical air emissions, but the issues surrounding resource consumption, vehicle range and cost will be very different. A qualitative discussion of further issues can be found in the full LCVA report[1].

The life-cycle air pollutant emissions of the EV1 compared with the NECAR 4a FCV rely, for the most part, on the relative electricity requirements for each system. Overall, the NECAR 4a requires 2.9 times more electricity for hydrogen produced through electrolysis than the EV1 needs to travel an equivalent distance. Since all of the air emissions for these two systems, except PM from brake and tire wear, are produced through electricity generation and its upstream activities, there is a direct correlation between the systems' life-cycle emissions and electricity generation requirements. This relationship applies to all of the electrolysis and electric vehicle systems, but is most evident in the natural gas and coal–based systems, as opposed to hydroelectricity, wind, and nuclear systems with almost no life-cycle air pollutant emissions.

WIND-BASED ELECTRICITY (FCV & EV)

Wind power generation is a zero-emission electricity generation technology. The only emission quantified from the operation of the entire system was the generation of particulate matter from brake and tire wear. PM emissions are lower for the EV1 than the NECAR 4a FCV because it was assumed that the EV1 would use 30% regenerative braking on average to recover and store energy from the wheels, and thus reduce the particulate matter generated from mechanical braking. Other zero-emission electricity generation technologies

HYDROELECTRICITY-BASED ELECTRICITY (FCV & EV)

Conventional hydroelectricity power generation with the use of large human-made reservoirs also resulted in extremely low life-cycle emissions. Emission of greenhouse gases, however, does occur from most natural water bodies and an effort was made to quantify these amounts. The resulting life-cycle GHG emissions from a hydroelectricity-based FCV or EV system are approximately 2% of the base case in Ontario and effectively zero in British Columbia.

NUCLEAR POWER-BASED ELECTRICITY (FCV & EV)

Nuclear power generation is another technology with very low life-cycle air emissions. Even though uranium needs to be mined, transported and processed, the emissions per unit of electricity generated are extremely small because the amount of uranium required is small. In comparison with the base case, emissions from a nuclear power–based vehicle system in Toronto are nearly zero, with the exception of PM emissions from vehicle brake and tire wear. The stressor category with the next largest relative emission was sulphur dioxide, with less than 2% of the base case emissions.

NATURAL GAS-BASED ELECTRICITY

Natural gas power can be used in all three provinces for incremental power generation. This means if additional electricity demand were to be placed on the grid, it may result in the increased output of natural gas power plants. In both British Columbia and Ontario, it was assumed that selective catalytic reduction (SCR) would be used in conjunction with low NO_X burners to reduce the plant's NO_X emissions. This is a result of recent trends in both provinces to reduce the emission of criteria air contaminants in densely populated areas. In Alberta, it is assumed only low NO_X burners will be used, as this is the easiest method of achieving the legislated requirements.

NATURAL GAS-BASED ELECTROLYSIS (FCV)

Using natural gas power plants for hydrogen generation through electrolysis shifts nearly all air emissions away from the vehicle tailpipe to the natural gas power plants and upstream natural gas supply system. In the regions of vehicle operation, there is a decrease in all stressor categories, as shown in Fig. 13, except PM, which increases in two of the regions due to power plant emissions. The remaining regions all demonstrate either an increase or little change in ADP, SO_2 and NO_X emissions. For the Calgary scenario, NO_X emissions in Alberta are over 2.5 times as large as the base case emissions due to the lack of selective catalytic reduction of NO_X at the power plants. The other stressor categories, GLOP, SPMP and CO, either decrease or have little change in most regions, and on a life-cycle basis, all decrease between 35% and 83%. Conversely, the life-cycle emissions of GHG, ADP, PM and NO_X generally increase with some small exceptions (ADP and PM emissions in the Toronto scenario decrease slightly due to fewer emissions for natural gas pipelining compared with crude oil pipelining).

Overall, fuel cell vehicles powered with natural gas–based electrolysis show considerable benefit to the cities where vehicle operation takes place, but on a life-cycle basis, the results show both benefits and disadvantages to choosing such a system.

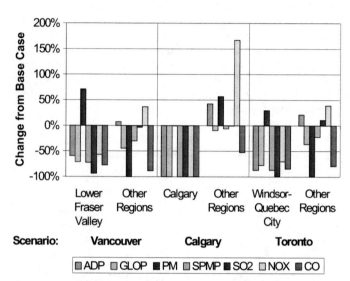

Figure 13 Change in Regional Emissions for the Hydrogen NECAR 4a Using Natural Gas-Based Electrolysis Compared with the Gasoline Focus

NATURAL GAS-BASED EV

Operation of a GM EV1 that is recharged from a natural gas power plant results in a significant improvement in air pollutant emissions over the base case in nearly all stressor categories in the various regions evaluated. There is an obvious shift in pollutant sources from the areas of vehicle operation and gasoline production to the power plant locations. However, both the life-cycle and regional emissions shown in Figs. 3 and 14 respectively demonstrate an overall benefit for the electric vehicle, due primarily to the increased fuel efficiency of the EV1 over the Focus and the superior emissions control capabilities of a large power plant compared with a vehicle.

One area of potential concern given the results is the location of natural gas electricity generation in Alberta. Many of the emissions from Calgary, Edmonton, and the oil and gas field are shifted, in part, to the power plants. This may result in a concentration of emissions in areas of large power plants or several smaller ones, thus having a more intense environmental influence in those areas.

Overall, however, the results show that the selection of an EV recharged from a natural gas power plant will have a large net benefit to the environment in all three of the scenarios. With the exception of NO_X emissions in the Calgary scenario, every stressor category demonstrates a reduction of life-cycle emissions of between 44% and 94%.

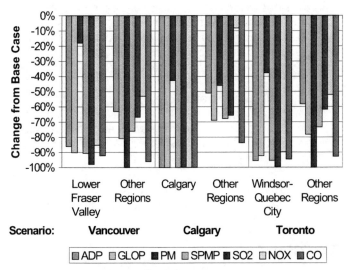

Figure 14 Change in Regional Emissions for the EV1 Using Natural Gas-Based Electricity Compared with the Gasoline Focus

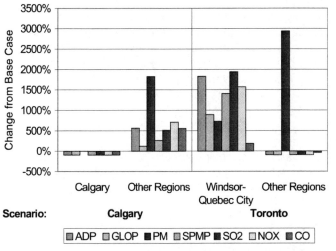

Figure 15 Change in Regional Emissions for the Hydrogen NECAR 4a Using Coal-Based Electrolysis Compared with the Gasoline Focus

COAL-BASED ELECTRICITY (FCV & EV)

The life-cycle results show that a transportation system powered by coal emits a very large amount of air pollutants when compared with the base case. On a life-cycle basis, stressor category emissions increase between 54% and 1400% for the FCV scenarios, and up to 440% in the EV scenarios. The EV scenarios do show some decrease in life-cycle emissions for GLOP, CO and GHG (Toronto scenario only) emissions, but the remainder of the stressor categories all increase. As shown in Figs. 15 and 16, these systems do shift many of the air pollutants almost entirely away from particular regions (the Calgary region, and the resource production regions for the Toronto scenario). However, these emissions are generally many times more intense at the coal power plants. Since there are only a handful of large coal power plants in Alberta and Ontario, the total

life-cycle emissions will be concentrated in the areas surrounding the plants. Particulate matter is emitted at high levels within areas of both coal mining (particularly the northeastern U.S. Appalachian region for the Toronto scenario) and electricity production.

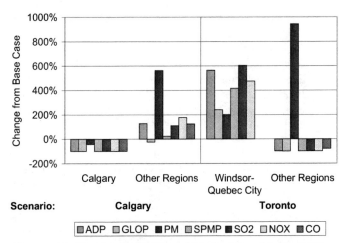

Figure 16 Change in Regional Emissions for the EV1 Using Coal-Based Electricity Compared with the Gasoline Focus

CONCLUSIONS

The conclusions reached within this LCVA based on the analysis of the environmental, economic, social and technical aspects of each system are listed below. Please refer to the full LCVA report[1] for further detail regarding the stressor categories and scenarios not discussed in detail within this paper.

- Fuel cell vehicles fuelled with hydrogen from renewable energy-based electrolysis show the greatest opportunity for minimizing negative environmental and social impacts of vehicle / fuel supply systems. However, at the current level of technology maturity and assuming mass production economies of scale, fuel costs are estimated to be higher for electrolysis-based systems than for conventional vehicles.

- Where renewable energy is not available, steam methane reforming (SMR) technology is the next most environmentally benign source of hydrogen, although distribution logistics for centralized plants and the operational issues of decentralized plants remain as hurdles. SMR-based FCV systems are also estimated to have fuel costs comparable to fuel costs for gasoline light-duty vehicles, but higher than diesel bus fuel costs at the current level of technology maturity.

- Liquid hydrogen requires a considerable increase in electricity compared with gaseous hydrogen, due to liquefaction, which adversely affects overall life-cycle environmental performance and fuel cost.

- Natural gas–based electrolysis for hydrogen FCVs has both advantages and disadvantages when considering their life-cycle air emissions. For most of the natural gas–based light-duty vehicle (LDV) scenarios, air emissions increase or show little change in as many stressor categories as they decrease in, whereas in the bus scenarios, life-cycle GHG and SO_2 emissions increase while the other air emissions decrease.

- Electric vehicles, both personal and trolley buses, use electricity more efficiently than producing hydrogen from electrolysis, but face major social, technical and economic challenges. Environmental performance is highly dependent on the source of electricity.

- Coal-based electricity for vehicle fuels demonstrates major disadvantages over conventional gasoline or diesel, on a life-cycle basis.

- Nuclear power plants have both positive and negative environmental and social attributes compared to other sources of transport energy. Nuclear power-based systems have near zero lifecycle air emissions, but create radioactive waste with negative long-term safety, security and environmental impacts.

- Currently, indirect methanol fuel cell vehicle technology demonstrates its ability to nearly eliminate vehicle tailpipe emissions, but has limited ability to significantly reduce lifecycle emissions in several categories. This is expected to improve as the technology matures.

- Natural gas vehicles have some of the lowest environmental impacts of any internal combustion engine vehicles and have the potential to be cost-competitive, but would face fuel infrastructure and storage challenges with wide adoption.

- Hybrid Electric Vehicles (HEVs) with internal combustion engines, battery storage and electric drive motors present an opportunity to reduce life-cycle impacts with no change to fuel infrastructure. HEV technology is an effective method of improving overall system performance and could be applied to most vehicle types. HEVs currently have higher vehicle costs, but lower fuel costs than conventional ICEVs.

- The benefits and disadvantages of selecting a diesel ICEV over a gasoline ICEV need to be evaluated on an individual basis as there are trade-offs in performance with no clear overall benefit of one over another.

- The use of oilsands instead of conventional oil sources has both benefits and disadvantages. Production of crude oil from oilsands requires more energy input, thus creating higher GHG and NO_x emissions. Conversely, oilsands operations create less volatile organic compounds and sulphur emissions than conventional oil production. Oilsands operations also concentrate the effects of oil production activities in particular areas.

- Regional considerations are very important to system performance. The shifting of environmental, social and economic burdens and benefits from one region to another are evident in a number of the systems analyzed.

- Full systems analysis and design is critical to ensuring that environmental, social and economic performance is optimized.

REFERENCES

1. Row, J., et. al. 2002. "Life-Cycle Value Assessment (LCVA) of Fuel Supply Options for Fuel Cell Vehicles in Canada." Pembina Institute for Appropriate Development. Drayton Valley, AB, Canada. www.pembina.org/publications_item.asp?id=131
2. Sullivan, J., et al. 1998. "Life Cycle Inventory of a Generic U.S. Family Sedan – Overview of Results USCAR AMP Project." Society of Automotive Engineers, Inc. Warrendale, PA, USA.

CONTACT

Jesse Row
Eco-Efficiency Technology Analyst
Pembina Institute
E-mail: jesser@pembina.org

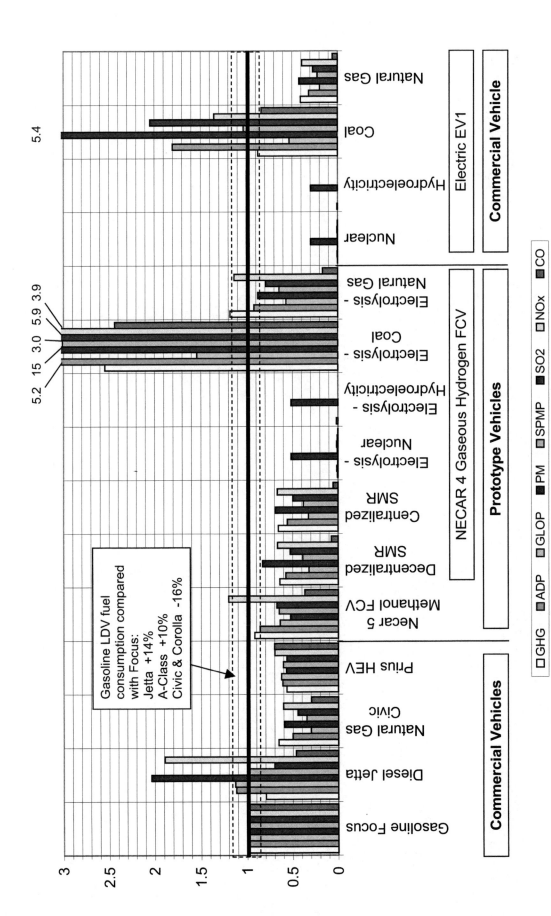

- Methanol FCV expected to reduce fuel consumption by up to 17% through weight optimization.
- PM emissions from vehicle brake and tire wear present for all vehicle types.

Figure 3   Normalized Life-Cycle Emissions for Light-Duty Vehicles in Toronto

57

2004-01-0788

Raison d'Être of Fuel Cells and Hydrogen Fuel for Automotive Powerplants

Antoni K. Oppenheim
University of California

Harold J. Schock
Michigan State University

ABSTRACT

The paper presents reportage of the debate on the topic expressed by its title that was held as a special session at the SAE 2003 Congress, supplemented by commentaries on its highlights. The debate brought to focus the fact that fuel cells are, indeed, superb powerplants for automobiles, while hydrogen is at the pinnacle of superiority as the most refined fuel. The problems that remained unresolved, are: (1) when fuel cells will be practically viable to replace internal combustion engines and (2) under what circumstances hydrogen, as the ultimate fuel, will be economically viable in view of its intrinsically high cost and hazards engendered by its extraordinary flammability and explosive tendency. This state of affairs provokes a fascinating question: is there another way to resolve the formidable problems imposed by the exorbitant demand for oil and excessive emission of pollutants imposed upon the automotive technology than abandoning internal combustion engines in favor of fuel cells and replacing hydrocarbons by hydrogen? Some answers to it are provided in the Commentaries.

FOCUS

Optimizing efficiency of automotive power plants, combined with their environmentally benign operation, is today at the forefront of interest to the society in our times. Major attention in this respect has been focused recently upon fuel cells, especially their low-temperature, PEM (Proton Exchange Membrane) version using hydrogen for fuel. The demand for hydrogen became therefore of prime concern. Hydrogen can be provided either by on-board reformation from a hydrocarbon, or, directly by distribution from refineries. Contemplation of hydrogen in a fuel-cell-powered vehicles, in comparison to its use in internal combustion engines as well as hybrid powerplants, be they piston engine and/or battery-powered, became thus of topical concern. The dilemma is, moreover, exacerbated by concomitant need for modification of infrastructure involving the high cost replacement of oil-based fuel distribution system by that for hydrogen, let alone the problems of its storage and handling due to hazards caused by its exceptional flammability and detonability as the most refined fuel with a well established reputation as an explosive material. It is consideration of the engineering aspects of these issues that provided the major purpose of the debate.

PANEL

The distinguished roster of panelists was comprised of Syed M. Shahed, President of SAE and Vice-President of Garrett-Honeywell, Rodica A. Baranescu, Past-President of SAE and Chief Engineer of International Truck and Engine Corporation, Stanford S. Penner, Professor of Engineering Physics at the University of California in San Diego, a highly respected authority in energy technology and the Founder Editor-in-Chief of the international journal on ENERGY, John R. Wallace, a well known Consultant in Alternative Fuel Vehicle Programs at Ford Motor Company, Richard Stobart, Professor of Engineering at the University of Sussex and chair of the Committee on Advanced Powerplants, the promoter and sponsor of this event.

The co-organizer of the debate was Harold J. Schock, Professor of Mechanical Engineering and Director of the Automotive Research Experiments Station at the Michigan State University, while the organizer and moderator of the debate was Antoni K. Oppenheim, Professor of Engineering at the University of California in Berkeley.

PRESENTATIONS

The session was started by **Shahed** who, upon outlining the mission of SAE as one "To advance the mobility community to serve humanity," and tracing the evolution of automobiles powered by internal combustion engines, identified the crucial issues facing us in terms of the following questions.

• In view of advanced micro-electronics, how much control can be exercised over combustion and with what results - at what cost?

• How vigorously should this be pursued before I.C. engines are "abandoned" in favor of alternatives?
• What is the right balance of effort and expectation between fuel cell and advanced IC engines ... with or without hydrogen?

Rodica **Baranescu** pointed out the distinguished position held today by the supply of crude oil that is plentiful for decades and the mature engine technology manifested especially by the phenomenal progress made since 1974 in reducing pollutant emissions of heavy-duty diesel engines. Thereupon she expressed the view that "engine only development approach is hitting a dead end" in-so-far as a tenfold reduction in both NOx and PM is expected – a true opinion shared widely through the world by the executives of automotive industry. After enumerating the extremely tough challenges facing the industry up to 2007, she concluded that "the internal combustion engine will continue to meet transportation needs as long as liquid fuels are available," by future developments in engine + catalytic processing of exhaust + ultra low sulfur fuel + vehicle optimization.

Sol **Penner** provided a well-documented, authoritative résumé of the steps leading to the production of hydrogen fuel and the operating principles of PEM (Proton Exchange Membrane) fuel cells for automotive applications. Besides providing a review of numerous studies of fuel cells and hydrogen fuel (vid. e.g. [1]), he recounted the following list of milestones in the evolution of fuel cell technology.

1839 W.R. Grove invented the "gaseous voltaic battery"
1889 L. Mond and C. Langer described a single cell FC
1894 W. Oswald called for replacement of inefficient steam engines
1986 & 1995 S.S. Penner et al, DOE studies on research needs and commercialization
1998 F.R. Kalhammer et al "Status and Prospect of Fuel Cells as Automobile Engines", Calif. Air Resources Board
2002 F.R. Kalhammer "FCs for Transportation Applications", Encyclopedia of Physical Science &Technology

A Schematic of an electrochemical cell is presented by Fig. 1. On the left, denoted by c, is a platinum-black electrode approximately 30 cm^2 in surface area. On the right, marked by a, is a titanium dioxide (TiO_2) semiconductor mounted on an indium plate, which serves as an electrode contact material. The TiO_2 is exposed to light (hν). Symbol d points out an external load across which the voltage was measured with a voltmeter V, while b describes a suitable electrolyte.

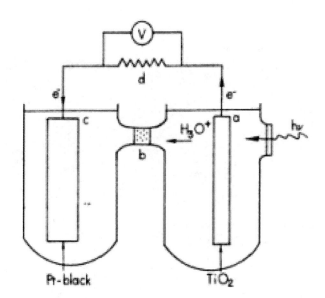

Fig. 1. Schematic of a Rudimentary PEM (Proton Exchange Membrane) Fuel Cell

The following reactions take place

$$TiO_2 + 2\,h\nu \rightarrow TiO_2 + 2\,e^- + 2\,p^+,$$
$$2\,p^+ + H_2O\,(l) \rightarrow (1/2)\,O_2\,(g)\uparrow + 2\,H^+,$$
$$2\,H^+ + 2\,H_2O\,(l) \rightarrow 2\,H_3O^+\,(l),$$
$$2\,e^- + 2\,H_3O^+\,(l) \rightarrow H_2\,(g)\uparrow + 2\,H_2O\,(l),$$
$$H_2O\,(l) + 2\,h\nu \rightarrow H_2\,(g) + (1/2)\,O_2\,(g).$$

Among the contemporary schemes considered for production of hydrogen, he presented two cycles studied recently in Japan, displayed in Fig. 2, and provided a comprehensive list of splitting processes for hydrogen production in Table 1, as well as schematics of hydrogen production from waste heat of a fission reactor displayed in Fig. 3.

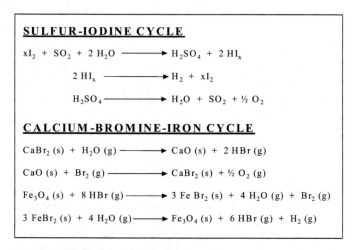

Fig. 2. Two Cycles Studied in Japan

Table 1: Characteristics of Splitting Process for Hydrogen Production

Splitting Process	Energy Required [kWh/N^3 of H$_2$]		Status	Efficiency [%]	Costs Relative to CH$_4$ Steam Reform	Fraction of Production [%]
	In Theory	In Practice				
Methane Steam Reforming	0.78	2 – 2.5	Mature	70 – 80	1	48
Methane Cracking			Mature	54	(0.9)	
Partial Oxidation of Heavy Oil	0.94	4.9	Mature	70	1.8 (1.6)	30
Naphtha Reformation			Mature			
Coal Gasification (TEXACO)	1.01	8.6	Mature	60	1.4 – 2.6 (2.3)	
Partial Oxidation of Coal			Mature	55		18
HYDROCRAB			R&D		(0.9)	
Steam Iron			R&D	46	(1.9)	
Chloralkali Electrolysis			Mature		By product	
Water Electrolysis	3.54	4.9	R&D Mature	27[3]	5 – 10 (3.0)	4
High Temperature Electrolysis			R&D	48	(2.2)	
Thermochemical Cycles			Early R&D	35 – 45	6	
Biomass Conversion			Early R&D		2 – 2.4	--
Photolysis			Early R&D		< 10	

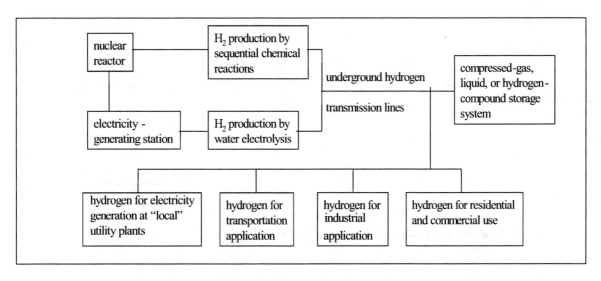

Fig. 3. Schematic of Hydrogen Production from Waste Heat of a Fission Reactor

John **Wallace** recounted the popularly accepted reasons for the introduction of fuel cell vehicles, citing such overwhelmingly attractive advantages as environmental cleanliness, low noise, comfort, low maintenance cost, reliability, and high efficiency. He outlined thereupon a long term vision offered by Freedom CAR within a relatively modest budget of $150 million. The goals and challenges he visualized to be addressed thereby are as follows.

Goals: Fuel Cells

- For direct hydrogen fuel cell system, achieve energy efficiency of 60%, and power density of 325 W/kg and 220W/L (including hydrogen storage).
- For reformer fuel cell systems, achieve peak energy efficiency of 45%, and meet emissions standards.
Cost targets of $45/kW by 2010 and $30/kW by 2015.
- Competitive performance with internal combustion engines (cold start-up time, reliability/durability, etc.).

Technical Challenges: Fuel Cells

- Low cost and durable electrolyte membranes extended for higher temperature operation and dry conditions.
- Electrode with low loading of precious metal catalyst.
- Manufacturing processes for high volume production of MEA and bipolar plates.

- Reduced complexity of system.
- High efficiency air compressors.
- Mature analytical tools.

Goals: Hydrogen Storage & Refueling

- Available capacity of 6 weight percent hydrogen, specific energy of 2000 Wh/kg and an energy density of 1100Wh/liter at a cost of $5/kW-hr.
- H2 refueling with developed commercial codes & standards and diverse renewable & non-renewable energy sources. Targets of 70% energy efficiency, well-to-pump, at a cost equivalent to gasoline at market prices.

Technical Challenges: Hydrogen Storage & Refueling

- Develop on-board hydrogen storage system capable of meeting efficiency goal of at least six (6) weight percent hydrogen.
- Many potential technological alternatives.
- Major challenge given weight and volume constraints for vehicles to meet customer expectations.
- Critical need to simultaneously develop hydrogen fuel infrastructure.

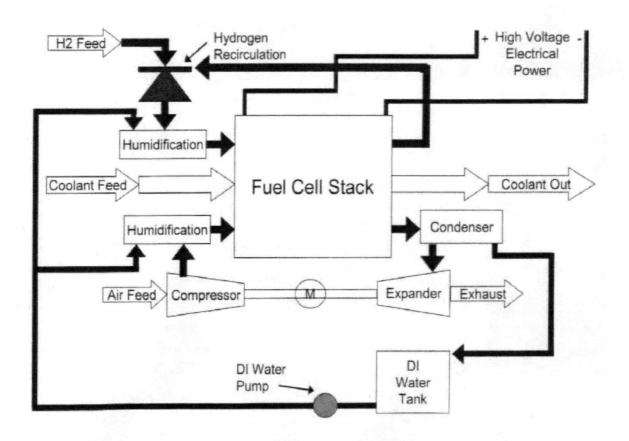

Fig. 4. Block Diagram of a Fuel Cell System

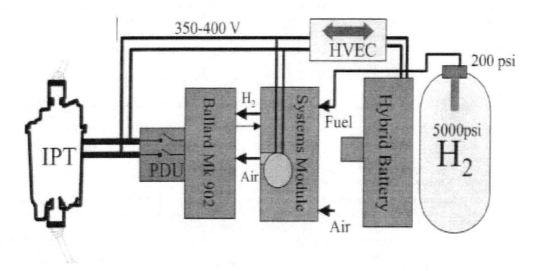

Fig. 5. Hybrid Fuel Cell electric Vehicle in Focus at Ford

A system diagram of a fuel cell is presented by Fig. 4, while Fig. 5 displays the architecture of a hybrid fuel cell electric vehicle in focus at Ford. The function of the hybrid battery is to execute the start up and shut down, as well as to provide peak power and the regenerative braking energy.

Richard **Stobart**, a world-renowned authority in the technology of fuel cells, presented a resumé of their fuel economy in comparison to other automotive powerplants illustrated by Fig. 6 taken from the paper of Stodolsky et al [3]. He concluded by the stunning pronouncement:

"Even with the most optimistic assumptions, the fuel cell powered vehicle offers only a marginal efficiency improvement over the advanced CI/hybrid and with no anticipation yet of future developments of IC engines. At $100/kW, the fuel cell does not offer a short term advantage even in a European market after Euro5."

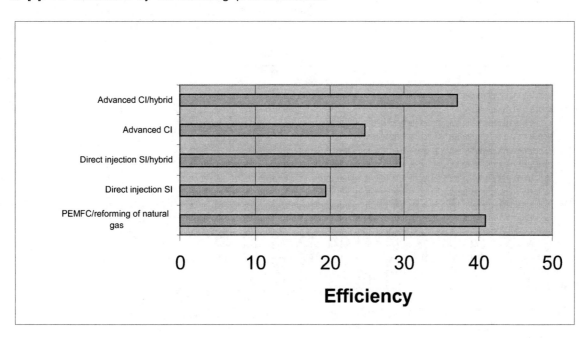

Fig. 6. Total Fuel Cycle Efficiency

In reviewing the prospective of fuel cells, **Oppenheim** pointed out that they are, indeed, supreme powerplants for all prime movers, let alone automotive vehicles. Their technology is well established and their usefulness impressively demonstrated in advanced engineering systems such as the space shuttle for which they were developed. To launch the debate, he brought forth the awkward problem of energy economy for automotive powerplants. An overall view is provided in Table 2.

Fuel cells are at the threshold between the economies of hydrocarbons and atoms. They will come into their own when the energy supply becomes inexhaustible, as it may at the age of atomic fusion. In the meantime we are faced with hydrocarbons as the major energy resource.

Table 2. Energy Economy for Automotive Powerplants

Resource	Technology	Century
Sun	Animal	<19th
Coal	Steam	~ 19th
Hydrocarbons	Combustion	~ 20th
Atoms	Electromagnetism	~2?th
Outer space	Antigravity	~??th

Upon the realization that the value of energy should be measured in energy units rather than money, consider the overall energy conversion system displayed by Fig. 7. The balance between its input and output is

$$C_nH_{2n+2} + (1.5n + 0.5)O_2 + X = nCO_2 + (n+1) H_2O + X$$

where X expresses any components of the system that do not participate in the molecular transformation, as in particular, nitrogen in air and all the impurities in the fuel. With M denoting the molar mass and h – the enthalpy of formation per unit mass at NTP, the enthalpy of the input:

$$H_R = M_{CnH2n+2}h_{CnH2n+2} + (1.5n + 0.5) M_{O2}h_{O2} + H_X$$

while enthalpy of the output:

$$H_P = nM_{CO2}h_{CO2} + (n + 1) M_{H2O} h_{H2O} + H_X$$

The calorific value of the transformation:

$$\Delta H_{CH} = H_R - H_P$$

Its part due to carbon:

$$\Delta H_C = [n/(2n + 1)] [M_{CnH2n+2}h_{CnH2n+2} + (1.5n + 0.5) M_{O2}h_{O2}]$$
$$- nM_{CO2}h_{CO2}$$

while its part due to hydrogen:

$$\Delta H_H = [(n + 1)/(2n+1)] [M_{CnH2n+2}h_{CnH2n+2} + (1.5n + 0.5)$$
$$M_{O2}h_{O2}] - (n+1) M_{H2O} h_{H2O}$$

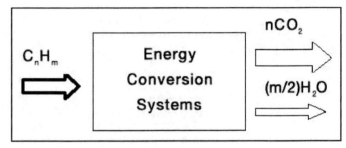

Fig. 7. System Diagram of Energy Balance for Energy Conversion Systems, Based on Oxidation of Paraffins, the Most Typical Hydrocarbons by Air

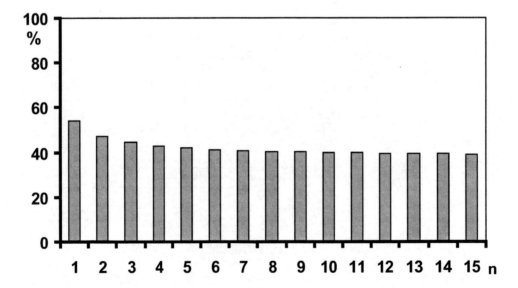

Fig. 8. Fraction of Calorific Value Derived from Oxidation of Hydrogen. ΔH_{H2}, with respect to that obtainable from the Paraffin Molecule, $\Delta H_{CnH2n+2}$, based on data of NIST Chemistry WebBook [4]

Table 3: Computations of Calorific Values Derived from Oxidation of Paraffins

n	h_{HC}(kJ/gmole)	$(n+1)h_{H2O}$(kJ)	nh_{CO2}(kJ)	CV_{HC}(kJ/gmole)	CV_H(kJ/gmole)	CV_C(kJ/gmole)	CV_H(%)
1	-74.87	-483.66	-393.52	802.31	433.75	368.56	54.06
2	-74.87	-725.49	-787.04	1437.66	680.57	757.09	47.34
3	-74.87	-967.32	-1180.56	2073.01	924.54	1148.47	44.60
4	-127.1	-1209.15	-1574.08	2656.13	1138.54	1517.59	42.86
5	-146.8	-1450.98	-1967.6	3271.78	1370.91	1900.87	41.90
6	-167.2	-1692.81	-2361.12	3886.73	1602.78	2283.95	41.24
7	-187.8	-1934.64	-2754.64	4501.48	1834.48	2667.00	40.75
8	-208.4	-2176.47	-3148.16	5116.23	2066.14	3050.09	40.38
9	-229.1	-2418.3	-3541.68	5730.88	2297.72	3433.16	40.09
10	-249.7	-2660.13	-3935.2	6345.63	2529.33	3816.30	39.86
11	-270.3	-2901.96	-4328.72	6960.38	2760.93	4199.45	39.67
12	-290.9	-3143.79	-4722.24	7575.13	2992.52	4582.61	39.50
13	-311.5	-3385.62	-5115.76	8189.88	3224.10	4965.78	39.37
14	-332.1	-3627.45	-5509.28	8804.63	3455.67	5348.96	39.25
15	-354.8	-3869.28	-5902.8	9417.28	3686.16	5731.12	39.14

The computations for paraffins, the essential ingredients of oil, in their mixtures are displayed in Table 3, while the calorific value of hydrogen with respect to the hydrocarbon used for its production, $[\Delta H_H/\Delta H_{CH}]$%, is depicted in Fig. 8, demonstrating that it is ~ 50% less than that provided by the. Using the rest of the rest of the energy to generate hydrogen by reformation with water makes the situation even worse, for it takes more available energy to get hydrogen from water than obtainable from its subsequent oxidation with air.

Thus, even if the efficiency of the energy conversion system fueled by hydrogen is 100%, ~50% of the calorific value of the hydrocarbon from which it is obtained is lost right at the outset. For comparison, indicated efficiencies of theoretical Otto and Diesel cycles fueled by paraffins are displayed in Figs. 9 and 10, where the compression ratio is denoted by r.

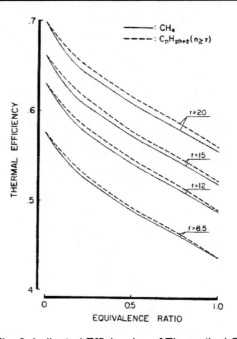

Fig. 9. Indicated Efficiencies of Theoretical Otto Cycles

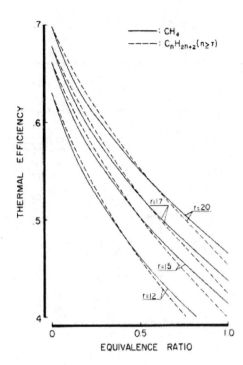

Fig. 10. Indicated Efficiencies of Theoretical Diesel Cycles

COMMENTARIES

Upon individual remarks made by the panelists, intended primarily to clarify their points of view, the debate raised by critical comments from the floor became quite lively to say the least.

Most of the criticisms were raised concerning the significance of Fig. 8, pointing out that the energy derivable from the remaining carbon molecules can be utilized to produced more hydrogen by steam reforming – a point whose futility for energy economy has been brought up in the previous section.

To elucidate this issue, provided here in Table 4 are thermodynamic data of formation for hydrocarbon molecules and the products of their oxidation. The values are derived from the NIST Chemistry Handbook [4], where, following the JANAF Tables [5], the standard reference states are H_2, O_2, N_2 and C (graphite) at normal temperature of 298K, so that their values in the table are 0. Beside the rise in enthalpies and entropies of formation per unit mass of hydrocarbon moles as the number of carbon atoms is diminished, the increase in available energies of products, expressed by Gibbs function, is quite striking.

To obtain hydrogen from water requires the expenditure of by far the largest amount of available energy per unit mass of H_2O, or even an order of magnitude more (a factor of 9 to be exact) per unit mass of H_2, as demonstrated in the last column of Table 4. Rather than providing service as a source of energy, hydrogen, in terms of the energy required for its production, is its considerable consumer.

Table 4. Thermodynamic Data of Formation for Hydrocarbon Fuels and Water with Respect to Hydrogen, Carbon(S), and Oxygen in Molecular Form

	$\Delta_f H^\circ$		$\Delta_f S^\circ$		$T^\circ \Delta_f S^\circ$		$\Delta_f G^\circ$		
	kJ/mole	kJ/gm	J/mole*K	J/gm*K	kJ/mole	kJ/gm	kJ/mole	kJ/gm	kJ/gmH2
$C_{12}H_{26}$	-332.1	-1.67	700.4	3.5373	208.7	1.954	-540.8	-2.73	-17.85
C_8H_{18}	-208.4	-1.62	467.1	4.0974	139.2	1.2211	-347.6	-3.04	-19.25
CH_4	-74.5	-4.66	188.7	11.794	56.2	3.5125	-130.7	-8.17	-32.68
H_2O	-241.8	-13.43	188.8	10.489	56.3	3.1278	-298.1	-16.56	-149.04

Table 5. Data on Combustion of Paraffins, C_nH_m, with Air for $\lambda = 1$ and $\lambda = 2$ at NTP

n	m	Fuel					Air							T_a	
		M_F gm/moleF	$Y_F(\%)$		M_A gm/moleF		M_A/M_F		$Y_A(\%)$		K	
			$\lambda = 1$	$\lambda = 2$	$\lambda = 1$	$\lambda = 2$	$\lambda = 1$	$\lambda = 2$	$\lambda = 1$	$\lambda = 2$	$\lambda = 1$	$\lambda = 2$
1	0	12	8	4	137	275	11	23	92	96	2303	1504
12	26	170	6	3	2541	5081	15	30	94	97	2299	1470
8	18	114	6	3	1717	3433	15	30	94	97	2275	1475
1	4	16	6	3	275	549	17	34	94	97	2226	1478
0	2	2	3	1	69	137	34	69	97	99	2383	1642

Table 6. Fuel Production Efficiencies for Automotive Powerplants

Engine	Fuel	NG Feedstock Production Efficiency	Conversion Efficiency	Fuel Storage, Transmission, & Distribution Efficiency	Efficiency Additional Compression	Overall Efficiency of Fuel Production
Conventional SI	NG	95%		97%	95%	87.5%
	H_2	95%	78.5%	97%	82%	59%
Conventional Diesel	F-T Diesel	95%	72%	97%		67%
	Dual-Fuel	95%	98%	97%	95%	86%
Hybrid SI	NG	95%		97%	95%	87.5%
Hybrid Diesel	F-T Diesel	95%	72%	97%		67%
	Dual-Fuel	95%	98%	97%	95%	86%
Battery All-Electric	Electricity	95%	42%	93%		37%
Fuel Cell	Methanol	95%	62.4%	97%		57.5%
	H_2	95%	78.5%	97%	82%	59%

The admiration of hydrogen as the most refined and vastly superior fuel is, nonetheless, well taken. Its technology is promoted by a great number of research studies, most of those conducted in our times published prominently by the International Journal of Hydrogen Energy, currently in its 28th volume. The superiority of hydrogen over hydrocarbon molecules is demonstrated by Table 5, from which it becomes apparent how a relatively small mass fraction of its mixture with air can generate significantly higher temperatures than hydrocarbons by combustion at initially normal (NTP) atmospheric conditions. The mass air/fuel ratio is expressed in terms of the air excess coefficient, λ, where $\lambda = 1$ corresponds to the stoichiometric proportion, while $\lambda = 2$ serves as a 'rule of thumb' indicator of LFL (Low Flammability Limit), at NTP.

The demand for hydrogen stems from the fact that it is the only admissible substance for Proton Exchange Membrane Fuel Cell (PEMFC) operating at a low room temperature. It is for this reason, that they are primarily considered for automotive powerplants, since carbon atoms, contained in any HC molecule, have a destructive influence upon the catalytic properties of the membrane.

Other kinds of fuel cells, like the Solid Oxygen Fuel Cell (SOFC), that admit carbon containing molecules, operate at relatively high temperatures of ~ 1000K.

At steady state conditions, this is far worse than operation at significantly higher peak temperatures of ~2000K, lasting intermittently for ~ 1 msec over intervals of ~30 msec in a four-stroke internal combustion engine operating at ~4000rpm.

Of direct relevance to the raison d'être of fuel cells and hydrogen are two recently published articles, one in Power, a supplement on "A New World of Power" to Mechanical Engineering, journal of the ASME, and the other in Science.

The first, authored by Kreith and West [6], was based on their previous paper published last year [7] that presented a comprehensive account of fuel production efficiencies for automotive powerplants displayed in Table 6. In [6], this account has been broken down into Well-to-Tank and Tank-to-Wheel with the overall Well-to-Wheel and presented in form of bar graphs portrayed here by Figs. 11, 12 and 13. Significantly enough, fuel cell using hydrogen is at the bottom of all the automotive powerplants in Well-to-Tank, as well as Well-to-Wheel, fuel production efficiency.

The conclusion reached on this basis is expressed succinctly by the subtitle of this article as follows:

"While the idea of using hydrogen to replace fossil fuels may seem like a dream come true, reality is more complex".

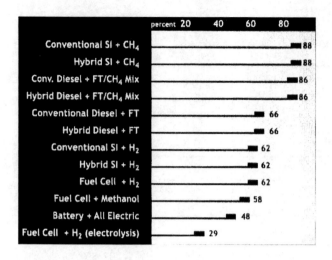

Fig. 11. Well-to-Tank Fuel Production Efficiency

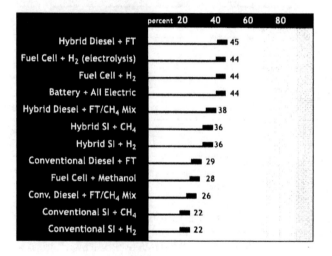

Fig. 12. Tank-to-Wheel Fuel Production Efficiency

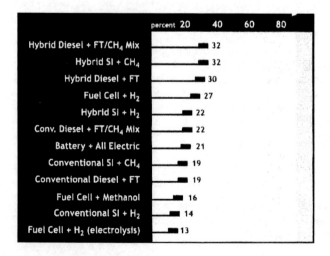

Fig. 13. Well-to-Wheel Fuel Production Efficiency

The second, authored by Keith and Farrell [8], following their earlier studies [9], stems from the recognition that "support for hydrogen cars has reached new heights, especially for fuel cell vehicles that use hydrogen directly." – the fact emphasized by the "Freedom Car and Fuel Initiative, which amounts to $1.7 billion over 5 years."

"Like electricity," they point out, "hydrogen is an energy carrier that must be produced from a primary energy source." It cannot be, therefore, of any assistance in replenishing our energy resources, but on the contrary, it imposes an additional burden upon its expenditure.

Taking "a longer term strategic view of energy policy" they provide critical arguments "against early adoption of hydrogen cars."

The most detrimental demonstration of the relatively minor influence of automobiles upon the protection of our environment is provided by the comparison between pollutant emissions from electric and internal combustion engine automotive powerplants in a form reproduced here in Table 7.

It becomes apparent from all of this that the panacea of an omnipotent solution offered by fuel cells and hydrogen fuel in preserving our energy resources and our environment is not as rosy as is purported judging from the lavish political and financial support it has engendered.

The most fascinating for us is the fact that in the frantic search for alternative powerplants and for alternative fuels, so little attention is given to the possibility of advancing the technology of combustion in piston engines by ushering into it the modern tools of Microtechnology. When the anti-pollution rules were enacted in late nineteen sixties, all the responsibility for execution and control of the exothermic process of combustion were mechanical in nature. Under such circumstances, any attempts to mitigate the formation of pollutants in the course of combustion, referred to as internal treatment, were bound to fail. The only way to comply with the strict anti-pollution measures was by resorting to the use of a chemical processing plant in the exhaust pipe, the catalytic reformer in the footpath of classical chemical engineering practice for cleaning the stack gases of industrial plants, known as external treatment.

In the meantime, great strides were made in technological progress promoted by the advent of Microtechnology that modernized the whole spectrum of engineering practice. As a consequence of its exceptionally fast and reliable action, the responsibility for operation and control of dynamic systems is nowadays undergoing a shift from mechanical to electronic. The effect of it is evident in most industrial products, including, of course, the automobiles and, in particular, most of their accessory equipment.

Table 7.Comparison between Pollutant Emissions from Electric and Internal Combustion Engine Automotive Powerplants

Sector	CO_2 (%)	Percentage (%)	SO₂ Emissions		Percentage (% total)	NOₓ Emissions	
			SO₂ per Gj of Fuel (kg SO_2/Gj)			NO₂ per Gj of Fuel (kg NO_2/Gj)	
			Current	Est. in 2010		Current	Est. in 2010
Cars and Light Trucks	19	1	0.02	<0.005	18	0.25	0.07
Other Transportation	14	5	0.08		39	0.70	
Fossil Fuel Electricity	41	70	0.04	0.28	32	0.24	0.12

The only technology left, as yet, untouched by this revolutionary progress is the execution and control of the exothermic process of combustion in piston engines. The specific techniques available for this purpose were described in a number of papers [10-14], and exposed comprehensively in a recently published book [15].

The principal concept for implementation of the conversion from mechanical to electronic technology involves, of course, the transition from external to internal treatment. Instead of letting the exothermic process of combustion be propagated in the form of FTC (Flame Traversing the Charge), as it occurs naturally whenever it is left by itself, have it executed by distributed exothermic centers combustion in form of FMC (Fireball Mode of Combustion). Instead of transforming the reactants into products in series typical of FTC, under distributed conditions created by turbulent FMC, they are processed in parallel. Today, practical feasibility of such a mode of combustion in an engine cylinder has been demonstrated by the highly popular HCCI engine.

The execution of distributed ('homogeneous') process of combustion can be realized by a Micro-Electronically Controlled Combustion (MECC) system featuring PJI&I (Pulsed Jet Injection and Ignition). The latter is an advanced version of the modern technology of direct injection [16], associated with pulsed jet ignition system of LAG*, and combined with RGR (Residual Gas Recirculation) [17, 18], constituting thus, de facto, a sequel to HCCI. In recognition of the fact that the sole purpose of combustion in piston engines is to create pressure for generating force to push the piston, a methodology was developed to assess the effectiveness of utilizing fuel for this purpose under the code name of

---

* Lavina Activatsia Gorenia, or Avalanche Activated Combustion, – a combustion system developed under the direction of M. M. Semenov, Nobel laureate for conceiving the chain reaction theory, upon an R&D program conducted for over half a century in Russia [10-13].

pressure diagnostics [14, 15].

In order to develop successfully a novel system like MECC, it should be carried out as a step-by step improvement of the state of art in current engines. Pressure diagnostics is then indispensable not only to provide an interface for robust control, but also to monitor the progress made at each step in predicting the gains in performance it may exert as a guideline for its implementation.

An example of such a service, presented here are the results of a study performed to evaluate the improvements that can be attained by the Renault F7P engine if, instead of being run in a conventional, throttled, Flame Traversing the Charge (FTC) manner, it were operated in a stratified charge, wide open throttle (WOT), Fireball Mode of Combustion (FMC) [14, 15].

The operating conditions under study are provided by in Table 8, where Q expresses the net energy loss incurred by heat transfer to the walls, while λ_o and λ_r are the air excess coefficients in the cylinder charge and in the chemically reacting zone, respectively.

Table 8. Operating Conditions Under Study

	$\theta_i°$	$\theta_f°$	Q	λ_o	λ_r	CASE
FTC	335	410	Q_H	1	1	0
FMC	335	410	Q_H	4.44	1.05	1
			$Q_H/2$	8.00	1.33	2
			0	9.09	2.00	3
	335	392.5	$Q_H/2$	5.41	1.67	4
			$Q_H/4$	6.90	2.00	5
			0	9.52	2.50	6

Identified here are six cases of FMC, besides the reference case of FTC, denoted by 0. Cases 1, 2, 3, are concerned with the effects of diminished heat transfer loss, achievable by reducing the contact of the reacting medium with the walls of the cylinder-piston enclosure, while the time interval within which fuel is consumed remains unchanged. Cases 4, 5, 6, take into account, moreover, the consequences of having the lifetime of the dynamic stage reduced by a factor of two, deemed attainable by increase in combustion rate due to turbulent mixing induced by jet injection and jet ignition.

In all the cases, the indicated power is maintained carefully at the same level, as displayed by the indicator diagrams of the FTC and FMC modes of combustion corresponding to initial and final states of the exothermic process of combustion, θ_i and θ_f adjusted for maximum IMEP, referred to as MBT (Maximum Brake Torque). The results are displayed in Fig.14.

Demonstrated thereby are the advantages accruable by just having the exothermic process executed in a distributed mode within a turbulent jet plume away from the walls.

By protecting it thus from excessive heat transfer losses, the exothermic reaction is sustained under large air excess conditions leading to significantly reduced peak temperatures, so that, as evident from Fig. 14, formation of NO can be practically annihilated and the concentration of CO reduced by an order in magnitude.

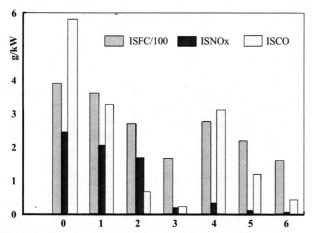

Fig. 14. Prognosis of Engine Performance Parameters

This example is by no means all that can be accomplished by MECC. Besides eliminating the formation of particulate by carrying the oxidation reaction in a highly diluted mixture, the concentrations of CO and of the unburned hydrocarbons can be drastically reduced by Residual Gas Recirculation (RGR) under control of Variable Valve Timing (VVT), let alone the expedient of oxygen enrichment of intake air by incorporating a semipermeable membrane in its filter, like, among others, that produced by Broken Hills Ltd. in Australia.

There is, of course, no free lunch. Executing the exothermic process of combustion in a highly diluted air/fuel mixture at a relatively low temperature leads, in principle, to less work delivered per cycle, as well as lower rate of burn. It is, in fact, this drawback that contributed toward the demise of the Russian LAG engine.

Today this drawback can be offset by taking advantage of the jet generated turbulence, whose technology has been greatly advanced over the last two decades, combined with explosively growing micro-electronic control technology, to attain significantly faster processing in parallel by distributed ('homogeneous') combustion, exemplified by HCCI and FMC, rather than flames traversing the charge (FTC), as is the case with current technology.

It will be advantageous to operate engines of this kind at a significantly higher speed in RPM. The deprecating problem of excessive friction can be alleviated by relieving the piston from the dual demand imposed upon it for dual service in transmitting the force exerted by cylinder pressure on the piston and in transmitting its axial displacement to the rotary motion of the crankshaft. This can be accomplished by segregating them into an axially moving piston rod and a crosshead mechanism in the crankcase (a la Scotch yoke). Taking then advantage of the absence of side thrust on the cylinder walls, piston rings can be replaced by gas lubricated labyrinth seals, yielding a considerable reduction in frictional losses at the high rotational speeds achievable by the rapid rate of burn of distributed, turbulent combustion.

Such prospects of progress in the technology of combustion in piston engines are, indeed, most attractive. In order to carry it out successfully, computational analytic means are required to assess specifically the effectiveness with which fuel is utilized in a piston engine and, thereupon, employ them to monitor and guide the design and development of the MECC system step by step in its evolution. These means are available today [15]. Particularly crucial to the implementation of the thee concepts is the availability of fast operating injectors whose stringent demands can be satisfied by Micro-Electronic Mechanical System (MEMS) valves that are, as yet, unavailable in spite of the well established MEMS technology.

CONCLUSIONS

1. Fuel Cells are the most sophisticated electric generators available today. Their Proton Exchange Membrane (PEM) version, operating at low temperatures, is particularly attractive because it is devoid of any irreversible effects due to heat transfer - the major drawbacks of heat engines. Their major disadvantage is the dependence on hydrogen as the only admissible fuel, according to current state of technology.

2. Fuel cells operating continuously at high temperatures, like the SOFC (Solid Oxide Fuel Cell), are handi-

capped by a significantly higher energy loss by heat transfer to the walls than the intermittently operating internal combustion engines

3.   Hydrogen is the most highly refined fuel, by far superior to hydrocarbons. It utilizes ~ 34 times its own mass of air (in contrast to ~15 for hydrocarbons) to generate the maximum calorific value. And, with that, its combustion with air attains an exceptionally high temperature of ~2300K, in contrast to ~ 2100K achievable by hydrocarbon fuels, while, for comparison, the melting temperature of iron is ~1800K.

4.   By the same token, hydrogen is the most flammable and explosive fuel – well known hazards posing serious problems of safety. At the same time, as a most refined product, hydrogen is most expensive in terms of calorific, rather than monetary, value.

5.   Albeit hydrogen is most abundant on earth, the only practical way to obtain energy out of it is by oxidation with air. For that reason, it is an energy source if, and only if, it is not in equilibrium with oxygen, as, indeed, is the case with hydrocarbon molecules in oil wells. But then, the calorific value of hydrogen is less than half of that obtainable by oxidation of the hydrocarbon.

6.   In water molecules, hydrogen is in equilibrium with oxygen. Hence, the energy expenditure for its extraction from them, irrespectively how it is accomplished, be it electrolysis or reformation, exceeds significantly that of the energy it may provide. In terms of available energy expressed by change in the Gibbs function, its level of ~150 kJ/gm is considerably higher than that of ~18 kJ/gm for heavy hydrocarbons up to ~33 kJ/gm for methane.

7.   In view of the energy expenditure required for its production and the safety issues it engenders, hydrogen is an energy drain, rather than a resource.

8.   In relatively modest amount, hydrogen can be produced at a reasonable cost by exploiting such energy resources as waste heat. In mass production, comparable in any way to that of hydrocarbon fuels, however, has to await an era of peaceful life on earth and infinite energy supply, in drastic contrast our times of terrorist thread and energy crisis.

9.   Instead of resorting to alternative powerplants and fuels, like fuel cell powered electric motors and hydrogen, in order to satisfy the demands for efficient and environmentally benign energy conversion systems, great strides can be made by advancing the technology of combustion in piston engines.

10. This can be accomplished by executing the exothermic process in piston engines by means of MECC (Micro-Electronically Controlled Combustion) - an advanced version of HCCI featuring PJI&I (Pulsed Jet Injection and Ignition) - to generate distributed FMC (Fire-ball Mode of Combustion) away from the walls, maximizing thus the effectiveness of fuel is utilization.

REFERENCES

1.   Bockris, J. O'M. and Scrinivasan, S., "Fuel Cells: Their Electrochemistry," xxxiii+659 p., McGraw-Hill BookCompany, New York, 1969.

2.   Norbeck, J., Heffel, J., Durbin, T., Montano, M., Tabbara, B., Bowden, J., "Hydrogen Fuel for Surface Transportation," ix+548 p., SAE Book Code R-160, 1996.

3.   Stodolsky et al, "Total Fuel Cycle Impact of Advanced Vehicles," SAE 1999-01-0322, 1999.

4.   4.   NIST Standard Reference Database Number 69, NIST Chemistry WebBook, March 2003, <ttp://webbook.nist.gov/chemistry>

5.   JANAF Thermochemical tables, edited by Stull D.R., Prophet H., National Bureau of Standards US Department of Commerce) Report NSRDS-NBS 37, 1141 pp., 1971.

6.   Kreith, F., West, R.E., Gauging Efficiency Well to Wheel Mechanical Engineering Power, pp. 20-23, 2003.

7.   Kreith, F., West, R.E., Isler, B.E., Efficiency of Advanced Ground Transportation Technologies, ASME Journal of Energy Resources Technologies, pp. 173-179, 2002.

8.   Keith, D.W., Farrell, A.E., Rethinking Hydrogen Cars, Science, v. 307, <www.sciencemag.org> 2003.

9.   Farrell, A. and Keith, D., Hydrogen as a Transportation Fuel. Environment, pp. 43-45, April 2001.

10. Oppenheim, A. K., "The Future of Combustion in Engines," Proceedings of the Institution of Mechanical Engineers, C448/022, IMech 1992-10, pp. 187-192, 1992.

11. Oppenheim A.K. and Kuhl A.L., "Paving the Way to Controlled Combustion Engines (CCE)," Futuristic Concepts in Engines and Components, SAE SP-1108, Paper 951961, pp.19-29, 1995.

12. Wolanski, P. and Oppenheim, A.K., "Controlled Combustion Engines (CCE)," SAE Paper 1999-01-0324, 5 pp., 1999.

13. Beck, N.J., Oppenheim, A.K. and Uyehara, O., "On the Future of Combustion in Piston Engines," Advanced Power-plant Concepts SP-1325, SAE 980117, pp. 1-10, 1998.

14. Oppenheim, A.K., Maxson, J.A., and Shahed, S.M., "Can the Maximization of Fuel Economy be Compatible with the Minimization of Pollutant Emissions?" SAE Paper 940479, pp. 12, 1994.

15. Oppenheim, A.K., "Combustion in Piston Engines," xv+174 pp, Springer-Verlag, Berlin Heidelberg, 2004.

16. Stan, C., "Direct Injection Systems – The Next Decade in Engine Technology," Society of Automotive Engineers, Warrendale, Pa. 2002.

17. Heitland, H., Rinnie, G., Wislocki, K., "Controlled Combustion in Direct Injection Stratified Charge En-

gines," SAE Paper 2000-01-0197 in Advances in Combustion, SP-1492, 63-69, 2000.

18. Heitland, H., Rinnie, Willmann, G., Vanhaelst, R., "IC Engines for 100 Miles/Gallon Cars," SAE Paper 2001-01-0258 in Advances in Combustion SP-1574, 99-105, 2001.

II. FUEL ISSUES

Fuel/Vehicle Analysis

2003-01-0412

Assessing Tank-to-Wheel Efficiencies of Advanced Technology Vehicles

Feng An and Danilo Santini
Argonne National Laboratory

ABSTRACT

This paper analyzes four recent major studies carried out by MIT, a GM-led team, Directed Technologies, Inc., and A. D. Little, Inc. to assess advanced technology vehicles. These analyses appear to differ greatly concerning their perception of the energy benefits of advanced technology vehicles, leading to great uncertainties in estimating full-fuel-cycle (or "well-to-wheel") greenhouse gas (GHG) emission reduction potentials and/or fuel feedstock requirements per mile of service. Advanced vehicles include, but are not limited to, advanced gasoline and diesel internal combustion engine (ICE) vehicles, hybrid electric vehicles (HEVs) with gasoline, diesel, and compressed natural gas (CNG) ICEs, and various kinds of fuel-cell based vehicles (FCVs), such as direct hydrogen FCVs and gasoline or methanol fuel-based FCVs. We focus on variations in estimates of vehicle gasoline-equivalent fuel energy use, glider and powertrain masses, and introduce powertrain effectiveness as a new surrogate measure for tank-to-wheel vehicle efficiency. We conclude that, while the degree of uncertainty across studies is considerable, it is not as great as a summary investigation and direct comparison implies. Our investigation suggests that there are logical and systematic reasons for variations among the studies. Further studies are required to improve both assessment and understanding of technical potentials of these advanced technologies, and to narrow the range of uncertainty currently present.

INTRODUCTION

In recent years, many studies have assessed the impacts of advanced technology vehicles (ATVs) on gallons of gasoline equivalent (gge) energy use, and some of them also include full-fuel-cycle (or well-to-wheel [WTW]) energy use estimates and greenhouse gas (GHG) potentials [1-9]. Recognizing the importance of this emerging area of research, and the quality of work presented to it, SAE awarded the 1999 full-fuel-cycle energy analysis paper by Stodolsky et al (7) an Arch T. Colwell award as one of the top papers of the year. Stodolsky et al provided a high and low estimate for each fuel/vehicle pathway examined. Studies since have either expanded in comprensiveness or added depth of detail. In general, these analyses demonstrate very different results for many ATVs and show a wide range of uncertainties. Similar to Stodolsky et al, both Weiss et al (1) and GM (2) recognize the importance of acknowledging uncertainty and include their own estimates of uncertainty ranges. Although uncertainty ranges can be sometimes quite wide, an evaluation of differences among technologies based on percentage change from the base vehicle's reference value reveals that the most probable values of one study are sometimes outside the uncertainty range of the other. Because the stakes are increasing for both industry and the federal government in choosing developmental paths, providing and allocating capital and technological investment, consistent and improving projections of future advanced technology vehicles are crucial for decision-makers at all levels. Advanced vehicles include, but are not limited to, advanced gasoline and diesel internal combustion engine (ICE) vehicles, hybrid electric vehicles (HEVs) with gasoline, diesel and CNG ICEs, and various kinds of fuel-cell based vehicles (FCVs) such as direct hydrogen (H2) FCVs and gasoline or methanol reforming FCVs. Like gasoline, diesel, and CNG ICE technology, FCVs may also be hybridized. Some studies include FCVs with and without hybridization.

In this paper, we examine results from four studies (cited by institutional or corporate affiliation of the authors) as follows [1-6]:

MIT = Weiss et al., 2000, *On the Road in 2020: A Life-Cycle Analysis of New Automobile Technologies,* MIT Energy Laboratory Report No. MIT EL 00-003, Energy Laboratory, Massachusetts Institute of Technology, Cambridge, Mass., Oct.

GM = General Motors Corp., et al., 2001. *Well-to-Wheel Energy Use and Greenhouse Gas Emissions of Advanced Fuel/Vehicle Systems — North American Analysis, Executive Summary Report.*

DTI = Thomas, C.E., B.D. James, F.D. Lomax, and I.F. Kuhn, 1998 "Societal Impacts of Fuel Options for Fuel Cell Vehicles," SAE paper 982496, Society of Automotive Engineers, Warrendale, Penn.

ADL = Arthur D. Little, Inc., 2002. Guidance for Transportation Technologies: Fuel Choice for Fuel Cell Vehicles. Final Report.

This paper focuses on the so-called "tank-to-wheel" (TTW) portion of the WTW process and introduces a new concept – powertrain effectiveness – to compare relative benefits of ATVs in these studies. The authors of this paper, on behalf of the U.S. Department of Energy's Office of Energy Efficiency and Renewable Energy, have been attempting to determine the primary causes for differences among these studies, and provide such information to the scientific and technical community. This is one of a series of papers that focuses on a subset of technologies and/or studies, and attempts to provide insights on the many potential intermediate causes for the different final estimates by authors of the various comprehensive studies [10-13]. Before starting our analysis, we would like first to address several critical issues associated with ATV analysis.

TECHNOLOGY TIMELINE ISSUE - "VERTICAL" VS. "DIAGONAL" ASSESSMENT

A careful examination reveals that these studies have assumed very different timelines for baseline conventional vehicles (CVs) and ATVs, creating confusion about which timeframe and baseline should be used for assessing the relative benefits of ATVs. For example, all of MIT's ATV analyses are based on a projected MY 2020 advanced conventional gasoline spark-ignition (SI) vehicle, instead of a current SI-ICE vehicle. However, GM has chosen a current Silverado pickup as a baseline vehicle, even though the study states that GM is "focusing on technologies that are expected to be implemented in 2005 and beyond" and "emissions targets for all vehicles were based on Federal Tier 2 standards, … for the 2010 timeframe"[2]. This creates a dilemma that can be illustrated by Figure 1, where the concepts of "vertical" vs. "diagonal" assessment are introduced. The vertical assessment can be defined as comparisons based on same timeframe, e.g., future ATVs vs. future CVs. The diagonal comparisons can be defined as comparisons based on different timeframes, e.g., future ATVs vs. current CVs.

While the diagonal comparisons are often used in assessing future potentials of the same types of vehicle technologies, as in studies recently conducted by the NRC and others [14-17], it can be argued here that the

vertical approach may better suit cross-comparisons of advanced powertrain technologies, because of the following reasons: 1) baseline CVs, as well as ATVs, are evolving, thus all of them are moving targets (not just ATVs); 2) we are most interested in cross-comparing relative benefits of various ATVs under the same timeframe; and 3) vertical assessment gives a more consistent and easier approach to cross-compare different studies.

Figure 1 – Vertical vs. Diagonal Assessment

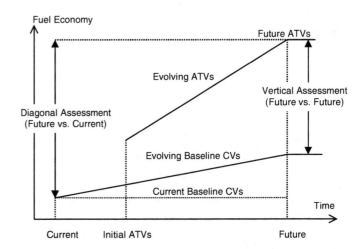

Nevertheless, the advantage of a diagonal assessment is obvious: it gives a clear picture of absolute benefits over current baselines. Of course, the diagonal approach merges with the vertical approach if the baseline CV is assumed not to be evolving, or in other words, the evolving baseline equals the current baseline. This appears to be the case for GM analysis, where no future evolving base vehicle is analyzed. The diagonal comparison scheme is explored further in the Appendix.

The variation in timelines and baseline vehicle attributes turns out to be a source of much confusion in comparing ATV benefits, which we will further discuss in the following sections. Another issue is how to compare vertical assessments under different future timeframes, e.g., GM under 2005-10 and MIT under 2020. This issue is unsolvable within the scope of this study and we will have to leave it for future analysis.

WELL-TO-WHEEL (WTW) VS. TANK-TO-WHEEL (TTW) ANALYSIS

Although this paper focuses on tank-to-wheel, or vehicle-level analysis, we believe that complete comparisons of advanced technology vehicles should be conducted on a life-cycle, or well-to-wheel (WTW) basis. WTW process includes both well-to-tank (WTT, or upstream) and tank-to-wheel (TTW or vehicle) steps. Both GM and MIT analyses include these steps and are subjects of this section.

These two studies are based on very different baseline vehicles – MIT is based on a midsize conventional car, and GM is based on a light-duty truck (Chevrolet Silverado). Each also includes somewhat different fuel and technology choices, but they share many common powertrain types within a very comprehensive analysis, and thus provide the basis for numerous comparisons. Table 1 lists the most probable WTW GHG emissions (in g/mi) of baseline and advanced technology vehicles, as well as GHG reductions associated with various ATVs. *Note that this paper focuses on the most probable values from each of these two studies, and does not discuss the uncertainty ranges within those studies.*

For vertical assessment purposes, we are interested only in GM assessments of ATVs and the baseline vehicle under the same 2005-2010 timeframe. Since GM provides only a current pickup truck as the base vehicle, we can assume that GM is allowing for no changes in the baseline vehicle in the 2005-2010 timeframe. Using this interpretation, we label the GM case "GM 2005-10." This implies that the GM and MIT (based on 2020) assessments are literately more than 10 years apart. Table 1 shows that these two studies demonstrate significant differences in almost all ATVs in terms of GHG emissions, rankings, and reductions.

Since the MIT baseline vehicle is much lighter and smaller than GM's, the GHG rates of both studies cannot be compared directly. Instead, we compare ranks and percentage of GHG reductions associated with each ATV. Note also that the MIT study provides two nominal 2020 evolving baselines. However, all ATVs are "spun off" the advanced 2020 baseline, so we use it as *the* 2020 baseline in Table 1.

Table 1 – Well-to-Wheel GHG Emissions of Advanced Vehicles in MIT and GM Studies

Technology	MIT 2020			GM 2005-10			MIT-GM
	g/mi	Rank	Reduction	g/mi	Rank	Reduction	Diffs
Current SI-ICE	424*			554			
2020 Base SI-ICE	278*	7					
2020 Adv. SI-ICE	**247**	6					
Advanced CI-ICE	218	4	12%	472	6	15%	-3%
Hybrid SI-ICE	176	2	29%	454	5	18%	11%
Hybrid CI-ICE	156	1	37%	384	4	31%	6%
Hybrid H2 FC[1]	201	3	19%	296	1	47%	-28%
Hybrid Gaso. FC	289	8	-17%	366	3	34%	-51%
Hybrid MeOH FC	222	5	10%	324	2	42%	-31%

* These MIT vehicles have different "gliders" (bodies) than remaining MIT cases, which all use the same low mass, low aero drag, low rolling resistance glider.

Bearing all the differences in mind, the GHG reductions for MIT ATVs range from -17% for gasoline FCVs to 37% for the diesel hybrid-electric vehicles (Hybrid CI-ICE). The GM study estimates that the benefits of ATVs range

from 15% for the conventional diesel vehicles (CI-ICE) to 47% for the direct hydrogen (H2) FCVs. The last column of Table 1 shows the differences between MIT and GM results. While MIT's results show higher benefits for hybrid ICE vehicles (Hybrid SI-ICE and Hybrid CI-ICE), and slightly lower benefits for the CI-ICE, the trend for fuel-cell based fuel cell vehicles (FCVs) is striking: MIT results indicate much lower benefits for all FCVs. One extreme result is the hybrid gasoline reformer FCVs. While MIT shows a negative GHG impact (comparing with the 2020 baseline case with a 17% increase), GM shows a strong positive impact (a 34% reduction over its baseline vehicle).

Please note that this conclusion is consistent with each team's publicized position on technology choice: MIT is generally more enthusiastic about the potential of conventional technologies and skeptical of fuel cell technologies. GM has made no secret that they prefer focusing research on fuel cell technologies, rather than attempt to dramatically improve fuel economy of conventional technology vehicles. GM's study is quite different from the Weiss et al study in that it does not include a case with improved conventional powertrains, even though research and development of such powertrains is being conducted. Another difference between the GM study and Weiss et al (as well as Graham et al (8)) is that the conventional hybrid powertrains do not include a higher efficiency engine than that found in the reference vehicle. Weiss et al assume that a gasoline direct injection engine will be successful in the U.S. in the future and will be used in both conventional and hybrid powertrains. Graham et al assume that the Toyota Atkinson cycle engine technology (not suitable for conventional powertrains) will be used in future hybrids.

There are also many other contradictory implications. Some examples are as follows:

➤ MIT projects that the hybrid gasoline reformer FC would have the highest GHG emissions rate among all of its 2020 vehicles (ranks 8th of nine in Table 1), including baseline evolving SI-ICE. However, GM projects that the hybrid gasoline reformer FC (ranks 3rd of seven in Table 1) has a lower GHG emissions rate than all of its ICE-based conventional or hybrid vehicles.

➤ MIT estimates that hybrid hydrogen (H2) FCV (ranks 3rd) compares unfavorably to two ICE-based hybrid vehicles (ranks 1st & 2nd) in terms of GHG emissions. GM estimates otherwise.

➤ MIT estimates that diesel hybrid vehicles have the lowest GHG rate and largest GHG reduction among all vehicles (ranks 1st). In the GM case, it is the hybrid direct hydrogen FCV.

---

[1] H2 from decentralized reforming NG stations

77

Figures 2 and 3 demonstrate the significantly different results of these two studies. These contradictory results lead us to ask hard questions regarding the root of the uncertainties. While it is convenient to divide the WTW process into Well-to-Tank (WTT) and Tank-to-Wheel (TTW) processes, one needs to remember that upstream WTT energy use and emissions are largely driven by TTW energy demands, thus the uncertainties and disagreements often come directly from the assessment of TTW, or net vehicle powertrain efficiencies. More specifically, for vehicles using petroleum fuels or fuels from other feedstock, WTT process would inherit and even expand the uncertainties from TTW process.

Figure 2 – GHG Reductions from Base Vehicles in MIT and GM Studies

Figure 3 – Differences in GHG Reductions Between MIT 2020 and GM 2010 Scenarios

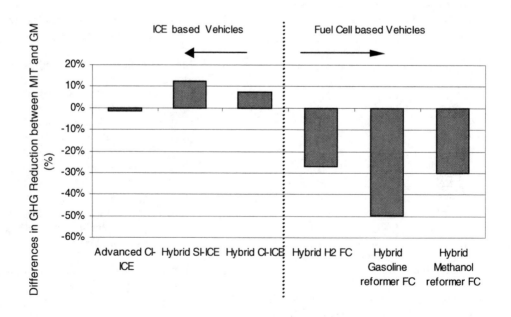

However, for our purposes, we are more interested in causes of variation among studies than within studies. Since the MIT and GM studies both relied significantly (MIT), or totally (GM), on different versions of the same upstream model (Argonne National Laboratory's Greenhouse Gas, Regulated Emissions, and Energy Use in Transportation [GREET] model), the upstream (WTT) variation across the two studies is not as large as the downstream (TTW) variation. Table 2 shows the Well-to-Tank (WTT), Tank-to-Wheel (TTW), and Well-to-Wheel (WTW) energy efficiencies of all ATVs in the MIT and GM studies. Figure 4 compares the WTW efficiencies of ATVs in both studies. Table 2 shows that the upstream WTT efficiencies are quite consistent among these two studies – about 83% for gasoline fuel, 86-88% for diesel fuel, 65% for methanol fuel based on NG pathway, and 56-57% for H2 fuel based on H2 production from decentralized NG stations. However, the vehicle TTW efficiencies differ greatly, as demonstrated by the "MIT-GM" columns. The last three columns of Table 2 under the "WTW Gains" show that the overall WTW efficiency gains of ATVs over the baseline vehicles, and the

differences between MIT and GM studies. It shows that the efficiency differences are even greater than the differences for GHG reductions.

While most vehicle-related studies focus on vehicle fuel economy (miles per gallon, or MPG), other measures such as fuel consumption (in MJ/km or L/km) and TTW efficiency (or vehicle powertrain efficiency) can also be used to analyze and compare ATVs. Vehicle TTW efficiency is defined as a ratio of vehicle useful work to fuel consumption. While the fuel economy and fuel consumption are most commonly used to measure vehicle energy consumption, they are greatly influenced by vehicle weight, aerodynamic characteristics of vehicle body design, and tire resistance, all of which are not exclusively related to specific powertrain technologies. Therefore, our investigation pays special attention to the relative advancement of different powertrain technology platforms. While TTW efficiencies are ideal measures to cross-compare these studies, only two of the four studies examined (MIT and GM) provided TTW efficiency figures. To overcome this, we introduced a new concept

Table 2 - Well-to-Tank (WTT), Tank-to-Wheel (TTW), and Well-to-Wheel (WTW) Efficiencies of ATVs in MIT and GM Studies

Energy Efficiency		MIT 2020			GM 2005-10			MIT-GM			WTW Gains		
Vehicle Type	Fuel	WTT	**TTW**	WTW	WTT	**TTW**	WTW	WTT	**TTW**	WTW	MIT 2020	GM 2010	MIT-GM
Current SI-ICE	Gasoline	82.6	**14.8**	12.2	82.5	**16.7**	13.8	0.1	**-1.9**	-1.6			
2020 evolv. SI-ICE	Gasoline	82.6	**18.0**	14.9									
2020 Adv. SI-ICE	Gasoline	82.6	**17.0**	14.1									
Advanced CI-ICE	Diesel	87.7	**20.2**	17.7	85.7	**19.4**	16.6	2.0	**0.8**	1.1	26%	21%	5%
Hybrid SI-ICE	Gasoline	82.6	**26.1**	21.5	82.5	**20.7**	17.1	0.1	**5.4**	4.5	53%	24%	29%
Hybrid CI-ICE	Diesel	87.7	**30.9**	27.1	85.7	**24.6**	21.1	2.0	**6.3**	6.0	92%	53%	39%
Hybrid H2 FC	H2 (CNG Station)	56.5	**36.0**	20.3	55.8	**41.4**	23.1	0.7	**-5.4**	-2.8	45%	68%	-23%
Hybrid gasoline FC	Gasoline	82.6	**17.6**	14.5	82.5	**27.3**	22.5	0.1	**-9.7**	-8.0	3%	63%	-60%
Hybrid Methanol FC	MeOH	64.9	**22.6**	14.7	64.5	**31.1**	20.1	0.4	**-8.5**	-5.4	4%	46%	-41%

Figure 4 – Well-to-Wheel Efficiencies in MIT and GM Studies

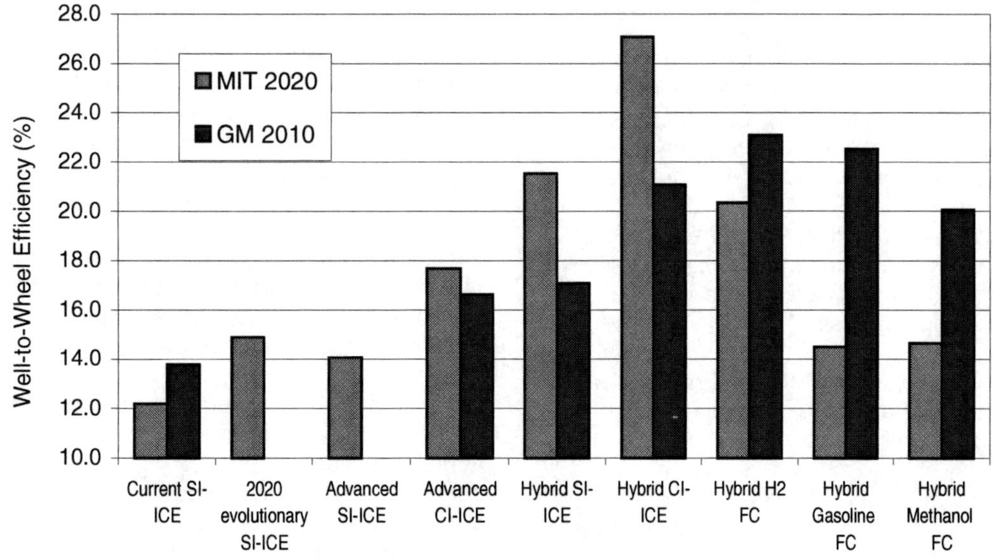

labeled "powertrain effectiveness" as a surrogate measure for the TTW efficiency. The powertrain effectiveness measure is unaffected by the failure of TTW efficiency to deal properly with influences of powertrain mass differences on fuel consumption. This approach would minimize problems in conducting cross-comparison among studies when different weight-class vehicles, including pickup trucks, midsize cars, and PNGV-type vehicles, are used.

GENERAL METHODOLOGICAL ISSUES

It is becoming increasingly clear that, in assessing technical potentials of various ATVs, a set of rules should be adopted to assure the assessment is fair and consistent. Santini et al. showed how violations of different sets of rules could influence the outcome of the technological prediction [10-13]. However, in many studies, standardized rules for comparisons are either not defined or not followed, intentionally or unintentionally. The following are some examples of these rules and related issues:

1. Carefully define and characterize baseline vehicle technologies.

In many studies, the baseline technology is not properly defined and characterized (e.g, NRC CAFE study [14]), resulting in questionable future projections for improvement potentials. It may be noted that in the NRC study, there is no difference in estimated percent fuel economy gain or cost by class of vehicle, despite the fact that separate tables are repeated with the same numbers for passenger cars (NRC Table 3-1), SUVs and Minivans (NRC Table 3-2), or Pickup Trucks (NRC Table 3-3).

2. Establish performance-equivalent criteria, such as 0-60 acceleration time, gradeability, range, interior volume, torque, and safety requirement.

Vehicle fuel efficiency is sensitive to its performance level. The 0-60 acceleration time is a most commonly used criterion. It is important that any comparison and evaluation be based on performance equivalence, and the estimates of performance should be published. However, this rule was violated in varying degrees by many analyses. One illustration of the difference that assumptions about peak powertrain kW per kg of vehicle mass (a surrogate for 0-60 time) can make is found in Santini, Vyas, Kumar, and Anderson [12]

3. Select and use common driving cycles. Comparison and evaluation should be conducted on the basis of the same driving cycles.

Vehicle fuel consumption is very sensitive to drive cycles. Meaningful comparisons can be derived only based on the same test cycles. Various studies worldwide have used FTP, HWY, CAFE, 1.25*CAFE, European and/or Japanese cycles as the basis for evaluation.

Comparisons can more easily be made successfully among studies that have used the same cycle(s). If the cycles are not identical, then very detailed information on the vehicles and powertrain components must be provided if evaluators (such as ourselves) are to successfully translate estimates from one driving cycle to another.

4. Base the comparison of vehicle fuel use on fuel economy (such as MPG), fuel consumption (such as MJ/km or L/km), or powertrain efficiency (or tank-to-wheel efficiency, vehicle efficiency). At this time there is no consistent approach agreed upon.

While the fuel economy and fuel consumption benchmarks are most commonly used, the advantage of vehicle and powertrain efficiency benchmarks is not always overlooked. Using TTW efficiency removes the influences of powertrain and vehicle weight and body structures. TTW has often been used to provide cross-comparison among a wide variety of vehicle and engine technologies and analyses, and the MIT and GM studies make use of it. However, we suggest that TTW is imperfect, and we develop and suggest an alternative in the closing section of this paper.

An advantage of using powertrain efficiency is that it gives insight regarding the technology advancement. While there is theoretically no limit to fuel economy potentials (which can in simulation be continuously improved by lowering vehicle weight and load), the limitation of powertrain efficiency is much better defined and understood, e.g., the SI gasoline engine peak thermodynamic efficiency limit is under 40%. Yet, peak efficiency alone is not enough, especially when low load efficiency varies. If idle fuel flow rate information is not available, fuel savings in hybrids by eliminating idling cannot be assessed.

5. When making comparisons, clearly define and synchronize technology scenarios.

The comparison of conventional and advanced technology vehicles should be conducted under well-defined technology aggressiveness or success scenarios. Technology potentials usually have a range defined by upper and lower boundaries. Adoption of upper boundary, middle point or lower-boundary values represents different scenarios often associated with names such as optimistic, moderate, or conservative. It is a mistake and creates confusion when scenarios are mixed but not clearly defined during comparisons. A detailed illustration is in Appendix A.

In many studies, technologies are too often compared on the basis of different scenarios, e.g., conventional vehicle technology in the aggressive upper-boundary scenarios against advanced technology vehicles under conservative lower-boundary scenarios, or vice versa. Distorted results and conclusions could occur under these circumstances.

In our view, a well-defined technology scenario cannot be replaced by simply assigning uncertainty ranges to individual technologies. Investigation of a preliminary draft of the NRC study [14] demonstrated that the multiplicative accumulation of savings at high ends of uncertainty ranges associated with multiple technologies could lead to clearly incorrect results. (For example, accumulation of upper-boundary potentials of all valve train technologies can lead to violation of thermodynamic laws). The problem is that the probability of one technology is often not independent of another technology, and there are very few technologies that are truly independent from others.

6. Conduct the comparison under clearly defined time periods. Confusion is created when projected benefits of distant future ATVs are compared with near-term benefits of different ATVs.

Since we all understand today's technologies better than that of the future, and conventional technologies better than emerging advanced ones, it is important to choose properly defined and comparable time periods to compare different technologies.

7. Advanced technologies are not static, but evolving, just like conventional technologies. For example, hybrid technologies can have many forms that result in different benefits [17-19]. It is not appropriate to designate one form of hybrid technology to be representative. The emergence of so called "minimum" or 42 V- based hybrids starts to blur the distinction between hybrid technology and evolving conventional technologies [15, 19].

8. The cost estimates of future technologies are highly uncertain. Cost considerations may involve ordering of technology sequence, current cost, future mass-production cost, and target cost. A more detailed discussion about technology cost is given by Santini, Vyas, Moore, and An, 2002 [10].

9. Conduct ATV comparisons based on same baseline reference vehicle. _A strongly preferred method for evaluating powertrain effectiveness is to have these ATVs sharing the same vehicle glider,_ which includes mostly vehicle body and chassis. If the effects of vehicle aerodynamic drag (Cd), tire rolling resistance (Cr), and body chassis design (glider mass) are to be isolated from powertrain effects, these values have to be held constant in the reference vehicle and the ATVs. Most studies do this, but MIT includes two conventional powertrain vehicle cases that do not.

10. Make the effects of cold start explicit. In addition to the complication that the change in "hot stabilized" idle fuel flow may not change by the same percentage as peak thermal efficiency, there are also problems that energy and time needed to warm the powertrain to operating temperature vary across technologies in ways very different from peak thermal efficiency. Ideally, studies would provide separate estimates of fuel consumption with and without cold start of a given driving cycle. None cited do this.

While we are aware of the complexity of these issues, it is not possible in this paper to address all of them. In the main body of this paper, we compare results of different studies directly, without adjusting them based on the above. So it is important to keep in mind that much of the presented differences come from several of the above sources. However, even with theoretically perfect adjustments according to the rules above, differences of opinion by experts will remain important, and uncertainty will remain. Nevertheless, nothing is more essential than judging the technology itself, isolated from the "noise" created when rules are not followed. Our analysis shows that even after some adjustments to reduce the above problems, there remain fundamental differences among these analyses in estimates of the technology potentials of various ATVs. These differences are the subject of our discussion in the following sections.

SUMMARY RESULTS

In the following sections, we compare results of Weiss et al (MIT), GM, Thomas et al (DTI), and ADL studies on the following three variables: vehicle mass, fuel economy in miles-per-gallon of gasoline equivalence (MPGGE), and powertrain effectiveness.

Largely on the basis of the MIT study, we have chosen vehicles listed in Table 3 as our basis for comparison. Please note that some advanced vehicles are not included in the other studies. Table 3 also lists the timeframe for baseline and ATV analysis under each study. Only DTI didn't explicitly provide a timeframe for their analysis, rather, their analyses are based largely on PNGV and DOE targets. On the basis of this, we interpret their timeframe to be around 2005.

Table 3 – ATVs Considered in the Four Studies

Vehicle Type	MIT	GM	DTI	ADL
Current SI-ICE	X	X		
Timeframe	2020	2005-2010	2005	2010
Evolutionary SI-ICE	X			
Advanced SI-ICE	X		X	X
Advanced CI-ICE	X	X		X
Hybrid SI-ICE	X	X		X
Hybrid CI-ICE	X	X	X	X
Hybrid CNG SI-ICE	X		X	
Hybrid H2 FC	X	X	X	X
Hybrid Gasoline FC	X	X	X	X
Hybrid MeOH FC	X	X	X	X

While all four studies provide MPG figures for various ATVs, only the MIT study provides complete figures of both vehicle mass and TTW efficiency. The GM study does not provide vehicle mass, and DTI and ADL studies do not provide TTW efficiencies. We estimate mass for the GM case, and develop a powertrain effectiveness measure for comparison among all studies that eliminates the need for TTW efficiency for the ADL and DTI studies.

FUEL ECONOMY BASED COMPARISION

The fuel economy in our analysis is defined as gasoline-equivalent miles per gallon: MPGGE or simply MPG. The baseline GM vehicle is a pickup truck and weighs about 1,900 kg; the baseline MIT and ADL vehicles are midsize cars and weigh about 1,300 kg; and the baseline ADL vehicle weighs about 1,170 kg. While it's not appropriate to compare the GM study directly with the other three for fuel economy values, it is more reasonable to compare directly MIT, DTI, and ADL results. As discussed earlier, we choose a vertical assessment approach, where the comparisons between baselines and ATVs are based on same timeframe for each study. Figure 5 shows the vertical assessment based on three different timeframes: DTI, GM, and ADL for 2005-10; and MIT for 2020, for hydrogen and gasoline FCVs and corresponding baseline vehicles.

Figure 5 shows that all three H2 FCVs (MIT, DTI, and ADL in solid squares) are grouped near the top of the map, demonstrating clear energy benefits of H2 FCVs on all scenarios. The three gasoline FCVs (in solid triangles) are grouped in the middle of the map, and the baseline vehicles (in solid diamonds) are at or near the bottom (connected by a dark dotted line). Since GM vehicles are all much heavier and should have lower MPGGE, all GM vehicles are in open symbols, so that they will not dominate the illustration. Among the baseline vehicles, the biggest distinction appears to be the obvious upswing of the MIT base vehicle in 2020, overtaking MIT's gasoline reformer FCV. The other studies estimate clear advantages of gasoline FCVs over the corresponding baseline vehicles. However, the figure also illustrates that if GM had an improved 2010 baseline, some of the advantage it estimates would disappear. This map clearly illustrates that MIT is bullish on the future baseline CV technologies.

Table 4 shows MPG gains over respective baseline vehicles for these studies, under different timeframes, as also shown in Figure 6. The MIT 2020 results show MPG gains range from -14% for the gasoline reformer FCV, to 92% for the direct H2 FCV. GM 2010 results show a range of 18% for diesel ICE vehicle, to 138% for H2 FCV. DTI 2005 results show up to 173% MPG gain for the direct H2 FCV. ADL 2010 study shows a 39% MPG gain for conventional diesel ICE vehicles. The "Gap"

Figure 5 – "Vertical" Assessments of ATVs and Baseline Vehicles for these Four Studies

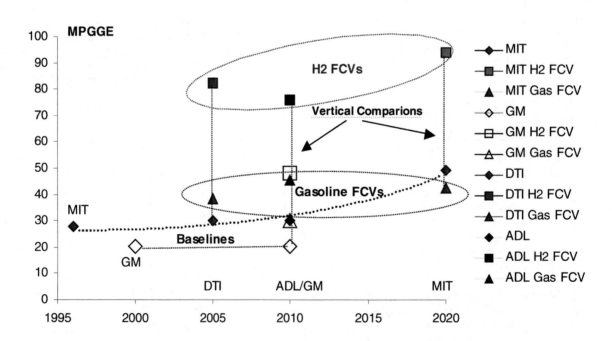

Table 4 – MPGGE and MPGGE Gains Over Baseline Vehicles for the Four Studies

MPG and Gains		MIT 2020		GM 2005-10		DTI 2005		ADL 2010		Gain
Vehicle Type	Fuel Type	MPGGE	Gains	MPGGE	Gains	MPGGE	Gains	MPGGE	Gains	Gap
Current SI-ICE	Gasoline	27.8		20.2		30.1		30.1		
Base evolutionary SI-ICE	Gasoline	43.2								
Advanced SI-ICE	Gasoline	49.1								
CI-ICE	Diesel	56.0	14%	23.8	18%			41.7	39%	25%
Hybrid SI-ICE	Gasoline	70.8	44%	24.4	21%			38.0	26%	23%
Hybrid CI-ICE	Diesel	82.3	68%	29.4	46%	58.0	93%	44.0	46%	47%
Hybrid CNG SI-ICE	CNG	73.4	49%			50.9	69%			20%
Hybrid H2 FC	Hydrogen	94.1	92%	48.1	138%	82.3	173%	76.0	152%	82%
Hybrid gasoline FC	Gasoline	42.3	-14%	30.2	50%	38.4	28%	45.6	51%	65%
Hybrid Methanol FC	Methanol	56.9	16%	34.5	71%	54.6	81%	48.5	61%	65%

Figure 6 – Fuel Economy Gains of Advanced Vehicles over Baseline Vehicles in the Four Studies

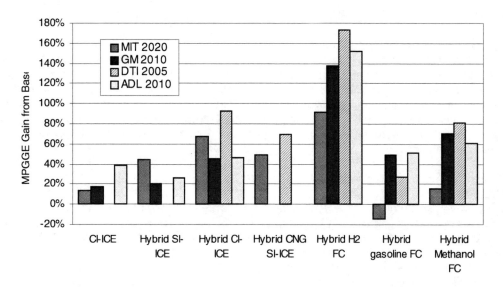

Figure 7 – Fuel Economy Gains and Gaps in the Four Studies

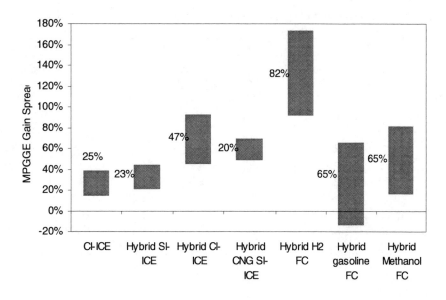

column in Table 4 represents the spread of the results over a different timeframe, or the largest differences of MPG gains among these studies, as shown in Figure 6. These figures clearly show that the gaps in MPG gain vary greatly among these studies, ranging from 23% to 82%. Not too surprisingly, the average gap rises from two-study comparisons to four-study comparisons. Figure 7 also shows that (by the measure printed in the figure), the largest uncertainty gaps among these three studies are associated with FC-based ATVs.

In the next two sections, we look separately at weight and a newly defined powertrain effectiveness measure. Both influence fuel efficiency. After first examining mass alone, we construct a powertrain effectiveness measure that combines effects of TTW efficiency and powertrain mass on fuel consumption required to move the glider and its passengers.

DECOMPOSITION PART A - MASS ANALYSIS

As we discussed earlier, vehicle fuel economy is significantly influenced by vehicle weight. While weight of conventional vehicles is well established, the assumptions associated with future technology vehicle weight are less certain. This uncertainty will spill over into uncertainty in vehicle fuel economy. In the following analysis, we will compare weight assumptions associated with these four studies.

Vehicle weight of each ATV is presented explicitly by the MIT, DTI, and ADL studies. GM did not present vehicle weight in its study. However, since the baseline vehicle is an MY 2000 Silverado pickup truck, and given information for both city/HWY MPGs, 0-60 time, and vehicle powertrain efficiencies, we are able to select the likely base Silverado model and estimate vehicle weight at each stage of advanced technology development.

Table 5 presents curb weight of baseline and advanced technology vehicles in these studies. Vehicle weight for the GM study is from our own estimates (in *italics*). The reference vehicle weight in each study is in **bold**. Table 5 also presents vehicle weight gains over baseline vehicles (in percentage). Again, these analyses show large differences, as demonstrated by Figures 8 and 9. It also appears that these differences do not show a consistent pattern that can be easily explained by timeframe differences. We can also separate the whole vehicle weight into "glider" and "powertrain" portions, and examine the weight increases of the powertrain portion only. More detailed discussion on this matter can be found in the Appendix. We consider the vehicle weight increment to be mostly from powertrain weight gain, and compare vehicle percentage weight gains as a first-order approximation analysis.

Table 5 shows that the weight-gain gaps among these studies range from 7% to 34%. The GM study estimates considerably higher weight gains associated with ICE-based hybrid vehicles. One of the large gaps is from diesel hybrid vehicles, where the DTI study shows a 5% decrease, MIT shows a 5% increase, and GM shows a 15% increase in mass. Both GM and MIT estimate much higher weight increases for FC-based vehicles. ADL assumes weight decreases for all ATVs with the exception of methanol FCVs. The largest gap is from hybrid gasoline FCVs, with MIT showing 32% mass increase and ADL a -2% increase.

In comparing MIT and GM studies, MIT assumes aggressive powertrain and vehicle weight reduction for ICE hybrid vehicles, while GM's estimates appear to be based on current technology. However, this appears not to be the case for fuel cell based advanced technology vehicles. The FC vehicle weight impacts are relatively close. Both MIT and GM comparisons indicate 15-17%

Table 5 – Curb Weight of Advanced Vehicles in the Four Studies

| Technology | MIT 2020 | | GM 2005-10 | | DTI 2005 | | ADL 2010 | | Wt. |
	Mass curb wt. (kg)	Increase over base	Mass curb wt. (kg)	Increase over base	Mass curb wt.	Increase over base	Mass curb wt.	Increase over base	Gain Gap
Current Reference, SI-ICE	1,322	31.3%	*2,134*		**1,168**		**1,304**		
Base evolutionary SI-ICE	1,108	10.0%							
Advanced SI-ICE	**1,007**								
CI-ICE	1,062	5.5%	*2,135*	0.5%			1,284	-1.5%	7%
Hybrid SI-ICE	1,023	1.6%	*2,430*	13.9%			1,289	-1.2%	15%
Hybrid CI-ICE	1,060	5.3%	*2,454*	15.0%	1,109	-5.1%	1,282	-1.7%	20%
Hybrid CNG SI-ICE	1,039	3.2%			1,083	-7.3%			10%
Hybrid H2 FC	1,179	17.1%	*2,457*	15.2%	1,155	-1.1%	1,219	-6.5%	24%
Hybrid gasoline FC	1,330	32.1%	*2,695*	26.3%	1,339	14.6%	1,284	-1.5%	34%
Hybrid Methanol FC	1,253	24.4%	*2,731*	28.0%	1,277	9.3%	1,314	0.8%	27%

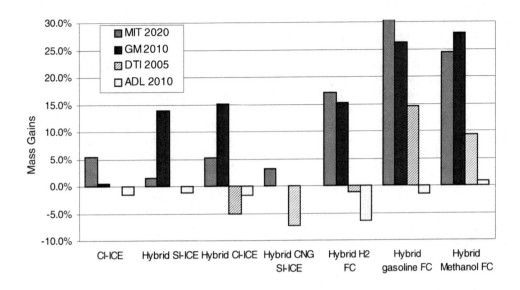

Figure 8 – Mass Gains of Advanced Vehicles over Baseline Vehicles in the Four Studies

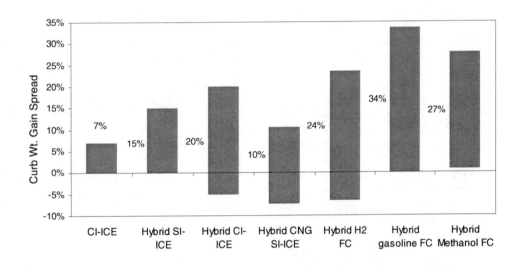

Figure 9 – Advanced Vehicle Weight Gains and Gaps in the Four Studies

weight increase in direct hydrogen fuel cell vehicles. MIT assumes a 32% increase in mass for gasoline FCVs and 24% for methanol FCVs; GM assumes a 26% increase for gasoline FCVs and 28% in methanol FCVs.

Among all these studies, the ADL and DTI analyses demonstrate consistently optimistic estimation of weight impacts for all advanced technology vehicles. This is a result of ADL and DTI's more optimistic assumption for both specific power of powertrains and application of lightweight materials for all advanced vehicles, while it appears that GM assumes current technology-based materials. In other words, much of ADL and DTI's MPG gains are the results of more aggressive mass reduction projections. In the Appendix, we illustrate the important effect of vehicle glider (body/chassis) mass in altering estimated changes in total vehicle mass when specific power of the powertrain is held constant.

DECOMPOSITION PART B – POWERTRAIN EFFECTIVENESS

While MIT and GM provide TTW efficiency estimates, DTI and ADL do not. TTW estimates should themselves be created by use of the specific vehicle simulation model that was used in the study. An external attempt to recover the TTW values with another vehicle simulation model would introduce another potential source of error in the estimate of what study authors intended. In order to compare the estimates of powertrain effectiveness across all four studies, we would like to have a relatively easy metric to construct that would allow us to circumvent the problem of missing TTW efficiency estimates.

TTW efficiency itself does not tell the entire story, since mass does matter, though some simple calculations imply that this is a secondary effect. For the most efficient conventional vs. FCV comparisons – the hybrid diesel vs. the hydrogen-fueled hybrid FCV – the MIT estimate of TTW efficiency gain shown in Table 2 is 16.5% (36% vs. 30.9%), and for GM it is 68% (41.4% vs. 24.6%). The gasoline equivalent fuel economy gains indicated in Table 4 are less, at 14.3% (94.1 vs. 82.3 mpgge) for MIT and 64% (48.1 vs. 29.4 mpgge), due to the fact that the hydrogen-fueled hybrid FCV powertrain is heavier than the diesel hybrid powertrain. So, although TTW is a good measure, and probably will capture most of the cause for difference in fuel economy gain, it does have the flaw that it does not completely address the net effects of both TTW efficiency and powertrain mass.

After deliberation, we suggest and compile a candidate standard measure of powertrain fuel consumption improvement effectiveness. First, although we elsewhere in this paper present fuel economy numbers that are consistent with the studies examined, we have recently argued that evaluations of advanced technology should be based on comparisons of fuel consumption, not fuel economy (10). Thus, the measure we suggest is based on fuel consumption (liters per km) rather than fuel economy (km per liter). We also convert here to metric units, though the papers cited consistently presented results in terms of miles per gallon of gasoline equivalent.

We ask the rhetorical question, what are the respective purposes of a vehicle and its powertrain? Think of the horse and buggy. The buggy is the object designed to carry passengers, and its motive power may be readily switched, and is clearly a separate part. In a modern automobile, the flexibility to switch motive power is far more limited. Nevertheless for the problem at hand, it is actually the glider (like the buggy) that is the object to be moved. Effectiveness of alternative powertrains in moving the glider and its passengers is the topic of interest. Ignoring economic issues here, we compare these technologies only on the basis of units of energy required to move a glider and passengers after a given powertrain has been incorporated into it. Since the question at hand is whether powertrain switches in a base glider should be made, we use the mass of the base (reference vehicle) glider and passenger load as the denominator in estimating energy use (liters of gasoline equivalent) per kg to estimate powertrain effectiveness. This corrects for the fact that TTW efficiency gives "credit" for moving the mass in a heavier powertrain, when no such credit is logically justifiable. This removes the vast majority of effects of different sizes of vehicle on the comparisons of powertrain effectiveness across studies.

Appendix A illustrates how much difference glider mass can make when constructing MPG change comparisons that do not standardize for glider characteristics.

The measure we present is "liters of gasoline equivalent consumed per 1000km of travel per 1000kg of glider plus passenger load." This metric provides two-digit values shown in Table 6. Glider mass used in the cases in Table 6 are as follows: MIT Current reference = 930 kg; MIT evolutionary 2020 = 845 kg; all other MIT 2020 = 756 kg; GM = 1451 kg; DTI = 852 kg; ADL = 900 kg. The GM glider mass is an educated guess. Glider mass share for the MIT 1996 reference passenger car was 29%. For DTI, the current reference car glider mass share was 27%, and for the ADL case, it was 31%. We assumed that the mass of the body of a pickup relative to its powertrain is less than that for a passenger car, and assumed 32% for the GM case.

Note that there is contradictory information in the DTI studies. Table 1 of Ref. 5 indicates glider mass is 852 kg. However, Table 4 of Ref. 4 indicates that the gliders of the Hybrid CI-ICE and Hybrid CNG SI-ICE are 694 and 723 kg respectively, with the same total vehicle mass as that we show in Table 5. However, according to our system of comparison, it is the mass of the base case glider that is to be used for the powertrain effectiveness measure. If the powertrain is heavier and causes modification of the glider that adds weight, this is a penalty assigned to the powertrain. Similarly, if the powertrain is lighter, allowing the glider mass to be less, then this provides a fuel saving properly credited to the lighter powertrain. Nevertheless, this degree of difference in reported glider mass calls into question the assumption that the same glider has been used in the DTI cases. MIT did have changes in glider mass consistent with the direction of change of powertrain mass, but these changes were *far* less relative to the base vehicle than reported by DTI.

Please note that since the powertrain effectiveness measure does rely on the underlying vehicle glider mass, the rule of selecting baseline vehicles still applies. The powertrain effectiveness measure has corrected for several comparison problems. These are: (1) the huge mass difference in the GM pickup truck and the remaining passenger cars; (2) the tendency of MPG values to "explode" for relatively small changes in actual fuel consumption; (3) the fact that the common glider in the MIT study was a 2020 glider, much lighter than the 1996 reference car characterized.

According to the estimated powertrain effectiveness measure, the net powertrain effectiveness of baseline vehicles for the 1996-2010 time frame are very similar to one another (ranging from 73 to 79 in Table 6), with the exception of the MIT 2020 baseline advanced SI-ICE (54 in Table 6). All studies are very similar in hydrogen H2 FCVs and methanol FCVs as well. Significant differences exist for diesel ICEs, hybrid ICE vehicles, and gasoline reformer FC vehicles among these studies. In the 2005-2010 timeframe, DTI is considerably more optimistic about the Hybrid CI-ICE powertrain than GM and ADL, and considerably more pessimistic about the hybrid gasoline FCV. To the extent that the hybrid ICE results of DTI are representative of optimism for relatively conventional powertrains, this optimism appears to be reflected in the MIT study as well, only to a greater

Table 6 - Liters of gasoline equivalent per 1000km per 1000kg of glider plus passenger load for four studies

Technology	MIT 2020		GM 2005-10		DTI 2005		ADL 2010		L/kg Range divided by Mean
	Liters/glider kg measure	Drop vs. current ref.	Liters/glider kg measure	Drop vs. current ref.	Liters/glider kg measure	Drop vs. current ref.	Liters/glider kg measure	Drop vs. current ref.	
Current Ref., SI-ICE	79		*73*		79		75		8%
Evolutionary SI-ICE	55								
Advanced SI-ICE	54								
CI-ICE	*47*	-13%	62	-15%			54	-28%	28%
Hybrid SI-ICE	*37*	-31%	61	-17%			60	-21%	45%
Hybrid CI-ICE	*32*	-41%	50	-31%	41	-48%	52	-32%	45%
Hybrid CNG SI-ICE	*36*	-33%			47	-41%			26%
Hybrid H2 FC	*28*	-48%	31	-58%	29	-63%	30	-60%	9%
Hybrid gasoline FC	62	15%	*49*	-33%	62	-22%	50	-34%	24%
Hybrid Methanol FC	46	-15%	*43*	-41%	44	-45%	47	-38%	9%

degree in the 2010 timeframe. The DTI studies were completed before the MIT studies (two 1998 DTI publications are cited by MIT) and both were completed before the GM and ADL studies. The DTI pessimism about the hybrid gasoline FCV is also reflected in the MIT results.

Generally, with the exception of gasoline and methanol FCVs, MIT characterizations of 2020 powertrain effectiveness indicate lower inherent fuel consumption than for the same technologies in the 2005-2010 time frame from the other three studies. The gasoline and methanol FCVs are exceptions. However, careful inspection reveals a pattern of greater incremental optimism for the variants of conventional drivetrains than for FCVs as a class of powertrains.

Thus, relative to the three studies characterizing 2005-2010 technology, MIT results imply greater gains from 2010 to 2020 for conventional drivetrain technology than for FCVs. In effect, MIT results imply that powertrain technology that has been in the market for over a century (ICEs and electric drive – though admittedly, electric drive has been in, out, and back in) will continue to improve from 2010-2020, while FCVs will not. However, another paper presented in this meeting suggests that there is potential for hydrogen FCVs to be even more efficient than characterized in these four studies (20).

Surprisingly, among the four studies, there is apparently considerable agreement based on this powertrain effectiveness measure. It should be stressed that this agreement masks considerable underlying differences. Study of Table 5 vs. Table 6 illustrates that some of the similarities are likely the result of differences in estimates of changes in powertrain mass and TTW efficiency that offset one another. For general purposes, it is comforting to see a degree of similarity in estimates of what advanced powertrains can achieve. However, if opinions differ considerably concerning the degree to which it is TTW gain or achievement of good specific power that contributes most to the powertrain's improved effectiveness, then the efficient allocation of research

dollars will be more difficult. Thus, we stress that those who prepare such studies in the future should present estimates of powertrain mass, kw, and TTW efficiency for each glider/powertrain pair. MIT's study is the model for documentation of assumptions and estimates.

While it is beyond the scope of this paper to delve into many of the more important details in these studies, we note that details can alter results considerably. Degree of hybridization is one example where differences can result (10, 11, 13, 17). For example, using numbers for the 10-second 0-60 time cases from Ref. 17, we compute powertrain effectiveness values of 43 for a "full" SI-ICE hybrid, and 47 for a "mild" hybrid (Table 7). The full hybrid provided 48% of peak power with a motor, while the mild hybrid provided 25%. The MIT case in Table 6 provides 33%, while the ADL cases that we cite in this study are for their "small battery" case, for which 9% of power is provided by a motor in the hybrid SI-ICE. GM's degree of hybridization is unknown, but we speculate that it would be between 9% and 25%. As discussed in Ref. 13, in the GM case, the hybridization also provides faster acceleration, a choice that works against fuel efficiency. So, while it warrants a separate study in itself, the degree of and approach toward hybridization can result in a wide range of fuel efficiency changes when a vehicle is hybridized. Some of the ADL and GM vs. MIT differences for hybrid ICEs are undoubtedly explainable by differences in degree of hybridization and performance requirements/changes.

Among the references cited, the only one to conduct sensitivity analysis with respect to effects of performance requirements was Plotkin et al [17]. This study only examined SI-ICE HEVs. Like ADL and DTI, this study failed to include TTW efficiency. However, taking advantage of the powertrain effectiveness measure, Table 7 illustrates that study's indication that acceleration performance requirements make a large difference in the effectiveness measure, especially for conventional vehicles. It can be seen that the performance effectiveness measure for a conventional vehicle from

Table 7 - L/1000km/1000kg-glider/pass [17]

0-98 km/h >	12 sec.	10 sec.	8 sec.
Conventional	57	65	79
Mild HEV	46	47	51
Full HEV	42	43	46

that study is consistent with the estimates in Table 6, since the acceleration capabilities of the conventional vehicles are in the 8-9 second range for 0-98 km/h (0-60 mph) time. Table 6 also illustrates how comparison of conventional vehicles with 8-9 second 0-98 km/h capability to those with 10-12 second capability can exaggerate estimated fuel consumption reduction estimates.

Another probable cause of differences in ICE results among the studies is the assumption of direct injection technology by MIT for the ICE engines. GM and ADL do not specify the ICE technology used, nor the peak efficiency of the ICE engines simulated. If they do use less inherently efficient ICE technology, then this contributes to some of the variation in powertrain effectiveness shown in Table 6.

Considering differences in hybridization approach, engine technology assumptions, and time frame, a good bit of the differences between the MIT study's projections of the potential for advanced ICE powertrains and those of the other three studies could be explained. However, the information given in the studies is not enough to readily develop an estimate of how much.

The large hybrid gasoline FCV differences also require explanation and evaluation. We hope to contribute to an explanation in a separate future paper.

CONCLUSIONS

This paper presented a comparative assessment of ATV energy consumption improvement potential from the "most probable" cases of four recent major studies carried out by MIT, GM, DTI, and ADL. Detailed analysis reveals that some of the "most probable case" estimates from these studies differ greatly on the implied MPGGE benefits of advanced technology vehicles at the tank-to-wheel level, leading to a superficial appearance of considerable uncertainties in estimating well-to-wheel greenhouse gas (GHG) emission reduction potentials.

Specifically, this paper compared these analyses on the basis of previously compiled fuel economy (MPGGE), mass and/or tank-to-wheel efficiency measures. Finally, we developed and discussed a suggested measure of powertrain efficiency that is meaningful across studies with widely varying light duty vehicle types. While the MPGGE and mass analyses provide results commonly

used, the superior (a) TTW efficiency and (b) liters of fuel consumed to move glider and passengers measures provide most of the intrinsic insight regarding the differences of powertrain effectiveness estimated among these studies.

This investigation indicated that while the degree of uncertainty across studies is considerable, it is not as great as a summary investigation and comparison implies. Our investigation suggests that there are logical and systematic reasons for variations among the studies. We note where more thorough documentation of results and input assumptions could help us to explain even more of the differences. Among the four studies discussed in detail here, the Weiss et al (MIT) group of analysts provided the most technical information and implicitly set up the best overall set of comparison rules. We have tried to build on that transparency. In addition to our assertion that differences can be understood with better documentation, we have also evaluated the methods used in the studies we have examined, and have suggested rules and tools for comparison that we hope could contribute to improvements in future studies.

ACKNOWLEDGMENTS

We gratefully acknowledge the support of Drs. Robert S. Kirk, Steven G. Chalk, Patrick Davis, and Peter R. Devlin of the Department of Energy Office of Energy Efficiency and Renewable Energy in supporting this and other cited efforts to evaluate studies comparing advanced technology vehicles. We also gratefully acknowledge the support of Dr. Larry Johnson, Director of the Argonne National Laboratory Transportation Technology Research and Development Center. We would like to thank John DeCicco and John German for comments provided. Errors are, of course, the responsibility of the authors, not the supporting institutions or reviewers. The manuscript was authored by a contractor of the U.S. Government under contract no. W-31-109-ENG-38. Accordingly, the U.S. Government retains a nonexclusive, royalty-free license to publish or reproduce the published form of this contribution, or allow others to do so, for U.S. Government purposes.

REFERENCES

1. Weiss et al., 2000, *On the Road in 2020: A Life-Cycle Analysis of New Automobile Technologies*, MIT Energy Laboratory Report No. MIT EL 00-003, Energy Laboratory, Massachusetts Institute of Technology, Cambridge, Mass., Oct.

2. General Motors Corp., et al., 2001, *Well-to-Wheel Energy Use and Greenhouse Gas Emissions of Advanced Fuel/Vehicle Systems — North American Analysis, Executive Summary Report,* available electronically at Argonne National Laboratory's

Transportation Technology Research and Development Center web site http://www.transportation.anl.gov/ at document address http://www.tis.anl.gov.8000/db1/ttrdc/document/DDD/126.PDF.

3. Thomas, C.E., B.D. James, F.D. Lomax, and I.F. Kuhn, 1998, *Integrated Analysis of Hydrogen Passenger Vehicle Transportation Pathways,* Draft Final Report prepared by Directed technologies, Inc. (Subcontract No. AXE-6-16685-01) for the National Renewable Energy Laboratory, Golden, Colo., March.

4. Thomas, C.E., B.D. James, F.D. Lomax, and I.F. Kuhn, 1998, "Societal Impacts of Fuel Options for Fuel Cell Vehicles," SAE paper 982496, Society of Automotive Engineers, Warrendale, Pa.

5. Thomas, C.E., 1999, *PNGV-Class Vehicle Analysis: Task 3 Final Report,* prepared by Directed Technologies, Inc. (Subcontract No. ACG-8-18012-01) for the National Renewable Energy Laboratory, Golden, Colo., March.

6. Arthur D. Little, Inc., 2002, *Guidance for Transportation Technologies: Fuel Choice for Fuel Cell Vehicles,* Final Report, Arthur D. Little, Inc. Cambridge, Mass., Feb. 6.

7. Stodolsky, F., L. Gaines, C. Marshall, and F. An. 1999, "Total Fuel Cycle Impacts of Advanced Vehicles." SAE Technical Paper 1999-01-0322.

8. Ford Motor Company Research Laboratory, 2000, *Direct-Hydrogen-Fueled Proton Exchange Membrane Fuel Cell System for Transportation Applications: Final Technical Report,* prepared for the U.S. Department of Energy's Office of Transportation Technologies under Contract No. DE-AC02-94CE50389, Dearborn, Mich., Dec.

9. Graham, R., et al., 2001, *Comparing the Benefits and Impacts of Hybrid Electric Vehicle Options*, Final Report, July 2000, Electric Power Research Institute, Palo Alto, Calif.

10. Santini, Danilo J., A.D. Vyas, J. Moore, and F. An, 2002, *Comparing Cost Estimates for U.S. Fuel Economy Improvement by Advanced Electric Drive Vehicles*, presentation at the 19th International Electric Vehicle Symposium and Exhibition "The Answer for the Clean mobility" Oct.19-23, BEXCO, Busan, Korea.

11. Santini, D.J., A.D. Vyas, and J.L. Anderson, 2002, "Fuel Economy Improvement via Hybridization vs. Vehicle Performance Level," Future Car Congress 2002 paper 02FCC-27, Arlington, Va., June 3–5.

12. Santini, D.J., A.D. Vyas, R. Kumar, and J.L. Anderson, 2002, "Comparing Estimates of Fuel Economy Improvement Via Fuel Cell Powertrains," Future Car Congress 2002, paper 02FCC-125, Arlington, Va., June 3–5.

13. Santini, D.J., et al., 2002, "Hybridizing with Engine Downsizing," paper number 02-4095, 81st Annual Meeting of the Transportation Research Board, Washington, D.C., Jan. 13–17.

14. National Research Council, 2002, *Effectiveness and Impact of Corporate Average Fuel Economy Standards*, National Academy Press, Washington D.C.

15. An, F., J. DeCicco, and M. Ross, 2001, "Assessing the Fuel Economy Potential of Light Duty Vehicles," SAE paper 2001-01FTT-31, Society of Automotive Engineers, Warrendale, Penn.

16. Santini, Danilo J., A.D. Vyas, J. L. Anderson, and F. An, 2001. *Partnership for a New Generation of Vehicles' Fuel Economy Goal: Evaluation of Trade-Offs Along the Path.* Transportation Research Record 1750, pp. 3-10.

17. Plotkin, S., et al., 2001, *Hybrid Vehicle Technology Assessment: Methodology, Analytical Issues, and Interim Results,* Argonne National Laboratory Report ANL/ESD/02-2, Argonne, Ill.

18. An, F., F. Stodolsky, and D. Santini, 1999, "Hybrid Options for Light-Duty Vehicles," SAE paper 1999-01-2929, Reprinted from *Electric and Hybrid Electric Vehicles and Fuel Cell Technology* (SAE SP-1466), Society of Automotive Engineers, Warrendale, Penn., presented at the Future Transportation Technology Conference, Costa Mesa, Calif., Aug.

19. An, F., A. Vyas, J. Anderson, and D. Santini. 2001. "Evaluating Commercial and Prototype HEVs," SAE Paper No. 2001-01-0951, Society of Automotive Engineers, Warrendale, Pa.

20. Rousseau, A., R. Ahluwahlia, B. Deville, and Q. Zhang, 2003, "Well-to-Wheels Analysis of Advanced SUV Fuel Cell Vehicles," SAE World Congress paper 2003-01-0374, Society of Automotive Engineers, Warrendale, Pa.

APPENDIX - ISSUES OF TECHNOLOGY TIMELINE AND SCENARIOS

In the section on Methodological Issues, we mentioned that ATV comparisons should use the same baseline glider (related to Issues 1 and 9) and be based on the same timeline (Issue 6) and technology scenarios (Issue 5). We have also discussed the differences between "vertical" and "diagonal" assessment (Figure 1). This section addresses for the first time in our own work how the effort to standardize the technology baseline, timeline, and scenario can affect the outcome.

We note that this effort is hardly the only way, as noted in our ten points made earlier. Another dramatic effect involves performance equivalence assumptions (Issue 2), particularly for the power of the powertrain chosen for each ATV in a set of comparisons. How this can affect results is illustrated in Santini, Vyas, Kumar and Anderson (12).

On the performance-equivalence issue, within these four studies, the DTI and ADL studies appear to deviate considerably from the GM and MIT studies. For ATVs, DTI and ADL allow much lower peak power per unit of vehicle mass than in the reference vehicle. The MIT study holds this constant. While the DTI study does not provide any explanation of the performance capabilities of the vehicles, ADL concludes from examination of prior examples that 11.5 seconds or less is acceptable 0-60 performance for ATVs, but compares such vehicles to a base conventional vehicle capable of considerably faster acceleration (≈ 0.076 kW/kg for base vehicle – about 9-10 seconds capability – see Ref. 17). The potential for use of unequal 0-60 time capability to significantly affect results for compared pairs is shown in Ref. 12, although the question of the "right" performance metric is not.

We note that the MIT and GM studies adopted constraints related to acceleration performance that should result in reasonably good comparability, so re-estimation to account for that effect should not significantly alter the readjusted MIT vs. GM results we present here. Nevertheless, the reader should keep in mind that many of the papers cited discuss different attempts to develop standard comparisons. A comprehensive analysis according to the ten points presented here (a challenge) has not yet been developed.

Previous MIT advanced vehicle technology analyses are based on the 2020 advanced SI-ICE because it shares the same vehicle glider with sequential ATVs. Now we would like to ask the following question: How can we assess MIT's ATV projection over current baseline vehicles (diagonal assessment)? What would happen if MIT's ATVs were based on a current baseline vehicle glider, as in the GM study? How would this change the

outcome? Here we would like to conduct a hypothetical experiment to answer these questions.

There are two fundamental reasons for choosing the MIT case in this Appendix: 1) MIT provides most complete data for detailed analysis; and 2) by establishing three baseline scenarios as current SI-ICE, 2020 evolving ICE, and 2020 advanced ICE, the MIT study provides a platform to conduct this "what if" analysis.

First, Table A1 presents the mass breakdown on all MIT vehicles. The glider mass includes predominantly vehicle chassis and body weight; propulsion system includes predominantly the engine (ICE or FC) and transmission systems; fuel mass is 2/3 of full tank. Load is a standard 136 kg two passenger weight (300 lb.) and is the difference between test weight and curb weight. Table A1 also lists rated power of the IC/FC engine, battery weight and rated power, total rated power of the vehicle. The last column lists power to weight ratio, which MIT held constant to maintain hypothetical "performance equivalence" throughout all ATVs. We note that vehicle simulation results in references 9 and 17 imply that a vehicle with a motor rather than ICE engine as a source of power can achieve the same 0-60 time with approximately 10% lower peak kw/kg requirements. However, the DTI and ADL studies far exceed this decline in kw/kg in their comparisons.

In MIT's assumptions, the glider mass decreases from the current SI-ICE to 2020 base ICE, and further to advanced SI-ICE. However, glider mass remains roughly the same afterwards. The small differences in different ATVs' glider mass are adjustments to support different weights of the propulsion system. Figure A1 shows the 1996-2020 evolution of the conventional powertrain specific power (defined as rated peak power/weight in kW/kg), and the specific power estimates associated with different 2020 ATV powertrain technologies. The higher number, the better. It shows that, with the exception of the big jump in specific power from current base ICE to 2020 evolving ICE, the hybrid and fuel cell based propulsion systems have diminished specific power rates relative to conventional ICE powertrains, thus increasing powertrain mass in the hybrids and FCVs. However, in comparison with the current SI-ICE (Table A1), all the ATVs have much smaller glider mass (using much lighter materials), so the total vehicle test weights of ATVs are lighter than the current SI-ICE, with the exception of the gasoline FCV.

In the following analysis, we assume all advanced ATVs use the same glider as MIT's 1996 SI-ICE with glider mass of 930 kg, as shown in Table A2. In the process, we increase mass of all other components accordingly, as well as rated power to keep the power-to-weight ratio of both the vehicle and propulsion system unchanged. It then becomes possible to project the impacts of ATVs as if they were installed in MIT's current SI-ICE (as GM did).

Table A1 – MIT Vehicle Mass Breakdown (based on Table 3.4 of Reference 1)

	Mass (kg)						Rated Power			
	Glider*	Propulsion System	Battery	Fuel	Load	Total Test Wt.	IC/FC Engine (kW)	Battery (kW)	Total kW	kW/kg
Current SI-ICE	930	340	12	27	136	1445	110		110	0.076
2020 Base SI-ICE	845	226	12	16	136	1235	93		93	0.075
Advanced SI-ICE	756	217	12	15	136	1136	85		85	0.075
Advanced CI-ICE	759	271	12	13	136	1191	89		89	0.075
Advanced SI-HEV	756	216	36	11	136	1155	58	29	87	0.075
Advanced CI-HEV	759	251	37	9	136	1192	60	30	89	0.075
Advanced H2 FCV	763	371	41	2.7	136	1314	66	33	99	0.075
Adv. gasoline FCV	794	465	46	17	136	1457	73	36	109	0.075
Adv. methanol FCV	778	390	43	28	136	1375	69	34	103	0.075

Figure A1 – Specific Power of Propulsion System in MIT Analysis

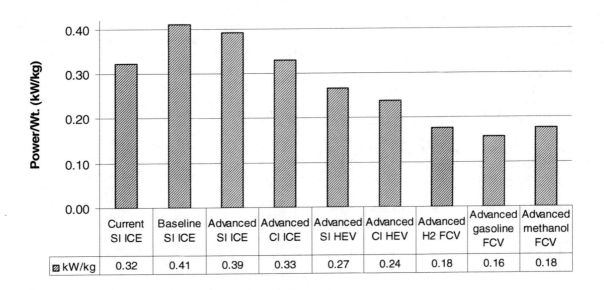

	Current SI ICE	Baseline SI ICE	Advanced SI ICE	Advanced CI ICE	Advanced SI HEV	Advanced CI HEV	Advanced H2 FCV	Advanced gasoline FCV	Advanced methanol FCV
kW/kg	0.32	0.41	0.39	0.33	0.27	0.24	0.18	0.16	0.18

Table A2 – Hypothetical MIT Vehicle Masses based on Current SI-ICE Vehicle Glider

Hypothesis	Mass (kg)						Rated Power			
	Glider*	Propulsion System	Battery	Fuel	Load	Total	IC/FC Engine (kW)	Battery (kW)	Total kW	kW/kg
Advanced SI-ICE	930	259	14	18	136	1357	102		102	0.075
Advanced CI-ICE	930	322	14	17	136	1419	106		106	0.075
Gasoline SI-HEV	930	258	43	13	136	1380	69	34	103	0.075
Diesel CI-HEV	930	298	44	11	136	1420	71	35	106	0.075
H2 FCV	930	439	48	3.3	136	1556	78	39	116	0.075
Gasoline FCV	930	528	52	20	136	1666	83	41	124	0.075
Mmethanol FCV	930	452	50	33	136	1602	80	40	120	0.075

This allows us a fairer comparison of the degree of relative optimism/pessimism for ATV powertrains in the two studies, as if those powertrains had been installed in contemporary vehicle gliders in both cases. The implications do change.

Table A3 shows the adjusted weight gain comparisons among MIT-current (based on the hypothetical case of Table A2), MIT 2020 (based on the original mass distribution of Table A1), and the GM case. It shows that compared to MIT-2020, the MIT-current case dramatically reduced percentage weight gains for all ATVs. This implies that, based on this adjusted technology scenario, MIT's assessment of mass impacts for all ATVs is greatly reduced, as clearly demonstrated by Figure A2, which shows that the ATV weight-gain gaps between the MIT and GM cases significantly increase. This is a logical reflection of MIT's assumption that its powertrain technology is based on the year 2020, which appears to be much more optimistic in comparison to GM's assumption.

The reduction in percent weight gains translates into increases in fuel economy gains, as shown in column 1 of Table A4. Here we use the approximation that the tank-to-wheel efficiencies of Table 4 will be unchanged by glider and powertrain masses. Thus the reductions in weight gain translate into decreases of percent gain of workload of the vehicles, resulting in increased percent gains in the revised MIT-current fuel economy gains for all ATVs. Under this assumption, Table A4 lists the adjusted MPG estimates for the MIT current cases, and compares the adjusted fuel economy gains among the MIT current case, MIT 2020 case, and GM case. The MIT current case shows significant jumps in MPGGE for all ATVs, as further demonstrated by Figure A3.

Table A3 – Comparisons of Original and Revised Curb Weight Gains of MIT-current, MIT-2020, and GM ATVs

Curb_Weight Gains	MIT current	MIT 2020	GM	MIT_current less GM
Advanced SI-ICE	-7.6%	0.0%		-8%
Advanced CI-ICE	-3.0%	5.5%	1%	-3%
Gasoline SI-HEV	-5.9%	1.6%	15%	-21%
Diesel CI-HEV	-2.9%	5.3%	17%	-19%
H2 FCV	7.4%	17.1%	17%	-9%
Gasoline FCV	15.7%	32.1%	29%	-13%
Methanol FCV	10.9%	24.4%	31%	-20%

There are two striking trends shown in Figure A3: while the gaps between MIT and GM cases have increased significantly for ICE-based ATVs, the gaps for gasoline and methanol based FCVs have narrowed significantly. The gap for the H2 FCV is about the same, but with the revised MIT result over-predicting, instead of under-predicting, the GM result.

Table A4 – Comparisons of Fuel Economy Gains of "Current Baseline" based ATVs and "2020 Baseline" based ATVs

MPGGE Gains	Adjusted MPG	MIT_ current	MIT_ 2020	GM	MIT_current less GM
Advanced1 SI-ICE	41.4	49%	0.0%		19%
Advanced CI-ICE	47.3	70%	14%	18%	18%
Gasoline SI-HEV	59.5	114%	44%	21%	51%
Diesel CI-HEV	69.3	150%	68%	46%	54%
H2 FCV	79.5	186%	92%	138%	-9%
Gasoline FCV	37.2	34%	-14%	50%	-42%
Methanol FCV	49.4	78%	16%	71%	-29%

Table A4 shows that advanced SI-HEV gains of 114% in MPG over the current SI-ICE baseline vehicles, and advanced CI-HEV gains of about 150% in MPG. Among these gains, the SI-ICE technology itself gains 49% in MPG over the current SI-ICE baseline vehicles, and hybridization itself gains another 44%. For diesel vehicles, advanced CI engine technology gains about 70% in MPG and hybridization gains another 47%. For the GM cases, the gasoline HEV gains 21% in MPG over its baseline ICE vehicle, while the diesel hybrid gains 46% over the baseline, with only 18% gain from diesel engine technology alone. These numbers have two implications: 1) MIT projects very significant potentials for further improvement in gasoline and diesel engine technologies; and 2) MIT also projects higher potential for hybridization than GM does.

In all MIT HEV and FCV cases, the battery power is 33% of peak power, which qualifies as full hybrids similar to Toyota Prius. The 40% plus MPG gain associated with full hybridization is consistent with our previous analysis on the Prius and other full hybrids [18]. Based on GM's estimate of 21% gain associated with hybridization, our prior research [16] implies that if GM is considering a so-called "mild" hybrid, with battery power fraction below 15%, then the estimates of percent gains via hybridization by MIT and GM are perfectly reasonable. So it is fair to say that the MIT result represents an upper bound potential of current hybrid technology, while GM assumes a less aggressive degree of hybridization. This demonstrates the importance of technology scenarios.

Since the MIT and GM hybridization percentage gains appear consistent with the few contemporary hybrids that exist, it appears that MIT is not overly optimistic about improvement in hybrid technology in the next 20 years. Clearly, MIT's very large gains for conventional powertrains imply an opinion that much improvement can be squeezed out of further refinement of today's dominant technology. For the FCV cases, the narrowed gaps are perhaps deceptive, because the narrowed MPG gaps are largely associated with widening weight gaps. In other words, MIT's optimistic assumption associated with mass of FC powertrains results in narrowed net MPGGE gain gaps vs. GM. The more intrinsic comparisons rely on tank-to-wheel efficiency comparisons as described in the text. Nevertheless, these comparisons demonstrate the importance of baseline, timeline, and scenarios of technological analysis.

Having made adjustments to make the MIT and GM studies more comparable, it appears that the largest difference between the MIT and GM studies is in the estimate of the net potential for improvement and refinement of the conventional powertrain over the next 20 years. In this regard, the GM study provides no competing estimate to the MIT study. In effect, GM did not provide an opinion concerning how great an improvement can be made over 20 years in the conventional powertrain. Notably, both studies imply great potential for the H2 FCV. Given the research that we have conducted on contemporary hybrids and simulated hybrids with contemporary component and engine efficiencies, these two studies also imply that hybrid technology has already reached a plateau.

Figure A2 – Comparisons of Curb Weight Gains of "Current Baseline" based ATVs and "2020 Baseline" based ATVs

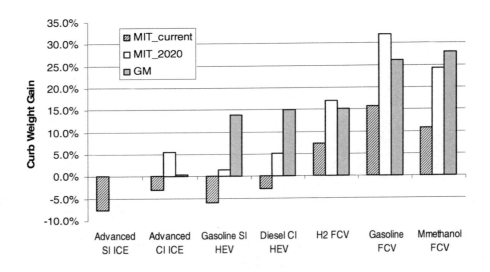

Figure A3 – Comparisons of Fuel Economy Gains of "Current Baseline" based ATVs and "2020 Baseline" based ATVs

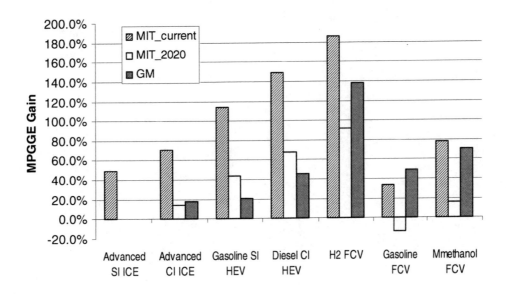

The Performance of Future ICE and Fuel Cell Powered Vehicles and Their Potential Fleet Impact

John B. Heywood, Malcolm A. Weiss, Andreas Schafer, Stephane A. Bassene and Vinod K. Natarajan
Laboratory for Energy and the Environment
Massachusetts Institute of Technology

ABSTRACT

A study at MIT of the energy consumption and greenhouse gas emissions from advanced technology future automobiles has compared fuel cell powered vehicles with equivalent gasoline and diesel internal combustion engine (ICE) powered vehicles [1][2]. Current data regarding IC engine and fuel cell vehicle performance were extrapolated to 2020 to provide optimistic but plausible forecasts of how these technologies might compare. The energy consumed by the vehicle and its corresponding CO_2 emissions, the fuel production and distribution energy and CO_2 emissions, and the vehicle manufacturing process requirements were all evaluated and combined to give a well-to-wheels coupled with a cradle-to-grave assessment.

The assessment results show that significant opportunities are available for improving the efficiency of mainstream gasoline and diesel engines and transmissions, and reducing vehicle resistances. Battery parallel hybrid systems with these improved engines and vehicles are more efficient still, but are significantly more costly. Vehicles with fuel cell systems, with gaseous hydrogen as fuel, are also significantly more efficient, but when the hydrogen fuel production energy is included in the assessment, no significant advantage remains.

The impacts of several of these vehicle technologies on US light-duty vehicle fleet fuel consumption were also assessed, using an empirical data-based model of the in-use fleet as it evolves over time. Fleet impacts are delayed due to both the time required for new and improved technologies to be mass produced and spread substantially across total new vehicle production, and due to the 15 year vehicle lifetime. These fleet calculations show that extrapolating the trends of the past 15 years will likely result in a 60% increase in US light-duty vehicle fleet fuel consumption by 2030. Effective ways to offset this are through efficiency improvements where implementation can start soon, and dealing with growth in fleet size and vehicle usage.

INTRODUCTION

Automotive manufacturers and suppliers around the world are investing heavily in the development of fuel cell systems (FCSs) for light duty vehicles. We at MIT have been assessing new automobile technologies that could be commercialized by 2020 with respect to life-cycle greenhouse gas (GHG) emissions, energy efficiency, and cost. As in all comparisons of future alternatives, the results depended on the assumptions made. The assumptions and methodologies used in our studies are given in detail in [1] and [2]. One purpose was to determine how competitive FCSs would be in comparison with internal combustion engine based vehicle systems (ICESs). The primary motivation for these assessments was to evaluate new automobile technologies which might lower emissions of GHGs believed to contribute to global warming. The GHG of most concern is the carbon dioxide (CO_2) in the exhaust of vehicles burning petroleum or other carbon-containing fuels. The transportation sector accounts for about 30% of all CO_2 emissions in OECD countries, and about 20% worldwide.

Any assessment of emissions from future vehicle technologies must consider the total system over its entire life cycle. The life cycle of an automotive technology is defined here to include all the steps required to provide the fuel, to manufacture the vehicle, and to operate and maintain the vehicle throughout its lifetime up to and including scrappage and recycling. Provision of the fuel from primary energy sources such as petroleum or natural gas must be considered from the point of resource recovery from underground reservoirs through transportation to refineries or plants where those resources are converted to fuels for vehicles. The fuel must then be distributed and deposited in the vehicle's fuel tank. The total of these steps is often called "well-to-tank". Analogously, the vehicle manufacture begins with ores or other raw and recycled materials necessary to make the parts included in a vehicle, fabrication and assembly of those parts, and distribution of the finished vehicle to the customer. The vehicle is then operated until the end of its lifetime when the vehicle is scrapped and recycled. Vehicle operation is often called "tank-to-wheels". "Well-to-wheels" normally means "well-to tank" plus "tank-to-wheels" but does not ordinarily include vehicle manufacture which should be included in any comprehensive life cycle analysis.

Part of our project included assessing the impact of both powertrain and vehicle improvements on total US light-duty-vehicle fleet fuel consumption and GHG emissions. Thus, this paper also compares the impact of the more promising of these future vehicle technologies on total US vehicle light-duty-vehicle fleet fuel consumption under various vehicle performance and penetration scenarios.

INDIVIDUAL VEHICLE TECHNOLOGIES AND PERFORMANCE

VEHICLE, POWERTRAIN, AND DRIVING PARAMETERS

All the vehicles examined in this study are functional equivalents of today's typical US mid-size family sedan. For the customer, this means that characteristics such as acceleration, range, passenger and trunk space, remain constant in future vehicles. All vehicles are designed to have the same ratio of peak power to vehicle mass, namely 75 W/kg, which is approximately today's average value and roughly equalizes the short-time acceleration performance of all vehicles.

The propulsion systems evaluated here consist of a) advanced spark-ignition (SI) and compression-ignition (CI) ICEs, fueled by gasoline and diesel respectively, as stand-alone engines and in parallel hybrid configurations, and b) fuel cell systems fueled by compressed hydrogen or by dilute hydrogen (about 40% by volume) in gas generated by processing gasoline on board, also in both non-hybrid and hybrid configurations. The systems are listed in Table 1. For all hybrid systems the battery and electric motor were sized to provide a ratio of peak electrical power to vehicle mass of 25 W/kg, and the power plant (ICE or FC) to provide 50 W/kg, giving the total of 75 W/kg cited above. All hybrid systems included regenerative braking.

The future advanced gasoline spark-ignition engine was assumed to have an 8% improvement in indicated efficiency (to 41%) and a 25% reduction in engine friction relative to current values, as a result of design changes, direct gasoline injection, variable valve control, and

increased compression ratio. The future advanced turbocharged diesel had a 7 percent higher indicated efficiency (52%) and a 15% reduction in friction, again relative to current values, through increased boosting, improved combustion control and other design changes. Allowance was made for a loss in efficiency due to the diesel aftertreatment technology that will be required in the future.

All vehicles, except the 2001 reference and the 2020 evolutionary base case, used the same type of advanced body with reduced vehicle mass (e.g. more extensive use of aluminum) and resistances (e.g. lower drag coefficient and rolling resistance) [1]. These vehicles are compared to the typical current (2001) US mid-size family sedan, for "reference", and to a 2020 evolutionary "baseline". Both the reference and the baseline are gasoline-fueled ICE cars with similar capacity and performance; the baseline has evolutionary improvements in powertrain and vehicle technology over the next 20 years or so similar to improvements achieved during the last 20 years.

Table 1. Propulsion Systems Asssessed

Propulsion System	Description
Gasoline ICE	Advanced SI engine and auto-clutch transmission
Gasoline ICE hybrid	Gasoline ICE engine with continuously variable transmission plus battery and electric motor in parallel
Diesel ICE	Advanced CI engine and auto-clutch transmission
Diesel ICE hybrid	Diesel ICE engine with continuously variable transmission plus battery and electric motor in parallel
Hydrogen FC	Fuel cell operating on 100% compressed hydrogen with electric drive train
Hydrogen FC hybrid	Hydrogen FC with addition of a battery
Gasoline FC	Like the Hydrogen FC, but fueled by hydrogen produced by processing gasoline on board
Gasoline FC hybrid	Gasoline FC with addition of a battery

The performance of each of the vehicles we assessed was calculated using computer simulations described in [1]. Originally developed by Guzzella and Amstutz [3] at the ETH, Zurich, these simulations back calculate the fuel consumed by the propulsion system by driving the vehicle through a specified cycle. Such simulations require performance models for each major propulsion system component as well as for each vehicle driving resistance.

FUEL CELL SYSTEM PERFORMANCE

Since one focus of this study is the comparative energy consumption of advanced fuel cell vehicles, our assumptions about the future performance of fuel cell systems (FCS) are critical, and were developed as follows. We define the FCS to include a fuel processor (for gasoline fuel) which converts the fuel chemically to hydrogen, hydrogen cleanup equipment, a fuel cell "stack" which converts the hydrogen energy electrochemically to electric power, associated equipment for heat, air, and water management, and auxiliary equipment such as pumps, blowers, and controls.

The overall efficiency of an FCS is defined here as the net DC energy output of the stack (obtained from the gross output by subtracting the electrical energy needed to operate FCS auxiliaries such as pumps and compressors) divided by the lower heating value (LHV) of the fuel consumed in the FCS—whether gasoline fed to a fuel processor or hydrogen gas from a high pressure tank or other on-board hydrogen storage system. That overall efficiency will vary with the load on the fuel cell and will generally increase as load decreases except at very low loads when parasitic power losses and/or fuel processor heat losses become comparatively high and overall efficiency declines.

We assume that all these FCSs include proton exchange membrane (PEM) stacks in which hydrogen, pure or dilute, fed to the anode side of the electrolyte reacts with oxygen in air at the cathode side of the electrolyte to produce water and electric power. The anode and cathode are porous electrodes impregnated with catalytic metals, mostly platinum. We assume the stacks operate at about 80ºC and a maximum pressure (at peak power) of about 3 atmospheres.

Fuel cell systems fueled by pure hydrogen iincur two types of efficiency losses: the losses in the stack (polarization loss during electrochemical conversion of the hydrogen's chemical energy to electrical energy) and the loss of generated electric power used to power the auxiliary equipment. FCSs fueled by dilute hydrogen from gasoline reformate have the same two types of losses in efficiency that pure-hydrogen FCSs suffer, but also have two additional types: 1) losses in the "fuel processor" during conversion of gasoline (by reaction with steam and air) to hydrogen and subsequent cleanup, and 2) incomplete hydrogen utilization in the

stack. We assume a hydrogen utilization of 85% as in [1]. That is, 15% of the hydrogen entering the stack from the fuel processor is purged and leaves the stack unreacted (but may be used to heat the reformer).

Our objective was to identify advances in FCS technology that were plausible—but not assured—with aggressive development, but not assume advances that depended on hoped-for but not yet demonstrated technical innovation. The new stack polarization data we used correspond to the current Ballard Mark 900 80 kW stack [4] with unit cell voltage increased by 0.05 V (about 5 to 8%) at all current densities to anticipate further improvements. We also assumed that operating a stack of given area on gasoline reformate rather than pure hydrogen would reduce peak power density and cell voltage by amounts consistent with the Ballard Mark 900 experience [5]. Table 2 lists the polarization data used. For our stack conditions, the ideal unit cell voltage is 1.22-1.23 V; this ideal voltage excludes all the losses found in an operating fuel cell. We defined peak power as the power level at which unit cell voltage drops to 0.6 V for both pure hydrogen and reformate fuels.

Table 2. Stack Polarization Data

Current Density mA/cm^2	Unit Cell Voltage, V	
	100% H_2	40% H_2 (reformate)
0	1.05	1.03
25	0.94	0.92
50	0.90	0.88
100	0.87	0.84
200	0.84	0.81
400	0.79	0.75
600	0.75	0.71
800	0.72	0.67
1000	0.68	0.61
1050	--	0.60
1200	0.63	--
1300	0.60	--

For FCSs fueled by processing gasoline to hydrogen, a customary expression of efficiency of the processor (including removal of CO from the gas stream) is equal to the LHV of the hydrogen in the gas stream leaving the processor divided by the LHV of the gasoline fed to the processor. This efficiency can be increased by supplying heat to the fuel processor by burning the hydrogen in the tail gas purged from the stack. Table 3 lists the efficiencies assumed in our most recent study [2] for gasoline fuel processors feeding a stack whose peak power output is about 60 kW. At high power, the efficiency is 0.81 LHV compared to 0.725 LHV assumed in our earlier study [1]. US DOE's current 2001 "baseline" (at peak power) is 0.76 [6]. Some reformers under development are claimed to have higher efficiencies but, according to a Ford authority quoted by DeCicco [7], "Effective reformers exist only in the laboratory".

Of the energy needed to drive the FCS auxiliaries, primarily pumps and blowers for water, air, and heat management, the largest load is the air compressor which delivers air to the cathode compartments of the stack; some of the air compressor load can be offset by an expander powered by the cathode exhaust gas. Table 4 shows our assumptions about total net requirements for auxiliary power expressed as a fraction of stack gross power.

Table 3. Efficiencies of Gasoline Fuel Processors

Stack Gross Power, % of Peak	Efficiency $LHV_{H2\,Out}/LHV_{Gasoline\,In}$	
	Previous Study [1]	This Study
0	0.725	0.60
5	0.725	0.73
10	0.725	0.79
20	0.725	0.81
30	0.725	0.81
100 (Peak)	0.725	0.81

Table 4. FCS Auxiliary Power Requirements

Stack Gross Power, % of Peak	Auxiliary Power as Percent of Gross Stack Power	
	Previous Study [1]	This Study
5	15	15
10	15	12
20	15	10
30	15	10
100 (Peak)	15	10

FUEL CELL AND VEHICLE COST AND WEIGHT

In addition to projecting future advances in FCS efficiency, we considered the prospects for reduction of both FCS cost and weight. Many projections of FCS costs reflect targets rather than an analysis of specific design and manufacturing steps that would directly determine FCS costs. An example is the FreedomCAR target [8] of $30/kW by 2015 (Table 5) for FCS systems fueled by hydrogen (including fuel tank) or by gasoline. For comparison, the ADL analysis carried out for DOE [6] estimates high volume manufacturing costs in 2001 for gasoline FCS to be $249-$324/kW. In another study [9], ADL did estimate potential costs of future FCS using results from analyses done with DOE and EPRI, and concluded that "factory costs of future FCVs would likely be 40-60% higher than conventional vehicles". Typical annual ownership costs for fuel cell vehicles would therefore be about $1200 to $1800 higher than for ICE vehicles. Long-term factory costs for the FCS were estimated at about $105/kW for hydrogen and about $130/kW for gasoline fuel processor FC systems.

Table 5. Unit Cost and Weight of Future Fuel Cell Systems - *Ex fuel and storage*

Source [Reference]	100% Hydrogen Fuel		Gasoline Reformate	
	$/kW	kg/kW	$/kW	kg/kW
Previous Study [1]	60	2.9	80	4.8
ADL DOE [6]	28	1.8	45	3
FreedomCAR [8]	30*	3.1*	30	--
ADL [10]	105	--	130	3.5

*   Includes hydrogen storage

The vehicle costs we used are those reported in [1]. Total vehicle costs were $18,000 for our 2020 baseline vehicle, 8 and 14% higher for advanced gasoline and diesel vehicles respectively, 17 and 23% higher for gasoline and diesel hybrids, and 23 and 30% higher for hydrogen and gasoline fuel cell hybrids. Our assumptions about fuel costs also are those developed in [1], since no new technologies have been identified that make a major change in the costs of the fuels we considered. ADL [10] notes that our fuel costs and fuel-chain energy use and GHG emissions are comparable to other studies. Our previous projections for FCS unit weights [1] still look optimistic but achievable and we have not changed them.

OVERALL FUEL CELL SYSTEM EFFICIENCY

Overall fuel cell efficiencies are listed in Table 6 under the heading "Components". These numbers combine the efficiencies (or losses) of the individual FCS components listed in Tables 2 to 4 with no allowance for performance degradation due to design compromises needed to obtain the best combination of characteristics of the total powerplant in the vehicle. Examples of such compromises—often to reduce cost, weight, or space or to provide for warm-up or transients—would be lower stack efficiency due to smaller stack area, lower processor efficiency due to simpler but less-effective processor heat management, or lower hydrogen utilization through changed stack design and operation.

For a total integrated system, we assumed an increase of 5% in the losses in each component. The column "Integrated" in Table 6 shows overall FCS efficiencies based on the component efficiencies column but additionally assuming: a) in the stack, unit cell voltage is reduced 5% (from, say, 0.8 V to 0.76 V) at any given power density, b) auxiliary power requirements are increased 5% (from, say, 10% of net output to 10.5%) at any given power, and c) all efficiencies in the reformer are decreased 5% (from, say, an efficiency of 0.80 to 0.76). Hydrogen utilization remained at 85%. These assumed losses due to integration result in significant increases in FCV fuel consumption relative to the "component" assumption. Consumption of on-board fuel per vehicle km traveled increases about 9 to 23% depending on the driving cycle, fuel, and hybridization.

ON THE ROAD RESULTS

ON-BOARD ENERGY USE

Table A1 in the Appendix lists the assumed characteristics, and the on-the road and life-cycle energy consumptions and GHG emissions of all the ICE vehicles we assessed. Table A2, also in the Appendix, does the same for all the fuel cell vehicles. Additional details can be found in [1] and [2].

Figure 1 shows the combined 55% urban/45% highway US Federal Test Procedure driving cycle results. All of the tank-to-wheels energy consumptions are compared on a relative scale where 100 is defined as the consumption of the "baseline" car—a gasoline-engine non-hybrid car—with lower-cost evolutionary improvements in engine, transmission, weight, and drag assumed to take place by 2020. The projected on-board fuel consumption of the baseline car in this combined driving cycle is 5.4 liters of gasoline/100 km which is equivalent to 43 miles per gallon or 1.75 MJ (LHV)/km. The 2001 predecessor of the baseline car had a fuel consumption of 7.7 l/100 km (30.6 mpg) or 2.48 MJ (LHV)/km.

The bar for each of the fuel cell vehicles in Fig. 1 (and also in Figs. 4 and 5) has a shaded area and a hatched area. The shaded area indicates the fuel consumption based on assuming that each of the components of the FCS can operate as efficiently as shown in Tables 2 to 4 with an overall FCS efficiency shown in the "Components" columns of Table 6. The hatched area shows the additional fuel consumption due to efficiency losses through integration as summarized in the "Integrated" columns of Table 6. In comparing different vehicles, modest differences are not meaningful due to uncertainties in the assumptions.

HYBRID BENEFITS FOR DIFFERENT DRIVING CYCLES

The advantage of hybrid systems relative to their non-hybrid equivalent depends on many factors: maximum power split between engine and electric motor; electrical power/ICE power transition thresholds; engine's efficiency variation over its load and speed map; transmission characteristics; capacity of the battery system to absorb regenerative power; characteristics of the vehicle driving cycle. While we explored several of the technical issues listed above to ensure that the details of the vehicle configurations we analyzed made sense, we examined the effects of different standard driving cycles on this hybrid non-hybrid comparison more extensively in [11].

The driving cycles used were the US Federal Urban and Highway Cycles, the US06 cycle, the New European Driving Cycle, and the Japanese 15-Mode Cycle. The characteristics of these different driving

Table 6. Overall Fuel Cell System Efficiencies

Net Output Energy, % of Peak	100 x Net DC Output Energy/Fuel LHV			
	100% Hydrogen Fuel		Gasoline Reformate Fuel	
	Components	Integrated	Components	Integrated
5	76	71	46	42
10	75	71	50	45
20	74	70	49	44
40	69	65	46	42
60	65	61	44	39
80	61	58	41	37
100	53	50	36	33

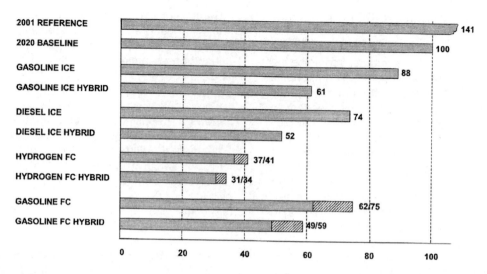

Figure 1: Relative on-board consumption of fuel energy for vehicle technology combinations. MJ(LHV)/km expressed as percentage of baseline vehicle fuel use. All vehicles (except 2001 reference and 2020 baseline) are advanced 2020 designs. Driving cycle assumed is combined Federal cycles (55% urban, 45% highway). Hatched areas for fuel cells show increase in energy use in integrated total system which requires real-world compromises in performance of individual system components.

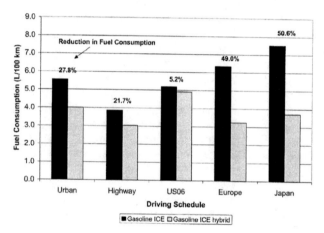

Figure 2. Effects of hybridization on gasoline SI ICE vehicles.

cycles are summarized in Table 7. The US Urban, European, and Japanese cycles are viewed as representing "urban" diriving in these three regions with low average speeds, substantial idle time, and repeated moderate acceleration and braking. The US Highway cycle represents "highway" driving, and average urban/suburban/highway fuel consumption is assumed to be 55% of the urban value added to 45% of the highway value. The US06 is a more recent high acceleration cycle intended to represent aggressive driving. Some auto companies use their own cycles (which can be roughly characterized as one-third of each of the US Urban, Highway, and US06 cycles) to represent modern light-duty vehicle driving.

Table 7. Characteristics of Driving Cycles Used

Driving Cycle	Duration(s)	Average speed (km/h)	Maximum Speed (km/h)	% Time at Idle	Maximum Acceleration (m,s^2)
US Urban	1877	34.1	91.2	19.2	1.6
US Highway	765	77.6	96.3	0.7	1.4
US06	601	77.2	129.2	7.5	3.24
European	1220	32.3	120	27.3	1.04
Japanese	660	22.7	70	32.4	0.77

Figure 2 shows the fuel consumption of the advanced gasoline ICE and gasoline ICE hybrid vehicles, and the reduction in fuel consumption (in %) the hybrid achieves for these five driving cycles. The greater the amount of "stop and go" driving (European and Japanese cycles) the greater the benefit (about 50% for these two cycles) from both increased average engine efficiency and regenerative braking. With higher speeds, and more aggressive accelerations, the hybrid benefit is much reduced (to 5% for the US06 cycle). (Note: modest differences in our assumptions for these technology combinations cause minor differences in calculated results. These are not significant.)

Figure 3 shows the results for the advanced fuel cell and fuel cell hybrid vehicles. The trends are similar, but the percentage hybrid fuel consumption benefits are significantly lower. Since the fuel cell is relatively more efficient at lighter loads, whereas the ICE exhibits the reverse trend, this part of the hybrid benefit (shifting the "engine" to higher loads) is much reduced.

These are specific illustrations of the impact of driving cycle characteristics on the fuel consumption advantages of the hybrid. Many other factors are important in this hybrid non-hybrid comparison, such as cost, towing capacity, performance on extended grades.

LIFE-CYCLE RESULTS

To estimate life-cycle energy consumption and GHG emissions, the energy use and GHG emissions for the fuel cycle, and the vehicle manufacturing cycle, were added to the tank-to-wheels estimates. The GHGs considered were CO_2 and methane from natural gas leakage: gC(eq) is equal to the carbon in the CO_2 released plus the carbon in a mass of CO_2 equal to 21 times the mass of methane leaked.

During the fuel cycle, gasoline and diesel fuels were assumed to be refined from crude petroleum and would have modest improvements in quality over the next 20 years. Hydrogen was assumed to be produced by the reforming of natural gas at local filling stations, and compressed to about 350 atmospheres for charging vehicle tanks. Energy consumptions during the manufacturing and distribution of these fuels were calculated to include energy from all sources required to produce and deliver the fuels to vehicle tanks. GHG emissions were calculated similarly. Results are given in Table 8 [1].

For vehicle "manufacturing" (which includes all materials, assembly, and distribution) we assumed, as in our previous report [1], intensive use of recycled

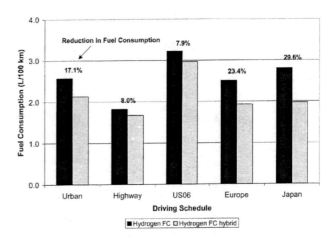

Figure 3: Effect of hybridization on fuel cell vehicles operating on direct hydrogen feed.

materials (95% of all metals and 50% of glass and plastics) and that manufacturing energy and GHGs were prorated over 300,000 km (vehicle life of 15 years driven 20,000 km/year). These manufacturing additions for the vehicles assessed ranged from 0.25 to 0.33 MJ/km in energy consumed and about 4.8 to 6.3 gC(eq)/km of GHGs released.

The full life-cycle results are shown for energy in Fig. 4, and for GHGs in Fig. 5. On a life-cycle basis, both energy consumption and GHG releases for the diesel ICE and hydrogen FC hybrid vehicles are closely comparable. The gasoline ICE and gasoline FC hybrids are not as efficient but, considering the uncertainties of the results, not significantly worse than the two other hybrids. Both life-cycle energy use and

Table 8. Fuel Cycle Energy Use and CO_2*

Fuel	Energy Use	Efficiency	GHG
	MJ/MJ		gC/MJ
Gasoline	0.21	83%	4.9
Diesel	0.14	88%	3.3
CNG	0.18	85%	4.2
F-T Diesel	0.93	52%	8.9
Methanol	0.54	65%	5.9
Hydrogen	0.77	56%	36
Electric Power	2.16	32%	54

*Per MJ of fuel energy in the tank.

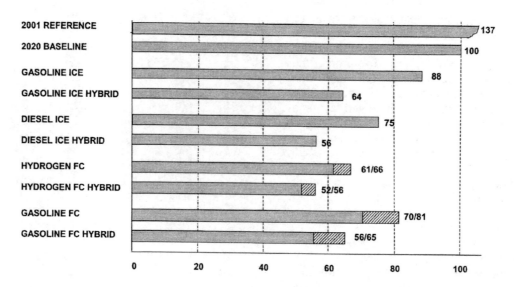

Figure 4: Relative life-cycle consumption of energy for vehicle technology combinations. Total energy (LHV) from all sources consumed during vehicle lifetime shown as percentage of baseline vehicle energy consumption. Total energy includes vehicle operation and production of both vehicle and fuel.

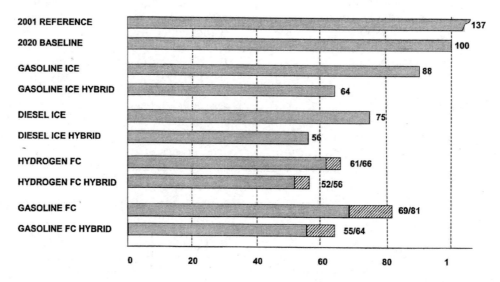

Figure 5: Relative emissions of life-cycle greenhouse gases for vehicle technology combinations. Total mass of carbon equivalent emitted during vehicle lifetime shown as percentage of baseline vehicle GHG emissions. Greenhouse gases include only CO_2 and CH_4 (assumed equivalent to 21 CO_2). Emissions include vehicle operation and production of both vehicle and fuel.

GHG releases from all four of these hybrids are between 52 and 65% of our 2020 baseline vehicle, and between 38 and 47% of our 2001 reference vehicle.

Table 9 breaks down life-cycle energy and GHG totals into the shares attributable to each of the three phases of the life cycle: operation of the vehicle on the road, production and distribution of fuel, and manufacture of the vehicle including embodied materials. The largest

single share of energy, ranging from 44 to 75% of the total, results from vehicle operation. The largest single share of GHGs, from 65 to 74%, is also attributable to operation except for hydrogen fuel where the fuel cycle accounts for about 80% of the total. Vehicle manufacturing increases its share of energy and GHGs for vehicles with higher on-the-road fuel economies, up to about 21%, comparable to several of the fuel cycle shares.

Table 9. Share of Life-Cycle Energy & GHG

Vehicle	Energy, % of Total			GHG, % of Total		
	Operation	Fuel Cycle	Vehicle Mfg.	Operation	Fuel Cycle	Vehicle Mfg.
2001 Reference	75	16	9	74	18	8
2020 Baseline	74	15	11	71	18	11
Gasoline ICE	73	15	12	72	18	10
Gasoline ICE Hybrid	69	14	17	67	17	16
Diesel ICE	75	10	15	74	12	14
Diesel ICE Hybrid	70	10	20	70	11	19
Hydrogen FC	45	34	21	0	81	19
Hydrogen FC Hybrid	44	35	21	0	79	21
Gasoline FC	67	14	19	66	16	18
Gasoline FC Hybrid	66	14	20	65	16	19

Note: Percentages for FCs are averages for "Component" and "Integrated" systems. Neither system varies more than about 1% from average.

POTENTIAL US FLEET IMPACTS

OVERVIEW

To this point, we have discussed individual vehicle characteristics. However, the impact of the above vehicle improvements in fuel consumption and GHG emissions that really matter are the resulting reduction in total US vehicle fleet fuel consumed and GHGs emitted. Due to both the long penetration times required for new technologies that do make it into mass production to grow in volume so they are used in a large fraction of each years' new vehicles, and the long lifetimes of vehicles in the in-use fleet (some 15 years), fleet impacts are significantly delayed.

We have examined these fleet impacts using a model of the U.S. car and light truck vehicle fleet [12]. The model calculates the effects of introduction of more efficient technology in new vehicles on fleet fuel consumption (and hence GHG emissions) over time. Historical data were used to check the validity of the fleet turnover calculations. The model is structured in three modules as follows:

(i) Fleet vehicle number and age distribution calculations, based on new vehicle sales each year, and vehicle retirement based on age-specific scrappage and removal rates and the fleet median age.

(ii) Annual vehicle usage distributions (km/year) for each major class of vehicles as a function of model year and vehicle age.

(iii) Annual fleet fuel consumption based on the fuel consumption characteristics of each vehicle technology and type, and model year, integrated over the vehicle usage and fleet make-up distributions.

US FLEET FUEL CONSUMPTION MODEL

New Vehicle Sales, Sales Mix, and Scrappage Rates - Projections were made of new passenger cars and light-duty trucks sales for each calendar year. Historical sales data were taken from [13]. In the reference case, the total light-duty vehicle sales were estimated to grow at the same rate as the U.S. population (0.8% per year on average from 2000 to 2030, according to the medium projection of the U.S. Bureau of Census). The light-duty truck share was modeled by extrapolating the historical data to a given 2030 market share by a second order polynomial curve. The reference case assumes that the current trend of increasing percentage of light trucks will increase from its current value of 50% of new vehicles and level off at 60% market share in 2030.

Historical data on vehicle scrappage rates were taken from [14] for model years 1970, 1980 and 1990. The vehicle survival rate data for each given model year were fitted using the following equation:

$$1 - \text{Survival Rate (t)} = \frac{1}{1 + e^{-\beta(t_0)}} ,$$

where, t_0 is the median age of the corresponding model year, t is the age on a given year, and β is a growth parameter defining how rapidly vehicles are retired around t_0.

The historical survival rate data for model years 1970, 1980 and 1990 show an increase in the median age of automobiles and a small decrease in the median age of light-duty trucks: see Table 10. The intermediate median age data were linearly interpolated for both fleets (passenger cars and light duty trucks). However, extrapolating this trend would lead to excessively high values for the median lifetime, so the median age was kept constant after the model year 2000.

Table 10: Median Age (years) [14]

	Model Year 1970	Model Year 1980	Model Year 1990
Cars	10.7	12.1	13.7
Light Trucks	16.0	15.7	15.2

Thus, the number of vehicles (passenger cars and light-duty trucks) in use for each model year and for any calendar year between 1960 and 2030 can be calculated. Since the calculation starts for model year 1960, the calculated total vehicle stock composition matches the data accurately only after 10 to 15 years, when the number of vehicles from model years prior to 1960 becomes negligible relative to the total stock.

Vehicle Kilometers Traveled (VKT) - Historically, vehicles have tended to drive less each year as they age. Data show that each calendar year, the annual distance traveled per vehicle for a given model year, decreases at a rate of 4.5% per year (Greene et al.,[15]). Thus, the usage degradation rate is kept constant in our model at 4.5% annual decrease; however, the distance traveled per year for new vehicles is allowed to evolve for each calendar year. The average annual growth rate of new vehicle kilometers traveled depends on economic conditions and the price of fuel. This rate has been 0.5% per year during the 1970-1998 period. The reference case assumes it remains at 0.5% per year from 2000 to 2030.

Vehicle Fuel Consumption - The fuel consumption of each model year was calculated as follows. For years before 2000, the historical data for average fuel consumption for new passenger car and light-duty truck fleets were used. For future model years, the performance characteristics of each considered technology were appropriately sales weighted to obtain the average new vehicle on-road fuel consumption for these two fleets. These projected "average" vehicle fuel consumptions for each model year serve as an input to the fuel use estimates. In all the scenarios considered, the future percentage improvement in light-duty truck fuel consumption was assumed to be the same as the improvement for passenger cars. A 17% increase was applied to US fuel consumption test procedure results to adjust these new vehicle fuel consumptions to on-the-road values. The 17% adjustment factor was also applied to ICE-hybrid vehicles since little data are yet available to calibrate on-road fuel consumption for this type of vehicle [12].

TECHNOLOGY PENETRATION SCENARIOS

In all the technology scenarios, the following input parameters remain constant: the average annual growth rate of new vehicle sales (0.8% per year); the annual growth rate of the average per-vehicle kilometers travelled (0.5% per year); the evolution of the share of light trucks in new light-duty vehicle sales (currently 50%, and rising to 60% market share in 2030). The five technology scenarios considered are following:

Reference Scenario (No Change) - The average new car and light-duty truck fuel consumptions remain at their 2000 levels until 2030 (on-road fuel consumption of 9.8 L/100 km for cars and 13.7 L/100 km for light trucks).

Baseline - The baseline scenario assumes a steadily decreasing fuel consumption for new vehicles as technologies for reducing vehicle fuel consumption are progressively rolled out by automakers into the light-duty fleet: see Fig. 6. Note this baseline inherently assumes that most of the realizable efficiency increase is not traded for larger heavier vehicles, higher performance, and other amenities. During the past decade or so, efficiency increases were fully traded for these attributes. Thus fuel consumption in all new 2005 vehicles decreases by 5% relative to new 2000 vehicles, and in new 2020 vehicles reaches the 35% reduction calculated in our technology assessment study "On the Road in 2020" [1] and reevaluated here. Further decreases in fuel consumption are assumed, to 50% of 2000 fuel consumption levels in new 2030 vehicles. These relative improvements are assumed to be the same for all light-duty vehicles.

Advanced Vehicles with Internal Combustion Engine Hybrids - To further reduce fuel consumption, more advanced technologies relative to those included in the baseline projection must come into production. In these fleet calculations, we considered ICE-hybrid vehicles as the incoming advanced technology. Again, the relative fuel consumption improvement for light-duty trucks is assumed to be the same as for cars. The current average fuel consumption of ICE hybrids was determined by scaling the fuel consumption of the Toyota Prius to a vehicle with the average mass of new passenger cars. The 2020 fuel consumption for the advanced gasoline ICE-hybrid vehicle is that calculated by our assessment here. Between these two levels, we assume a linear decrease. Beyond 2020, we assumed a less steep slope, leading to a 66% fuel consumption improvement in new 2030 hybrid vehicles (5% better than the 2020 value) relative to the 2000 baseline fuel consumption. These relative ICE hybrid fuel consumption improvements are also shown in Fig. 6.

The baseline fuel consumption assumptions (solid line in Fig. 6) apply to all the vehicles produced in a given model year. For hybrids, a production penetration scenario is needed. Three cases were considered (see Table 11):

- Low penetration scenario with a 2030 market share of 25%,
- Medium penetration scenario with a 2030 market share of 50%,
- High penetration scenario with a 2030 market share of 75%.

With these parameters, the sales-weighted fuel consumption was calculated for each calendar year, for both passenger car and light-duty truck fleets. These data are the input to the total fleet fuel use calculations.

FLEET SCENARIO RESULTS

Reference Scenario - This scenario assumes that light-duty vehicle fuel consumption is not reduced over the next 30 years, continuing the trend witnessed during the last 10-15 years, when improved vehicle efficiency was traded for performance, power, size, weight and other amenities while the CAFE standards remained unchanged. This scenario can be thought of as "business as usual." Table 12 shows this reference scenario light-duty vehicle fleet fuel use. Total fuel use grows steadily because of the fleet and vehicle kilometers traveled growth. The 2030 level (774 billion liters of gasoline per year) is 63% higher than the 2000 level. Light trucks account for about two thirds of the total fuel use in 2030.

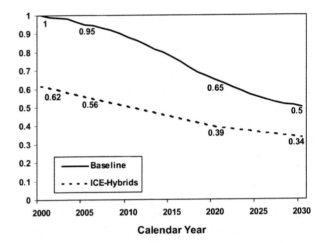

Figure 6: Relative Improvement in fuel consumption (vertical axis) relative to the 2000 new car average fuel consumption.

Table 11: Market Penetration Scenarios for New ICE-Hybrid Vehicles

Year	LOW %	LOW Thousand Vehicles	MEDIUM %	MEDIUM Thousand Vehicles	HIGH %	HIGH Thousand Vehicles
2005	0.5%	82	1.0%	163	1.5%	245
2010	2.1%	357	4.2%	713	6.2%	1,053
2015	7.2%	1,273	14.5%	2,563	22%	3,836
2020	16%	2,962	32%	5,942	48%	8,904
2030	24%	4,841	48%	9,702	73%	14,543

Table 12. Fleet Fuel Use for the Reference Case

	Billion Liters	Million Barrels per Day (Mbd)
1990	390	6.7
2000	475	8.2
2010	580	10.0
2020	680	11.7
2030	774	13.3

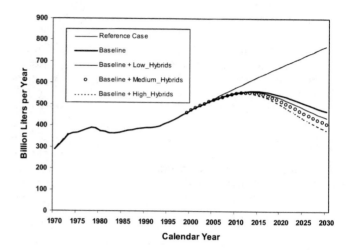

Baseline Scenario - This scenario assumes that fuel economy is no longer largely traded for increased performance, vehicle size/weight, and amenities, and the technologies progressively rolled out into the fleets result in significant vehicle fuel consumption improvements. As a result, average new car fuel consumption decreases steadily as defined by the solid curve in Fig. 6. This same percentage improvement in fuel consumption is assumed for new light trucks. In 2020, the average estimated new car and new light truck on-road fuel consumptions are 6.4 L/100 km and 8.9 L/100 km, respectively, as compared to the 2000 values of 9.8 L/100 km and 13.7 L/100 km.

The cumulative effect of these less-fuel-consuming vehicles results in significant fleet fuel savings compared to the reference case. Around 2015, the fuel consumption reduction offsets the growth in the fleet size and VKT, and total fuel use begins to decrease. The maximum fleet fuel use under the baseline scenario is 562 billion liters of gasoline per year in 2015, a 20% reduction over the reference case (in 2030 a 40% reduction is projected). Figure 7 shows the total fleet fuel use for these two cases.

Baseline + Advanced ICE-Hybrids - Here, ICE-hybrid vehicles, with the advanced body design, are substituted progressively for the baseline vehicles defined above. According to the three penetration rates, Low, Medium and High, the hybrid vehicles' market share gradually increases to 25%, 50% and 75% of the light-duty vehicle market share by 2030. Again, light trucks are assumed to gain the same percentage improvement in fuel consumption, and the fraction of hybrids in new light-duty vehicles is assumed to be identical for cars and light trucks. The fleet fuel consumption, and average vehicle fuel consumption, are shown in Fig. 7 and Fig. 8. Until about 2013, the impact of hybrids is negligibly small due to low (though growing) production numbers. Beyond about 2015, these hybrid fuel consumption improvements decrease baseline fleet fuel use by 2.6%, 5.2%, and 7.9% for the low, medium and high market share cases in 2020, and by 6.2%, 12.4% and 18.6% in 2030.

Note that to continue the decrease in the fleet energy use requires a continuing fuel consumption reduction for new vehicles to counterbalance the effects of growth in vehicle fleet size and increasing VKT.

Figure 7: Light-duty fleet fuel use for various technology scenarios.

Sensitivity to Fleet Growth and VKT - We also examined the effects of changes in (1) sales mix, (2) new vehicles sales growth rate, and (3) average annual VKT growth rate. The reference case assumption of 60% market share of light trucks in 2030, was changed to 50%, 40% and 30% of the 2030 light-duty vehicle market. The results are presented in Table 13.

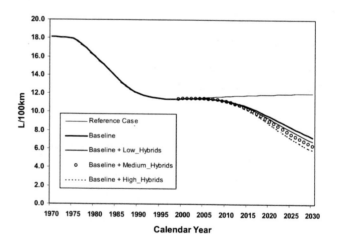

Figure 8: Average light-duty vehicle fuel consumption for various technology scenarios.

Table 13: Sensitivity Analysis of Light Truck Share of New Vehicle Sales

| | 2030 Light Truck Market Share | | | |
	60%*	50%	40%	30%
	Fuel Use (billion liters)	Percent fuel use change		
Reference				
2020	679	-0.7%	-1.3%	-2.0%
2030	774	-1.6%	-3.3%	-4.9%
Baseline				
2020	541	-0.6%	-1.2%	-1.8%
2030	467	-1.5%	-3.1%	-4.6%

* Reference case

The maximum reduction in fleet energy use due to changes in light truck market share, relative to the reference case, is 2% in 2020 and 5% in 2030 for a decrease in the share of light trucks to 30% of new sales in 2030. This percentage reduction, relative to the baseline fuel use level, is less.

The reference case assumes a 0.8% annual growth rate for the new light-duty vehicle sales. We analyzed the case where this average annual growth rate is halved to 0.4%. The effects are surprisingly significant. Half the reference case growth rate leads to an additional 6% fleet fuel savings in 2020, and 9% in 2030. It is plausible that a slow down in new light-duty vehicle sales might occur, due to approaching saturation in vehicles per licensed driver.

In the reference case, the average annual per-vehicle kilometers traveled grows at an annual rate of 0.5% from 2000 to 2030. The effect of reducing this increase was examined. The results show that a 0% growth of annual per-vehicle travel can lead to fuel savings of 8% of the baseline case level in 2020 and nearly 12% in 2030. Thus, successful travel reduction strategies can have a significant impact on the fleet energy use.

All these individual fuel conserving strategies are illustrated in Fig. 9, relative to the baseline technology scenario. A composite scenario was then examined. Relative to the baseline, this considers the introduction of advanced ICE-hybrids under the medium market share assumption (50% market share in 2030), concurrently with the improving baseline vehicles. In addition, the annual new vehicle sales growth rate is halved to (0.4%), while the annual per-vehicle kilometers traveled is assumed to remain constant (0% growth). This scenario also assumes a decline in the market share of light trucks to 40% in 2030. Such a composite scenario illustrates the potential impacts that a series of measures can have on the fleet fuel consumption. The

composite scenario is also shown in Fig. 9, and the quantitative benefits of each individual strategy and the composite strategy added to the baseline, are quantified in Table 14.

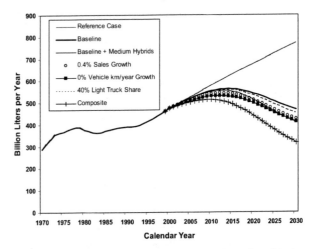

Figure 9: Light Duty Fleet Fuel Use for Various Scenarios

DISCUSSION

Several conclusions can be derived form the above fleet impact analysis. First, the projected reduction of new vehicle fuel consumption through improvements in mainstream technology (the baseline) provides the most significant savings in fleet energy use over the next 20 or so years because these improvements are substantial, and can be implemented in large volume most rapidly. This underlines the benefits of early action to improve vehicle fuel consumption. Changes in the share of light trucks in the new vehicle sales mix will have only a modest effect on fleet energy use. Measures like travel reduction and slowing down of the growth in fleet size, over many years, could have a significant impact on fleetwide fuel savings due to compounding. Considering the baseline scenario as a reference, the effect of the latter measures is comparable in magnitude to the introduction of advanced ICE-hybrids vehicles into the fleet. Also, travel and fleet growth reduction strategies have a more immediate effect on fleet fuel consumption.

As shown in Fig. 9, the total fleet energy use for the composite scenario peaks in 2020, five years earlier than what would be achieved if only technology improvements were implemented.

It is important to note that, with the assumptions of the reference scenario (on sales mix, sales and VKT growth), the model predicts that a minimum annual rate of reduction of average new vehicle fuel consumption of 1.3% is needed to offset the effects of stock and VKT growth and stabilize the total light duty vehicle fleet fuel use as shown in Fig. 10. This number is sensitive to new vehicle sales growth rate and per-vehicle annual

VKT growth rate. A continuing decrease in new vehicle fuel consumption is needed to limit the growth of light-duty vehicle fleet fuel use and GHG emissions.

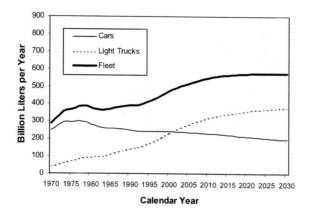

Fig. 10: Stabilization in light-duty vehicle fleet fuel use due to a steady 1.3% annual decrease in average new vehicle fuel consumption.

ICE hybrids (see Fig. 7) have limited impact relative to the baseline before about 2015 because production volumes have not become substantial enough. However, beyond that point their beneficial impact steadily increases as their relative production volume grows. Beyond about 2030, their increasing penetration into the in-use vehicle fleet steadily drops the total fleet fuel consumption below the baseline vehicle technology fleet consumption levels.

We have not examined the impact of fuel cell vehicles on fleet energy consumption. One reason is that with hydrogen as the fuel-cell fuel, how that hydrogen is produced and the energy consumed and GHG emissions released in its production and distribution are critical questions. With all vehicles using petroleum-derived fuels, the fuel production and distribution energy consumption is smaller (about 15%) and the relative penalty that is added to fuel use on the vehicle is constant. Note also that the lead times for fuel cell technology to enter large scale mass production and penetrate across a significant fraction of new car production will, at best, be much longer than ICE-hybrid technology which is already in limited mass production.

Fuel cell technologies and hydrogen are not likely to be available at acceptable cost, scale, and robustness to make significant contributions to petroleum reduction within this 20 or so year timescale. In any case their estimated well-to-wheels benefits are not significantly better than those achieved through ICE-based technology improvements unless the hydrogen used is produced without releasing significant CO_2.

CONCLUSIONS

Our assessment shows that substantial reductions in energy use and GHG emissions over the next 20 years can be achieved through improvements in mainstream vehicle technologies (ICEs, transmissions, and vehicles). Use of ICE hybrids would increase these reductions, but at significantly higher cost. However, judging solely by lowest life-cycle energy use and greenhouse gas releases, there is no current basis for preferring either fuel cell (FC) or internal combustion engine (ICE) hybrid powerplants for mid-size automobiles over the next 20 years or so using fuels derived from petroleum or natural gas. That conclusion applies even with optimistic assumptions about the pace of future fuel cell development.

All hybrid vehicles are superior to their non-hybrid counterparts, but their relative benefits are greater for ICE than for FC powertrains. Hybrids can reduce both life-cycle energy use and GHGs to between about 37 to 47% of current comparable vehicles, and to between about 52 to 65% of what might be expected in 2020 as a result of normal evolution of conventional technology.

These reductions in energy use and GHG releases result from not only advances in powertrains but also from reduction of both vehicle weight and the driving resistances of aerodynamic drag and tire rolling resistance.

If automobile systems with GHG emissions much lower than the lowest estimated here are required in the very long run future (perhaps in 30 to 50 years or more), hydrogen appears the most promising fuel option identified to date. But the hydrogen must be produced from non-fossil sources of primary energy (such as nuclear or renewables) or from fossil primary energy with effective carbon sequestration. Biofuels may also increase their currently limited role. A comparison of the on-the-road and life-cycle energy and GHG results for hydrogen—superior in the former but about the same in the latter—illustrates why a valid comparison of future technologies for light-duty vehicles must be based on life-cycle analysis for the total fuel and vehicle system.

The effects of new vehicle technologies such as hybrids on US fleet fuel consumption are significantly delayed due to both the time for these new technologies to achieve large-scale mass production and the 15 year in-use vehicle lifetime. Growth in total vehicle fleet size and annual kilometers traveled, and increasing percentage of light trucks in the new vehicle sales mix, all counter these individual vehicle improvements. About a 1.3% annual decrease in average new vehicle fuel consumption is required to offset these growth

Table 14: Savings in Light-Duty Vehicle Fleet Fuel Use for Chosen Actions

	Fuel Use: Billion Liters or Percent Change						
Year	Reference Case	Baseline	Medium Hybrids	0.4% Sales Growth	0% VKT Growth	40% Light Truck	Composite
2020	679	-20.3%	-24.5%	-24.8%	-26.8%	-21.3%	-35.2%
2030	774	-39.6%	-47.1%	-45.1%	-46.7%	-41.5%	-58.8%
2020		541	-5.2%	-5.6%	-8.1%	-1.2%	-18.7%
2030		467	-12.4%	-9.1%	-11.7%	-3.1%	-31.7%

trends and stabilize the total US light-duty vehicle fleet fuel consumption around 2015 and beyond. Implementing the baseline technology improvements discussed in this paper would produce fleet fuel savings of 20% in 2020 relative to the no-change reference scenario. ICE-hybrid vehicle penetration into the market relative to this baseline case, even with their much lower than baseline vehicle fuel consumption, has limited impact before 2015 but does usefully improve fleet performance beyond about 2020.

A sobering overall conclusion is that it requires combining all potentially plausible technology, growth, and sales mix options together—clearly tasks requiring a major national commitment—to reduce US light-duty vehicle fleet annual fuel consumption over the next 20 years to levels below today's value of about 500 billion liters per year.

REFERENCES

1. Weiss, M. A., Heywood, J.B., Drake, E., Schafer, A., and AuYeung, F., "On the Road in 2020: A life-cycle analysis of new automobile technologies". Energy Laboratory Report # MIT EL-00-003, Massachusetts Institute of Technology, October 2000. http://lfee.mit.edu/publications/reports

2. Weiss, M.A., Heywood, J.B., Schafer, A., and Natarajan, V.K., "Comparative Assessment of Fuel Cell Cars," MIT LFEE 2003-001 RP, February 2003. http://lfee.mit.edu/publications/reports

3. Guzzella, L., and A. Amstutz. "Quasi-Stationaren-Simulations". Matlab programs and text Benutzeranleitung, Laboratorium fur Motorsysteme, Institute fur Energietechnik, ETH-Zurich, 1998.

4. Ballard Power Systems. "Ballard Fuel Cell Power Module, Mark 900 Series, Mk 900 Polarization". Burnaby, BC, Canada, November 2001.

5. Barbir, Frano, et al. "Design and Operational Characteristics of Automotive PEM Fuel Cell Stacks". Energy Partners, LC, West Palm Beach, FL. SAE 2000-01-0011.

6. Arthur D. Little, Inc. "Cost Analysis of Fuel Cell System for Transportation, 2001 System Cost Estimate, Task 3 Report to: Department of Energy", Ref. 49739 SFAA No. DE-SCO2-98EE50526, August 2001.

7. DeCicco, John M. "Fuel Cell Vehicles: Technology, Market, and Policy Issues". SAE Research Report RR-010, November 2001.

8. US Department of Energy. "FreedomCAR Goals". Office of Advanced Automotive Technologies.

9. Arthur D. Little, Inc. "Guidance for Transportation Technologies: Fuel Choice for Fuel Cell Vehicles, Main Report, Phase II Final Report to DOE", 35340-00, December 14, 2001.

10. Arthur D. Little, Inc. "Guidance for Transportation Technologies: Fuel Choice for Fuel Cell Vehicles, ADL Phase II Results Comparison to MIT Study, Revised Phase 3 Deliverable to DOE", 75111-00, February 26, 2002.

11. Natarajan, V.K., "The Performance of IC Engine and Fuel Cell Hybrid Propulsion Systems in Light-Duty Vehicles," MIT Mech. Eng. S.M. Thesis, February 2002.

12. Bassene, S.A., "Potential for Reducing Fuel Consumption and Greenhouse Gas Emissions from the US Light-Duty Vehicle Fleet," MIT S.M. in Technology and Policy Thesis, September, 2001.

13. Davis, S.C., Transportation Energy Data book: Edition 20. Center for Transportation Analysis, Oak Ridge National Laboratory, Oak Ridge, TN, 2000.

14. Davis, S.C., Transportation Energy Data Book: Edition 19. Center for Transportation Analysis, Oak Ridge National Laboratory, Oak Ridge, TN, 1990.

15. Greene, D.L., and Hu, P.S., "Influence of the Price of Gasoline on Vehicle Use in Multivehicle Households," Transportation Research Record, pp. 19-24, 1984.

APPENDIX

Table A1: Vehicles Using Internal Combustion Engines [2]

	Gasoline				Diesel	
	2001	2020	2020	2020	2020	2020
	Reference	Baseline	Advanced	Hybrid	Advanced	Hybrid
Mass (kg)						
Body & Chassis	930	845	746	750	757	758
Propulsion System (3)	392	264	252	269	293	297
Total (Incl. 136 kg payload)	1458	1245	1134	1155	1186	1191
Vehicle Characteristics						
Rolling Res. Coeff	0.009	0.008	0.006	0.006	0.006	0.006
Drag Coeff.	0.33	0.27	0.22	0.22	0.22	0.22
Frontal Area (m^2)	2.0	1.8	1.8	1.8	1.8	1.8
Power for Auxiliaries (W)	700	1000	1000	1000	1000	1000
Engine						
Displacement (L)	2.50	1.79	1.65	1.11	1.75	1.16
Indicated Eff. (%)	38	41	41	41	51	51
Frictional ME Pressure (kPa)	165	124	124	124	153	153
Max. Engine Power (kW)	110	93	85	58	89	59
Max. Motor Power (kW)				29		30
Use of On-Board Fuel						
Driving Cycle						
US Urban (MJ/km)	2.82	2.00	1.78	1.20	1.53	1.03
US Highway (MJ/km)	2.06	1.45	1.25	0.91	1.04	0.78
US06 (MJ/km)	2.81	1.94	1.67	1.49	1.39	1.29
Combined (MJ/km) (4)	2.48	1.75	1.54	1.07	1.30	0.92
Combined (mpg) (8)	30.6	43.2	49.2	70.7	58.1	82.5
Combined as % Baseline	141	100	88	61	74	52
Life-Cycle Combined Energy						
Vehicle Operation (MJ/km)	2.47	1.75	1.55	1.07	1.31	0.92
Fuel Cycle (MJ/km) (5)	0.52	0.37	0.32	0.22	0.18	0.13
Vehicle Manufacturing (MJ/km)	0.29	0.25	0.25	0.26	0.26	0.26
Total (MJ/km)	3.28	2.37	2.12	1.55	1.75	1.31
Total as % Baseline	138	100	89	65	74	55
Life-Cycle Combined GHG Emissions						
Vehicle Operation (gC/km) (7)	48.5	34.4	30.2	21.0	27.1	19.1
Fuel Cycle (gC/km) (6)	12.1	8.6	7.6	5.2	4.3	3.0
Vehicle Manufacturing (gC/km)	5.5	4.8	4.8	5.0	5.0	5.1
Total (gC/km) (9)	66.1	47.8	42.6	31.2	36.4	27.2
Total as % of Baseline	138	100	89	65	76	57

Notes:
(1) 1 liter (0.737 kg) gasoline = 32.2 MJ (LHV)
(2) 1 liter (0.856 kg) diesel = 35.8 MJ (LHV)
(3) Propulsion system mass includes ICE, drive train, motors, battery, fuel (2/3 full), and tank
(4) Combined cycle is 55% urban/45% highway
(5) Fuel cycle energy, MJ per MJ fuel in tank: gasoline 0.21, diesel 0.14
(6) Fuel cycle gC per MJ fuel in tank = gasoline 4.9, diesel 3.3
(7) Vehicle operation gC per MJ burned = gasoline 19.6, diesel 20.8
(8) Gasoline equivalent miles per gallon calculated as equal fuel LHV
(9) gC of GHG calculated as C in CO_2 released plus carbon in CO_2 equal to 21 times mass of methane leaked

Table A2: Vehicles Using Fuel Cell Systems [2]

	Hydrogen				Gasoline			
	Non-hybrid	Non-hybrid	Hybrid	Hybrid	Non-hybrid	Non-hybrid	Hybrid	Hybrid
	Comp.	Integrated	Comp.	Integrated	Comp.	Integrated	Comp.	Integrated
Mass (kg)								
Body & Chassis	776	780	752	754	821	822	775	776
Propulsion System (3)	465	479	372	378	638	640	460	463
Total (Incl. 136 kg payload)	1377	1395	1260	1268	1595	1598	1371	1375
Vehicle Characteristics								
Rolling Res. Coeff	0.006	0.006	0.006	0.006	0.006	0.006	0.006	0.006
Drag Coeff.	0.22	0.22	0.22	0.22	0.22	0.22	0.22	0.22
Frontal Area (m^2)	1.8	1.8	1.8	1.8	1.8	1.8	1.8	1.8
Power for Auxiliaries (W)	1000	1000	1000	1000	1000	1000	1000	1000
Propulsion System								
Max. Net Stack Power (kW)	103	105	63	63	120	120	69	69
Max. Motor Power (kW)	103	105	95	95	120	120	103	103
Use of On-Board Fuel								
Driving Cycle								
US Urban (MJ/km)	0.75	0.82	0.60	0.66	1.29	1.56	0.96	1.16
US Highway (MJ/km)	0.52	0.57	0.47	0.51	0.85	1.03	0.73	0.88
US06 (MJ/km)	0.92	1.00	0.78	0.87	1.51	1.83	1.27	1.56
Combined (MJ/km) (4)	0.65	0.71	0.54	0.59	1.10	1.32	0.86	1.04
Combined (mpg) (8)	117.3	106.5	140.3	128.1	69.2	57.4	88.4	73.1
Combined as % Baseline	37	41	31	34	62	75	49	59
Life-Cycle Combined Energy								
Vehicle Operation (MJ/km)	0.65	0.71	0.54	0.59	1.10	1.32	0.86	1.04
Fuel Cycle (MJ/km) (5)	0.50	0.55	0.42	0.46	0.23	0.28	0.18	0.22
Vehicle Mfg. (MJ/km)	0.31	0.32	0.28	0.28	0.33	0.33	0.28	0.28
Total (MJ/km)	1.46	1.58	1.24	1.33	1.66	1.93	1.32	1.54
Total as % Baseline	61	66	52	56	70	81	56	65
Life-Cycle Combined GHG Emissions								
Vehicle Operation (gC/km) (7)	0	0	0	0	21.5	26.0	16.8	20.3
Fuel Cycle (gC/km) (6)	23.3	25.6	19.4	21.3	5.4	6.5	4.2	5.1
Vehicle Mfg. (gC/km)	5.8	5.9	5.3	5.3	6.2	6.3	5.4	5.4
Total (gC/km) (9)	29.1	31.5	24.7	26.6	33.1	38.6	26.4	30.8
Total as % of Baseline	61	66	52	56	69	81	55	64

Notes:
(1)   1 liter (0.737 kg) gasoline = 32.2 MJ (LHV)
(2)   1 kg hydrogen = 120.0 MJ (LHV)
(3)   Propulsion system mass includes fuel cell system, drive train, motors, battery, fuel (2/3 full), and tank
(4)   Combined cycle is 55% urban/45% highway
(5)   Fuel cycle energy, MJ per MJ fuel in tank: gasoline 0.21, hydrogen 0.77
(6)   Fuel cycle gC per MJ fuel in tank = gasoline 4.9, hydrogen 36
(7)   Vehicle operation gC per MJ burned = gasoline 19.6, hydrogen 0
(8)   Gasoline equivalent miles per gallon calculated as equal fuel LHV
(9)   gC of GHG calculated as C in CO_2 released plus carbon in CO_2 equal to 21 times mass of methane leaked

III. SYSTEMS DESIGN AND EVALUATION

III.   Systems Design and Evaluation

It is in the field of systems integration that there has been the most notable advance in fuel cell technology in recent years. Discussions about hybridization are illustrative of the discussion about vehicle architecture that is widespread in the industry and that includes fuel cell technology as well. Hybridization continues to be debated. While vehicle efficiency depends on the duty cycle and the para,eters of the hybridization, there may be circumstances under which a simpler and lighter vehicle simply outweighs the efficiency benefits of hybridization.

The most prominent development in new vehicle architectures is in the application of auxiliary power units (APUs). The APU is a small power generating device that works with the main power unit of a vehicle to offer an overall performance that is energy efficient and pollutes less. The efficiency of the APU is less important than the operational flexibility and its potential to use the vehicles host fuel.

Architectures

The paper 2003-01-0809 describes how a combination of a fuel cell and conventional engine technology can be used to achieve the performance of current vehicles with a significant fuel economy benefit. Such hybrid approaches use the best features of both types of powerplant.

In 2003-01-0418 the authors present the design of a passenger car where the primary power source is a hydrogen-fuelled PEM fuel cell stack. An ultra-capacitor is used to buffer electrical energy generated during braking of the vehicle. The architecture of the system is selected for good dynamic behaviour, volume of the powertrain components as well as overall efficiency. Experience in operating the vehicle demonstrated the importance of water management.

The paper 2001-01-0538 discusses the merits of operation of a fuel cell system under pressure. A pressurized system operates at higher stack efficiencies, is smaller in size for a given power output but incurs losses elsewhere, notably in the management of air flow. The authors study the nature of the trade-off questions involved in the choice of operating pressure.

In 2002-01-1959 the authors analyse the design requirements for both a hybrid and non-hybrid fuel cell powered vehicle. Their goal was to develop the best combination of component choices to meet the vehicle performance requirements.

The paper 2004-01-1302 describes a study conducted using the software simulation tools PSAT (for vehicle) and GCTool (for fuel cell system). The authors consider a system solution in which both the degree of hybridization and the energy management strategy are varied to meet performance targets.

Development of Auxiliary Power Units (APU)

In 2004-01-1479 the authors describe the results of a study into the motivation for solid oxide fuel (SOFC) cell based APUs in line-haul trucks. The analysis includes the utilization of modern day trucks in the United States and describes the fuel consumption gains likely to accrue from the widespread use of APUs. The study identifies that idling is responsible for between 3 and 8% of total vehicle energy consumption.

The authors of 2002-01-0411 present a progress report on a project whose aim is to demonstrate a passenger car APU based on an SOFC. The particular appeal of SOFC technology is its requirement for just a simple reformer to convert the host vehicle's liquid fuel into a reformate suitable for the fuel cell stack.

Paper 2004-01-1477 is a design study into the application of a PEM (proton exchange membrane) fuel cell and partial oxidation reformer to a military truck. The fuel cell system supplements a diesel propulsion system and both use JP8 fuel. The fuel cell as considerably more efficient than the engine at idle conditions, and the use of the fuel cell offers a 20% fuel economy improvement in a representative duty cycle.

The authors of paper 2004-01-1479 report on a detailed study into the energy usage of on-highway trucks in the United States. The study's objective was to develop requirements for an APU to meet the power supply needs when a truck is stationary. In the majority of current trucks those loads are met by idling the engine. The authors conclude that a 4kW APU based on solid oxide technology and using diesel fuel would generate total fuel consumption savings of between 3 and 8%.

Vehicle Development and Test

In 2003-01-0422 the authors investigate the cold start performance of the propulsion system in a fuel cell powered vehicle.

Paper 2003-01-0417 covers the design and test of the conversion of a passenger bus from diesel propulsion to a fuel cell based system. The converted bus, using Toyota components was tested and the test results reported.

Paper 2002-01-0408 is a progress report on a European Union funded programme. The paper reports the design, build and first tests of a fuel cell vehicle based on the PSA Peugeot Citröen Partner van and describes both the vehicle's characteristics and plans for further work.

Paper 2002-01-1930 is the result of a cost analysis of a PEM fuel cell system. Preliminary results were presented in 2000, and this paper includes the reaction to those comments and subsequent modelling and analysis work. The analysis work was based on the assumption of high production volumes appropriate to mass produced systems for passenger vehicles.

In paper 2002-01-1945 the authors consider a range of hybridization from a pure electric vehicle through to a load following fuel cell stack. With other parameters such as vehicle mass held constant, the authors concluded that there is benefit in a small degree of hybridization. Control system strategy is the result of an interplay of performance objectives and component efficiencies.

The paper of Anders Folkesson and his colleagues was presented at the meeting, Scientific Advances in Fuel Cell Systems held in Amsterdam in September 2002. The authors set out the case for the application of fuel cell power to public transport buses and report on the energy flows that were measured during the operation of a test vehicle. The net efficiency of the fuel cell system is reported to be 40% while energy usage can be reduced by up to 28% through regenerative braking.

In paper 2002-01-0414, the authors comment on the current status of the two main branches of fuel cell technology: proton exchange membrane (PEM) and solid oxide (SO). The authors consider each of the main challenges facing fuel cell developers and demonstrates the status of each of the two technologies stands against those challenges.

The authors of paper 2004-01-1003 describe the conceptual design of a fuel cell vehicle using the freedoms offered by fuel cell devices. The authors describe a design based on a low drag body and the use of electrically powered sub-systems. Improvements to the architecture of fuel cell stacks allow faster warm up and more efficient operation.

III. SYSTEMS DESIGN AND EVALUATION

Architectures

2003-01-0809

Cost Competitiveness of Fuel Cell Vehicles Through Novel Hybridization Approaches

Jan H. J. S. Thijssen, J. P. Mello and J. R. Linna
TIAX, LLC

ABSTRACT

A combination of fuel cells with internal combustion engines (FC/ICE hybrids) for light duty vehicle propulsion has the potential to offer significant benefits over conventional battery electric hybrid architectures. Specifically, compared to conventional battery / ICE hybrids the FC/ICE hybrid concept may offer improvements in vehicle utility, performance, and emissions. Furthermore, the FC/ICE hybrid can be implemented at a substantially lower cost than that associated with full fuel cell vehicles (FCVs).

Numerous studies have indicated significant emission and energy efficiency benefits from FCVs, especially those based on proton exchange membrane fuel cells (PEMFCs). However, recent more detailed studies indicate that achieving competitiveness with IC engine-based powertrains, especially with respect to cost, would require significant additional improvements in fundamental PEMFC technology. The power density of IC engine powertrains, especially high-powered ones, also would be hard to match for full fuel cell powertrains.

Still, the promise of energy efficiency and emissions benefits that fuel cells offer may be difficult to match with advanced engine technology, especially in city traffic. Thus, we believe it is time to consider integrating fuel cells with engines, rather than hybridizing either with batteries. In fact, recent TIAX analysis suggests that much of the emissions and efficiency benefits promised by "conventional" FCVs could be achieved with FC/ICE hybrids, at a much lower cost and technical risk.

Preliminary analysis conducted by TIAX indicate that such novel FC/ICE hybrids could cut fuel consumption of high-end light-duty vehicles such as full-size SUVs in half, while potentially providing significant emissions benefits as well. The cost for achieving this is estimated at about $4,100 to the manufactured vehicle cost.

INTRODUCTION

The period from now until 2020 is very likely to be the beginning of a transition period to the next generation of automotive powertrains. The automotive industry and governments in Japan, Western Europe, and North America are facing the challenges of mitigating global warming and improving air quality while meeting the public's growing mobility needs. To meet these conflicting requirements, considerable improvements in the efficiency and emissions of current powertrain technology will be needed. The engineering challenges involved in achieving these improvements are formidable to the point that entirely new powertrain technologies and even entirely new fuel infrastructures are being considered.

It is against this background that the FC/ICE configuration emerges as an interesting option. The FC/ICE concept has the potential to offer significant improvements in vehicle fuel efficiency and emissions, at a substantially lower cost than that associated with full fuel cell vehicles (FCVs), even in the longer term.

A DIFFERENT TYPE OF HYBRID VEHICLE: THE FC/ICE HYBRID.

We believe that along with improvements in ICE technology, hybrid electric vehicles (HEVs), and FCVs, another hybridization approach may deserve consideration. FC/ICE hybrids could have a relatively small fuel cell that is operated for low and idling loads while a more or less conventional IC engine would provide acceleration and peak power.

---

[1] Formerly the Technology and Innovation Division of Arthur D. Little, Inc.

Such an FC/ICE hybrid would bear some similarity to a battery-electric micro-hybrid[2], in which the battery is replaced with a fuel cell. Such a hybrid could provide several key advantages over a conventional ICEs and HEVs:

High efficiency - By using the high-efficiency fuel cell to minimize low-efficiency low-load / idling ICE operation, and to avoid round-trip efficiency penalties associated with batteries and alternators;

Low emissions – By virtually eliminating emissions during low and idling loads significant emissions reductions are achieved. In addition, closer integration of the reformer of the fuel cell system with the engine might eventually substantially reduce start-up emissions, possibly leading to PZEV emission levels. Accepting a temporary performance penalty, operation on the fuel cell only could allow occasional zero emissions city driving;

Added functionality - By offering auxiliary power unit (APU) function and possibly 4WD (electric motor works on the other wheels) user functionality is increased. In addition, none of the performance limitations associated with conventional HEVs are expected.

Compared with FC/battery hybrids***, the FC/ICE hybrid would offer substantial benefits as well. These include the following:

Modest cost – The use of a fuel cell that is more than five times smaller than in a full FCV reduces cost almost proportionally. Because the required fuel cell size is decoupled from desired vehicle peak performance, this impact is particularly high-powered vehicles. Battery power requirements are also minimal, compared with a substantial battery cost in full FCVs;

No limitations in performance – Because the fuel cell is small and most of the power comes from the ICE, the system carries only a modest weight penalty even if very high fuel cell power densities are not achieved. The IC engine will allow rapid start-up and load-following as required.

Finally, there are some other benefits that could be quite appealing to the driver of the vehicle, especially for high-end vehicles:

- The fuel cell can provide an APU function to support electrical amenities such as air conditioning during parking;
- By powering the other wheels, the system can provide four wheel drive, at least at low speed;

- The fuel economy for very high-powered engines and vehicles can be substantially improved;
- The low emissions that could be achieved even for upscale SUVs and luxury cars could allow a combination of high performance with a green profile.

In some ways the FC/ICE hybrid has some similarities with the combination of FC APUs and ICE propulsion systems discussed in the literature [1,2]. In these systems the FC APU would power electrical engine accessories, and perhaps even provide reformate to the ICE during start-up to reduce emissions, but it would not provide power for propulsion under any circumstances. In the FC/ICE described here the latter is the primary function of the FC, whereas the APU functions are secondary.

In summary, FC/ICE hybrids would offer many of the societal benefits of a full fuel cell vehicle, but at a lower cost. The FC/ICE hybrid concept also provides additional functionality that is relevant primarily to the upscale and profitable market segments. Consequently, these market segments offer a more likely path to commercialization than implementation in environmentally friendly mid-size cars would.

This paper provides some preliminary insight into the potential performance of FC/ICE hybrids in comparison with conventional vehicles and FCVs. But first we would like to provide some background on those more conventional approaches to achieving the necessary reductions in emissions and fuel consumption.

WHY DO WE NEED TO AUGMENT THE IC-ENGINE?

Significant improvements in vehicle fuel economy and emissions have been accomplished over the past 25 years. Figure 1 shows that in the US, average acceleration performance of passenger cars has improved by over 20%, while average fuel consumption has been reduced by over 25%.

---

[2] We are adopting the definition of battery electric micro-hybrids from Ref. 4: i.e. 42-volt electrical system, electrically powered ancillaries and stop-start ICE operation

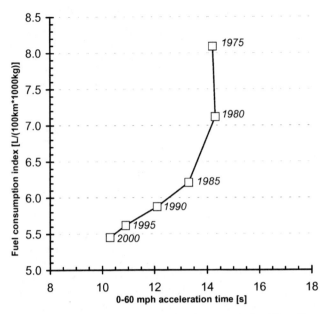

Figure 1   Average Fuel Consumption/performance trade-off trend for the U.S. passenger car new fleet 1975–2000 [3]

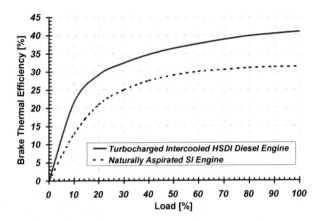

Figure 2   Typical thermal efficiency for naturally aspirated stoichiometric SI and turbocharged intercooled HSDI diesel

Naturally, the data are an aggregate of improvements in both engine and transmission efficiency and in vehicle weight reduction and/or downsizing. Nevertheless, the data show that from 1980 until today, powertrain engineers have repeatedly managed to increase both efficiency and maximum thrust produced in a stepwise manner. Despite this history of successful improvement, and the potential for further enhancements, there are reasons to consider alternative powertrain configurations. Specifically, further fuel economy improvements are becoming increasingly costly since the "low-hanging-fruit" has already been picked. Consequently, we are entering an era in which the continued need to improve vehicle efficiency without diminishing consumer expectations will necessitate augmentation of the IC engine.

In particular, efficiency at low loads is the Achilles heel of IC engines. For stoichiometric, naturally aspirated, SI engines, the brake efficiency normally peaks in the mid-30% range at high load and roughly half the rated engine speed. For compression ignition (diesel) engines, the corresponding peak efficiency falls in the low to mid-40% range. However, as the engine load is reduced, the brake thermal efficiency falls off rapidly for both engine types and becomes 0% by definition at idle (no brake load), see Figure 2. To date, several alternative powertrains have been proposed to address this issue, including the hybrid electric powertrain, and fuel cell powertrain, respectively coupling IC engines and fuel cells with batteries and electric motors.

HYBRID ELECTRIC VEHICLE CHALLENGES

HEV powertrains are technically viable options for the auto-industry today, and initial commercial HEV models have been introduced in several markets. Despite this, a number of technical and market challenges must be met if HEVs are to become a mainstream powertrain technology. Such challenges include battery technology, HEV cost, and duty cycle management.

First, battery energy and power density, cycle and calendar life, and charge/discharge efficiencies are key technical issues that are limiting the penetration rate of HEV. Reasonable and cost-effective battery life need to be commensurate with current regulatory guidelines targeting lifetimes of 150,000 miles or 15 years, whichever comes first. Current battery technology offers compromises on the part of many of these performance factors, thereby limiting the competitiveness of HEVs in many market segments.

Market-side challenges for HEVs are largely related to battery cost. With current fuel prices, the fuel cost savings achieved, at least in North America, by HEVs do not offset the significantly higher purchased vehicle prices. In Europe, modern high-speed turbocharged diesel engines provide a more cost effective way for the consumer to improve fuel economy. It also appears that profit margins on HEVs are currently well below the average in comparable conventional vehicles.

In addition, duty cycle management challenges have shown to lead to significantly reduced fuel consumption benefits and even to reduced vehicle performance. This is especially noticeable during deviations from the expected mission duty cycle, such as sustained hill climbing or trailer towing. Such excursions from the design duty cycle significantly increase the rate of energy consumption, increasing fuel consumption or decreasing the maximum speed.

The major market barrier for HEV, then, becomes one of consumer acceptance of a vehicle with a higher initial price and reduced versatility. Advances in battery storage capacity and energy density can overcome these limitations only to a certain degree. In the end, however, the consumer will have to purchase these vehicles with a specific duty cycle in mind. Achieving significant market success with HEVs is likely to require a considerable significant amount of consumer re-education and revised expectations.

FUEL CELL VEHICLE HURDLES

After about a decade of intensive research and development of fuel cells for vehicles, initial system and vehicle prototypes have started to demonstrate some of the purported benefits: clean and quiet operation and high efficiency [6,7,8]. While early pre-commercial vehicles are expected to reach markets in 2002-2004 [9, 10, 11], most automotive OEMs do not expect full commercialization until after.

These FCVs are generally expected to be hybrid vehicles, in order to be able to meet the dynamic power requirements of the vehicle, and in order to limit the size and cost of the fuel cell required. Almost uniformly these hybrids are fuel cell / battery combinations, in which the battery provides from 10% - 50% of peak power [12]. Some vehicles incorporate supercapacitors. Unfortunately, both fuel cells and traction power batteries are expensive and heavy compared with ICEs.

Given the early stage of development of fuel cell vehicles (FCVs) then, there are significant hurdles to be overcome before fuel cells could mount a serious challenge to the dominance of internal combustion engines (ICEs) for automotive propulsion. These hurdles include the following:

Powertrain Cost - The most serious hurdle appears to be achieving an acceptable manufactured powertrain cost, which is estimated to be four to five times higher than is considered acceptable [13], even if high production volumes were achieved. Recent TIAX analysis indicates that even with rather aggressive assumptions for fuel cell technology development, the manufactured cost for a typical light SUV would increase by 40 – 60 % over that of a current conventional vehicle [12] (Figure 3)

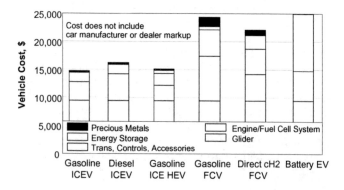

Figure 3   Projected cost for conventional, hybrid, and fuel cell [12]

Hydrogen Fuel Infrastructure - In addition, many of the current fuel cell vehicles require hydrogen as a fuel, but a hydrogen infrastructure to supply such vehicles is not available and likely expensive to implement. Recent TIAX analysis conducted for the US DOE has shown that eventually, with a mature and depreciated hydrogen infrastructure, the fuel cost for the owner of a fuel cell vehicle may be similar to that for owners of current gasoline-ICE-powered vehicles because the higher energy cost is mostly off-set by a higher fuel efficiency. However, the transition costs associated with the introduction of the new fuel infrastructure may require a transitional hydrogen price that is more than three times higher than this break even point.

Power Density - Another considerable hurdle is power density, which is critical to develop compelling fuel cell vehicles (FCVs) without sacrificing vehicle utility or performance. Through dramatic improvements in power density of hydrogen-fueled PEMFC over the past ten years, even system power density of hydrogen-fueled FCVs is now starting to approach that of conventional powertrains. However, for reformer-based FCVs significant weight-reduction is still required to achieve acceptable power densities, as is shown in Table 1.

System Weight and Specific Power Estimates			
Sub-System	2000 DOE Targets (W/kg)	2001 Status	
		(kg)	(W/kg)
Fuel Cell	350	292	171
Fuel Processor	600	173	289
Total w/o start-up batteries[1]	250	489	102
Total[1]	250	543	92

[1] Total includes Balance-of-Plant components.

Table 1   Estimated automotive fuel cell system weights based on 2001 status of technology for reformer-based fuel cell systems.

System Life and Reliability - Finally, proving and improving system life and reliability still requires significant additional attention.

Interestingly, to date most attention has been devoted to finding ways to reduce the cost of fuel cell vehicles so that their environmental and energy efficiency benefits can be brought into reach of consumers. However, relatively little attention has been devoted to finding ways in which fuel cells could be used to make vehicles more appealing by offering superior performance or additional vehicle utility, and doing so at a modest cost.

Given the considerations above, we suggest that rethinking of the basic approach to FCV architecture may be required. The approach we discuss here, the FC/ICE hybrid, might help overcome some of the hurdles to FCV development and commercialization without requiring significant additional technological breakthroughs.

POSSIBLE SYSTEM CONFIGURATIONS

When considering FC/ICE hybrids, numerous combinations of fuel cell types, ICE types, and fuels are possible. However, many of these possible combinations were not included in the present analysis. Instead, this study is focused on one of the most likely combinations.

Given the small size of the fuel cell system in these configurations, reformer-based PEMFCs may not be attractive. Combinations that would require two separate vehicle fuels would also appear unlikely to be attractive. Some of the following combinations may be worthwhile considering.

Hydrogen-fueled SIE or CIE with PEMFC – the improved efficiency of this hybrid may extend the range of the vehicle enough to enable more practical fleet applications of hydrogen as a vehicle fuel (e.g. taxis and police cars).

Gasoline-fueled SIE with SOFC – the SOFC could improve efficiency and emissions while providing four wheel drive for high-end SUVs.

Diesel-fueled CIDI with SOFC – for somewhat heavier duty vehicles, and army vehicles, the combination of a CIDI with an SOFC may provide advantages in efficiency by reducing idling and emissions.

In the following paragraphs we will focus on the gasoline-fueled SIE with SOFC as an example.

GASOLINE FUELED SUV WITH SIE/SOFC HYBRID

A possible example of a FC/ICE hybrid is that of a high-end, full-size SUV with a SIE combined with a SOFC. To avoid the infrastructure issues associated with hydrogen fuel, a vehicle that could use a conventional fuel such as gasoline would avoid one significant barrier to commercial success. This type of vehicle typically has a 200-250kW V8 spark-ignition engine, coupled with an automatic transmission. Many of these vehicles are outfitted with 4WD or AWD, primarily to provide traction during winter months. Average vehicle power consumption over the federal city/highway drivecycle is around 10 to 20 kW, or less than 10% of peak power.

One option for FC hybridization of this gasoline platform would be the use of a SOFC. Small-scale gasoline fueled SOFC have been designed by TIAX and have been built by Delphi and others. These SOFC prototypes have demonstrated efficient operation on gasoline over thousands of hours [1]. The power density of such devices is also rapidly approaching the acceptable range. Thermal cycling remains a major challenge, and

drastic further improvements in start-up time would be required. Assuming these issues can be solved without significantly compromising performance, the SOFC would be an interesting match with the SIE.

In the hybrid system studied, a small SOFC would be installed near the fuel tank to provide power to the rear wheels, while the SIE drives the front wheels. The powertrain would have the following characteristics:

- 10 kW fuel cell, 75 l / 75 kg, efficiency ~37% at peak power, 45% at 10% load, fuel cell start-up in 5 min; Reformer provides fuel for SIE during start-up to reduce emissions. Fuel cell is operated for any loads < 10kW, and idled / switched off if load exceeds this number;
- 0.2 kWh lead acid battery;
- 250 kW naturally aspirated, V8, SIE

As indicated above, such an FC/ICE hybrid would offer several key benefits to the SUV owner:

- 4 WD for low-power operation
- APU function to power up to 10 kW of appliances, or perhaps to provide emergency power to up to three homes
- Lower fuel consumption
- Low emissions, especially during city driving and idling, reducing emissions to PZEV levels.

The FC/ICE hybrid will be examined below in more detail, and a preliminary characterization of possible performance and cost implications will be reviewed.

PERFORMANCE ASSESSMENT

A powertrain model, built using the commercially available software package GT-Drive™, was used to estimate the fuel consumption performance over the U.S. Federal Test Procedure City-Cycle (Bag 1 and 2 of the FTP-75 drive cycle). The average speed (excluding idling) of this cycle is 24 mph, with a maximum speed of 57 mph and a total duration of 1372s. Key characteristics of the modeled vehicle are summarized in Table 2.

Curb Weight	1,853	kg
Frontal Area	3.5	m^2
Drag Coefficient	0.33	-
Engine Type	N/A S.I. V8	
Engine Rated Power	250	kW
Engine Displacement	4.2	L

Table 2   Model Vehicle Key Characteristics

In the baseline configuration, (dedicated ICE configuration - no APU), over the City-Cycle, the IC-engine produces on average 10kW (idling and deceleration excluded) with a resulting fuel consumption of 13.2 L/100km, see Figure 4.

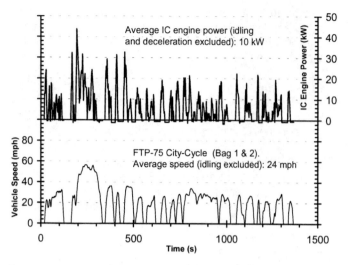

Figure 4   Baseline IC-engine power delivered during test-cycle

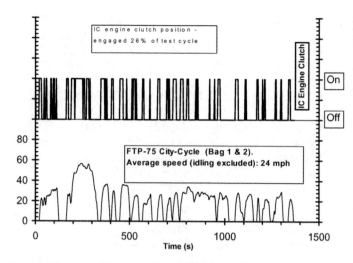

Figure 5   IC-engine duty-cycle with a 10 kW threshold

Next, the FC/ICE concept was modeled by combining the IC-engine with a fuel-cell APU and an electric drivetrain. The fuel-cell capacity was sized to correspond to the average IC-engine power over the test cycle, (i.e. 10kW). A basic operating strategy was modeled consisting of two mutually exclusive operating modes:

- The fuel-cell provides all power, with the IC-engine stopped (default mode).
- The IC-engine provides all power, with the fuel-cell being turned-off (high-power mode).

An electromechanical clutch was used in the model to switch between the two operating modes and transfer traction power from either the fuel-cell/electric drivetrain (default) or the IC-engine (high-power). In the model, selection of traction power source defaults to fuel-cell power. The high-power operating mode (IC-engine) is only activated if the traction power required exceeds a pre-selected switchover threshold.

With a 10kW switchover threshold the IC-engine is engaged a total of 26% of the time, with fuel-cell providing power for the balance (74%) of the city-cycle. As a result, the average power from the IC-engine (off-time excluded) is 16kW. For the fuel-cell the corresponding average power is 2kW, see Figure 5 and Figure 6.

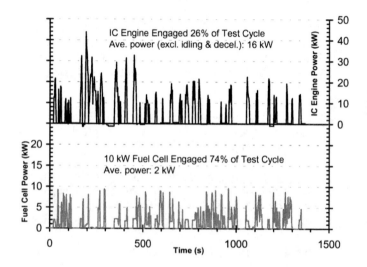

Figure 6   IC-engine and Fuel-Cell power delivered during the City-Cycle with a 10 kW switchover threshold

FUEL CONSUMPTION

The simulation model was used to assess the impact of some key FC/ICE design parameters on fuel economy over the chosen test cycle. At this point it must be noted that several simplifications and idealizations were used in building the model. In particular, the transition between operating modes (i.e. between fuel-cell and IC-engine power) is modeled as an instantaneous event with no power-up or shut-down losses. Also, at the start of the test cycle, both power sources (fuel-cell and ICE) were assumed to be available without any start-up delay. Consequently, the objective is not to predict absolute fuel consumption numbers. Instead, the results should be viewed and understood in the context of a sensitivity assessment.

To this end, Figure 7 shows the impact of varying the switchover threshold for a 10 kW fuel-cell – 250 kW IC-engine combination.

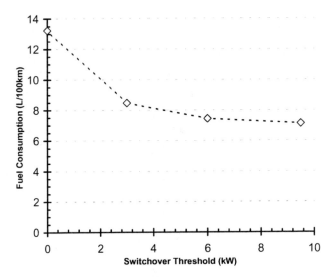

Figure 7   Total fuel consumption (IC-engine and Fuel-cell) at different switchover thresholds for the model vehicle over the city-cycle with a 10 kW fuel-cell.

The results shown in Fig. 6 lead to the following observations:

- With a fuel-cell capacity sized for the average traction power requirement of the vehicle, (10kW), the FC/ICE hybrid concept holds potential to significantly reduce the fuel consumption, (-45%), over the city-cycle.
- Most of this fuel consumption improvement can be realized even if the switchover threshold is set much lower than 10kW. That is, even if the fuel-cell is only managing the power supply when the requirement is at or less than 3kW, the fuel consumption improvement is still on the order of 30%.

Consequently, the data in Figure 7 suggest that a significant fraction of the fuel economy benefit associated with the FC/ICE concept is related to the long recognized advantages of shutting-off the IC-engine during coasting and idling, and powering accessories electrically (in this case via a fuel-cell). To test this hypothesis, model runs were performed with four different configurations:

- Conventional ICE (250kW)
- ICE (250kW) with engine shut-off during idle and deceleration.
- FC/ICE hybrid, 10kW F/C – 250kW ICE
- Dedicated Fuel Cell Vehicle (60kWl) (the fuel cell is 50 kW, so clearly peak vehicle performance for this configuration can not match that of the baseline and FC/ICE vehicles)

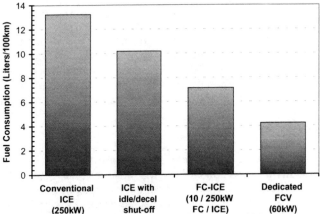

Figure 8   City-Cycle fuel consumption for the model vehicle with different drivetrain configurations.

As clearly shown by Figure 8, much of the efficiency gain promised by FCVs can be realized with the FC/ICE hybrid and a relatively modest fuel-cell size (20% of the peak power required over the FTP-75 city cycle, or in this case less than 10% of engine capacity). Also, the numbers suggest that roughly half of the fuel consumption benefit associated with the FC/ICE concept is the result of providing high-efficiency fuel-cell traction power during measured driving.

Equally important (i.e. the other half of the fuel consumption benefit) is ICE shut-off at idle and decelerations. The predicted reduction shown in Figure 8 is entirely consistent with previously reported results [14], in which the elimination of standby and driveline losses (as assumed herein) was reported to reduce fuel consumption by approximately 23% (17.2% from eliminating standby and 5.6% from the elimination of driveline losses) over the FTP-75 city cycle.

EMISSIONS

The greenhouse gas emissions from FC/ICE hybrids will be reduced in proportion to fuel consumption.

The impact of the FC/ICE hybrid on criteria pollutants will likely strongly depend on the precise manner in which the system is integrated. We discuss the potential emissions impact by part of the drive cycle.

Start-Up

Start-up emissions from the FC/ICE hybrid will depend on the degree to which the FC and the SIE will be integrated. Without any special measures, they will likely be similar to those from SIEs, since start-up power would be mainly derived from the SIE. However, more sophisticated integration approaches have been suggested by some [1] in which the reformer provides hydrogen-rich gaseous fuel to the SIE during start-up. This would significantly decrease engine-out emissions during start-up and perhaps accelerate engine heat-up. However, it would require an extremely fast-starting reformer and a very tight integration between engine and fuel cell. While we believe based on some of our own experience that this may be achievable, we also think of this integration as a longer-term improvement.

Low-Power Driving

During low-power driving, when only the fuel cell is operating, emissions of criteria pollutants will likely be close to zero. As this represents a significant portion of the drive-cycle, this too is expected to have a significant impact on overall emissions.

High-Power Driving.

During high-power operation, when only the SIE is operating, emissions would likely be similar to those from conventional SIE-powered vehicles. Conceivably, by using the SOFC prudently during transients emissions from such events may also be reduced.

However, in order to achieve any of the criteria pollutant emissions, significant technology development and integration hurdles must be overcome. In particular the SOFC start-up and turn-down must be much improved, and appropriate controls technology will be required to achieve smooth switching between the SOFC and the SIE.

COST EFFECTIVENESS

Based on a high-level preliminary analysis, we expect that the added cost of the FC/ICE hybrid would be about $4000 (excluding value chain mark-up). The SOFC system would be responsible for about $3100 out of this number, as shown in Figure 9. The fuel cell cost analysis was based on a detailed conceptual design and cost estimate of a 5 kW gasoline-fueled APU, reported to SAE in 2000 [15]. The basis for the analysis is 2001 SOFC technology, but assuming that thermal cycling, life, and start-up issues can be overcome without a significant cost-penalty. Appropriate scaling considerations were made for each sub-system and component identified in that analysis. The results of this analysis are shown in Figure 9. For the electric drivetrain, cost correlations were used that were developed in a study of hybrid electric vehicles performed for EPRI in 2000 [16].

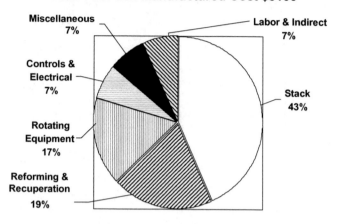

Total Fuel Cell Manufactured Cost $3100

Figure 9   Cost-Structure of 10 kW SOFC System

As can be seen from Figure 9, the stack represents the largest cost item in the 10 kW SOFC system, but reforming & recuperation and air and fuel handling also contribute significantly.

The 10 kW electric drivetrain required would add another approximately $1000 to the manufactured powertrain cost. While this may seem high, it is very modest compared with the cost of a full FCV powertrain (Figure 10). For the analysis of the full FCV powertrain, we extrapolated our analysis of gasoline-based PEMFC fuel cell system cost as a function of power rating, as performed for DOE in 2001 [12]. Especially for high-powered vehicles, the disadvantage of fuel cells in achieving economy of scale is clearly apparent. Most FCV cost analyses and most FCV prototypes have focused on relatively under-powered vehicles with fuel cell engines with capacities ranging from 40 – 90 kW. Such a low power level would clearly be inadequate to compete with a 250 kW SIE.

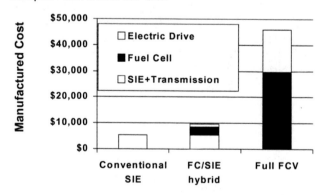

Figure 10   Cost-Structure of 10 kW SOFC System

In addition, unlike the conventional FC/battery hybrid, the FC/ICE hybrid would provide significant additional functionality to the driver. Considering for example the cost of conventional 4WD/AWD, and the benefits of the APU function of the FC/ICE hybrid, the added cost may be acceptable to the driver without considering the trade-off against fuel and maintenance costs. Such benefits are also more likely to translate into a higher vehicle market value than improved fuel economy, especially in higher-end markets.

CONCLUSIONS AND RECOMMENDATIONS

Considering radically different options for the use of fuel cells in light duty vehicle propulsion, we have come to the conclusion that hybridization of IC engines with fuel cells may provide most of the benefits that could be derived from full FCV operation, but at a cost premium that is about ten times lower than that projected today for full FCVs. In addition, these vehicles could well provide additional functionality compared with conventional vehicles and FCVs. Combined with the relatively modest cost premium this may make FC/ICE hybrids commercially viable with reasonable advances in fuel cell technology. By contrast, viability of full FCVs depends on spectacular further advances in fuel cell technology, likely combined with a strong government regulatory driver.

Based on our preliminary analysis of the platform, we have identified several key technical hurdles that must still be overcome:

- Rapid system start-up in less than ten minutes must be achieved, and low-power idling must be proven;
- Reliability with respect to both operating life and number of thermal cycles must be improved and proven
- Low emissions over the drive-cycle must be proven.
- Controls and vehicle integration of the hybrid system must be proven. The switching between fuel cell and engine power must occur smoothly, and without undue impacts on emissions, fuel consumption or maintenance cost for the concept to have merit.

Overcoming these barriers will require the concerted efforts of industry and government alike. However, if it can be done, very appealing and cost-effective fuel cell-powered vehicles will result, while providing significant benefits to the environment and to energy efficiency.

ACKNOWLEDGMENTS

The authors would like to thank the DOE's Office of Fuel Cell and Hydrogen Technology and the National Energy Technology Laboratory's Solid State Energy Conversion Alliance for their support of the studies that were cited as underpinnings for the findings, reviewed here. We also would like to thank several colleagues at TIAX for the contribution of expertise in the wide range of topics spanned in this study. In particular, we would like to thank Masha Stratanova of TIAX for providing fuel cell efficiency estimates.

In addition, we would like to acknowledge the direct and indirect support from our strategic partner in the recently completed study "Future Powertrain Technologies, the Next Generation", DRI-WEFA, and our strategic partner for engine and driveline simulations, Gamma Technologies. Finally we would like to thank the many members of the fuel cell community and automotive industry, too numerous to mention, that have been willing to discuss and refine the thinking outlined in this paper.

REFERENCES

1. Zizelman, J., S. Shaffer, and S. Mukerjee. Solid Oxide Fuel Cell Auxiliary Power Unit - A Development Update. in SAE 2002 World Congress. 2002. Detroit, MI, USA.
2. Crosbie, G.M., et al. Solid Oxide Fuel Cells for Direct Oxidation of Liquid Hydrocarbon Fuels in Automotive Auxiliary Power Units: Sulfur Tolerance and Operation on Gasoline. in SAE 2002 World Congress. 2002. Detroit, MI USA.
3. Light-Duty Automotive Technology and Fuel Economy Trends 1975 Through 2000, . 2000, EPA.
4. Future Powertrain Technologies, the Next Generation, DRI-WEFA and TIAX, December, 2001
5. AuYeung, F., Heywood, J.B., and Schafer, A., "Future light-duty vehicles: predicting their fuel consumption and carbon reduction potential," SAE Paper No. 2001-01-1081, 2001.
6. Le, K.-C., et al. Hyundai Santa Fe Fcv Powered By Hydrogen Fuel Cell Power Plant Operating Near Ambient Pressure. in SAE 2002 World Congress. 2002. Detroit, MI, USA: SAE.
7. Michalak, F., J. Beretta, and J.-P. Lisse. The Hydro-Gen Project: 2nd-Generation Pem Fuel Cell System With High Pressure Hydrogen Tanks for An Electric Vehicle. in SAE 2002 World Congress. 2002. Detroit, MI, US: USA.
8. Matsumoto, T., et al. Development of Fuel-Cell Hybrid Vehicle. in SAE 2002 World Congress. 2002. Detroit, MI USA: SAE.
9. Okuda, H., Chairman's Message, Toyota Motor Company, . 2002, Toyota: Tokyo.
10. Honda, Honda Unveils the Prototype of its FCX Fuel Cell Vehicle, Planned for Commercial Release This Year, . 2002: Tokyo
11. DaimlerChrysler, The Fuel Cell on Its Way to the Customer: "F-Cell": The World's First Fleet of Fuel-cell-Powered Cars Enters the Practical Test Stage, . 2002, DaimlerChrysler.
12. Thijssen, J.L., Steve; Unnasch, Steffan, Fuel Choice for Fuel Cell Vehicles, . 2002, US Department of Energy, Energy Efficiency and Renewable Energy, Office of Advanced Transportation Technology: Washington DC.
13. Carlson, E.J., et al. Cost Modelling of Pem Fuel Cell Systems for Automobiles. in Future Car Congress. 2002. Arlington, VA.
14. Amann, C. A., "The stretch for better passenger-car fuel economy," SAE 972658.

15. Read, C.J., J.H.J.S. Thijssen, and E.J. Carlson. Fuel Cell Auxiliary Power Systems:Design and Cost Implications. in SAE 2001 World Congress. 2001. Detroit, MI.
16. Arthur D. Little, Inc., Comparing the Benefits and Impacts of Hybrid Electric Vehicle Options, . 2001, EPRI: Palo Alto.

DEFINITIONS, ACRONYMS, ABBREVIATIONS

4WD	Four wheel drive
APU	Auxiliary Power Unit
AWD	All wheel drive
CIDI	Compression Ignition Direct Injection
CIE	Compression ignition engine
FC	Fuel Cell
FCV	Fuel Cell Vehicle
FTP	Federal Test Procedure
HEV	Hybrid Electric Vehicle
HSDI	High-speed direct injection
IC	Internal combustion
ICE	Internal Combustion Engine
kW	Kilowatt
OEM	Original equipment manufacturer
PEMFC	Proton exchange membrane fuel cell
PZEV	Partial zero emissions vehicle
SIE	Spark Ignition Engine
SOFC	Solid Oxide Fuel Cell
SUV	Sport Utility Vehicle

CONTACT

Jan Thijssen
TIAX, LLC
Acorn Park
Cambridge, MA 02140
(617) 498-6084
thijssen.j@tiax.biz

J.P. Mello
TIAX, LLC
Acorn Park
Cambridge, MA 02140
(617) 498-6130
mello.john@tiax.biz

J.R Linna
TIAX, LLC
Acorn Park
Cambridge, MA 02140
(617) 498-6191
linna.janroger@tiax.biz

Performance and Operational Characteristics of a Hybrid Vehicle Powered by Fuel Cells and Supercapacitors

Paul Rodatz, Olivier Garcia and Lino Guzzella
Swiss Federal Institute of Technology (ETH)

Felix Büchi, Martin Bärtschi, Akinori Tsukada, Philipp Dietrich, Rüdiger Kötz, Günther Scherer and Alexander Wokaun
Paul Scherrer Institute

ABSTRACT

The paper presents experimental results of a fuel cell powered electric vehicle equipped with supercapacitors. This hybrid vehicle is part of an ongoing collaboration between the Paul Scherrer Institute (PSI, Switzerland), the Swiss Federal Institute of Technology (ETHZ), and several industrial partners. It is equipped with a fuel cell system with a nominal power of 48 kW and with supercapacitors that have a storage capacity of 360 Wh. Extensive tests have been performed on a dynamometer and on the road to investigate the operating ability. The highlights of these tests were the successful trial runs across the Simplon Pass in the Swiss Alps in January 2002.

The fuel cell system consists of an array of six stacks with 125 cells each and an active area of 200 cm^2. The stacks are electrically connected as two parallel strings of three stacks each in series in order to match the voltage requirement of the powertrain. The reactant gases and the cooling liquid are fed in parallel through a manifold.

The supercapacitors are sized for peak power levelling to assist the fuel cell during hard acceleration. Moreover, the supercapacitors are used to store the energy obtained from regenerative braking and serve to optimize the vehicle efficiency.

Polarization curves, efficiency data of the fuel cell system and fuel consumption data from the New European Driving Cycle are presented. The transient behavior of the fuel cell system and its influence on the performance of the vehicle are analyzed.

INTRODUCTION

In the past Polymer Electrolyte Fuel Cells (PEFC) were exclusively used in aerospace or military applications. The demands on PEFCs in those fields differ greatly from those in automotive applications, where the pressure of costs is much more severe. Furthermore, the reduction of volume and weight are at the center of ongoing research. Most often these demands can only be satisfied at the expense of efficiency or other parameters. In the automotive application the PEFC has to compete against the internal combustion engine, which has a power density in the range of 1 kW/kg. This value is still hard to meet with a fuel cell propulsion system.

PEFCs have a substantially higher part-load efficiency than internal combustion engines. As passenger vehicles are mostly operated under part-load conditions the use of PEFCs permits significantly lower fuel consumption. The high torque output of electric motors represents a further advantage of fuel cell propulsion systems. As electric motors offer a high torque output even at low speeds, fuel cell vehicle have a remarkably good acceleration behavior. Combined with an electrical storage device (such as batteries or supercapacitors) at least part of the braking energy can be recuperated, which leads to a further reduction in fuel consumption.

In the last decade a variety of demonstration vehicles have been presented by the major car manufacturers. DaimlerChyrsler has been playing an important role in the research of fuel cell powered vehicles, as their prototype series has already reached the fifth generation with the Necar 5. But manufacturers such as Ford, GM, Toyota, Nissan, Honda and many others as well have made considerable progress in their efforts toward mass producing fuel cell cars in the near future (e.g. [1]).

Nevertheless, numerous problems have to be solved yet before an attractive fuel cell powered vehicle will be available on the market. The objective of this work was to build an experimental fuel cell vehicle that can be used to explore the performance of new materials and system architectures, hence giving insights for further developments.

In the two sections following this introduction the vehicle concept and the fuel cell system setup, respectively, are presented. Some general observations concerning the fuel cell system and results from experiments performed on a dynamic test bench are discussed in the fourth section. The fifth section titled vehicle performance is divided into three subsections. First, the transient vehicle behavior is analysed. Next the in-vehicle performance of the fuel cell system is presented and, finally, results from drive cycle tests are shown. The conclusion section sums up the insights gained in course of this project.

VEHICLE CONCEPT

With the knowledge gained from a small pilot system (less than 10 kW), the test vehicle (see Fig. 1) was equipped in early 2002. It is based on a Volkswagen Bora (a 5-seat sedan vehicle known in the U.S. as VW Jetta) and was modified to accommodate the new powertrain. The test vehicle is equipped with a hybrid powertrain in which the fuel cell acts as the primary power source. The supercapacitors supplement the fuel cell with enough instantaneous power to achieve good driving performance and allow the capture of regenerative braking energy.

Fig. 1: Test vehicle "Hy.Power"

Supercapacitors are electrical storage devices with a high power density. Their energy density is up to one hundred times higher than that of conventional capacitors, and their power density is up to ten times higher than that of batteries. With their wide operating temperature range and their long lifetime, supercapacitors are the short-term storage elements of choice. Jointly developed by PSI and Montena SA, the supercapacitors installed in the vehicle each have a rated capacitance of 1600 F with a rated voltage of 2.5V. Altogether, the 282 pair-wise connected supercapacitors provide a storage capacity of 360 Wh and are able to provide 50 kW for a duration of roughly 15 s. The maximum voltage of the supercapacitor module is 360 V. An active voltage-balancing unit mounted on the supercapacitors serves to balance the cell voltage inside the module and to avoid the overcharging of specific cells.

Fig. 2 shows the configuration of the powertrain. The vehicle is powered by an AC motor with a permanent power output of 45 kW, a peak power output of 75 kW, and a maximum torque of 255 Nm (all values apply to an input voltage of 280 V). The input voltage of the motor inverter is kept at a constant high voltage, thereby assuring the highest possible motor torque and good efficiency over the whole speed range. Fuel cell and supercapacitors are connected to a DC link by means of DC-DC converters [2]. A supervisory controller is used to actively regulate the power flow between motor inverter, fuel cell and supercapacitors. A one-step transmission without a clutch completes the powertrain.

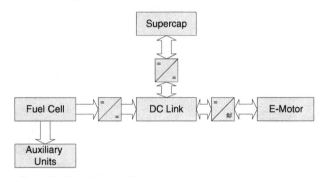

Fig. 2: Outline of powertrain

Two 26-liter tanks store the hydrogen at a pressure of up to 350 bar. This amounts to roughly 1.05 kg of hydrogen, which is equivalent to about 4 liters of gasoline. The vehicle has a range of 50 to 100 km depending on the drive cycle. The total weight of the test vehicle amounts to roughly 2000 kg.

Supervisory Control–A predefined energy management is used to control the energy flows between fuel cells, supercapacitors and electric motor. It serves to optimize these flows with regard to the consumption of hydrogen. The requirements on an ideal energy management are numerous, with the main three being the following:

1. Fulfill all power demands
2. Recuperate as much braking energy as possible
3. Ensure minimum fuel consumption

Under certain circumstances these demands are mutually conflicting. To achieve good acceleration, the fuel cells have to be assisted by the supercapacitors, which therefore should be fully charged and thus cannot absorb any recuperated energy. The conflicts are resolved with the help of different priorities assigned to the various demands, the highest priority being the satisfaction of the power demand at all times.

A number of security aspects have to be considered at the highest priority level. These include:

- supercapacitors must not be overcharged
- supercapacitors may not be discharged below 25% of

maximum energy storage capacity (because of deteriorating efficiency)

- charging and discharging currents of supercapacitors may not exceed certain limits to avoid overload of power electronics
- the current drawn from the fuel cells may not exceed the current that the fuel cells are able to supply (important during warm-up and fast load changes)

Any energy management will result in a compromise between good acceleration and yielding high recuperation of braking energy. Based on the assumption that with increasing vehicle speed the need for strong acceleration decreases and the amount of energy that can be recuperated increases, the following energy management may be designed. At low vehicle speeds (e.g., below 60 km/h) the supercapacitors should be charged to a high degree, whereas with increasing speed they are discharged gradually to allow the recuperation of as much braking energy as possible. The exact gradients are a function of the power drawn by the electric motor.

Only the desired fuel cell power calculated by the supervisory controller is sent to the DC-DC converter as a reference value. The fuel cell DC-DC converter transfers the requested power P_{FC} from the fuel cell to the DC link. The supercapacitor DC-DC converter controls the voltage of the DC link to its nominal value. Hence the difference between the motor inverter power P_D and the fuel cell power P_{FC} is transferred to/from the supercapacitors. The inverter power P_D may be restricted by the supervisory controller if it exceeds the momentary maximum combined output of the fuel cell and the supercapacitors. Similarly, during braking the amount of regenerative braking may be limited, depending on the state of charge of the supercapacitors. This mode of operation ensures the power balance between all components while keeping the DC link voltage very stable.

For further information about the supervisory controller, the interested reader is referred to reference [3].

The control algorithms are centrally managed by a dSpace™ MicroAutoBox (MABX) 1401/1504. It has a Motorola™ PowerPC 603e running at 200 MHz with a slave processor to handle the digital I/O units. Matlab/Simulink™ is used to program the control algorithms. Logical sequences were programmed with the help of the Matlab toolbox Stateflow™. Euler's method was chosen as solver with a fixed step size of 5 ms.

The MABX has only a limited number of I/O units. Therefore, the communication between the sensors and the MABX, as well as the communication to the actuators is handled by a CAN-Bus. For this purpose the highly flexible WAGOTM 750 CANopen series is installed, which is a modular I/O system. Modules are available for almost every type of sensor signal or actuator output and the configuration can be expanded easily by adding additional elements.

FUEL CELL SYSTEM

The direct-hydrogen fuel cell system can be divided into three main subsystems according to the fluid that is handled:

- air subsystem: supply of the process air at the required pressure, flow rate, temperature, and humidity
- H_2 subsystem: supply of hydrogen at the required pressure and flow rate
- cooling subsystem: guarantee adequate cooling of fuel cell stacks and ensure small temperature gradient across stacks

System Components–The air subsystem consists of a compressor, a humidifier, and a pressure control valve. The compressor is integrated into a feedback loop containing a flow meter and a PI controller which regulates the air flow. Water is injected into the pressure side of the compressor to cool down the hot air to cell temperature. As a desirable side effect, the air humidity is also increased to the benefit of the stack operation [4].

The hydrogen subsystem has to supply enough fuel to the stacks under all operating conditions. Furthermore, water droplets from the anode side of the cells have to be removed so as not to block the reaction sites. The hydrogen supply subsystem is a closed circuit. It is pressure regulated and the hydrogen influx depends only on the pressure drop in the stacks caused by the depletion of hydrogen due to the electrochemical reaction.

Satisfactory dynamics of the fuel cell power output require a supply of excess hydrogen to the stacks. An ejector is used to recirculate the excess hydrogen and thus to prevent wasting any hydrogen to the surrounding. Experiments have shown that a shock wave generated across a ventilation valve induces a temporarily increased influx of hydrogen in the fuel cell flow field. An additional benefit is that the diffusion layer situated between the flow channel and the membrane electrode assembly (MEA) is dynamically inflected by the shock wave which causes any liquid water droplets that may have formed inside the fuel cell to be dispersed. Furthermore, the liquid particles are blown out from the fuel cell, allowing for the delivery of additional hydrogen and thus preventing reactant starvation in parts of the cell. The shock waves are generated with a solenoid valve and a vacuum vessel. Similarly, if the pressure inside the vessel is higher than the pressure in the fuel cell, shock waves above the system pressure are induced. A detailed description of this procedure may be found in [5].

The closed cooling loop includes a variable-speed electric pump and an air-water heat exchanger with two stepwise variable-speed fans. Since the cooling media is de-ionized water, special precautions have to be taken to prevent freezing. Additionally, the low operating temperature of the fuel cell stacks and the limited frontal area of the vehicle create major problems for heat rejection, especially at higher ambient temperatures.

<u>Fuel Cell Stack</u>–Adequate performance under conditions of an automotive application, low degradation over time and the possibility for optimized preparation procedures are the aims that have guided the selection process of the electrochemical components. Commercially available membranes (Nafion 112, DuPont) and electrodes (ELAT, E-Tek) were evaluated and the respective preparation and assembly procedures were developed. The bipolar plate (BIP) is a multifunctional part, which represents the largest volumetric part of the stack. The BIP has to distribute the air and hydrogen to the membrane-electrode-assembly (MEA), support the cooling of the MEA, prevent the mixing or leaking of the different media and conduct the current between electrochemical cells. A new bipolar plate was developed in order to reduce volume and weight. During the design phase, special attention was paid to the optimization of the manufacturing process of the bipolar plates.

The new components were tested in a single cell with less than 100 W output. Single cells were then stacked in series through the optimized bipolar plate to form multi-kW stacks. The scale-up from a single cell to a 125-cell stack was realized with only small deviations among the stacks. Six stacks were assembled to an array in order to generate a power output of several tens of kW. This modularity of a fuel cell system, however, also has certain disadvantages. The main disadvantage is the large number of parts that are needed for powerful systems. The array of six stacks shown in Fig. 3, contains more than 5000 parts. Most of these parts had to be handled, prepared, checked and finally assembled individually. Except for a few standard parts such as screws, springs, and tie rods, all parts were individually designed.

The six stacks are electrically connected as two parallel strings of three stacks each in series in order to match the voltage requirement of the powertrain. The reactant gases and the cooling liquid are fed in parallel through a manifold plate. The parallel arrangement of six stacks requires the equal distribution of the reactant flow across 19500 channels (6 stacks * 125 cells per stack * 26 flow channels per cell). The challenge of this task is demonstrated by the measurement results of a 100-cell stack shown in Fig. 4. At approximately 35 min. the hydrogen stochiometry was reduced to 1.1. Even though 10% excess hydrogen was fed to the stack, a number of cells experience a steep voltage drop. At 39 min. the hydrogen flow was set to the original high stochiometry and all cells recovered.

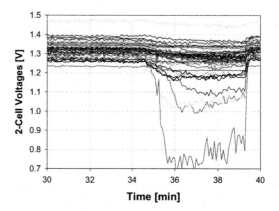

Fig. 4: Effect of low hydrogen stochiometry on cell voltages. Every line shows the voltage over two cells of a 100 cell stack. At about 35 min. the stochiometry was reduced to 1.1.

RESULTS AND DISCUSSION

The standard test conditions were as follows: stack temperature 70 °C, gradient of cooling water across stacks ΔT 5-8 °C, gas pressures 2 bar (absolute) at exit, stoichiometries of 2 for both reactant gases, and dew points of 55 and 50 °C for hydrogen and air, respectively. Under these conditions each stack generated an output of 8 kW at a voltage of 0.6 V (an efficiency based on the lower heating value (LHV) of about 50%). Therefore, the array of six stacks should have a combined power output of 48 kW. This power is generated if the process gas flow into the array is distributed equally across all cells and flow channels. Especially on the hydrogen side this is difficult to accomplish as the hydrogen flow is considerably smaller than the air flow.

The results from the single-stack experiments were irreproducible with the array of six stacks due to difficulties in the gas supply. When the power output was increased beyond 20 kW, the voltage became unstable (see Fig. 5). Larger than expected amounts of liquid water accumulated at the anode. Although the shock waves were very effective

Fig. 3: Array of six stacks. The manifold plate behind the stacks supplies media to the different stacks

in the single-stack arrangement, the complexity of the piping caused the effect to be insufficient to remove all the water from the anode in the array setup. As a consequence, the anode is increasingly flooded by water, hydrogen is prevented from reaching the reaction site, the local reactant is starved and the reaction breaks down. The magnitude of this phenomenon rises with increasing current as more water is produced in the course of the reaction. In one attempt to improve matters a purge valve was installed through which hydrogen was continuously purged to the surrounding. This led to an increased hydrogen flow in the fuel cell and hence to a easing of the anode flooding. Of course this improvement was gained at the cost of efficiency as hydrogen was wasted to the surrounding.

Fig. 5 demonstrates the beneficial effect of purging. For these experiments the net power output of the fuel cell was increased in steps of 2 kW. A minimum of 20 seconds was allowed for the voltage to settle down after each step. The setup without the purge valve is represented by gray lines. Also shown is the setup with the purge valve (black lines). Every stack voltage is plotted separately, therefore each case is displayed as a family of curves. The voltage of the setup without purge valve is noticeably lower and the difference between the two cases grows with increasing power. Furthermore, the voltage starts to oscillate above 10 kW, becoming unstable at 22 kW and forcing the system to be turned off.

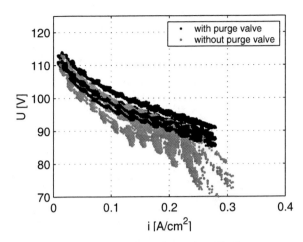

Fig. 6: Current-Voltage plot of experiments from Fig. 5.
Experiments with purge valve show a more stable voltage.

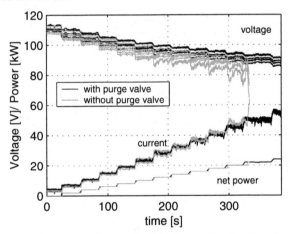

Fig. 5: Comparison of voltage behavior for experiments with and without purge valve. Net power was increased in steps of 2 kW. At least 20 s was allowed for the voltage to settle down

The same data is presented in Fig. 6 as a voltage-current graph. It confirms that with a purge valve the oscillation spectrum is reduced and mass transport limitations are avoided. However, the purge valve can only be a temporary solution since hydrogen is wasted. In a future setup this hydrogen should be recirculated by means of a pump.

Test bench–A dynamic test bench at the Measurement and Control Laboratory of the Swiss Federal Institute of Technology Zurich was used for extensive testing of the powertrain. The test bench by APICOM, type SM L-4, consists of an asynchronous electric motor/generator, power electronics, and feedback controllers. The main test bench signals are the brake torque and speed. The torque was measured with a torque meter placed between the brake and the electric vehicle motor. No gearbox was mounted between the vehicle motor and the brake. The test bench was operated in a speed controlled mode, causing any power limitations to appear in the torque profile rather than in the speed profile.

The main objective of those tests was to demonstrate the operativeness of the powertrain and to eliminate any problems before its final assembly into the vehicle. Ordinary drive cycles (NEDC, ftp, etc.) are not sufficiently demanding to include all the operating conditions of the powertrain. Therefore, a ride across the Simplon-Pass between Brig and Gondo in the Swiss Alps was recorded using an electric vehicle with conventional batteries. Fig. 7 depicts the speed profile of the first 6 km recorded on the beginning slope of the Simplon Pass, whereas Fig. 8 shows the torque profile. Figs. 8 to 12 show measurement results from the test bench for various net power values of the fuel cell. Fig. 9 shows the power drawn by the motor inverter from the DC link as well as the reference value recorded during the drive across the Simplon Pass. Fig. 10 depicts the total power output of the fuel cell. Since the total includes the power consumed by the air compressor, the values are higher than those of the net output. Fig. 11 shows the fuel cell voltage averaged over the two strings. The supercapacitor voltage is shown in Fig. 12. For the purpose of achieving a satisfactory charge/discharge efficiency, the supercapacitors are not discharged below 25% of the maximum energy storage capacity. This corresponds to 50% of the maximum voltage, hence at 180 V the supercapacitors are considered empty.

This series of measurements was conducted with the purge valve installed. Nevertheless, at a net power output of the fuel cell of 30 kW, the fuel cell voltage again started

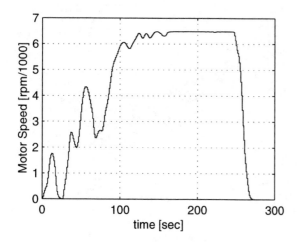

Fig. 7: Speed profile of the first 6 km of the ride across the Simplon Pass

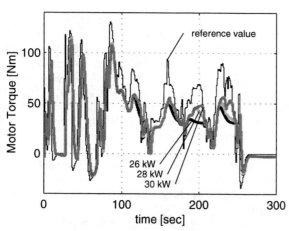

Fig. 8: Torque profile of the first 6 km up the Simplon Pass and the test bench measurements for different net fuel cell power levels

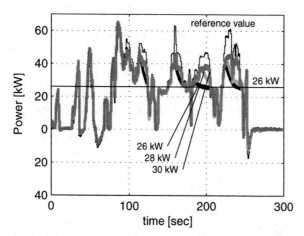

Fig. 9: Reference value and measurement of the power drawn by the motor inverter from the DC link for different net fuel cell power levels

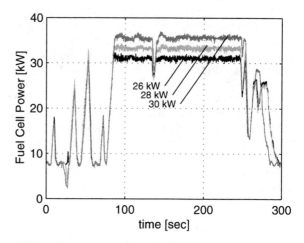

Fig. 10: Total fuel cell power (incl. compressor power) for different net fuel cell power levels on Simplon Cycle

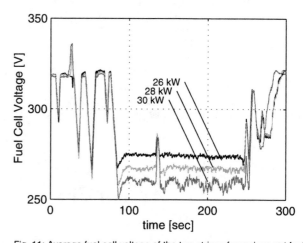

Fig. 11: Average fuel cell voltage of the two strings for various net fuel cell power levels on Simplon Cycle

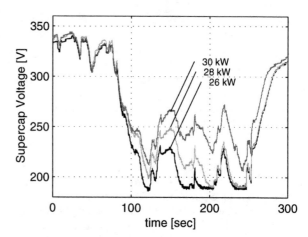

Fig. 12: Total supercapacitor voltage for different net fuel cell power levels on Simplon Cycle

to oscillate. Therefore, no experiments were performed beyond 30 kW. Until approximately 100 sec after start of experiment no or only small differences between the experiments are visible and the torque reference is quite well matched. All three experiments have in common that as soon as top speed is achieved the supercapacitors are near depletion. This is consistent with the strategy described earlier in which the supercapacitors are discharged at high velocity to accommodate for any recuperation energy. However, at subsequent accelerations the supercapacitors are no longer able to assist the fuel cell. Therefore, the requested torque is limited by the supervisory controller and the power drawn by the inverter is restricted to the current fuel cell power output. Large deviation from the torque reference thus cannot be avoided. Obviously, higher levels of fuel cell net power attenuate this behavior.

VEHICLE PERFORMANCE

Transient Vehicle Behavior–The transient response of the powertrain to a fast acceleration demand from standstill is shown in Figs. 13 through 15. For this response experiment the maximum fuel cell net power was limited to 24 kW, the maximum gradient on fuel cell power was 1 kW/s, and the supercapacitor current was restricted to 125 A. The fuel cell power gradient is very conservative since under normal operation it is set to 2.5 kW/s.

As evident in Fig. 15 the torque response to the driver's input is very fast; maximum torque is achieved in about one second. With increasing motor speed the power drawn by the inverter continues to rise up to the limit given by the supervisory controller (see Fig. 13). The power is mainly provided by the supercapacitors, as the power response from the fuel cell is much slower. The supercapacitors are continuously discharged, thus causing the voltage to drop. As the supercapacitor current is already at the allowed limit, it cannot be increased to compensate the decreasing voltage and to achieve a constant power output. The inverter limit remains unchanged for a long time, as the drop in supercapacitor power is made up by the fuel cell power which is slowly building up. However, to achieve a constant torque output, the inverter power would need to be increased. Consequently, the torque is reduced. Once the supercapacitors are nearly depleted, the power output is gradually brought down to zero and the vehicle is supplied only from the fuel cell. Fig. 14 shows the responses of the fuel cell voltage and current. Clearly visible are the undershoot of the voltage and the overshoot of the current. The voltage undershoot results from a temporary shortage in hydrogen due to a slow response of the supply system. The current overshoots to compensate the voltage drop as the fuel cell is power controlled. Slight deviations in stack resistance lead to varying string resistance and therefore to somewhat different string currents after the voltage has settled down.

The acceleration performance of the vehicle (with a maximum fuel cell power of 27 kW and a gradient of 1

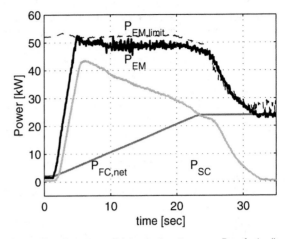

Fig. 13: Transient response of motor inverter power P_{EM}, fuel cell net power P_{FC} and supercapacitor P_{SC}. For this experiment the fuel cell net power was limited to 24 kW, the maximum fuel cell net power gradient was set to 1 kW/s.

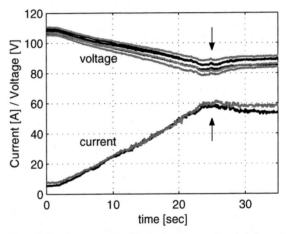

Fig. 14: Transient response of fuel cell voltage and current. Arrows indicate voltage undershoot and current overshoot.

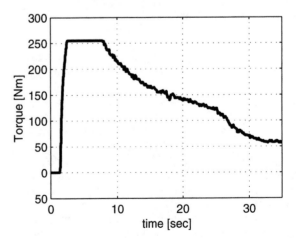

Fig. 15: Transient response of motor torque

kW/s) was measured to be 15 sec for 0-80 km/h and 50 sec for 0-100 km/h. The large difference between these two values is due to the limited capacity of the supercapacitors which does not allow for faster accelerations up to a speed of 100 km/h. A top speed of 115 km/h was achieved.

In-Vehicle Performance–In this section, stack performance and cell ageing are discussed. Plots (a) through (f) in Fig. 16 shows the polarization curves for all stacks after different operating stages. The cell performance and the obtained data was divided into four stages, namely (1) experiments on the dynamic test bench, (2) operation in the vehicle after final assembly, (3) operation in the vehicle after some cells in stack 3 were damaged due to exposure to negative voltage and (4) operation in vehicle after damaged cells were replaced. All markers represent operating points under stationary conditions and a stack temperature of at least 50 °C. Plot (g) shows the polarization curve averaged over all six stacks. The following equation [6] was used to fit the data:

$$E = E_0 - b \log i - R i$$
$$E_0 = E_r + b \log i_0 \tag{1}$$

where E is the cell voltage, b and i_0 are the Tafel slope and the exchange current density respectively, i is the current density, R is the ohmic resistance and E_r is the reversible potential.

Table 1: Parameters of equation 1 derived from averaged stack data

	E_0 V	b V decade^{-1}	R Ohm cm^{-2}
Testbench experiments	115.6	2.964	0.225
vehicle operation	115.5	2.698	0.232
vehicle operation with damaged stack	113.9	2.142	0.251
vehicle operation after repair	116.6	4.259	0.180

As mentioned above the stacks are ordered as follows: stacks 1 to 3 as well as stacks 4 to 6 are connected in series. The two strings are connected in parallel resulting in an equal string voltage. Stacks 3 and 6 are located at the bottom of the array, 2 and 5 in the middle and 1 and 4 at the top. A manifold plate is attached to one side of the array, as shown in Fig. 3. The reactant gases and the cooling medium are supplied to each stack from this manifold. The reactant gases and the cooling are fed at the bottom of the manifold and have to rise the height of the array to reach the two uppermost stacks. Therefore, any flooding is most likely to start at stacks 3 and 6. This was confirmed by the observation that the cells located near the entrances of stacks 3 and 6 were always among the first cells to fail.

The transition from the test bench to the vehicle was accomplished without any voltage degradation, on the contrary even slight improvements were observed. The fuel cell array is located in the rear of the vehicle in a closed and sealed compartment which under operating conditions is heated up by convection from the fuel cell to the stack temperature. The pipes and fittings leading from the compressor and humidifier to the fuel cell are thus heated up to the stack temperature. This is in contrast to the test bench setup where the stacks are surrounded by ambient air at room temperature leading to a constant heat convection from the pipes to the ambient air and a decrease in the temperature of the air flow along the length of the pipe. Therefore, oversaturated air enters the fuel cell and liquid water blocks part of the reaction site.

In the course of the vehicle testing a number of cells in stack 3 were severely damaged due to longtime exposure to negative voltage. This failure prevented that the initial stack voltage was reached. The stack voltage was lower by as much as 5 V. Nevertheless, due to the tight testing schedule, the drive tests had to be continued and the failed cells were ignored. Subsequent investigations revealed that several hotspots had formed in the failed cells leading to holes in the membrane and even to local melting of the bipolar plates.

After the replacement of the damaged cells, the initial performance of stack 3 was restored. However, in the course of the experiments the stack voltages began to deteriorate. The extent of the degradation varied from stack to stack, with stacks 3 and 6 most heavily effected. Explanations for this phenomenon must remain tentative since very little information on the effects of excessive dryness or flooding of the MEA is found in the literature. On our testbench only a coarse filter was installed in the air intake fitting. Therefore, dirt may have been sucked into the air system and consequently into the fuel cell where it may lead to a contamination of sensitive areas of the MEA. Even though precautions have been taken to avoid corrosion in the humidification system, the possibility of corrosion can never be ruled out completely. Again, any particles washed into the fuel cell cause contaminations of the fuel cell. The longtime exposure to excess water or, worse yet, flooding helps to distribute the poison to extended areas in the MEA. St-Pierre et al. reported that contamination of the MEA by impurities led to a replacement of H$^+$ions by foreign cations and a reduced conductivity proportional to the ionic charge [7]. The degradation may also result from a reduction of the catalyst surface area, as reported by Wilson et al. [8].

Stacks 3 and 6, which are located at the bottom of the array, are exposed to larger amounts of water than the other stacks. Therefore, the degradation is most visible here. However, contrarily to the observations by St-Pierre et al. and Wilson et al. this degradation seems to stem mostly from a reduced kinetic performance. During the last operating stage the Tafel slope b has increased considerably (see Table 1). Furthermore, Fig. 16 shows that the deviation from the initial polarization curve builds up in the region of low current density and then stays

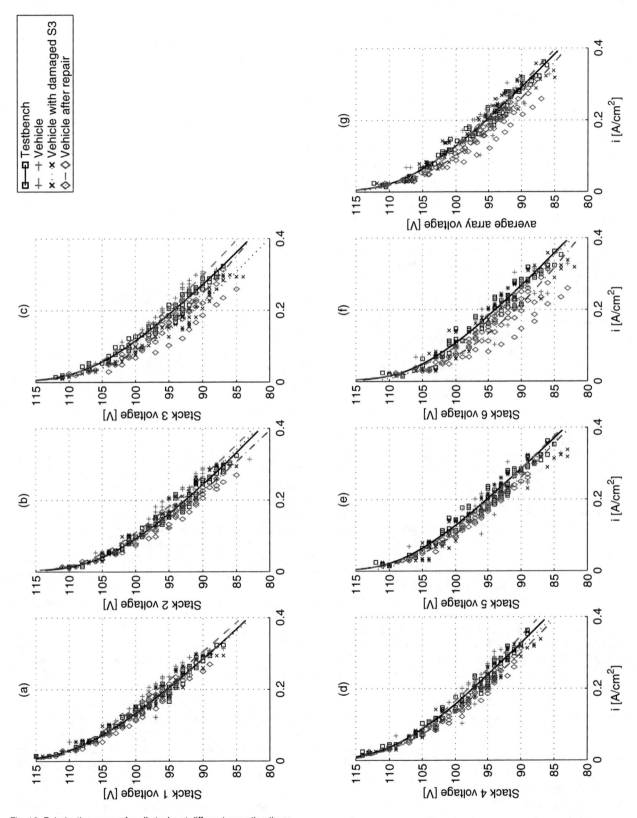

Fig. 16: Polarization curves for all stacks at different operating times

unchanged with increasing current levels. No mass transport limitations are observed in the region of high current density. As the humidification setup is identical to that of the other operating stages, the degradation cannot result from increased flooding. These observations thus point to a contamination of the MEA.

Drive Cycle Tests–Fig. 17 shows measurement data from the New European Driving Cycle (NEDC) obtained on the dynamometer of the Swiss Federal Laboratories for Materials Testing and Research, EMPA. The experiments were conducted with a maximum fuel cell net power of 22 kW and a maximum gradient of 2.5 kW/s. The solid line in plot (a) shows the measured vehicle speed, whereas the target speed of the cycle is given by the thin dashed line. The speed profile clearly shows that the NEDC is a succession of four identical urban sections of about 200 seconds each, followed by a highway section. As expected, the measurement data show a recurrent system behavior for the four urban sections. For ease of reading, only the first urban section (from 40 to 240 s) and the highway section (from 820 to 1220 s) are depicted in plot (b) as well as in the subsequent plots. Plot (c) shows the total DC power drawn by the vehicle (electric motor and other power consumers within the vehicle). The bold dashed lines in plot (c) show the upper and lower dynamic saturation points which limit the power that may be requested by the driver. The net power output of the fuel cell system is indicated on plot (d). The net power refers to the power fed into the DC link. The power that is needed to supply the auxiliary components of the fuel cell is not included in this value. The sum of the net power and the power drawn by the auxiliary components is the total fuel cell power, which is the product of voltage and current. The fuel cell voltage and the current density are shown in plot (e) and (f), respectively. The supercapacitor power is shown in plot (g). Negative supercapacitors power values indicate a power flow into the cells, which means that they are being charged. Finally, the state of charge (SOC) of the supercapacitors is depicted in plot (h).

As evident from plot (a) the vehicle is able to match the speed setpoint well, except for the final acceleration to 120 km/h. The supercapacitors were completely depleted by that time and the vehicle is solely powered by the fuel cells. (Remember that 180 V is the lower operating limit of the supercapacitors, which corresponds to a 25% SOC.) The 22 kW supplied by the fuel cells are insufficient to accelerate the vehicle to top speed within a reasonable time. Clearly, the high transient and negative power capabilities of the supercapacitors allow to smooth the power profile of the fuel cell.

Table 2 lists the fuel consumption data of the vehicle during the NEDC. Clearly visible is the deteriorating effect of purging on the fuel consumption, as roughly 40% is wasted hereby. It seems possible that in future experiments the purged hydrogen can be recirculated by a pump with a minimal additional parasitic power loss.

Table 2: Fuel Consumption for NEDC

	kg H_2	l_{ge}*/100km
total consumption	0.309	10.8
consumption (excluding purging loss)	0.176	6.16
consumption (excluding purging loss and auxiliary components)	0.127	4.45

* l_{ge} = gasoline equivalent and SOC corrected
For comparison:
BMW 7 Series Weight 1935 kg: 10.7 l/100km
Bora Original 55 kW Weight 1300 kg: 6.7 l/100km

CONCLUSION

The fuel cell vehicle was intended as a demonstration platform to investigate new components in a realistic surrounding. Complications with the peripheral equipment were unavoidable, since many prototypes as well as devices that were not designed for automotive usage were used in the vehicle. Nevertheless, the experimental vehicle was a major success. During the operation time a total distance of about 1000 km was travelled and the vehicle was demonstrated at numerous occasion, such as the Geneva Motor Show or Earth Summit in Johannesburg in 2002.

Water management proved to be a critical aspect for the efficient and stable operation of the fuel cells. Especially the removal of liquid water was a limiting factor of the fuel cell system. Condensation in the fuel cell channels, in the manifold plate and in the piping resulted in the accumulation of liquid in the fuel cell. For the present experiment the removal of liquid water from the anode could only be accomplished by feeding the fuel cell with additional hydrogen, which was purged to the atmosphere.

The experiments showed that poisoning of the active area of the electrode is facilitated by any extended time of contact with liquid water. For one, impurities in the water can effectively be absorbed on the electrode and secondly, water promotes the distribution of poisons through extended areas in the cells.

The hybrid powertrain presented here allows to keep the fuel cell system comparably small, while achieving a high peak power output for acceleration. Owing to the hybridization with the supercapacitor, the powertrain shows a good dynamic behavior. If the additional hydrogen flow needed for the removal of liquid water from the anode is recirculated effectively, an impressive total fuel consumption in the region of 6 to 7 l/100km for the NEDC cycle can be achieved.

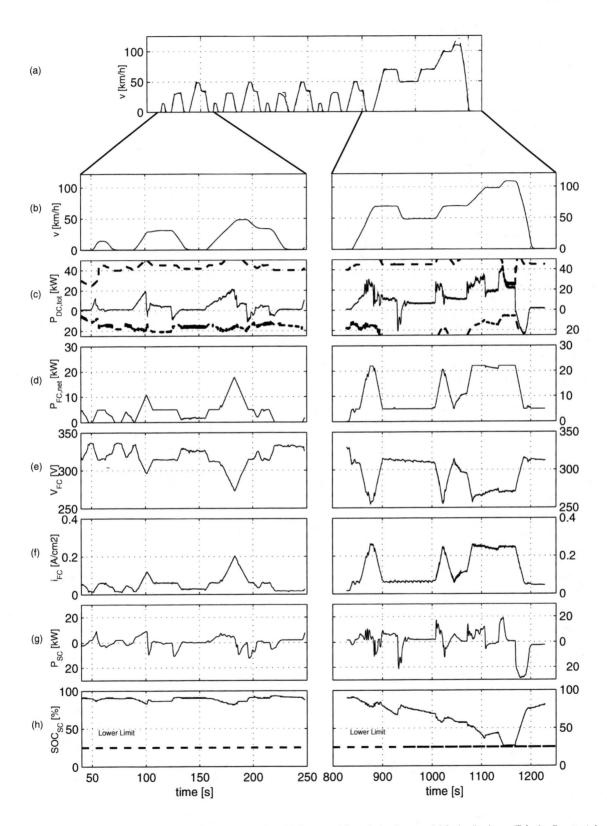

Fig. 17: Dynamometer measurements (a) and (b) vehicle, (c) total DC power, (d) net fuel cell power, (e) fuel cell voltage, (f) fuel cell current density, (g) supercapacitor power, (h) supercapacitor state-of.charge SOC

ACKNOWLEDGMENTS

Financial support by the Swiss Federal Energy Office and by AMAG Schweiz AG is gratefully acknowledged.

REFERENCES

[1] S. Ashley; Fuel cells start to look real; Automotive Engineering International, Vol.64, March 2001

[2] H. Stemmler, O. Garcia; A simple 6-way DC-DC converter for power flow control in an electric vehicle with fuel cells and supercapacitors; Proceedings EVS-16, Beijing, (1999)

[3] P. Rodatz, G. Paganelli, A Sciarretta, L. Guzzella; Optimal power management of an experimental fuel cell/ supercapacitor powered hybrid vehicle; Submitted to Control Engineering Practice

[4] S. Pischinger, C. Schönfelder, O. Lang, H. Kindl; Development of fuel cell system air management utilizing HIL tools; SAE Paper 2002-01-0409, SAE 2002 Congress, Detroit, MI (2002)

[5] P. Rodatz, A. Tsukada, M. Mladek, L. Guzzella; Efficiency Improvements by Pulsed Hydrogen Supply in PEM Fuel Cell Systems; Proceedings of the XVth IFAC World Congress, Barcelona 2002

[6] D. Chu, R. Jiang; Comparative studies of polymer electrolyte membrane fuel cell stack and single cell; Journal of Power Sources 80, pp. 226-234 (1999)

[7] J. St-Pierre, D.P. Wilkinson, S. Knights, M.L. Bos; Relationships between water management, contamination and lifetime degradation in PEFC, Journal of Materials for Electrochemical Systems, 3, pp. 99-106 (2000)

[8] M.S. Wilson, F.H. Garzon, K.E. Sickafus, S. Gottesfeld; Journal of the Electrochemical Society, 140, p. 2872 (1993)

Vehicle System Impacts of Fuel Cell System Power Response Capability

Tony Markel and Keith Wipke
National Renewable Energy Laboratory

Doug Nelson
Virginia Polytechnic University and State Institute

ABSTRACT

The impacts of fuel cell system power response capability on optimal hybrid and neat fuel cell vehicle configurations have been explored. Vehicle system optimization was performed with the goal of maximizing fuel economy over a drive cycle. Optimal hybrid vehicle design scenarios were derived for fuel cell systems with 10 to 90% power transient response times of 0, 2, 5, 10, 20, and 40 seconds. Optimal neat fuel cell vehicles where generated for responses times of 0, 2, 5, and 7 seconds. DIRECT, a derivative-free optimization algorithm, was used in conjunction with ADVISOR, a vehicle systems analysis tool, to systematically change both powertrain component sizes and the vehicle energy management strategy parameters to provide optimal vehicle system configurations for the range of response capabilities.

Results indicate that the power response capability of the fuel cell system significantly influences the preferred powertrain component characteristics and the resulting fuel economy in a neat fuel cell vehicle. Slower transient capability leads to larger component sizes and lower fuel economy. For a hybrid fuel cell vehicle, optimal combinations of component sizes and energy management strategy parameters can be found that lead to only a minor variation in vehicle fuel economy with respect to fuel cell system power response capability.

INTRODUCTION

ADVISOR is a vehicle simulator capable of simulating conventional, hybrid electric, electric, and fuel cell vehicles [1, 2]. It uses drivetrain component characteristics to estimate vehicle fuel economy and emissions over defined drive cycles as well as other quantitative performance metrics (i.e., maximum-effort acceleration, gradeability). Roughly 30 different drive cycles and numerous complex test procedures can be used to assess the vehicle fuel economy, emissions, and performance under various simulated test conditions.

Because of the complexity of hybrid electric vehicles (HEVs), including issues such as component sizing, energy management strategy, and battery state-of-charge (SOC) balancing, optimization becomes necessary to give results that can be accurately compared with other vehicles. ADVISOR executes quickly, allowing a single drive cycle (~1000 seconds) to be run on the order of 20 seconds on a standard PC. It is well suited to being linked to optimization routines that may need to evaluate several thousand designs to determine the best.

ADVISOR v3.1 was used in this study as the 'objective function' call within the MATLAB environment. Various optimization algorithms have been linked to ADVISOR to both understand the differences in their approach and their effectiveness in satisfying the needs of finding optimal solutions in the challenging design space of hybrid electric vehicles [3]. It was concluded that the non-gradient based methods of the DIRECT algorithm were extremely effective in finding the regions likely to contain local and possibly global optimum solutions, but required considerable amount of solution time to converge onto the answer within a small tolerance.

Previous work relating to optimization of hybrid vehicles has included efforts at University of California, Davis [4] that examined whether, and under which conditions, hybridization of a fuel cell vehicle would be beneficial. More recently, collaborative work was performed with Virginia Polytechnic Institute and State University (Virginia Tech) on sizing of components for fuel cell hybrids, which this work will build upon [5]. Atwood, et. al. concluded that some level of fuel cell system hybridization was beneficial to the design. Based on a revision to their published results, it was shown that the best City/Highway fuel economy could be obtained at a fuel cell to total power (fuel cell + battery) ratio of 0.26. This translated to a fuel cell providing a net peak power of 52 kW and a 150 kW battery pack for a GMC Suburban vehicle (a total of 202 kW was required to meet the performance requirements). An assumption of [5] was that the vehicle mass will remain constant for all cases. As a result, there was no mass penalty associated with a larger fuel cell system.

In a recent study completed at NREL, optimization tools were applied to the design of a hybrid fuel cell sport utility vehicle (SUV) for a variety of driving schedules [6]. It was assumed that the hydrogen fuel cell system would have 10 to 90% power transient response capability of 2 seconds. This correlates to the U.S. Department of Energy (DOE) Fuel Cells for Transportation Program technical target for a fuel cell stack in 2004 [7]. This performance target has been set such that, in a vehicle application, there would be no performance degradation.

A fuel cell stack itself can respond to changes in flowrate and pressure quickly. However, for a complete fuel cell system (including fuel, air, water, and thermal management subsystems) it may be difficult to provide fast transient load-following capability. For example, the rotational inertia of an air compressor may limit how quickly the inlet air flowrate and pressure provided to the fuel cell can be changed. It may not be possible to provide inlet flow conditioning (flowrate, pressure, temperature, and humidity level) within a short time-frame to provide pure load-following capability. This will be especially true for a system that includes a fuel reformer composed of additional hardware and support systems that all have their own individual transient limitations.

The DOE has set a technical target to provide an integrated fuel cell system with reformer operating on gasoline with a 10 to 90% power transient response time of 5 seconds by 2004. While in [7] the current status of such technology is denoted as having a 15 second transient response capability. Mays, et. al. notes that the transient response capability of existing fuel cell systems is on the order of 20 seconds to reach maximum power [8]. They suggest that this should be sufficient for hybrid vehicle applications. Additionally, Adams, et. al. provides test results in their paper that indicate a fuel cell system dynamic response on the order of 0.8 seconds [9]. Clearly, there is variability in this fuel cell system attribute. This paper will address how the rest of the vehicle powertrain system may be designed around a fuel cell system with specific operating characteristics.

Potter and Reinkingh evaluated the impacts of the fuel cell system dynamic response in neat and hybrid transit bus applications [10]. Of the cases they evaluated, a system with a response rate of 5-10 kW/s (10-20% per second) provided a system that could satisfy the drive cycle demands and offer a good balance between cost, mass and volume impacts. They varied component sizes in discrete steps based on existing technology, while in this study we will allow the component attributes to vary continuously to provide optimal configurations.

As was eluded to in [10], it may be possible to reduce system mass, volume, and/or cost with alternative fuel cell system designs if the transient response capability is not a critical parameter with respect to the vehicle fuel

economy. For example, a fuel cell system air compressor would need to be sized to handle rapid changes in mass flow to satisfy the needs of the fuel cell stack if the fuel cell system is to have full fast transient capability. However, if other systems in the vehicle (i.e. batteries in a hybrid vehicle) can filter the fast transient and allow the fuel cell system to follow a slower transient then it may be possible to downsize the air compressor or various other support subsystems. The fuel cell system design trade-off will not be discussed in this paper but would have significant cost, volume, and mass implications.

In this study, the impacts of fuel cell system power response capability on the optimal vehicle design will be explored for both hybrid and neat[1] fuel cell SUV. Both component sizing and energy management strategy parameters will be varied to provide the best possible vehicle in each design case. Derivative-free optimization algorithms will be employed to provide sufficient confidence that the design is the best possible design within a large non-linear and discontinuous design space. The operating behavior of the optimal vehicles will be discussed and conclusions will be drawn based on the results provided. The specific fuel cell system design necessary to provide the indicated transient response capability or the subsystem that would induce such a transient response limitation is not considered in this study.

OPTIMIZATION PROBLEM DEFINITION AND CONFIGURATION

VEHICLE MODELING ASSUMPTIONS

For this study, the HEV vehicle characteristics are assumed to be based on a current production baseline conventional mid-size SUV similar to a Jeep Grand Cherokee. Table 1 and Table 2 outline the vehicle assumptions and the components used.

Table 1: Vehicle Assumptions

Vehicle Type	Rear wheel drive mid-size SUV (i.e. Jeep Grand Cherokee) VEH_SUV_RWD.m
Baseline conventional vehicle mass	1788 kg
HEV glider mass (no powertrain)	1202 kg
Rolling Resistance	0.012
Wheel Radius	0.343 m
Frontal Area	2.66 m^2
Coefficient of Aerodynamic Drag	0.44

The HEV components are based on the current state of the art technology for which data is available. The fuel cell and motor/controller data files are based on proprietary data, and have not been included in a public release of ADVISOR at this time.

---

[1] In a neat fuel cell vehicle the fuel cell system is the only source of energy. This vehicle does not have an energy storage system to supplement the fuel cell power or to capture regenerative braking energy.

Table 2: Baseline Components

Component (ADVISOR filename)	Description
Fuel Converter (FC_HH52_Honeywell.m)	Efficiency vs. net power performance data for 52 kW net pressurized fuel cell system based on Honeywell stacks
Motor/Controller (MC_AC119_VPT.m)	AC induction motor developed by Virginia Power Technologies 83 kW @ 275 Vmin
Energy Storage System (ESS_NIMH45_Ovonic.m)	Ovonic 45 Ah nickel metal hydride battery modules

In this analysis, the total mass of the vehicle will vary with respect to the size and the mass of the powertrain components. The optimizer will have control over both the component sizes and the energy management strategy parameters. All parameters will be allowed to vary over a continuous range. There will be no delay or fuel penalty associated with fuel cell system start-up and shutdown during the cycle. A fuel consumption penalty is included, however, to account for inefficiencies during warm-up of the fuel cell system.

The fuel cell data and model used in this analysis represents a pressurized hydrogen fuel cell system. The transient response time range evaluated is appropriate for either a hydrogen or a gasoline reformed fuel cell system. The absolute fuel economy results would be somewhat different for a reformed system since the system efficiencies will be different. However, the trends with respect to transient response capability will be applicable to either a hydrogen or gasoline reformed fuel cell system.

For this study, it will be assumed that the power response capability of the fuel cell system will be the same during both ramp-up and ramp-down events.

We will assume that the fuel storage system will be sized to provide 563 km (350 miles) of range. The mass associated with storing a sufficient amount of hydrogen to provide the desired range was calculated and will remain constant for all analyses (average fuel consumption of 3.62 L/100 km {65 mpg} gasoline equivalent and a fuel storage specific energy of 2000 Wh/kg was assumed).

All component masses are assumed to scale linearly with respect to peak power except for the energy storage system. The energy storage system mass is assumed to scale linearly with respect to the number of modules in the pack and by the following relationship with respect to capacity:

$$scaled\ mass = base\ mass * (C1 * ess\_cap\_scale + C2) \quad (1)$$

Where C1 and C2 are 0.9832 and 0.01602, respectively and ess_cap_scale is the factor by which the capacity of the battery pack has been increased relative to the baseline component characteristics. The coefficients, C1 and C2, were derived from nickel metal hydride battery technology data in the ADVISOR data file library.

INTERFACING BETWEEN ADVISOR AND OPTIMIZATION TOOLS

The ability to use ADVISOR in a "GUI-free" or batch mode was introduced and documented with the release of ADVISOR v3.1. This mode was specifically developed to make it easy to use ADVISOR as an automated function or response-generating tool to be connected to optimization routines as shown in Figure 1. As currently configured, this functionality provides the user with nearly all of the functionality available from the GUI, and in some instances even more functionality.

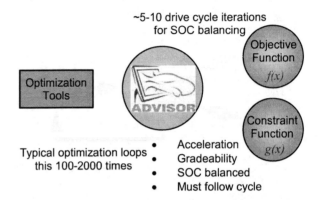

Figure 1: ADVISOR Linkage with Optimization Tools for Vehicle Systems Analysis

The general approach for linking the optimization tools to ADVISOR includes three primary files and five basic steps. The files include a main function routine for configuring the workspace and performing post-processing operations, a function for generating the objective response value, and a function for generating the constraint response values. Most optimization software tools will require minor variations to this implementation process but use the same general approach. As a result, it requires minimal effort to apply multiple algorithms to the solution of the same problem.

The basic optimization process, using ADVISOR, can be summarized as follows,

1. Initialize the MATLAB workspace
2. Modify the design variable values in the workspace with input from optimizer
3. Run simulation to generate objective responses
4. Run simulation to generate constraint responses
5. Process results with optimization tool and return to step 2 until convergence criteria is satisfied

Each of the first four operations is achieved using the unique options as input to the adv_no_gui function as defined in the ADVISOR documentation [11].

OPTIMIZATION PROBLEM DEFINITION

The design variables for this study consist of 8 variables: 4 defining sizing of the fuel cell, motor, and batteries, and 4 defining energy management strategy variables including the maximum and minimum fuel cell power,

Table 3: Design Variables

ADVISOR name	Description	Units	Lower Bound	Upper Bound
fc_pwr_scale	Fuel Cell System Peak Power Scale	--	1 (52 kW)	3 (156 kW)
mc_trq_scale	Motor/ Controller Peak Power Scale	--	0.8 (66 kW)	2.5 (207kW)
ess_module_num	Battery Pack Number of Modules	#	11 (143 V)	35 (455V)
ess_cap_scale	Battery Module Maximum Ah Capacity Scale	--	0.333 (15 Ah)	2 (90 Ah)
cs_min_pwr	Minimum Power Setting (% of Peak Power)	%	0	50
cs_max_pwr	Maximum Power Setting (% of Peak Power)	%	50	100
cs_charge_pwr	Charge Power Setting (% of Peak Power)	%	0	50
cs_min_off_time	Minimum Off Time Setting	s	10	1000

Table 4: Vehicle Performance Constraints

Type	Description	Condition
Acceleration	0-96.5 km/h (0-60 mph)	<= 11.2 s
	64-96.5 km/h (40-60 mph)	<= 4.4 s
	0-137 km/h (0-85 mph)	<= 20.0 s
Gradeability	@ 88.5 km/h (50 mph) for 20 min. at Curb Weight + 5 passengers and cargo (408 kg)	>= 6.5%
Drive Cycle	Difference between drive cycle requested speed and vehicle achieved speed at every second during the drive cycle	<= 3.2 km/h (2 mph)
SOC Balancing	Difference between final and initial battery state of charge	<= 0.5%

battery charge power, and the minimum fuel cell off-time. The ADVISOR variables and their upper and lower bounds are listed in Table 3.

In order to ensure performance equivalence, constraints were established that ensure that the vehicle will have the same acceleration, gradeability, and charge-neutrality of the baseline conventional SUV upon which this fuel cell vehicle is based. Table 4 includes all of the vehicle performance constraints used in the optimization problem. Finally, the objective of this study will be to maximize fuel economy (i.e. minimize fuel consumption).

In a conventional internal combustion engine-powered hybrid, a manufacturer would need to consider cost and emissions in addition to fuel economy when optimizing the vehicle. Since a hydrogen fuel cell-powered hybrid is essentially a zero emissions vehicle, we have eliminated emissions from the problem. At this time, cost models are under development by other U.S. Federal National Labs that will allow us to include vehicle cost in the equation for future studies. For more detailed descriptions of the ADVISOR design variables and data files mentioned in Tables 1-4, please refer to the documentation for ADVISOR v3.1 [12].

As a result of the constraints and objectives defining this problem, the evaluation of a single design point in ADVISOR will include an iterative zero-delta SOC-balanced fuel economy calculation for the applicable drive cycle(s) individually, a 20 minute gradeabilty test, and a 0-137 km/h (0-85 mph) acceleration test. On a Pentium 4 1500 MHz machine a single evaluation will, on average, take ~75 seconds of processing time.

PARAMETRIC SWEEP OF POWER RESPONSE CAPABILITY

The optimization problem defined in the previous section was performed using ADVISOR and DIRECT for a range of fuel cell system transient response times. For hybrid vehicles, optimal vehicle configurations were derived for 10% to 90% power transient response times of 0, 2, 5, 10, 20, and 40 seconds over 3 different driving schedules including the combined City/Highway test, the US06 cycle, and the New European Drive Cycle (NEDC). For neat fuel cell vehicles, optimal designs were generated for response times of 0, 2, 5, and 7 seconds for 2 driving schedules including the combined City/Highway test and the US06 cycle. Response times greater than 7 seconds in a neat fuel cell vehicle required unreasonably large systems to satisfy the performance constraints.

DISCUSSION OF SIMULATION RESULTS

In all, 26 different optimal vehicles were derived in this study. The optimizer called on ADVISOR to evaluate more than 75,000 vehicle designs and was able to complete these analyses within a 10 day period by distributing the analysis among 3 to 7 available desktop PCs in parallel.

The results of this study will be presented in two parts. First, the characteristics of the optimal vehicle configurations will be compared. Then the detailed operating characteristics of some of these vehicles will be explored.

CONFIGURATION CHARACTERISTICS

A review of the results of the optimal hybrid vehicle configurations provides insight into how the vehicle system may adapt to or take advantage of the performance characteristics of one of the components within the system. Figure 2, provides a summary of the optimal vehicle component characteristics with respect to the transient response capability of the fuel cell system.

As the transient response capability decreases (longer response time) the battery energy content will increase. The battery is relied upon to provide propulsion power more often during the drive cycle for a slow responding fuel cell system. Also, the fuel cell power and the battery power capabilities are, in general, inversely related, meaning that the system needs a specific amount of total power and will trade fuel cell power for battery power and vice versa. Note that some variation in the component sizes is also due to the variation of the vehicle mass. Specifically, we see that the motor power increases with increasing response time. The motor size must increase directly with the vehicle mass to provide equivalent vehicle performance.

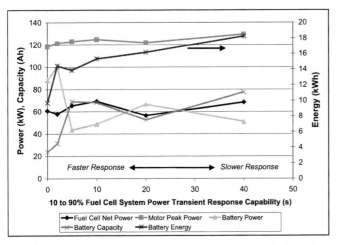

Figure 2: Component Characteristics of Hybrid Vehicles Optimized for Maximum Fuel Economy on Combined City/Highway Driving

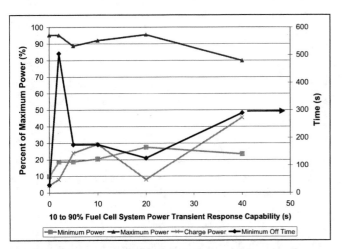

Figure 3: Energy Management Strategy Parameters of Hybrid Vehicles Optimized for Maximum Fuel Economy on Combined City/Highway Driving

Along with the component sizes, the optimizer also varied the energy management strategy parameters. Figure 3 provides a summary of the energy management strategy parameters for the optimal hybrid vehicle configurations. As the power response time grows longer, the control will try to close down the width of the fuel cell operating range. Notice how the minimum power setting increases with increasing response time and the maximum power setting decreases with increasing response time. Since the fuel cell will take longer to respond both on the ramp-up and the ramp-down events in cases with longer response times, it is desirable to stay closer to the middle of the operating zone. This is also a very efficient operating area for the fuel cell system.

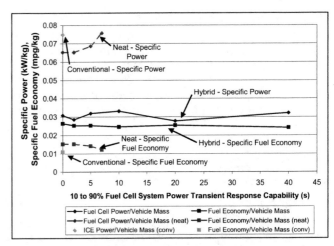

Figure 4: Fuel Economy Impacts of Transient Response Capability for Hybrid and Neat Fuel Cell Vehicles on Combined City/Highway Driving

The corresponding fuel economy results for these combinations of component sizes and energy management strategy parameters are provided in Figure 4. For comparison, the neat fuel cell cases and a baseline conventional vehicle are also included. The fuel economy results have been normalized by the mass of the vehicle such that we can isolate the fuel economy impacts of the powertrain system variation with respect to the transient response capability. Likewise, the fuel cell power capability has been normalized by the vehicle mass and is provided for each case.

The first conclusion that can be drawn from Figure 4 is the fuel economy impacts of the fuel cell system and that of hybridization. With a 0 second response time, the neat fuel cell vehicle normalized fuel economy (dashed line with squares) is 38% better than that of the conventional vehicle (square). Likewise, the hybrid fuel cell vehicle normalized fuel economy (solid line with squares) is 73% better than that of the neat fuel cell vehicle. The mass differential between these vehicles has a significant impact on these relative improvements. On an absolute fuel economy basis the step from conventional to neat fuel cell provides a 65% gain while the step from neat fuel cell to hybrid fuel cell offers a 50% gain.

From Figure 4, we can also conclude that for a neat fuel cell vehicle configuration, the fuel economy begins to drop significantly with increasing transient response time. This is in part because the fuel cell size (peak power) is growing rapidly with respect to transient response time. It must do so in order to provide equivalent performance.

On the other hand, for the hybrid cases, the transient response time has almost no impact on either the fuel economy or the fuel cell power requirement. Basically, in a hybrid system we have the flexibility to optimize the combination of component characteristics and energy management strategy parameters to nullify the fuel economy impacts of a slow responding fuel cell system.

OPERATIONAL CHARACTERISTICS

Once all of the optimal vehicle configurations had been derived and saved, it was possible to review the detailed operating characteristics of each vehicle independently within ADVISOR. This provides some insight into the reasons for and the impacts of the choices made by the optimization tools.

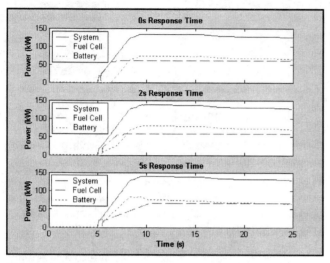

Figure 5: Distribution of Power During Acceleration Event for Hybrid Fuel Cell SUV

First, it is important that we review what the transient response time really means. The impacts of the power response capability on vehicle operation are clearly apparent during a maximum effort acceleration test. Figure 5 provides the component and system power output during an acceleration test for three different hybrid vehicles. These vehicles have 10 to 90% power transient response times of 0 seconds, 2 seconds, and 5 seconds from top to bottom, respectively. The system power is simply the sum of the fuel cell power and the battery power outputs. It is nearly the same in all three cases because the total power delivery to the wheels is determined by the traction characteristics of the vehicle. In comparing the three plots we see that with a shorter response time (0 second case) the fuel cell initially provides the entire vehicle power demand up to its maximum capability. The fuel cell initially provides a majority of the power demand in the 2 second case. In the 2 second case, the battery power is used to supplement the system power while the fuel cell power is ramping up at a rate of ~22 kW/s. Likewise, it is apparent that the battery provides a majority of the power demand in the 5 second case while the fuel cell power ramps at a rate of ~10 kW/s.

Figure 6 shows only the fuel cell system power output during the acceleration test for the same three hybrid configurations (0, 2, and 5 seconds). The differences in ramp rates are more clearly defined. Likewise, Figure 7 provides the same set of fuel cell performance data for the neat fuel cell vehicle cases during an acceleration test.

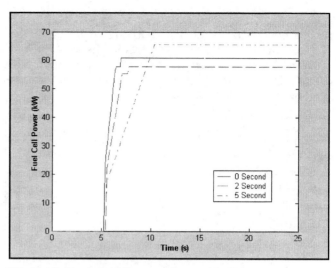

Figure 6: Fuel Cell System Power Delivery during Acceleration for <u>Hybrid</u> Fuel Cell SUV

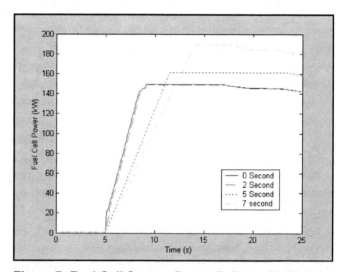

Figure 7: Fuel Cell System Power Delivery During Acceleration for <u>Neat</u> Fuel Cell SUV

For a neat fuel cell vehicle the battery is not present to supplement the capability of the fuel cell. Therefore, as was seen in Figure 4 the power capability of the fuel cell must be significantly larger so that it can compensate for its slow response during a fast transient. Notice the difference in the scales between Figure 6 and Figure 7.

Although the fuel cell is the only power source in a neat fuel cell vehicle, in Figure 7, the actual power delivered to the driveline from the fuel cell is different in each case. As a result, the acceleration performance of each vehicle is slightly different. We have enforced three acceleration time constraints (0-60 mph, 40-60 mph, and 0-85 mph) on these vehicles and all of the constraints must be satisfied. Typically only one constraint will be active while the performance is better than required for the other two acceleration constraints. In the 7 second case, the 0-60 mph constraint is active while in the 2 second case, the 0-85 mph constraint is active.

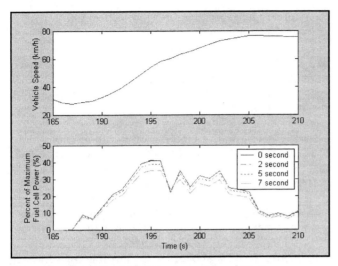

Figure 8: Neat Fuel Cell System Power Delivery Limitations During a Portion of the FTP

In Figure 8, the actual operating point as a percent of the maximum capability of the fuel cell system is shown for each of the neat fuel cell vehicles during a short acceleration event within the Federal Test Procedure (FTP) driving schedule. A longer transient response time leads to an optimal vehicle with a larger fuel cell system such that it can satisfy all of the performance constraints. As a result, we see that the fuel cell itself operates at a lower percentage of its peak capability on normal drive cycles with respect to increasing transient response times.

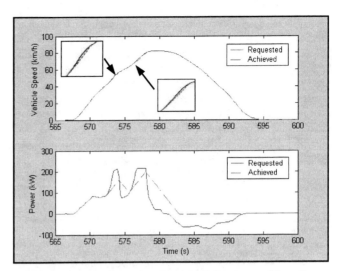

Figure 9: Operation of a Fuel Cell System with a 5 Second Transient Response Time in a Neat Fuel Cell Vehicle on a Portion of the US06 Cycle

Sections of the US06 cycle have extreme power transients. Figure 9 highlights the operational impacts of the power response capability of the fuel cell system in a neat fuel cell vehicle. The top plot shows the vehicle speed trace from a portion of the US06 cycle. The actual trace value (solid line - Requested) and the achieved speed (dashed line - Achieved) are nearly identical. However, the vehicle speed falls away from the trace near 573s and 576s. As the vehicle begins to miss the

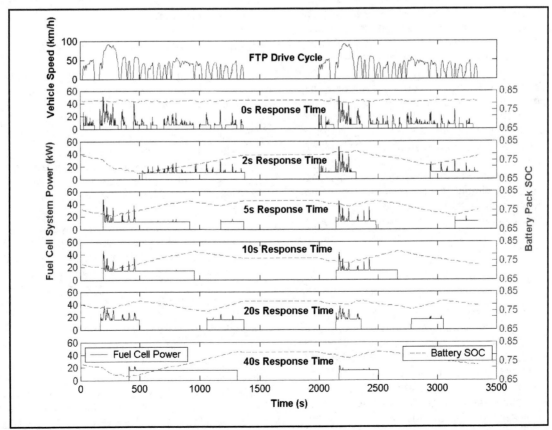

Figure 10: Fuel Cell and Battery Operating Characteristics of Hybrid Vehicles Optimized For Combined City/Highway Driving Vary with Respect to Fuel Cell System Transient Response Capability

requested speed more power is requested of the fuel cell. The lower plot shows the power requested (solid line) by the drivetrain of the fuel cell system such that the vehicle can follow the speed trace. The dashed line defines the power actually delivered by the fuel cell system. Clearly the fuel cell system in this case with a transient response rate of ~40 kW/s is unable to respond to the fast power transients. Both on the rising side and the falling side the fuel cell power achieved lags what is requested. It is unclear how the excess power during a falling event may impact the vehicle performance in a real vehicle.

Thus far we have looked at the vehicle details and their operation during small portions of driving cycles. It is also interesting to compare the vehicle system operation of the various vehicles on a complete drive cycle. In Figure 10, the fuel cell power output and the battery pack state of charge over an FTP (2 urban dynamometer driving schedules) is provided for the 6 different hybrid vehicles optimized for operation on the City/Highway test procedure. The top portion of the figure provides the actual driving schedule. The six plots below provide the component operating conditions for the 0, 2, 5, 10, 20 and 40 second response time cases.

As one may expect, the vehicle with a fuel cell system with the 0 second response time operates almost entirely under a load following (i.e. when on, the power varies proportional to the drivetrain load) approach. While the vehicle with a slow response fuel cell system (40 second response time) operates almost entirely with a thermostatic-type control (i.e. when on, the system operates at a single power level). The vehicles between the 0 second and 40 second cases show some combination of the behavior of the two boundary cases. It is also clear that the 0 second vehicle case uses its battery pack very little with the SOC remaining within a very narrow band while the other vehicles exercise the battery pack significantly through an ~15% SOC window. It should be noted that the total capacity of the battery packs vary from case to case but that in general the longer the response time the larger (more capacity) the battery pack. SOC is simply a relative measure of the available capacity in a battery pack therefore a 10% variation of SOC in a large pack represents significantly more energy throughput than the same SOC variation for a smaller battery pack.

CONCLUSION

In this study, the impacts of fuel cell system power response capability on optimal hybrid and neat fuel cell vehicle configurations have been explored. Optimal hybrid vehicle design scenarios for an SUV were derived for fuel cell systems with 10 to 90% power transient response times of 0, 2, 5, 10, 20, and 40 seconds. A derivative-free optimization algorithm was used with ADVISOR to systematically change both powertrain component sizes and the vehicle energy management strategy parameters to provide optimal vehicle system configurations for the range of transient response times.

Several conclusions can be drawn from the results collected and analyzed.

For a hybrid fuel cell vehicle, as the transient response capability of the fuel cell system increases (shorter response times) when trying to maximize fuel economy, the optimal vehicle configuration will shift from one with a large capacity battery pack and a thermostatic control strategy to one with a smaller battery pack and a more load following strategy. System costs and packaging considerations were not considered in this analysis and would likely influence the optimal design characteristics.

For a neat fuel cell vehicle, a small fuel cell transient response time is critical to satisfy vehicle performance constraints and to provide significant fuel economy improvement over that of a typical conventional vehicle.

In contrast, it was demonstrated that system optimization could be used to find combinations of component sizes and energy management strategy parameter settings for a hybrid fuel cell vehicle that will nearly nullify the effects of transient response time with respect to fuel economy.

The results presented in this study were derived assuming that a hydrogen fuel cell system would be used in an SUV. Both hydrogen and gasoline reformed fuel cell systems will have unique power response capabilities. The trends highlighted above will apply similarly to a gasoline reformed fuel cell system in a vehicle application.

Future areas of exploration may include,

- Resolve the details of the actual fuel cell system configuration and control to provide systems with the transient response characteristics described in this paper. This would help identify the design advantages (mass, volume, and/or cost) of alternative fuel cell system designs.
- Resolve the current optimal designs using gradient-based optimization tools within a small design space centered around the current best case designs.
- Generate optimized vehicle designs based on parametric sweeps of other system attributes such as fuel cell system specific power, specific cost, power density, and efficiency. These attributes will impact the balance between fuel cell and battery in addition to the energy management strategy employed in the vehicle.

ACKNOWLEDGMENTS

The authors would like to acknowledge the support of the U.S. Department of Energy, specifically Robert Kost and Patrick Sutton, for their support of ADVISOR. In addition we would like to acknowledge JoAnn Milliken and Steve Chalk for their support of our fuel cell vehicle systems analysis efforts.

REFERENCES

1. NREL's Vehicle Systems Analysis web site. http://www.nrel.gov/transportation/analysis.
2. Wipke, K.; Cuddy, M.; Burch, S. "ADVISOR 2.1: A User-Friendly Advanced Powertrain Simulation Using a Combined Backward/Forward Approach." *IEEE Transactions on Vehicular Technology*, v. 48, n. 6, ISSN 0018-9545, Nov. 1999.
3. Markel, T.; Wipke, K. "Optimization Techniques For Hybrid Electric Vehicle Analysis Using ADVISOR." *Proceedings of the ASME International Mechanical Engineering Congress and Exposition.* New York, New York. November 11-16, 2001.
4. Friedman, D. "Maximizing Direct-Hydrogen PEM Fuel Cell Vehicle Efficiency – Is Hybridization Necessary?" SAE Publication 1999-01-0530. *Proceedings of 1999 SAE Congress.* Detroit, Michigan. March 1999.
5. Atwood, P.; Gurski, S.; Nelson D.J.; Wipke, K.B. "Degree of Hybridization Modeling of a Fuel Cell Hybrid Electric Sport Utility Vehicle." SAE Publication 2001-01-0236. *Proceedings of SAE Congress 2001.* Detroit, Michigan Jan. 2001. Reprinted in SAE SP-1589, *Fuel Cell Power for Transportation 2001*, pp. 23-30.
6. Wipke, K.; Markel, T.; Nelson, D. "Optimizing Energy Management Strategy and Degree of Hybridization for a Hydrogen Fuel Cell SUV." *Proceedings of 18th Electric Vehicle Symposium.* Berlin, Germany. October 2001.
7. Davis, P.; Milliken, J.; Ho, D.; Garland, N. "DOE Fuel Cell Activities Overview." Presentation. October 30, 2001. <http://www-db.research.anl.gov/db1/cartech/document/DDD/99.pdf>.
8. Mays, C.R.; Campbell, A.B.; Fengler, W. A.; Rowe, S.A. "Control System Development for Automotive PEM Fuel Cell Vehicles." SAE Publication 2001-01-2548.
9. Adams, J.A.; Yang, W.-C.; Oglesby, K.A.; Osborne, K.D. "The Development of Ford's P2000 Fuel Cell Vehicle." SAE Publication 2000-01-1061.
10. Potter, L. and Reinkingh, J. "SPFC Bus Design Studies." ETSU F/02/00134/REP. 1999.
11. ADVISOR v3.1 Documentation. "Section 2.3: Using ADVISOR without the GUI." http://www.ctts.nrel.gov/analysis/advisor_doc/advisor_ch2.htm#2.3.
12. ADVISOR v3.1 Documentation. http://www.ctts.nrel.gov/analysis/advisor_doc/.

CONTACT

Tony Markel, Engineer II, National Renewable Energy Laboratory, 1617 Cole Blvd., Golden, CO 80401, USA. Phone: 303.275.4478, Fax: 303.275.4415, E-mail: tony_markel@nrel.gov.

2001-01-0538

A Comparison of High-Pressure and Low-Pressure Operation of PEM Fuel Cell Systems

Joshua M. Cunningham, Myron A. Hoffman and David J. Friedman
University of California, Davis

ABSTRACT

This paper compares the merits of operating a direct-hydrogen fuel cell (DHFC) system using a high-pressure air supply (compressor) versus one using a low-pressure air supply (blower). Overall, for the system modeled, it is shown that there is no inherent *performance* advantage for either mode of operation at the DHFC <u>stack</u> level. However, in practical applications, as will be shown in this paper, a systems analysis (stack and air supply) of power and efficiency needs to be performed.

Equivalent PEM DHFC stack peak power values can be obtained using both high-pressure and low-pressure air supply systems. For each air supply configuration, air mass flow and pressure operating conditions can be found that result in an equal value of the oxygen partial pressure at the cathode catalyst layer surface.

However, at the system level, the required air supply power needed to achieve the same DHFC stack performance values can be drastically different for high and low pressure operation. In order to *compare* the two *systems*, an optimal air supply control strategy is first developed to obtain the desired stack operating conditions with minimal parasitic loads based on each air supply configuration. Second, the resulting air supply parasitic loads are compared directly between the two configurations – both comparisons are set in the context of the system performance. In other words, the systems are sized such that the peak <u>net</u> power values are equal while the stack gross power may be different.

The results of the study demonstrate the well-known fact that equivalent DHFC *peak net system* power values (86kW) can be obtained with both types of air supply configurations but require different stack sizes. For the blower application, the stack size had to be increased by 16.3% (500 vs. 430 cells in this example) for a peak net power of 86kW. Differences are also apparent with the WTM sub-system. Quantitative results will be presented for both the high pressure and the low pressure applications.

INTRODUCTION

Currently, fuel cell vehicles (FCVs) are being developed as a potential alternative to the conventional internal combustion engine vehicle (ICEVs). FCVs offer the potential of higher fuel efficiency and lower vehicle emissions compared to the ICEVs. However, an area needing further development is a realistic systems analysis combining the various sub-systems and components for FCVs.

Depending on the output electrical power required from the fuel cell stack, the air mass flow rate and, in many stack designs, the pressure of the air will need to be varied. This acts to control the oxygen partial pressure at the cell reaction sites in order to produce a specific or desired "gross power" output from the fuel cell stack. Various air supply technologies are available, with varying performance capabilities, to provide the oxygen to the cathode sites. In general, the components can be separated into two categories: those of relatively high pressure ratio (r_c) ability (compressors), and those of low pressure ratio ability (blowers). By operating the system at higher pressures, significantly more power is needed for the compressor operation. However, benefits can include higher stack efficiencies and smaller stack sizes/costs.

This paper seeks to compare the performance results and geometry of the system (fuel cell stack and air supply componentry) using both the compressor and the blower by scaling the fuel cell stack itself to obtain an equivalent peak net system power. The system modeled was that of a direct hydrogen fuel cell configuration.

MODELING SETUP

The system performance results used for this paper's comparison were obtained by using the UCDavis computer simulation program created with Matlab's Simulink visual modeling software. The model uses air system performance data provided by air component developers from actual lab tests. The data are included in the model in the form of two-dimensional performance maps for shaft power, shaft speed, and exit air temperature. The PEM fuel cell stack characteristics are directly modeled and have been validated against lab performance data from Los Alamos National Laboratory. Additionally, a water/thermal management (WTM) model was incorporated to account for condenser and radiator parasitic loads. The WTM model is based on fundamental relationships and data tables.

For the simulations, a twin-screw, positive displacement compressor from Vairex Corp. was chosen that has a maximum r_c capability of 2.5 and a maximum air mass flow of 105 g/s at an r_c of 1.8 (at STP conditions). For the low-pressure application, a "regenerative" blower from Siemens-Airtech was chosen which has a maximum r_c capability of 1.4 and a maximum air mass flow of 93 g/s at ambient pressure. The performance maps used in the model adequately account for the associated limitations of the particular technology (maximum and minimum performance regions). Additionally, a variable speed motor and controller map was utilized to determine the electric efficiency for the corresponding shaft speed and torque.

One unique feature of the model is the optimization procedure between the fuel cell stack performance, the parasitic load of the air supply technology utilized, and the parasitic loads of the condenser and radiator for the WTM system. The optimization in the model determines the air system operating scheme such that the net system electric power is maximized for each value of the stack current density. A full description of this optimization procedure can be found in references [8, 9].

The defined net electric power is simply the stack gross electric power minus the parasitic loads of the air system electric motor (calculated from the air system model during the optimization process) and the WTM radiator and condenser loads. Equations 1 and 2 below specifically define these relationships.

$$P_{stack} = (VI)_{stack} \quad ; \quad P_{as\_motor} = \frac{(P_{sh\_comp})}{\eta_{as\_motor}} \quad (1)$$

$$P_{net} = P_{stack} - P_{as\_motor} - P_{rad} - P_{cond} \quad (2)$$

Note that the air system power in equation 1 is a function of the compressor load and does not include power recovered from an expander. Most pressurized systems will, however, incorporate an energy recovery device such as an expander. Performance data was not available to include the expander in this analysis. However, it has been shown that the use of an expander can reduce the air system power and stack size/cost [5].

The gross power of the fuel cell stack is directly dependent on the partial pressure of oxygen (p_{oxygen}) at the cathode catalyst reaction sites. Each single value of p_{oxygen} corresponds to a single cell voltage value at each particular current density. However, p_{oxygen} is a function of both the total air pressure and the air mass flow rate,

and can be achieved through different combinations of the two. Figure 1 shows this relationship for an *example* simulation (discussion purposes only). The figure shows a contour of constant p_{oxygen} values versus values for air pressure and air mass flow rate at a fixed current density of 400 mA/cm². It can be seen that a wide range of pressures and air mass flows can achieve the exact same oxygen partial pressure, and therefore the exact same cell voltage (and therefore the exact same stack gross power). Thus, a high pressure system can achieve the same p_{oxygen} at low air flow rates that a low pressure system can achieve at high flow rates. It is therefore quite possible for both the compressor and the blower applications to produce the same partial pressure of oxygen and same stack gross output. For example, if the blower application has a limit in maximum output pressure, an increased amount of air mass flow compared to the compressor application can result in equal values of p_{oxygen}.

Equal stack gross powers do not, however, necessarily produce equal net system output powers. Instead, the proper r_c and air mass flow combination must be found to maximize the net system power at any given current density, compressor/blower choice, and specified p_{oxygen}. This will likely result in different air mass flow and r_c conditions for the blower and compressor applications. The resulting optimum choice takes into consideration the various component performance characteristics. (Note: the relationship between the current density and the mass flow is dependent on the air stoichiometric ratio)

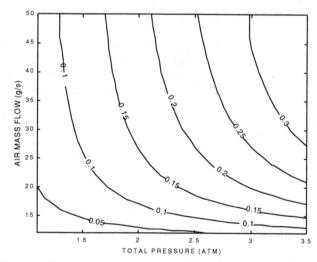

Figure 1: Lines of constant Partial Pressure of Oxygen for a range of air mass flow (g/s)and total pressure (atm) Note: fixed current density, J=400mA/cm²

In addition to the stack / air supply interactions, the water and thermal management systems need to be considered. As described in reference [5], water is introduced to the cathode from several sources. It is dragged across the membrane from the anode (drag) along with the hydrogen ions, it is formed at the cathode reaction sites (form), and in some configurations it is directly injected into the air stream for cooling purposes, humidifying the stream in the process (hum). Two factors are important to study here. First, different r_c / air mass flow combinations require different total amounts of water for the <u>same net power level</u>. This can occur if the corresponding stack current is different for the various control schemes, and/or if the humidification needs are different depending on the exhaust air temperature from the compressor or blower. Second, different r_c / air mass flow combinations lead to different water states (%vapor vs. %liquid) in the fuel cell exhaust. Both of these factors have ramifications on condenser and radiator loads and will affect pump and fan (parasitic) electric loads. Again, refer to the following references for modeling the WTM sub-systems and the related performance optimization procedure [1, 7, 9].

SIMULATION DESCRIPTION: HIGH-PRESSURE VS. LOW-PRESSURE

The following table outlines the air system used along with the corresponding fuel cell stack characteristics modeled. Notice that in order to achieve the same net power, the blower application corresponds to a larger stack size, modeled as an increase in the number of cells while maintaining the same cell active area.

Table 1: Simulation Input Parameters

Configuration	Net Power	#Cells	AA/cell
Twinscrew Comp.	86 kW	430	490 cm^2
Regenerative Blow.	86 kW	500	490 cm^2

The specific stack size chosen was partially dependent on the voltage restrictions of the vehicle drive motor. In the UCDavis vehicle model, the drive motor operated properly if the voltage supply is held between approximately 200 – 400 volts during normal operation. With this restriction, a cell active area was chosen such that the number of cells for both the compressor and the blower applications resulted in adequate stack voltages.

It should be noted that to truly optimize a low pressure or high pressure system, the fuel cell stack for each application could be significantly different in physical design (i.e. cathode flow fields carefully designed to minimize total pressure losses for the blower application). This simulation was limited to simply varying the number of cells and therefore, conclusions from the study are somewhat limited as a result.

Figure 2 shows the system configuration including the primary components involved and the distribution of water in the system. More specific details regarding the WTM components can be found in reference [9].

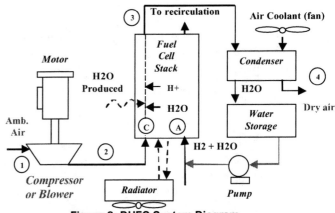

Figure 2: DHFC System Diagram

SIMULATION RESULTS

AIR SYSTEM AND FUEL CELL STACK INTERACTION-
In Figure 3, the net system and air supply electric powers are graphed for the range of stack current density (mA/cm^2). Both the blower and the compressor modeled provided the same peak P_{net} of 86 kW ($P_{stack} - P_{as\_motor} - (P_{cond}+P_{rad})$). For the blower application, this range of net power occurred with lower parasitic loads (blower electric power) due to its lower pressure ratio operation (as shown in later figures). As a result, the peak net system power occurred at a lower gross stack power compared to the high pressure application. The major difference was that the blower required a stack with 16.3% more cells resulting in a costlier stack. Also note that the P_{net} values occurred at different stack current densities.

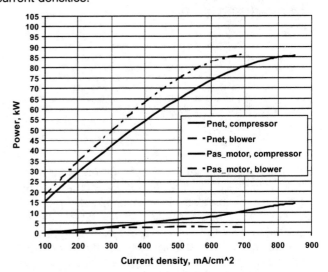

Figure 3: Fuel Cell Stack Gross and Air System Electric Power

Figure 4 shows the various power characteristics for the peak load condition (WTM parasitic loads not shown). It is evident that the blower's parasitic load is significantly less than that of the compressor at peak power.

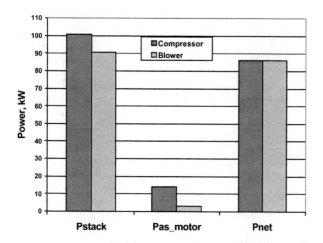

Figure 4: Fuel Cell Stack peak gross and net power, and air system electric power

Table 2 describes the efficiency relationships of the system.

Table 2: System Power and Efficiency

Parameter	Compressor	Blower
Pnet, kW	85.7 (18.0)	86.6 (18.4)
Pas_motor, kW	14.2 (0.8)	2.9 (0.4)
Ratio P_{as_motor}/P_{stack}	0.14 (0.04)	0.03 (0.02)
Efficiency_stack(LHV) %	45.7 (62.3)	42.9 (62.4)
Efficiency_net, %	38.9 (59.6)	41.1 (61.0)

* *Values in ( ) are for part load performance, approximately 21% of peak net power*

The efficiencies are defined as follows:

$$\eta_{stack} = \frac{P_{stack}}{\overset{\bullet}{m}_{H2} * LHV_{H2}} \quad (5)$$

$$\eta_{net} = \frac{P_{net}}{\overset{\bullet}{m}_{H2} * LHV_{H2}} = \eta_{stack} * (1 - \frac{(P_{as\_motor} + P_{rad} + P_{cond})}{P_{stack}}) \quad (6)$$

For both technologies, the net efficiency for the P_{net} range was similar while achieving the same peak Pnet. The blower application maintained a η_{net} 1.5-2.0 percentage points higher than the compressor application over the entire P_{net} range. Both applications resulted in similar net efficiency variations of a typical fuel cell application where the peak efficiency occurs at the lower end of the P_{net} range and slowly tapers off as power is increased.

The small difference in η_{net} can be understood by looking at Equation 6 and Figure 3. For the same P_{net}, if $\overset{\bullet}{m}_{H2}$ is lower, η_{net} increases. This was the case for the blower application. Figure 3 shows that current density (directly proportional to $\overset{\bullet}{m}_{H2}$ for a fixed utilization of hydrogen at the anode) was lower for nearly all stack power levels,

and it can be shown that this relationship holds for the net system power as well.

But why was the current density lower? For a lower pressure in the fuel cell, the concentration of oxygen at the cathode catalyst is reduced. The low pressure application (given a fixed cell active area) makes up for this by increasing the number of cells per stack and thus total stack voltage.

Considering the similarity in η_{net}, both of these technologies would perform in a similar manner over a vehicular driving cycle where vehicle fuel economy over the range of engine power is a key performance parameter. However, other important differences may lead to dissimilar vehicle fuel economy. For instance, stack voltage differences at the same system net power and efficiency may result in different vehicle motor efficiencies and overall performance. Additionally, the transient response of the compressor or blower to vehicle load demands may differ impacting vehicle performance. Physical mass differences of the two systems will also impact vehicle response for a given system net power capability.

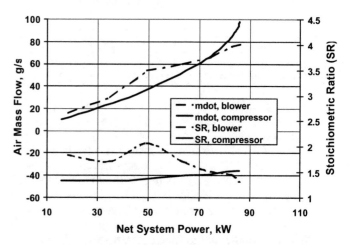

Figure 5: Air mass flow rate and SR for the compressor and Blower

Figure 5 shows the air mass flow rate as a function of the net power. For both applications, the air mass flow increases with power. However, for the blower, the air mass flow requirements are higher over most of the power range to provide the optimum partial pressure of oxygen at the cathode catalyst sites with a corresponding lower total air pressure.

Also shown in the figure is the cathode air stoichiometric ratio (SR). This is a quantitative measure of the excess air mass flow in the stack required from the air supply. Considering the blower is limited in output pressure and compensates for this with sufficiently higher mass flows, the SR is higher for much of the net power range. The compressor, on the other hand, shows a relatively constant SR of 1.5.

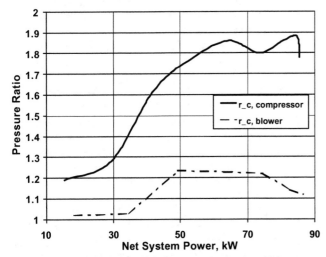

Figure 6: Exit pressure of the compressor and blower

Figure 6 shows the resulting pressure ratio operation for the air supply systems. The Blower operated near ambient pressure part of the time with a slight increase in pressure between 35 and 75kW(net). The compressor, on the other hand, shows a steadily increasing r_c, operating between 1.2 and 1.9 pressure ratios.

Also modeled in the simulations were pressure drops. At the high power levels, the air mass flow was larger leading to an increased pressure loss across the fuel cell. The pressure loss was much more restrictive to the blower application which already operated *near* ambient pressure. In these simulations, if the pressure loss characteristic had been larger, the blower system would not have obtained a net power of 86kW. This is an important consideration when performance matching a fuel cell stack and an air supply and why geometry would likely change.

In general, the non-linear performance curves in Figures 5 and 6 of both the air mass flow and r_c are a direct result of the optimization model which searches for the air supply air mass flow and r_c combination that maximizes the net power for a given current density. The resulting non-linear curve is unique to the specific compressor or blower and is dependent on the component's operating efficiencies.

It is important to note that all of the high pressure application results would differ if an expander were to be included. P_{net} would be achieved at reduced P_{stack} powers and thus different air pressure and mass flow schemes. Stack size would be further reduced, potentially increasing overall power density. Net system efficiency *may* improve as well.

WATER AND THERMAL MANAGEMENT INTERACTIONS - As mentioned in the modeling section above, management of the water and thermal loads is critical in fuel cell systems. Ensuring that the fuel cell stack is adequately humidified is necessary, requiring the recovery of liquid water from the fuel cell stack

exhaust by using water traps and condensers. Table 3 below shows the amount of water involved at the maximum load condition for both the compressor and blower application. The data presents the conditions at the stack cathode exit.

Table 3: Conditions at stack exit – Comparison at peak load condition

Parameter	Compr.	Blower
Pnet, kW	85.7	86.6
Pstack, kW	100.8	90.4
Current density, mA/cm^2	850	700
Total current, A	417	343
r_c, at the exit of the stack	1.8	1.01
Air stoichiometric ratio	1.5	1.3
$\dot{m}_{h2o-form}$, g/s	19.4	16.0
$\dot{m}_{h2o-drag}$, g/s	4.0	3.2
$\dot{m}_{h2o-hum}$, g/s	1.1	0.8
$\dot{m}_{h2o-total}$, g/s vapor and liquid	24.5	20.0
% of exit water mass flow in vapor form	88 %	100 %

* Note: Assumes dry air into compressor inlet and a stack operating temperature of 80°C

Several trends can be seen in the results. The total water involved was lower for the low pressure, blower system. This is largely because $\dot{m}_{h2o-form}$ and $\dot{m}_{h2o-drag}$ are reduced due to lower current densities at the same net power. Additionally, the actual air mass flow utilized in the stack at the peak condition of 86kW(net) was less for the blower application in these simulations (total air mass flow divided by SR). However, since the pressure was lower at the exit of the fuel cell stack where the gas temperature was approximately 80°C, the percentage of water in the vapor form was higher for the blower application (100% vs. 88% for the compressor). Note that 100% of the water was in vapor form for the blower application at peak load because the gas temperature of 80°C was higher than the saturation temperature.

In these simulations, the actual amount of water that was <u>needed</u> for system operation (operating self-sufficiency) was simply $\dot{m}_{h2o-drag}$ (4.01g/s for the compressor and 3.24g/s for the blower applications). Consequently, even though the amount of system water required for the blower application was lower in these simulations, all of the water must be condensed at the condenser (as compared to the stack) and at a lower pressure. This has the effect of increasing the condenser load, resulting in either a larger condenser area and/or an increased cooling load on the condenser fan compared to the high pressure application. Additional details of the effects of cathode pressure on the condenser and radiator loads, and the associated tradeoffs, are discussed in reference [9].

Though not modeled, one possible way to increase the percentage of liquid water exiting the stack at the low pressure is to operate the fuel cell stack at lower temperatures. This has the effect of shifting the condensation load from the condenser to the radiator but also reducing stack efficiency.

PHYSICAL SIZE/GEOMETRY CONSIDERATIONS

As mentioned previously, the load on the condenser is dependent on several parameters, including the total quantity of vapor to condense and the gas stream pressure and temperature conditions. Comparing the low vs. high pressure systems at peak system load, the blower application operates with more water vapor exiting the stack at a lower gas pressure, both of which increase the condenser loads. To compensate for this, the condenser size and/or the fan load must be increased to extract the required amount of liquid water.

As shown, the fuel cell stack size was different as well. With the blower application, the total number of cells necessary in the stack was 16.3% larger than that needed for the pressurized system (500 vs. 430cells), resulting in a larger stack size and <u>cost</u>. These stack size differentials are based on the same active area of 490cm^2 per cell for both applications.

The actual size of the compressor and the blower are assumed to be similar. If an expander were to be added to the pressurized application, the compressor/expander module may be larger. However, with the use of an expander, the stack size could be further reduced while still maintaining the same net power [5], and the condenser load may also be reduced if a significant portion of the vapor in the gas stream condenses in the expander (assuming the condenser is placed after the expander).

It was not possible, based on the current data, to determine which system has the larger volumetric power density (86kW / (total system volume, m^3)). However, it is anticipated that the low pressure system will be larger in physical size due to the stack size differential and potentially larger WTM components. This would result in a lower volumetric power density.

CONCLUSION

The following conclusions can be made from these specific simulations:

- The same peak P_{net} can be achieved with both a blower (low pressure) and a compressor (high pressure), but the required fuel cell stack sizes are different. For the same peak P_{net} of 86kW, 16.3% more operating PEM cells were needed in the stack for the blower application (500 vs. 430 cells with a constant active area of 490cm^2).
- The blower system was able to obtain the same net power by operating just above ambient pressure at

the stack and providing sufficiently higher air mass flow rates compared to that of the compressor for much of the P_{net} range.
- The parasitic loads for the blower are significantly less than that of the compressor at the high P_{net} region. The ratio of P_{as_motor}/P_{stack} was 14.1% for the compressor vs. 3.2% for the blower at a peak P_{net} of 86kW (though these occur at different P_{stack} values).
- Overall, the net system efficiencies over the P_{net} range were very similar for both the blower and the compressor. However, the blower system did maintain a net efficiency 1.5 – 2.0 percentage points higher than the compressor system over most of the net power range.
- High pressure application results would differ if an expander were to be included. P_{net} would be achieved at reduced P_{stack} powers and thus different air pressure and mass flow schemes. Stack size would be further reduced, potentially increasing overall power density. Net system efficiency *may* improve as well.

ACKNOWLEDGMENTS

The work presented in this paper was supported by the Fuel Cell Vehicle Modeling Project at the Institute of Transportation Studies, University of California, Davis. The modeling process utilizes the Matlab/Simulink programs. The authors would also like to acknowledge the support received from Vairex Corp. and Siemens-Airtech in the development of the air supply performance. Acknowledgements are given to team members P Badrinarayanan and Anthony Eggert for their assistance in the water and thermal management work.

CONTACT

Joshua M. Cunningham
Graduate Research Assistant,
Fuel Cell Vehicle Modeling Program,
Institute of Transportation Studies
University of California, Davis
jmcunningham@ucdavis.edu

REFERENCES

1. Badrinarayanan, P., A.Eggert, K.H. Hauer, "Implications of Water and Thermal Management Parameters in the Optimization of an Indirect Methanol Fuel Cell System," Presented at the 35th Intersociety Energy Conversion Engineering Conference, 24-28, July 2000, Paper # 2000-3046.
2. Barbir, F., "Air, Water and Heat Management in Automotive Fuel Cell Systems," Presented at the Commercializing Fuel Cell Vehicles 2000 conference, 12-14 April, 2000, Berlin, Germany.
3. Barbir, F., et al., "Trade-off Design Analysis of Operating Pressure and Temperature in PEM Fuel Cell Systems," Proceedings of the ASME Advanced Energy Systems Division, AES-Vol. 39, ASME 1999

4. Cunningham, J.M., D.J.Friedman, M.A.Hoffman, R.M.Moore, "Requirements for a Flexible and Realistic Air Supply Model for Incorporation into a Fuel Cell Vehicle (FCV) System Simulation," Presented at the SAE Future Transportation Technology Conference and Exposition, 17-19 August, 1999, SAE #1999-01-2912, SAE International, Warrendale, PA, 1999

5. Cunningham, J.M., M.A.Hoffman, "The Implications of Using an Expander (Turbine) in an Air System of a PEM Fuel Cell Engine," Presented at the 17th International Electric Vehicle Symposium & Exposition, 15-18 October, 2000, Montreal Quebec.

6. Doss, E.D., et al., "Pressurized and Atmospheric Pressure Gasoline-Fueled Polymer Electrolyte Fuel Cell System Performance," Presented at the SAE 2000 World Congress, March 6-9, 2000, SAE # 2000-01-2574, SAE International, Warrendale, PA, 2000

7. Eggert, A., et al., "Water and Thermal Management of an Indirect Methanol Fuel Cell system for Automotive Applications," Proceedings of the 2000 International Mechanical Engineering Congress and Exposition, Orlando, FL, November 5-10, 2000 (pending)

8. Friedman, D.J., R.M.Moore, "PEM Fuel Cell System Optimization", Proceedings of the Second International Symposium on Proton Conducting Membrane Fuel Cells II, Edited by S. Gottesfeld et al., Electrochemical Society, Pennington, NJ, 1999

9. Friedman, D.J., A.Eggert, P.Badrinarayanan, J.M.Cunningham, "Maximizing the Power Output of an Indirect Methanol PEM Fuel Cell System: Balancing Stack Output, Air Supply, and Water and Thermal Management Demands," Presented at the SAE 2001 World Congress, March 5-8, 2001, SAE # 2001-01-0535, SAE International, Warrendale, PA, 2001

NOMENCLATURE

I = stack total current, A

J = stack current density, A/cm^2

$\dot{m}$ or mdot = mass flow rate, g/s

P = power, kW

p = pressure, atm

p_{oxygen} = partial pressure of oxygen

r_c = compressor pressure ratio = p_{c2} / p_{c1}

SR = stoichiometric ratio of air: ratio of **moles of O_2 in the air per second** supplied to fuel cell stack vs. **moles of O_2 per second** utilized at the corresponding stack power level (or fuel consumption rate)

T_1 = ambient and compressor inlet temperature, 25°C (298K)

T_2 = compressor exit temperature, Kelvins

T_3 = fuel cell stack exit and turbine inlet temperature, Kelvins

V = stack total voltage, volts

η = efficiency

Subscripts:

as_motor = air supply motor characteristic

c = compressor

cond = condenser

drag = H_2O transported across the membrane

form = H_2O formed at from the catalyst reactions

hum = H_2O for humidification of air into the fuel cell

rad = radiator

sh = shaft

t = turbine/expander

1 = inlet conditions to the compressor –atmosphere

2 = exit conditions from the compressor and inlet to the fuel cell stack

3 = exit conditions from the fuel cell stack and inlet to the turbine

4 = exit conditions from the turbine - atmosphere

2004-01-1302

Energy Storage Requirements for Fuel Cell Vehicles

A. Rousseau, P. Sharer and R. Ahluwalia
Argonne National Laboratory

ABSTRACT

Because of their high efficiency and low emissions, fuel-cell vehicles are undergoing extensive research and development. As the entire powertrain system needs to be optimized, the requirements of each component to achieve FreedomCAR goals need to be determined. With the collaboration of FreedomCAR fuel cell, energy storage, and vehicle Technical Teams, Argonne National Laboratory (ANL) used several modeling tools to define the energy storage requirements for fuel cell vehicles. For example, the Powertrain System Analysis Toolkit (PSAT), which is a transient vehicle simulation software, was used with a transient fuel cell model derived from the General Computational Toolkit (GCtool). This paper describes the impact of degree of hybridization, control strategy, and energy storage technology on energy storage requirements for a fuel cell SUV vehicle platform.

INTRODUCTION

Hydrogen fuel cell vehicles are undergoing extensive research and development as a means to address both environmental and oil dependency issues in the United States. Considering the current status of fuel cell technologies, it is very likely that the first fuel cell vehicles will be hybrids. To define the most appropriate energy storage technology for such an application, several FreedomCAR Technical Teams worked to define the future electrochemical energy storage requirements for fuel cell vehicles by using advanced simulation tools.

Over several decades, Argonne National Laboratory (ANL) has developed and used a number of computer models in support of the U.S. Department of Energy's (DOE's) advanced automotive R&D program, which has addressed aspects of vehicular life cycles, ranging from design and manufacturing through recycling. Advanced batteries, fuel cells, engines, and vehicle configurations have been developed, tested, and modeled in DOE's facilities at ANL. This combination of analytical, developmental, and testing experience has been applied to several types of advanced vehicle powertrains at the vehicle (PSAT) and fuel cell system (GCtool) levels.

In this paper, we describe how GCtool was first used to define fuel cell system characteristics representative of mid-term technologies and the process used to define the impact of hybridization degree, control strategy, and energy storage technology on the requirements by using PSAT.

GCTOOL FUEL CELL SYSTEM MODELING

GCTool was developed at ANL for steady-state and dynamic analysis of fuel cell systems. It allows users to establish realistic system constraints and conduct constrained optimization studies. The analyses are typically conducted in design or off-design modes, but mixed modes are also permitted. In the design mode, the components are sized to meet specified performance targets. In the off-design mode, GCTool determines the performance of components of a given size and their physical attributes. GCTool has an extensive library of model classes for components and devices that appear in practical energy conversion systems. In particular, the library includes various types of fuel cells (polymer electrolyte, solid oxide, phosphoric acid, and molten carbonate), hydrogen storage devices (compressed gas, liquid hydrogen, metal hydrides, glass microspheres, etc.), catalytic reactors (such as for auto-thermal reforming, steam reforming, water-gas shift, preferential oxidation, and sulfur removal), and heat exchangers (counterflow, air-cooled condenser, finned radiator, etc.). Several thermodynamic codes are available in GCTool for equations of state of mixtures of gases, liquids, and condensables, which can be used for gaseous (e.g., hydrogen and methane), liquid (methanol, ethanol, octane, etc.), and synthetic fuels (gasoline and diesel).

GCTool is focused on design and searches for optimum configurations. The detailed algorithms in GCtool (thermodynamic and chemical transport) are generally inappropriate for use in vehicle studies because of the greatly increased computer run time. For this reason, engineering models of fuel cell systems and components using the GCTool architecture have been developed for vehicle analysis, as has a procedure to automate the linkage to MATLAB-based vehicle codes (i.e., PSAT).

PSAT VEHICLE MODELING

PSAT is a powerful modeling tool that allows users to evaluate fuel consumption, exhaust emissions, and vehicle driving performance [1]. ANL developed PSAT to study transient effects in future vehicles and the interactions among components with accurate control commands. For this reason, PSAT is a forward-looking model, as it allows users to model with commands.

PSAT is also called a command-based model: developed under MATLAB/Simulink, PSAT allows users to realistically estimate the wheel torque needed to achieve a desired speed by sending commands to the different components, such as the throttle for the engine, displacement for the clutch, gear number for the transmission, or mechanical braking for the wheels. In this way, we model a driver who follows a predefined speed cycle. Moreover, as components react to commands as in reality, we can implement advanced component models, take transient effects into account (such as engine starting, clutch engagement/ disengagement or shifting), or develop realistic control strategies. Finally, PSAT has been validated by using several vehicles [2–4].

To automate the GCTool link with PSAT, a translator has been developed to produce a MATLAB/Simulink that is executable from the GCTool model. The GCTool model is written in a C-like language that is interpreted by GCTool. The executable then becomes a member of the drivetrain library in PSAT using an S-function, which can be used for analyzing transient fuel cell system responses during drive-cycle simulations of hybrid vehicles. The executable is specific to the fuel and the system configuration setup in the GCTool model, and a new one must be produced if there is any change in system attributes The methodology has been demonstrated by using direct hydrogen fuel cell systems.

VEHICLES DEFINITION

Three different vehicle platforms have been selected for this study: compact, midsize, and sport utility vehicle (SUV). Only the results with the SUV platform will be presented in this paper. The vehicle's characteristics are provided in Table 1.

Table 1. Reference Vehicle Validation

	Units	Test	PSAT
Vehicle Assumptions			
Vehicle Mass	kg	2104	
Glider Mass	kg	1290	
Engine		VL, V6, SOHC, 210hp	
Frontal Area	m²	2.46	
Drag Coefficient		0.41	
Rolling Resistance		0.0084	
Wheel Radius	m	0.368	
Model Validation			
Acceleration (0–60 mph)	s	10.5	10.5
Combined Fuel Economy	mpg	20	21

Both fuel economies mentioned in Table 1 are EPA-unadjusted values. The fuel economy obtained with PSAT is higher than the reference because the effect of cold start was not taken into account.

Several fuel cell vehicles have been defined to provide performance similar to that of the reference vehicle, including 0–60mph acceleration (10.5 s), sustained grade of 6.5% at 55mph, and maximum speed above 100 mph. The goal is to design vehicles with characteristics similar to those of the reference vehicle for customer acceptance. The Federal Urban Driving Schedule (FUDS) and the Federal Highway Driving Schedule (FHDS) have been used in this study.

The fuel cell system powertrain, described in Figure 1, includes a fixed ratio in addition to the final drive, as well as DC/DC converters for the high-voltage battery and the 12-V accessories.

The fuel cell systems have been designed to provide power for top speed and grade performance and to have 1-s transient response time for a power

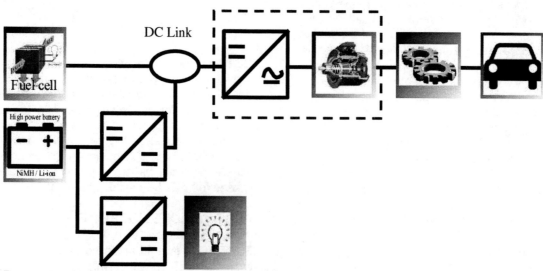

Figure 1. Fuel Cell Powertrain Representation (without vehicle)

request change of 10– 90% of the maximum power. Moreover, they should reach maximum power in 15 s for cold start from 20°C ambient temperature and in 30 s for cold start from -20°C ambient temperature.

As the fuel cell systems defined with GCTool were based upon mid-term technology (2005), the Li-ion technology was selected as the reference energy storage technology. The Saft Li-ion HP6 was used as it was recently tested at ANL, and industry considers it to be state-of-the-art.

DEFAULT CONTROL STRATEGY

Because of the high efficiency of fuel cell systems, it appears natural not to use the energy storage as the primary power source. Indeed, when comparing the fuel cell system efficiency to the internal combustion engine (ICE), as shown in Figure 2, note that the fuel cell system has high efficiency at low power. For a hybrid ICE, it is interesting to use the battery at low and medium power levels and the ICE at high power levels; that is not the case for fuel cell vehicles. Consequently, the control strategy has been developed so that the main function of the battery is to store the regenerative braking energy from the wheel and return it to the system when the vehicle operates at low power demand (low vehicle speed). The battery also provides power during transient operations when the fuel cell is unable to meet driver demand.

Component limits, such as maximum speed or torque, are taken into account to ensure the proper behavior of each component. Battery state-of-charge (SOC) is monitored and regulated so that the battery stays in the defined operating range. The three controller outputs are fuel cell ON/OFF, fuel cell power, and motor torque. A battery state-of-charge equalization algorithm has been used to ensure a fair comparison.

To minimize the impact of the variation of SOC, the same values were selected for both the initial conditions and the goal. As shown in Figure 3, the consequence is that the battery will supply the system

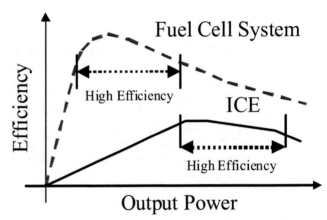

Figure 2. Fuel Cell System Efficiency vs. Internal Combustion Engine Efficiency

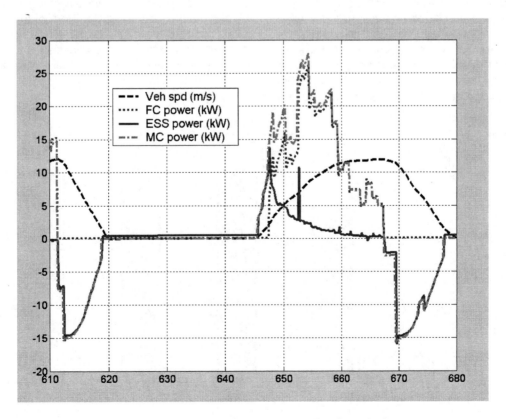

Figure 3. Default Control Strategy – Part of FUDS Cycle – 100-kW Fuel Cell – Hot Start

163

with the energy that it had just recovered from regenerative braking. For instance, the SOC will go up after regenerative braking, and this recuperated energy will be returned to the vehicle during the next acceleration, thus returning the SOC back to its goal value. In other words, to maintain the SOC goal, the battery does not store any net energy over the cycle. The energy that is recovered during braking is immediately returned to the vehicle during the next acceleration.

To implement this aspect of the control strategy, we compare the total power required by the vehicle to a threshold: the minimum power demand needed to use the fuel cell. This control strategy parameter was set by using the PSAT graphic user interface (GUI). More specifically, this control parameter is defined as the sum of the wheel power demand from the driver model (set to zero in the default control strategy) plus an additional power, depending upon the SOC value. If the SOC is above its goal, the additional power will be negative, and, consequently, the fuel cell will be used later. For example, if the SOC is 70%, the value will be zero, but with a higher SOC (71%), the minimum power might be 3 kW, allowing the energy storage to be discharged and return to the SOC goal.

IMPACT OF HYBRIDIZATION DEGREES

The first step in defining the energy storage requirements consists of selecting the proper hybridization degree. As it has been defined that 160 kW peak electrical power should be provided to the electric motor to obtain performance characteristics similar to those of the reference vehicle, several vehicles were defined for each hybridization degree selected. As shown in Figure 4, four options were selected, from 20 kW energy storage and 140 kW fuel cell (on the left)

to 80 kW energy storage and 80 kW fuel cell (on the right). We did not use fuel cell systems with a lower power than 80 kW because it is the minimum power necessary to sustain a 6.5% grade at 55 mph, which was one of the vehicle requirements.

For the Li-ion technology and the default control strategy used, the most significant increase in fuel economy is obtained at the lowest hybridization degree (140 kW fuel cell). This large fuel economy increase is mostly due to regenerative braking energy, as shown in Figure 5.

A further increase in the degree of hybridization still provides some improvements in fuel economy until we reach the optimum of a 100-kW fuel cell. Fuel economy then starts to decrease. At this point, the decrease in fuel cell system efficiency on the driving cycle is greater than the gain due to regenerative braking.

Referring back to figure 2 and the efficiency curve of the fuel cell system, this result is in agreement with the expectations arising from this figure. The fuel cell has a "sweet spot" at relatively low power. If the average operating point of the cycle falls in this "sweet spot," the maximum fuel economy is attained. Downsizing the fuel cell will cause the average operating point to shift to the right. If the initial operating point is before the "sweet spot," then downsizing will be advantageous. The operating point will move to the right and enter the "sweet spot." (The fuel economy trend in Figure 4 is from 140 kW to 100 kW.) However, additional downsizing will push the operating point farther to the right and out of the optimal efficiency region. (The fuel economy trend in Figure 4 is from 100 kW to 80 kW.) At this point, one observes that the fuel cell has been over-downsized — or, to state it another way, the fuel cell vehicle has been over-hybridized.

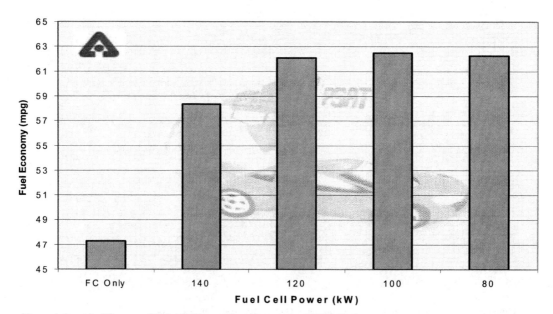

Figure 4. Impact of Degree of Hybridization on Fuel Economy — FUDS Cycle

164

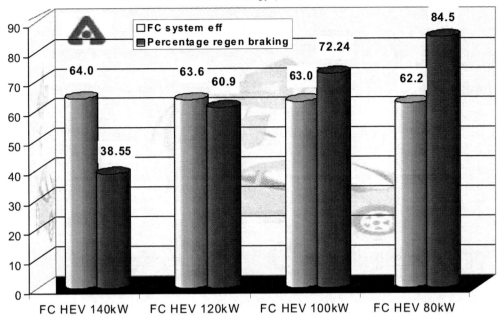

Figure 5. Impact of Degree of Hybridization on Fuel Cell System Efficiency and Regenerative Braking

For the component technologies considered, we conclude that a small hybridization degree is the most suitable solution to optimize the regenerative braking gains while maintaining a high fuel-cell-system efficiency. We can conclude that the degree of hybridization has a significant impact on component behavior and, consequently, will be a determining factor of the energy storage requirements.

IMPACT OF TEMPERATURE

As GCtool allows users to evaluate the influence of temperature, we studied the impact of cold (-20°C), ambient (20°C) and hot (80°C) starts. As shown in Figure 6, initial temperature mostly affects the energy storage requirements during the first 200 s of the cycle. Moreover, because of lower efficiencies and, consequently, a higher amount of heat rejected, the cold start temperature of the fuel cell increases faster than that for the ambient condition.

IMPACT OF CONTROL STRATEGY

As previously mentioned, the battery SOC and the minimum fuel cell power demand threshold are key parameters to the control strategy. To evaluate the impact of control strategies options on the energy storage requirements, we modified both of these parameters.

Figure 7 illustrates the impact of the minimum fuel cell power demand threshold. Using 15 kW instead of zero leads to more use of the energy storage, as shown on the right. Since the battery provides more energy to help propel the vehicle and we want to closely monitor the SOC, it is logical that the fuel cell provides more power to recharge the battery. For example, the fuel cell peak power is 25 kW for the default control and 30 kW when using 15 kW for the minimum power demand threshold.

Table 2 provides the results of modifying the minimum threshold for fuel cell power demand. As expected on the basis of the efficiency curves of the fuel cell system, increasing the minimum demand threshold (and, consequently, using the energy storage more) leads to a decrease in fuel economy as a result of an increase in powertrain losses — even if the amount of regenerative braking increases. As previously discussed, regenerative braking energy and fuel cell system efficiencies are key to the system optimization. However, in this case, an increase in regenerative braking energy does not lead to an increase in fuel economy because a larger increase in fuel cell system energy loss nullifies the benefit associated with regenerative braking.

The other parameter of interest is the energy storage SOC target. Figure 8 compares fuel economy results when the SOC is 0.7 and 0.5 (both for initial conditions and goal). Note that an increase in fuel economy of up to 4% can be achieved just by selecting a lower energy-storage SOC.

The main reason for this improvement in fuel economy is an increase in regenerative braking energy combined with a small increase in fuel cell system efficiency, as shown in Figure 9.

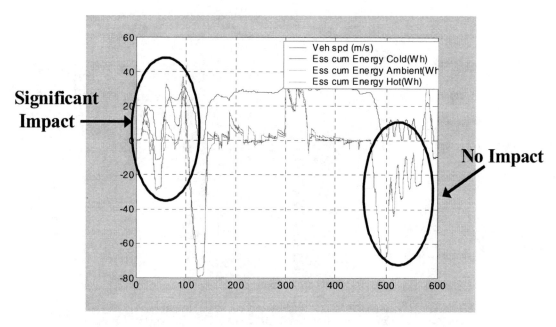

Figure 6. Impact of Initial Temperature on Energy Storage Requirements

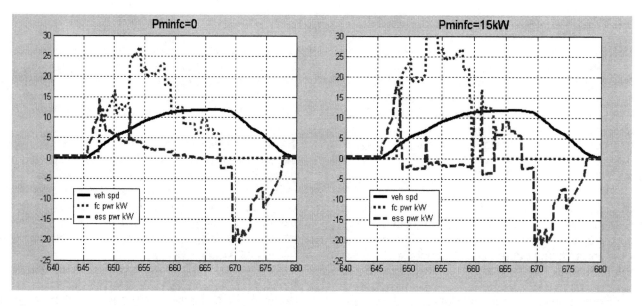

Figure7. Impact of the Minimum Power Demand to Use Fuel Cell

Table 2. Effect of Increasing Minimum Power Demand on Powertrain Losses

	Units	0 kW	5 kW	15 kW
Mech. Braking Energy Loss	W•h	106	100	76
Fuel Cell Energy Loss	W•h	1818	1839	1906
Difference	W•h		**14.9**	**57.7**

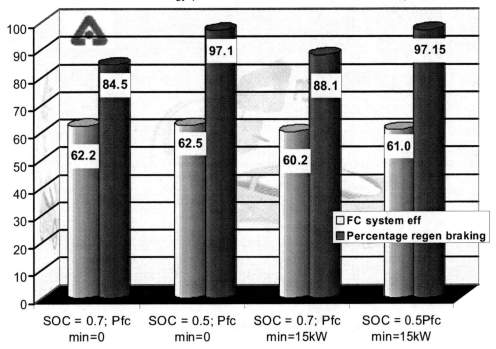

Figure 9. Reasons for Increase in Fuel Economy when Using Lower SOC Target

We noticed that modifying the parameters of the default control strategy had a significant impact on the behavior of the powertrain and, consequently, on how the components are used and their requirements. Figure 10 illustrates the impact of another control strategy philosophy, where the energy storage will be used as the main energy source rather than the fuel cell ("large ess" case). For both the FUDS and FHDS cycles, the fuel cell system efficiency significantly decreases when the use of the energy storage increases. However, for the US06, a larger SOC window may be desirable because by allowing the battery to be more discharged during acceleration, more regenerative braking energy can be recovered during deceleration.

In summary, we conclude that control strategy philosophies and their parameters have a significant impact on energy storage requirements. Several options to increase the energy storage usage were investigated by increasing the minimum wheel power demand to use the fuel cell and by changing the control strategy philosophy by using energy storage as the first choice. The results demonstrated increasing energy storage usage resulted in a decrease in fuel economy. A better option to increase regenerative braking would be to decrease the SOC goal. We chose a a value of 50% because there is still enough available energy to start the vehicle at very low temperatures.

IMPACT OF ENERGY STORAGE SYSTEM TECHNOLOGIES

In the previous example, the Saft Li-ion HP6 battery has been used. To properly define the energy storage requirements for fuel cell vehicles, NiMH and ultracapacitor technologies were investigated. The NiMH battery used had a capacity of 28 Amp-h and was manufactured by Ovonic. The ultra-capacitor had a capacitance of 2,700°F and was manufactured by Maxwell. As shown in Figure 11, the best fuel economy is obtained for different hybridization degrees for each technology. Where, in this example, the Li-ion is optimum with a 100-kW fuel cell and a 60-kW battery, both NiMH and ultracapacitor achieve best performance at low hybridization degrees.

These differences are explained both by the difference in power density, as shown in Figure 12, and in physical characteristics. At a low degree of hybridization (i.e., a140-kW fuel cell), the potential regenerative energy capability is the main reason for achieving better fuel economy, whereas at a high degree of hybridization, the mass increase from NiMH and ultracapacitor technologies is significant.

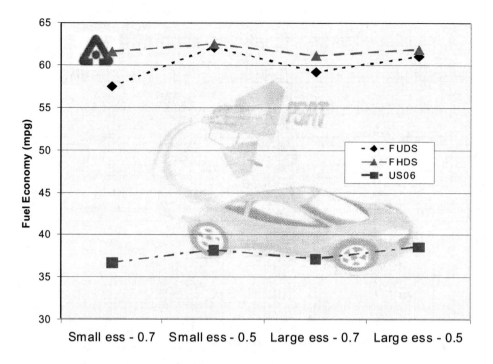

Figure 10. Effect of Using more Energy Storage on Fuel Cell System Efficiency

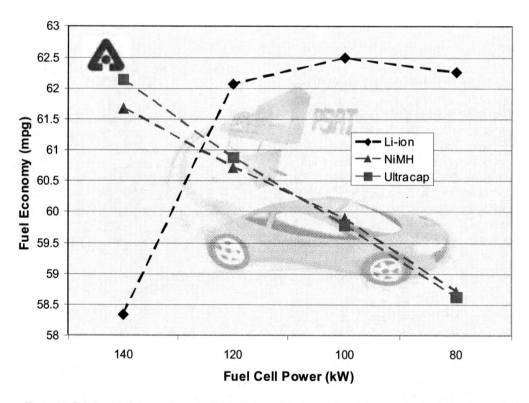

Figure 11. Relationship between Degree of Hybridization Chosen and Energy Storage Technology – FUDS Cycle

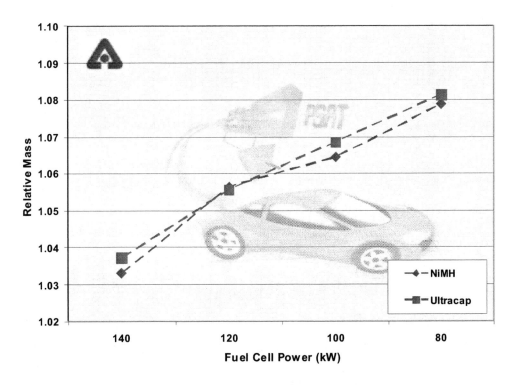

Figure 12. Relative Comparison of Vehicle Test Mass for Each Energy Storage Technology (reference: Li-ion)

CONCLUSION

By using GCTool and PSAT, specific direct hydrogen fuel cell systems and powertrain were developed to achieve performance characteristics similar to those of conventional vehicles. For a specific vehicle platform, we demonstrated that, to define the energy storage requirements of fuel cell vehicles, a system approach was needed. On the basis of mid-term component technologies, we demonstrated that the degree of hybridization should be chosen to optimize the regenerative braking and yet minimize the fuel cell system's losses. Moreover, selecting a lower battery-SOC target allows an increase in regenerative braking and thus can contribute to further lowering of the degree of hybridization. The control strategy should be oriented toward optimizing regenerative braking energy by using a narrow SOC range for low transient cycles (FUDS and FHDS) and a large one for high transient cycles (US06). This study allowed us to narrow the scope of the study for the other vehicle platforms and component technologies. The results will be used to define the energy storage requirements for each case.

ACKNOWLEDGMENTS

This work was supported by the U.S. Department of Energy, under contract W-31-109-Eng-38. The authors would like to thank Bob Kost and Lee Slezak from DOE, who sponsored this activity, as well as the FreedomCAR vehicle, battery, and fuel cell technical team members for their support and guidance.

REFERENCES

1. A. Rousseau; Sharer, P.; Besnier, F. 2004. *"Feasibility of Reusable Vehicle Modeling: Application to Hybrid Vehicles,"* SAE04-454, SAE World Congress, Detroit.
2. B. Deville, and Rousseau, A. 2001. *"Validation of the Honda Insight Using PSAT"*, DOE report, September.
3. A. Rousseauand Pasquier, M. 2001. *"Validation Process of a System Analysis Model: PSAT,"* SAE paper 01P-183, SAE World Congress, Detroit, March 4–8.
4. P. Sharer and Rousseau, A. 2001. *"Validation of the Japan Toyota Prius Using PSAT,"* DOE report, March.

CONTACT

Aymeric Rousseau
(630) 252-7261
E-mail: arousseau@anl.gov

III. SYSTEMS DESIGN AND EVALUATION

Development of Auxiliary Power Units (APU)

2002-01-0411

Solid Oxide Fuel Cell Auxiliary Power Unit – A Development Update

James Zizelman, Steven Shaffer and Subhasish Mukerjee
Delphi Automotive Systems

ABSTRACT

Delphi Automotive Systems and BMW are jointly developing Solid Oxide Fuel Cell (SOFC) technology for application in the transportation industry primarily as an on-board Auxiliary Power Unit (APU). In the first application of this joint program, the APU will be used to power an electric air conditioning system without the need for operating the vehicle engine.

The SOFC based APU technology has the potential to provide a paradigm shift in the supply of electric power for passenger cars. Furthermore, by supplementing the conventional fuel with reformate in the internal combustion engine, extremely low emissions and high system efficiencies are possible. This is consistent with the increasing power demands in automobiles in the new era of more comfort and safety along with environmental friendliness.

Delphi Automotive Systems and BMW were successful in demonstrating an Auxiliary Power Unit (APU) based on Solid Oxide Fuel Cell (SOFC) technology in February, 2001. A SOFC APU generates power using hydrogen and carbon monoxide reformed from fuels such as gasoline, diesel, or natural gas. The proof-of-concept unit and the advantages of using a SOFC APU will be discussed. This paper will also describe our most recent activities in the development of a second generation APU. This development has been targeted towards resolving the fundamental issues with the following key subsystems: fuel cell stack, fuel reformers, and energy and thermal management. Major focus has also been directed at system integration challenges to make a more robust and efficient product.

INTRODUCTION

Delphi Automotive Systems is developing Solid Oxide Fuel Cell (SOFC) systems for automotive applications.

This program, started in a joint effort with BMW, has demonstrated the basic viability of using a SOFC system as an automotive APU (1-5). This paper briefly reviews the trend to high power and high efficiency electrical systems and the role which fuel cells may have in this trend to vehicle electrification. It continues with an overview of the SOFC APU mechanization with discussion of several of the key subsystems. Further detailed information is provided summarizing the actual execution of the APU including system mechanization, vehicle installation and a performance summary at the system and subsystem level. A second focal point of this paper is the overview of the Generation 2 gasoline SOFC APU system. This overview includes the mechanization and analytical results illustrating performance and efficiency improvements that are expected.

In addition, the stack, reformer, and thermal management subsystem for Generation 2 will also be described. It concludes with some vision on target applications (such as a diesel APU).

The application of the fuel cell APU for transportation comes in the context of a significant trend to electrification of vehicle accessories. Many automotive systems are being converted to electric power to support cost, weight, and packaging objectives. Intermittent accessories, such as heated seats, heated windshield, power steering and brakes have been adopted first.

Higher power and more continuous accessories, such as electric air conditioning and electromagnetic valve train systems have been developed more slowly due in part to the limited on-board electrical capacity and its relatively low efficiency.

Higher power and higher efficiency electrical machines are being introduced at 42V. This will provide the capacity to continue the trend to electrification with new features such as starter-generators. Fuel cell systems have significantly higher fuel to electric conversion efficiency (>50%) because of the direct chemical to electrical conversion. This will allow the completion of the transition to electric accessories and allow independence of the engine from accessory operation - an important step for the introduction of mild hybrid vehicles.

This paper discusses the advantages of applying such a fuel cell power unit to the dual-voltage 42V/14V automotive electrical system meeting the evolving 42V PowerNet specifications. Other key transportation applications include APUs for work trucks, recreational vehicles, fire-rescue vehicles, military vehicles, ships, and aircraft. These vehicles require the APU to operate on fuels other than gasoline, such as diesel or jet fuel. These applications are expected to play an important role in the introduction of this technology to the market place.

The Solid Oxide Fuel Cell technology is also easily applicable to stationary power. Residential primary and back-up power, distributed power, and co-generation are achievable utilizing essentially any liquid or gaseous hydrocarbon fuel.

WHY A SOLID OXIDE FUEL CELL?

The Proton Exchange Membrane (PEM) fuel cell is widely regarded as the likely fuel cell of choice for the automotive propulsion application. Although PEM has been widely accepted as an attractive fuel cell for propulsion, it has several clear challenges. Firstly, it is seeking to replace the internal combustion engine, which, after a century of development is highly competitive in the key attributes of cost density <$50/kW and specific power. Secondly, it requires relatively pure hydrogen as its fuel. On-board reformer systems may be used for fuels such as methanol and gasoline but these systems are inherently large and complex (Figure 1).

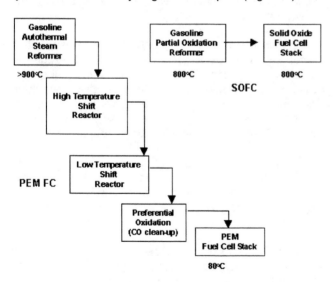

Figure 1: Comparison of gasoline reforming for PEM vs. SOFC

Thirdly, the PEM fuel cell is typically applied to an electric vehicle - requiring power electronics, an electric drivetrain, and some amount of battery storage. All of these subsystems have been developed in electric vehicle applications, but they face similar cost and performance challenges compared to mechanical alternatives.

A fuel cell APU is a high efficiency generator that runs with the engine on or off. In this sense, it is not a threat to the internal combustion engine and does not have to reach the extremely low cost/kW levels for propulsion. It can be applied in conventional or mild hybrid configurations - and is not linked to a fully electric drivetrain.

SOFC is attractive for use as an APU. It is compatible with conventional petroleum fuels - with a simple partial oxidation reforming process. It has less stringent requirements for reformate quality (using carbon monoxide directly as a fuel) and has less sensitivity to contaminants such as sulfur.

PEM fuel cells may also be used for application in an APU (1), but require a complex reforming process for conventional fuels (or stored hydrogen) and need on-board water management.

DESCRIPTION OF THE SOLID OXIDE FUEL CELL

The SOFC is a solid state energy conversion device that produces electricity by electrochemically combining fuel and oxidant gases across an ion-conducting ceramic membrane. As with other fuel cells, SOFC consists of two electrodes (anode and cathode) separated by a solid electrolyte. Reformate is fed to the anode, and air is fed to the cathode.

Some of the advantages of using Solid Oxide Fuel Cells are:

- High cell electrical efficiency: SOFC can attain fuel to electric efficiencies of >50% with a hydrocarbon fuel such as natural gas.
- Ability to use CO along with H_2 as fuel. Furthermore, the SOFC demonstrates a high tolerance to fuel impurities such as sulfur.
- Temperature and water conditions permit internal reforming at the anode of SOFC.
- No noble catalysts.
- Relatively simple reformer technology and compatibility with hydrocarbon fuels.
- No humidification of reactants necessary because of the dry electrolyte.
- Water management is not required.

"PROOF OF CONCEPT" SOFC APU

Delphi Automotive Systems and BMW successfully demonstrated a Proof of Concept SOFC APU in February, 2001. This was a major step towards showing the feasibility of using SOFC technology for automotive APU applications. Figure 2 shows a picture of the Delphi APU system integrated in the trunk of a BMW 7- series sedan. The system used gasoline as a fuel and was sized to run the HVAC in the car independent of the engine.

Figure 2 : Delphi SOFC APU mounted on a BMW 7-series sedan.

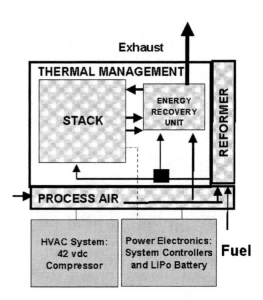

Figure 3b : Mechanization of SOFC APU System.

The Proof of Concept SOFC APU system was a discrete mechanization that used a simplified arrangement of sub-systems which were independently engineered and then integrated (Figure 3a,b). The APU system was a stand-alone packaged unit integrated in an insulated box as shown in Figure 3a. The system can be divided into several sub-systems:

- Solid Oxide Fuel Cell Stack
- Fuel Reformer system: micro reformer and main reformer
- Thermal Management system
- Energy Recovery Unit (ERU)
- SOFC Control system and Power Electronics system
- LiPoTek™ plastic lithium ion battery pack

The development of the sub-systems have been reported in some detail earlier (2). The integrated APU system in the box can be divided into three separate zones – the stack assembly zone; the reformer zone and the low temperature zone (for automotive valves and sensors). The stack assembly (Figure 4) is comprised of four modular stacks mounted on a manifold base. The stack modules were manufactured and supplied by Global Thermoelectric of Canada. The stack technology was based on anode supported cells and metallic interconnects using compressive seals. The stacks produced nominal power at 750°C (6,7). The Energy Recovery Unit (ERU) also resides in the stack assembly zone. The ERU acted as a heat exchanger for the input gases to the stack and provided a post combustor function for the anode tailgas.

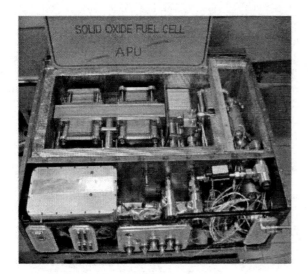

Figure 3a: Picture of SOFC APU System

Figure 4 : Stack plate with a 60 cell modular stack and ERU.

Hot cathode air was utilized to bring the stack to operating temperature. This process was executed in a controlled fashion, such that the allowable temperature difference in the stack was not exceeded. The startup procedure was an automated and controlled process that included catalytic combustion at the ERU to produce the energy to heat up the stack. The reformer system consisted of a gasoline partial oxidation reformer and a micro-reformer. The reformer was capable of producing good quality reformate within minutes of start-up. A three way reformate control valve was used in the hot zone to control flow of reformate into the stack or through the ERU. Figure 5 shows the fast cycling of the reformer to follow the load cycling at the stack.

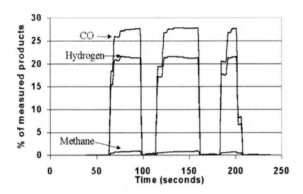

Figure 5: Reformer cycling.

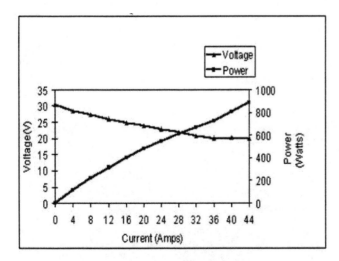

Figure 6: Data from 2 15-cell stacks integrated in the APU system running on gasoline POx reformate. (Temp 750°C)

Extensive sub-system and system based testing has been carried out to understand the validity of the proof of concept mechanization. To offer an illustration of the integrated system performance, test data from two 15-cell modules completely integrated with the rest of the APU system is shown in Figure 6. The POx reformate (gasoline) from the reformer consisted of H_2(20%), CO (23%), CO_2 (2%), HC (< 2%), H_2O (3%), with the remaining gas consisting primarily of N_2. The power density was 0.37 W/cm^2.

GENERATION 2 SOFC APU SYSTEM

The Proof of Concept APU system was the first step in implementing and demonstrating the usefulness of SOFC technology as applied to an Auxiliary Power Unit for automotive applications. Delphi is currently focussed on developing a second generation Solid Oxide Fuel Cell APU (Generation 2 APU) that is more consistent with the requirements of a production unit. The design is being optimized based on customer requirements and manufacturability assessments to increase robustness and reduce cost.

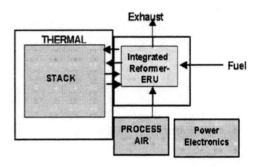

Figure 7 : Generic mechanization of Generation 2 SOFC APU .

The Generation 2 APU system is being designed to have a more functionally integrated mechanization in order to reduce weight and volume of the overall unit and improve overall gravimetric and volumetric power densities. The mechanization in Figure 7 shows an integrated reformer and ERU as one of the options.

The second generation SOFC APU is being tailored for increased efficiency by lowering parasitic losses, increasing fuel utilization and optimizing thermal losses. Also, the mechanization is being optimized for improved reformer efficiency by adding water from recirculated stack anode exhaust gas (there is no external water being used). The Generation-2 APU targets are shown below.

System	Targets
Power (Net)	3-10 kW
Fuel	Gasoline
Durability (continuous)	5000 - 10000 hrs.
Durability (thermal cycles)	>5000
Efficiency	> 35%
Start up time	< 20 minutes
Weight	10 - 20 kg/kW
Volume	10 - 20 L/kW

Table 1: Generation 2 APU targets.

A key to the development of the Generation 2 APU is the research and engineering being undertaken to address the technical challenges in two of the key sub-systems of the APU: the stack and the reformer.

GENERATION 2 STACK

Detailed analysis and testing is being carried out to optimize the stack design and materials to develop a robust stack that meets the requirements of the automotive industry. The stack design is comprised of anode supported cells capable of providing high power densities at 750-800°C. Other key components for stack design are metallic interconnects and seals. Multiple concepts are being analyzed and tested to develop a robust stack.

Multiple design features are under consideration as shown in Table 2.

SOFC Cells	Seals	Interconnect (metallic)	Loading mechanism	Cell Support	Manifolds air/fuel
Square	Compressive	Thin foil with mesh	None (if bonded)	Picture frame	Rectangular chimney
Other shapes	Bonded - Glass - Brazing	Embossed features	Load frame with compliant member	Edge Cell	Multi holes chimney
		Etched features	Load frame without compliant member		Open Manifolds

Table 2 : Design features under investigation for Generation 2 stack.

Each of these design characteristics play a critical role in optimizing the performance of the stack. They are being investigated in detail to develop the most robust design to meet the targets specified below:

Stack	Targets
Power	3.5 -11 kW
Fuel	Gasoline "steam" (with water from anode exhaust recycled) reformate
Durability (continuous)	5000 - 10000 hrs
Durability (thermal cycles)	> 5000
Fuel utilization	> 60%
Start up time	< 20 minutes
Weight	< 3 kg/kW
Volume	< 3 L/kW

Table 3 : Generation 2 stack targets

Detailed analysis using Computational Fluid Dynamics (CFD) is being used to study flow distribution in the plane of the cell and in the manifolds to understand distribution of fuel and air from cell to cell (Figure 8a,b). This has consequences in improving fuel utilization, heat transfer and efficiency. Figure 8a shows an example of CFD analysis done on the plane of the cell to understand the distribution of reformate gas or air. The holes in the figure are openings through which reformate or air flows into and out of the cell. CFD analysis is also being used to understand the effect of flow distribution in the manifolds for cell to cell distribution through the height of the stack (Figure 8).

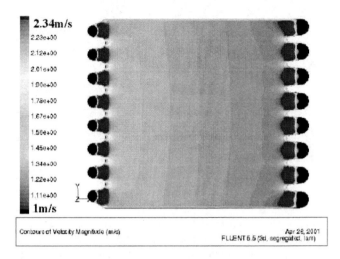

Figure 8a: CFD analysis on a stack concept to understand flow distribution in the plane of the cell

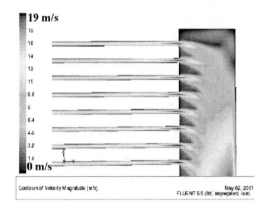

Figure 8b: CFD of analysis on cell to cell flow distribution in a stack.

CFD and Finite Element Analysis (FEA) is being used to understand the thermal stresses during fast start-up and to optimize stack designs to meet the start-up requirements for the APU system. Delphi is working with Pacific Northwest National Laboratory (PNNL) to address this challenge with analysis and testing. A sequentially coupled thermal–structural analysis is being carried out on stack concepts to evaluate the thermal gradients and stresses developed during fast start-up (8, 9). They are then correlated to the acceptable stresses in the components of the stack. This modeling tool has proven to be very useful in understanding and optimizing the stack design to reduce stresses during fast start-ups. Figure 9 gives an example of thermal gradients observed during start-up in the plane of the cell. In this view, the cathode air used for heating enters at the lower left, flows upward then on the plane of the cell and flows out downward on the right of the cell.

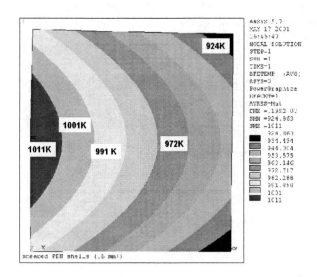

Figure 9: Thermal gradient predicted from analysis during start-up in the plane of the cell on a stack concept.

Work is in progress to address the key challenges for stack development. Durability and fast start-up are key to development of this technology and extensive effort is being put towards addressing these two fundamental issues.

GENERATION 2 REFORMER

Multiple Generation 2 reformer concepts are being evaluated that address the key issues learned during the Proof of Concept development. Some of these issues are reformer efficiency, durability, start up time, weight and volume. Each of the concepts are evaluated against the targets shown below.

Reformer	Targets
Power	Sized for 3-10 kWe (net)
Fuel	Gasoline
Durability (continuous)	5000 - 10000 hrs
Durability (thermal cycles)	>5000
Start up time	< 1 minute
Weight	0.7 Kg/kWe
Volume	0.7L/kWe

Table 4: Generation 2 reformer targets

The two main reformer designs being evaluated are tubular and planar. Figure 10 a,b shows examples of these reformers.

Figure 10a: Tubular reformer concept

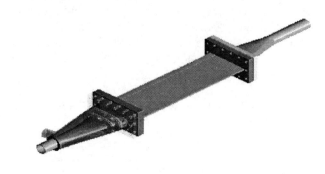

Figure 10b: Planar reformer concept

The reformer efficiency is being increased by developing strategies that allow the reformer to operate in partial steam reforming mode rather than just catalytic partial oxidation. By recycling a percentage of the anode tailgas, the water necessary for steam reformation is made available. This recycling also increases efficiency by utilizing the unused hydrogen and carbon monoxide in the anode tailgas. By using recycle, the system efficiency can be improved by approximately 80%.

The key attributes for reformer durability are temperature control and uniformity in the catalyst bed and homogeneity and fuel vaporization and mixing of the air/fuel mixture entering the catalyst bed. The reformer must be controlled in very tight operating windows so as not to be in modes that would create extreme temperature spikes or carbon formation. Delphi's experience in the research, development, and manufacturing of durable and robust automotive catalytic systems is being used to advance the reformer toward a commercial device.

By taking advantage of the geometry allowed in the planar reformer concept, an integration of the reformer and ERU functions can be accomplished. The weight and volume of the integrated unit will be substantially less than the separate components. This integration also allows for highly efficient thermal management which is required to accommodate high percentages of anode tailgas recycle. Figure 11 is a rendering of what an integrated reformer and ERU may look like.

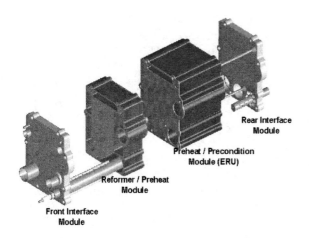

Figure 11: Integrated Reformer and ERU

FUTURE PROGRAM DIRECTION

The SOFC APU system described in this paper may also be applicable to many other transportation and residential applications.

Examples of other transportation applications are:
- APU for Class 8 heavy duty trucks (diesel)
- APU for work trucks and recreational vehicles (gasoline and diesel)
- APU for military vehicles (gasoline and diesel)
- Ships, boats (gasoline and diesel)

DIESEL APU SYSTEM

Concurrent with the gasoline APU development, Delphi is also developing a diesel APU system focussed mainly for trucks and military applications. The targets for the diesel APU are shown in the table below:

System	Targets
Power (Net)	3-10 kW
Fuel	diesel "partial steam" (with water from anode exhaust recycled)
Durability (continuous)	10000 - 20000 hrs
Durability (thermal cycles)	5000
Efficiency	>35%
Start up time	< 20 minutes
Weight	10- 20 kg/kW
Volume	10-20 L/kW

Table 5 : Diesel APU targets

A major thrust of our effort is in the development of a reformer that can reform diesel fuel to a hydrogen and CO rich gas. System mechanizations with optimized integration for maximum efficiency are being developed. Different design concepts as well as catalysts are being developed for the diesel reformer. Delphi is working with key partners like TotalFinaElf as well as Los Alamos

National Laboratory to develop a robust diesel reforming technology. Initial results on Swedish diesel (low sulfur content) have been very encouraging. Table 6 below shows some of our initial results from dodecane and low sulfur Swedish diesel. The data also shows the enhanced H_2 and CO composition in the reformate expected from the addition of water from recirculated anode exhaust.

Fuel In	% H_2	% CO	% CH_4	% H_2O	% CO_2	% N_2
Dodecane (POx)	20.5	20.0	0.9	1.3	1.6	56
Swedish Diesel (POx)	19	18.3	0.43	1.1	1.58	59.5
Diesel * ("steam reformed" from recirculated anode exhaust)	35	44	1	13	7	0

* Calculated using Aspen model

Table 6 : Results from reformation of diesel fuel.

RESIDENTIAL POWER

Delphi is also very focussed on applying SOFC technology for stationary and residential applications. Some key applications of this technology in the residential markets are:
- Residential back-up generator (5-10 kW)
- Distributed generation (5-25kW)
  - Residential primary power: independent or connected to the grid
  - For continuous power in hospitals, restaurants etc.
  - Community power source independent of the grid
- Cogeneration

As part of the Solid State Energy Conversion Alliance (SECA) led by the US Department of Energy, Delphi's goals are consistent with SECA's vision of mass producing 5 kW SOFC modules that can be used for both residential and transportation applications. These modules can be combined as required for each type of application.

SUMMARY AND CONCLUSIONS

Solid Oxide Fuel Cell based APU is a paradigm shift in the supply of electric power for transportation. Its applications can include premium class automobiles, work trucks, recreational vehicles, fire-rescue vehicles, military vehicles, ships, and aircraft. This trend toward separating the on-board electricity production from the vehicle's internal combustion engine is consistent with the new era of increased comfort and being able to provide it to the customer in an efficient, environmentally friendly way.

Delphi is currently optimizing its sub-systems and system integration to develop a more robust and manufacturable product. Delphi has formed key collaborations and is working with both suppliers and customers to take this technology to market faster. Delphi is interested in using this technology for other applications that are complementary to this development in the non-transportation arena. Clearly, this technology is applicable to many forms of stationary power.

This paper describes our "Proof of Concept SOFC APU" system in a BMW sedan and the continuation of our efforts in developing a Generation-2 SOFC APU. It describes some of the key challenges and our approach towards solving them. Current development is focussed on:

- Reducing cost
- Improving cycle durability
- Improving start-up characteristics
- Improving power density
- Developing reformation technology for a broad range of hydrocarbon fuels.

If the current challenges of the SOFC are overcome, this technology offers the potential for high market penetration in the next decade.

ACKNOWLEDGMENTS

The authors would like to acknowledge the rest of the SOFC APU team for their significant contributions.

REFERENCES

1) J. Tachtler, T. Dietsch, G. Götz, "Fuel Cell Auxiliary power unit-Innovation for the electric supply of passenger cars?" SAE SP-1505, Detroit, March, 2000.

2) J. Zizelman, Dr. J. Botti, J. Tachtler, W. Strobl, "Solid Oxide Fuel Cell Auxiliary Power Unit-A Paradigm shift in electric supply for transportation", Convergence 2000-Paper 2000-01-C070, Detroit, 2000.

3) C. DeMinco, S. Mukerjee, J. Grieve, M. Faville & J. Noetzel, M. Perry, A. Horvath, D. Prediger, M. Pastula, R. Boersma & D. Ghosh, "Development of a Solid Oxide Fuel Cell (SOFC) Automotive Auxiliary Power Unit (APU) Fueled by Gasoline", 10th Canadian Hydrogen Conference, May, 2000.

4) S. Mukerjee, M. J. Grieve, K. Haltiner, M. Faville, J. Noetzel, K. Keegan, D. Schumann, D. Armstrong, D. England, J. Haller and C. DeMinco, "Solid Oxide Fuel Cell Auxiliary Power Unit-A New Paradigm in Electric Supply for Transportation" SOFC VII, PV 2001-16, p173, The Electrochemical Society Proceedings Series, Pennington, NJ 2001.

5) S. Mukerjee, M. Grieve, C.DeMinco, M. Perry, A. Horvath, D. Prediger, M. Pastula, R. Boersma, D. Ghosh in Fuel Cell 2000 Abstracts, Portland, Oregon; p-530 , 2000.

6) H. Buchkremer, U. Diekmann, L.G.J. de Haart, H. Kabs, D. Stolten, D. Stöver, I.C. Vinke, "Status of the Development of the Anode Supported Planar SOFC system", Proceedings / 1998 Fuel Cell Seminar, Palm Springs, 1998.

7) D. Ghosh, G. Wang, R. Brule, E. Tang and P. Huang, "Performance of Anode Supported Planar SOFC Cells", SOFC VI, Honolulu, Hawaii, 1999.

8) K. Keegan, S.M. Kelly, M. Khaleel, L. Chick, K. Recknagle "Analysis of a Planar Solid Oxide Fuel Cell based Automotive Auxiliary Power Unit" SAE02P-484, Detroit, March, 2002.

9) M.A. Khaleel, K.P. Recknagle, Z. Lin, J. E. Deibler, L. A. Chick and J. W. Stevenson, "Thermomechanical and Electrochemistry modeling of planar SOFC stacks" ,SOFC VII, PV 2001-16, p1032, The Electrochemical Society Proceedings Series, Pennington, NJ 2001.

2004-01-1477

Fuel Cell APU for Silent Watch and Mild Electrification of a Medium Tactical Truck

Zoran Filipi, Loucas Louca, Anna Stefanopoulou, Jay Pukrushpan, Burit Kittirungsi and Huei Peng
Automotive Research Center
University of Michigan

ABSTRACT

This paper investigates the opportunities for improving truck fuel economy through the use of a Fuel Cell Auxiliary Power Unit (FC APU) for silent watch, as well as for powering electrified engine accessories during driving. The particular vehicle selected as the platform for this study is a prototype of the Family of Medium Tactical Vehicles (FMTV) capable of carrying a 5 ton payload. Peak stand-by power requirements for on-board power are determined from the projected future digitized battlefield vehicle requirements. Strategic selection of electrified engine accessories enables engine shutdowns when the vehicle is stopped, thus providing additional fuel savings. Proton Exchange Membrane (PEM) fuel cell is integrated with a partial oxidation reformer in order to allow the use of the same fuel (JP8) as for the propulsion diesel engine. The APU system is modeled and linked with the complete vehicle system simulation, and accessory duty cycles are derived for both silent watch and driving. The results indicate six-fold improvements of the silent watch fuel economy with the FC APU compared to main-engine idling, and relatively modest improvements from the mild electrification and FC APU use during the driving cycle. Combined fuel economy benefits calculated over the hypothetical daily military mission with a combined 10 hour highway/local/off-road driving and 10 hour silent watch are 20.1%.

INTRODUCTION

Current truck related research initiatives such as the 21st Century Truck, DOE's Program on Reduction in Parasitic Energy Losses for Class 3-8 Trucks, Future Tactical Trucks Systems and Future Combat Systems call for significantly improved fuel efficiency, while preserving or improving truck mobility and freight carrying capacity. Total fuel consumption of commercial trucks in U.S. is more than 42 billion gallons per year, and in recent years it has been steadily growing. The strong impact of the price of fuel on economy and operational cost of freight fleets stimulates development of fuel efficient technologies. In military applications, the impact of the fuel economy is amplified, due to the fact that much of the present logistics support is devoted to moving fuel. The Army transformation requires ability to operate with an agile, more deployable force and a smaller logistics tail. This translates into a technical goal of improving the fuel economy. Given the complex mission of military trucks, the fuel economy needs to be considered not only during driving, but also for silent watch, i.e. when the truck is parked and the crew performs surveillance or communication tasks. Similarly, about one million heavy duty line haulers have sleeper cabins that require energy for ventilation, climate control and various accessories while the driver is resting. Hence, fuel economy, as well as noise and gaseous emissions generated while providing power for cabin auxiliary (hotel) loads during stops are receiving increasingly more attention in the commercial sector.

The effectiveness of some of the approaches for increasing vehicle fuel economy, previously developed for passenger cars, is limited when applied to heavier vehicles. Specifically, opportunities for truck mass reduction are limited due to structural constraints, as well as the fact that reduction of truck's weight is typically viewed as one of the avenues for possible increases in payload rating. In addition, medium and heavy trucks are normally equipped with highly-efficient diesel engines. Consequently, advanced hybrid propulsion systems and reduction of parasitic losses are identified as critical enablers on the technology roadmap to future ultra-efficient truck systems. Electrification of accessories and use of an Auxiliary Power Unit is seen as being particularly effective in reducing parasitics and fuel consumption associated with hotel loads. While fuel cells are considered not to be ready for heavy propulsion, the technology has already been demonstrated on a smaller scale suitable for APU [1, 6].

Both military and commercial sectors are keenly interested in APUs providing stand-by electric power for on-board devices. Examples of these advanced technology devices in military trucks include advanced radio communication equipment, navigation systems, GPS, movement tracking systems, night vision systems, cabin ventilation, Air Conditioning (A/C) and Nuclear Biological and Chemical (NBC) protection systems, computers, and displays. The US Army projects the need for up to 10 kW of electric power for various battlefield equipment per Dobbs et al. [1], while maintaining minimal thermal and acoustic signature. Power for these devices needs to be available during driving as well as when the vehicle is parked, i.e. during a silent watch. Traditionally, the silent watch for military trucks is provided by propulsion engine idling and running the alternator. This forces the engine to operate at a very inefficient regime for extended periods of time, and creates undesirable noise and gaseous emissions. The option of a small ICE or gas turbine APUs allows much more efficient operation, but noise and thermal signature still require very careful management. Hence, the Fuel Cell APU is a very attractive alternative, offering a well rounded trade-off among fuel economy, energy density and low thermal/acoustic signature.

The focus of this work is the feasibility analysis and evaluation of fuel economy benefits of using the Fuel Cell APU to power the electrified engine and vehicle accessories. The particular vehicle selected as the platform for this study is the prototype version of the Family of Medium Tactical Vehicles (FMTV) with a 6x6 drivetrain, capable of carrying a 5 ton payload over a smooth or rough terrain. Fuel economy potential of hybridization of Class VI trucks was addressed in previous simulation studies – see [2, 3, 4]. In particular, work published by Filipi et al. [4], addressed the selection of the hybrid architecture, presented modeling of the complete vehicle system, and proposed the methodology for sequential optimization of hybrid propulsion system design and power management. The results demonstrated the effectiveness of the hydraulic hybrid (HH) propulsion system in regenerating and reusing the braking energy. Hence, the HH FMTV prototype vehicle is the starting point for the study of the additional benefits provided by the APU. Very high number of FMTVs in army fleets implies high impact of possible future insertion of new technologies into the medium tactical trucks.

The option of using a FC APU has recently attracted a lot of attention and initial studies have shown the viability of 5 kW units for cabin auxiliaries [6, 7, 8, 9]. In parallel, researchers considered opportunities for reducing engine parasitic losses (parasitics) by using controllable, electric engine accessories, e.g. Hnatzuk et al. [10] evaluated the potential for savings through electrification of coolant and water pumps, while Hendricks et al. [11] performed a comprehensive study of the benefits of removing belt-driven mechanical loads, but without addressing the impact of replacing the mechanical loads with electric loads. The work presented here introduces a combined approach, i.e. in addition to using the FC APU for supporting the silent watch requirements, we propose to electrify some of the engine accessories and use the FC APU to power them during driving. This enables accessory control for reduced power consumption and efficient power generation, thus contributing to improved driving fuel economy. The

concept is illustrated in Figure 1. Strategic selection of electrified accessories enables engine shut-down when the vehicle is stopped, thus providing additional fuel savings. The Fuel cell APU system consists of a JP8 Fuel Processor (FP) system, and a Proton Exchange Membrane (PEM) Fuel Cell (FC) system that includes the air delivery, humidification and cooling sub-systems. The reformer of the FP system is assumed to be based on a catalytic partial oxidation process [12]. The PEMFC is sized based on the estimated combined silent watch and engine electrification requirements. The original FMTV battery pack (four 14V 6MTF Lead-Acid batteries) is used to supplement the FC APU power output during fast changes in load demand.

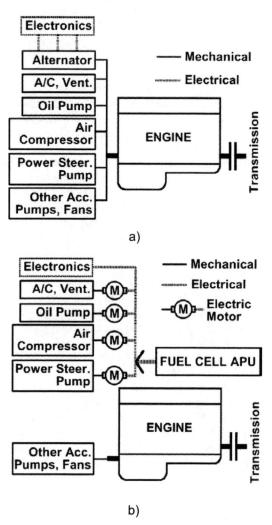

Figure 1: Schematic of the engine with accessories: a) original configuration and b) proposed accessory electrification and the use of FC APU

The study first surveys the electric and electronic loads from typical cabin devices and equipment aboard the tactical vehicle. Then, we identify a set of accessories that hold a promise of tangible energy savings if decoupled from the engine crankshaft and electrified. Selecting engine oil pump, power steering pump and air brake compressor allows shut-downs of the engine during vehicle stops. Next, modeling and integration of the complete vehicle system is presented, followed by the details of FC APU modeling and control. Silent watch mission profile that captures auxiliary power requirements during a ten hour period is constructed. Duty cycles for engine/drivetrain accessories are devised based on their function and in conjunction with the actions taken during the driving schedule. Then, a sequence of simulations is performed for the silent watch and the vehicle following the representative driving schedule. Finally, conclusions about the fuel economy potential of the proposed concept are given in the last section.

STAND-BY POWER REQUIREMENTS

Auxiliary power requirements on modern trucks are rapidly increasing due to growing demand for electronic equipment, driver protection and comfort. The projections for future military trucks are based on the plans for army transformation and battlefield digitization. Equipment necessary for digitization of the tactical vehicle includes digital radio systems, navigation systems (GPS, 3D mapping), encryption systems, computer and displays, movement tracking system (MTS), identification friend-or-foe, and driver vision enhancement (DVE) [1]. In addition, Nuclear, Biological and Chemical (NBC) protection requires high-capacity ventilation system for over pressurizing the cabin combined with air/conditioning. Power for this equipment needs to be available at any time, during driving as well as during extended stand-by periods. The stand-by requirements should be preferably met with minimal noise and thermal emission, hence the term silent watch.

For tactical trucks, such as the FMTV, the estimated peak power requirement for the ventilation, NBC protection and air-conditioning is 3.4 kW. The communication equipment's maximum power consumption is 0.6 kW. It is estimated that the rest of the electronics, such as navigation, MTS, DVE, displays etc., will require another 0.6 kW. This brings the total peak auxiliary power requirement to 4.6 kW. Duty cycles for electronics and other auxiliaries are discussed in the dedicated section given in the latter part of the paper.

There are generally three ways of providing power for auxiliaries during stand-by: discharging batteries, engine idling and running the alternator, and using an Auxiliary Power Unit. The APU can be based on a generator-set powered by a small IC engine, micro-turbine, or a fuel cell.

Batteries found on the FMTV truck (four military batteries) exceed the average size, due to cold start requirements. Nevertheless, their ability to sustain stand-by loads for extended period of time is very limited. The main determining factors are the state-of-charge lower limit that guarantees safe restarting of the engine, and the effect of the depth of recycling on battery life. Dobbs [1] estimates the practical stand-by operation, using only electronics (no A/C), to be up to 30 minutes with two military batteries. This would increase to one hour on the FMTV equipped with four batteries – still far less than the expected length of the silent watch of 4 – 10 hours. Hence, on most military vehicles engine idling is used as the source of energy for extended stand-by. Unfortunately, while heavy-duty diesels provide very good efficiency at high loads, their fuel economy is much poorer at idling. Extended idling in cold weather often leads to deteriorated diesel combustion and over-fueling, which can lead to diluting of oil and increased wear. In addition, engine noise emission can not provide the truly silent operation, and its exhaust emission provides unwanted heat signature.

Using an APU with a small IC Engine can dramatically improve the stand-by fuel consumption, but is still accompanied by the noise and thermal signature. The micro-turbine APU is typically less economical and more costly than the ICE APU. This makes the Fuel Cell APU worth exploring.

While this paper focuses on a typical tactical truck, it is worth noting that the total power requirement for the commercial Class 8 vehicle with a sleeper cabin would be very similar. The ventilation and A/C system are comparable, and cabin comfort and entertainment features, e.g. electric coffee maker, refrigerator, hair-dryer, TV and video equipment, replace the extensive use of battlefield electronics. Consequently, total stand-by requirement for a commercial truck is approximately 5 kW [6, 7, 9]. It is estimated that roughly half a million trucks idle between 3.3 and 16.5 hours a day, and 18 states already have anti-idling regulations. Using a Fuel Cell APU would offer similar benefits as in the military application. In addition, availability of the Fuel Cell APUs on dual-use vehicles would enable on-site generation and increase mission flexibility in case of disaster and relief efforts.

ELECTRIFICATION OF ACCESSORIES

The engine/drivetrain accessories enable functioning of the systems critical for safe engine and vehicle operation, such as cooling, lubrication, steering and braking. In the conventional engine/drivetrain configuration all of the main accessories are mechanically driven by the engine. The direct mechanical coupling makes accessory speed dependant on engine speed and severely limits our ability to control its operating conditions. Electrification offers potential fuel savings through accurate control of accessories independent of instantaneous engine speed.

The concept is illustrated using the power steering pump as an example. When the steering wheel is turned, the power steering system provides assist using the pressurized fluid supplied by the pump. The pump is sized so that it can satisfy the required fluid pressure and flow even at low engine speeds. This determines the nominal torque needed to drive the pump. However, as the engine speeds up, the relief valve opens in order to limit the pressure, the torque remains roughly constant, but the consumed power increases almost linearly with speed – see Figure 2. In addition, even when the pump is unloaded, internal friction losses will cause small, but non-negligible power consumption. If the pump is decoupled from the engine, it can be operated only when needed and always at the most efficient point. In case of the FMTV vehicle, this reduces the peak power requirement from 12.9 kW to only 4.2 kW, as shown in Figure 2. In addition, the power consumption is reduced from 0.7 kW to zero when there is no steering. Similar reasoning regarding the nominal power consumption can be applied to the engine oil pump, as shown by Hnatczuk et al. [10].

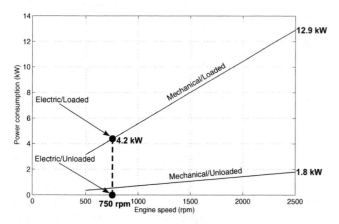

Figure 2: Power consumption of the mechanically driven power steering pump and the desired operating point achievable with the electrified version

The accessories of the FMTV vehicle that are considered in the analysis and their peak power requirements are as follows: engine fan 26 kW, transmission fluid pump 15 kW, power steering pump 12.9 kW, air brake system compressor 3.7 kW, engine oil pump 4 kW, and engine cooling pump 2 kW. The biggest consumers, e.g. engine fan, are too large for a practical FC APU design. Prime candidates for electrification were selected considering the ease of implementation and savings that can be provided by allowing engine shut-downs when the vehicle is stopped. Thus, the engine oil pump, power steering pump and air compressor are electrified in the virtual vehicle equipped with the FC APU.

The peak power requirement for the original, mechanically driven components is 4+12.9+3.7 = 20.6 kW. However, electrifying the same three components reduces the peak requirement to 1.4+4.2+1.2 = 6.8 kW for the reasons explained earlier in this section and depicted in Figure 2. The APU needs to fulfill this requirement during driving, in addition to the power needed for cabin auxiliaries described in the previous section. Consequently, the total peak power requirement is 11.4 kW. In reality, it is unlikely that all of the accessories will operate at their peak at any given instant in time, and even if this occurs the battery on the vehicle is fully capable of acting as a buffer and covering the short term deficit. Thus, the practical sizing target for the FC APU is set at 10 kW.

VEHICLE DESCRIPTION AND MODELING

The vehicle system is based on the 5-ton standard cargo vehicle from the Family of Medium Tactical Vehicles with a gross vehicle weight of 15,300 kg. It's a 6X6, full time all wheel drive truck, powered by a 246 kW six-cylinder, turbocharged, intercooled, direct injection diesel engine coupled to a 7-speed automatic transmission. For the modeling purposes of this study, the FMTV truck is decomposed into the engine, drivetrain, hydraulics and vehicle dynamics. The schematic of the propulsion of this hybrid vehicle is shown in Figure 3. Since the emphasis of the work is fuel economy the torque look-up table based engine model is used, rather than a high-fidelity thermodynamic model previously used for studies of engine transient response [21]. In the conventional drivetrain, diesel engine drives the vehicle through the torque converter (TC), whose output shaft is then coupled to the automatic transmission (AT), which drives the transfer case (TrC) that equally splits the torque to the front and rear wheels. The front prop-shaft delivers the torque to the front differential (DF). The other output of the transfer case, through the rear-front prop-shaft, delivers power to the inter-axle transfer case (TrCA) that further splits the torque to the two rear axles. The first output of the inter-axle transfer case drives the rear-front differential (DRF) and the other the rear-rear differential (DRR) through the rear-rear prop-shaft. Finally, the torque is delivered to the wheels through the half-shafts.

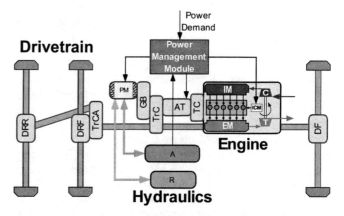

Figure 3: Schematic of the 6x6 truck with the parallel hydraulic hybrid propulsion system

In addition to the conventional components, the drivetrain includes a hydraulic pump/motor (PM), accumulator (A), and reservoir (R) to provide the hybrid functionality. The pump-motor is connected to the transfer case after a fixed speed reduction provided by a gearbox (GB). The pump/motor is connected to the accumulator that stores a high-pressure hydraulic fluid by means of compressed gas. On the other hand, the reservoir stores or supplies the hydraulic fluid used by the pump/motor. The operation of the hybrid drivetrain is controlled by the power management module that determines the operating points of the engine, transmission, and pump/motor based on the power demand and the available hydraulic fluid in the accumulator.

The drivetrain drives the three solid axles of the FMTV truck, which are connected to the chassis with leaf spring suspensions and shock absorbers. Due to the particular focus on fuel economy, the lateral motion of the vehicle is neglected. Therefore, the model assumes left-right symmetry that results in a planar pitch plane model with only three degrees of freedom for the vehicle motion.

The basis for the FMTV component modeling is the previously developed high-fidelity Vehicle-Engine Simulation (VESIM) environment [20, 21, 22]. VESIM has been validated against measurements and proven a very versatile tool for mobility, fuel economy and drivability studies [21]. VESIM emphasizes the high degree of model fidelity and feed-forward logic, thus enabling studies of transient conditions and easy implementation and evaluation of controllers. The model is developed in the 20SIM system modeling and simulation environment that supports hierarchical modeling and allows the physical modeling of subsystems and components using the bond graph formulation [23, 24]. This environment allows easy modification (addition or removal) of model complexity, which makes model development a straightforward task. The model can be easily developed by including only the necessary physical phenomena that contribute to the response of interest, i.e. fuel economy. Model complexity is systematically identified using an energy-based model reduction technique [25].

At the top level of the model hierarchy (see Figure 4), the engine, drivetrain, hydraulics, and vehicle dynamics are excited by the environment. One source of excitation is the driver who controls the vehicle velocity through the gas and brake pedal. The cyber driver is modeled as a controller that attempts to follow the driving cycle. Due to the off-road driving of the truck, the other excitation comes from the road, which is usually uneven and prescribes a vertical velocity to the tires at the contact point. The road excitation is applied to the front, rear-front, and rear-rear tires as a function of their longitudinal position. A more detailed description of the complete FMTV model can be found in [4].

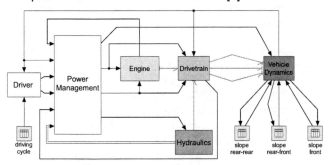

Figure 4: VESIM model of the FMTV truck

ACCESSORIES MODELING

As previously described, engine accessories will be considered for electrification due to the potential reduction of parasitic losses and fuel economy

improvements. Alternative design configurations will be considered; therefore, models of the selected accessories (air-brake compressor, oil pump, AC compressor, and power steering pump) are developed in order to calculate the additional fuel consumption. The flexibility of the VESIM environment allows easy integration within the vehicle system simulation.

The complexity of the engine accessory models depends on their interactions with the system and operating conditions. The dynamics of accessories are orders of magnitude faster from the rest of the vehicle system and their inertial effects are much smaller compared to the engine inertia. Therefore, only the steady-state behavior of these devices is necessary in order to capture the dominant power interactions between the engine and accessories. Such a model will accurately predict the engine response and fuel economy.

Under steady state conditions the power steering pump and air conditioning compressor produce a constant torque load that needs to be provided by the engine (see Figure 5). The actual values of this torque depend on the operation of the accessory and they will be defined later in the paper. These accessories have a one-way coupling with the engine and vehicle, in contrast with the alternator and oil pump loads that are coupled with the engine speed. All accessories are driven by the engine via belts or gears that are represented in the model by transformer elements (TF). The accessories are connected to the engine through the crankshaft port that provides the total torque (T_{ACC}) to drive the accessories.

Given the electric power, $P_{electric}$, needed by the electronics and AC fan, the required torque at the shaft of the alternator is calculated based on engine speed and alternator efficiency. This is done in the "Alternator" block in Figure 5 and the alternator torque (T_A) is applied to the engine. The electric accessory loads are defined as power requirements, which will also be defined later. The oil pump torque (T_{OP}) is a function of engine speed and is calculated from a lookup table obtained from experimental data.

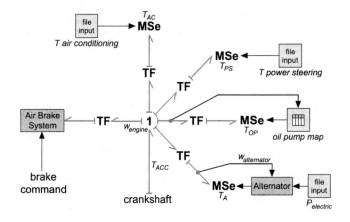

Figure 5: Model of mechanically driven accessories

For the air brake compressor torque, the brake command during driving is used to determine the operation of the compressor. Calculations are performed by the "Air Brake System" block using a dynamic model of the air brake system. This model is depicted in Figure 6. It consists of the compressor and air tanks (wet, primary and secondary). The compressor model provides the airflow (mass and energy) to fill the air tanks based on engine speed and backpressure from air tanks. In addition, it provides the torque load to the engine. This simplified model is derived from a detailed thermodynamic model with complete cycle dynamics and flow through inlet/exhaust valves. The air consumption is calculated based on the brake command where for each brake event an amount of air equivalent to the volume of the six brake chambers is removed from the tanks. The model also includes heat transfer effects between the air tanks and the environment. Finally, a governor is used to maintain the air pressure within the working limits for a safe operation of the brakes. In contrast with the other accessories, there is two-way coupling between the air compressor and engine due to the brake command feedback.

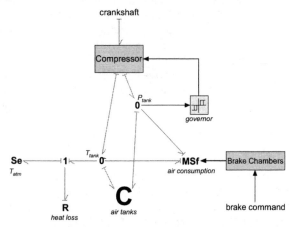

Figure 6: Air-brake system model

For electrically driven accessories the model is modified since the accessories are now decoupled from the engine and are driven by the electric power of the FC APU. Therefore, electric motors are used to convert the electric to mechanical power. Design of accessories is unchanged, but their operating speed is assumed to be equivalent to the engine idle speed, which is the minimum speed at which they can have acceptable performance. The alternator is eliminated since the electric loads are directly provided by the FC APU and their electric power is just added to the other power requirements. This model is given in Figure 7 where the GY element represents the motor, which includes the motor constant and mechanical efficiency. The total power (P_{APU}) that the FC APU must supply to drive these accessories is the sum of power of the electrically driven accessories and direct electric loads ($P_{electric}$) given by:

$$P_{APU} = V_{BUS} \cdot i_{ACC} + P_{electric} \qquad (1)$$

where V_{BUS} is the electric bus voltage, i_{ACC} is the total current drawn by the accessories.

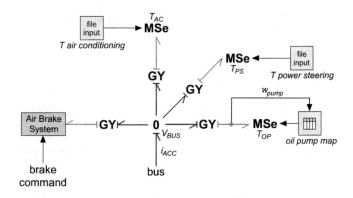

Figure 7: Electrically driven engine accessories

FUEL CELL APU SYSTEM

SELECTION OF COMPONENTS

Among all fuel cell types, the front-runners in APU applications are the Solid Oxide Fuel Cell (SOFC) and Proton Exchange Membrane Fuel Cell (PEMFC). PEMFCs are favored over SOFCs due to their more advanced stage of development, material strength and corrosion characteristics, lower operating temperature, manufacturing process, and relatively lower cost. One of the key issues of PEMFC stacks is that they require hydrogen (H_2) rich gas with almost-zero carbon monoxide (CO) and sulfur concentration. Since storing hydrogen is generally not desired in military applications, a fuel processor (FP) system is adopted to extract hydrogen from the same fuel that is currently used for propulsion, namely JP8. The fuel processor, also known as fuel reformer, is integrated with the PEMFC to form a self-contained FC APU. Venturi et al. demonstrated the use of a FC APU running on methanol [6] and a synthetic fuel that is similar in chemical composition to JP8 [7]. An APU based on fuel processor plus PEMFC is expected to have lower overall efficiency than an APU based on SOFC due to the multiple processes required to develop the H_2 rich gas [14, 19]. We choose to use a PEM Fuel Cell Auxiliary Power Unit with an on-board fuel processor due to their anticipated early adoption in military applications.

A typical fuel processor (FP) consists of four main reactors shown schematically in Figure 8. The liquid fuel is first vaporized and supplied to the first reactor, the hydro-desulfurizer (HDS) for Sulfur removal. The cleaned fuel gas is then supplied to the main reformer, which is responsible for the majority of the H_2 conversion. The gas exiting the main reformer is rich in Hydrogen but contains large quantities of CO_2 and CO. Due to the detrimental effects of CO on the PEMFC, additional reactors are needed to remove CO from the gas stream before it is directed to the fuel cell. Carbon monoxide removal is achieved in three separate reactors: the two water gas shift (WGS) and the preferential oxidation (PROX) reactors.

186

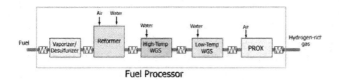

Figure 8: Schematic diagram of a typical Fuel Processor

The two most popular chemical processes to convert hydrocarbon fuel to hydrogen include Steam Reforming (SR) and Partial Oxidation (POX). The most common method, steam reforming, is endothermic, well suited for steady-state operation and delivers relatively high conversion efficiency [16]. However, it suffers from poor transient operation [14]. The partial oxidation process offers several advantages such as compactness, rapid-startup, and responsiveness to load changes [17], but delivers lower efficiency as shown in Table 1.

Table 1: Main H_2 Reactor Efficiency for different Hydrocarbon Fuel and Conversion Principle

Fuel	Energy Efficiency (%)	
	POX	SR
Methanol	Data n/a	83.2
Natural Gas	77.5	85.5
Gasoline	55.8	81.2
Diesel	55.7	81.2
Jet fuel	54.9	81.2

The values of Table 1 are derived from Table 6 in [14] using theoretical input energies for producing a mole of H_2. In particular, the SR and POX efficiencies are calculated from:

$$\eta_{SR} = \frac{Q_{LHV_{H_2}}}{Q_{tTH_{C_xH_y}}} \quad \text{and} \quad \eta_{POX} = \frac{Q_{LHV_{H_2}}}{Q_{TH_{C_xH_y}}} \quad (2)$$

where:

$Q_{LHV_{H_2}}$ is the H_2 low heating value,

$Q_{tTH_{C_xH_y}}$ is the total theoretical C_xH_y fuel heat input per mole of usable H_2, and

$Q_{TH_{C_xH_y}}$ is the theoretical C_xH_y fuel heat input per mole of usable H_2.

Note here that in the SR case the total Theoretical C_xH_y Fuel Heat input per mole of usable H_2 includes the heat required for the endothermic reaction. The efficiencies in the POX-based fuel processor are calculated without assuming utilization of the exothermic heat. This choice leads to low efficiency assumptions for the POX-based FP system. In all cases, the system

efficiency is calculated assuming that the gas leaving the FC anode has 8% hydrogen partial pressure. The molar composition of the fuel affects the Partial Oxidation reactor efficiency. The lowest efficiency (54.9%) is observed when Jet fuel, chemically similar to JP8, is used. Despite of the low efficiency, a POX-based fuel processing system is adopted due to its faster start-up and compact characteristics.

FUEL CELL STACK SIZING AND MODELING

The fuel cell stack (FCS) is sized to satisfy the 10 kW peak power identified in the previous sections. The sizing process has many degrees of freedoms and has to satisfy many constraints. Design degrees of freedom include the number (n) of cells stacked in series, the cell active area (A), the reactant supply pressure, and the cooling method. Typical constraints are the desired vehicle operating voltage, the volume, the weight, and the hydrogen supply method. The PEMFC APU is sized using conservative published data, in order to better evaluate its short term benefits instead of their long-term potential.

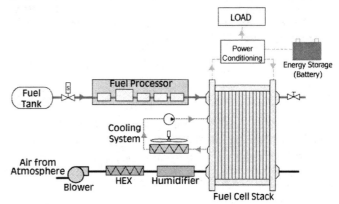

Figure 9: Schematic diagram of the FC APU system

A liquid-cooled low pressure FC system design is selected because it can operate seamlessly with a fuel processing system that typically operates in near-ambient pressure conditions [1]. Its configuration is shown in Figure 9. Non-equilibrium conditions are neglected and a single polarization curve is utilized to describe the electrical output of a single fuel cell. Specifically, Figure 10 shows the single cell voltage (V_{fc}) versus current density ($i=I_{fc}/A$ in A/cm^2) drawn from the fuel cell. The specific characteristic is obtained from [15] under low reactant pressure. We consider only one steady-state polarization curve for all cells within the stack because the current drawn from the fuel cell stack is controlled so that quasi-steady FC conditions are maintained (see FC APU Control section). Due to the electrical connection of n cells in series the stack current is equal to the single cell current ($I_{fsc} = I_{fc}$), whereas, the stack voltage is given by ($V_{fsc} = nV_{fc}$).

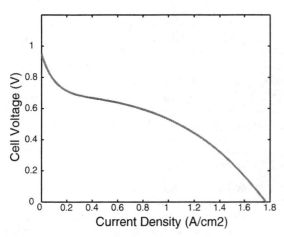

Figure 10: Fuel Cell Polarization Curve

A typical active cell area (250 cm^2 < A < 400 cm^2) and an output stack voltage (V_{fcs} = 42 Volts) are considered as target values for the FC stack. The parasitic losses (P_{aux}) due to the FCS auxiliary components such as the air blower are captured by a linear function. The minimum auxiliary load is 200 W and it is assumed to increase linearly by 0.57 W/Amp with the load current drawn from the stack. The net fuel cell stack power is then given by:

$$P_{fcs}(I_{fcs}) = nV_{fc}I_{fc} + \max\{0.2, 0.57I_{fc}\} \qquad (3)$$

A simple search procedure shows that n = 65 and A = 300 cm^2 satisfies the system requirements of 42 Volts voltage output and 10 kW power output. The FC stack net power (P_{fcs}) and efficiency values versus the current (I_{fc}) are shown in Figure 11.

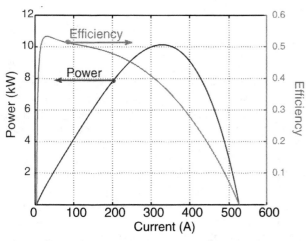

Figure 11: Fuel Cell Stack Power and Efficiency versus Current

The peak FC power $P_{fcs}^{\max}$ = 10.05 kW is attained at stack current I_{fc} = 350 Amps or at the current density value of i = 1.17 A/cm^2. The stack efficiency (η_{fcs}) is defined as:

$$\eta_{fcs} = \frac{P_{fcs}}{nV_0I_{fc}} \qquad (4)$$

where V_0 is the theoretical maximum single cell voltage also known as the open circuit voltage (from lower heating value).

FUEL PROCESSOR MODELING

The fuel consumption from the Fuel Cell Auxiliary Power Unit is calculated using basic electrochemistry and the conversion efficiency of the fuel processor. Specifically, the hydrogen consumption rate (moles per second) depends on the stack current drawn from the fuel cell system:

$$N_{H_2} = \frac{nI_{fc}}{2F} \qquad (5)$$

where F is the Faraday number (96485 Coulombs). The fuel consumption rate (moles/sec) is then calculated from the POX FP efficiency:

$$N_{fuel} = \frac{Q_{TH_{C_xH_y}}}{Q_{LHV_{C_xH_y}}} N_{H_2} \qquad (6)$$

where $Q_{LHV_{C_xH_y}}$ is the low heating value of the C_xH_y hydrocarbon fuel and $Q_{TH_{C_xH_y}}$ is defined in the previous section. Finally, the mass flow rate (g/s) of the C_xH_y hydrocarbon fuel is calculated from:

$$\dot{m}_{fuel} = M_{C_xH_y} N_{fuel} \qquad (7)$$

where $M_{C_xH_y}$ is the molar mass of the C_xH_y hydrocarbon fuel (in our case JP8). The instantaneous mass flow rate is integrated throughout the driving cycle and the silent watch in order to calculate its impact on the overall fuel consumption.

The dynamic response of the POX-based fuel processor during transient conditions is modeled using a first order lag with a time constant $\tau_{fp} = 1$ sec. This time constant is chosen based on our experience in transient control of a 200 kW natural gas reformer and other published data [12, 18]. Although the system time constant and efficiency actually vary with fuel flow rate (hydrogen flow demand), using a simple first order dynamics with fixed fuel processor efficiency is deemed sufficient for the vehicle system level fuel economy calculations.

FC APU CONTROL

An overview of the FC APU controller that allows integration with the vehicle batteries is shown in Figure 12. The key objectives of the FC APU controller are to:

- Ensure that the requested power is delivered
- Protect the fuel cell stack and fuel processor system from sudden current loads
- Supplement fuel cell system power deficit from batteries without depleting them below a threshold state of charge

The fuel consumption calculations are also included in the model to allow direct implementation within VESIM. The input to the FC APU controller is the desired APU power (P_{APU}). Any power request from the Battery (P_{bt}^{req}) is added to the FC APU power requests to provide the total power demand given by:

$$P_{fcs}^{dem} = P_{apu} + P_{bt}^{req} \qquad (8)$$

The overall fuel cell system power demand P_{fcs}^{dem} is then used as the input to a lookup table to calculate the current demand from the FCS (I_{fcs}^{dem}). The lookup table essentially inverts the FCS power function found in Eq.(3) for the stack current calculation.

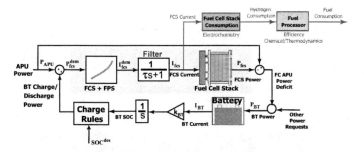

Figure 12: Schematic of the FC APU controller

CURRENT CONTROL AND HYDROGEN STARVATION CONSTRAINTS

The responsiveness of H$_2$ generation imposes a limitation on how fast stack current can be drawn from the FCS. As discussed before, the FP responds to a step change in hydrogen command with a first order lag in H$_2$ generation. To avoid H$_2$ starvation in the FCS anode the current drawn from the fuel cell is always filtered using the FP time constant:

$$I_{fc}(s) = \frac{1}{\tau_{fp}s+1} I_{fc}^{dem}(s) \qquad (9)$$

The current drawn from the FCS is then used to calculate the FCS power P_{fcs} using Eq.(3). The FP time constant thus dominates the fuel cell system dynamics. To support the instantaneous current demands associated with the steering requests or other time-critical functionalities, we employ the existing vehicle battery and integrate it with the FC APU. The deficit between the demanded power from the APU (P_{apu}) and the power delivered by the FCS is supplemented by the vehicle batteries. However, to avoid battery depletion, a separate battery controller is designed which is explained in the next section.

BATTERY STATE OF CHARGE CONTROLLER

The original FMTV battery pack (four 14V 6MTF Lead-Acid batteries) is used to supplement the FC APU power output during pulsed load demand. The power requested and drawn from the battery is:

$$P_{bt} = P_{apu} - P_{fcs} + P_{other} \qquad (10)$$

where P_{other} is the additional electrical power demand from the vehicle. A lookup table with the battery efficiency is used to calculate the current drawn from or supplied to the battery during discharging and charging, respectively ($I_{bt} = f_{bt}(P_{bt})$). Note that the power request for the battery is positive when discharging and negative when charging. The battery state of charge (SOC) depends on the battery current (I_{bt}) and the battery rated Amp-Hours ($1/k_{bt}$):

$$SOC(t) = SOC(t_0) - k_{bt} \int_{t_0}^{t} I_{bt} d\tau . \qquad (11)$$

A simple proportional controller defines the power request (P_{bt}^{req}) and consequently regulates the battery state of charge at a desired level (SOC^{des}) selected to be 0.55 :

$$P_{bt}^{req} = k_c \left(SOC^{des} - SOC \right). \qquad (12)$$

The controller gain k_c is adjusted based on the desired speed for charging the battery.

MISSION PROFILE/ACCESSORY DUTY CYCLE

Fuel consumption is heavily dependent on the operation of the vehicle. For accurate and representative predictions, a mission profile based on the typical use of the FMTV truck is needed. The mission is separated into driving and silent watch as described below.

DRIVING

The design study is focusing on fuel economy improvement as the FMTV truck carries out a typical mission. This mission is described by a speed profile (driving cycle) that the truck must maintain as it drives over an uneven road profile. The specific road and speed profiles, which consist of city, highway, and cross-country roads, shown in Figure 13, represent a typical mission of the FMTV vehicle. Note that the beginning of the cycle represents the primary and secondary roads with a flat surface and frequent accelerations and decelerations. In the third, cross-country part, driver attempts to maintain constant speed on roads with uneven profile. The driver model adjusts the throttle and brake signal in order for the vehicle to follow the prescribed speed profile. In addition, the engine needs to provide power for the air conditioner and on board electronics according to the duty-cycle described in the next section.

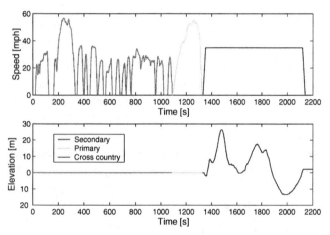

Figure 13: Vehicle driving schedule (50% secondary roads, 20% primary roads, and 30% cross-country)

A power steering model is also included in the mission for more realistic fuel economy predictions. The selected model is very simple with a predefined torque load applied to the engine. The torque varies from a high value when the vehicle steers, to low when it drives straight with no steering. This results in a steering torque pulses when the vehicle steers. The duration of the pulses is set to 2 seconds and the low and high torque values are determined from measurements.

The timing of the pulses (steering event) is defined using a statistical approach. A steering event is initiated based on a steering probability that represents typical driver/vehicle steering behavior as a function of vehicle speed (see Figure 14). At zero speed the steering probability is zero; however, it rapidly increases as the vehicle starts to move, reaching its maximum at a vehicle speed of 7 mph. After that point, the steering probability slowly decreases in order to return to zero at 100 mph (theoretical). The shape of this probability is based on a gamma distribution.

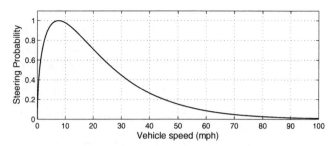

Figure 14: Probability distribution of power steering usage as a function of vehicle speed

The probability distribution is used along with the total usage of steering to generate the steering cycle. The overall steering usage depends on the driving conditions and it can be determined from previous SAE studies/standards [26]. For the driving cycle considered in this study, the steering usage is 60% for secondary roads, 10% for primary roads, and 35% for cross-country. Using this information and the probability distribution for steering the time history of the power steering pump torque is generated as shown in Figure 15. For example, in the case of primary roads we see frequent steering at low speeds due to the high probability, while higher speeds show sporadic usage.

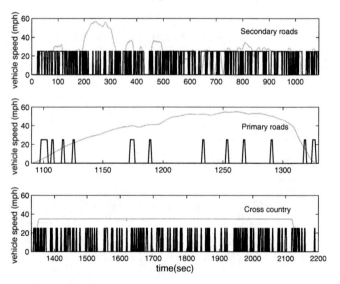

Figure 15: Power steering usage

SILENT WATCH LOADS

The cabin auxiliary loads come from electronics, radio, and ventilation/air conditioning that the driver and soldiers use in the cabin during driving and standby conditions. These loads appear in the form of electric power and are assumed to be the same for both driving and standby conditions, as long as the truck is manned and performing a mission.

More specifically, the electronics load is assumed constant at 0.6 kW since all these devices are turned on throughout the entire time that the vehicle is on a mission. The radio load is assumed to have a peak value of 0.6 kW (zero when not used), and it occurs during 6% of the mission time. The duration of each interval of usage (pulse) is 40 seconds, which is randomly distributed over time. Finally, the air conditioning is assumed to have a periodic use with 3.4 kW peak power consumption and a period of 8 minutes. In addition, the air conditioning stays on for 2 minutes each time it is turned on, which is representative operation of commercial air conditioning systems. To determine the total power consumption, the power requirements for the three devices are added. This produces the duty cycle shown in Figure 16, with a peak electric load of 4.6 kW and a minimum of 0.6 kW.

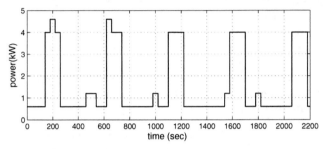

Figure 16: Combined accessory duty cycle during silent watch conditions

SIMULATION RESULTS

The study of the silent watch was performed first. Fuel consumption predictions were obtained for the ten-hour silent watch mission, using the accessory loads and duty cycles described previously. Two vehicle configurations were compared: one relying on the main propulsions diesel engine to generate power, and the other using a 10 kW FC APU. In the baseline vehicle, the 246 kW engine was operated at near-idle conditions to support both electrical and mechanical loads. Summary of results is shown in Table 2. The effect of avoiding inefficient engine operation at near-idle conditions and replacing it with the FC APU is impressive. The fuel consumption decreases from 8.6 gallons to 1.5 gallons with the APU, an 82.6% reduction.

The distribution of visitation points across the FCS load range during 10 hours of silent watch shown in Figure 17 helps explain the benefits. The fuel cell is relatively oversized for silent watch, since it has to accommodate the additional loads during driving. Consequently, the operating points are concentrated in the low current region. Contrary to engines that obtain their peak efficiency at high loads, the fuel cell achieves its best efficiency at very low loads, hence the big impact of replacing engine idling with a FC APU. The FCS average efficiency is 50% and the FP conversion efficiency is 55%, thus the overall FC APU efficiency is 28%. It is clear that reforming of the complex fuel significantly diminishes the overall efficiency to a level attainable with a small diesel powered APU. Nevertheless, the low thermal signature and noise level of the fuel cell APU makes it attractive. The efficiency results rely on the current level of technology and represent a realistic short-term outlook, while expected advances of the fuel cell and reforming technology could significantly improve the long-term perspective.

Table 2: Predicted fuel consumption during the ten hour silent watch for the baseline vehicle, and the one equipped with the Fuel Cell APU.

Energy Source	Fuel consumed [gallon]	Improvement
Diesel engine @ idle speed	8.6	-
Fuel Cell APU	1.5	575%

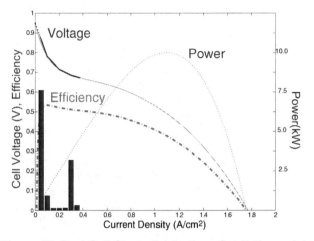

Figure 17: Fuel Cell Stack distribution of visitation points during silent watch

The effect of FC APU during driving was evaluated through simulations over the previously described vehicle duty cycle. Four vehicle system configurations were considered: baseline diesel hydraulic hybrid truck, vehicle with the FC APU added, vehicle with the FC APU and power management strategy allowing engine shut-downs when the vehicle is stopped, and finally the configuration that includes engine downsizing (or de-rating) to 95% of the original power plant. The overall fuel economy results are summarized in Table 3.

The first of the three proposed vehicle designs with the FC APU improves fuel economy over a typical driving schedule by 5.8%. The savings are in part due to the reduction of the diesel propulsion fuel consumption, and in part due to the reduction of fuel consumed for powering accessories. The latter can be attributed to two factors: reduced energy consumption of the controllable, electrified accessories and relatively efficient energy conversion in the APU. Normally the hydraulic hybrid powertrain does not allow engine shut-downs, since engine needs to run the essential accessories at all times. However, electrification of the power steering and air brake compressor allows safe shut-downs, thus increasing the fuel economy benefit to 9.2%. Downsizing the engine by 5% provides only an incremental improvement. Obviously the majority of engine visitation points are in the mid-load range of the engine where small changes of relative load do not have a dramatic impact on specific fuel consumption.

Table 3: Predicted fuel consumption over the FMTV driving schedule and the percent improvement for three configurations of the vehicle equipped with the FC APU.

Energy Source	Propulsion Fuel Cons. [gallon]	Accessory Fuel Cons. [gallon]	Improvement
Diesel engine only	2.28	0.369	-
Engine + Fuel Cell APU	2.22	0.306	5.8%
Engine + Fuel Cell APU (Engine shutdown)	2.14	0.306	9.2%
Engine + Fuel Cell APU (Engine shutdown & downsizing to 95%)	2.11	0.306	10.6%

Figure 18 shows the engine visitation points as function of speed and load, and illustrates the effect of electrifying accessories and power management strategies on the diesel engine. First of the enlarged sections from right-to-left shows a large number of operating points at low load and idle speed, corresponding to the engine in the baseline vehicle overcoming losses in the torque converter and running the accessories while the vehicle is stopped. The second enlarged section shows changes brought about by electrification of accessories and connecting them to the APU: there are much fewer engine operating points at near idle condition, with a single large spike associated with the engine overcoming the loss in the torque converter during idling. The last insert, far left, indicates further dramatic reduction in the number of near-idle points in case when the engine is being shut-down during extended vehicle stops. Since the very low load engine operation represents the least efficient region, elimination of near-idle points directly and significantly benefits the overall propulsion system efficiency.

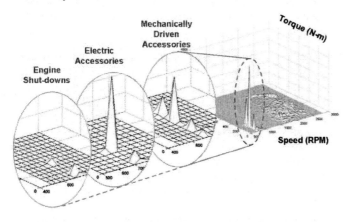

Figure 18: Changes of the distribution of engine visitation points shown as a function of engine speed and load. Magnified portions of the 3-D plot illustrate reduction of near-idle times due to electrification of accessories and engine shut-downs.

The savings realized through the dramatic reduction of the time the engine spends at near-idle conditions are in part offset by the effect of decoupling the accessories from the engine and thus changing the relative position of operating points on the Break Specific Fuel Consumption (BSFC) map. In short, the engine operating point in the baseline system is determined by the sum of the propulsion power and the accessory power. If the accessory power is removed, the relative load of the engine decreases. This means that the point moves downward on the engine BSFC map as shown in Figure 19. If this happens in the low-to-mid speed and part load region, the effect can be relatively significant – see example "A" in Figure 19. However, at higher speeds and loads, the effect might be non-existent – example "B" on Figure 19. In the particular study, the overall effect was not very large, given the very small change between runs with the regular engine and the downsized (95%) engine. However, in case larger accessories are decoupled and connected to the Auxiliary Power Unit, engine downsizing would be highly recommended in order to avoid any offsets of fuel economy gains.

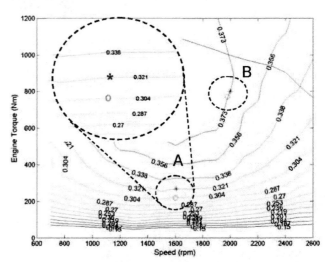

Figure 19: The effect of unloading the engine due to decoupling of the accessories on its efficiency: 'stars' mark operating points with accessories, and 'circles' the corresponding points without accessories.

During driving, the simultaneous operation of cabin accessory devices and engine/drivetrain accessories causes very dynamic APU load changes. Figure 20 shows the instantaneous desired FC APU power (solid), the power delivered by the fuel cell stack (dashed), and the corresponding battery power during a portion of the driving cycle. Whenever the rate of change of load exceeds the ability of the FC system to respond, e.g. when driver suddenly turns the steering wheel, the battery compensates as indicated by the spikes in battery power plot in Figure 20. Note that at the end of the event, e.g. when the steering wheel returns to neutral position, the FC system does not return instantaneously to zero load but slowly ramps down as dictated by the first order lag response. During the power-down events, the FC automatically recharges the battery by using the generated hydrogen from the FP system instead of purging it – see the spikes with the opposite sign in Figure 20. This eliminates the need for a high gain controller for battery recharging.

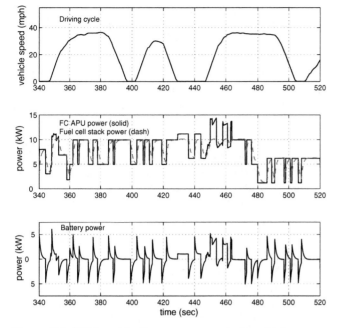

Figure 20: APU and battery response during driving

The distribution of FC visitation points during the driving cycle is shown in Figure 21. In this case, the increased power demand due to the addition of electrified engine/drivetrain accessories moves a number of points towards the high current region, with detrimental effects on average FC efficiency that drops to 47%. The overall FC APU efficiency is as low as 26%.

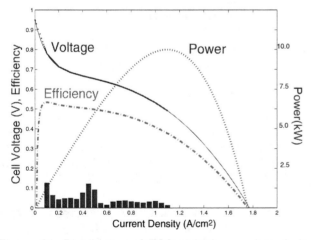

Figure 21: Distribution of FCS visitations points during the FMTV driving schedule

Finally, an attempt is made to assess the realistic impact of the electrification and the use of FC APU from the point of view of the fleet operator and logistics planner. To this end, a hypothetical FMTV truck work day is considered, consisting of 10 hours of driving, 10 hours of silent watch and 4 hours of rest. In this scenario, the daily fuel consumption is reduced from 51.9 to 41.5 gallons, which represents a 20.1% improvement of fuel economy and potential extension of vehicle range by roughly 70 miles.

CONCLUSIONS

The simulation study quantifies the fuel economy benefits of a combined approach of using the Fuel Cell Auxiliary Power Unit (FC APU) for supporting the silent watch requirements, and for powering the selected electrified engine accessories during driving. Models of the low pressure PEM fuel cell stack and the JP8 partial oxidation reformer were derived, accounting for the efficiency of main energy/chemical conversion devices, as well as the effect of fuel choice on overall system efficiency. The virtual vehicle is based on the comprehensive model of the 5-ton, Family of Medium Tactical Vehicles (FMTV) 6x6, Class VI truck with hydraulic hybrid propulsion system, previously modeled in the Automotive Research Center. The silent watch mission defines the use of advanced communication, navigation, computer and Air Conditioning (A/C) and Nuclear Biological and Chemical (NBC) protection systems. The engine/drivetrain accessories decoupled from the engine were the engine oil pump, the air brake compressor and the power steering pump. The peak power requirements of all electrified accessories were estimated at 11.4 kW, out of which 4.6 kW is the peak silent watch power. The fuel cell APU was sized for 10 kW output, assuming that the battery will provide a buffer in case of rare maximum power peaks.

Simulation of the ten-hour silent watch mission indicates the potential for a six-fold improvement of fuel economy compared to the option of idling the diesel engine normally used for propulsion. The results emphasize the fact that the APU eliminates the most inefficient engine operation at very low loads, and replaces it with efficient part load operation of the relatively oversized Fuel Cell system. Fuel savings over the FMTV's combined on-off-road driving schedule were relatively modest at ~6%. The benefits come from operating the controllable electrified components at the optimum conditions, independently of engine speed, and efficient power generation by the FC APU. However, this is somewhat offset by the fact that unloading the accessories from the main-engine moves its operating points to regions of slightly higher specific fuel consumption. Strategically selected engine and drivetrain accessories allow engine shut-downs when the vehicle is stopped, increasing the fuel economy improvement to ~9%.

Compounded fuel economy benefits calculated over the hypothetical daily military mission with a combined highway/local/off-road driving for 10 hours and silent watch for 10 hours, are ~20%. All results were obtained assuming conservative estimates of fuel cell efficiencies attainable in the short term. Reformer efficiency is also low at ~50% due to the fact that it processes a complex fuel (JP8). Fuel choice was dictated by the application of the APU on a military truck. Projected advances in the fuel cell and reformer technology would improve the long term fuel economy potential, and reduce the space and packaging burden.

ACKNOWLEDGMENTS

The authors would like to acknowledge the technical and financial support of the Automotive Research Center (ARC) by the National Automotive Center (NAC) located within the US Army Tank-Automotive Research, Development and Engineering Center (TARDEC). The ARC is a U.S. Army Center of Excellence for Automotive Research at the University of Michigan, currently in partnership with University of Alaska-Fairbanks, Clemson University, University of Iowa, Oakland University, University of Tennessee, Wayne State University, and University of Wisconsin-Madison. In particular, Jim Yakel, Don Szkubiel, Ron Chapp, and Ken Deylami provided technical guidance related to the FMTV Program goals. Herb Dobbs, Erik Kallio (Alternative Fuels & Fuel Cell Team), Jim Miodek and Fred Krestik (Team Power) provided valuable information about hotel loads and battery characteristics. Finally, the authors would like to thank Dave Allen and Bob Page (Engineered Machined Products Inc.) for providing the FMTV engine accessory load data, Bob Julian (Stewart and Stevenson) for air compressor data and John Bennett (Arvin Meritor) for air-brake system data.

REFERENCES

1. Dobbs, H.H, Kallio, E.T., Pechacek, J.M., "U.S. Army Strategy for Utilizing Fuel Cells as Auxiliary Power Units," SAE paper 2001-01-2792.

2. Lin, C. C., Filipi, Z., Wang Y. S., Louca L., et al, "Integrated, Feed-forward Hybrid Electric Vehicle Simulation in SIMULINK and Its Use for Power Management Studies", SAE Paper 2001-01-1334, Warrendale, PA, 2001

3. Wu, B., Lin, C., Filipi, Z., Peng, H., Assanis, D., "Optimization of Power Management Strategies for a Hydraulic Hybrid Medium Truck," 6th International Symposium on Advanced Vehicle Control, Hiroshima, Japan, September 2002

4. Filipi, Z.S., L.S. Louca, B. Daran, Chan-Chiao Lin, U. Yildir, B. Wu, M. Kokkolaras, D.N. Assanis, H. Peng, P.Y. Papalambros, and J.L. Stein, 2003. "Combined Optimization of Design and Power Management of the Hydraulic Hybrid Propulsion System for the 6x6 Medium Truck"; to appear in the Special Issue on Advanced Vehicle Design and Simulation of the International Journal of Heavy Vehicle Systems, Inderscience Publishers, Geneve, 2004

5. Kepner, R.D.,"Hydraulic power assist ~a demonstration of hydraulic hybrid vehicle regenerative braking in a road vehicle application", SAE paper 2002-01-3128, Warrendale, PA, 2002

6. Venturi, M., Martin, A.,"Liquid-fueled APU fuel cell system for truck application", SAE paper 2001-01-2716, Warrendale, PA, 2001

7. Venturi, M., Kallio, E., Smith, S., Baker, J., Dhand, P.,"Recent results on liquid-fuelled APU for truck application", SAE paper 2003-01-0266, Warrendale, PA, 2003

8. Zizelman , J., Botti, J., Tachtler , J., Strobl, W.,"Solid oxide fuel cell auxiliary power unit ~ A paradigm shift in electric supply for transportation", SAE paper 2000-01-C070, Warrendale, PA, 2000

9. Brodrick, C. J., Farshchi, M., Dwyer, H.A., Gouse, S.W., Martin, J., Von Mayenburg, M.,, "Demonstration of a proton exchange membrane fuel cell as an auxiliary power source for heavy trucks", SAE paper 2000-01-3488, Warrendale, PA, 2000

10. Hnatczuk, W., Lasecki, M. P., Bishop, J., Goodell, J., "Parasitic loss reduction for 21st century trucks", SAE paper 2000-01-3423, Warrendale, PA, 2000

11. Hendricks, T., O'Keefe, M., Laboratory "Heavy vehicle auxiliary load electrification for the essential power system program: Benefits, tradeoffs, and remaining challenges", SAE paper 2002-01-3135, Warrendale, PA, 2002

12. Pukrushpan, J.T., Stefanopoulou, A.G., Varigonda, S., Pedersen, L. and Ghosh, S. "Multivariable Control of Natural Gas Reformer for Fuel Cell Power Plant," Proceedings of the American Control Conference, 2003

13. Gelfi, S., Pukrushpan, J., Stefanopoulou, A.G., Peng, H. "Dynamics and Control of Low and High Pressure Fuel Cells," Proceedings of the American Control Conference 2003

14. Brown, L.F. "A Comparative Study of Fuels for On-Board Hydrogen Production for Fuel-Cell-Powered Automobiles," International Journal of Hydrogen Energy, v.26, pp. 381-397, 2001

15. Pukrushpan, J.T., Peng, H., Stefanopoulou, A.G. "Simulation and Analysis of Transient Fuel Cell System Performance Based on a Dynamic Reactant Flow Model," Proceedings of the ASME International Mechanical Engineering Congress & Exposition, 2002

16. Ahmed, S., Krumpelt, M. "Hydrogen from Hydrocarbon Fuels for Fuel Cells," International Journal of Hydrogen Energy, v.26, pp. 291-301, 2001

17. Dicks, A.L., "Hydrogen Generation from Natural Gas for the Fuel Cell Systems of Tomorrow," Journal of Power Sources, v.61, pp.113-124, 1996

18. Ballard Power System Inc. "Xcellsis HY-80 Light Duty Fuel Cell Engine Brochure," http://www.ballard.com/pdfs/XCS-HY-80_Trans.pdf, 2003

19. Steele, B., Heinzel, A., "Materials for Fuel Cell Technologies," Nature, v.414, pp.345 – 352, 2001

20. Louca, L.S., Yildir, U.B.,"Modeling and Reduction Techniques for Studies of Integrated Hybrid Vehicle Systems". Proceedings of the 4th International Symposium on Mathematical Modeling, Vienna, Austria. Published in the series ARGESIM-Reports, ISBN 3-901608-24-9, Vienna, Austria, 2003.

21. Assanis, D.N., Filipi Z.S., Gravante S., Grohnke D., Gui X.,Louca L.S., Rideout G.D., Stein J.L., Wang Y., "Validation and Use of SIMULINK Integrated, High Fidelity, Engine-In-Vehicle Simulation of the International Class VI Truck", SAE Paper 2000-01-0288, Warrendale, PA, 2000.

22. Louca, L.S., J.L. Stein, and D.G. Rideout, 2001. "Integrated Proper Vehicle Modeling and Simulation Using a Bond Graph Formulation". Proceedings of the 2001 International Conference on Bond Graph Modeling, Vol. 33, No. 1, pp. 339-345, Phoenix, AZ. Published by the Society for Computer Simulation, ISBN 1-56555-221-0, San Diego, CA.

23. Karnopp, D.C., D.L. Margolis, and R.C. Rosenberg, "System Dynamics: A Unified Approach", Wiley-Interscience, ISBN 0471-62171-4, New York, 1990.

24. Brown, F.T., "Engineering System Dynamics: A Unified Graph-Centered Approach", Marcel Dekker, ISBN 0-8247-0616-1, New York, 2001.

25. Louca, L.S., J.L. Stein, G.M. Hulbert, and J.K. Sprague, "Proper Model Generation: An Energy-Based Methodology". Proceedings of the 1997 International Conference on Bond Graph Modeling, Vol. 29, No.1, pp. 44-49, Phoenix, AZ. Published by the Society for Computer Simulation, ISBN 1-56555-103-6, San Diego, CA, 1997

26. SAE Standard 1343:" Information Relating to Duty Cycles and Average Power Requirements of Truck and Bus Engine Accessories"

2004-01-1479

Modeling Stationary Power for Heavy-Duty Trucks: Engine Idling vs. Fuel Cell APUs

Nicholas Lutsey, John Wallace, C. J. Brodrick, Harry A. Dwyer and Daniel Sperling
Institute of Transportation Studies, University of California, Davis

ABSTRACT

Line-haul truck engines are frequently idled to power hotel loads (i.e. heating, air conditioning, and lighting) during rest periods. Comfortable cabin climate conditions are required in order for mandatory driver rests periods to effectively enhance safety; however, the main diesel engine is an inefficient source of power for this conditioning. During idle, the diesel engine operates at less than 10% efficiency, consuming excess diesel fuel, generating emissions, and accelerating engine wear. One promising alternative is the use of small auxiliary power units (APUs), particularly fuel cell-based APUs. The Institute of Transportation Studies (ITS-Davis) developed an ADVanced VehIcle SimulatOR (ADVISOR)-based model to quantify the costs and benefits of truck fuel cell APUs. Differences in accessories, power electronics, and control strategy between the conventional engine idling and APU-equipped systems are analyzed and incorporated into the model. A case study of a diesel-reformer solid oxide fuel cell (SOFC) APU system retrofit on a Class 8 line-haul truck is presented here. We sized a SOFC for heavy duty truck APUs to have a power output of 4 kW (net) electric. The modeled SOFC system was found to have an efficiency of about 30-35% efficient and consume about 0.14-0.17 gallons of diesel fuel per hour. The fuel savings resulting from replacement of avoidable idling are estimated at 3-8% of total vehicle energy use. For trucks that idle 10 hours or more daily, overall fuel consumption would be reduced by about 5-12%.

INTRODUCTION

Approximately 400,000 heavy-duty trucks in the U.S. primarily drive over 500 miles from their home base [1]. Federal regulations mandate rest periods in an effort to increase driver safety. Line-haul drivers commonly rest in their truck cabs, which are generally equipped similar to small mobile homes with beds, electrical appliances, heating, and cooling. To power cab appliances and maintain cab climate in seasonal weather, the trucks' main propulsion engine is often utilized. These engines were developed and optimized to run at 300+ horsepower to haul a loaded 40-ton vehicle; they are inefficient when powering very light loads, such as air conditioners and light bulbs.

Idling engines consume relatively large amounts of fuel, generate pollution, are noisy, and accelerate engine wear [2]. These negative consequences are widely acknowledged and a variety of steps are being taken to curb the negative effects. Regional organizations are rapidly restricting idling in order to attain ambient air standards [3,4]. Localities have begun to restrict engine idling to minimize the nuisance of noise, smoke, and emissions. A national program is targeting reduced fuel consumption, long a concern of truck operators for whom fuel is a large operating cost [5].

In order to assure continued driver safety, truck operators will need to find alternatives to engine idling. The principal on-board technologies currently available are diesel-fired heaters, thermal storage, and gen-sets [2]. Parking areas with shore power hook-up installations, commonly referred to as electrification, are increasing [6].

To be accepted by drivers and truck owners, any technology should efficiently and economically provide hotel amenities for vehicles at truckstops, loading docks, and remote locations. To date, market penetration of on-board technologies is estimated at about 5% [7]. Although electrification is spreading rapidly, there is shortage of truck stop spaces, and it is uncertain how often trucks stop at truck stops, opposed to other locations [2]. Several cost-benefit studies have been conducted on existing on-board technologies and truck stop electrification [8] based on existing, average emissions factors. However, sophisticated analyses of driver behavior, truck operation, and emissions are necessary to definitively quantify the costs and benefits for modern vehicles.

RESEARCH METHOD

In order to asses the potential energy benefits of fuel cell APUs in heavy duty trucks, ITS-Davis developed a physical model that simulates the energy use, emissions, and performance of trucks with and without fuel cell APUs. The model is modified from the National Renewable Energy Laboratory's (NREL's) Advanced VehIcle SimulatOR (ADVISOR 2002 edition) a MATLAB/Simulink based program [9]. The majority of our enhancement effort was focused on developing realistic load profiles of tractor operation during idling.

Similar to driving cycles, which are vehicle speed-time traces, idle load profiles depict the change in power demand over time. We refer to these load-profiles simply as "cycles" and store them along side driving cycles in the model. The profiles were developed based on a literature search, supplemental survey data collected from line-haul truck drivers in California, load measurements, and discussions with truck component suppliers. For APUs, the cycles were then adjusted to account for the differences between the current mechanical power requirements of some conventional accessories and the electrified accessories likely to be used in combination with APUs. Next, updated diesel engine maps were incorporated to improve the baseline estimates of fuel consumption and emissions during diesel idling. Unlike the default maps, these engine maps included low torque, low engine rpm operations typical of idling. Last, APU modules with control strategies were built into the ADVISOR model. A prominent reason for using ADVISOR as our APU model framework is its ability to accept data and simulate a wide variety of systems, including fuel cells.

Cycle Development

Developing the cycle for line-haul truck idling involved several steps. First, we defined the distribution of driving and idling time. Then we identified the accessories that are used during idle and the percent of idle time that each accessory was engaged. Next, power requirements were assigned to these accessory devices. Modifications from the idling engine to the APU scenario were made to reflect the differences in mechanical and electrical power requirements of the two systems. Finally, power cycles, i.e. accessory load vs. time traces, were created for the two scenarios.

Idle Duration

The American Trucking Associations' Technology and Maintenance Council (TMC) reports 6 hours per day for its daily idling duration for long-distance, freight-hauling, heavy duty trucks [10]. Stodolsky et al, pointing out the seasonality of idling, use 10 hours per day during winter (85 days) and 4.5 hours per day the rest of the working year (218 days); this equates to a base case of 6 hours per year [2]. Other estimates are in line with these estimates. A California Air Resources Board-sponsored

study logged total hours where heavy-duty trucks were at rest, as a percentage of total engine run time. With 84 trucks logged over a total of 1,600 hours, the average idling time was found to be about 42% of total engine-on time [11]. This equates to approximately 7 hours of idling for every 9 – 10 hours day of driving; although no distinction is made regarding which stops were necessary, unavoidable stops (e.g., red lights) versus which ones were unnecessary, avoidable (e.g., idling while resting). Another source reported in-truck sleeping time of about 5 hours per day [12]. Based on these sources, a driver who idles about 6 hours per day appears to be "typical."

A small pilot survey was used to obtain quantitative data on idling characteristics. Two student researchers went to a public rest area off Interstate 80, near Sacramento, CA, to solicit information from drivers of Class 8 trucks. The survey was verbally administered to the drivers, with driver responses filled in by researchers. In order to reduce memory bias, researchers asked drivers to recall behavior in the last 24 hours, as opposed to referring to a "typical day". The response rate was about ten percent. Twenty-nine surveys were completed in one workday.

The small sample size led to high variability, as well as, a higher chance for biases. The single geographic location for one day jeopardizes the generalizability of the data to drivers in other geographic and climate conditions. Because the survey was only used for general guidance in association with all other available data, these problems are not thought to be substantial. The preliminary analysis from a recent nationwide survey appears consistent with the pilot survey results presented here.

The resulting data indicated disparate truck operating patterns as was expected. For example, 17% of drivers in our survey reported that they rarely idled their engines when resting, while another 17% idled their engines over 10 hours that day. The survey average was approximately 5 hours idling per truck per day, which is consistent with the literature.

Table 1 presents the idling data obtained from the literature and the pilot study. It is important to note that none of the available studies offers a rigorous statistical collection of data that can inform conclusively about line-haul trucking in the US for this study. Some are based on industry estimates that may not be generalizable for different fleet sizes and independent owner-operators. One paper is based on average time slept in truck cabins, regardless of how often engine is at idle or accessories are in use. The study involving datalogging does not offer adequate distinction between short and long term idling. In the absence of comprehensive data, final results for this analysis (e.g., diesel savings) are reported as a range that reflects the variation in idling time reported by truck drivers (~0-12 hours/day).

TABLE 1 Idling Estimates for HD Trucks from Available Studies

Study	Estimated average idling duration [a]		Comments
	Hours per day	Percent of engine run-time	
TMC, 1995 [10]	6	40	Estimation used in calculations
Stodolsky et al, 2000 [2] (basecase)	6	40	Informal estimates from fleets (Given here is the "base case" for driver with 10 hrs/day in 85 winter days, 4.5 hours/day for 218 days)
Webasto, 2001 [12]	5	36	Based on estimated average sleeping time in truck, *not* actual time with engine idling
Maldonado, 2002 [11]	6.5	42	Datalogs of 84 trucks over 1600 total hours in California fleets, without distinction between nondiscretionary and avoidable resting idling
Pilot Survey	5.0	35	Small sample (n=29) of Class 8 tractor-trailers in northern California
"Typical"	6	38	Assumed "typical" line-haul HD truck driver for this analysis

[a] unless otherwise stated in study, 9 hours driving per day is assumed

Accessory Use and Duration

Next, we determined the duration and power required for truck cabin accessory use during idle. Our survey revealed the likelihood of drivers to have various accessories (Table 2). This is compared to our results from the our previous survey of truck drivers in California [13]. The results show that the two driver samples appear to have similarly equipped cabins.

TABLE 2 Accessories in truck cabins, reported in two pilot surveys

Accessory	Percentage of trucks with the following accessories	
	2001 Pilot Survey[a]	2002 Mini-Pilot Survey
Stereo	96%	86%
TV	60%	21%
Computer	35%[b]	28%[c]
CB radio	90%	86%
lamp (built-in)	84%	66%
AC light bulb	N/A	41%
Refrigerator	52%	48%
Coffee maker	14%	7%
Microwave	12%	10%
A/C powered by engine	92%	93%
Electric A/C	7%	0%
Heat from engine	94%	N/A
Heat from other source	2%	N/A
Stove using battery	9%	N/A
Stove using other source	3%	N/A
VCR	9%	N/A
Cell phones	N/A	28%
"Other"	5%	N/A
Power-take-off	13%	N/A

[a] from Brodrick et al., 2001 [13]
[b] no distinction between PC and dash-readout/company computer was made
[c] dash-readout/company computer percentage is given; 10.3% of trucks had personal computers

Accessory Power Requirements

Next, the power draw for each of these accessories was estimated. Existing files in ADVISOR utilize the duty cycles in SAE standards report J1343 [14] on accessory power requirement, and standard auxiliaries like the engine cooling fan, the air brakes, and the alternator are included. However, many of the stationary aftermarket accessories (TV, microwave, etc.) are not included. The missing data were derived from several sources, including some field measurements of voltage and current from idling trucks at a nearby truck dealership and a review of widely available appliances. like televisions, coffee makers, and light bulbs.

Table 3 shows the power draws assigned for the accessories on the idling truck. Several clarifications must be emphasized. The power of two of the devices, cabin air conditioning and the engine cooling fan, vary with engine speed. The current built-in ADVISOR files for cabin air conditioning dictate variable shaft energy, from about 1.3 to 3.0 kW at 600 to 1200 rpm. Similarly, the engine cooling fan draws between 700 W and 3 kW over the multiple idling speed ranges. The fan run time is likewise variable, and we assume that it runs about 40% of idle time. The rest of the loads for the idling case are electrical and are either "on" with the given power draw or "off" with zero electric load.

Electrical power for all accessories is shown in Table 3. These are the loads "seen" at the accessory and do not account for the alternator efficiency losses. For example, the ADVISOR model has look-up tables to reflect that alternator operation varies with electric power delivered and engine speed. When idling the main diesel engine, heating the cabin draws excess engine heat in addition to drawing up to 300 W of electrical energy to power the fans that transport this heated air into the cabin. In total, the base case (i.e. idling) truck had an estimated average power draw of about 2.1 kW for stationary power at idle.

TABLE 3 Estimations for Key Characteristics for Average Truck Idling

Accessory used during idle time	Fraction of idle time	Estimated power for idling base case (W)	Estimated power for fuel cell APU case (W)
Stereo (dashboard)	0.31	30	30
CB radio	0.39	10	10
Television	0.05	300	300
Dash-read/company comp.	0.19	50	50
Personal computer	0.01	50	50
Microwave	0.01	1200	1200
Refrigerator/Electric Cooler	0.26	300	300
Overhead lamp (built-in DC)	0.15	30	30
Light Bulb (AC)	0.04	60	60
Coffee maker	0.01	900	900
Electric Blanket(Other)	0.06	100	100
Cell Phone(Other)	0.32	10	10
Cabin air conditioning	0.32	2100*	1700
Cabin heating	0.32	300	2400
Engine cooling fan	0.40	1800*	0

*These are taken from ADVISOR model, for an engine speed = 850 rpm. In reality and in the model these power magnitudes vary with idling rpm

199

The key differences in hotel loads between the idling base case and the APU case are in the power requirements for heating and cooling in the cabin (See Figure 1). The heating power requirement for the APU case will differ markedly from the idling engine because there would be no engine waste heat available. For air conditioning, the APU electric system will have a different power requirement than the idling belt-driven system. For heating and cooling in the APU case, we assume the specifications of off-the-shelf electric heat pump technology. We used specifications for a 115 V AC heat pump: 10,000 Btu/hr for cooling from 14.8 Amps (~1.7 KW) of electricity and 6,825 Btu/h for heating from 20.3 amps (~2.4 kW). An initial spike of about 4.4 kW is required at startup [15].

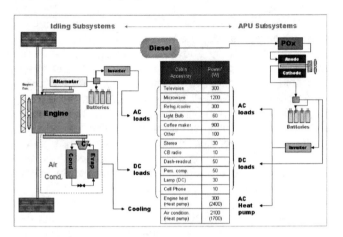

FIGURE 1 Systems Diagram for Idling and APU Subsystems

Cycle Generation

The procedure used to transform the data in Tables 1, 2 and 3 into a power vs. time cycle is similar to that utilized by ADVISOR. Accessories shift on and off, simulating real-world user operation. As in ADVISOR, accessories cycle on at full-power and off at zero power according to a user-determined frequency and duration. Hence, multiple appliances may draw power at any given time, and the combination of appliances changes over time.

The exception to the on-off operation of the other loads is the air conditioning and engine cooling fan systems whose power level is variable as a function of engine speed (rpm), as mentioned above. Engine rpm varies substantially in the field [13] and has a profound effect on idling fuel consumption [16,17]. Data from the Brodrick et al [13] pilot survey revealed that drivers generally set higher engine speeds for accessory loads. To account for this variation, we ultimately ran the model under a realistic distribution of engine speeds.

Figure 2 shows the created power-time traces for the model. The reported percentages of idle time that accessories are engaged (Table 3) are incorporated into the cycle. For example, the television is on for 5% of the

time, and the air conditioning is one for about one-third of the time. Inconsistencies, such as running a heater and air conditioning simultaneously, are prohibited. The scenario depicted is a desert condition with air conditioning, running in the first portion of the cycle, and the heat pump running on the later portion of the same cycle.

The "Estimated power for idling base case" in Figure 2 reflects the power idling diesel engine is expected to draw. This trace is the sum of the electrical and mechanical loads. As discussed above, this cycle changes with engine speed, due to rpm-variable accessories like air conditioning.

The "Estimated power for fuel cell APU case" reflects what power the fuel cell APU is expected draw. For this trace, the heat pump values have been divided by 0.85, the assumed DC-AC inverter efficiency, in order to determine the amount of net power requested from the fuel cell. The average power of this "typical" fuel cell APU cycle shown in Figure 2 is about 1.8 kW. A second "high load" APU cycle (not shown) was created to represent higher loading situations (e.g. long periods of desert or frigid weather, or more powerful accessories) by scaling up the entire power trace. This cycle was created to further test the fuel cell system in light of the prevailing uncertainties and will be further discussed in the "results" section of the paper.

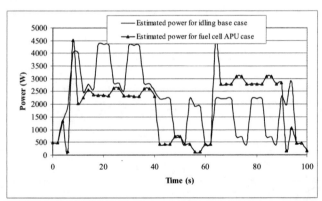

FIGURE 2 "Typical" Estimated Accessory Power Profiles for Idling Scenario (Electrical and Mechanical) and APU Electric Load over Duration of Stationary Period

Diesel Engine Fuel Consumption and Emissions

The default ADVISOR engine maps did not adequately characterize both fuel and emissions at the low speed and low torque regions typical of idling. New maps were added. The ADVISOR heavy-duty truck model is equipped with a 330-kW engine component with a fuel consumption map, but excludes the accompanying emissions data. As a result, available maps from a 1999 engine with a peak power of 209 kW were scaled up to be utilized in the model.

As is normally the case, there are data gaps in the maps, particularly for low torque and rpm values of engine operation – those points that are especially relevant for our idling situations. As a result, a variety of methods were employed to approximate emissions and fuel rates in these areas, and calibration was done to ensure a baseline performance that is comparable to existing empirical data. In general the fuel mass flow rate of an engine decreases linearly with torque for a given rpm, and as a result, extrapolating the grams/second maps to lower torque values was deemed to be the most logical and accurate method. Other extrapolating techniques, including keeping grams/kWh and grams/second constant with engine rpm and grams/second linear with torque at constant rpm, were also explored. The engine maps for fuel consumption, before and after extrapolating, are shown in Figure 3.

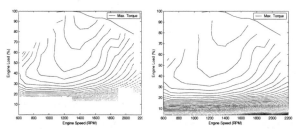

FIGURE 3 Engine Fuel Maps - Before and After Extrapolating for Missing Data

Baseline Fuel Consumption

Although there is not comprehensive statistical data on how line-haul trucks consume diesel fuel at idle, the key variables are known. Those key variables are engine speed (rpm) and accessory brake horsepower (bhp). The American Trucking Associations' Truck Maintenance Council estimates the relationship between RPM, bhp, and gallons of diesel consumed per hour, shown in Figure 4 [10]. These curves are similar to current truck models, but may be too high because they are based on pre-1995 truck data. As a result, available data on newer trucks was used to validate and calibrate the fuel consumption results of our model.

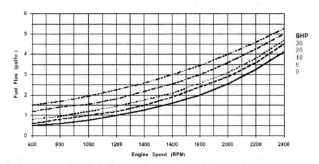

FIGURE 4. Estimated fuel consumption with varying engine speed (rpm) accessory loading (bhp) (TMC, 1995)

Much of the testing done on idling engines intentionally involves disengaging all nonessential accessories for reasons of replicablility and consistency. There are only two known sources that specifically tested idling engines over a range of engine speeds and accessory loadings (including "hotel loads"). Looking at data from the Irick et al [17] and Brodrick et al [16], strong positive correlations emerge between idling fuel consumption (in gallons of diesel per hour) and engine rpm and accessory loading. As seen in Figure 5, there are near linear increases (R^2 values of 0.72 to 0.98 for each accessory loading line) of fuel consumption with rpm, with higher trendlines for increased accessory loading.

Along with empirical data points, ADVISOR outputs for idling fuel consumption are shown on Figure 5. These datavalidate that our baseline model for fuel consumption is accurate in capturing the correct relationship for fuel consumption rate with respect to engine speed. The model was than calibrated by adjusting the points in the engine data maps in the extrapolated regions for low torques. In a previous survey, respondents reported engine speed during idling for accessory use [13]. The typical range of 600 to 1200 rpm corresponds to 0.53 to 1.25 gallons of diesel per hour.

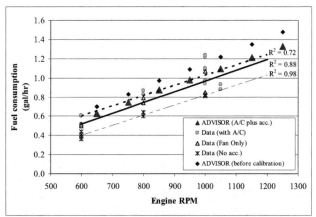

FIGURE 5 Fuel Consumption at Idle: ADVISOR Results with Empirical Data (fitted to linear regression lines)

Baseline Emissions

Several impediments lie in the way of developing an accurate, interactive emissions modeling capability for varying idling situations. There are some studies that suggest that engine speed and accessory loading may play a large role in emissions as for fuel consumption [16,17,18]. However, the bulk of emissions testing has been conducted without these conditions and ample data was not available to show well defined relationships such as those in Figure 5. Steady-state engine emissions maps (and especially ones with very low torque points) are very difficult to obtain. As a result, analysis of idling emissions will be addressed in later model iterations.

APU Module

ITS-Davis altered ADVISOR's engine/accessory control strategy to accommodate APU operation by adding an APU 'block'. The APU block checks the state of the engine (on/off), takes the stationary accessory load as an input, uses fuel cell performance, and outputs the power achieved, fuel consumed, and emissions produced during the process. The model has been created to be flexible enough to use several different types of APUs, including a fuel cell, diesel generator, or large battery pack, as well as predict their performance over any combined driving/idling cycle. The case of a SOFC APU is used here for illustration purposes.

Control Strategy

Another key issue is how the APU and the existing electrical system (primarily the battery) interact, or when each one supplies power to the accessories.

Characteristics of the strategy include–
- APU is off when vehicle is driving, on when vehicle is at rest.
- When vehicle is at rest, the APU-battery system acts out the following subroutine:
  - APU delivers all accessory power up to its peak
  - When APU cannot deliver all required power, the battery delivers the difference.
  - If the battery state-of-charge (SOC) is below its initial state of charge, and the accessory loading is below the APU peak power, the APU increases its output to charge the battery. The excess power from the APU is proportional to the difference of initial SOC and current SOC.

The control strategy characteristics, particularly with respect to start-up time, of the SOFC system may require modification from this simple strategy. Even if there is a long start-up, it may be a relatively small problem for this application. It may be reasonable to assume drivers, knowing roughly when they will take their rest, could press the "warm-up" button a half-hour in advance. For this start-up period, a certain minimum power draw would be required. In light of uncertainties about the still developing SOFC technology and how such issues will ultimately be resolved, the strategy above was retained. Similarly, modifications to the control strategy may be required to more properly address the discharge and cycling of the batteries. The chosen control strategy has the ability to utilize the battery to buffer on occasions where the APU cannot supply power to the demanded load. Though this adds flexibility for the fuel cell, issues with respect to the batteries' state of charge, cycling characteristics, and lifetime under a variety of different power cycles warrant more thorough investigation than provided here.

Fuel cell performance map

Our ADVISOR APU model inputs fuel cell performance data from TIAX, LLC for a diesel-fueled solid oxide fuel cell (SOFC) system with a partial oxidation (POx) reformer [19]. The fuel cell system is represented by an efficiency vs. load relationship. As shown in Figure 6, this relationship accounts for efficiency losses due to parasitic loads, heat loss, and auxiliary subsystems (reformer, air compressor, etc.), as well as the stack efficiency. The bottommost line for overall system efficiency is used to calculate, for a given fuel cell electric power draw, the amount of diesel fuel that is consumed.

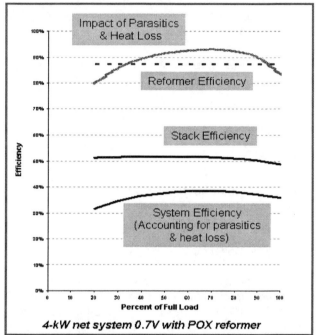

4-kW net system 0.7V with POX reformer

FIGURE 6 Fuel Efficiency Curves for 4-kW SOFC System with POx Reformer (from [19])

It must be noted here that there are many uncertainties about the actual performance characteristics of future solid oxide fuel cells and their accompanying reformers. For example, the reformer characteristics of a SOFC system may be a key factor in determining the ability of the SOFC system to react quickly to transient changes in demand, but to what extent is relatively unknown. Perhaps a slightly more realistic model would limit the rate of change of power output of the SOFC unit; however, compensation by the battery pack during these times could make the effect of this change on system performance and efficiency a small one.

RESULTS

The first step in determining the fuel consumption of the base case idling engine and the APU systems was to determine the optimal fuel cell size. A parametric study was conducted to find the optimal fuel cell size in terms of energy use. The model was run seven times with a fuel cell to supply power for the "typical" accessory power cycle and fuel cell peak power of 2 through 8 kW. Optimal fuel consumption for this "typical" cycle was found to be a 4-kW (net) peak power SOFC system. However, the model can easily be run with other battery-fuel cell combinations. For example, the smallest and least expensive fuel cell that could provide the average power requirement (1.8 kW) would subject the battery to frequent battery discharge and recharging. If instead the fuel cell is sized to peak follow perfectly (up to 4.4 kW), the battery is never discharged but the fuel cell stack operates in a lower efficiency mode and its initial cost increases. The design selection will depend on the relative cost, lifetimes, and cycling capabilities of the fuel cells and battery.

The key results presented here are the modeled fuel consumption of the two systems (main diesel engine idling and 4-kW SOFC APU) and an estimation of total diesel savings over total truck duty cycle (i.e. drive cycle plus idling). Figure 7 depicts the second-by-second diesel fuel consumption over for the "typical" stationary truck cycle ("Estimated power for idling base case" in Figure 2) from the modified ADVISOR model. The engine speed at idle has a much larger effect on fuel consumption than does toggling on and off individual accessories. The average diesel consumption rates for the 600, 900, and 1200 rpm model runs are 0.53, 0.95, and 1.25 gal/hr, respectively.

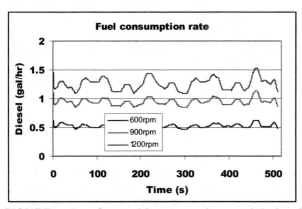

FIGURE 7 Second-by-second modeled fuel consumption of idling base case for "typical" stationary cycle at different engine speeds

Similarly, second-by-second traces of fuel consumption for the 4-kW SOFC APU system are shown in Figure 8. Here, two cycles, the "typical" and a "high" load, were used. This high load was introduced because the electric load has much more effect on fuel consumption in this case, and the exact specifications of the necessary electric air conditioners and inverters for the APU system involve more uncertainty. The higher load is a scaled-up version of the "typical" load, with an average of 2.7 kW (compared with 1.8 kW) and maximum of 6.5 kW (compared with 4.4 kW). While the "typical" load required an average of 0.14 gal/hr, the higher load required 0.17 gal/hr. The average SOFC system efficiency over the cycles was about 33%.

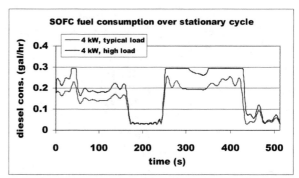

FIGURE 8 Second-by-second modeled fuel consumption of 4-kW SOFC APU for the "typical" and "high" cycles

The two systems are compared in Table 4. Considering that high engine speeds are normally set for higher accessory loads, the higher load on the APU is used to compare against the results for the 1200 RPM idling condition. Taking into account the high degree of uncertainty for the SOFC efficiency, ±20% efficiency bars were used for the system efficiency of Figure 6, and these results shown in parentheses in Table 4. According to the model results, drivers who idle at low engine speeds may expect 70-81% savings in fuel used during idle, while those at a higher rpm may save 83-89% of fuel used during idle.

TABLE 4 Comparison of Idling and SOFC APU systems to provide power for stationary power loads

Accessory load (engine speed)	Fuel Consumption (gal/hr)		Percent of Idled Fuel Saved with SOFC APU
	Idling engine	SOFC APU (in lieu of idling)	
"typical" (600 rpm)	0.53	0.14 (0.10 – 0.16)	74% (70-81%)
"typical" (900 rpm)	0.95	0.14 (0.10 – 0.16)	85% (83-89%)
"high" (1200 rpm)	1.25	0.17 (0.14 – 0.21)	86% (83-89%)

Typical cycle: average 2 kW, max 3.7 kW; High load: average 2.7 kW, max 4.7 kW; ( )'s denote 20% SOFC error bars in system efficiency curve

Assuming 300 annual days in service per truck, the annual potential diesel savings were calculated and are shown in Figure 9. The figure shows large differences in potential diesel savings for trucks that are idled at various engine speeds (from 600 to 1200 rpm) and different daily idle durations (from 0-12 hours per day). For example, a typical truck that is idled 6 hours per day at 900 rpm is estimated to save about 1300 gallons per year with a SOFC APU, whereas a heavier idler (1200 rpm, 10 hrs/day idling) is estimated to save about 3000 gallons per year.

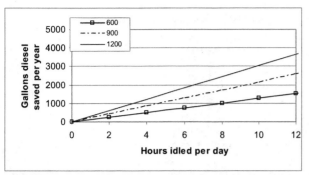

FIGURE 9 Annual diesel saved replacing SOFC for engine at idle for given engine idle speed (rpm)

In estimating the percentage of potential fuel savings at idle compared to the total consumed diesel (including driving), 9.1 hour/day and 300 days/yr driving on the U.S. highway cycle are assumed. This equates to about 110,000 miles driven per year on the highway driving cycle. Shown in Figure 10, the typical driver (~6 hrs idle/day, 900 rpm) could save about 5 percent of total fuel consumption. Heavier idlers (>10 hrs idled/day), which could represent 15% of line-haul trucks, could save 5-12% with a SOFC APU in place of idling.

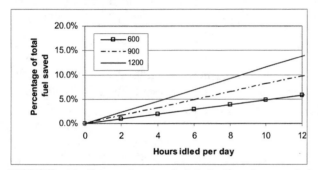

FIGURE 10 Percentage of total diesel saved by replacing SOFC for engine at idle for given engine idle speed (rpm)

CONCLUSION

This research work, characterizing existing data on truck driver behavior, utilizing a modified ADVISOR vehicle platform, and comparing heavy-duty truck idling to fuel cell APU operation, allows for a number of key conclusions.

- The oft-reported industry average of one gallon per hour diesel consumption at idle appears to be accurate. Using our model, engine speeds of 600 to 1200 rpm yielded approximately linear fuel consumption from 0.53 to 1.25 gal/hr. The average survey-reported idle engine speed of 900 rpm corresponded to 0.95 gal/hr.
- Using several criteria, the optimal size of a fuel cell was determined to be 4-kW (net) rated peak power. This decision consisted of an estimation of the average power required for sleeper cabs, an estimation of the demands for harsher climates, a minimization of fuel consumption, and an attempt to

sparingly utilize the battery to peak-shave. In particular, a more comprehensive understanding of driver accessories from survey data and a more thorough look at battery discharge could further refine this choice for fuel cell rated power.

- The modeled 4-kW (net) solid oxide fuel cell APU, operating over varying sleeper cab accessory power cycles, averages about 33% efficiency (LHV to net fuel cell power), consumes 0.14-0.17 diesel gal/hr, and reduces diesel consumption 74-86% compared to baseline idling for sleeper cabs while providing the same in-cab services (appliances, heating, cooling, etc.).
- Potential total fuel savings resulting from the replacement of avoidable idling with the SOFC system for a "typical" line-haul driver are estimated at 3-8% of total vehicle energy use. For heavy-idling trucks (~10 hours idling per day), fuel consumption could be reduced by 5-12%.
- Uncertainties in the vehicle and engine data exist. Fuel and emissions engine maps with more detail at low torque and RPM would eliminate the need for extrapolating into these regions. Vehicle fuel and emissions testing of tractors with variation of engine idle speed and accessory power required at idle for different model year tractors would allow for a better picture of how fuel and emissions truly vary in the fleet.
- Uncertainties in driver behavior characteristics exist. Variables and their distributions here are estimated primarily based on small surveys in northern California. The extent to which the analyses here are valid nationwide is not clear, and a well-crafted nationwide survey with input from individual drivers and from fleet managers could minimize uncertainties of this report.

ACKNOWLEDGEMENTS

We are grateful to the U.S. Department of Energy and the California Air Resources Board for their funding of our research. The National Science Foundation's Integrative Graduate Education and Research Traineeship (IGERT) program provided student researcher support. Fellow researchers at TIAX LLC, Drs. Jan Thijssen and Masha Stratanova, were important collaborators.

REFERENCES

1. U.S. Department of Commerce. *Vehicle Inventory & Use Survey.* CD-ROM CD-EC97-VIUS. February 2000.

2. Stodolsky, F., L. Gaines, and A. Vyas. *Analysis of Technology Options to Reduce the Fuel Consumption of Idling Trucks.* Argonne National Laboratory, Argonne, IL, ANL/ESD-43. 2000.

3. Levinson, T. "Anti-Idling Laws and Regulations." The Maintenance Council Annual Meeting. Nashville, TN. March 15, 2001.

4.   Texas Administrative Code. *Control Requirements for Motor Vehicle Idling, Rule §114.502.* January 18, 2001.

5.   National Energy Policy Development Group. *National Energy Policy: Reliable, Affordable, and Environmentally Sound Energy for America's Future.* U.S. Office of the Vice President, Washington, DC. 2001

6.   IdleAire. "Idleaire Deployment." http://www.idleaire.com/about/deployment. July 1, 2003

7.   Jones, Ruth. "Powering Up for Comfort." *Landline Magazine.* February 1999.

8.   Van den Berg, A. Joseph. "Truckstop Electrification: Reducing CO_2 Emissions From Mobile Sources While They Are Stationary." *Energy Conversion Management.* Vol. 37, Nos. 6-8, pp. 879-884. 1996.

9.   National Renewable Energy Laboratory. Center for Transportation Technologies and Systems, 2002. ADVISOR Download site http://www.ctts.nrel.gov/analysis/download.html. U.S. Department of Energy. 2002.

10. Truck Maintenance Council. *Analysis of Costs from Idling and Parasitic Devices for Heavy Duty Trucks*, Truck Maintenance Council Recommended Procedure 1108, American Trucking Associations, Alexandria, Va., March 1995.

11. Maldonado, Hector. "Development of Heavy Duty Truck Chassis Dynamometer Driving Cycles for Source Testing for Emissions Modeling." California Air Resources Board. 12th CRC On-Road Vehicle Emissions Workshop. April 15-17, 2002.

12. Webasto. "Essential Power Systems Workshop: Cab/Engine Heaters." December 12, 2002. http://www.trucks.doe.gov/pdfs/V/108.pdf. Accessed August 6, 2002.

13. Brodrick, C. J., N.P. Lutsey, Q.A. Keen, D.I. Rubins, J.P. Wallace, H.A. Dwyer, D. Sperling, D., and S.W. Gouse III. "Truck Idling Trends: Results of a Northern California Pilot Study." *Society of Automotive Engineers Technical Paper Series.* 2001-01-2828.

14. Society for Automotive Engineers. "Information Relating to Duty Cycles and Average Power Requirements of Truck and Bus Engine Accessories." SAE J1343. August 2000.

15. Allen, K.. Personal communication. Engineering Project Manager, Refrigeration and Air-Cooled Products, Taylor Made Environmental, Inc. Discussion with David Grupp, July 18, 2002.

16. Brodrick, C.J., M. Farshchi, H.A. Dwyer, D.B. Harris, F.G. King, Jr. "Gaseous Emissions from Idling of Heavy-Duty Diesel Truck Engines" *Journal of the Air & Waste Management Association.* **52**: 174-185. September 2002.

17. Irick, D.K., B. Wilson, D.C. Lambert, M. Vojtisek-Lom, P.J. Wilson. "Emissions and Fuel Consumption of HDDVs During Extended Idle." (unpublished). 2002.

18. Gautum, M., N.N. Clark, W.S. Wayne, G. Thompson, D.W. Lyons, W.C. Riddle, and R.D. Nine. "Qualification of the Heavy Heavy-Duty Diesel Truck Schedule and Development of Test Procedures." Final Report, CRC Project No. E-55-2. West Virginia University Research Corporation. Prepared for California Environmental Protection Agency, Air Resources Board and Coordinating Reasearch Council, Inc. March 2002.

19. Stratanova, M. and J. Thijssen. *TIAX SOFC Fuel Cell Model.* TIAX, LLC. Cambridge, MA. 2002.

CONTACT

Nic Lutsey (corr. author) nplutsey@ucdavis.edu

III. SYSTEMS DESIGN AND EVALUATION

Vehicle Development and Test

2003-01-0422

Cold Start Fuel Economy and Power Limitations for a PEM Fuel Cell Vehicle

Stephen D. Gurski and Douglas J. Nelson
Mechanical Engineering, Virginia Tech

ABSTRACT

Fuel cells are being considered for transportation primarily because they have the ability to increase vehicle energy efficiency and significantly reduce or eliminate tailpipe emissions. A proton exchange membrane fuel cell is an electrochemical device for which the operational characteristics depend heavily upon temperature. Thus, it is important to know how the thermal design of the system affects the performance and efficiency of a fuel cell vehicle. More specifically, this work addresses issues of the initial thermal transient known to the automotive community as "cold start" effects for a direct hydrogen fuel cell system. Cold start effects play a significant role in power limitations in a fuel cell vehicle, and may require hybridization (batteries) to supplement available power. The results include a comparison of cold-start and hot-start fuel cell power, efficiency and fuel economy for a hybrid fuel cell vehicle.

Fuel cell system design can significantly affect the cold start performance of a fuel cell system. Through modeling, it is possible to quantify the impact of thermal mass on warm up time to operating temperature of a fuel cell system. As expected, performance reduction is seen during cold start that affects both available power and fuel use. The overall cold start energy use penalty is relatively small (~ 5% difference) for the combination of component sizes and control strategy presented here.

INTRODUCTION

Fuel Cell System Model

Understanding the complex interaction between the thermal fluid systems of a fuel cell is needed to quantify the impact that cold start has on vehicle efficiency and performance. A vehicle fuel cell system developed at Virginia Tech is used as a practical example to follow for the system model. The Virginia Tech 50 kW fuel cell system uses a design approach where practicality, simplicity and safety were key requirements. A diagram of the system, shown below in Figure 1, identifies the major system components that are in the physical fuel cell system, and are accounted for in the fuel cell system model.

A fuel cell system can be broken down into three major sub-systems; air supply, coolant loop, and fuel supply. Energy and mass balances for each component are used to model the system. The details and equations used are available in Gurski (2002).

MODEL CAPABILITIES AND OPERATIONAL STRATEGY

To make a fuel cell system useful in a vehicle system, some amount of control over the system is necessary to implement requests from the operating strategy. With laboratory development, the fuel cell system can be characterized and a controller used to change inputs to the system to achieve the desired power generation. However, for this model, a single input iterative approach was chosen for simplicity. In the actual vehicle development, the fuel cell control strategy was based upon a current request to the fuel cell stack. This allows the air compressor speed and humidification for the fuel and air to be set to the appropriate levels.

Safe and proper operation of the fuel cell has constraints such as maximum power and minimum cell voltage; these parameters are usually set by the fuel cell manufacturer and are included in the model. In an effort to increase system efficiency, previous work done by Kulp and Nelson (2001) suggests that a minimum power request be placed upon the system. As part of the implementation into a vehicle simulation tool called ADVISOR, the model needs to handle a net power request. To account for the safe operation requirements and minimum system power, the model uses a goal seeking function to determine a system operating point, given a net power request. The model takes into account the following system electrical parasitics to determine a net power operating point: air compressor, radiator fan, condenser fan and coolant pump.

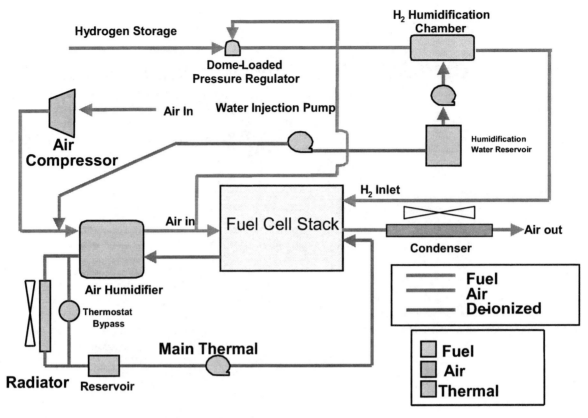

Figure 1. Fuel Cell System Components

The model does the following at each time step:

1. Net power request from vehicle control strategy
2. Guess current density that generates a gross power to meet net power
3. Determine net power operating point
4. Enforce one of the three following limits, through iteration
   A. Generate the net power request, if possible
   B. Impose either a minimum cell voltage or a maximum current density limit that results in limited power output
   C. Generate no power, per minimum power limits from control strategy

The goal of this work is to understand the transient thermal effects that result from limits placed upon fuel cell stack and vehicle operation. Requirement B above can occur for the following reasons:

1. The net power request is much higher than the fuel cell system peak power capability (due to stack sizing and hybridization design).
2. The fuel cell is cold which reduces the stack output voltage
3. Low oxygen content due to air flow and pressurization from the air compressor control

FUEL SYSTEM

A pressurized hydrogen gas storage and delivery system is utilized on-board the vehicle. Hydrogen is humidified and any unused excess is recirculated back into the stack inlet to be reused. Since the hydrogen and water vapor in hydrogen carry comparatively little energy, they have been neglected in the thermal model. From experience, proper design of a direct hydrogen fuel system achieves flow and pressure response that do not limit the fuel cell system response. A dome loaded pressure regulator in the vehicle fuel cell system matches the pressure of the hydrogen to the pressure of the air system. Modeling the dome loaded regulator required the assumption that the hydrogen flow and pressure could be met at all times, and the system model did not utilize recirculation, but rather "dead headed" the fuel cell stacks (the same net effect of recirculation). Purging of hydrogen from the fuel system is not included in the model for the current results.

AIR SYSTEM

Air System - Air Compressor

The air compressor in the system is based on an Opcon 1050 twin screw compressor with an internal compression ratio of 1.44. Empirical data in a 2D lookup table yields the volumetric efficiency, temperature rise and adiabatic efficiency of the compressor as a function of mass flow and outlet pressure. In the vehicle fuel cell system, the compressor works against a fixed orifice that increases the fuel cell operating pressure with an increase in air mass flow. The model control strategy has the ability to change the cathode operating pressure with respect to fuel cell stack current density and fuel cell temperature.

Air System – Humidifier

A liquid-to-air heat exchanger based on an automotive intercooler is utilized as a humidifier in the pressurized air stream. Water is directly injected into the air inlet in the humidifier, and the heat required for vaporization is obtained from the fuel cell coolant running in adjacent flow channels. A simple energy balance between the coolant and the air with humidity water injection is performed. Each of the temperatures and mass flows are obtained from the current operating conditions.

Air System - Condenser

Water balance (water used to humidify inlet air versus water collected from exhaust air) is an issue in practical use and operation of fuel cells for transportation. A simplified air-air heat exchanger with condensation model has been implemented into the system to evaluate water balance. A curve fit based upon empirical data (Kroger, 1984) relates exterior air mass flow to heat transfer capability. The pressure drop across the air side of the condenser is a function of air mass flow, which allows the calculation of power required for the condenser fan. Fan work associated with the condenser is accounted for in the net power calculation of the system.

THERMAL SYSTEM

Thermal System - Coolant Reservoir

In the vehicle system, the pump draws directly from the coolant reservoir. In the model, the coolant reservoir is responsible for the heat lost to the ambient from all of the plumbing, and the lumped capacitance of all the coolant in the system. A backward looking finite difference method and a simple mixed tank model is used to account for the thermal transient of the coolant in the reservoir.

Thermal System - Radiator

The radiator in the system is modeled using a technique similar to that of the condenser in the air system. An energy balance between the coolant and the heat removal capacity of the air flow across the radiator yields the outlet coolant temperature.

The pressure drop across the air side of the radiator is a function of air mass flow, which allows the calculation of power required for the radiator fan. A simple thermostat is part of the radiator model that will not allow heat to be rejected form the radiator below the desired fuel cell operating temperature.

FUEL CELL STACK

The most complex device in the system model is the fuel cell stack. For air reactant flows that are not saturated at the cathode inlet, the system may generate water in vapor form inside the stack. Since the lower heating value of hydrogen is used in the energy balance, the heat generation term assumes that all water product is in vapor form. The water that is generated after the cathode stream is saturated is condensed in the stack, and gives an addition internal heat load term for the stack.

In the model, three streams flow into and two flow out of the stack; the hydrogen outlet has been "dead headed". To make the system model simpler, the water vapor in air and hydrogen are separated in to flows into the fuel cell stack. Now there are five flows streams in (air, water vapor in air, hydrogen, water vapor in hydrogen, coolant) and three flows out (air, water vapor in air, coolant). All the water vapor in the hydrogen stream is assumed to diffuse through the membrane and exit via the cathode. Making the assumption that a fuel cell stack is an excellent heat and mass exchanger, all of the outlet flows and the stack thermal mass are at the same temperature.

Again, a backward looking finite difference model using a lumped capacitance is used to evaluate the thermal transients in the stack temperature.

OPERATING CONDITIONS

Operating conditions of the fuel cell stack are very important to the performance and thermal response of the system. Listed below are the pertinent base operating conditions for the fuel cell model.

FUEL CELL STACK	
Anode inlet humidity	80% Rh
Cathode inlet humidity	60% Rh
Min cell voltage	0.6 V/cell
Max coolant inlet temperature	80 deg C
Max coolant temp. rise in stack	10 deg C
System Operating Conditions	
Min system power	5000 W
Max fuel cell power	50 kW Net

CHARACTERIZING THE SYSTEM

Before exercising the model in a dynamic vehicle environment, some tests have been performed that describe the system during steady state and simple transients. The first test performed is a steady state characterization of the system parasitic power used to run the fuel cell stack. In Figure 2 below, the parasitic power of the system is determined over the range of net system power up to 50 kW, operating at the base conditions.

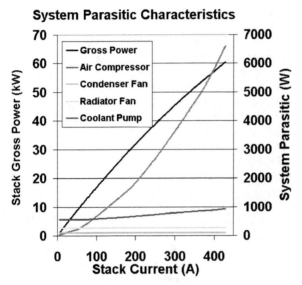

Figure 2. Fuel Cell System Parasitic Power Characteristics

This test was performed at normal operating temperature of 80 deg C. The maximum cathode pressure is 1.8 atm, which corresponds to a stack current of 430 A, or 60 kW of gross power. The minimum stack pressure is 1.05 atm. Note that system component sizing and selection, as well as operating strategy for air flow and pressure as a function of fuel cell current can strongly influence these characteristics.

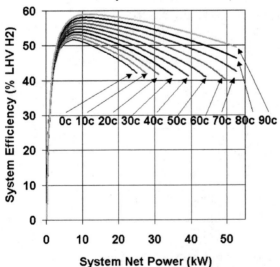

Figure 3. Fuel Cell System Power and Efficiency Variation with Stack Temperature

Characterizing system efficiency and temperature dependence is performed next. Figure 3 shows the system efficiency based on net system power output. Each of the curves represents a different operating temperature and the remaining operating conditions are the same at each power level. For example, at each of the different temperatures operating at 20 kW would have the same cathode pressure, mass flow etc. Each of the efficiency lines end when a minimum cell voltage is encountered. From this figure, the temperature effects are evident in decreasing the available system power and maximum fuel cell system efficiency. In both Figure 2 and Figure 3, the air compressor speed and power were allowed to go to zero; in practice this is difficult to achieve and still have acceptable dynamic response of the system. Figure 3 shows that a minimum power operating strategy can eliminate the low system efficiency region of operation at low loads, which also sets a minimum speed and flow for the air compressor (Kulp et al., 2002).

THERMAL TRANSIENT RESPONSE TO STEP INPUT

Since it is desirable to operate the system at its maximum available efficiency and power, characterizing the thermal transient or cold start performance is necessary. A step power input to the system while requesting the maximum system design power is shown in Figure 4.

When requesting a 50 kW load initially from the fuel cell system, there is a period where net system power is less than requested due to the low stack temperature. For this system, it takes 350 seconds for the fuel cell system to initially warm up to the steady state operating temperature. After 300 seconds, the system has the capability to produce the full rated net power. The temperature limiting effect on power seen in Figure 4 is also seen in the steady state results of Figure 3.

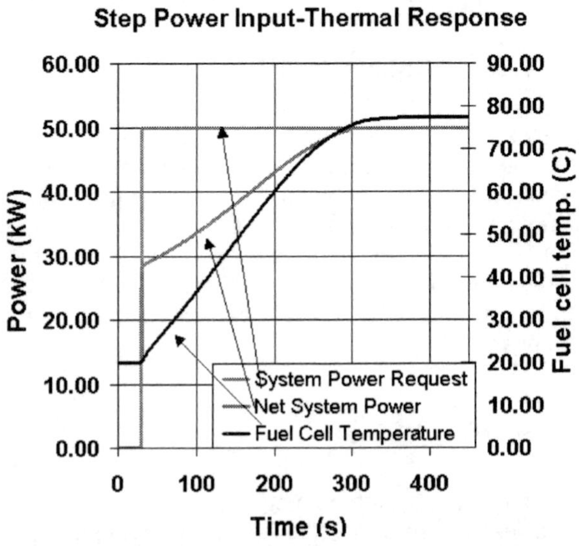

Step Power Input-Thermal Response

Figure 4. Thermal and Power Response to Step Power Request

VEHICLE MODELING AND ENERGY IMPACT

To understand how the initial cold start transient affects the performance and fuel economy of a vehicle during a drive cycle, the fuel cell system model is incorporated into a vehicle model. The vehicle model is based on a 2002 Ford Explorer that has been converted to a hybrid fuel cell vehicle. Table 1 lists the characteristics of the vehicle used for the drive cycle simulations.

Table 1. Hybrid Fuel Cell Vehicle attributes

Attribute	Value
Mass	2400 kg
Cd	.41
Fa	2.8 m^2
Drivetrain	83 kW GE EV2000
Batteries	16 Ahr 336 V Lead Acid
Fuel Cell	50 kW Net

Table 2 contains performance results of the vehicle model that were obtained using a hot fuel cell system and the performance capability of the 2WD vehicle with a single electric drive axle. Adding a second electric drive axle could improve the performance of the vehicle to levels close to a conventional sport utility vehicle (see Atwood, et al. 2001).

Table 2. Vehicle Performance

Attribute	Model
0-97 kph	18 sec
1/8 mile	15 sec
Gradeability	4.8% @ 55mph

The drive cycle that is used for the hot and cold start comparison is a standard EPA FTP cycle. Using vehicle simulation software called ADVISOR™, the fuel cell vehicle model is used to compare the energy impact for a hot and cold start. Both runs use the same control strategy, and have the same characteristics with the exception of the initial starting temperature. The initial temperature for the cold start is 20 degrees C, and the hot start temperature is 80 degrees C. Also both of the runs are battery state of charge (SOC) corrected for each drive cycle within ½%.

The results show that the cold driving cycle consumed 4.35% more fuel than the hot driving cycle. Taking a closer look at the losses in the system, Figure 5 shows that the majority of the system losses are incurred by the fuel cell system. What is difficult to see in Figure 5 is where the majority of the changes occurred. Table 3 below details how much energy is lost in the system for the duration of the drive cycles. As expected, the majority of the difference comes from the fuel cell operation.

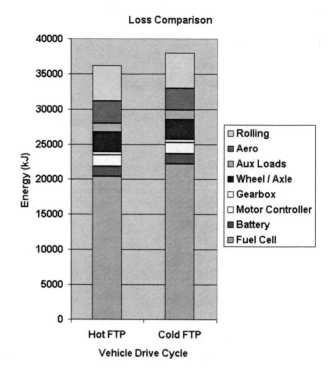

Figure 5. Hot Start and Cold Start Energy Loss Comparison

Table 3. Component Energy Loss Comparison

Component Loss	Hot FTP	Cold FTP	Difference	% of Total Loss
Fuel Cell	20355.00	22145.00	1790.00	94.46
Battery	1455.00	1508.00	53.00	2.8
Motor Controller	1617.00	1598.00	19.00	1.00
Gearbox	432.00	426.00	6.00	0.32
Wheel / Axle	2835.00	2862.00	27.00	1.42
Aux Loads	1314.00	1314.00	0	0
Aero	3130.00	3130.00	0	0
Rolling	5020.00	5020.00	0	0
Total kJ Loss	36158.00	38003.00	1895.00	

Taking a closer look at the fuel cell systems losses, Figure 6 shows that the majority of the losses are accounted for in inefficiencies associated with the fuel cell stack, specifically the energy conversion that generates electrical power.

The fuel cell stack is responsible for the majority of the difference in losses for the hot and cold drive cycles. Figure 7 compares the operating points of the fuel cell stack and the relative efficiencies. Each of the points in Figure 7 represents one second of the drive cycle. The noticeable difference between the two graphs is that the cold FTP has many more operating points occurring at lower efficiencies as the stack warms up.

213

Fuel Cell System Losses

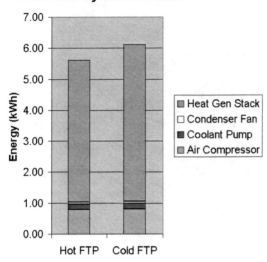

Figure 6. Fuel Cell System Loss Comparison

Thermal effects that change fuel cell stack overpotentials are the dominant cause for the reduction in efficiency and available power. Operating points below the solid line in the figure are not possible because of the minimum cell voltage limitation imposed by the control strategy. This strategy also prevents operation of the stack at lower efficiencies, and so tends to make the cold start energy penalty relatively small.

Another operating limitation of the fuel cell is also seen as a minimum power limit. The fuel cell system incorporates a minimum power level below which the fuel cell does not generate power. As stated before, a

minimum system power request can increase the overall system efficiency because the majority of the power generated is allocated to offset parasitics at very low power. This strategy does increase the amount of energy that must be processed through the battery energy storage system, and may increase battery losses as well as sizing requirements (Atwood, et al. 2001).

STANDARD EPA FUEL ECONOMY TEST

To gather an understanding of how this model vehicle compares to that of current production vehicles, the standard EPA fuel economy test has been performed. This is the fuel economy test that yields window sticker (corrected) fuel economy numbers and the published EPA fuel economy results. The test consists of two driving cycles, the FTP-75 and the HWFET.

The EPA test uses a cold start FTP-75 cycle and a hot start HWFET cycle. Fuel economy from the tests are reported here as uncorrected and combined fuel economy. The uncorrected fuel economy figures are simply the raw fuel economy numbers from the FTP-75 and HWFET simulations.

The combined fuel economy number, typically used for annual fuel cost calculations, is a weighted percentage of city and highway fuel economy. For the combined fuel economy, the city accounts for 55% of the total and the remaining 45% comes from the highway results. Table 4 contains the results in miles per gallon of gasoline equivalent energy (MPGGE) from the simulations to yield the fuel economy results for the fuel cell model and the stock vehicle.

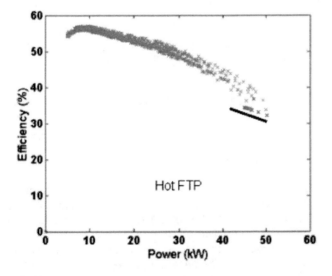

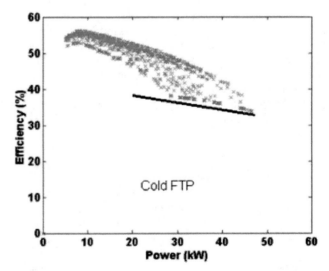

Figure 7. Comparison of Hot and Cold Efficiency Points over a FTP Cycle

Table 4. EPA fuel economy results, MPGGE

Cycle	Raw	Combined
Model: Fuel Cell Vehicle		
FTP-75	34.3	
HWFET	39.2	
Combined		36.3
Production Vehicle: Internal Combustion		
FTP-75	16.9	
HWFET	25.2	
Combined		20.6

The fuel economy results of the fuel cell vehicle are significantly better than that of the production vehicle. Using the combined results, a 77% increase in fuel economy over the production vehicle is accomplished. The performance level of the production vehicle is significantly higher, however, and this accounts for some of the difference in fuel economy.

IMPACT OF POWER LIMITING ON PERFORMANCE

Fuel cell power limiting impacts not only vehicle fuel economy, but also performance metrics such as acceleration and driveablility. To better understand the impact that temperature has on vehicle performance, a full power acceleration 0 to 97 kph (0 to 60 mph) test is performed. The test is performed with the system initially started at 20 deg C (cold) and 80 deg C (hot). At time equal 20 seconds the vehicle performs a maximum power available acceleration. In Figure 8, the results show that the cold start vehicle 0-97 kph time was 8 seconds longer than that of the hot start vehicle. This performance difference could be reduced by increasing the size and power available from the battery, but then this extra battery capacity would go unused most of the time once the vehicle is warmed up.

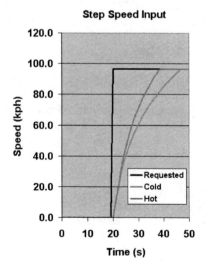

Figure 8. Hot and Cold Acceleration Performance

Figure 9 shows a more dynamic drive cycle example. A US06 drive cycle is used as a comparison because of the more dynamic acceleration, speed and overall aggressiveness relative to that of the standard EPA city and highway drive cycles. A trace miss comparison in Figure 9 shows the degree that the vehicle was unable to maintain the requested speed trace.

The trace miss is the difference between the requested speed during the cycle and the actual speed achieved; taller peaks are larger differences between the requested speed and actual speed. During the 600 second cycle, the first 200 seconds are of most interest because the vehicle had the largest and most frequent trace misses. After 200 seconds both vehicles equally fell short of the request for the drive cycle. Another point to note, during the US06 drive cycle it takes the system approximately 350 seconds to arrive at the desired system operating temperature.

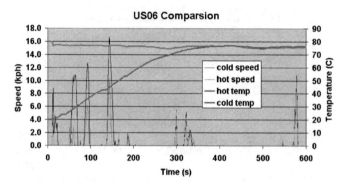

Figure 9. US06 Drive Cycle Hot and Cold

Illustrating the limiting effects of temperature on performance is the intent of evaluating the vehicle on the US06 drive cycle. On an absolute scale, the model vehicle was not able to meet the speed trace criteria of the US06 drive cycle for either hot or cold conditions due to drivetrain component sizing.

The performance limitation of this vehicle during a cold start cycle (inability to meet the speed trace) is attributed to the fuel cell system interaction with the hybridization batteries. To meet the criteria of a SOC corrected drive cycle, the model required a very low initial battery SOC (~45%). The cold start performance of the vehicle further suffers because of power limitations of the batteries at low SOC.

During a hot start US06 drive cycle the vehicle also exhibits performance limitations, although not as extreme as the cold start scenario. During this case, the speed trace miss is not the result of a reduction of performance of the fuel cell system due to temperature. The performance limit is the result of under sizing of the vehicle propulsion system with respect to the vehicle size, weight and performance requirements of the US06 drive cycle.

Several vehicle design modifications can be modeled that would result in a higher performance vehicle capable of meeting the US06 speed trace: increased electric drive power, and increased fuel cell plus battery power. Atwood et al. (2001) present results for a full performance sport utility vehicle for the hot start case.

Energy use summary of EPA and US06 Drive Cycles

Two different drive cycles were used to compare the cold and hot start fuel economy and performance of the modeled fuel cell vehicle. The FTP-75 drive cycle is used to illustrate the differences in a fuel cell vehicle and the US06 is used to compare the performance limitations. Table 5 below is a summary of fuel economy in mpgge.

Table 5. Fuel Economy Comparison of fuel cell vehicle model

	Hot Start	Cold Start
FTP – 75	36.4	34.3
US06	15.5	14.1

CONCLUSIONS

Fuel cell system characteristics and design can significantly affect the cold start performance of a fuel cell system. Through modeling, it is possible to quantify the impact of thermal mass and control strategy on warm up time to operating temperature of a fuel cell system. As expected, a performance reduction is seen during cold start that affects both available power and fuel use. The overall cold start energy use penalty is relatively small for the combination of component sizes and control strategy presented here. Changing the fuel cell system design and operating control strategy could reduce the effects of cold start. One of the challenges in fuel cell vehicle design is the fuel cell stack size necessary to achieve a power level useful in a vehicle. In this model, the majority of the thermal capacitance of the system is tied up in the fuel cell stack and bipolar plates. Industry goals include increasing the power density, which would result in a decreased stack size with subsequent reduction in thermal mass of a fuel cell system.

Future work to reduce cold start effects using this model would be to explore the operating control strategies. Use of additional fuel cell stack power that potentially could be generated may increase the system operating temperature more quickly. Active control of fuel cell stack operating pressure may also decrease the cold start transients. Electric or fuel-fired heaters could be evaluated to see if the heater energy used is offset by increased fuel cell efficiency.

Finally as seen in the performance limitations with the US06 drive cycle, fuel cell system size could be increased to decrease the cold start power limitations in the system. However, this modeling effort cannot predict the practical limitations (weight, size, cost) that may plague such a scenario.

REFERENCES

P. Atwood, S. Gurski, D. J. Nelson, and K. B. Wipke (2001), "Degree of Hybridization Modeling of a Fuel Cell Hybrid Electric Sport Utility Vehicle," SAE Paper 2001-01-0236, in *Fuel Cell Power for Transportation* 2001, SP-1589, pp. 23-30.

M. Fronk, Matthew H., David L. Wetter, David A. Masten, Transportation," SAE paper 2000-01-0373, in *Fuel Cell Power for Transportation* 2000, SAE SP 1505, pp.101 – 108.

Gurski, Stephen D. (2002), "Cold-Start Effects on Performance and Efficiency for Vehicle Fuel Cell Systems," MS Thesis, VPI&SU, Blacksburg, VA. Available on http://scholar.lib.vt.edu/theses/available/etd-12192002-162600/ .

D.G. Kroger, "Radiator Characterization and Optimization," SAE paper 840380, Vol. 93 *SAE Transactions*, 1984 pp. 2.984-2.990

G. Kulp and D.J. Nelson (2001), "A Comparison of Two Fuel Cell Air Compression Systems at Low Load", SAE paper 2001-01-2547 in Fuel Cells and Alternative Fuels/Energy Systems, SP-1635, *Proceedings of the 2001 SAE Future Transportation Technology Conference*, Costa Mesa, Ca., pp. 81-90

G. Kulp, S. Gurski, and D.J. Nelson (2002), "PEM Fuel Cell Air Management Efficiency at Part Load", *Proceedings of the 2002 Future Car Congress*, June 3-5, Arlington, VA, SAE paper 2002-01-1912, 14 pgs.

M. Sadler, R.P.G Heath and R.H. Thring "Warm Up Strategies for a Methanol Reformer Fuel Cell Vehicle," SAE paper 2000-01-0371, in *Fuel Cell Power for Transportation* 2000, SAE SP 1505, pp. 95 – 100.

ACKNOWLEDGEMENTS

We would like to thank CTTS analysis team especially Keith Wipke, Tony Markel, Kristina Harraaldson and Bill Kramer at the National Renewable Energy Laboratories for supporting this work under contract XCL-1-3116-01.

CONTACT INFORMATION

Dr. Douglas J. Nelson
Virginia Polytechnic Institute and State University
Mechanical Engineering Department
Blacksburg, Virginia 24061-0238
(540) 231-4324 Doug.Nelson@vt.edu

2003-01-0417

Development of Fuel-Cell Hybrid Bus

Tatsuaki Yokoyama, Yoshiaki Naganuma, Katsushi Kuriyama and Makoto Arimoto
Toyota Motor Corporation

ABSTRACT

In order to improve air quality and to reduce urban noise, Toyota Motor Corporation has developed a fuel cell hybrid bus, FCHV-BUS2, in cooperation with HINO Motors, Ltd. The FCHV-BUS2 is based on a HINO low floor city bus model, and powered by a hydrogen fuel cell hybrid system. Hydrogen is stored in high pressure tanks on the bus roof.

Based on the Toyota fuel cell hybrid technology for passenger cars, this fuel cell hybrid bus is equipped with two fuel cell stacks, two traction motors and four secondary batteries, making its vehicle efficiency approximately 1.7 times better than the diesel engine powered bus. The vehicle efficiency is boosted by charging the secondary batteries with regenerated energy while deceleration and by stopping the fuel cell stack(s) power generation during low fuel cell power modes.

INTRODUCTION

Amid the worldwide call for environmental protection, demand for the development of cleaner vehicles with higher energy efficiency has grown[2]. A fuel cell uses hydrogen and oxygen to generate electricity and does not emit CO_2, NO_X and PM. It is a clean and highly efficient energy conversion device that raises the expectations for the commercialization of a vehicle that uses a fuel cell as the power source.

While Toyota has actively been pursuing the development of fuel cell hybrid vehicles in the form of passenger cars[1,3], it is also developing fuel cell hybrid systems for public transport buses in an effort to find ways of improving air quality and reducing urban noise. The FCHV-BUS2 is based on a Hino low-floor city bus model that can hold 60 passengers.

The bus features roof-mounted high-pressure hydrogen storage tanks and high-performance Toyota FC Stacks. The use of a unique hybrid system, which includes secondary batteries to store energy regenerated during deceleration, gives the FCHV-BUS2 more-efficient operation and a cruising range of about 250 km.

This paper gives the overview of the FCHV-BUS2 and describes the fuel cell hybrid system newly developed for the FCHV-BUS2. Additionally, a new power distribution algorithm which maximizes the fuel cell hybrid system efficiency is discussed in detail. Finally, experimental results are shown and discussed.

OVERVIEW

Figure 1 and 2 show the appearance and a schematic of the FCHV-BUS2, respectively.

The FCHV-BUS2 uses a Hino low-floor city bus as its base with two Toyota fuel cell stacks, four secondary batteries and two permanent magnetic motors installed in place of a conventional diesel engine and transmission. Fuel cell stacks convert hydrogen gas, which is supplied by roof-mounted high pressurized tanks, into electric power, and the permanent magnetic motors drive the wheels of FCHV-BUS2.

Figure 1. Toyota FCHV-BUS2

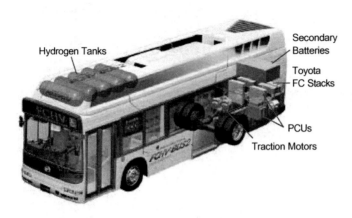

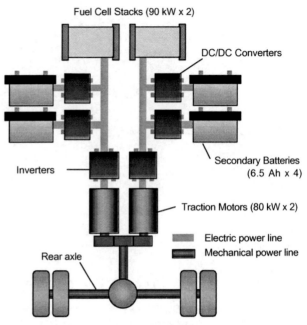

Fuel Cell Stacks (90 kW x 2)

DC/DC Converters

Secondary Batteries
(6.5 Ah x 4)

Traction Motors (80 kW x 2)

Inverters

Rear axle

■ Electric power line
■ Mechanical power line

Figure 3. Fuel Cell Hybrid System Schematic

Figure 2. Schematic of FCHV-BUS 2

Table 1 and 2 show the vehicle specifications and main components of the fuel cell hybrid system. Even though the FCHV-BUS2 is equipped with a fuel cell hybrid system, it maintains almost the same occupant capacity as the base low-floor bus.

Table1. Vehicle Specifications

Base platform	Hino low-floor city bus HU2PMEE
Dimensions (L/W/H)	10,515/2,490/3,360 mm
Maximum speed	80 km/h
Driving range	250 km
Occupant capacity	60 persons

Table 2. Components Specifications

Fuel Cell Stacks

| Type | Polymer electrolyte |
| Output | 90 kW x 2 |

Motors

Type	Permanent magnet
Maximum output	80 kw x 2
Maximum torque	260 Nm x 2

Secondary Batteries

| Type | Nickel-metal hydride |
| Capacity | 6.5 Ah x 4 |

Fuel

Type	Pure hydrogen
Storing method	High pressure tank
Storage pressure	35 Mpa

FUEL CELL HYBRID SYSTEM

Since the FCHV-BUS2 is aimed at urban transportation, on which buses stop and go at frequent intervals and their average speed is relatively slow, a fuel cell hybrid system is adopted for the FCHV-BUS2. Figure 3 shows a schematic of the FC hybrid system.

This FC hybrid system has been developed based on the Toyota fuel cell hybrid system for passenger cars[3] and redesigned for a bus. As a result, the system mainly divided into two parts, each part consists of a fuel cell stack, two secondary batteries and a traction motor.

In each part, the fuel cell stack and the traction motor/inverter are connected directly in order to avoid loss of energy through a converter. On the other hand, each secondary battery is connected through a DC/DC converter, respectively.

The DC/DC converters manage not only the battery input/output power, but also fuel cell output power by means of controlling its output voltage at desired operational point. Additionally, DC/DC converters keep two secondary batteries' state of charge at the almost same level.

Each fuel cell stack is electrically separated from each other in order to control the stacks independently. Two traction motors are mechanically connected and jointly drive the rear axle of the vehicle.

FUEL CELL SYSTEM

Figure 4 shows a schematic of the system which consists of two fuel cell systems. Each system has one fuel cell stack, a hydrogen delivery system, and an air supply system.

Hydrogen is supplied to the fuel cell stack from compressed storage tanks through a pressure regulator. Surplus hydrogen from the stack is re-circulated through a hydrogen pump in order to improve the fuel cell performance.

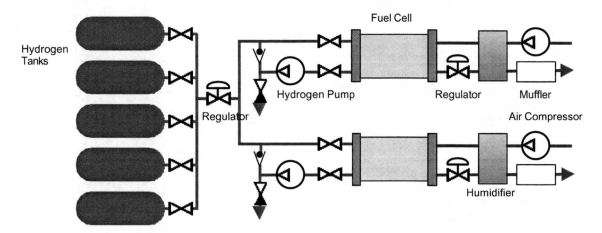

Figure 4. Fuel Cell System Configuration

Air is compressed and pumped into the stack through a humidifier. The humidifier takes water vapor from stack exhausted air and uses it to humidify the incoming compressed air. A regulator between the stack and the humidifier is electrically controlled to keep air pressure at a certain value.

IMPROVING VEHICLE EFFICIENCY

Power Distribution System

As Figure 5 shows, electrical power is supplied by the stacks and batteries in the hybrid system. Vehicle demanded power is a sum of motor power and auxiliary power, such as air compressors, hydrogen pumps and other bus equipment. The power distribution system has been designed to divide the demanded power into the fuel cell system power and the battery power with minimum loss of energy.

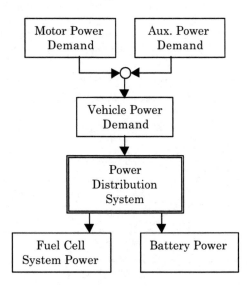

Figure 5. Fuel Cell System/Battery Power Distribution

In order to derive the optimum power distribution, it is necessary to define the power loss of the fuel cell system (FCS) and secondary batteries as a function of output power.

Power Loss of the Fuel Cell System

The FCS power loss, L_{FCS}, is the sum of power loss of the fuel cell itself, L_{FC}, and power loss of auxiliary parts, L_{AUX}, such as the air compressor and the hydrogen pump. In addition, the FCS net power, P_{FCS}, is derived as follows:

$$L_{FCS} = L_{FC}(P_{FC}) + L_{AUX}(P_{FC}) \qquad (1)$$

$$P_{FCS} = P_{FC} - L_{AUX}(P_{FC}) \qquad (2)$$

where

L_{FCS} :	FCS power loss
P_{FCS} :	FCS net power
P_{FC} :	Fuel cell output power
L_{FC} :	Power loss of fuel cell (I-V loss)
L_{AUX} :	Auxiliary power

Figure 6 shows the FCS power loss characteristics of the FCHV-BUS2 that increases exponentially as P_{FCS} increases.

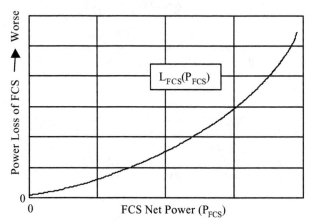

Figure 6. Power Loss of Fuel Cell System

In order to compare the power loss of the FCS and secondary batteries, power loss of secondary batteries is defined in the following equations:

Power Loss of Secondary Batteries

Once secondary batteries are used to output power (discharging), it is necessary to charge the batteries with the FCS output power. Consequently, the power loss of secondary batteries is defined as follows:

$$L_{BAT} = L_{BAT\_DIS}(P_{BAT}) + \frac{P_{BAT}}{P_{CHG}}\{L_{BAT\_CHG}(P_{CHG}) + L_{FCS}(P_{CHG})\}$$
(3)

where

P_{BAT} : Battery output power
P_{CHG} : Battery charging power
L_{BAT_DIS} : Power loss while discharging
L_{BAT_CHG}: Power loss while charging

The first term of Eq.3 represents the power loss of battery itself and the DC/DC converter while discharging at P_{BAT}. The second term is that of battery and DC/DC converter while charging at P_{CHG}. The third term represents the FCS power loss at the FCS output power P_{CHG}. Figure 7 shows the L_{BAT} characteristics of the FCHV-BUS2.

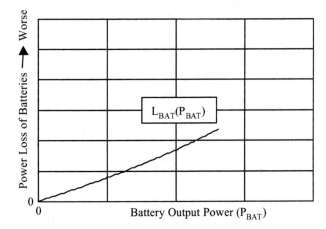

Figure 7. Power Loss of Secondary Batteries

Figure 8 shows the comparison of the power loss of the FCS and batteries. It is possible to compare L_{BAT} with L_{FCS} directly because L_{BAT} includes not only the power loss while discharging, but also the power loss of batteries and the FCS while charging.

In range A, for example, all of demanded power is supplied from the FCS because L_{FCS} is smaller than L_{BAT}. On the other hand, in range B, since L_{FCS} increases exponentially, battery output power should be increased in order to minimize the total power loss.

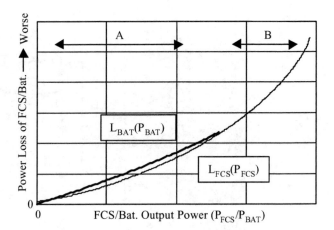

Figure 8. Comparison of L_{FCS} and L_{BAT}

Power Distribution Algorithm

To derive the optimum power distribution between the FCS and secondary batteries, the total power loss of the FCHV system, L_{SYS}, is defined as follows:

$$L_{SYS}(P_{FCS}, P_{BAT}) = L_{FCS}(P_{FCS}) + L_{BAT}(P_{BAT})$$
(4)

The optimum power distribution is defined to be where P_{FCS} and P_{BAT} give a minimum value of L_{SYS}.

$$\min_{P_{FCS}, P_{BAT}} \{L_{SYS}(P_{FCS}, P_{BAT})\}$$
(5)

$$\text{where} \quad P_{DEMAND} = P_{FCS} + P_{BAT}$$
(6)

Figure 9 shows the relation between L_{SYS}, P_{FCS} and P_{BAT}. Firstly, the power distribution, (P_{FCS}, P_{BAT}), is on a line P-Q which satisfies the vehicle demand power, P_{DEMAND}. Secondly, the curved lines represent contours of the total power loss, i.e. the FCHV system power loss, L_{SYS}, is the same value on each curved line. Finally, the optimum power distribution is the point R.

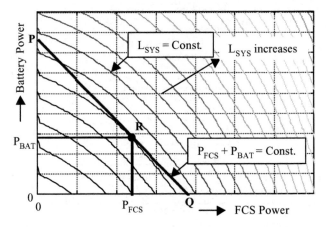

Figure 9. System Power Loss and Power Distribution

Figure 10 shows the resulted FCS/Battery power distribution.

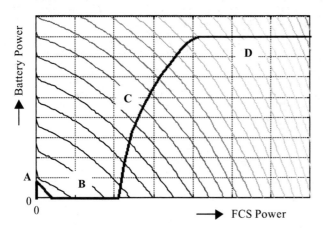

Figure 10. FCS/Battery Power Distribution

The power distribution is mainly divided into four ranges:

Range A:
The FCS stops operating and batteries supply all of demanded vehicle power at low-load conditions, because the battery power loss is lower than that of the FCS (see Fig.8).

Range B:
The FCS supplies the demanded power alone with good efficiency.

Range C:
Batteries assist the FCS in order to maximize the efficiency, because the FCS power loss increases exponentially in this power range.

Range D:
Batteries supply the maximum power and the FCS makes up the difference.

In the following section, we discuss simulation results with the proposed power distribution algorithm.

Simulation Results

In order to investigate the advantage of the proposed power distribution algorithm in city driving conditions, Tokyo transient pattern[6] No.5 was selected as the simulation driving cycle (Figure 11).

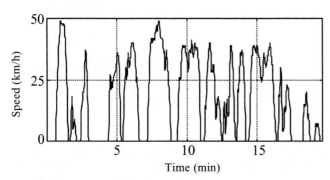

Figure 11. Driving Cycle of Simulation

With this driving cycle, following two power distribution methods are simulated and compared with each other.

Case I:
The vehicle power demand is supplied by the FCS, except when the FCS power exceeds its maximum limit or in a transition state.

Case II:
The vehicle power demand is divided into the FCS and battery power according to the proposed algorithm. Batteries are charged under high FCS efficiency range.

In both cases, batteries are charged with regenerated energy during deceleration.

Power Distribution Results

Figures 12-a and 12-b show the resulting power distribution of Case I and II, respectively, at the beginning of the driving cycle.

Case I:
The demanded power is mainly supplied by the FCS. Batteries only assist the FCS in transient state and are charged during deceleration. As a result, the FCS output range is from zero to the maximum power.

Case II:
Only the FCS supplies power while the demanded vehicle power is within range B. If the demanded power increases to range C, then batteries start to assist in order to prevent the FCS efficiency from becoming worse.

In range D, the battery output power reaches the maximum power and FCS supplies the remainder. FCS stops operating and batteries are charged during deceleration in range A.

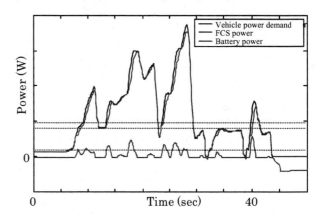

Figure 12-a. Power Distribution Result of Case I

221

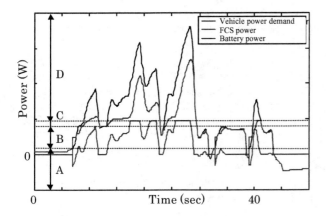

Figure 12-b. Power Distribution Result of Case II

FCS power frequency and power loss

Figure 13-a shows power frequencies and accumulated power losses of the FCS as a function of output power.

a) In low-power range ("a" in Fig.13-a), the FCS power frequency is decreased in comparison with Case I, because the FCS efficiency is relatively low. In this range, as the FCS is off most of the time, the FCS power loss becomes nearly equal to zero.

b) The power frequency in "b" mid-power range is increased because the power distribution algorithm keeps the FCS operation in this range most of the time.

c) As high-power FCS operation causes more power loss than that of mid-power operation, the power frequency in range "c" is explicitly decreased in comparison with Case I.

Although the power loss increases in range "b", the loss explicitly decreased in range "a" and "c". Consequently, the total amount of the FCS power loss decreased by about 21% with the distribution algorithm

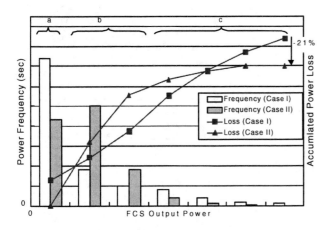

Figure 13-a. Histogram of FCS Power/Loss

Battery power frequency and power loss

Figure 13-b shows power frequencies and accumulated power losses of the batteries as a function of output power.

As the result of keeping the FCS operation in mid-power range, batteries assist the FCS power more frequently and are charged under the FCS high efficiency condition. This means that the battery power loss slightly increases but is negligible compared with that of the FCS power loss.

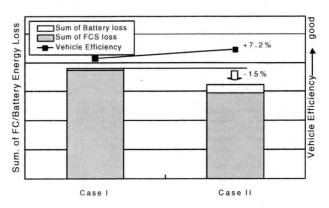

Figure 13-b. Histogram of Battery Power/Loss

Simulation results show the effectiveness of the proposed algorithm, Case II, which decreases the total amount of power loss about 15% and improves the vehicle efficiency about 7.2% (Figure 14)

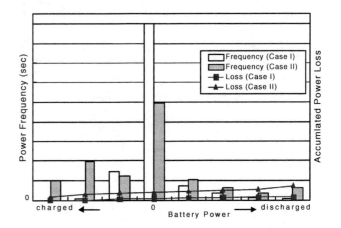

Figure 14. Comparison of FCS/Battery Energy Loss

EXPERIMENTAL RESULTS

Vehicle Efficiency

Figure 15 shows the experimental results of FCHV-BUS2 and the conventional diesel engine bus (ICE). The vehicle efficiency of FCHV-BUS2 is about 66% better than that of the ICE bus. In addition to the high efficiency of the Fuel Cell System itself, this high vehicle efficiency is realized by the power regeneration during deceleration and the proposed power distribution algorithm. The power regeneration and the power distribution algorithm improved the vehicle efficiency by 17.6% and 5.0%, respectively.

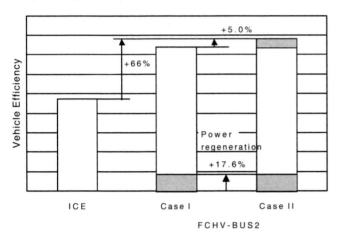

Figure 15. Vehicle Efficiency Comparison

Driving Performance

FCHV-BUS2 also has been evaluated for the driving performance. Evaluation results show that the FCHV-BUS2 driving performance is better than that of the ICE bus (See Table 3).

Table 3. Driving Performance

	0-40 km/h Acceleration
ICE bus	10.5 sec
FCHV-BUS2	9.3 sec

CONCLUSION

1. Toyota Motor Corporation has developed the fuel cell hybrid bus, FCHV-BUS2, in cooperation with HINO Motors, Ltd. in order to realize clean and efficient urban transportation.
2. Based on the Toyota fuel cell hybrid technology for passenger cars, a fuel cell hybrid system has been redesigned for the FCHV-BUS2. This system consists of two fuel cell stacks, two traction motors and four secondary batteries.
3. Fuel cell stacks and secondary batteries are controlled based on the newly developed power distribution algorithm with the following characteristics:
   a. At low-load conditions, the fuel cell system stops operating and the batteries supply all of demanded power.
   b. The FCS supplies all of the demanded power at mid-load conditions.
   c. As demanded power increases, the batteries start to assist the FCS in order to keep the total efficiency high.
4. Experimental results show:
   a. The vehicle efficiency of FCHV-BUS2 is about 66% better than the base diesel engine powered bus.
   b. Driving performance is better than that of the base ICE bus.

References

1. Y. Nonobe, Y. Kimura, S. Ogino. "A Fuel Cell Electric Vehicle with Methanol Reformer," The 14th International Electric Vehicle Symposium, (1997).
2. S. Sasaki, T. Takaoka, H. Matsui, T. Kotani. "Toyota's Newly Developed Electric-Gasoline Engine Hybrid Power Train System," The 14th International Electric Vehicle Symposium, (1997).
3. T. Matsumoto, N. Watanabe, H. Sugiura, T. Ishikawa, "Development of Fuel-cell Hybrid Vehicle," SAE World congress 2002,(2002).
4. J. Reinkingh, J.Matthey, "Dynamic Requirements for a Fuel Cell System in a Hybrid Bus," Proceedings. International Symposium on Automotive Technology & Automation, (1999)
5. R. Wimmer, "Fuel Cell Transit Bus Testing & Development at Georgetown University," Intersociety Energy Conversion Engineering Conference, (1997)
6. Tokyo Metropolitan Research Institute for Environmental Protection.

2002-01-0408

The HYDRO-GEN Project : 2nd Generation PEM Fuel Cell System with High Pressure Hydrogen Tanks for an Electric Vehicle

Franck Michalak, Joseph Beretta and Jean-Pierre Lisse
PSA Peugeot Citroën, Division of Research and Automobile Innovation

ABSTRACT

The objective of the European Hydro-Gen project is to develop an innovative PEM Fuel Cell (PEMFC) energy system, based on high pressure gaseous hydrogen tanks. The developed system would result in a large increase in performance compared to the state of the art in terms of components, PEMFC stack and energy system, with an emphasis placed on cost reduction for on board applications in fuel cell electric vehicles. This is achieved by developing a low temperature (80°C) and low pressure (1.5 Bar abs) 30 kW PEMFC stack based on innovative core components, 700 bars H_2 tanks with optimized weight and volume, and integrating the fuel cell drive train in a compact electric van Peugeot Partner Electric.

The vehicle has a driving range of 300 km for a 350 bar fill-up, and overall performances comparable to those of the Partner Electric version. The efficiency of the FC stack alone is 56% at 1.5 bar abs. The energy consumption of the vehicle is around 1.2 MJ/km that is 1 kg of H_2 per 100 km.

INTRODUCTION

In a world where fossil energy resources are becoming scarce, and where emissions of green house effect gases are presented as responsible for the global warming, car manufacturers are developing engines with increasingly better fuel economy. Another possibility is changing the fuel. Hydrogen "burning" fuel cells are more efficient than internal combustion engines, and only produce water (locally). In 1996 PSA Peugeot Citroën, together with Air Liquide, CEA, Nuvera Fuel Cells Europe, Renault and Solvay, helped by the European Commission, started the HYDRO-GEN research project on this technology.

STATE OF THE ART

The state of the art at the start of the project, regarding the fuel cell as well as the hydrogen storage made it impossible to built a vehicle having an acceptable level of performance, because the components were too bulky or heavy.

PROBLEMS TO SOLVE

To be able to reach an acceptable level of performance (speed, range), the consortium had to develop new high pressure tanks, a fuel cell having good performances with enough power, and then a system around it with all the fuel cell auxiliaries, and integrate everything in the passenger vehicle. This work was split in seven work-packages (see Figure 1)

The work-package 1 deals with the fuel cell, power module (fuel cell and auxiliaries), and hydrogen tanks specifications

The work-package 2 is related to research activities on innovative core components for the fuel cell, such as low platinum loading MEAs and low cost membrane alternative to Nafion®.

In work-package 3, Nuvera Fuel Cells Europe task consists in developing and delivering a 5 kW fuel cell prototype for study, and the final 30 kW stack, including all necessary core components like MEAs and bipolar plates.

The work-package 4 is concerned with hydrogen storage, and in particular the definition and manufacturing of the high pressure hydrogen cylinders.

The work-package 5 deals with the power module built around the 30 kW stack, notably the air and hydrogen lines, cooling, and electronic control of the power module system.

In the frame of work-package 6, PSA Peugeot Citroën studies the mechanical modifications that are necessary to integrate the main components such as the hydrogen tanks and the power module. The electronic control of the vehicle taking into account the energy management is also studied. Finally the vehicle can be built and tested.

The work-package 7 deals with all safety aspects during the vehicle life.

The role of the six partners is summarized in Table 1.

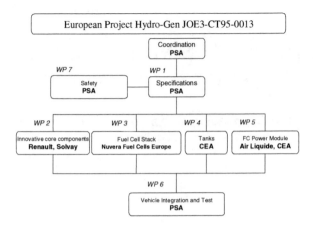

Figure 1: Hydro-Gen project organization.

Partners	Tasks in the project
PSA Peugeot Citroën	Co-ordination of the project; Specifications; Innovative fuel cell core components; Safety; Vehicle integration and tests.
Air Liquide	Study and fabrication of the power module (Fuel cell system)
CEA	Fuel cell stack and system modeling; Study and fabrication of the high pressure hydrogen tanks .
Nuvera Fuel Cells Europe	Study and fabrication of the 30 kW fuel cell stack
Renault	Innovative fuel cell core components (Electrodes)
Solvay	Innovative fuel cell core components (proton conducting membranes)

Table 1: role of the 6 partners in the Hydro-Gen project

RESULTS

Figure 2 : Drawing of the initial implantation of the fuel cell drive train in the Peugeot Partner. 1) : Fuel cell. 2) DC/DC

converter. 3) Water management. 4) Compressed hydrogen tanks.

CHOICE OF INSTALLATION OF THE MAIN COMPONENTS - The demonstrator vehicle is built from a Peugeot Partner Electric model which is produced on the line and sold on the market. The original under-hood installation (electric motor and transmission) has been kept. But the floor and the rear part had to be modified to accept the tanks, the DC/DC converter and the fuel cell system (Figure 2). The vehicle seats 5 persons

THE FUEL CELL DRIVE TRAIN - The specific components in the fuel cell drive train are presented on Figure 3.

It shows the fuel cell system at the back, the DC/DC converter, the hydrogen tanks which are under the floor, and the traction motor and the power battery.

The NiMH power battery is mounted in the front, under the hood, in place of one traction battery pack. The other traction battery pack is removed. The power battery is used together with the FC stack during the acceleration phases, and for regenerative braking.

The DC/DC converter is used to raise the cell stack voltage up to the voltage of the drive train, e.g. 162 V. The electric drive system is the standard one used for PSA Peugeot Citroën EVs. Figure 4 presents the connections within the fuel cell drive train

Following an installation study of the fuel cell drive train by CAD, the structure of the vehicle was frozen (Figure 5), and the project entered in an active prototype components manufacturing phase.

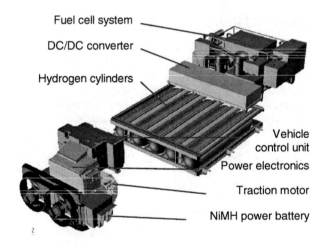

Figure 3 : Numerical mock-up of the fuel cell drive train

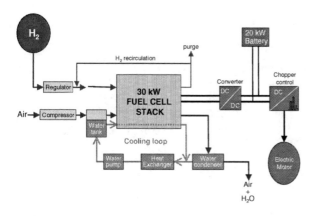

Figure 4: Schematic of the Hydro-Gen fuel cell drive system

Figure 5 : Global view of the vehicle after integration study by CAD

THE FUEL CELL STACK - The MEAs developed in the frame of WP2 reached a remarkable platinum loading of 0.16 mg/cm^2 both sides added, but could not be installed in the prototype stack unit because of lack of time. Nuvera used standard automated electrodes instead. A new internal membrane humidification by direct water injection was developed. Water injection is also used to cool the stack. The characteristics of the 30 kW stack are listed in Table 2. A picture of the stack can be seen on Figure 6.

Table 2: characteristics of the 30 kW fuel cell stack

Weight	130 kg
Power	30 kW, at 70V
Efficiency (LHV)	56%
Air pressure	1.5 Bar (abs)
H2 pressure	1.8 Bar
Humidification - Cooling	Direct water injection
Platinum loading (Total)	0.3 mg/cm^2(45 g)

Figure 6: 30 kW fuel cell stack on test bench

THE HIGH PRESSURE HYDROGEN TANKS - The hydrogen tanks are made by carbon fiber / epoxy coiling on a thin wall aluminum alloy liner. One cylinder is shown on Figure 7. The characteristics of the cylinders are given in Table 3:

Table 3 : Characteristics of the 700 bars approved hydrogen tanks.

Max working pressure	700 bars
Burst pressure	> 1400 bars
Internal volume (5 cylinders)	140 dm^3
Weight 5 cylinders	110 kg
H$_2$ content 5 cylinders	3.3 kg at 350 bars 6.0 kg at 700 bars*

* estimated

The cylinders are filled up to 350 bars with a hydrogen compressor from commercial 200 bars cylinders

Figure 7 : picture of a 700 bars hydrogen cylinder made of composite materials : carbon fiber / epoxy on a thin wall aluminum alloy liner.

THE FUEL CELL SYSTEM - The fuel cell system built around the stack has good performances and dynamics as can be seen on Figure 8. It takes less than 1 second to go from 50 A to 550 A net, and around 0.7s from 50 A to 450 A net. This is better than expected .We did not know how the air compressor would behave, and so it is possible that the power battery will have a too big capacity. (see 2.9 Future actions).

Transient record from 50 to 550 A

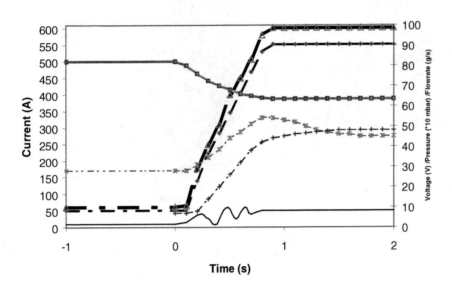
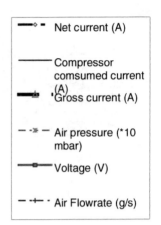

Legend:
- Net current (A)
- Compressor comsumed current (A)
- Gross current (A)
- Air pressure (*10 mbar)
- Voltage (V)
- Air Flowrate (g/s)

Figure 8: transient record of the fuel cell system after a 50 A to 550 A step.

BUILDING OF THE VEHICLE - The vehicle was assembled in two phases. During the first one, all components of the electric drive train (without the fuel cell) were installed in a functional mock-up.

Electronic electric architecture validation - Thanks to the mock-up, the whole electrical architecture and the energy management software was validated. This software was implemented in the vehicle control unit so that the correct balance between the power batteries and the fuel cell system is maintained whatever the power battery state of charge, and the power set-point. CAN protocol was used for the dialog between the FC electronic control and the vehicle supervisor.

Fuel cell components integration - In a second phase, the hydrogen cylinders and the fuel cell power module were integrated in the vehicle, under the floor and in the boot as shown on Figure 9.

VEHICLE TESTING - The performance of Hydro-Gen is summarized in Table 4. It is similar to those of the Partner Electric model, except for the range which is 300 km for a fill-up at 350 bars.

Table 4: Performance of the Hydro-Gen Peugeot Partner

0-50 km/h.	8 s
Max speed	95 km/h
Fuel economy on CEN cycle	100 km / kg H_2
Range (350 b)	300 km
Weight	1800 kg
Max power	33 kW

Figure 9: picture of the fuel cell power module in the car.

Figure 10 : Hydro-Gen on the road.

CHARACTERISTICS OF THE HYDRO-GEN VEHICLE

Table 5: characteristics of the Hydro-Gen Peugeot Partner

Motor	Type	DC, separate excitation
	Supplier	Leroy Somer
	Nominal power rating	20 kW
	Maximum power rating	33 kW
	Maximum torque	210 Nm
	Maximum operating conditions	6,500 rpm
	Maximum traction current	230 A
	Maximum excitation current	10 A
	Cooling method	pulsed air
Transmission	Wheel	Front-wheel drive
	Type of reduction gear	epicycloidal
	Gear ratio	0.1396
	Speed per 1,000 rpm	15.11 km/h
Batteries	Type	Nickel-metal hydride
	Supplier	Varta
	Number of 1.2 V cells	135
	Location	front compartment in front of electric motor
	Nominal voltage	162 V
	Capacity	10 Ah
	Weight	80 kg
	Cooling method	air
	Battery charging	Fuel cell or regenerative braking
DC/DC converter	Supplier	CIRTEM
	Input	60-100 V / 540 A
	Output	162V / 250 A
Fuel cell	Supplier	Nuvera Fuel Cells Europe
	Power rating	30 kW net at 70 V
	Supplier of Fuel cell system	Air Liquide
	Efficiency of fuel cell alone	56%
	Air pressure	1.4 bar
	Hydrogen pressure	1.8 bar
	Humidification	direct injection
	Cooling method	forced water circulation
	Location	in trunk
Tank	Design	CEA
	Fuel	compressed hydrogen
	Pressure	350 bar (validated at 700 bar)
	Weight	110 kg (5 cylinders)
	Capacity	140 liters (i.e. 3.3 kg of hydrogen)
	Materials	carbon fiber composites, epoxy resin and aluminium alloy liner
	Pressure relief device	Thermal fuse
Vehicle	Type	Peugeot Partner minivan
	Dimensions	
	Length	4,110 m
	Width	1,960 m
	Height	1,800 m
	SCx	0.9
	Weight without load	1,800 kg

- **FUTURE ACTIONS** – PSA Peugeot Citroën will continue testing in various conditions. We will also work on the vehicle supervisor to determine the best battery for the system taking into account the better than expected dynamics of the fuel cell system. In parallel, PSA Peugeot Citroen will work on fuel cells as a secondary energy source, to recharge batteries, as illustrated by Fuel Cell Cab concept. Depending on the mission profile of this vehicle and the cost of the fuel cell, this may be an interesting solution.

ACKNOWLEDGEMENTS

This work was funded in part by the European Commission under the Non-Nuclear Energy Joule III program (Contract JOE3-CT95-0013). The partners of the project are Air Liquide, CEA, De Nora Fuel Cells (Nuvera Fuel Cells Europe), PSA Peugeot Citroën, Renault and Solvay. PSA Peugeot Citroën is the coordinator of the project.

CONTACT

Franck Michalak
PSA Peugeot Citroën, Division of Research and Automobile Innovation, Centre Technique de Vélizy, Route de Gisy, 78943 Vélizy-Villacoublay Cedex – France.
Tel : +33-1 41 36 21 79
Fax : +33-1 41 36 30 66
E-mail: **franck.michalak@mpsa.com**

2002-01-1930

Cost Modeling of PEM Fuel Cell Systems for Automobiles

Eric J. Carlson, Johannes H. Thijssen, Stephen Lasher, Suresh Sriramulu and Graham C. Stevens
Arthur D. Little, Inc.

Nancy Garland
Office of Transportation Technologies, U.S. Department of Energy

ABSTRACT

Cost is one of the critical factors in the commercialization of PEM fuel cells in automotive markets. Arthur D. Little has been working with the U.S. Department of Energy, Office of Transportation Technologies to assess the cost of fuel-flexible reformer proton exchange membrane (PEM) fuel cell systems based on near-term technology but cost modeled at high production volumes and to assess future technology scenarios. Integral to this effort has been the development of a system configuration (in conjunction with Argonne National Laboratories), specification of performance parameters and catalyst requirements, development of representative component designs and manufacturing processes for these components, and development of a comprehensive bill of materials and list of purchased components. The model, data, and component designs have been refined based on comments from the Freedom Car Technical Team and fuel cell system and component developers. The baseline results from the project were reported in the first Future Car Congress in 2000.

In this presentation, we provide an overview of changes resulting from comments from the fuel cell development community. A sensitivity and Monte Carlo analysis on the revised baseline model is presented. The results of this analysis include a ranking of the most important cost factors and the minimum and maximum fuel cell system cost given the uncertainty in our model assumptions. The impact of various system scenarios on the fuel cell system cost ($ / kW_e) and power density were assessed, including the effect of unit cell voltage at rated power (i.e., power density versus efficiency tradeoff) and system power rating (i.e., hybridization). The results of these analyses are presented and their implications discussed.

BACKGROUND

Any new technology faces substantial barriers to commercialization; however, transportation markets present unique challenges including the size of the market, cost drivers, consumer expectations, product requirements (e.g., performance, safety, and reliability), well-entrenched technology and other alternative technologies, infrastructure issues, OEM profits, and societal (government) requirements. The latter includes emissions of pollutants and greenhouse gases, efficiency goals, and energy policy. Within this forest of challenges, developers and automotive companies are attempting to develop cost-effective electric fuel cell powertrains for passenger vehicles. The current Department of Energy (DOE) goal for a fuel cell powertrain cost is $45/$kW_e$[1]. For some of us, the battery electric vehicle (BEV) story starting in the mid-1970s vividly illustrates the significant hurdles radical new powertrains face in passenger vehicles. Even though the most recent industry/government BEV initiative, the U.S. Automotive Battery Consortium (USABC), failed to produce a viable mass-market EV battery, this effort led to significant advances in powertrain components, including electric motors and power electronics, that have benefited the fuel cell and hybrid electric vehicle efforts.

Starting in 1998, Arthur D. Little began a five-year project[2] to model and project fuel cell system costs for the DOE, Office of Advanced Transportation Technology. The project started with the characterization of the status of fuel cell system technology, development of a system layout and thermodynamic model, and development of an activities-based cost model for production of this system at high production volumes. The initial cost results were presented at the first Future Car Congress[3]. In addition to projection of fuel cell system costs, the objectives of the study include:

- Technology assessment
- Solicitation of feedback from industry on assumptions concerning the system, performance parameter, manufacturing, and cost assumptions
- Identification of technology development needs
- Analysis of the impact of system design parameters on cost and performance
- Inputs for establishment of DOE program cost targets

In this paper, we present (1) an updated cost projection for near term technology, (2) an analysis of several system scenarios (i.e., impact of stack operating voltage on cost and impact of system-rated power on cost per kilowatt), and (3) a long-term projection of reformate and hydrogen fuel cell system costs.

COST MODELING APPROACH

We began cost modeling by developing a system layout for the technology in question with the level of detail dictated by the project objectives. Using this system configuration, a model was developed to determine the state parameters (temperature, pressure, and mass and molar flows) of the various streams and components by thermodynamic modeling. The individual reactor beds and the fuel cell stack were then scaled using available kinetic or polarization data for the assumed catalyst materials and their loadings.

Our manufacturing group developed production process options for key subsystems and components, and they obtained raw material costs from potential suppliers. Purchased components

were listed separately and suppliers were contacted regarding the availability and cost of outsourced components. An activity-based process model in Microsoft® Excel was then developed to estimate manufacturing costs using capital equipment and raw material costs, labor rates, and throughput for 500,000 units per year. The model yields cost per kilowatt, cost by subsystems and components, and a breakdown of material and process costs. With additional software, sensitivity and Monte Carlo analyses were performed on the base model. The key assumptions, manufacturing process flows, and model results were presented to developers and OEMs as part of our effort to obtain industry feedback. As the system and manufacturing technologies evolve during the project, the cost projection will be updated using revised assumptions or process flows.

As shown in Figure 1, the estimated cost includes factory costs (e.g., direct materials and labor, factory expenses, capital equipment) but excludes corporate charges for profit, sales expenses, and general services and administration. Table 1 lists the components included in the baseline fuel flexible reformer fueled fuel cell system. For purposes of this project, the design of the fuel processor was based on gasoline. The level of component outsourcing assumed is characteristic of a vertically integrated manufacturer. Alternatively, an OEM could also assemble purchased subsystems and accept the higher cost associated with outsourcing. The cost of a 50 kW_e system produced at a volume of 500,000 units per year was modeled based on guidance from the DOE and the Freedom Car fuel cell technical team. A 50 kW_e electric system would be sufficient to power a family sedan and would lead to a $/$kW_e$ estimate representative of systems in this power range.

BASELINE ASSUMPTIONS

Linking the system and cost models allows one to assess the impact of performance and material assumptions on the overall system cost. Tables 2 and 3, respectively, show the assumptions used to size the reformer and fuel cell stack, including reaction rates and catalyst loadings. The selection of 0.8 volts for the fuel cell voltage at rated power was driven by overall system-efficiency targets rather than operation of the stack at or near its high power point.

Figure 1: Definition of Factory Cost

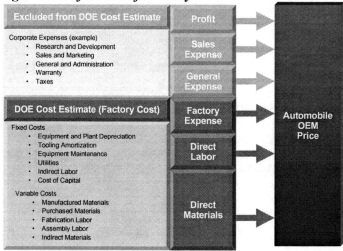

Table 1: Component Breakdown Between Subsystems

Fuel Processor Subsystem		Fuel Cell Subsystem	Balance-of-Plant
• **Reformate Generator** • ATR • HTS • Sulfur Removal • LTS • Steam Generator • Air Preheater • Steam Superheater • Reformate Humidifier	• **Fuel Supply** • Fuel Pump • Fuel Vaporizer	• Fuel Cell Stack (Unit Cells) • Stack Hardware • Fuel Cell Heat Exchanger • Compressor/Expander • Anode Tailgas Burner • Sensors & Control Valves	• Startup Battery • System Controller • System Packaging • Electrical • Safety
• **Reformate Conditioner** • NH₃ Removal • PROX • Anode Gas Cooler • Economizers (2) • Anode Inlet Knockout Drum	• **Water Supply** • Water Separators (2) • Heat Exchanger • Steam Drum • Process Water Reservoir		
• Sensors & Control Valves for each section			

Table 2: Fuel Processor Subsystem Performance Assumptions

Parameter	Catalyst Beds				Clean-up Beds	
	ATR	HTS	LTS	PROX	Sulfur Removal	NH₃ Removal
Temperature (°C)	1030	430	230	205	490	80
Catalyst	Pt/Ni	Fe₃O₄/CrO₃	Cu/ZnO	Pt	ZnO	Activated Carbon
Support	Alumina	Alumina	Alumina	Alumina	None	None
GSHV (1/hour)	15,000	10,000	5,000	10,000	NA	None

Table 3: System and Fuel Cell Stack Operating Assumptions

Operating Assumptions	Units	2001 Baseline
Net Power	kW	50
Ambient Temperature	°F (°C)	**120** (49)
System Pressure	atm	3
Compressor/Expander Efficiency	%	70/80
Unit Cell Voltage	volts	0.8
Power Density	mW/cm²	250
Fuel Utilization	%	85
Cathode Stoichiometry		2.0
Operating Temperature	°C	80
Percent Anode Air Bleed	%	1

After discussions with developers, we concluded that the assumptions made were representative of generally available technology. Additionally, the system configuration modeled in this project does not portend to solve all of the technical challenges fuel cell systems face (e.g., fast startup and low-temperature operation). For the purposes of this project, we assumed these represent engineering challenges rather than major cost additions.

UPDATED COST PROJECTIONS FOR AVAILABLE TECHNOLOGY

In the presentation made at the 2000 Future Car Congress, the cost given for a fuel flexible 50 kW$_e$ system was $295 per kW$_e$. Following discussions with component and system developers, we revised the initial estimate upward by 10% to $324 per kilowatt. Increases in catalyst and PEM material costs increased the fuel cell subsystem cost by 25%, while revisions to the reformer bed sizing calculation and assembly time assumption reduced costs of these areas. The fuel cell subsystem continues to dominate the system cost as shown in Figure 2. For this Figure and all following Figures, the cost estimate was based on production volumes of 500,000 per year and 50kW$_e$ systems.

Figure 2: Breakdown of 50 kW_e Fuel Flexible Reformate Fuel Cell System Cost

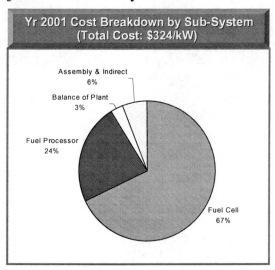

Figure 3 shows the breakdown of costs in the fuel cell subsystem. We continued to use a turbo-compressor expander for pressurization of the system to 3 atmospheres. Based on developer input, we used a volume cost of $630 for the turbo-compressor expander and its controller. The balance of the air supply system included a cathode gas humidifier, air filter, valves and piping.

Figure 3: Fuel Cell Subsystem Cost by Major Components

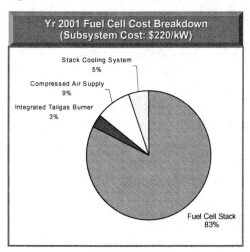

Figure 4 further breaks down the cost of the fuel cell stack and then the membrane electrode assembly (MEA). Platinum content of the electrodes represent approximately 50% of the MEA cost, hence the polarization curve (current density versus

voltage) and design criteria (high efficiency versus high power) have a significant impact on the total platinum and cost in the system. Later in our presentation, we detail the factors influencing MEA costs and the potential for cost reduction. On a gram per kilowatt basis, this system uses 3.6 g/kW_e of platinum in the stack with a total cost of $3,270 ($18/gram) including a London Metals Exchange (LME) price of $15 per gram and processing costs.

Figure 4: Breakdown of Fuel Cell Stack and MEA Costs

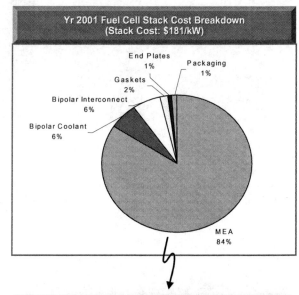

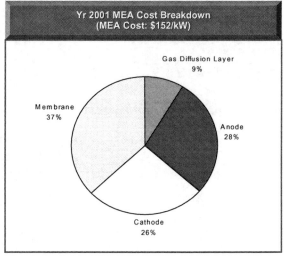

In a similar manner, Figure 5 shows the cost of the fuel processor components. The reformate generator consists of an autothermal reformer (ATR), shift beds, desulfurization bed, and heat exchangers (i.e., preheaters and steam generator). A two-stage

234

preferential oxidation reactor (PrOX) and ammonia removal bed makeup the reformate conditioner.

Figure 5: Breakdown of Fuel Processor Costs

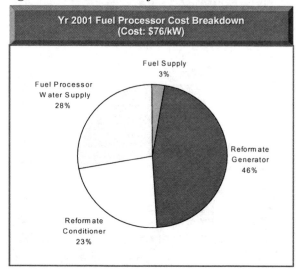

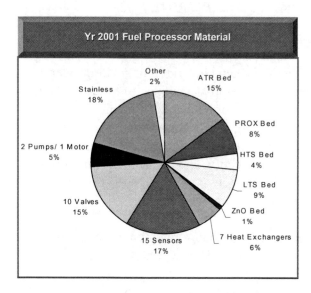

A Monte Carlo simulation was performed on the cost model to quantify the uncertainty in system cost arising from uncertainty in the values of the model parameters. Figure 6 shows the results of this analysis. The difference between the $324/kW_e$ cost projection and the mean cost in the simulation arises from the skewed ranges assumed for the input variables in the simulation. In the simulation each model parameter is assigned a base value, upper and lower limits, and a probability distribution (e.g., triangular or normal) between the limits. In this simulation, the base values tended to be closer to the upper limits leading to difference between the

baseline cost estimate and the mean value resulting from the simulation.

Figure 6: Monte Carlo Simulation on Technical Cost Model

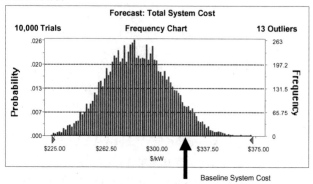

SYSTEM SCENARIOS

OPERATION OF THE STACK AT HIGHER POWER DENSITY

In our discussions with developers, the DOE, and review committees, we were asked to define the impact of operating the fuel stack at a higher power point (e.g., lower voltage on the system cost and weight). Obviously, the stack could be made smaller and less costly for a given rated power using a lower cell voltage. For purposes of the high power design analysis, we doubled the power density from 250 milliwatts/cm^2 (at 0.8 volts) to 500 milliwatts/cm^2 (at 0.65 volts). For this change in stack operating point, the calculated overall system efficiency decreased by 20% from 37% to 29%.

Due to the large contribution of the fuel cell stack to the system cost, decreasing its size led to an overall 20% reduction in system cost to $265/kW_e$. The reduction in stack cost is partially offset by increases in the stack cooling system and the fuel processor size to accommodate the lower stack efficiency. The expected decrease in total system weight was offset due to increases in fuel processor catalyst beds, larger fuel cell stack heat exchanger, and larger startup battery.

In summary, the decision to operate the stack at higher power densities becomes a trade-off between efficiency and cost, but not necessarily weight because of the increases in the weight of other components to accommodate the lower fuel cell efficiency.

HYBRID SCENARIOS

As with internal combustion engine powertrains, hybridization of fuel cell systems is being considered to reduce the total power needed from the primary power unit and improve fuel economy through regenerative braking. An analysis was performed to estimate how fuel cell system cost ($/kW$_e$) might scale with rated power of the stack. We did not consider the added cost of the hybrid electric vehicle battery in this particular analysis. We started by dividing the components into two groups, those with fixed costs and those where cost scaled with power. For the 50 kW$_e$ system, "fixed cost components" represent approximately 20% of the system cost. Table 4 shows how the components segregated into these categories. Many of the low cost but high count components were assumed to be fixed in cost, while the larger component costs were allowed to scale with power.

Table 4: Breakdown of Fixed- and Scaled-Cost Components

Fixed Cost Components	Scaled Cost Components
◆ Sensors (temperature, oxygen, sulfur, CO, pH) ◆ Fittings ◆ Injectors ◆ Valves ◆ Pumps ◆ Fans ◆ Filters ◆ Control and Electrical Systems ◆ Piping System ◆ Compressor/Expander	◆ Fuel Cell Module ◆ Heat Exchangers ◆ Catalyst Beds ◆ Drums (knockout, steam, water separators) ◆ Tanks ◆ Start-up Batteries

As shown in Figure 7, the cost per kilowatt increases with decreasing rated power and decreases with increasing rated power from our baseline of 50 kW$_e$ due to the shifting proportion of fixed and scaleable cost components. On a kilowatt basis, the 25 kW$_e$ system is 30% more expensive and the 100 kW$_e$ is 15% less expensive. Hence, the 25 kW$_e$ system, instead of halving the cost is 66% of the 50 kW$_e$ system cost. To complete the comparison between the hybrid and baseline system, the cost of the battery needs to be factored into the cost/kW$_e$ of the system. A complete vehicle integration analysis

is being performed as part of another ongoing project for the DOE, with results expected in early 2002[3].

Figure 7: Fuel Cell System Cost ($/kW$_e$) versus Rated Power

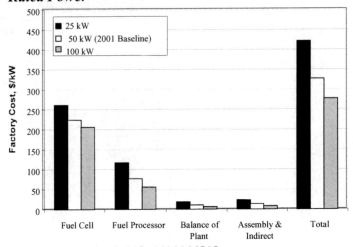

PLATINUM LOADING ANALYSIS

APPROACH/ASSUMPTIONS

The factors influencing platinum loadings were assessed to develop an estimate of future fuel cell stack cost for reformate and hydrogen systems. Platinum loading per kilowatt is a critical issue as indicated by the cost contribution of platinum in the $324/kW$_e$ projection and by the DOE funding resources allocated to platinum reduction and performance improvements.

Projections of minimum platinum requirements were estimated based on an analysis that considered:

- Impact of catalyst particle size and catalyst activity on kinetics
- Impact of electrolyte adsorption on catalyst kinetics
- Development of polarization curves based on electrochemical kinetics
- Impact of ohmic resistance losses on polarization curve

In this paper, we provide an overview of the analysis, key assumptions, and the results. A technical paper focused solely on this analysis will be submitted for publication in the near future.

We started with the assumption that the fuel cell will operate at relatively high unit cell voltages (e.g., 0.8 volts) to achieve overall system efficiency targets. However, analyses that consider overall vehicle efficiency and reward higher fuel cell system power density will allow some relaxation of this stack efficiency target. At high cell voltages, cathode kinetics control the voltage losses within the stack. Tafel kinetics were used to assess the effect of operating conditions and cathode voltage losses. Voltage drops were then assigned to the anode and ohmic components based on experimental data. Analysis of a series of scenarios involving catalyst loadings, pressures, and temperatures led to the conclusion that the stack would have to operate at high temperature (e.g., 160°C) and elevated pressure (e.g., 3 atm) to satisfy DOE fuel cell power density (milliwatt/cm^2) goals. Compressor/expander technology has been and continues to be a development area for DOE, while the resources allocated to high temperature membranes increased significantly in the 2002 period.

The polarization curves and functional relationships between platinum loading and performance were then combined with the stack cost model to assess the relationship between platinum loading and materials cost. Several overall assumptions are important to understanding the basis of the analysis. First, we assumed the performance of the catalyst is not limited by diffusion, the structure of the electrode, or the dispersion of the catalyst. In the early days of PEM fuel cell development, significant performance gains were made while significantly decreasing the catalyst loadings. In our analysis, reduction in platinum loading leads to lower power density. Second, introduction of ohmic resistance losses into the analysis reduces the power density that can be achieved by increasing catalyst loading or by decreasing the cell voltage. Voltage losses across the ohmic resistances (I^2R) in the system reduce the voltage available to perform useful work at the electrodes. Third, in the baseline cost estimate, the fuel stack materials represent over 90% of the stack cost. Consequently, in the following analysis, we only consider the material cost in the stack results (i.e., MEA, gas diffusion layer, and bipolar plate).

Figure 8 shows the material cost ($/kW$_e$) versus cathode platinum loading for stacks operating at 3 atm, 160°C, and 0.8 volts with direct hydrogen and reformate. Assumptions in this analysis include use of an alloy catalyst having a kinetic activity two times that of platinum, a unit cell resistance of 0.1 ohm cm^2, and an anode catalyst loadings one half the cathode loading.

Figure 8: Fuel Cell Stack Material Cost versus Cathode Platinum Loading

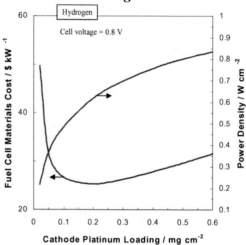

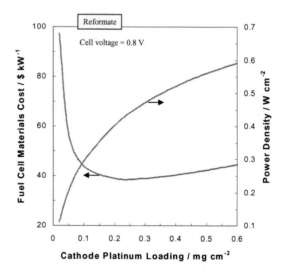

The analysis produced an "L" shaped curve where material costs rise sharply at low platinum loadings (low power density) and then show weak dependence with increasing platinum loading. Figure 9 shows the flat portion of the curve for various values of ohmic resistance. Independent of the resistance value, all of the curves have a minimum cost in the platinum loading region of

0.1 to 0.3 mg/cm^2. Insights into the factors influencing minimum platinum loading in addition to the value (mg/cm^2) and the resulting "L" shaped curve include:

- Reduction of platinum loadings beyond some minimum value negatively impacts cost due to the combination of increasing cost of the non-platinum materials in the stack (e.g., membrane, gas diffusion layer, and bipolar plate) relative to the platinum cost per unit area and the greatly reduced power density.

- Cost goes through a minimum with increasing platinum loading due to the negative impact of ohmic resistance on power density. The kinetic benefit of the increased platinum loading is not realized due to the voltage losses arising from ohmic resistances.

- This analysis shows the critical cost implications of ohmic losses within the stack.

Assumption of different material costs will shift the curves but not change their nature or general shape.

Figure 9: Stack Cost versus Ohmic Resistance

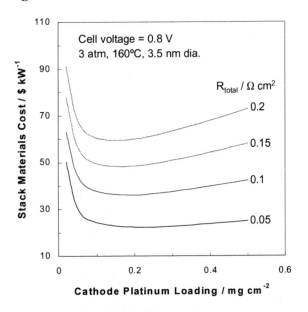

FUTURE SCENARIOS

Two future scenarios were considered; a direct hydrogen system with compressed hydrogen storage and a modified reformer system. The changes to the baseline reformate system included replacement of the catalyzed nickel autothermal reformer bed with a short contact time (high space velocity) reactor based on rhodium and the use of a single shift reactor bed. Furthermore, use of a higher temperature stack and assumption of no sulfur in the fuel allowed elimination of the PrOX and sulfur removal beds. For the hydrogen system projection, a cost of \$24/kW$_e$ was assumed for the compressed hydrogen storage system. For both scenarios, increased current densities were selected on the basis of the high temperature kinetic analysis. These assumptions are summarized in Table 5.

Table 5: Future Scenario Parameters and System Changes (Stack Operation at 0.8 volts and 160°C)

Parameter	Baseline	Future Reformate	Future Hydrogen
Stack Improvements			
◆ Current Density (mA/cm^2)	310	500	750
◆ Cathode Pt (mg/cm^2)	0.4	0.2	0.2
◆ Anode Pt (mg/cm^2)	0.4	0.2	0.1
◆ Anode Ru (mg/cm^2)	0.2	0.0	0.0
Fuel Processor Improvements	See Table 1	◆ Short contact time reactor ◆ Improved shift catalysts ◆ No sulfur bed ◆ No PrOX	◆ No Fuel Processor ◆ Compressed H$_2$ storage ◆ Simpler tailgas burner
System and Material Cost Reduction		Reduced Sensor, CEM, and Membrane costs	

Figure 10 shows the cost projections for the future system scenarios. The contribution of precious metal cost to each system is broken out separately to highlight their impact on system cost. This analysis compares the cost of the reformate and direct hydrogen systems on a 50 kW$_e$ basis. In practice, the lighter weight of direct hydrogen systems will produce equivalent vehicle performance (e.g., acceleration, hill climb) at lower net system power. The higher efficiency of the hydrogen system may also lead to operation of the stack at lower voltages (higher power density) leading to lighter, less costly stacks. Hence, the cost of the hydrogen system could be reduced further relative to the reformate system.

Figure 10: *Projections of Future Reformate and Direct Hydrogen System Costs*

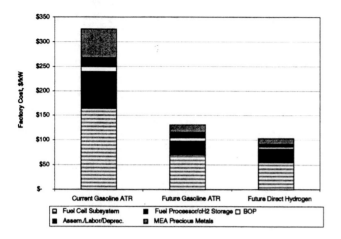

CONCLUSIONS

The future fuel system costs presented here are significantly lower than earlier cost estimates based on near-term technology. However, the projected costs for reformate and hydrogen systems are still higher than Freedom Car goals of "$45/kW_e." The direct hydrogen system is closest to the Freedom Car target, but will need technology advances in high temperature membranes and further reduction in material and component costs to achieve the cost goals.

ACKNOWLEDGEMENTS

Romesh Kumar and Daniel Doss of Argonne National Laboratories worked with us to develop the system configuration and the thermodynamic model. We gratefully acknowledge the financial support of DOE (SFAA No. DE-SC02-98EE50526) and the encouragement of the Fuel Cell Team within the Office of Advanced Automotive Technologies.

(1) U.S. Department of Energy, 2001 Annual Progress Report, "Fuel Cells for Transportation."

(2) Contract Number DE-AM26-99FT40465, Subcontract Number 000300008.

(3) Chalk, Steven G., Hale, H. Jackson, Wagner, Fred W., April 2-6, 2000. "Advanced Power Sources for a New Generation of Vehicles," 2000 Future Car Congress Proceedings, Hyatt Regency Crystal City, Arlington, VA.

Available online at www.sciencedirect.com

SCIENCE @ DIRECT®

JOURNAL OF
POWER
SOURCES

ELSEVIER

Journal of Power Sources 118 (2003) 349–357

www.elsevier.com/locate/jpowsour

Real life testing of a Hybrid PEM Fuel Cell Bus

Anders Folkesson[a,*], Christian Andersson[b], Per Alvfors[a], Mats Alaküla[b], Lars Overgaard[c]

[a]*Energy Processes, Department of Chemical Engineering and Technology, Royal Institute of Technology, Teknikringen 50, SE-10044 Stockholm, Sweden*
[b]*Department of Industrial Electrical Engineering and Automation, Lund University, Lund, Sweden*
[c]*Bus Chassis Pre-Development Department, Scania, Sweden*

Abstract

Fuel cells produce low quantities of local emissions, if any, and are therefore one of the most promising alternatives to internal combustion engines as the main power source in future vehicles. It is likely that urban buses will be among the first commercial applications for fuel cells in vehicles. This is due to the fact that urban buses are highly visible for the public, they contribute significantly to air pollution in urban areas, they have small limitations in weight and volume and fuelling is handled via a centralised infrastructure.

Results and experiences from real life measurements of energy flows in a Scania Hybrid PEM Fuel Cell Concept Bus are presented in this paper. The tests consist of measurements during several standard duty cycles. The efficiency of the fuel cell system and of the complete vehicle are presented and discussed. The net efficiency of the fuel cell system was approximately 40% and the fuel consumption of the concept bus is between 42 and 48% lower compared to a standard Scania bus. Energy recovery by regenerative braking saves up 28% energy. Bus subsystems such as the pneumatic system for door opening, suspension and brakes, the hydraulic power steering, the 24 V grid, the water pump and the cooling fans consume approximately 7% of the energy in the fuel input or 17% of the net power output from the fuel cell system.

The bus was built by a number of companies in a project partly financed by the European Commission's Joule programme. The comprehensive testing is partly financed by the Swedish programme "Den Gröna Bilen" (The Green Car). A 50 kW$_{el}$ fuel cell system is the power source and a high voltage battery pack works as an energy buffer and power booster. The fuel, compressed hydrogen, is stored in two high-pressure stainless steel vessels mounted on the roof of the bus. The bus has a series hybrid electric driveline with wheel hub motors with a maximum power of 100 kW.

Hybrid Fuel Cell Buses have a big potential, but there are still many issues to consider prior to full-scale commercialisation of the technology. These are related to durability, lifetime, costs, vehicle and system optimisation and subsystem design. A very important factor is to implement an automotive design policy in the design and construction of all components, both in the propulsion system as well as in the subsystems.
© 2003 Elsevier Science B.V. All rights reserved.

Keywords: PEM fuel cell system; Hybrid bus; Test; Hydrogen

1. Introduction

Even though only a small percentage of all vehicles in the world are urban buses, they impact disproportionately on public health. The reason for this is that they are concentrated to urban areas. The most important emissions from ordinary city transit buses with Compression Ignition (CI, i.e. Diesel) engines are particulates and nitrogen oxides (NO_x). Particles especially are believed to cause severe health problems and are possibly carcinogenic. Nitrogen oxides are important components in the formation of smog, contribute to acid rain, the eutrofication of lakes and seas and cause health problems.

Urban buses are, for several reasons, one of the best applications for commercialisation and testing of new, alternative fuels and advanced propulsion technologies:

1. They operate in urban areas where air pollution is considered a problem.
2. They are usually co-ordinated and fuelled centrally.
3. They are highly visible for the public.
4. They are often subsidised by government funds.
5. There is space available for the fuel cell system (both in volume and weight).

Fuel cell technology enables Zero or Ultra Low Emission Vehicles (ZEV or ULEV) with increased comfort due to very low noise levels and the possibility to use fully electric, stepless, drivelines. Fuel cells systems have high efficiencies, especially at part loads and have the potential for further

* Corresponding author. Tel.: +46-8-7906531; fax: +46-8-7230858.
E-mail address: anders.folkesson@ket.kth.se (A. Folkesson).

0378-7753/03/$ – see front matter © 2003 Elsevier Science B.V. All rights reserved.
doi:10.1016/S0378-7753(03)00086-7

development in building simple systems with long lifetime due to very few moving parts. Drawbacks with the fuel cell technology today are high costs, limited lifetime and the need of a new fuel infrastructure.

There are other types of environmentally friendly bus concepts. For example, busses with conventional Internal Combustion Engines (ICEs) fuelled with different kinds of alternative fuels, such as ethanol, methanol, Liquefied Petroleum Gas (LPG), Compressed Natural Gas (CNG), biogas, improved diesel, synthetic diesel or hydrogen. Different ICE/Electric hybrid configurations are also possible. Comparisons, evaluations and modelling studies of different alternative propulsion technologies can be found in the literature [1–9]. Studies of fuel cell systems and fuel cell electric hybrid propulsion systems for automotive applications have also been published [10–12]. New ICE technologies as Homogenous Charge Compression Ignition (HCCI) engines are under development as well as different ICE-based hybrid electric configurations. All those techniques have environmental benefits compared to conventional IC engines running on regular diesel or petrol, and are good environmentally friendly alternatives in the short perspective. However, given the limited resources of fossil fuels, fuel cells are, from a long-term perspective, a very competitive alternative for powering environmentally friendly vehicles, at least in urban areas. For long range heavy transports will the constantly improving CI-engine technology remain the best alternative for the foreseeable future. Fuel cells might be relevant for non-propulsion purposes such as cabin heating and electricity supply.

Several Fuel Cell Bus prototypes have been presented during the last decade and more are to come. However, there is very little information published about fuel cell vehicle testing, concerning both cars and buses [13–15]. Some new projects like the Clean Urban Transport for Europe (CUTE) project, now targets not only the evaluation of the vehicle technology but also the question of the necessary infrastructure [16].

2. Background/Hybrid Fuel Cell Concept Bus project

The objective of the Hybrid Fuel Cell Concept Bus project was to design and build a demonstration vehicle in the shape of a hybrid Fuel Cell Bus. The project was supported by funds from EU's Non-nuclear energy (Joule) programme. Several companies and institutes were involved as partners or participants in the project:

- Air Liquide (France)
  Project management.
  Fuel cell module (FCM) design and construction.
  Hydrogen storage system design and construction.
- SCANIA (Denmark/Sweden)
  Bus construction.
  Battery supply and construction.

SCANIA has ZF (Germany) as electric driveline supplier.
- SAR (Germany)
  Power bus controller and dc/dc electrical converter design.
- Nuvera Fuel Cells Europe (Italy)
  Fuel cell design and construction.
- Universita di Genova (Italy)
  Air compressor module design.
  Compressor from Opcon Autorotor (Sweden).
- Commissariat à l'Energie Atomique (France)
  Fuel cell tests.

The project started in 1996 and SCANIA was involved from 1999. The project was originally to be closed by the end of year 2000, but due to technical problems it was delayed to the summer of 2001. The concept bus study continues within a project supported by the Swedish National Research Programme for Green Car Research and Development (Sweden: Den Gröna Bilen). The aim of this project is to gather knowledge and experience in using fuel cells and hybrid technology in heavy vehicles. The project involves SCANIA, The Royal Institute of Technology—KTH (Stockholm) and Lund University of Technology—LTH (Lund). Results from the first part of that project; Testing of the Hybrid Fuel Cell Concept Bus is presented in this paper. Test results obtained in this part will be used in part two of the project, in simulation studies of future bus concepts.

3. The bus

The bus type is a construction from the 1990s for inner city and airport traffic, called SCANIA Service Bus. The bus is 9.2 m long, 2.5 m wide and 3.2 m high and has capacity for 15 seated and 37 standing passengers (Table 1). It is a true low floor bus, which means that the passenger compartment has a completely flat and low floor.

A diesel–electric hybrid bus, developed by SCANIA and the German company ZF in the 1990s, is partly used as base for the Fuel Cell Bus (Fig. 1), but it has been fundamentally reconstructed for the new fuel cell and hydrogen technology.

Fig. 1. The SCANIA Hybrid Fuel Cell Concept Bus.

Table 1
General technical description of the bus and the propulsion system

Bus type: SCANIA Service Bus	Description
Technical	
Dimension ($L \times W \times H$)	9.2 m × 2.5 m × 3.2 m
Max weight	13 t
Passenger capacity	52
Propulsion system type	Series hybrid with regenerative braking
Fuel cell system	
PEM FC stacks (×2)	2 × 105 cells
Power output (gross)	0–50 kW
Cooling	Water
Hydrogen storage	
Material	Stainless steel
Maximum pressure	200 bar
Capacity	13.2 kg H_2
Driveline battery	
Lead acid VR	44 × 12 V
Nominal voltage	528 V
Energy density	35 Wh/kg
Power density	380 W/kg
Cooling	Air
Driveline	
Motors (×2)	Wheel hub
Power output	2 × 50 kW
Cooling	Water

The bus is equipped with a series hybrid driveline (Fig. 2), which means that the driveline is completely electric and uses energy that is supplied from more than one source.

In this configuration, the driveline receives energy from both the fuel cell system and a battery. As the battery serves as a high power energy reservoir, it enables the use of a rather small, and therefore less expensive, fuel cell system.

The fuel cell system has a designed maximum power output of 50 kW. The fuel is compressed hydrogen and the oxygen for the fuel cell is compressed ambient air. An integrated dc/dc converter adjusts the fuel cell output voltage with the voltage of the common power bus (600 V).

The propulsion system is located in the rear end of the bus (see Figs. 1 and 3). The whole system, including the fuel cell system, battery, wheel motors and power electronics and auxiliaries can easily be removed from the rest of the bus. This simplifies service and other work on the system.

The propulsion system consists of

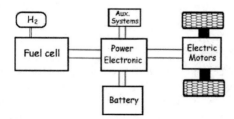

Fig. 2. A series hybrid system.

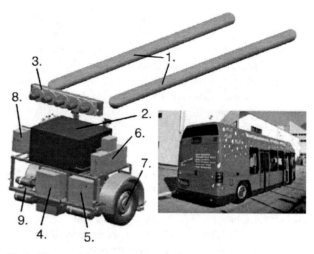

Fig. 3. The propulsion system located in the rear of the rear end of the bus.

(1) 200 bar hydrogen storage vessels;
(2) fuel cell system;
(3) radiator and fans for the secondary cooling system;
(4) high voltage battery module;
(5) power electronics;
(6) inverter and control system for wheel hub motors;
(7) wheel hub motors;
(8) auxiliary inverters; and
(9) auxiliary systems: air compressor and hydraulic system.

3.1. The fuel cell system design

In a fuel cell, the chemical energy in a fuel (i.e. hydrogen) is directly, without combustion, converted to electricity in an electrochemical reaction. If hydrogen is stored onboard a vehicle, the only local emission from the vehicle will be water. The fundamentals of fuel cells can be found in several previous published publications [17,18].

The fuel cell system (Fig. 4) consists of fuel cell stacks, a hydrogen circuit, an air circuit, a primary and a secondary

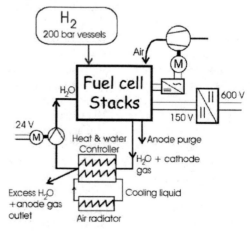

Fig. 4. The fuel cell system.

cooling water loop. The output voltage is harmonised with the high voltage system in the bus via a dc/dc converter.

The heart of the fuel cell system is the stack module with two PEM fuel cell stacks. Each stack contains 105 cells. The stack module has a maximum power output of 50 kW in the SCANIA Fuel Cell Bus configuration. The stack supplier is the Italian company 'Nuvera Fuel Cells Europe' and the two stacks were integrated into a complete fuel cell module designed and constructed by the French company 'Air Liquide'.

The stacks assembly components are metallic. Its dimensions are 58 cm height, 42 cm width and 57 cm length, giving a total volume of 139 l. This results in a power density of approximately 0.2 kW/l. It has to be noted that the stack design is from 1997/1998 if compared with today's state of the art stacks where power densities of >1.5 kW/l have been demonstrated [19,20]. The new generations of stacks that Nuvera works with today have an improved performance, with power densities of over 1 kW/l. With power densities of >1 kW/l, the stack size and weight is not a key problem in bus applications. More critical problems are the size and weight of the auxiliary systems, the fuel storage systems and the batteries.

Excess heat has to be removed from the fuel cell system in order to keep the temperature in the fuel cell stacks within the desired temperature interval. A lot of heat is transferred to the surroundings with the exhaust gases but the rest has to be cooled away. In the design tested, pure water is directly injected to the stack. This, the primary cooling circuit is heat-exchanged with a secondary cooling circuit, integrated with the heating system for the bus cabin. The secondary cooling systems are connected to a fan assisted radiator system, located on the roof of the bus. Some heat is also removed through a ventilation system that ventilates the fuel cell/engine section of the bus and via heat radiation.

A water management system is integrated with the thermal management system. There is no system for pre-humidification of the reactant gases. This reduces the cost, complexity, and size of the whole system. Instead of pre-humidification, the pure cooling water is directly injected into the stack where it both humidifies the membrane and controls the temperature of the stack, i.e. cools it down.

The outgoing airflow from the stack contains great amounts of water. In order to keep the global water balance, some of that water is collected in an exhaust gas condenser.

Air (i.e. oxygen) is supplied to the fuel cell via a twin-screw, oil-free, compressor. An air filter and silencer system is mounted prior to the compressor. The compressor motor is direct-powered by one of the fuel cell stacks, without conversion via the main dc/dc converter, but via conversion to ac electric power in a separate inverter for the compressor motor.

The fuel, hydrogen, is stored as compressed gas in two 200 bar stainless steel pressure vessels, located on the roof of the bus. The total amount of stored hydrogen gas is 875 l or 13.2 kg. The pressure is lowered in two stages before the gas enters the stacks. It is first lowered to <10 bar on the roof,

then transferred to the fuel cell where the pressure is lowered to the working pressure of the stacks.

4. Testing and test results

The performance of the bus has been tested thoroughly at IDIADA in northern Spain [21]. IDIADA is a commercial proving ground that vehicle producers from all over the world use for different kinds of vehicle testing. The facilities include:

1. high-speed circuit;
2. external noise test track;
3. dynamic platforms;
4. handling track;
5. general road circuit;
6. accelerated fatigue track;
7. test hills;
8. straight line braking surfaces;
9. comfort track;
10. customer workshops.

All tests were performed with the bus loaded with external weights so that the total weight was $12,500 \pm 25$ kg.

4.1. Data acquisition equipment

The data acquisition system used in the bus comes from IPETRONIK and is called SIM. The system is compact, with different kinds of modules for different sensor types. All the modules have in-built power supply (0–60 V) for the sensors. This makes the system suitable for automotive applications.

Software from National Instruments, called DIAdem, is installed on the notebook and used to control the data acquisition and to store the data on disc. This measurement data becomes available for further processing in Matlab. The programming in DIAdem is graphically similar to Lab-VIEW. An overview of the measuring system layout is shown in Fig. 5.

4.2. Aerodynamics and roll-resistance

To test the aerodynamics and roll-resistance of the bus, a roll-out test was done. The bus was accelerated to a certain speed, in wind-free conditions, and then allowed to roll on a flat road until it stopped. At a roll-out test from 60 km/h, the bus rolled 1300 m.

4.3. Acceleration

The acceleration performance of a bus is an important factor since it must be able to follow the traffic pace in a city. Acceleration tests were performed on a flat road in wind-free conditions and the acceleration performance is plotted in Fig. 6. The bus reaches 30 km/h in 7 s and 60 km/h in 25 s.

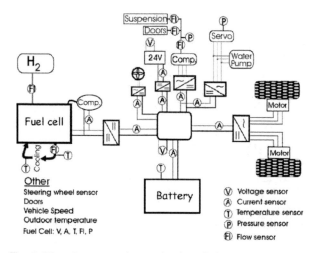

Fig. 5. Measuring system layout showing all important sensors and subsystems.

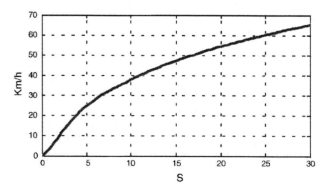

Fig. 6. Acceleration performance.

4.4. Hill climbing

The hill climbing capability of a bus is important in hilly cities, if the bus shall manage to drive certain routes. This requests a minimum of toque of the electric drive motors. The bus managed to climb an 18% steep hill for 75 m and a long 12% steep hill without problems.

4.5. Duty cycles

The bus was tested in accordance to several standardised duty cycles in order to evaluate the performance of the bus. Two duty cycles are discussed in this paper; the Braunschweig city duty cycle and the FTP 75 duty cycle. The Braunschweig city duty cycle is a recorded urban cycle, with high accelerations and many stops. It represents a typical driving pattern for a bus or a distribution truck travelling in urban areas. The duty cycle is used by MTC AB, a subsidiary of the Swedish Motor Vehicle Inspection Company (ASB) for testing of buses and distribution trucks. FTP 72 is also a recorded cycle. It is an American cycle from the beginning of the 1970s. The full version is called FTP 75, which is the FTP 72 cycle extended with 500 s. The bus was also tested in accordance with other cycles, such as artificially made duty cycle ECE 15. Those test results are not presented in this paper, due to the fact that they do not represent real city driving as FTP and Braunschweig do.

The bus ability to follow the standardised duty cycles is satisfactory. A comparison of the reference speed and the actual speed is shown for the Braunschweig and the FTP 75 duty cycles in Figs. 7 and 8.

A segment (55 s) showing a single acceleration and retardation in the Braunschweig cycle is shown in Fig. 9. The power from/to the electric drive motors are shown in the same figure. When the bus accelerates, the electric motors consume the power that the battery and the fuel cell supplies. Consequently, when the bus brakes the electric motors start to generate power, which is used to charge the battery. The efficiency of the regenerative system, i.e. the storage of energy into the battery during braking and the utilisation of energy from the battery during acceleration, was typically 85% in the tests.

The maximum total power from the battery and the fuel cell system is approximately 135 and >40 kW (net), respectively and the maximum power to the driveline motors during acceleration and from the driveline motors during regenerative braking are 130 and 100 kW, respectively.

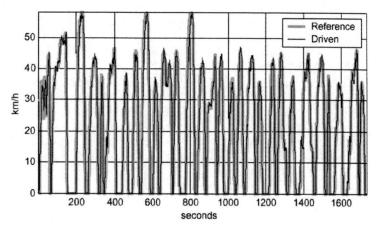

Fig. 7. The Braunschweig duty cycle. Presentation of reference speed and actual (driven) speed.

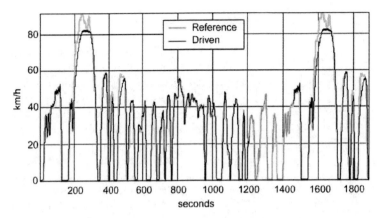

Fig. 8. The FTP 75 duty cycle. Presentation of reference speed and actual (driven) speed. Note that the maximum speed is electronically limited to 80 km/h.

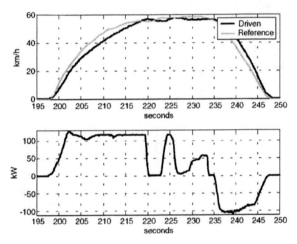

Fig. 9. Actual (driven) and reference speed and power from/to the electric motors for a segment of the Braunschweig duty cycle.

A comparison of the energy consumption for the Fuel Cell Bus at the two duty cycles and a standard SCANIA Omni City bus with a CI-engine is shown in Fig. 10. Energy consumption values are converted to diesel equivalents per 100 km. The energy consumption is between 42 and 48% lower for the Fuel Cell Bus than for the conventional bus. The figure clearly shows the advantages with a hybrid vehicle, in which the regenerated energy can be stored in a

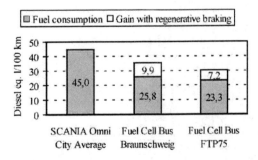

Fig. 10. The energy consumption of the Fuel Cell Bus, as diesel equivalents for the Braunschweig and FTP 75 duty cycles, in comparison with a SCANIA standard Omni City bus.

battery. The regeneration extends the range of the bus with 24–28% in these city duty cycles. Also, without the regenerative braking the bus would have been 21–32% more efficient than the standard bus.

4.6. Subsystems

The energy consumption of different subsystems in the bus was measured in order to map the energy flow in the bus and to find optimisation potentials. The subsystem consumes approximately 7% of the total energy input. The mean power consumption for the subsystems is 3–4 kW depending on the duty cycle, which is approximately 7% of the lower heating value of the consumed hydrogen. This includes ordinary bus stops (door openings and vertical adjustment of bus) at city driving but no air condition system. An air condition system for a 9 m bus, consumes up to 15 kW [13].

4.7. Fuel cell tests

The fuel cell module was specifically tested at the Air Liquide facility in Sassenage in France. Due to technical problems during the final fuel cell test, detailed test results are only presented up to 13 kW gross output power. Nominal power was achieved and tested at an earlier stage, hydrogen fuel consumption was not measured at that time, though. Results are presented for the tests and a prediction is made for higher loads in Fig. 11.

The temperature level in the stack during the tests was in the 50–75 °C range. The pressure on the air side was approximately 1.3 bars and on the hydrogen side approximately 1.5 bars. The stoichiometric factor (excess air factor) was approximately 1.5 during normal operation and as high as 4 at very low power output. The voltage and current levels of the systems are 145–180 V and 0–350 A, respectively.

4.8. Energy flow visualisation

It is interesting to study the different energy flows within the bus. In Fig. 12, the time average power flows during the

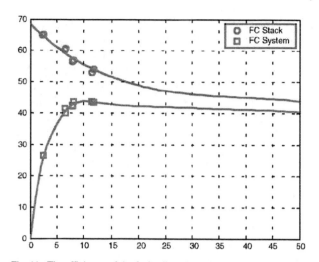

Fig. 11. The efficiency of the fuel cell stacks and the fuel cell system as a function of load. Rings and squares mark measured values and the lines predicted values.

Braunschweig cycle is shown. It is important to stress that the figure represents average values and that the situation shown in the diagram does not correspond to an actual situation.

4.9. Noise tests

External noise emissions of the bus were measured at IDIADA. The test was performed as an accelerated passage test, defined by the European regulation 70/157/EEC. In the test, the bus is accelerated to a speed of 50 km/h, which is held constant when it approaches the measuring area. As it enters, a full acceleration is performed during a 20 m long test (measuring) strip while the noise is measured at a specified distance on both side of the bus. The noise level

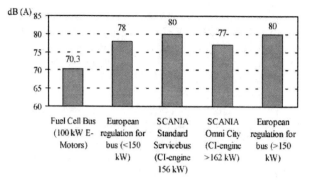

Fig. 13. Results from noise measurements, defined by the European regulation 70/157/EEC.

is compared with the same bus type, but with its standard engine and driveline configuration in Fig. 13. A comparison is also made with the present noise regulations. There is a breakpoint in the regulations for buses at 150 kW but it is not clear, though, which noise level the Hybrid Fuel Cell Bus shall be compared with. This is due to the fact that power sources together deliver more than 150 kW, but the driveline only consumes a maximum of 130 kW.

5. Experiences and conclusions

Hybrid Fuel Cell Buses have a big potential, not only with their high efficiencies and ZEV potential, they also offer other values such as very low noise levels and high comfort levels.

Regenerative braking gives a higher efficiency. Gains of between 24 and 28% can be made in urban duty cycles with many "stop and go" situations. The gain with a regenerative braking system is also achieved in ICE hybrid electric driveline systems.

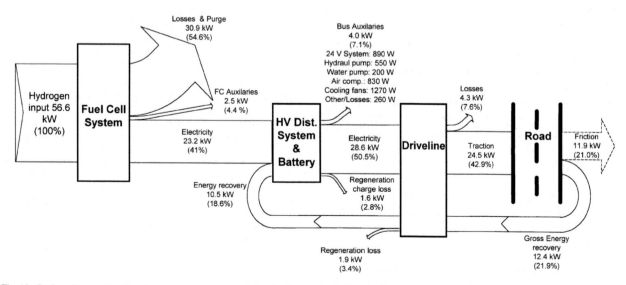

Fig. 12. Sankey diagram showing the average energy flows during the Braunschweig duty cycle and the proportion of all energy flows in relation of the total hydrogen input (based on LHV).

A. Folkesson et al. / Journal of Power Sources 118 (2003) 349–357

The mean power consumption for the 12.5 t concept bus is approximately 17–24 kW during the tested duty cycles. This means that a fuel cell system with a nominal power output of approximately 35–50 kW would be enough for a full size (12 m) hybrid electric city bus, even with a 20–25 kW air condition system installed. The energy buffer and power booster system, consisting of batteries, supercapacitors, or a mix of both, will then handle power peaks.

Many of the installations in the Fuel Cell Concept Bus are not optimised for automotive use concerning weight, size and lifetime. Nor is the bus designed for gaseous fuels from the beginning. Consequently, there is an optimisation potential in the general bus concept design. Electric drivelines in general enable completely new vehicle designs, with no limitations imposed by large mechanical transmission systems.

6. Future work

Several tasks or problems with the fuel cell technology and hybrid electric drivelines have to be solved prior to a mass-introduction of fuel cells on the vehicle market in general and on the urban bus market in particular.

6.1. Durability and lifetime

All propulsion system components as well as all subsystem components must be designed following automotive design rules for heavy-duty vehicles. The lifetime must be improved, in particular for the fuel cell stacks.

6.2. Cost reduction

Cost reduction can be achieved by lowering the complexity of the fuel cell system and by implementing industrial standards and better manufacturing methods. Development of new materials or a different choice of materials is also an important matter of concern. For instance, the noble metals that are used as catalyst in the fuel cells stacks and in fuel reformer systems are both a limited resource and expensive. Also the cost for other components in the propulsion system must be reduced. For example, driveline components such as power electronics and motors, or energy buffer systems such as batteries or supercapacitors. The cost of a new fuel infrastructure is also an important issue.

6.3. Fuel storage systems

Safe, light, energy efficient and inexpensive fuel storage systems for onboard storage of hydrogen must be developed.

6.4. Cooling system

A problem with this generation of PEM fuel cells is the relatively low working temperature of the stacks (70–85 °C). This demands large cooling systems due to the low temperature difference between the warm medium and the surroundings. The temperature difference is the driving force in the heat exchange process and a low difference requires a high flow of the cooling media and/or large heat exchanger surfaces. Another problem is that the cooling liquid will have a temperature that is too low for use in heating systems designed for CI-engines even though the amount of heat energy available is enough. This in turn leads to special and expensive systems only suitable for fuel cell powered vehicles. The temperature of the fuel cell systems have to increase to match or at least be closer to the temperature in CI-engine cooling systems.

6.5. Voltage

Higher voltage and consequently lower currents in the fuel cell system will lead to an easier handling of the power conversion and therefore lower cost, volume and weight of the power conversion system.

6.6. Optimisation

The specification of the vehicle, i.e. the relation in capacity of the primary energy source (fuel cell) contra the energy buffer (battery, supercapacitors) must be optimised. It is also necessary to design the whole vehicle for gaseous fuels from the beginning.

6.7. Subsystems

All subsystems that are not electrical today must be replaced by similar optimised electrical subsystems.

Acknowledgements

This work was financed by VINNOVA (Swedish Agency for Innovation Systems) and SCANIA.

References

[1] P. Mizsey, Comparison of different vehicle power trains, J. Power Sources 102 (2001) 205–209.
[2] F. Kreith, R.E. West, B.E. Isler, Efficiency of advanced ground transportation technologies, J. Energy Res. Technol. 124 (2002) 173–179.
[3] L. Lave, H. MacLean, C. Hendrickson, R. Lankey, Life-cycle analysis of alternative automobile fuel/propulsion technologies, Environ. Sci. Technol. 34 (17) (2000) 3598–3605.
[4] K. Jonasson, P. Strandh, M. Alaküla, Comparative study of generic hybrid topologies, in: Proceedings of the 18th International Electric Vehicle Symposium, Berlin, Germany, October 20–24, 2001.
[5] G. Rizzoni, L. Guzzella, B.M. Baumann, Unified modeling of hybrid electric vehicle drivetrains, IEEE/ASME Trans. Mechatron. 4 (3) (1999) 246–257.
[6] A.K. Shukla, A.S. Aricò, V. Antonucci, An appraisal of electric automobile power sources, Renew. Sustain. Energy Rev. 5 (2001) 137–155.

[7] L.U. Gökdere, K. Benlyazid, R.A. Dougal, E. Santi, C.W. Brice, A virtual prototype for a hybrid electric vehicle, Mechatronics 12 (2002) 575–593.

[8] P. Van den Bossche, Power sources for hybrid buses: comparative evaluation of the state of the art, J. Power Sources 80 (1999) 213–216.

[9] A.F. Burke, M. Miller, Fuel efficiency comparisons of advanced transit buses using fuel cell and engine hybrid electric drivelines, in: Proceedings of the 35th Intersociety Energy Conversion Engineering Conference, Las Vegas, NA, USA, July 24–28, 2000, vol. 2, 2000, pp. 1333–1340.

[10] H.-G. Jeong, B.-M. Jung, S.-B. Han, S. Park, S.-H. Choi, Modeling and performance simulation of power systems in fuel cell vehicle, in: Proceedings of the Third International Power Electronics and Motion Control Conference, Beijing, China, August 15–18, 2000, vol. 2, 2000, pp. 671–675.

[11] S.G. Chalk, J.F. Miller, F.W. Wagner, Challenges for fuel cells in transportation applications, J. Power Sources 86 (2000) 40–51.

[12] K.S. Jeong, B.S. Oh, Fuel economy and life-cycle cost analysis of a fuel cell hybrid vehicle, J. Power Sources 105 (2002) 58–65.

[13] R. Wimmer, Fuel cell transit bus testing and development at Georgetown University, in: Proceedings of the 32nd Intersociety Energy Conversion Engineering Conference, Honolulu, HI, USA, July 27 to August 1, 1997, vol. 2, 1997, pp. 825–830.

[14] G. Friedlmeier, J. Friedrich, F. Panik, Test experiences with the DaimlerChrysler fuel cell electric vehicle NECAR 4, Fuel Cells 1 (2) (2001) 92–96.

[15] J.T. Larkins, R.R. Wimmer, Fuel cell powered transit bus demonstration and test program, in: Proceedings of the 1998 Fuel Cell Seminar, Palm Springs, CA, USA, November, 1998, pp. 734–737 (Abstracts).

[16] Fuel Cells Bull. March (42) 2002.

[17] J. Larminie, A. Dicks, Fuel Cell Systems Explained, Wiley, England, 2000.

[18] L. Carrette, K.A. Friedrich, U. Stimming, Fuel cells—fundamentals and applications, Fuel Cells 1 (1) (2001) 5–39.

[19] Ballard Fuel Cell Power Module Mark 900 Series, Brochure, 100-0015-PA, Ballard Power Systems, 2000.

[20] A.P. Meyer, M.E. Gorman, D.M. Flanagan, D.R. Boudreau, Progress in the development of PEM fuel cell engines for transportation, SAE Technical Paper 2001-01-0540, 2001.

[21] http://www.idiada.es.

249

2002-01-0414

Long-Term Prospects for PEMFC and SOFC in Vehicle Applications

Jan H.J.S. Thijssen and W. Peter Teagan
Arthur D. Little, Inc.

ABSTRACT

After about a decade of considerable investments in polymer electrolyte fuel cell (PEMFC) and in solid oxide fuel cell (SOFC) technology, both are being actively considered for vehicle applications. The two vehicle applications being most actively considered for fuel cells are propulsion (mainly for PEMFC) and auxiliary power (for both PEMFC and SOFC). For all transportation applications, fuel cells promise the benefits of clean and quiet operation, potentially low maintenance and high efficiency, and ultimately greater utility to drivers and passengers.

Initial system and vehicle prototypes have started to demonstrate some of these benefits, but much technology development is still needed before commercialization can occur. Not surprisingly then, there are serious hurdles to be overcome if fuel cells are to become true competitors for internal combustion engines (ICEs) in automotive applications. The most serious hurdle appears to be manufactured cost, which would appear to be several times too high for propulsion applications to be competitive, based on today's technology. Another considerable hurdle is power density, which is critical to be able to develop compelling fuel cell vehicles (FCVs) without sacrificing vehicle utility or performance. Finally, proving and improving system life should receive far more attention than has been the case so far.

While the barriers may seem daunting, the automotive industry and other stakeholders are bringing many of the critical scientific and engineering resources to bear on solving the problem.

INTRODUCTION

Over the past decades, significant progress has been made in the improvement of both performance (notably power density and compatibility with environmental conditions) and cost of proton exchange membrane fuel cell (PEMFC, sometimes also referred to as polymer electrolyte fuel cells) and solid oxide fuel cell (SOFC) technology. These improvements have led to the consideration of both technologies for automotive applications. The rapid pace of technology development and a barrage of often contradictory and confusing press announcements make it difficult for the automotive engineer to understand how these intriguing technologies may come to affect powertrain and vehicle design in the next decades. Although our crystal ball is probably no more accurate than yours, we have developed a unique understanding of both the requirements of likely vehicle applications and the emerging capabilities and cost of fuel cell power systems which we would like to share with you. This knowledge is based on over a decade of intimate involvement with both the fuel cell industry and with IC engine developers on in-depth studies of power system and vehicle architecture as well our well-to-wheels comparisons[1,2]. In addition, our involvement in both fuel cell and ICE technology development programs provides us with a unique perspective on the technology development challenges that are faced by the emerging fuel cell industry as they apply to automotive applications.

In this paper we first review the two leading potential applications of fuel cells and the benefits fuel cells could provide in each of these applications. Next we will discuss the key technical and cost barriers to market success for fuel cells. We will conclude with an outlook on the opportunities for fuel cells in the vehicles of the next decade.

CURRENT FUEL CELL DEVELOPMENT STATUS

STAGE OF DEVELOPMENT - Over the past ten years, power densities of both PEMFC and SOFC stack technology have improved to exceed 500 mW/cm² under typical operating conditions relevant for automotive applications, with peak power densities under idealized conditions of greater than 1000 mW/cm² reported for both technologies[3,4,5,6,7,8]. Such power densities translate to high volumetric stack power densities in excess of 1 kW/L. for PEMFC. SOFC have lower volumetric stack power densities because of the relatively thick interconnects required and because of the volume of the system insulation required to support the high operating temperature.

Nevertheless, fuel cell systems for vehicles are still at an early stage of development as reflected in characteristics of current prototype vehicles as shown in Figure 1. Several prototype FCVs have been built over the past years. While these vehicles provide valuable experience with fuel cell / vehicle integration and with the operation of fuel cells in vehicles, the vehicles hardly provide the functionality commercial FCVs will have to provide. The systems take up too much space, are too heavy, and the system life and reliability are still a question mark. Hence we see current FCV prototypes as initial vehicle prototypes.

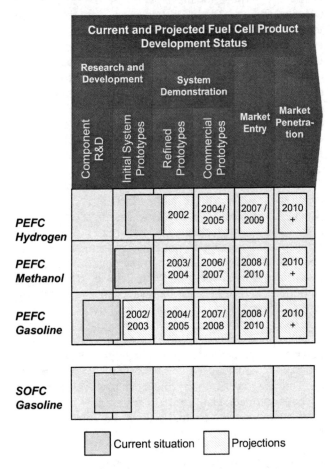

Figure 1. Overview of Current Development Status of Automotive Fuel Cell Technology (source, Arthur D. Little Analysis).

Some of the OEMs have projected commercial vehicle production starting anywhere in the 2003 - 2010 timeframe[9,10]. However, we expect that such vehicles will in reality be sold into "controlled" markets such as fleets before then and that their manufacturing cost may be subsidized by the manufacturer, the owner, or a third party such as a

government agency. In addition, such vehicles will likely be direct-hydrogen fueled. Because we do not expect hydrogen fuel to be widely available at that time, special arrangements must be made to ensure access to fuel for these vehicles in the local areas they drive in. In general, the auto industry and fuel cell developers have recognized that considerable further technology improvement is required and with it considerable R&D spending, in order that vehicles can be produced that are economically viable under normal economic circumstances. Our current industry projections for development and commercialization are also summarized in Figure 1. The projections are based on our experience with technology and product development times.

Another observation that can be made from Figure 1 is that PEMFC is only a little bit ahead of SOFC in the stages of technology development. This is somewhat counter to conventional wisdom in the fuel cell community (which has held that PEMFC are considerably closer to commercial viability than SOFC). However, while this may be true for some of the SOFC players, a few SOFC developers have steadily accumulated system operating experience that likely exceeds the operating experience with PEMFC systems.

Below we first summarize some of the vehicle prototype efforts on-going and then the public APU development efforts.

CURRENT FUEL CELL VEHICLE PROTOTYPES - Several manufacturers have built prototype fuel cell vehicles. Most claim fuel cell vehicles will be available for purchase before the end of the decade, several in the next few years in limited markets. To aid in the commercialization of fuel cell vehicles, meet these commercialization plans, and contribute to improvements in air quality, the California Fuel Cell Partnership was formed. This joint project is aimed at demonstrating the everyday practicality of fuel cell vehicles and preparing the California market for this new technology. The partnership plans to test more than 70 cars and buses between 2001 and 2003, incorporating innovative drive technologies under everyday operating conditions fueled by hydrogen, methanol, and a pure form of gasoline. Table 1 provides an overview of current FCV prototypes. The list is not intended to be complete, and some of the European manufacturer prototypes are omitted. Still, it provides an idea of the key characteristics of current systems.

In understanding Table 1, it is important to recognize that prototype FCVs are based on conventional vehicle bodies, modified to fit the fuel cell and electric drivetrain. Optimization of the integration of the fuel cell system with the vehicle body and other components is clearly far from ideal in these vehicles. One would expect that significant improvements could be made, especially once the definition of fuel cell system architecture and characteristics become clearer over the next few years. Nevertheless, some important lessons about fuel cell performance today, and the challenges for the coming years can be learned from the vehicle characteristics.

Table 1. Current Concept Light-Duty Fuel Cell Vehicles

Features	Daimler-Chrysler	Daimler-Chrysler	General Motors
Vehicle	NECAR 4A	NECAR 5	HydroGen1
Platform	A-Class	A-Class	Opel Zafira
Body Style	4-Door	4-Door	Van
Overall Length	3.57 m	3.57 m	4.32 m
Curb Weight	1750 kg	1430 kg	1570 kg
Comparable ICE weight	1190 kg	1190 kg	1393 kg
Fuel	Compressed H_2	Methanol	Liquid H_2
Fuel Pressure	35 MPa	--	--
Range	190 km	480 km	400 km
Top Speed	145 km/h	150 km/h	135 km/h
Fuel Cell	Ballard Mark 900	Ballard Mark 900	GM 60 kW PEMFC
Electric Motor	55 kW	55 kW	56 kW
Features	**Ford**	**Ford**	**Mazda**
Vehicle	Focus FCV	Focus FCV	Premacy
Platform	Ford Focus	Ford Focus	Premacy
Body Style	4-Door Sedan	4-Door Sedan	5-Door Sedan
Overall Length	4.34 m	4.34 m	4.35 m
Curb Weight	1727 kg	1769 kg	1850 kg
Comparable ICE weight	1180 kg	1180 kg	1630 kg*
Fuel	Compressed H_2	Methanol	Methanol
Fuel Pressure	24 MPa	N/A	N/A
Range	160 km	-	-
Top Speed	128 km/h	128+km/h	124 km/h
Fuel Cell	Ballard Mark 900	Ballard Mark 901	65 kW Ballard Mark 901
Electric Motor	67 kW	65 kW	65 kW
Features	**Volkswagen**	**Honda**	**Toyota**
Vehicle	Bora Hymotion	FCX-V3	FCHV-4
Platform	Volkswagen Jetta	EV Plus	Highlander
Body Style	4-Door Sedan	2-Door	SUV
Overall Length	4.38 m	4.05 m	2.72 m
Curb Weight	N/A	1750 kg	N/A
Comparable ICE weight	1355 kg	1630 kg*	1636 kg
Fuel	Liquid H_2	Compressed H_2	Compressed H_2
Fuel Pressure	--	25 MPa	25 MPa
Range	355 km	180 km	250 km
Top Speed	145 km/h	130 km/h	150 km/h
Fuel Cell	Ballard Mark 900	Ballard Mark 900	90 kW PEMFC
Electric Motor	67 kW	75 kW	80 kW

Features	Hyundai	Nissan
Vehicle	Santa Fe FCV	Xterra FCV
Platform	Hyundai Santa Fe	Nissan Xterra
Body Style	SUV	SUV
Wheelbase	2.62 m	2.65 m
Curb Weight	1615 kg	N/A
Comparable ICE weight	1588 kg	1622 kg
Fuel	Compressed H_2	Compressed H_2
Fuel Pressure	35 MPa	25 MPa
Range	200 km	--
Top Speed	128 km/h	120 km/h
Fuel Cell	IFC S300	Ballard Mark 900
Electric Motor	65 kW	N/A

*EV plus is an electric vehicle, not ICE
Source: reference 11,12.

Fuel Choice – Current FCV prototypes are predominantly fueled by hydrogen, either compressed hydrogen or liquid hydrogen. A few methanol-fueled vehicles were built (including one by Toyota not listed here) and other developers are working on methanol-fueled vehicles. Due to the relative simplicity of the on-board hydrogen systems, hydrogen-fueled vehicles carry less technical risk, and are therefore moving along the technology development timeline more rapidly. So far, no gasoline- or hydrocarbon-fueled FCVs have been tested, but GM has announced the construction of a S10 pick-up truck, of which a mock-up was shown at several occasions[13]. Other developers have announced that they are working towards prototypes. While the challenges of reforming are phenomenal in the early development of FCVs, the choice of liquid fuels facilitates the fuel infrastructure aspects of FCV commercialization[1].

Operating Pressure – Though not listed in Table 1, virtually all of the prototype vehicles operate with pressurized stacks. In principle, this helps the oxygen reduction kinetics on the cathode, and could lead to higher stack power density. However, other system factors can offset this, at least partially. In fact, the Hyundai vehicle, which is equipped with an atmospheric pressure IFC stack is relatively less heavy than the others vehicles.

Vehicle Weight – Table 1 shows clearly that most of the current FCV prototypes are considerably heavier than their conventional ICE-based counterparts. The Honda FCX-V3 is also heavier than the all-electric EV+ that it is based on[14]. Although some of this is related to the limited amount of vehicle integration, much of the difference is the result of the higher weight of the fuel cell, and the fuel storage or conversion system. This heavier weight is part of the reasons that the performance of the vehicles, in terms of acceleration and top speed is not as good as the ICE counterparts. The fuel cell system power is also lower, to avoid having even higher weights, which compounds the system. In other words, the fuel cell power density (on a mass basis) must be significantly improved to achieve better weight / performance characteristics. Both Ballard / Xcellsis, GM, and IFC have announced stack peak power densities in the 1.5 to 2.2 kW/L range. Although no weight numbers were provided, this high volumetric power density indicates that the weight is also significantly reduced, compared with the weights of the stacks as they are used in the prototype vehicles listed in Table 1. Still, combined with the balance of plant weight for the fuel cell systems, additional weight reduction will be required in all parts of the fuel cell system to achieve the desired system weight.

Vehicle Range - The range of many of the FCV prototypes is not adequate for current consumers, who have come to expect a range of well over 500 km for gasoline vehicles in the classes represented here (and about a 1000 km if the comparison is with state-of-the-art diesels). The key to this problem is the energy storage density of the fuels used in the FCV prototypes listed. The energy density of gasoline is around 35 MJ/L (HHV). Methanol has an energy density of about 18 MJ/kg. The difference with gasoline is largely offset by the higher efficiency of the fuel cell powerplant compared with conventional engines, even in these prototype vehicles. The energy density of liquid hydrogen is around 10 MJ/L, and that of compressed hydrogen at 35 MPa is a little more than 4 MJ/l. Due to the low energy density of stored hydrogen, a trade-off between vehicle range and the available trunk space arises, even though the direct hydrogen vehicles have even higher powerplant efficiencies than methanol-fueled FCVs. In the case of the current hydrogen-fueled FCV prototypes, both range and trunk space are compromised compared with conventional vehicles. Ultimately, this problem could be alleviated the use of higher hydrogen storage pressures, or by storing the hydrogen on-board as a metal or chemical hydride. Gasoline-based FCVs would have particularly compact fuel storage, combining high efficiency with a high-energy density fuel. However, the volume other fuel cell system components take up must also be reduced to achieve an attractive overall package.

Thermal Management – Despite the high efficiency of the fuel cell systems (around 30-60% depending on fuel choice, system integration, and operating conditions), a significant amount of heat must be rejected to the environment. Because of the rather low operating temperature (typically 75-90°C for automotive applications) of the stack, which is where most of the losses occur, only a small portion of the heat generated

is carried away by the exhaust gases. As consequence, to avoid overheating the stack the stack must be cooled, and the heat removed by the cooling fluid must be rejected to the environment, typically through a radiator. Because of the modest temperature, the driving temperature difference between the coolant temperature and the environment can be quite small if the outside temperature is high. This then necessitates a large amount of radiator area, several times more than for conventional ICEs. Of course this adds additional volume and weight to the vehicle, and it lowers the power density of the overall fuel cell systems.

Trunk Space – Due to the low energy density of compressed hydrogen, and the size of the fuel cell system itself, most of the FCV prototypes have severely compromised trunk or passenger space.

System Life – Because these vehicles are early prototypes, they were not specifically meant for long-term testing. However, the limited number of miles that appear to have been driven with most of these vehicles along with the limited long-term testing experience with stationary PEMFC systems give reason to believe Stationary PEMFC systems have achieved operation of hundreds to a few thousand hours. Because the stack is the most expensive part of the system, it appears critical that long-term test experience is developed with fuel cell systems on vehicles; periodic replacement of the stack would be unacceptably expensive.

Vehicle Performance – The high weight and modest power of current FCV prototype vehicles limits their acceleration and top speed. However, as electric-drive vehicles (i.e. all power to the wheels comes directly from electric motors, unlike in a parallel HEV in which part of it comes from the ICE via the gearbox), FCVs already provide reasonably good initial acceleration from standstill.

Quiet Operation – Fuel cell vehicles are generally reasonably quiet in operation. The only thing that currently somewhat interferes with this benefit is the air compressor, which makes some of the FCV prototypes fairly noisy. However, we believe that in future vehicles this noise will be reduced, and that with appropriate sound insulation, FCVs can become truly quiet vehicles. APU PROTOTYPE DEVELOPMENTS – The potential application of fuel cells as vehicle APUs has more recently become the subject of fuel cell system development. Nevertheless, several fuel cell APU prototypes were constructed or are under development. Table 2 provides an overview of those efforts. These are clearly initial system prototypes that demonstrated some of the concept but not exactly the feasibility of incorporating fuel cells into vehicles as APUs.

Table 2. Overview of APU-Related Fuel Cell Activities

Participants	Application	Size range	Fuel /Fuel Cell type	Nature of Activity
BMW, International Fuel Cells[15]	Passenger Car, BMW 7-series	5kW net	Hydrogen, Atmospheric PEM	Demonstration
Ballard, Daimler-Chrysler[16]	Class 8 Freightliner heavy-duty Century Class S/T truck cab	1.4 kW net for 8000 BTU/h A/C unit	Hydrogen, PEM	Demonstration
Ford[17]	Passenger Cars	Around 5 kW	SOFC, direct hydrocarbon SOFC	R&D effort
BMW, Delphi, Global Thermo-electric[15]	Passenger car	1-5kW net	Gasoline, SOFC	Technology development program

There are three noteworthy observations that may be made in relation to the fuel cell APU development programs:

- System capacity for APU applications is more than ten times smaller than for propulsion
- Overall volumetric power density of fuel cells may be a serious challenge for APU applications, perhaps even more so than cost. The power density for the smaller systems is typically lower than for the larger systems, as certain components do not scale down favorably in volume (e.g. insulation, valves, blowers, control boards, etc.)
- Unlike the fuel cells for propulsion programs, fuel cell APU development programs have been largely commercially driven, with limited government funding.

Technology status issues surrounding PEMFC technology as it could be applied in APU applications more or less mirror those that apply in propulsion applications, and are not repeated here. The issues facing SOFC are worth describing briefly here.

SOFC technology has been under development for several decades, focussing mostly on its applications to (industrial-scale) stationary power generation. A myriad of SOFC stack architectures has been under development, but until recently the most notable progress in SOFC technology was made by Siemens-Westinghouse with its tubular SOFC technology. The Siemens technology has been mainly focussed on stationary applications with capacities greater than 100 kW, and into the MW-range. This application range is partly determined by the relatively high operating temperature of around 900 – 1000°C[18].

In recent years a new generation planar SOFC stack has evolved, based on a cell design with a very thin (as thin as single-digit microns) electrolyte. This cell design allows operation at much reduced operating temperatures (around 750-850°C) and much higher power densities[7, 8]. The lower operating temperature significantly reduces the challenges associated with sealing the SOFC stack, and with the thermal management within the system. This development has led directly to the development of an APU system prototype by Delphi and Global Thermoelectric.

However, for the SOFC technology, achieving acceptable start-up times may pose a significant challenge. Arthur D. Little analyses have determined that the thermal mass of a stack together with expected limitations on internal temperature gradients within the geometry of a stack would likely necessitate cold-start times of more than ten minutes to more than half an hour[19]. Although system insulation can help to keep systems warm for some time, they are still expected to cool down overnight. Even if the half-hour start-up time is realized, this would only be acceptable in some of the APU applications. For others, such a long time would not be acceptable.

TYPICAL FUEL CELL SYSTEM ARCHITECTURE AND PERFORMANCE CHARACTERISTICS

Before we consider the benefits and disadvantages of fuel cells, we must discuss the fuel cell system's architecture and general characteristics as fuel cells represent the most radically different of the alternative power technologies for vehicles currently under consideration.

a	Gasoline-Fueled PEMFC Power Unit

Fuel Processor Subsystem

Reformate Generator	Reformate Conditioner
♦ Autothermal Reformer	♦ Ammonia Removal
♦ Shift Reactors	♦ CO-Clean-Up
♦ Sulfur Removal	♦ Anode Gas Cooler
♦ Air Preheater	♦ Economizers (2)
♦ Steam Superheater	♦ Anode Inlet Knockout Drum
♦ Reformate Humidifier	

Water & Fuel Supply	
♦ Fuel Pump	♦ Heat Exchanger
♦ Fuel Vaporizer	♦ Steam Generator & Drum
♦ Water Separators (2)	♦ Process Water Reservoir

Fuel Cell Subsystem

♦ Fuel Cell Stack (Unit Cells)	♦ Compressor/Expander
♦ Balance of Stack Hardware	♦ Anode Tailgas Burner
♦ Fuel Cell Heat Exchanger	♦ Air Purification System

Balance-of-Plant

♦ Startup Battery	♦ System Controller
♦ System Controller	♦ Sensors & Control Valves
♦ System Packaging	♦ Startup Battery

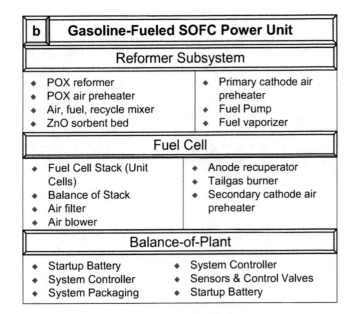

b	Gasoline-Fueled SOFC Power Unit	
Reformer Subsystem		
• POX reformer • POX air preheater • Air, fuel, recycle mixer • ZnO sorbent bed	• Primary cathode air preheater • Fuel Pump • Fuel vaporizer	
Fuel Cell		
• Fuel Cell Stack (Unit Cells) • Balance of Stack • Air filter • Air blower	• Anode recuperator • Tailgas burner • Secondary cathode air preheater	
Balance-of-Plant		
• Startup Battery • System Controller • System Packaging	• System Controller • Sensors & Control Valves • Startup Battery	

Figure 2. Typical System Components for PEMFC (a) and SOFC (b) Systems

Fuel cell systems are built around fuel cell stack, which is a combination of single electrochemical fuel cells in which air and fuel are electrochemically reacted, producing electric power and heat in the process. Most commonly these single cells are planar, although tubular geometries enjoy considerably popularity in SOFC systems, especially for larger stationary applications. Any of these stacks must be supported by several other subsystems, which are described in Fig. 2 for gasoline-fueled SOFC and PEMFC systems. These equipment lists are based on detailed studies Arthur D. Little has carried out of a 50 kW automotive propulsion PEMFC power unit and of a 5 kW POX-planar SOFC truck APU[19, 20].

PEMFC – PEMFC stacks operate at low temperature and must be operated moist, thus requiring humidification. While these operating parameters allow the use of relatively low-cost construction materials, and simplify sealing. However, the low operating temperature makes the stack catalysts very sensitive to contaminants, such as carbon monoxide and sulfur. These characteristics require the use of pure hydrogen as a fuel, or the use of a reformer to convert other fuels into a humidified, hydrogen-rich mixture of acceptable quality.

Hydrogen is neither widely available nor inherently convenient to store on-board and reformers can add significant volume and cost to the system, as illustrated by the current FCV prototypes. The alternative, reformers can considerably complicate the system because they add a large number of necessary components (Fig. 2). Although methanol reformers can be somewhat simpler than gasoline reformers, the increased system complexity, compared with direct hydrogen systems, is still significant.

With direct methanol fuel cells (DMFCs) reformers could be avoided as well as the challenges of on-board hydrogen storage. Although most automotive PEMFC fuel cell efforts have been focused on the development of hydrogen-fed PEMFCs (either from on-board stored hydrogen or via a reformer), some are currently developing DMFCs, which could use methanol as a fuel without reforming. DMFCs are still in a very early stage of development and, based on their current performance, are not yet suitable for incorporation into a vehicle platform.

A significant challenge in the development of DMFCs is overcoming the methanol cross-over, which leads to oxidation of fuel without the production of electric power. It significantly reduces the power density and efficiency of the stack.

SOFC – Although conceptually the stack design of SOFC systems considered for automotive applications (the so-called planar bi-polar design) is similar to that of PEMFCs, the operating conditions are considerably different. The ceramic electrolyte can and must be operated at high temperatures (600 – 850°C for current technology) to achieve acceptable conductivity. This high operating temperature brings with it considerable challenges in thermomechanical design and materials development as well as requiring significant heat recovery and insulation for thermal management. However, it provides the benefit that a variety of fuels can be used in the stack directly, including hydrogen and carbon monoxide. This would require only a very simple reformer to produce a carbon monoxide and hydrogen containing gas from gasoline or other simple hydrocarbons. In addition, some hydrocarbons can be converted inside the fuel cell itself (internal reforming), further simplifying the reformer design.

If efforts aimed at the complete direct oxidation of or internal reforming of hydrocarbons within the stack are successful, the SOFC system could be further simplified. However, similar to DMFC technology, this technology is currently not sufficiently far developed for SOFC and is facing significant development challenges. Also, because the reformers in SOFC systems are rather simple, the benefit of eliminating them is not as significant.

FUEL CELLS FOR VEHICLE APPLICATIONS: THE BENEFITS AND THE REQUIREMENTS

The two principal vehicle applications for which fuel cells are being considered are propulsion and auxiliary power. Both the benefits of fuel cells in each application and the requirements of the applications must be considered.

POTENTIAL BENEFITS OF FUEL CELLS AS AUTOMOTIVE POWERTRAINS - The interest fuel cells generate in the automotive engineering community stems mostly from four benefits that fuel cells promise for these two applications.

Low Emissions Potential – One of the strongest drivers currently for the investment in fuel cells for vehicles are ever-tightening local air quality regulations. Probably the most stringent proposed local air pollution regulation, California's Zero Emission Vehicle (ZEV) mandates, may effectively mandate some percentage of vehicles sold in California to be all-electric vehicles. Because FCVs could well provide that ZEV environmental performance while still providing good vehicle utility (e.g., they could provide better range and more rapid refueling than battery electric vehicles (BEVs). ZEV mandates are scheduled to be introduced within the next few years in California and some of the Northeastern states. Though the metrics by which different fuel cell vehicles are being counted are still being developed, the mandates could create a market large enough to support true mass manufacturing of FCVs, provided the number of models introduced is not too large. Similar but less drastic regulations are being considered in other regions.

Fuel Efficiency and Greenhouse Gas Emissions Reduction Potential - Fuel cells have the potential to improve the energy efficiency of vehicles. This aligns with the desires to conserve resources and reduce greenhouse gas. This effect could be strengthened if fuel cells were to use a renewable fuel. However, other powertrain technologies such as HEV technology and the use of renewable fuels (such as ethanol) in conventional vehicles could well achieve a similar impact.

Figure 3, based on a recent in-depth Arthur D. Little study of well-to-wheels energy use of fuel cell vehicles illustrates these trends[1, 2]. This analysis is based on a technology scenario for the 2008 – 2010 timeframe, and includes high temperature PEMFC membranes (see below). It shows that gasoline-based fuel cell vehicles could provide a roughly 30-35% improvement in efficiency over conventional vehicles. This is roughly the same as that of ICE-based HEVs. Hydrogen-fueled FCVs could be even more efficient, with a fuel consumption about 60% lower than that of conventional ICE vehicles. This is substantially better than that of gasoline-based FCVs because the hydrogen-fueled FCVs do not suffer the losses from the reformer, which are estimated to be especially significant under turn-down conditions (e.g. when operating at twenty percent of maximum system capacity), due to heat losses amongst other factors. The turn-down efficiency of hydrogen-fueled FCVs on the other hand is excellent,

taking full advantage of the fact that stack efficiency increases at turn-down.

Vehicle Drivability and Utility – As all-electric vehicles, FCVs can benefit from the favorable low-end torque characteristics of some electric motors and the low noise the drivetrains produce. In addition, as electric vehicles there is ample electric power available on-board for operating new amenities and on-board systems. In addition, FCVs are expected to be relatively quiet in operation.

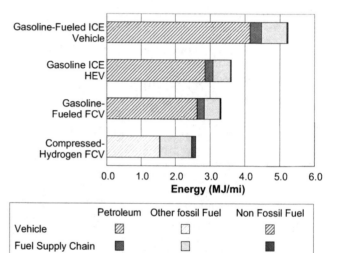

Figure 3 Well-to-Wheels Primary Energy Consumption of FCVs and Other Vehicles[1]

The latter is precisely the reason for the use of APUs in the first place, and fuel cell APUs are thought to possibly provide just this functionality with minimal noise and emissions.

Competitive Positioning - Several market positioning-related considerations are important to consider when assessing the support for FCV development in light-duty vehicle propulsion. For example, fuel cells are playing an increasing role in the competitive positioning of both automotive OEMs and energy companies.

POTENTIAL BENEFITS OF FUEL CELL APUs

One of the main potential benefits of fuel cell APU application especially for heavy and light-duty trucks is the reduction of engine idling or very low load operation. Such conditions can occur when the engine is used to provide power for a variety of non-propulsion uses. A summary of such uses (and potential uses) is provided in Table 3. Avoiding engine idling under such conditions could bring a range of benefits which are also listed in Table 3. The alternative power sources available to reduce the need for idling (i.e. battery packs, auxiliary generators, direct-fired heaters and absorption coolers) all have severe economic and technical drawbacks that have limited their market acceptance.

Fuel cell systems may be a good fit for APU requirements. Fuel cells are typically quiet, efficient, clean and potentially will have low maintenance, which are typically important factors for APU applications. On the other hand, in comparison with propulsion power applications, APU applications have requirements that are easier to meet for fuel cells as they may not have the load following requirements and physical size and weight constraints and because their cost targets are somewhat higher.

Table 3. Potential APU Loads and Benefits of APUs

APU Loads		Potential Benefits
Hotel loads	• Power truck sleeper compartment • Provide AC when parked	• Can operate when main engine unavailable • Reduce emissions and noise while parked • Extend life of main engine • Improve power generation efficiency when parked • Fuel and operating cost reductions
Amenities	• Power accessories such as TVs, refrigerators, telecommunications, equipment during non-driving operation	
Start-up	• Avoid cold-start problems	
General Practice	• e.g., maintain air system pressure during delivery operation	
Light-duty idling	• Power accessories during driving cycle itself	

CHALLENGES FOR FUEL CELL DEVELOPERS

Despite the demonstrated and potential benefits of fuel cells for vehicle applications, a number of significant challenges remain for fuel cell developers, which must be overcome before fuel cells can be used widely and commercially in vehicles.

Despite the differences in technology the challenges that SOFC and PEMFC technologies face are similar:

• Cost must be significantly reduced, especially for propulsion applications.
• Long system life must be achieved and demonstrated.
• Heavy weight and large volume must be reduced to allow practical integration into vehicles.

We will now describe the implications of these challenges for each technology.

PEMFC - For propulsion applications, the U.S. Department of Energy has established goals for gasoline-based fuel cell performance and cost that provide a starting point guide for the discussion of the challenges. Fig. 4 shows the difference between current technology and long-term DOE system targets, as well as a future scenario for the 2008 – 2010 timeframe.

These scenarios were developed for both reformer- and direct-hydrogen-based FCVs as a part of several assignments for the DOE, and assume that R&D efforts currently under way are successful within reasonable bounds. The projection is meant to provide an idea of what fuel cell powertrain technology may look like if current R&D efforts are successfully carried through until around 2010. Figure 4 shows that with these improvements fuel cells can overcome the barriers that fuel cells currently face, but that it will not lead to meeting the DOE cost targets.

The projection assumes some evolutionary progress in technology and one set of more-radical changes in system performance. The most important improvement is the development of high-temperature electrolytes that could operate at temperatures in excess of 120°C, more or less independently of the level of humidification.

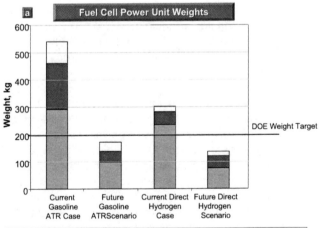

a — Fuel Cell Power Unit Weights

Legend:
▨ Fuel Cell Module ☐ MEA Precious Metals ■ Fuel Processor /cH2 Storage
☐ BOP ■ Assembly/Labor/Depreciation

* The Fuel Processor Total for the Future Hydrogen scenario includes a compressed hydrogen storage system.

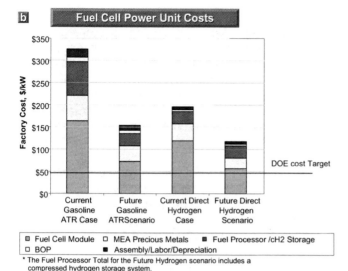

Figure 4   Current and Projected Weight (4a) and Cost (4b) of 50 kW Fuel Cell Power Units[20] (Cost projected based on currently demonstrated performance and assuming high-volume 500,000 units/year production rates). Future scenarios are for 2010.

While such membranes are not currently available with good performance, numerous organizations are working towards their development[21]. However, the implications for FCV power units are profound and include:

- Improved CO tolerance, reducing or obviating the need for a preferential oxidation reactor and for air-bleed (thus improving system efficiency by 5–10%) and considerably reducing start-up time. The remaining CO will be combusted in the catalytic tailgas burner to prevent emissions of CO.
- Facilitated Stack Cooling, reducing considerably the radiator area and the stack cooling plate requirements and consequently reducing system weight considerably because greater driving temperature differences will be available for each of these systems.
- Humidity-Independent Operation, which will be necessary to allow the use of high temperature membranes, reducing the need for humidifiers and water recovery, again reducing weight and cost and making FCV operation consistent with typical automotive environmental conditions

Key observations regarding technology challenges and to what degree they may be addressed by projected potential improvements are described below.

Power Density - Powertrain power density has been considerably improved since the early 1990s but requires significant additional improvement for vehicle integration. Generally, increased power density tends to

almost proportionally reduce system cost, so improving power density is doubly important because of the continuing need for cost reduction. Specifically:

- Fuel cell stack power densities over 1 kW/L, and peak power densities of over 2 kW/L[13] have been demonstrated, but further improvements especially under high-efficiency operation are necessary.
- The weight and volume associated with thermal and humidity management in the fuel cell subsystem is currently quite high. The development of high-temperature, humidity-independent membrane and stack technology would address this issue.
- Additional increases in fuel processor power density mainly through improved catalyst space velocities will be required to achieve acceptable system power density. In addition, a simplification of the system is important. The development of high-temperature membranes would significantly simplify or allow the elimination of the CO polishing step before the stack (the so-called preferential oxidation reactor) leading to considerable weight and cost savings.

Although the weight of compressed hydrogen storage systems is approaching acceptable levels, their volume is still too large. A breakthrough in hydride storage or other storage methods could help, but the technical uncertainties associated with each of these options are significant. Additional FCV prototypes incorporating these alternative hydrogen storage technologies should produce data that will help identify the preferred storage option. In that choice too the implications for the infrastructure should be a key consideration.

Water Management - Simplification of stack water management is a key hurdle to further improvements in stack power density and stack performance. Development of high-temperature membranes would remove the difficulties of handling liquid water in the stack and the necessity of recovering and recycling large amounts of water from the stack effluents.

Precious Metal Content - Additional reductions in precious metal use for FCVs will be required to achieve transportation market cost competitiveness. Currently, the overall platinum content of a fuel cell powertrain is around 4 g/kW, which, at current prices, represents a cost of $60/kW. Based on fundamental electrocatalysis experiments and analysis, there are certain limits to the reduction of the platinum content[2, 3, 4, 5]. Nevertheless, a further reduction by a factor of 5 to 10 appears both possible and necessary to allow fuel cell technology to approach competitive costs, compared with alternative advanced powertrains.

Compatibility with Environmental Conditions - Compatibility of the fuel cell system with vehicle environmental conditions is needed to enable vehicle operation under everyday conditions. Specifically:

- Tolerance of the fuel cell system to airborne sulfur, ammonia, and heavy hydrocarbons must be improved. Most likely, reliable traps will be needed to address most of these sensitivities.
- Although stack tolerance of freezing conditions has been improved, incompatibility with high ambient temperatures and low humidity still limits the operating conditions for fuel cells. The introduction of high-temperature membranes would largely solve this problem.

Start-Up Time - System start-up time of reformer-based fuel cell power units must be improved to allow practical operation of fuel cell powertrains and to achieve better system efficiency[22, 23]. Currently, start-up times for reformers of between three and seven minutes have been reported. Such long times would require large battery capacities, as well as consuming a large amount of energy. Reducing this start-up time would require a combination of:

- Increased fuel processor power density, reducing the amount of material to be heated,
- Widening of the temperature windows of operation for each of the system components, and
- Improved automatic system and temperature controls.

System Life - Component life, in particular stack life, must be improved to achieve acceptable system life. Most of the concerns about component life involve:

- membrane stability and life,
- catalyst deterioration (leading to a loss of operating cell voltage and hence system efficiency), and
- delamination of the membrane electrode assembly.

Although increasing catalyst loadings can compensate for the second issue, this does not constitute an acceptable solution because it would increase cost.

Despite the many tough challenges, fundamental technology limits do not appear to present absolute barriers for fuel cell application to powertrains.

Fuel Choice and Infrastructure - There are two basically different strategies being pursued by the industry for operation of automotive fuel cell systems.

- On-Board Reforming. Use of an on-board fuel processing system that converts gasoline (or some other easily stored fuel, such as methanol) into a hydrogen-rich gas stream for fuel cell operation.
- On-Board Stored Hydrogen. Generating the hydrogen, which is subsequently stored on the vehicle in specially designed storage tanks (compressed, liquified, metal hydrides, etc.), externally to the vehicle.

A third option is to use methanol or DME directly in the fuel cell (a so-called direct methanol fuel cell or DMFC), but this technology is currently in an early stage of development. Consequently the system performance is not good enough for integration into a vehicle.

There are major industry participants pursuing each of these strategies and several pursuing both. This reflects the current reality that both have major pros and cons (Fig. 5), which makes a firm selection of one over the other difficult at this time.

The fundamental problem has to do with the allocation of technical and commercial risk to the two strategies. The on-board reforming option results in considerably more complexity in the vehicle with associated risk and cost implications. It has the advantage, however, of eliminating fuel availability risk because it can use the existing fueling infrastructure (or modest variation thereof). The use of on-board stored hydrogen greatly reduces system complexity, development risk, and costs from the point of view of the automotive companies. Much of the development risk in this case is transferred to the fueling industry (oil companies, etc.) that would be called upon to develop a very costly and still unproven infrastructure for the generation, storage, and dispensing of hydrogen at service stations on a national scale. The automotive industry and energy industries and the government are in on-going discussions on this issue, which could be critical in determining the timing and extent of the market for FCVs. Several projects to demonstrate key aspects of such a hydrogen infrastructure are on-going, including those associated with the California Fuel Cell Partnership and the European Cube Project for hydrogen city buses.

	On-Board Reformer	Direct Hydrogen Fuel Cell Vehicle
Fuel	Efficient production	Moderately efficient production
	Infrastructure exists (gasoline)	New infrastructure required
	Moderate Fuel Cost (~ $7/GJ for gasoline, exclusive of taxes)	High fuel cost (more than $20/GJ for compressed hydrogen)
Fuel Cell Power Unit	Large stack: due to reformate quality	Compact stack
	Complex: primarily because of fuel processing system	Simple: Low: pressurized hydrogen Complex: metal hydrides
	Heavy: due to larger stack and fuel processor	Lighter: no fuel processor and compact light stack
	Good Efficiency	Excellent Efficiency
Vehicle	Established safety standards	Safety standards yet to be established
	Compact storage (energy dense fuel)	Bulky storage (low energy density)
	Requires sizable battery needed to bridge cold-start	Requires small battery for start-up & transients

Figure 5. Comparison of On-Board Reforming vs Direct Hydrogen FCVs.

The manner in which this critical issue is resolved will have major implications for companies interested in potential opportunities in fuel cell–based automotive power trains. For example, with on-board reforming there are many opportunities in a wide range of heat exchangers, controls, pumps, catalysts, etc. associated with the complex fuel-processing system. With on-board hydrogen storage there would be new opportunities associated with the sophisticated hydrogen storage vessels.

Most importantly, if one or more major energy companies commit to establishing the core of a hydrogen-fueling infrastructure (even if initially for fleet vehicles which would likely be the case), the timing of significant markets for FCVs might be accelerated.

Currently, the automotive industry and the fuel industries are working together in several collaborations to tackle the tough problems surrounding the introduction of a hydrogen-fueling infrastructure and issues related to risk (including financial risk), environment, and safety.

SOFC BARRIERS – At a high level, many of the same barriers to the successful commercialization of fuel cells listed for PEMFC also apply to SOFC in automotive applications. However, the details are drastically different. The key barriers to success currently are:

- Improving volumetric power density. Currently, the systems are quite bulky due to the large volume of the insulation of hot system components as well as the large volume of air that must be handled for stack cooling.

- High cost is an issue for SOFC too and increases in power density and broadening of the stack operating temperature range are seen as key to lowering the cost. Broadening the operating temperature range reduces the required recuperator size and the airflow requirement. Of course all improvements must be consistent with low-cost manufacturing techniques, such as tapecasting and screenprinting.
- Achieving long system life

In a recent study we projected forward a plausible technology development path for SOFC. They path would achieve most of the desired system characteristics by 2008. Key features of the projection are:

- Stack operation at 650°C with stack active from 500°C
- On-anode reforming of most of the fuel, requiring only prereformer
- Power density of 800 mW/cm^2, leading to volumetric power density of 0.15 kW/L
- Cost reduced to below $500/kW for a 5 kW system
- System efficiency around 40% (based on LHV)
- Start-up time less than half an hour (obviously, such a long start-up time will only be acceptable for certain applications).

Several designs being currently developed by SOFC players would lead to this type of performance. In SOFC development the challenge is arguably more a technical challenge, as the fuel choice is a bit more obvious (no significant penalty for using gasoline) and the cost would likely meet APU requirements if the technology objectives can be achieved.

IMPLICATIONS FOR THE FUTURE

TECHNOLOGY PATHS - Technology developments are key to fuel cell success in automotive applications. The previous section provided a description of potential developments that could bring technical performance of FCVs in line with competing technology. In addition, further improvements in cost will be required to achieve cost parity with conventional vehicle powertrains. For both PEMFC and SOFC fuel cell technologies, key elements of the technology path to achieve this include:

- Fundamental stack improvements, notably the development of high-temperature membranes with high current density at high operating cell voltage, humidity-independent operation, and low catalyst loading for PEMFC, and the development of robust ceramic structures and interconnect materials that can handle high current densities and reasonable thermal gradients.

- Improvements in and simplification of fuel processor technology, in particular the elimination of CO polishing steps and anode humidification systems for PEMFC systems, and the reduction in the amount of excess cooling air required for SOFC
- Simplification of system components and development of low-cost, compact, automotive-style combination components for both technologies.

The time schedule according to which these developments may take place is a subject for considerable discussion. We believe that technical performance targets could be met by both technologies 2008/2010, leading to fully functional FCVs and APUs, provided a high degree of success in the technology development programs is achieved. APUs are likely to lead the way together with hydrogen-based vehicles, with gasoline-based systems 1–2 years behind. Cost for powertrain power units could have been reduced to $100–150/kW (complete powertrain cost) which may be sufficient for regulated and niche markets (certainly to compete with battery-electric vehicle cost). It is clear that achieving cost parity with conventional power sources will require additional technology breakthroughs and is not likely to occur until after 2010. For APU applications the cost of systems may have been reduced to around $500 per kW, which may be sufficient for sizeable initial markets.

It is important to note that to achieve these improvements considerable additional R&D spending from the entire future value chain will be required. Currently that support exists, but challenges will undoubtedly arise with industry cycles in both the automotive and oil industries.

MARKET PROJECTION SCENARIOS - Because fuel cell technology is still in an early stage of development, estimates of market share over time are fraught with uncertainty. First, technology performance and cost uncertainty for fuel cells is considerably greater than that for other technologies. Second, planned environmental regulation is expected to have a major impact on the level of acceptance of fuel cell technology. Third, the fact that fuel cells would represent the most radical shift in powertrain technology makes it more difficult to anticipate customer and industry response to fuel cells.

Because the core fuel cell technology has such a big impact on FCV performance and cost, the level of market success will likely depend strongly on the degree of success in fuel cell technology development. Recognizing this, we can identify a couple of different possible scenarios for the success of fuel cell vehicles.

Fleet Applications Scenario – Even if technology development does not lead to a viable technology for general light-duty vehicle applications, niche applications of FCVs may occur in local fleet vehicle markets (e.g., buses or delivery vehicles). Annuals sales of thousands or tens of thousands of vehicles may be reached but sales will have to be subsidized.

Niche Applications Scenario - In this scenario, fuel cell technology will be sufficiently developed to provide solutions for small light duty vehicle (LDV) markets, and be attractive in markets where favorable regulations or requirements help overcome the cost differences with conventional vehicles. In this scenario FCV weight and cost will be reduced by 2008 to make FCVs superior to battery-electric vehicles and thus, in many regulated markets, fuel cell powertrains will capture increasingly most of the market. In such a case, sales could grow to the hundreds of thousands of LDVs by 2015, even if fuel cell powertrain costs would hover around $100/kW. This technology scenario is consistent with the future technology scenario described above. We do not expect that APU applications would be motivated much by such regulations, but under these cost and performance considerations, fuel cells may very well still be extremely competitive in APU applications, and find widespread application. For APU applications it does not appear that additional technology breakthroughs are a precondition for this scenario with APUs, though breakthroughs such as direct hydrocarbon fuel cells could further broaden the attractiveness of fuel cell APUs.

Technology Success Scenario - In the technology success scenario, fuel cell technology will achieve cost and performance levels that make it broadly competitive with other powertrain technologies. For propulsion applications, the technology victory scenario would require significant additional improvement in technology principally to achieve a cost approaching $50/kW. Additional technology breakthroughs will be required to achieve this; we do not expect these to come to fruition until after 2010. In the meantime, the projections for this scenario would mimic the Niche Applications Scenario. Transition from the demonstration-only scenario would appear to be unlikely because the development time and expenses required could probably not be supported without significant intermediate sales.

CONCLUSION

Fuel cells offer the most radical potential shift in powertrain technology and could offer tremendous benefits to customers and to society, but to achieve such impacts fuel cell technology developments must overcome serious challenges in technology performance, weight reduction, and cost. If this is achieved and the economic and regulatory environments are favorable, fuel cells could have a significant impact as the future propulsion power source for LDVs.

Finally, fuel cells could have considerable impact in light-duty vehicle markets (and heavy-duty and special-purpose vehicle markets as well) as APUs. The requirements for the APU markets appear to be more easily met by fuel cells than those in the propulsion markets, and the fuel cells provide tangible benefits to the vehicle users.

However, significant steps need to be undertaken to reduce cost and increase power density of the systems without any compromises in system level reliability and life characteristics. Demonstrating useful stack and system lifetimes ranging from over 5,000 hours for propulsion applications to over 20,000 hours or longer for most APU applications in order to achieve acceptable operating and maintenance cost structures will be a critical test for success in the coming years.

It is our judgment that given the current investment levels the requirements for some of these applications may be achieved, even if achieving manufactured cost levels that are broadly competitive with conventional ICEs appears to require significant additional technology breakthroughs.

ACKNOWLEDGMENTS

The authors would like to thank the DOE's Office of Transportation Technology and National Energy Technology Laboratory for their support of the various studies that were cited as underpinning the findings reviewed here. In addition we would like to thank the many members of the fuel cell community and automotive industry, too numerous to mention, that have been willing to discuss and refine the thinking outlined in this paper.

REFERENCES

1. Arthur D. Little Draft Report to DOE, "Fuel Choice for Fuel Cell Vehicles", November 2001.
2. Arthur D. Little, "Pathways to Low Cost", Presentation to DOE, August 2001.
3. Bahar, B., C. Cavalca, et al., in Journal of New Materials for Electrochemical Systems, Vol 2, 1999, p. 179-182.
4. Barbir, F., M. Fuchs, et al., in Society of Automotive Engineers, Detroit MI, SAE World Congress, 2001.
5. Cleghorn, S. J. C., X. Ren, et al., in Int. J. Hydrogen Energy, Vol. 22(12), 1997, 1137-1144.
6. Ralph, T. R., G. A. Hards, et al., in J. Electrochem. Soc. Volume 144(11), 1997, p. 3845-3857.
7. E. T. D. Ghosh, M. Perry, D. Prediger, M. Pastula, and R. Boersma, in Electrochemical Society, Vol. 2001-16, 2001, p. 100.
8. J.-F. J. A. Virkar, N. S. Kapur, D. W. Prouse, G-Y. Lin, Y. Jiang, P. Smith, D. M. England, D. K. Shetty, in 2000 Fuel Cell Seminar, Portland, Oregon, 2000
9. DaimlerChrysler website: http//www.daimlerchrysler.com/index_e.htm.
10. Toyota press announcement , January 2001 .
11. AEI, "Fuel Cells Start to Look Real," March 2001
12. Various Ford, GM, Toyota, Honda Company press releases 1999 – 2001.
13. GM Company press release, August 2001, http://www.gm.com/company/gmability/environment/products/fuel_cells/pickup_080701.html.
14. Honda press release, April 2001.
15. BMW Company press releases, 1999.
16. "Freightliner unveils prototype fuel cell to power cab amenities", O. B. Patten, Roadstaronline.com news, July 20, 2000.
17. Crosbie, G. M., Perry Murray, E., Bauer, D. R., Kim, H., Park, S., Vohs, J. M., and Gorte, R. J., "Application of Direct Oxidation of Liquid Hydrocarbon Fuels in Solid Oxide Fuel Cells to Automotive Auxiliary Power Units," SAE Paper No. 2001-01-2545, Society of Automotive Engineers, Intl., Sept. 2001.
18. A. M. H. Yokoyama, S. E. Veyo, in Solid Oxide Fuel Cells V (SOFC-V) (S. C. S. U. Stimming, H. Tagawa, W. Lehnert, ed.), The Electrochemical Society, Aachen, Germany, 1997, p. 94.
19. J. H. J. S. Thijssen, C. J. Read, W. P. Teagan, Conceptual Design of POX /SOFC 5kW net System, Final Report to US DOE, NETL SECA , January 8, 2001, http://www.seca.doe.gov/Events/Arlington/ADLCOST.pdf.
20. E. Carlson, Cost Analysis of Fuel Cell System for Transportation, Task 1 and 2 Final Report to DOE, March 2000, SFAA No. DE-SCO2-98EE50526, www.ott.doe.gov/pdfs/baseline_cost_model.pdf.
21. Savodogo, J., Journal of New Materials for Electrochemical Systems, Vol. 1: 47, 1998.
22. USDOE, "Fuel Cells for Transportation, FY98 Annual Contractors Report", 1998, p. 8 and 26.
23. USDOE, 1999 Annual Progress Report Energy Conversion Team, US Department of Energy, Energy Efficiency and Renewable Energy, Office of Transportation Technology .

CONTACT

Dr Thijssen is a Director of Alternative Fuel and Fuel Cell Technology in Arthur D. Little's Technology & Innovation Business.

Dr. J.H.J. Thijssen
Arthur D. Little, Inc.
Acorn Park, 20/434
Cambridge, MA 02140-2390
Thijssen.j@adlittle.com

DEFINITIONS, ACRONYMS, ABBREVIATIONS

APU: Auxiliary Power Unit

BEV: Battery-Electric Vehicle

DMFC: Direct Methanol Fuel Cell

FCV: Fuel Cell Vehicle

HEV: Hybrid Electric Vehicle (ICE-based)

HHV :   Higher Heating Value

ICE: Internal Combustion Engine

ICEV: Internal Combustion Engine Vehicle

LDV: Light Duty Vehicle

LHV: Lower Heating Value

PEMFC: Proton Exchange Membrane Fuel Cell

PROX: Preferential Oxidation

SOFC: Solid Oxide Fuel Cell

High Performance Fuel Cell Sedan

Codrin-Gruie Cantemir, Chris Hubert and Giorgio Rizzoni
CAR, The Ohio State University

Bogdan Demetrescu
Ford Motor Co.

ABSTRACT

New vehicle technologies open up a vast number of new options for the designer, removing traditional constraints. Some recent conceptual designs, such as GM's Hy-wire, have recognized this and offered innovative new architectures. Unfortunately, many other new technology concept cars do not exploit the freedoms of the new technologies, hampering themselves with traditional design cues developed for conventional powertrains. This paper will present the conceptual design of a high-power, high-speed fuel cell luxury sedan. One of the main motivations of this case study was to explore what could happen when a vehicle was designed from the ground up as a fuel cell vehicle, optimized at the overall system level as well as at the individual component level. The paper will discuss innovations in vehicle architecture and novel concepts for the electrical transmission, fuel cell system and electromagnetic suspension.

Vehicle architecture introduces new topology based upon new criteria which are needed for a fuel cell electric vehicle. The main elements of the vehicle— such as the hydrogen fuel tanks, super-capacitors, electric machines, fuel cell stacks, fuel cell auxiliaries, radiators, brake resistor and power electronics—are located in positions allowing maximum active and passive safety with the same or enhanced functioning performances. These goals can be met, without punishing passenger comfort or trunk volume by using the added freedom in component placement that the electric powertrain provides. General styling introduces a very intimate amphitheater concept together with a low drag body. An adjustable trunk maximizes storage space when necessary and improves drag when the storage is not needed.

The fuel cell design contains new high efficiency concepts for the stack itself as well as the auxiliary systems. Stack design eliminates the need for graphite plates and de-ionized cooling water, improving both the cold startability and the pressure drop across the stack. Integration of a fast auxiliary warm-up system further improves cold startability. This system allows for a 2-4 minute transition time when operating at $-20\,°C$, and permits the fuel cell to start at temperatures as low as $-40\,°C$. A continuously controlled pressure wave compressor (Comprex®) offers substantial efficiency gains over a classical electrically driven compressor, with energy savings in excess of 50%.

The electric transmission introduces some aspects of power electronics and energy management for low cost implementation. The fuel cell and the multi-level inverter have been designed together to eliminate any intermediary DC to DC converters. Virtual electromagnetic differentials distribute torque between the front and rear axles. Two such twin rotor electric machines allow for active traction control on all four wheels, while minimizing costs and increasing machine efficiency. A zero-power active levitation system provides the basis for the electro-magnetic suspension, which allows real-time camber control and recovery of some of the energy absorbed by suspension damping.

INTRODUCTION

The high performance fuel cell sedan represents a concept design case study developed by the Center for Automotive Research (CAR) at the Ohio State University. Investigators working on the case study strove to achieve a unified design at the system and sub-system level. Project researchers believe that alternative propulsion technologies, such as electric traction, powertrain hybridization, and fuel cells can have a very strong future in automotive design. However, the authors feel that the tremendous amount of past research, development, and experience with more "conventional" powertrain technologies makes it very difficult to simply "swap" new technologies for old and try to sell the change solely using efficiency arguments. Instead, new designs must be developed to fully exploit all the benefits of the new technologies.

Moreover, some thought must be given to the platforms themselves. In the future, these technologies will not be so "new" and costs will be reduced to the point where they can be used in very budget conscious vehicles, where their fuel savings potentials can be exploited. However, in the nearer term, the initial costs are too high to make such a plan feasible. New technologies are rarely introduced in any automaker's "economy" lines of vehicles, and when discussing the "core" of the vehicle—its powertrain—it hardly seems reasonable to break the trend. Thus, the performance and luxury markets have been targeted with a very high performance sedan. While economy and emissions are very important design considerations, this paper looks to present a vehicle that is possible—both technically and commercially—in the very near future.

GENERAL OVERVIEW

Figures 1 and 2 show computer aided renderings of the concept vehicle. In Fig. 2, a transparent shell reveals the component packaging. A very streamlined profile suggests the high performance; however ample seating for five adults provides the spacious feel expected of a luxury sedan. Two seats in the front are positioned low in the vehicle, giving the driver a very sporty feel.

Figure 1: High performance fuel cell sedan

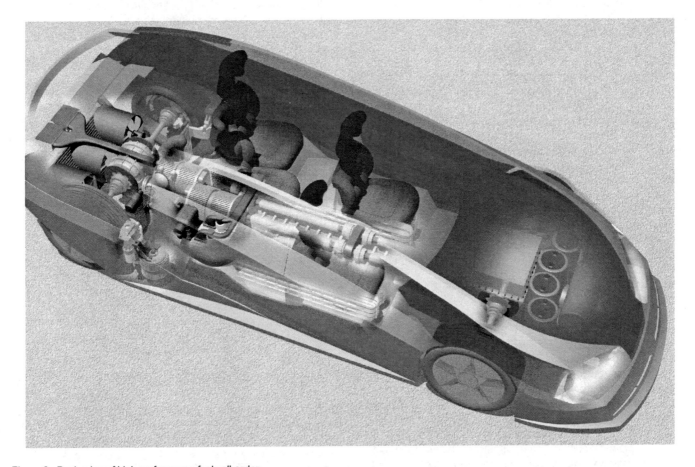

Figure 2: Packaging of high performance fuel cell sedan

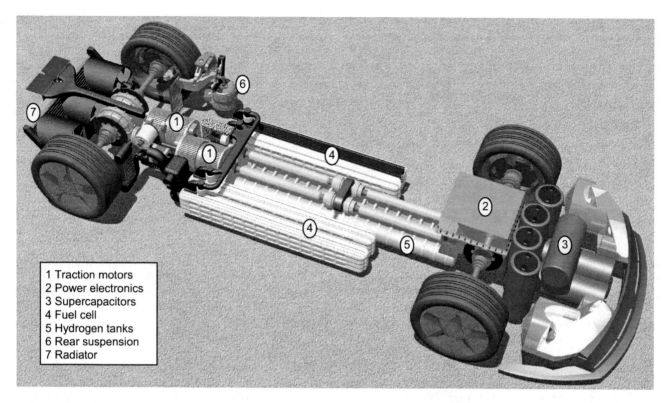

Figure 3: Powertrain layout

Table 1: Performance specifications of vehicle

Vehicle mass	2500 kg	5500 lb$_m$
Continuous power	160 kW	210 hp
0 to 60 mph time	3.8 s	
Top speed (continuous power)	250 kph	155 mph

The second row consists of three distinct seats, with the middle seat placed slightly forward of the outer two, and all three being elevated above the front seats. This amphitheater seating concept helps to maximize the passenger's view, similar to the seating in EDAG's Cinema 7 concept.

Table 1 reveals the various performance characteristics of the vehicle. The fuel cell system produces 160kW of power, allowing for a top speed of 250 kph (155 mph). Much higher power levels can be achieved using the ultracapacitor pack, allowing for very impressive accelerations (0 to 60 mph in under 4s).

POWERTRAIN & CHASSIS TOPOLOGY

Figure 3 presents a CAD rendering of the powertrain layout. Prior to developing the layout, designers worked to develop a clear methodology for the design of fuel cell hybrid electric vehicles. Hydrogen tank safety lies at the center of the powertrain design—both literally and figuratively. Six tanks (with a total capacity of 4kg H$_2$)

store the vehicle's hydrogen fuel. Safety must be the primary concern when storing such a combustible gas at high pressures and the safest place is the center of the vehicle—where the tanks are located. A shroud placed around the tanks allows for air flow through the center of the vehicle, venting any possible hydrogen leaks, keeping the passenger compartment safe.

Surveying the rest of the powertrain, one will notice that the fuel converter no longer sits at the front of the vehicle. The fuel cell sits in the middle of the vehicle, with a total of four stacks—two on each side of the hydrogen tanks. Accessories needed for fuel cell functioning—including the compressors—and the traction motors have been placed at the rear of the vehicle. Placing the elements with the strongest acoustic signals in the rear of the vehicle helps to mitigate noise in the passenger compartment. Two traction motors power the vehicle—one for the right side and one for the left. Electric traction makes it very simple to have the two driveshafts rotate in opposite directions, potentially improving the NVH characteristics. Drivability and performance have also been enhanced with the use of inboard mechanical brakes. This allows for the same dive characteristics whether braking is regenerative (via electric machines), mechanical, or some mix of the two.

The radiator has been placed at the rear of the vehicle eliminating the aerodynamic distortions caused by open front grills. Exhausting the air at the rear of the vehicle further reduces the coefficient of drag. GM's Precept concept vehicle used similar radiator positioning along with some typical protuberance reductions—wheel skirts and external mirror removal—to help achieve a coefficient of drag of ~0.16 in a family sedan. [1] In this concept, prismatic mirrors, such as those made by Bendinglight [2] allow for a very low profile, which can be nested inside the body's shell as the profile tapers back from its widest point near the front wheels. Flow separation around the rear wheels has been reduced by partially shrouding the wheels. Air tunnels on the inside of either wheel provide much of the needed inlet air. Referring to Fig. 1, one sees that the shrouds have been integrated into the body styling; while Fig. 3 shows the forward mounted rear suspension, needed to maximize air flow through the tunnels. Use of solid wheels further reduces drag; again this has been done while maintaining a stylish appearance. "Inverse spokes" filled with a transparent material, give a more traditional and aesthetically appealing look than a homogeneous solid rim. Even the trunk enhances aerodynamics, with a collapsible floor. When the trunk is empty the bottom pivots up, providing an upward tapering underbody profile.

Packaging flexibility afforded by electric traction offers further enhancement of the front/rear load balance. Supercapacitors and power electronics have been placed at the front, beneath what would customarily be the vehicle's hood. However, the new component packaging has eliminated the need for a hood—the components have high reliability with no required maintenance (no oil or coolant to add, no belts or hoses to change). Instead, a single transparent surface covers the front vehicle. This styling cue recalls the revolutionary use of a single piece of glass to cover all the headlamps in the Citroen DS and logically extends this idea into modern automotive design. Fast acting circuit breakers and short circuit elements help to enhance vehicle safety, disconnecting the power bus and dumping the supercapacitor charge when the front end begins to crumple.

TRACTION MOTORS

Two electric machines provide tractive power (and regenerative braking capability) to the vehicle. However, the motors are not used in a motor per axle configuration, but rather as an independent "right and left side" drive system. Splitting motors across the front and rear of the vehicle can result in an inefficient use of the motor's capabilities. During strong accelerations, weight transfers to the rear axle, requiring a larger motor to maximize acceleration capability. Similarly, sharp braking transfers weight to the front, increasing the size of the front motor for maximal regeneration. The net result is that one must either over-design the electric machines on a motor-per-axle vehicle, or lose some performance relative to the theoretical limits of an all-wheel drive vehicle. Separate motors on the left and right wheels eliminate the need for a mechanical differential. Instead, a virtual electric differential operates between the left and right side.

An all-wheel drive (AWD) vehicle requires a differential between the front and rear wheels as well as across the left and right (typically a transfer case serves this function). Again, the concept vehicle provides the "differential action" without a mechanical differential. Each of the electric traction machines has two rotors in a single stator. Here electromagnetic forces tend to keep wheels turning at the same rate, but, just as with a good limited slip mechanical differential, the absence of any rigid constraints allows for slip. Moreover, when slip occurs (one rotor turns at a different speed than the other) the electromagnetic differential offers a higher level of control. Dynamic control of the high voltage waveform allows active control of the torque distribution when the rotors move at different speeds.

Conceptual motor design leverages the success of the Ohio State Formula Lightning electric vehicle racing team. Figure 4 shows the traction motor which powers the "Smokin' Buckeye" while Fig. 5 shows the vehicle in action. Multiple Formula Lightning series championships and the series speed record of 144 mph have validated the motor design. The baseline design for the fuel cell sedan uses the rotor (two per motor) and stator laminate design from this machine. Some additional power electronics (discussed later) allow the number of poles to be doubled while maintaining approximately the same range of speeds at peak torque.

Figure 4: Smokin' Buckeye electric machine

Fig 5: OSU's "Smokin' Buckeye" Formula Lightning Electric race car

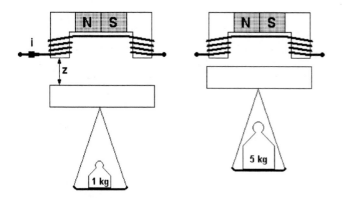

Figure 6: Zero power magnetic suspension principle

SUSPENSION

Electromagnetic suspensions, also known as magnetic levitation (maglev) systems, have traditionally been associated with and developed for the next generation of high-speed trains. However, with an adequate design electromagnetic suspensions can prove equally useful in automotive applications. The systems never leak and provide the opportunity for advanced active ride control. Several different system designs exist, based on differing principles. A promising candidate is the "zero power system" shown in Fig. 6. A permanent magnet (considered fixed) attracts a vertical motion reaction armature. For any given armature load (conceptually represented by the mass in Fig. 6) there exists an air gap distance, z, such that the magnetic force exactly equals the load. Larger loads must be balanced with shorter gaps (larger magnetic forces), as shown in Fig. 6.

Of course, the equilibrium—while it must exist—will always be unstable. If the armature is perturbed so as to decrease the gap, the magnetic forces will increase, further decreasing the gap. Similarly, perturbations which increase the gap lead to reduced magnetic forces and thus a further increasing gap. An additional electric coil, wound around the same magnetic circuit provides stability. Current flowing through the coil generates an additional flux which will either add to or subtract from the flux created by the permanent magnetic (depending on the current sense). With proper control of the current, the correct net attractive force can be maintained, stabilizing the system. A four-quadrant controller supplies the coil based on sensor information providing the air-gap distance, the armature's vertical speed and the armature's vertical acceleration. Figure 7 shows a CAD representation of the rear suspension based on a trailing arm suspension design. The design uses a single composite-material arm to balance the forces from the electromagnets (3) and the wheel (5) across a ball joint pivot (4).

A servomotor (labeled with a 1 in Fig. 7) mechanically adjusts the position of the permanent magnet to provide the same ground clearance at different vehicle loads (which require different "gap distances"). When the vehicle is at rest, the system can then calibrate itself to provide constant ground clearance regardless of passengers or cargo. A second servomotor (labeled as 2 in Fig. 7) provides active camber regulation. Larger camber angles can enhance stability in extreme situations (high speed operation, particularly when cornering). Figure 8 shows the suspension operating with $0°$ (left of Fig. 8) and $12°$ (right of Fig. 8) camber angle. An additional linkage provides near zero scrub regardless of camber angle. (Scrub will occur during camber adjustment—however the additional stability warrants the extra tire wear.)

Some regulating current must be continuously maintained when the vehicle is stopped (and in stand-by mode). However, the regulation power for a 2.5 ton vehicle will be around 250 W, the equivalent of the headlights. Additionally, the electromagnetic suspension can recover some of the energy it dissipates when absorbing shocks. This recovered energy further mitigates the cost of the regulation power.

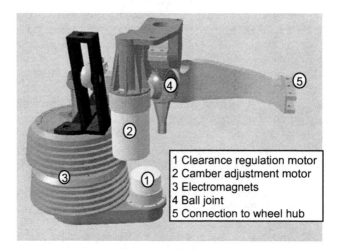

1 Clearance regulation motor
2 Camber adjustment motor
3 Electromagnets
4 Ball joint
5 Connection to wheel hub

Figure 7: Electromagnetic suspension

Figure 8: Active camber regulation: left side shows $0°$ camber angle, right side shows a $12°$ angle

FUEL CELL SYSTEM

STACK DESIGN AND CONFIGURATION

The stack design uses only existing proton exchange membrane, PEM, technologies. With these currently available technologies, conceptual design has focused on systems level interactions, tuning the design towards more efficient operating regimes, and improving the packaging and manufacturability. From a systems level perspective, use of graphite plates and de-ionized cooling water presents serious constraints in automotive applications. Thus, the new stack design concept avoids these elements. Larminie and Dicks [3] present several alternate PEM stack designs to the traditional method of using machined graphite for the bipolar plates. One alternative is the use of foamed or perforated metal plates in place of the carbon. The concept presented here uses the perforated metal plates with a geometry designed to maximize surface area as well as provide ample space for the flow of air and hydrogen.

Figure 9 shows the basic electrode design. Electrodes can be formed using a stamping process. Two electrodes are displayed independently in the upper right of the picture, while the lower left of Fig. 9 shows a pair of electrodes assembled together. Using the basic shapes and assembly shown in Fig. 9 allows for the successive combination of an unlimited number of electrodes. A sketch (not to scale) of the basic cell configuration is shown in Fig. 10. The PEM lies between each set of "mating" electrodes, as seen in Fig. 10 and hydrogen and air (oxygen) are vented into alternative channels. Looking at the central electrode (in Fig. 10), one can see that the "top" half functions as an anode, while the "bottom" half functions as a cathode. Each electrode in the stack (except the ends) functions as an anode in one cell and a cathode in the adjacent cell. This provides a natural series connection between successive cells, eliminating the need for additional electrical connections and increasing the internal efficiency.

Figure 11 shows a cross section through a stack assembly. Two identical end caps hold the electrodes and PEM in place (the top and bottom pieces in Fig. 11). As with the electrodes, cap design considerations included manufacturability: either an extrusion or an injection molding process can produce the end caps. Within each cap, a set of longitudinal channels provide hydrogen and air to the stack (the small round channels supply hydrogen, while the larger rectangular channels supply the air). The main supply/exhaust channels in the end caps connect to the various channels between the electrodes with a diagonal input-output configuration, giving each cell the same gas flow length. Cooling water circulates through tubing in either "side" (front or back as shown in Fig. 11) of the stack. Containing the cooling water in tubes eliminates the need for de-ionized water that hampers many of the bipolar graphite plate designs.

Using the electrode design presented in Figs. 9–11 results in an equivalent active surface area of 500 cm^2 across each cathode and anode (i.e. the total equivalent active surface area of an electrode is 1000 cm^2), with a total electrode length exceeding 60 m. Fuel cell stack design used a nominal current of 220 A (400 A max) at 180 V. This comparatively low current density—Larminie and Dicks note ~1A/cm2 as a realizable limit [3]—results in stack efficiencies in excess of 55% and allows for use of much thinner and lighter electrodes. The CAD model estimates the mass of a single stack to be less than 25 kg. To match the vehicle's power requirements four stacks are used in a parallel gas flow configuration. As shall be discussed later, the power electronics allows for different series/parallel connections of the four stacks, to better match the fuel cell characteristics to the load and inverter.

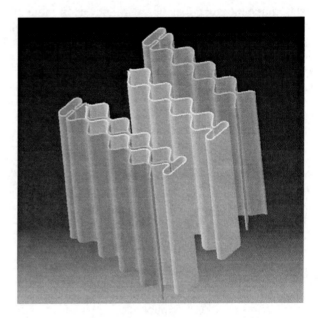

Figure 9: Fuel cell electrode configuration

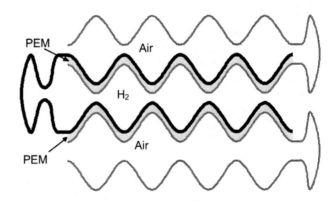

Figure 10: Schematic of basic cell configuration

270

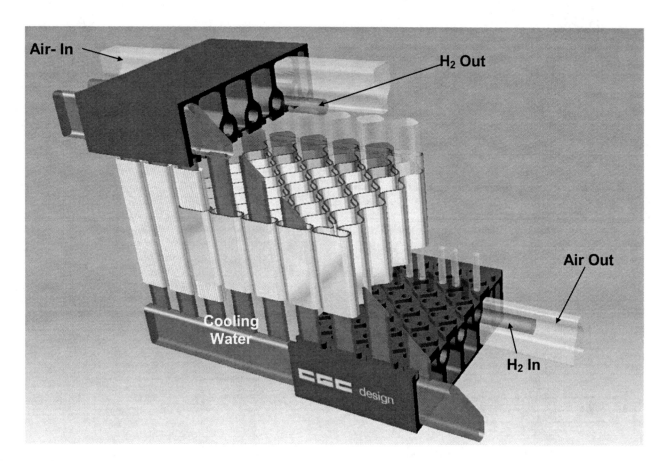

Figure 11: CAD rendering of fuel cell stack

AIR COMPRESSOR SYSTEM

Figure 12 shows the basic layout of the air system for the fuel cell. A pressure wave supercharger (PWS) serves as the primary compressor for the system. [4] Much of the energy needed to compress the air comes from the exhaust—offering a considerable improvement in system efficiency. Exhaust gasses will often not have sufficient energy to compress all the air required by the fuel cell stacks: a catalytic burner prior to the PWS directly burns hydrogen to increase the energy in the exhaust. Driving the compressor using chemical energy instead of electrical energy increases the peak available electricity—only small amounts of electricity are needed to drive the PWS rotor. Use of the PWS also offers the possibility of using the water in the exhaust stream to (at least partially) humidify the inlet air.

At start-up, a roots type supercharger provides the needed boost; it also provides faster transient response when needed. During start-up, the system is still cold and must come up to temperature—thus the roots blower has been placed after the intercooler. In "standard" operation, the bypass valve on the roots blower (not shown in Fig. 12) allows the already pressurized flow to be diverted. Flow control valves (FCVs) are placed at the outlet of the fuel stacks, allowing some of the stacks to be shut down when full power is not needed. Controlling the flow at the outlet

end (as opposed to the inlet) keeps the stacks pressurized, further improving transient response.

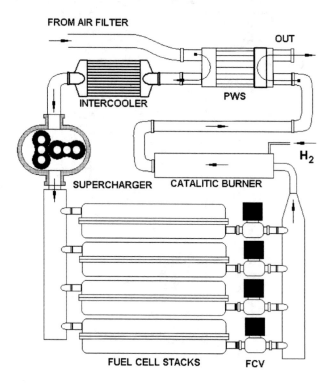

Figure 12: Air compressor system

POWER ELECTRONICS

Figure 13 shows the circuit schematic for the power electronics between the fuel cell stacks and the traction motors. The boxed-in area (labeled "full bridge") shows the inverter topology for a single phase. (The circuitry below the box presents identical topologies for the remaining two phases.) Looking at the stack connections (top of Fig. 13) in more detail one can see that the four stacks have been grouped into two pairs (in the figure stacks 1 and 2 are grouped together and stacks 3 and 4 are grouped together). A three-level inverter configuration has been chosen; thus the stacks have been tapped across the series connection of the two pairs and between the two pairs (providing the full and half voltage sources). Within each stack pair, several diodes and a switch allow for either series or

parallel connections. When large supply voltages are needed, the switches can be closed, connecting the two 180V (nominal) stacks in series, for a total voltage of 360V. (Thus, the three-level inverter is supplied with 360V and 720V.) However, at low speed operation the switch can be opened, giving a parallel connection with lower voltage and higher current—more suited to the needs of the traction motor.

The power electronics shown in Fig. 13 differ from a classic three level inverter. To increase the power which can be developed from the traction machine, a voltage boosting method has been adopted. This concept, which has been discussed by several of the authors (and others) [5], is based on the observation that in a classic wye configuration, as shown in Fig. 14, the neutral point is kept at half of the DC bus potential. Thus the maximum voltage that can be supplied to any phase will be one half of the DC bus potential, yet the insulation will be rated for the full potential. Figure 15 shows two voltage boosting configurations, the first (top)

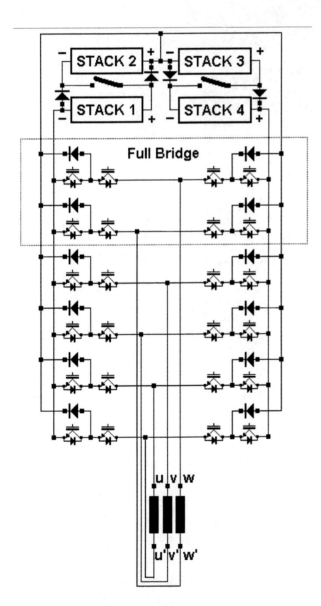

Figure 13: Power electronics

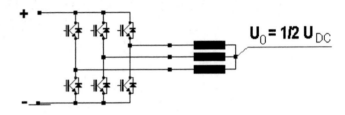

Figure 14: Traditional IGBT inverter with wye connection

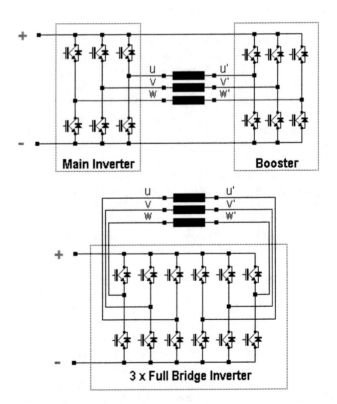

Figure 15: Two voltage boosting variants

272

using the same "traditional" IGBT with an additional booster, and the second with the boosting electronics integrated into the IGBT. Figure 13 further extends the integrated circuitry to a three-level inverter. Using this power electronics topology allows for twice the voltage to be applied to a single phase without any modifications to the motor. As a result, the motor can produce the same levels of torque at nearly twice the speed—nearly doubling the motor's power. (Theoretically, the speed and power be doubled, practically, some additional losses will occur.)

MARKET ANALYSIS

As previously mentioned, growth of hybrid, fuel cell, and electric traction technology in the automotive markets depends not only on the proper use and development of technology, but also on the proper choice of platforms for technology implementation. Attempts to introduce such technologies based purely on efficiency concerns are always faced with a classic catch 22. Much development remains to bring the cost of the new technologies to a point where they can compete in the "economy" market with the fully developed conventional technologies. However, this development effort requires incoming funds as well as practical experience—both of which come from use of the technologies in a product line. So, to use the components in economy vehicles requires more development, but development requires that the components be used.

This riddle actually has a very simple answer—one that has been known and used for many years in the automotive industry. New technologies are not introduced into economy lines, but instead are developed in other platforms, where aspects in addition to economy are exploited. In fact, quite often new automotive technologies begin in racing, the military, or have been transplanted from aerospace. With that in mind, a very high-performance luxury sedan has been developed. Electric traction offers considerable flexibility in component packaging (wires are much easier to route than driveshafts) and this flexibility has been used to enhance the abilities and the comfort of the vehicle. A new electro-magnetic suspension has been adopted, offering highly adjustable riding characteristics—ranging from the soft ride of a luxury automobile to the stiff tuning needed in a high performance sports car.

The continuous power (~160 kW) of the concept is fairly low for the targeted market segment. However, the addition of a supercapacitor bank gives the vehicle more than sufficient power to match the performance levels needed for the segment. Along with this power, the all-wheel drive system, with the electric and electro-magnetic differentials, allow the vehicle to maintain a tractive effort equal to 80% of its weight (near the edge of the adhesive limits of the tires and most road surfaces) at speeds exceeding 90 mph (145 kph). Table 2 summarizes a range of high performance luxury vehicles and their pricing.

At the top of the table, one sees some of the more "conservative" performance sedans (or 4 passenger coupes). With nearly 500 hp, the Mercedes sedans begin to approach the acceleration possible with the electric traction and super capacitor buffer. Moving down the table, one sees some more drastic performance cars. Here one can see the kind of pricing that the combination of very high luxury with very high performance (and of course, very high power) can obtain. The Bugatti, represents an extreme of performance, with ~1000 hp of power providing extreme acceleration. The Veyron is a clear coupe, but it seems that some sedans (perhaps the Arnage) may use the 64-valve quad-turbo W-16 powerplant.

Table 2 highlights a clear trend in the performance/luxury market: an increasing demand for more power. Horsepowers approaching 500 are not unreasonable, and some even reach 1000. In light of this increasing demand for power, the overload capacity of the electric traction motors along with the battery or super capacitor buffering in a hybrid, becomes even more attractive.

CONCLUSION

A conceptual design for a high performance fuel cell sedan has been completed. Design efforts have focused on developing new ideas and concepts to use existing technology to create a fuel cell vehicle which

Table 2: Prices in target market segment

Vehicle	Power (hp)	0-60 mph (s)	Price USD
Jaguar XK8	294	6.4	69,995
Jaguar XKR	390	5.4	81,995
Mercedes S430	275	6.9	74,320
Mercedes S500	302	6.1	82,770
Mercedes SL500	302	6.1	89,220
Mercedes S55AMG	493	4.6	110,170
Mercedes S600	493	4.6	122,820
BMW 745i	325	5.9	68,500
BMW 745Li	325	5.9	73,195
BMW 760Li	438	5.4	116,495
Aston Martin DB7 Vantage	420	5.0	180,000
Aston Martin DB9	450	4.7	TBA
Bentley Arnage	450	5.5	229,000
Maybach 62	550	5.3	360,000
Bugatti Veyron	1001	0-62: 2.9 / 0-186: 14	1,000,000

can be successful in the marketplace in the near future. Several new ideas or new uses for existing ideas have resulted from this approach, including the following.

- Fuel cell vehicles allow and sometimes even demand new vehicle topologies.
  - Safety concerns suggest that tanks should be centered.
  - Added freedoms allow elements with potentially large acoustic signatures to be placed at the rear of the vehicle.
  - Maintenance no longer requires a conventional hood. This allows for a very modern styling, with a seamless front end.
- Some other new styling ideas have further enhanced the performance and luxury.
  - An amphitheater style seating concepts allows comfortable seating for five, and greatly enhances the riding experience for rear seat passengers.
  - The floor of the trunk collapses, to enhance underbody aerodynamics when empty.
  - Flat wheel hubs enhance aerodynamics, with the use of clear plastic still giving the look of a traditional wheel.
- A unique suspension design incorporates several new technologies and ideas.
  - An electro-magnetic suspension uses zero-power active levitation system based on those developed for high speed trains. The system provides active ride control, allowing a full gamut of ride characteristics between stiff performance suspensions and soft luxury rides.
  - Additional servomotors and controls allow a constant ride height to be maintained regardless of vehicle load.
  - Further mechanisms provide active camber adjustment, enhancing stability at high speeds.
- Several non-conventional ideas have been developed to improve the net efficiency of the fuel cell system.
  - Perforated or foamed metal electrodes have been selected over the graphite electrodes more commonly used in PEM fuel cells.
  - A new electrode design has been developed around the perforated/foamed technology. Each electrode serves as an anode in one cell and a cathode in the next cell—reducing the system losses created by additional series connections.

- A pressure wave supercharger reduces the system losses, using some of the enthalpy still present in the exhaust gas to compress the supply air.
- Power electronics have been designed around the fuel cell and traction motors to maximize efficiency and performance.
  - New voltage boosting circuitry allows the traction motors to produce comparable torques at nearly double the speed (essentially doubling the power).
  - Additional switching circuitry within the stack allows for series or parallel connections: better matching the DC supply with the needs of the traction machines and three-level inverter. The three-level inverter further enhances the load to source matching and thus the net efficiency.

The design case study strove to integrate the new technologies to create a unified product. Subsystems have been designed to work with one another and promote maximal system efficiency and performance. What has been presented here as a final effort, is in fact an initial proposal, intended to show what can be both technologically possible and economically viable with recent propulsion advances and a clean sheet approach.

REFERENCES

1. J. Dunne, "This is a dupe," Popular Mechanics, http://popularmechanics.com/automotive/concept_cars/2001/1/GM_hybrid_gets_80_mpg/print.phtml accessed Aug. 26, 2003.
2. Bendinglight website. http://www.bendinglight.co.uk Accessed Aug. 28, 2003.
3. J. Larminie and A. Dicks. *Fuel Cell Systems Explained.* John Wiley & Sons, Ltd., Chichester, England, 2000.
4. G. Gyarmathy, "How does the Comprex® pressure-wave supercharger work?" 1983 SAE International Congress and Exposition, SAE 830234, March 1983.
5. C.G. Cantemir, C. Hubert, G. Rizzoni, G. Ursescu, C. Yakes, and K. Yasuda, "High-power High-speed Road Train System," 2003 SAE International Truck and Bus Meeting and Exhibition, SAE 2003-01-3380, November 2003.

IV. COMPONENT DEVELOPMENT

IV. Component Development

The fuel cell system is made up of a number of components apart from the fuel cell stack itself. As in the case of the internal combustion (IC) engine, the supply of air and fuel and the removal of exhaust products is required. Like the IC engine, the fuel cell system requires a control system to ensure that the processes that make up the power generation process are managed in an optimal way that does not compromise the overall system efficiency.

The choice of components depends critically on the nature of the fuel cell system and its design parameters. The most fundamental distinction is in the temperature of operation. A PEMFC stack runs at 80ºC, while an SOFC operates at temperatures in the range 500ºC and upwards. Fuel supply may vary from hydrogen or aqueous solutions of methanol with PEMFC to methane and syn-gas (carbon monoxide and hydrogen) for SOFC. A secondary and important design parameter is the pressure of operation. In general PEMFC systems may operate under pressure (up to 3bar) and require a compressor-expander system to supply air and recover energy.

The choice of fuel supply dictates whether a fuel processor is included in the system. A fuel processor will convert a primary fuel source into a reformate gas that after treatment to remove potential contaminants is supplied to the stack. The degree of processing depends on the type of stack. For a PEMFC that requires high purity hydrogen, the fuel processor must produce gas whose carbon monoxide content is low. For an SOFC, the carbon monoxide will be converted in the stack, but for both types of stack, sulfur must be removed as part of the reforming process.

The fuel cell stack delivers a variable voltage DC current. It is a demand device where the oxidation of the fuel gas varies according to the load placed on the stack. The system will include electrical power components that will convert the stack electrical output into a form that may be used by traction motor inverters and ancillary electrical devices. Most work reported in the vehicle sector is focused on the specific aspects of managing low voltage systems alongside traction voltages. The conversion between 12 or 42V and the traction voltage must be both cost effective and efficient.

The principal components that make up the system are:

- **Air supply**– either a fan for a low pressure ventilated system, or a compressor expander for pressurized systems. Compressor units may be positive displacement devices (such as scroll compressor, screw compressor) or radial flow compressors.
- **Electrical systems** that manage the flow of electrical power from the stack to both storage and traction systems. Power systems include the energy stores needed to buffer excess output or regenerated energy from braking.
- **Fuel cell stacks** that must be supplied with both fuel and oxidant and in the case of PEM be cooled on a separate circuit for which there will be component choices including pump and radiator.
- **Water and thermal management** components required for the maintenance of operating temperatures, the management of humidity of gas passing into the fuel cell stack and the removal of enough water from the exhaust streams to make the system self-sufficient in water.
- **Fuel processors** that converts a primary fuel into a reformate gas that is compatible with the fuel cell stack.

Control systems including sensors, actuators and the computer control system are covered in a later Chapter, but form a substantial part of the system complexity and cost. Estimates vary but point to about 30% of manufacturing cost is likely to be attributed to control and monitoring.

Other component topics for which references are listed in the bibliography section include energy storage and the choice of materials for system components.

Component issues include durability and life cycle issues. Work is already published on the durability of fuel cell stacks in the face of realistic cooling strategies. Long term chemical interactions of the coolant and the internal surfaces of the stack can lead to damage and a shortened lifespan of the components. Specific measures are required to control the chemical composition.

Air Supply

Paper 2002-01-0409 covers the investigation of part load air supply strategies for PEM fuel cell systems. Air supply is considered together with the need for humidification and management of the temperatures and heat flux in the intake system. The development process uses a Hardware-in-the-loop (HIL) method to simulate the stack and vehicle systems while the compressor-expander device is run on a test-bench.

The paper 2004-01-1009 describes the development of a model based control strategy to manage both air mass flow and supply pressure. The authors show experimental data to verify the performance of the proposed control system.

Electrical Systems

In paper 2002-01-1902 the author presents a design for a dc/dc converter that is developed from the system requirements for the vehicle. Of particular interest is the starting process where the dc/dc converter must start the compressor (for air supply) and the cooling pumps. Once start is complete the dc/dc converter must change direction and convert the voltage of current flowing from the fuel cell stack.

In paper 2004-01-1006 the author describes the application of a VCU (voltage and current control unit) to manage the flow of power from the fuel cell stack and to manage the receptivity of the capacitor. The management of the output of the fuel cell stack allows the rated voltage of the traction equipment to be reduced thereby offering a cost and size reduction.

Paper 2004-01-1303 explores the power system architecture that meets the operation requirements of a fuel cell powered vehicles. The paper describes a power control strategy and the use of an ultra-capacitor to control the battery current during load transients.

The authors of paper 2004-01-1304 describe a modeling exercise in which different configurations of energy store are used in a hybrid fuel cell vehicle. The modeling exercise is supported by experimental results from the Toyota FCHV. Lithium ion batteries show significant promise offering a substantial gain in overall efficiency as a result of lower operating voltage and lower battery specific mass.

Fuel Cell Stacks

In 2002-01-0407 the authors report on developments in the manufacture and use of materials in the construction of the membrane-electrode assembly (MEA), the active element of the PEM fuel cell.

In paper 2002-01-0410, the authors offer a progress report on a project in which the internal reforming capacity of solid oxide fuel cells has been investigated. The authors report on the ability of the fuel cell to reform and then oxidize complex hydrocarbons including some of the components that make up diesel fuel.

The paper from Ronald Mallant was presented at the meeting, Scientific Advances in Fuel Cell Systems held in Amsterdam in September 2002, presents an analysis of the thermodynamics of a PEMFC stack operating at the normal temperature of 80ºC. This temperature is imposed by the stability and function of currently available polymer electrolyte material. The author presents a case for a polymer electrolyte able to operate at high temperatures with the implication that the need for water to support proton conduction is significantly reduced or eliminated.

Water and Thermal Management

The paper 2003-01-0802 is concerned with the long term behavior of coolant in a PEMFC system. The coolant, consisting of a mixture of de-ionized water and ethylene glycol can leach material from the containment vessels and pipe-work and to create issues for durability and thermal performance. The authors present the results of a study to investigate coolant treatment.

The authors of paper 2001-01-0537 present a design study into the thermal and water management requirements in a PEMFC stack. The performance of both stack and system is investigated as anode saturation temperature is varied. The results show that a high anode saturation temperature leads to good stack efficiency but because of the changes external requirements, there is not necessarily a gain in system efficiency.

Fuel Processors

Paper by Detlef zur Megede was first presented at the Grove Fuel Cell Symposium held in London in September 2001. The author describes the development of fuel processor technology in the context of first Daimler-Benz and later DaimlerChrysler's fuel cell research and demonstration programme.

In 2003-01-1366, the authors present developments in the reformer design in which reacting flows and heat exchange are managed in narrow channels.

The authors of paper 2003-01-0810 investigate the steady state and dynamic behaviour of a methanol fuel processor and show that the operation of the fuel processor interacts strongly with other systems components including the stack and heat exchangers.

In 2003-01-0806 the operation of the reforming process is supplemented by acoustic energy applied to the catyalust bed of a methanol steam reformer. Improvements include an enhancement of the conversion efficiency.

The authors of paper 2004-01-1473 describe progress in a three year program to develop a fuel processor for a vehicle application that will supply hydrogen rich reformate to a PEM fuel cell stack. The authors describe the demonstration of a 75 liter unit that has been operated at 200kWth on gasoline fuel.

IV. COMPONENT DEVELOPMENT

Air Supply

2002-01-0409

Development of Fuel Cell System Air Management Utilizing HIL Tools

Stefan Pischinger and Carsten Schönfelder
Institute for Combustion Engines (VKA)

Oliver Lang and Helmut Kindl
FEV Motorentechnik GmbH

ABSTRACT

In this paper, boosting strategies are investigated for part load operation of typical fuel-cell-systems. The optimal strategy can mainly be obtained by simulation.

The boosting strategy is one of the most essential parameters for design and operation of a fuel-cell-system. High pressure ratios enable high power densities, low size and weight. Simultaneously, the demands in humidification and water recovery for today's systems are reduced. But power consumption and design effort of the system increases strongly with the pressure level. Therefore, the main focus must be on the system efficiencies at part load. In addition, certain boundary conditions like the inlet temperature of the fuel-cell stack must be maintained. With high pressure levels the humidification of the intake air before, within or after the compressor is not sufficient to dissipate enough heat. Vaporization during the compression process shows efficiency advantages while the needs in heat dissipation decreases.

With the simulation tool MATLAB/SIMULINK, supplemented by detailed self-developed design tools for compressor and expander units and hardware-interfaces for hardware-in-the-loop simulations, fuel-cell-systems can be modeled, simulated and verified accurately. The access to a wide database from compressor and fuel cell test benches allows precise simulation results with adjustable model complexity.

INTRODUCTION

In order that fuel cell systems can be used for propulsion, the system components have to be optimized regarding design and behavior. Many of them are not commercially available for the specific range of fuel cell systems. A main focus must further be taken on the strategies for system operation and control, because a defective mode of operation can dramatically deteriorate the efficiencies, or the system can even be damaged.

The pressure level in the fuel cell that is needed to obtain a maximum of efficiency is still discussed controversially. Concepts range from low pressure systems, that operate only slightly above the pressure loss of the system to reduce volume and weight of the air-supply, to high pressure designs of up to 4 bar, to maximize the power density of the fuel cell and to guarantee a balanced water circuit [1]. Such systems need a power of up to 25% of the fuel cell for the compression of the air-supply. The recovery of the pressure energy utilizing expander units is therefore recommended to maintain high efficiencies.

To compare different compressors concerning efficiency behavior, weight and cost, an extensive database of measurements is necessary. Furthermore, detailed knowledge about the layout of compressors and expanders as well as suitable simulation models are essential. A strong experience about super- and turbocharging systems from applications on internal combustion engines is available at the team [1-13].

The development of air supply concepts is closely related to the humidification of the cathode inlet air. Humidification is done in order to increase the efficiency of the fuel cell stack and to prevent cracking of its membrane. Therefore, an optimized integration of the humidification into the air supply system is extremely important with regard to high efficient, small size and low weight fuel cell systems. A promising approach for the humidification of the cathode inlet air is the usage of intermittent injection valves, which are derived from internal combustion engine applications or burner technology and can directly be integrated into the compressor.

AIR MANAGEMENT

MODELING CONCEPT - Complex systems like fuel cell systems for propulsion or electricity generation can be calculated in different ways. For the assessment of stationary designs it is sufficient to use simple calculation software, even spreadsheets. A detailed investigation demands a more sophisticated software, that can express and solve (partial) differential equation systems in time (and space variables). MATLAB/SIMULINK with specific user extensions is such a software that gains increasing currency.

Depending on the task, there are different strategies to simulate a technical system. E.g. it is possible to build up a simplified system using constant efficiencies or heat transfer coefficients. But the expressiveness of the results may be limited in this case. Detailed calculations, e.g. taking into account the flow patterns or the local diffusive transport, make the results much more accurate but increase in the same manner the simulation times. In addition, such calculations need information of the geometry and the used materials, so that the models are less flexible with regard to quick variations.

On the one hand a compromise must be found in prediction accuracy and fast and flexible calculations on the other. General assemblies can be designed by means of simplified calculations. To develop new components it is necessary to use a detailed description based on physical models. For the optimization of existing systems during the design process and the development of control strategies, measured maps can be included to obtain fast and accurate results. In this case the measured data must also comprise the dynamic system behavior.

A comprehensive database is therefore very useful. The team can look back on numerous measurements from their compressor and fuel cell test benches. The self-developed code CATS, that runs as a part in the SIMULINK environment, is able to calculate the compressor and expander units in detail and enables scaling of existing measurement data to different compressor designs based on similarity models. So it is possible to develop air-supply systems that are matched to the particular demands.

PART LOAD STRATEGY - Fuel cell systems are often calculated only at one point, for example at rated power output. But the system is driven mainly at part load, where the efficiencies of motor, compressor and other components change. It has also to be considered, whether the efficiency at rated power has to be optimized or if it is more useful to design a smaller system with less efficiency for maximum power. Increasing part load efficiency while saving volume and weight can improve the entire performance of the vehicle.

Concerning the boosting strategy of a fuel cell system a compromise in the pressure level has to be found. High levels enable less water injection and water recovery. With higher pressure, less water content causes the same humidity, so that the efficiency of the fuel cell rises. On the other hand the compression needs more power - up to 25% of the electric power of the fuel cell has to be used for the compressor drive. It has also to be considered, if the power density has to be maximized in order to save volume and weight.

The optimal strategy can only be calculated by simulating the entire fuel cell system.

The assembly of a simulation model for fuel cell systems can be described as follows: The compressor is realized by measured maps for efficiency and mass flow. Concerning dynamics, the mass moment of inertia and the heat capacity has to be included. The current-voltage characteristic, depending on pressure and humidity, is the main part of the fuel cell model. Pipes and other components influence the pressure and mass flow dy-

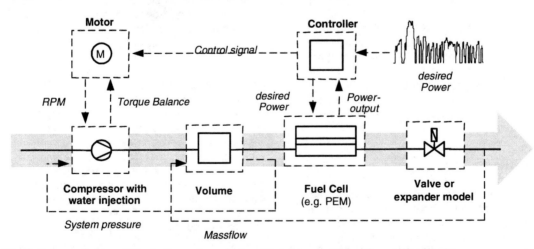

Fig. 1: Schematic signal flow in the main part of a simulation model for air supply systems

namics. The motor can be modeled with maps for efficiency and characteristic curves for constant and maximum torque.

Figure 1 shows a schematic overview of the main part of the simulation model for the air-supply system. The high pressure pipe to the stack is reduced to one volume. Depending on the system concept the pressure is regulated by an expander model or a pressure holding valve. A controller realizes the compressor speed that is necessary for the air-supply depending on the power demand. Furthermore, the valve position and the water injection is controlled. Other parameters like compressor outlet temperature can also be implemented in the control system.

HUMIDIFICATION CONCEPTS - The humidification of the intake air can be realized by mixing steam into the air-flow, using bubblers, etc. For mobile applications the injection of liquid water is a promising method, that is compact and energy efficient. The vaporization of the water cools the air, so that the high temperatures of the compression (up to 180°C) can be reduced to meet the desired gas temperature of about 80°C for the fuel cell stack.

The water can be injected, using an intermittent electro-magnetic valve, before, after or also into the compressor. Intermittent injection valves have advantages regarding turn-down ratio and control. The injection before or into the compressor has the advantage, that the gases are internally cooled during the compression. Thereby, theoretical benefits in efficiency can be achieved. Figure 2 shows the water injection into the housing of a screw type compressor (injection into the first part of compressor, near the inlet).

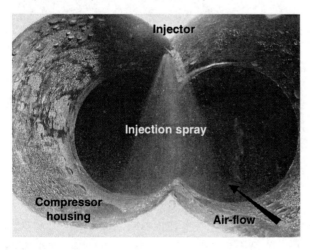

Fig. 2: Water injection into a screw-type compressor

The effects of the internal cooling can be calculated by a simple model. The compression is divided into stages of dry compression and aftercooling by internal water vaporization. A calculation of one stage means injection

after compression, infinite stages simulate continuous internal cooling by continuous vaporization. Figure 3 shows a temperature-entropy (T-s) diagram of the compression line. Dry compression with water injection after the compressor is shown as 1-2-2', while compression with internal cooling by water vaporization is represented by the curve 1-2'. There is also a compression process divided into two stages of dry compression and after-cooling shown.

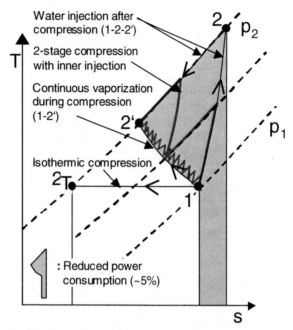

Fig. 3: Simulation of internally cooled compression

The cooling causes a decrease in pressure, while the vaporized water increases the gas mass and therefore the pressure. In total the pressure declines, so that the compression work decreases. This causes also a change in the internal compression ratio. The gases are compressed with a new polytropic exponent, lower than the isentropic exponent. The benefit in power consumption is shown in figure 3 as grey area. Assuming a real screw type compressor with already high efficiencies the power consumption (and the injected water mass, for constant output temperatures) decreases by about 5%. With external continuous cooling and without water vaporization, there is a potential of about 6.5%. A compressor with lower efficiency may have an even higher efficiency potential.

For the simulation of a compressor with water injection in a fuel cell system it is helpful to calculate injection maps. These maps contain the injection masses for each parameter like compressor speed, pressure ratio, inlet temperature and humidity. Also the outlet temperature and the dewpoint for each operation point can be determined. These maps can be used as a pre-control for the injection mass and as boundary conditions for simulations.

The calculation of these maps is based on the measured dry compression maps containing torque and isentropic and effective efficiency depending on compressor speed and pressure ratio. Using the model, the injection masses can be evaluated according to the desired boundary conditions like dewpoint and gas temperature.

An example for an injection map is shown in Figure 4. For a specific inlet temperature and humidity the injection quantity, the outlet and dew point temperature are presented. The chosen outlet conditions are: gas temperature below 80°C and dewpoint above 50°C. The desired dewpoint can only be reached with a pressure higher than 2 bar. At 2 bar boost pressure dewpoint and gas temperature are similar, so that an additional heater is needed in order to reach 80°C gas temperature. With high inlet temperatures and humidity there can also be too much heat energy in the system, so that water vaporization up to a humidity of 100% is not enough to cool the gases to 80°C. This means, that the area, where the desired outlet conditions can be achieved, is limited to both sides.

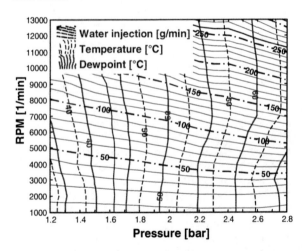

Fig. 4: Calculated water injection maps to meet desired output temperature and dewpoint for a screwtype compressor

An important result for the injection control is, that it is not possible to use a fixed nominal dewpoint, because it is not always possible to reach this value. So the control has to take into consideration the theoretical limits.

RESULTS OF SIMULATION - Using the simulation models - as described above - for the compressor, the motor (with maps for efficiencies of motor, controller and gear box), fuel cell with pressure and humidity dependency and an optimized controller unit for dynamic tests, the optimal part load strategy for each system can be investigated. It is possible to analyze system configurations with different components like the boosting system.

Interesting questions are for example, if an expander unit is still useful for lower system pressures or what kind of control concept is useful (e.g. wastegate, afterburner, etc.). With this method it is also possible to determine the optimal layout of the compressor, i.e. highest efficiencies must correlate to highest dwell times in the vehicle test cycle, etc. the team has a wide range of data of investigated compressor and expander units, that makes it easier to develop the optimal fuel cell system for the particular demand.

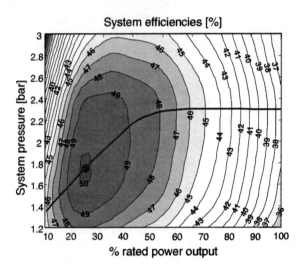

Fig. 5: System efficiencies of a hydrogen fuel cell system depending on power output and system pressure (screwtype compressor)

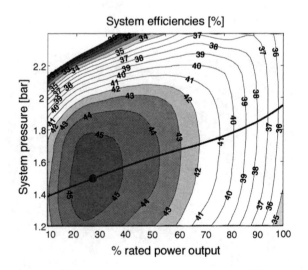

Fig. 6: System efficiencies of a hydrogen fuel cell system depending on power output and system pressure (turbo compressor)

Figure 5 and 6 show the results of simulation for different compressor types. A system with a screwtype compressor shows an optimal efficiency for a wide range

286

from full load to 50% part load at constant pressure levels in the fuel cell of 2.1 to 2.3 bar (corresponding to the pressure losses the pressure ratio of the compressor is higher). Below 50% the optimal pressure decreases continuously to slightly above 1 bar. This behavior, typical for this type of compressor, can be approximated by two straight lines.

The turbocharged fuel cell system shows a different behavior. Optimal efficiencies are reached with continuous ascent of the fuel cell pressure with the system power. Also, at full load the maximal pressure stays below 2 bar. The characteristics of these curves depend, among others, on the design of the compressor. Simulation results for different designs are shown in figure 7. Turbo compressors have up to 10% higher efficiencies in their optimal point of operation than screw types. In the examples of Figure 5 and 6 the specific screwtype compressor used leads to a 5% higher efficiency at part load.

For the general assessment of these concepts it must be emphasized, that a different stack characteristic can lead to a different optimized boosting device. Moreover, a turbo-compressor can easily be combined with a turbine as expander unit, so that further advantages are expected.

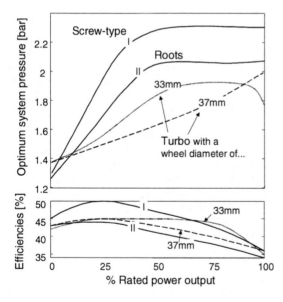

Fig. 7: Boosting strategy for part load operation of different compressor designs and system efficiencies

The behavior of the screwtype compressor can mainly be explained with the efficiency characteristic of the compressor itself. Figure 8 presents a measured "humid" map of a compressor with water injection. The effective efficiency of this compressor has its maximum at higher pressure ratios from 50% part load to full load. Below

50% part load and at lower air flow rates the efficiency decreases rapidly.

While the effective efficiency slightly rises with water injection, also an increase in volumetric efficiency can be expected. The water that deposits on the compressor screws reduces the gaps between screws and housing, resulting in an increased volumetric efficiency, while durability is not influenced by this measure.

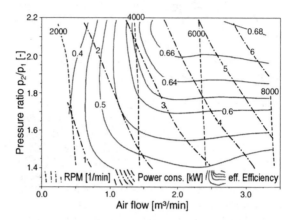

Fig. 8: Measured map of a screwtype compressor with water injection into first phase of compression (effective efficiency)

DEVELOPMENT OF FUEL CELL SYSTEMS

In order to develop complete fuel cell systems, simulation, testing and validation of single system components is not sufficient. The system and the operation strategies, like the part load strategy, has to be verified under static and dynamic conditions.

Results and data from measurements and simulations have to be used as directly as possible for calculations of the system behavior to save time and costs.

STRATEGIES FOR OPTIMIZED LAYOUT OF AIR SUPPLY WITH THE FUEL CELL SYSTEM - For optimization and fine-tuning of the system's components with regard to transient operation there are different assembly strategies possible. The classical procedure is to build up the complete system in hardware and to perform the fine-tuning on the test bench. Calculations are often only used for rough estimations and selection of the system components, e.g. matching of the compressor size.

The quality of a pure simulation-based prediction of the complete fuel cell system behavior is always linked with the experience of the software user. Nevertheless, there are always effects, that cannot be foreseen. In most cases the time to set-up a software model is limited, so that the simulation has to be simplified. In particular, the aspects of control design are sensitive to changes in the measured data and the closed-loop controlled system. The system can become unstable, which can influence the operation of the whole system.

For a reduction of development time and cost the fuel cell at the test bench is replaced by a dummy in the first step. This has also the advantage, that stack independent system data will be collected. Otherwise, the results will not be transferable and only valid for the tested system.

With the described HIL methodology a flexible exchange between soft- and hardware modules is possible.

Humidity measurement - As an example the measurement of humidity or dewpoint can be mentioned - actually a minor consideration. If the fuel cell needs high values of humidity near 100%, the most sensors fail. The reaction times are in the range of minutes to hours and the transfer characteristics are extremely non-linear. Therefore, a control strategy is difficult to develop. Utilizing high-order dewpoint transmitters, that are heated in order to prevent water condensation at the sensor head, these effects can be minimized, but not generally avoided. In addition, some unsteady behavior is influencing the pipe characteristics: Depending on the quality of atomization and the conditions of the water injection, not the entire injection mass can be vaporized in the compressor, but causes wall film effects in the outlet pipe, changing the dynamic behavior of the system.

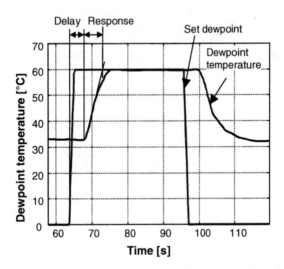

Fig. 9: Responses of a down-stream heated dewpoint transmitter to water injection turn-on and turn-off into a screwtype compressor

Figure 9 shows the response times of a high-resolution dewpoint transmitter. A PI-control for the dewpoint temperature was used. The delay and response times are

caused by the dead times of the sensor (conditions were used, which guarantee complete vaporization of the water injection; the delay time due to the flow speed was 0.3 s). The dead times can be explained with the diffusion of the water molecules into the surface material of the sensor head. In this example the delay is in the range of 5 to 10 seconds. The asymptotic characteristic of the dewpoint when turning off the water injection is also effected by the vaporization of some water droplets on the screws and the outlet walls.

This system behavior is hard to take into account while setting-up a simulation model. Therefore, measurements on the test bench with the relevant system components like the compressor with water injection, the piping and the dewpoint measurement is essential. Furthermore, a control strategy for the water injection must be developed, that enables a transient operation mode. Reaction times of 5 seconds or more are too slow to use only a simple control, a pre-control based on steady injection maps, as described above, is necessary.

Hardware-in-the-loop (HIL) - The direct design of the complete fuel cell system based on test bench measurements (e.g. compressor combined with fuel cell test bench) is difficult. Feedbacks between the system components have to be taken into consideration and fine-tuning is essential. To identify possible effects and problems of the complete fuel cell system in advance, the integration of the test benches for the system components (like air-supply and fuel cell) and the simulation model into one common environment has clear advantages. This is possible using a hardware-in-the-loop system, where the separated component test benches can be operated like software submodules in a real-time simulation environment.

the team has built up a hardware-in-the-loop system with the simulation software MATLAB/SIMULINK and a conventional personal computer with I/O cards as hardware interface. The models created in SIMULINK are compiled to real-time codes, which can be uploaded to the target PC and run in real-time. The driver for the analog or digital multi-channel I/O PCI-cards and the real-time operating system on the target PC are provided by the SIMULINK extension xPC-target. Using an ordinary low-cost 1 GHz Athlon single CPU with 256 MB memory sample rates of 5000 Hz or more are possible (with 16 channel input, 8 channel output, running complex models). Conventional operating systems like MS Windows are not able to provide comparable sample rates for input and output simultaneously.

This system can be used on all different types of test benches. Utilizing notebook PCs mobile application are also possible without the need of special and expensive hardware solutions. Even an upload and control of the model on the target PC via Internet and mobile phone is possible, so that the running system, e.g. in a prototype vehicle, can be influenced on remote operation.

Figure 10 shows an application of such a HIL-system. The air-supply for a fuel cell system to drive a passenger car can be designed and optimized from the very beginning as if operating in the real vehicle. Among others the following questions can be answered:

- Feedback of the fuel cell on the air-supply and effects of changing air temperature and humidity can be taken into account,

- Faults in the water injection control can cause (virtual) damage in the fuel cell stack and power loss because of decreasing fuel cell efficiency.

- The dynamic behavior of the speed control can be evaluated in order to guarantee in each case the desired stoichiometry with an accuracy of about 10%.

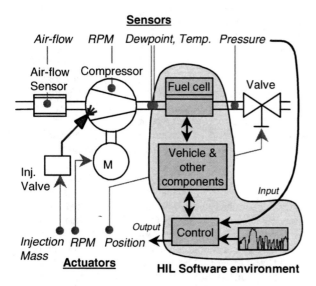

Fig. 10: Use of a hardware-in-the-loop system for the development of air-supply units for fuel cell systems

Therefore, the measurement data at the (virtual) fuel cell inlet like dewpoint, gas temperature and air flow is transferred through the I/O cards to the simulation model. Depending on the power demand the output power of the fuel cell, the electric drive and the vehicle dynamics can be calculated. A reformer model is integrated for methanol, gasoline or diesel fuel systems, that provides the gases for the anode. Corresponding to the power demand the control values for compressor speed, water injection and system pressure are chosen. The injection masses are controlled by 4-dimensional maps, depending on inlet temperature, humidity, compressor speed and pressure ratio. The control variables for injection mass (or pulse-width for the electromagnetic valve), motor speed, position of the throttle (pressure control) or the stator vane using an appropriate expander, are transferred via interface cards with high sample ratios.

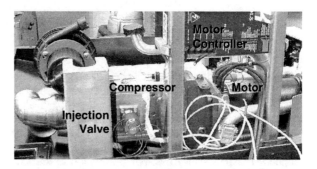

Fig. 11: Air supply unit with motor, controller, compressor and injection valve

Figure 11 shows an exemplary air-supply system developed by the team consisting of an extremely small and light electric motor with controller, compressor with integrated water injection and the injection valve. The prototype fuel cell system is designed for use in a production-type passenger car. Before assembly, the control strategy for the water injection and the boosting strategy for part-load operation was developed at the compressor and the fuel cell test bench.

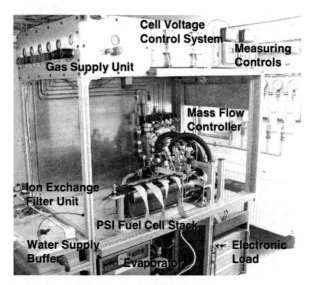

Fig. 12: Testing of a PSI stack at the fuel cell test bench

In order to verify the validity of the fuel cell model it is also possible to make use of the hardware-in-the-loop system at the fuel cell test bench (see figure 12). The fuel cell test bench is pressurized directly from gas cylinders. Humidification is realized by a steam generator. The conditions at the intake are stable and all parameters can be modified dynamically. The test bench control system REDLINE ADAPT was specifically adapted to the stack test bench requirements. The hardware-in-the-loop system can exchange data with the process control system using analog and digital interface units. Using the measured data from the compressor test bench, the system can actually be operated as if this air-supply system would be attached to the fuel cell system. With the measurements at the fuel cell test bench the software model of the fuel cell can be optimized and used for operation at the compressor test bench.

Dynamic operation. The developed air-supply system was finally checked under dynamic operation using the HIL-methods described above. For this, the calculated boosting strategy for part load and the humidification control was used. In addition, the used simulation model comprises the complete control for the compressor test bench. This allows fully automatic operation of the test bench controlled by the SIMULINK model right after start of the driving cycle. Further manual interventions are not necessary.

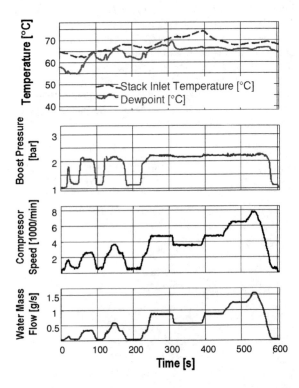

Fig. 13: Measured data of a dynamic HIL-test of an air-supply system using a defined boosting strategy for part load similar to NEDC

Figure 13 shows the results for a test cycle following the NEDC: Temperature at the (virtual) inlet of the fuel cell stack, dewpoint, pressure level, compressor speed (as indicator for power demands) and the amount of water injected into the compressor. A constant pressure of about 2.2 bar between 50% and 100% load was chosen. The amount of water injected into the air supply is controlled in the same manner, because the other parameters, like intake temperature and humidity remain constant. A dewpoint temperature of 60°C can not always be realized: At low pressures there is not enough energy available to evaporate sufficient quantities of water. The gas temperature would further decrease. Hence, the

optimization goal of the transient investigation is to maintain an approximately constant stack inlet temperature.

As shown in Fig. 13, the effects of stepwise change of rotational speed and pressure on gas temperature and dewpoint are minor, all parameters deviate in a small range. At part load, when the conditions for evaporating the injected water deteriorate due to low flow speeds, a still sufficient humidity can be kept. At high pressures, the stack inlet temperature does not exceed the maximum allowable temperature.

CONCLUSION

Different concepts of air-supply have been investigated to reach an optimum in system efficiency and transient behavior of a fuel cell vehicle. Finally, an air-supply system was built, which will be part of a prototype fuel cell vehicle.

First, the system components have been analyzed and optimized by detailed physical calculations. Water injection into the compressor shows benefits in efficiency of up to 5% and improves the controllability. Injection maps can be generated in order to meet desired inlet conditions of the fuel cell concerning gas temperature and dewpoint.

The control strategies were developed by detailed simulations, essentially based on measurement data from the compressor and the fuel cell test bench. Water injection for dynamic operation was mainly realized by a precontrol, based on detailed injection maps. Depending on the utilized compressor design, the optimal system efficiencies of about 50% can be reached from part load to full load at pressure levels in the fuel cell of 2.1 to 2.3 bar for a screwtype compressor. The obtained boosting strategy for part load operation was verified under transient conditions.

To develop, test and optimize the air-supply system, a hardware-in-the-loop system based on the modeling and simulation environment MATLAB/SIMULINK was used. It is possible to control the compressor and the fuel cell test bench from the beginning as a part of the software model of the complete passenger vehicle. In this way, the feedback of the other system components could be taken into consideration.

The complexity of a fuel cell air-supply system with water injection, especially regarding dynamic control, is a demanding challenge in development. Hardware-in-the-loop concepts are an essential key for the process of development. Time and costs can be saved, the outcome is improved.

The work presented in this paper is planned to be continued by detailed investigations regarding selection and control of expander concepts. Also different types of applications like stationary systems for indoor installation or auxiliary power units for mobile needs will be analyzed. For future hybrid applications this development strategy can also be used. Additional powertrain components then must be applied to the HIL system.

REFERENCES

1. Kalhammer, P.R. Prokopius, V.P. Roan, G.E.
   Status and Prospects of Fuel Cells as Automobile Engines
   A Report of the Fuel Cell Technical Advisory Panel, July 1998

2. A. Wiartalla, S. Pischinger, W. Bornscheuer, K. Fieweger, J. Ogrzewalla
   Compressor/Expander Units for Fuel Cell Systems
   SAE Paper 2000-01-0380, SAE 2000 Congress, Detroit, Michigan, March 6-9, 2000

3. Amphlett J.C. et. al.
   Simulation of a 250 kW diesel fuel processor/PEM fuel cell system
   Journal of Power Sources 71 (1998) S. 179 - 184

4. M.B.V. Virji, P.L. Adcock, P.J. Mitchell, G. Cooley
   Effect of Operating Pressure on the System Efficiency of a Methane-fuelled Solid Polymer Fuel Cell Power Source
   Journal of Power Sources 71 (1998) S. 337-347

5. A. Selimovic, J. Pálsson, L. Sjunnesson
   Integration of a Solid Oxide Fuel Cell into a Gas Turbine Process
   1998 Fuel Cell Seminar, Palm Springs, November 16th – 19th, 1998

6. Rouveyre, Luc
   Optimization of a Fuel Cell System for Electric Vehicles
   Ph.D. thesis École de Mines, Paris, 1998

7. O. Hild, K. Fieweger, S. Pischinger, H. Rake, A. Schloßer
   The Control System of a Direct Injection Diesel Engine for Passenger Vehicles regarding Boost-Pressure and EGR Control
   MTZ 60, March 1999, page 186-192

8. H. Paffrath
   Investigations on the Potential of a Controlled Two-stage Supercharging System for Heavy Duty Engines
   Ph.D. thesis, RWTH Aachen, 1995

9. K. Habermann, K. Fieweger, M. Rauscher
   Supercharging of SI-Engines as a Method for Fuel Consumption Improvement
   Conference "Downsizing Concepts for Gasoline and Diesel Engines"
   Haus der Technik e.V., 26. - 27.06.2000, Munich

10. A. Wiartalla, W. Bornscheuer
    Fuel Processing for Automotive Fuel Cell Systems - Applications of Components from Internal Combustion Engines
    ISAF XIII, Stockholm, Sweden, 3-6 July 2000

11. A. Wiartalla, W. Bornscheuer
    Fuel Cell Systems for Automotive Applications – Fuels and Fuel Processing
    2nd International Symposium on Fuels and Lubricants ISFL-2000, 10-12 March 2000, New Delhi, India

12. K. Fieweger, M. Rauscher, O. Lang
    Potential of an Electrically Assisted Turbocharger in Modern DI-Diesel Engines for Passenger Vehicles 7th Aufladetechnische Konferenz, Technische Universität Dresden, 28. - 29.09.2000

13. S. Pischinger, C. Schönfelder, W. Bornscheuer, H. Kindl, A. Wiartalla,
    Integrated Air Supply and Humidification Concepts for Fuel Cell Systems
    SAE Paper 2001-01-0233

CONTACT

S. Pischinger; Institute for Combustion Engines, RWTH Aachen, Germany; ☎ ++49 241 80-96200, email: office@vka.rwth-aachen.de

O. Lang, FEV Motorentechnik GmbH, Aachen, Germany; ☎ ++49 241 5689-652; email: Lang_o@fev.de

Control of a Fuel Cell Air Supply Module (ASM)

Johannes Reuter, Utz-Jens Beister, Ning Liu and Dave Reuter
IAV Automotive Engineering Inc.

Bill Eybergen, Mohan Radhamohan and Alan Hutchenreuther
EATON Corporation

ABSTRACT

Fuel cell systems emerge as a new technology, which is expected to play an important role for future powertrain applications. To enable this technology's entrance into the market, new developments to improve robustness, cost efficiency and maintainability are necessary. Besides the stack itself, several subsystems are required to operate a fuel cell system. The technical challenges for developing and optimizing these subsystems are comparable to the challenges in the stack development itself. The air delivery system is considered to have a major impact, subject to overall efficiency, noise emissions and costs. These properties are determined not only by the system hardware, but also by the chosen control strategy.

This paper describes an intelligent model based control strategy, which enables the system to use optimal operation points of compressor and motor. The quantities to be controlled are air mass flow and pressure. The precise achievement of the desired values is crucial for proper stack functionality and stack lifetime. The controlled parameters are the compressor speed/torque and the opening of the return manifold pressure valve. The paper considers all aspects of the controller development and the implementation of such a system including controller hardware and hardware-in-the-loop simulation.

The results presented in this paper were achieved in simulations and on a real system.

INTRODUCTION

Generating electricity in fuel cells is one of the most efficient processes to convert energy based on a hydrogen/oxygen (air) mixture. The basic technology is well known and has been used by the aerospace industry for many years. Due to this fact, many expect that hydrogen fueled fuel cells will become the future propulsion system for vehicles. However, the use in automobiles is a new application and still far away from mass production. Critical criteria for the success are costs, reliability, drivability, robustness, hydrogen storage (range), hydrogen supply (infrastructure) and well to wheel efficiency.

As some of these criteria pertain to infrastructure, technical development will be the main driver for bringing fuel cells into mass production. Due to these projects importance for the economy and the high risk involved (long term development) the development of hydrogen propelled vehicle systems is funded by the US government (DOE). The FreedomCar program is one example.

One focus of the fuel cell development is the fuel cell stack itself to increase power density, material cost effectiveness and reliability. Besides this, subsystems such as an air delivery module, humidifier, battery, hydrogen storage systems etc. need to be developed to fulfill today's automotive standards. While a lot of development tasks are more hardware related, a very important part to overcome current obstacles is a proper and sophisticated control system for fuel cell systems and subsystems. It has been proved possible to apply powertrain control systems expertise to fuel cell applications using proven development methods and quality standards to achieve the targeted performance. The controller discussed in this paper is controlling the air supply module (ASM) for a fuel cell stack to ensure the proper air mass flow and pressure over the entire operating range (important for performance but also to avoid stack damage due to unequal pressure on both sides of the MEA (membrane electrolyte assembly)). The air supply module is a high-pressure system, which uses a compressor instead of a blower to allow for a higher power density. It consists of a compressor, an electric motor including a motor controller (driving the compressor), an intercooler to limit the intake air temperature of the stack and a backpressure valve to control the pressure.

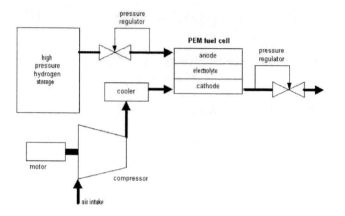

Figure 1 ASM Overview

The ASM controller developed by IAV is controlling the motor speed/torque and the pressure valve and is interfacing with the motor controller and with the overseeing fuel cell controller. In addition, the ASM controller is performing diagnostic functions. The hardware platform used is IAV's rapid prototyping platform UCU 5 (Universal Control Unit), with an MPC 555 microcontroller.

Hereby it is important to understand that the control system cannot be seen independent of the system hardware or the physical behavior. The success of the unit can only be guaranteed by considering the overall system. One good example is the noise issue. High-pressure fuel cell systems normally suffer from the noise of the compressor. Using an intelligent control strategy can reduce the noise factor depending on the parameters used to optimize the operating strategy.

MODELLING

For this project we have focused on a simplified model suitable for use in controls application rather than system validation. In order to serve as a control systems design guide, the model needs to reflect the system dynamics in a qualitative rather than quantitative manner. In this paper we follow the modeling approach similar to papers [1] and [2], however avoiding the linearization step. Please refer to these papers for further details. Nevertheless, for the sake of readability of this paper, we outline the basic results obtained there, as far as they are relevant for our approach. As mentioned earlier, in this paper we are focusing on the air supply system rather than the complete fuel cell stack.

In principle, the oxygen (air) path of a fuel cell can be seen as a mass flow through 3 different volumes. The first volume is the intake manifold, the second the volume of the stack and the third one the return manifold, please refer to Figure 1.

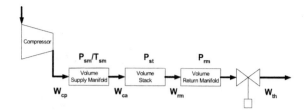

Figure 2 Model of the ASM

The quantities to be controlled are the flow through the stack volume and concurrently the pressure in the air path. The actuators used here are a compressor, which mainly controls the flow and a valve at the return manifold that in turn controls the pressure. However, since flow and pressure are coupled quantities, and changing the compressor speed affects both flow and the pressure and likewise changing the valve opening has influence on both pressure and flow, the system represents a nonlinear MIMO system. The measured quantities are the flow into the supply manifold, the pressure in the supply and return manifold as well as the temperature in the supply manifold. With T_{sm}, p_{sm}, p_{st} and p_{rm} being the temperature and pressure in the supply manifold, the stack and return manifold respectively, the state space equations can be written as follows:

$$
\begin{aligned}
\dot{p}_{sm} &= \frac{\kappa R}{V_{sm}} \left(W_{cp} T_{cp} - W_{ca} T_{sm} \right) \\
\dot{p}_{st} &= \frac{R_{st} T_{st}}{V_{st}} \left(W_{ca_{in}} - W_{ca_{out}} \right) \qquad (1) \\
\dot{p}_{rm} &= \frac{R_{rm} T_{rm}}{V_{rm}} \left(W_{ca_{out}} - W_{rm} \right)
\end{aligned}
$$

Where W_{cp} is the flow generated by the compressor, into the supply manifold, W_{cain} is the flow into the stack and W_{caout} the flow out of the stack. Please note, that humidifying is not considered at this point. Nevertheless, vaporization and humidity can be seamlessly considered, by using inherent fuel cell parameters, and state quantities such as current drawn from the fuel cell etc.

The flow into the supply manifold W_{cp} is a function of the ratio of supply manifold pressure P_{sm} to ambient pressure P_{am} and the compressor speed. A map that reflects this relationship has been generated from some stationary points as shown in Figure 3.

$$M_{cp} = c_p \frac{T_a}{n\ \eta(n, W_{cp})} \left(p_r^{\frac{\gamma-1}{\gamma}} - 1\right) W_{cp}(n, P_r)$$

$$(3)$$

Where ΔT is the difference in temperature between ambient and compressor outlet, M_{cp} is the required torque, η is the efficiency, n the compressor speed and P_r is the pressure ratio between the supply manifold and the ambient pressure.

For the pressure control valve, a nozzle equation is used, as can be found in many textbooks.

$$W_{rm} = \frac{p_{rm}}{\sqrt{RT_{rm}}} A(\nu)\Psi(p_{amb}/p_{rm})C_d \ (4)$$

Where $A(\nu)$ represents the effective opening area as a function of the control input ν, C_d a discharge factor and Ψ is the flow function

$$\Psi(p_r) = \begin{cases} \sqrt{\frac{2\gamma}{\gamma-1}\left(p_r^{\frac{2}{\gamma}} - p_r^{\frac{\gamma+1}{\gamma}}\right)} & , \text{for } p_r > \left(\frac{2}{\gamma+1}\right)^{\frac{\gamma}{\gamma-1}} \\ \sqrt{\frac{2\gamma}{\gamma-1}\left(\left(\frac{2}{\gamma+1}\right)^{\frac{2}{\gamma-1}} - \left(\frac{2}{\gamma+1}\right)^{\frac{\gamma+1}{\gamma-1}}\right)} & , \text{otherwise} \end{cases} \ (5)$$

The effective opening area as a function of the control input $\nu[\%]$ is shown in Figure 5.

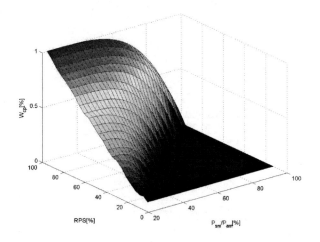

Figure 3 Compressor flow map

Likewise, a map representing the efficiency has been generated, based on two 2-dimensional exponential functions.

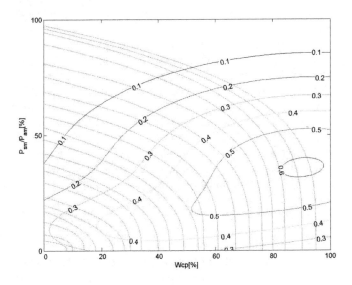

Figure 4 Compressor efficiency map

The state change in the compressor system is assumed to be adiabatic, thus both the temperature and the required torque have been computed by [3].

$$\Delta T = \frac{T_a}{\eta(n, W_{cp})} \left(p_r^{\frac{\gamma-1}{\gamma}} - 1\right) \qquad (2)$$

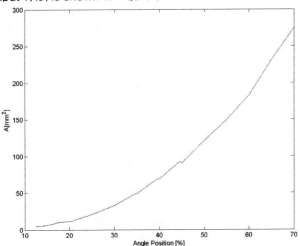

Figure 5 Effective area of the BPV

Since only stationary points have been available, the model validation was not possible for the dynamic behavior of the system.

CONTROL SYSTEM ARCHITECTURE

It is possible to linearize the system around certain stationary points [Stefanopoulou ACC 2002], however we have used a different approach here, which enables us to better take into account transients. Therefore, an exact I/O linearization approach is used. The control inputs are the compressor speed n and the opening area A. The I/O linearizing approach is obvious, since the first and third state equation contains both, a control input and the controls variables. Solving the first equation for W_{ca} yields

$$W_{ca} = \frac{1}{T_{sm}}\left(W_{cp}T_{cp} - \frac{\dot{p}_{sm}V_{sm}}{\kappa R}\right) \quad (6)$$

The first summand on the right hand side of this equation expresses mainly the stationary part while the second summand takes the dynamics into account. Selecting W_{cp} to

$$W_{cp} = \frac{1}{T_{cp}}\left((W_{ca_{des}} + u_2)T_{sm} + \frac{\dot{p}_{sm}V_{sm}}{\kappa R}\right) \quad (7)$$

with new feedback control input u provides a feedforward control law, that linearizes the dynamics between u and W_{ca}. In the implementation, T_{cp} is computed using equation 2, T_{sm} is measured and dp_{sm}/dt is directly computed from filtered measurements of prm. Please note that the inverse dynamics is analytical and differentiable everywhere.

In addition to the feed forward portion, a PID controller, generating the feedback control signal u is used to compensate for modeling errors. To generate the actual speed command out of W_{cp}, an inverse map n(W_{cp}, Pr) is used, as shown in Figure 6.

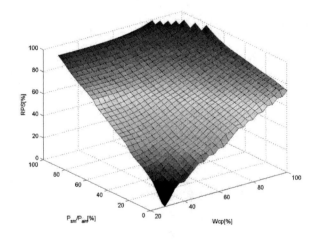

Figure 6 Inverted flow map

Further, the equation expressing the requested torque is used as a feed forward control command for the inner current control loop of the compressor motor controller.

The control loop for regulating the pressure in the stack, the return manifold respectively is designed using the same concept. Putting equation (3) into (1c) yields

$$\dot{p}_{rm} = \frac{R_{rm}T_{rm}}{V_{rm}}\left(W_{ca_{out}} - \frac{p_{rm}}{\sqrt{RT_{rm}}}A(\nu)\Psi(p_{amb}/p_{rm})C_d\right) \quad (8)$$

Selecting the control law

$$A(\nu) = \frac{V_{rm}}{R_{rm}T_{rm}}\left((W_{ca_{out}} + u_2)\frac{\sqrt{RT_{rm}}}{p_{rm}\Psi(p_{amb}/p_{rm})C_d}\right) \quad (9)$$

results in the exact linear dynamics

$$\dot{p}_{rm} = u_2 \quad (10)$$

Here u_2 is the output of a PID controller as well, which compensates for modeling errors.

It has to be taken into account, that Ψ is 0 for pressure ratio 1, that is: no flow. However, since this is only relevant for startup conditions, this does not affect the controls performance. Another difficulty results from the fact, that W_{ca} is not measured. To deal with this it is assumed that the flow control loop can hold the flow close to the desired flow, and therefore the desired mass flow $W_{ca_{des}}$ can replace $W_{ca_{out}}$ and deviations are considered external disturbances. To ensure feasible trajectories for $W_{ca_{des}}$, a profile generator has been implemented. This takes into account that the real physical system can only follow trajectories, which are at least twice continuously differentiable.

At the current state, the zero dynamics of the system (if there is any) has not been considered. In [2] it has been shown that the overall fuel cell system has an instable zero. However, this instable zero results from the drop of power provided by the stack to the outlet, during transients when major parts of the produced power are consumed by compressor acceleration. Since we are dealing with the air delivery system only, this does not affect the control system.

During testing the control architecture on the real systems, it turned out, the signal from the mass-flow sensor was found to be extremely noisy, as can be seen from Figure 18. It was planned to use this signal for the overlaying PID-T1 control loop, however increasing proportional and differential gains had the effect, that the motor speed is directly modulated by that noise. Heavy filtering of this signal would result in lack of performance and could not be applied here. To overcome the problem, the control error signals for the P- and D-portions were taken from a flow estimation module according to equation 5, while the error for the I- portion uses the actually measured signal.

Besides the actual control, a safety layer has been implemented. This layer enforces all sensor readings and communication data subject to signal range and plausibility. This was done, to give the controller a certain robustness by making sure it always operates under defined conditions, as well as for protecting the system.

IMPLEMENTATION

The complete control system is designed as a 3-layer architecture. The top most level encompasses the application layer and was designed using MATLAB/Simulink and Stateflow only. The functions are distributed between 3 tasks: The valve opening control loop is executed in a 1ms task, the overlaying pressure control loop, as well as the flow control loop is implemented in the 5ms task. The overall operation mode manager is implemented in a 10ms task. The second layer consists of the RTOS and the interface between the application layer and the hardware abstraction layer. This layer is also implemented under MATLAB/Simulink. A complete graphical interface has been created to incorporate features of the RTOS into Simulink and to configure it.

The hardware abstraction layer finally controls and communicates with the peripherals of the microcontroller. The architecture, as well as style considerations for including Simulink/RTW models in an OSEK compliant RTOS is described in more detail in [3].

The hardware platform used for this project was the UCU 5, a rapid prototyping control unit developed by IAV. This unit is equipped with 2 optically isolated CAN interfaces, 2 H-Bridges, low- and high-side switches, analog inputs and outputs, etc. Figure 7 gives an overview of the complete toolchain.

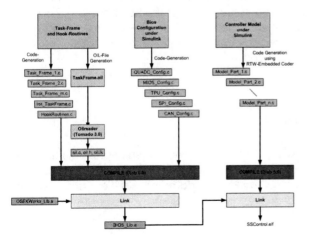

Figure 7 Make-process overview

In this project we used the RTW/Embedded Coder as code generator for the MPC555 target platform. It was necessary to customize this tool in several ways to enable a code generation process suitable for embedded targets as well as to generate an ASAP 2 file, which has been used for the calibration process.

A GUI for Simulink has been implemented, to control and execute the complete process from Code generation, building the executable, Asap2 File generation to code download from inside the Simulink environment. A blockset for interfacing and configuring the peripherals of the microcontroller has been developed by IAV. Furthermore Simulink blocks, that are used to generate code for OSEK tasks and the OSEK implementation language description file (OIL) has been developed.

HIL - SETUP

Hardware-in-the-Loop (HIL) has been used to perform the initial evaluation of the software algorithms, the control system as well as for basic calibration

This system consists of a dSPACE Micro AutoBox a PC laptop computer running dSPACE Control Desk, variable power supplies, and the IAV UCU controller running the application software.

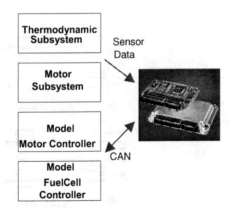

Figure 8 HIL Setup

The plant model according to the equations given above has been implemented in Simulink, and is executed on the Micro AutoBox in realtime, The analog outputs have been used to provide the system controller with emulated sensor signals. Concurrently a model of the motor and motor controller including the CAN communication interface as well as the communication layer of the fuel cell controller has been implemented, to provide suitable interfaces to the IAS controller. With this environment it was possible, to completely test the communication and the command execution. It was further possible to evaluate the control strategy on the HIL system.

The Control Desk block interpreted user input from the laptop into information that was outputted to the UCU embedded controller as either fuel cell CAN commands, analog sensor responses, or motor CAN responses.

Motor CAN responses could include any of the error responses that the motor controller would send via CAN such as over temperatures, over currents, and other alarms or errors. This allows the user to generate virtually any condition one might think could happen during the operation of the fuel cell system.

The MicroAutoBox receives back CAN responses from the UCU controller, the motor CAN commands, any indicator lamps that are customer specific, and hardware analog lines that are customer specific, and the commanded position of the backpressure valve.

From the inputs received, the HIL system breaks them down internally into status responses that are sent to the use via the Control Desk link and plant related inputs such as motor commands, and backpressure position commands.

The plant model (ASM) also receives ambient sensor values from the user via the Control Desk link. With in the ASM block the sensor information is broken down into inlet pressure, inlet temperature, and other controlled user variables such as compressor oil temperature. Motor commands are broken down into torque feed forward, velocity command, and run enable. These inputs are then used in the motor controller model and the main core plant model of the air system. This is shown in the diagram below.

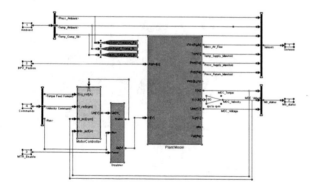

Figure 9 Top layer of the implemented HIL system

This plant model then feeds back as analog voltage sensors outputs of mass flow, temperature of the supply manifold, the pressure of the supply manifold, and the pressure of the return manifold to the UCU embedded controller. Also compressor motor actual torque and velocity and voltage are sent to the UCU embedded controller as motor CAN responses.

Inside the core air system model block are model blocks for the compressor, supply manifold, return manifold, backpressure valve and the fuel cell stack. These model behaviors have already been previously discussed within this document. A typical layout of these blocks in the model environment is shown in Figure 10.

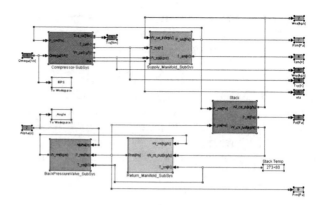

Figure 10 Simplified ADM model

Requested speed of the compressor and the angle of the backpressure valve are the only required inputs and the return values have been mentioned in the discussion of the outputs of the ASM.

This HIL system allows the engineers to test the CAN communications between the UCU embedded controller, basic functionality, command execution, response of system, basic calibrations, and provides a rapid development environment for software changes and trails. Using such a system the engineers can move to the dyno and physical hardware with a high degree of confidence that the embedded control system is ready to go.

COMMISIONING AND CALIBRATION

For calibrating the system a 3-phase approach has been performed. First those characteristics that are converting voltage to the physical value have been calibrated using sensor data sheets if possible and a reference measurement system. At this time the safety relevant features such as signal range check and controlled shut down behavior could be validated. The subsequent phase included the tuning of the backpressure valve and motor controller loops. Finally the overall control loops have been adjusted. For calibration of the system CCP has been used.

RESULTS

Here we first show simulation results utilizing the model and control architecture described above. Figure 11 shows the tracking performance of flow and pressure over the full range of operation.

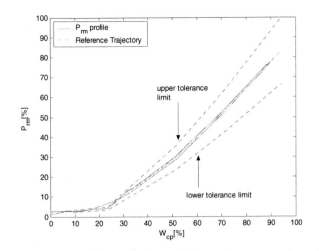

Figure 11 Reference profile tracking

The system was steered from zero flow to full flow and back to zero flow. The associated desired pressure was given by a characteristic. It can be seen that the tracking accuracy is excellent. The flow profile created by the controller is shown in Figure 12. While the flow into the stack accurately follows the desired profile, the flow out of the stack as well as out of the return manifold is much more affected by the pressure controller.

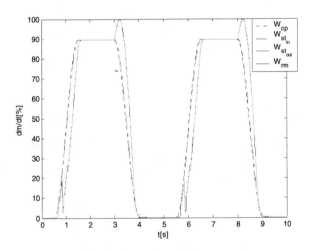

Figure 12 Flow control simulation

The large overshoot of the flow out of the stack and the return manifold is caused by the opening of the pressure valve to decrease the pressure according to the desired trajectory. The pressure is much more directly controlled as it can be seen from FIGURE 13.

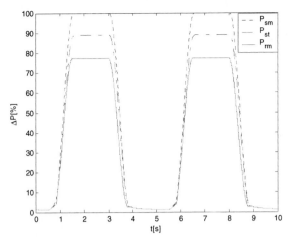

Figure 13 Pressure control simulation

FIGURE 13 shows the pressure profiles for the supply manifold pressure, the stack pressure as well as for the back pressure. It is remarkable that although there is a large pressure drop over the stack volume, the rise time for all pressures is in the same range.

The following figures show results achieved during extensive testing of the system.

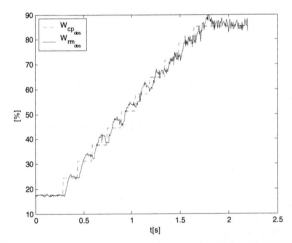

Figure 14 shows a sequence of step responses. Pressure and flow steps have been commanded concurrently. Each plateau has a duration of 150ms. It can be observed that in lower ranges the flow has a slight overshoot. This could have been eliminated by an adjustment of the feed forward part of the flow controller. Pressure control has very good performance throughout the complete range of operation.

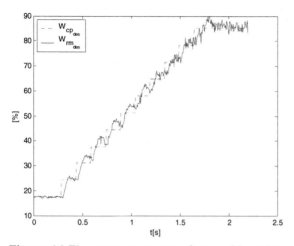

Figure 14 Flow step response for positive slope

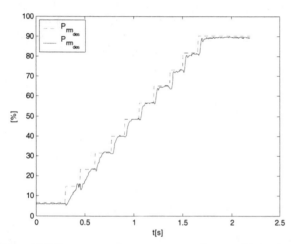

Figure 15 Pressure step response for positive slope

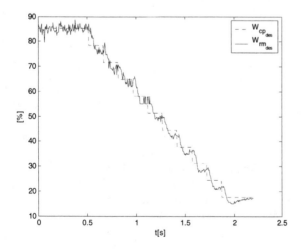

Figure 16 Flow step response for negative slope

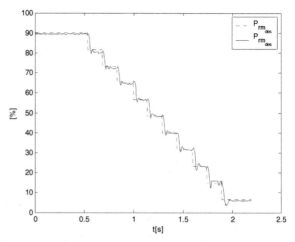

Figure 17 Pressure step response for negative slope

It is remarkable, that the performance of the controller is almost identical in the complete range of operation. This proves, that the approach, using exact I/O linearization is a proper choice for these kinds of systems.

The dynamic performance of the system is required to be suitable for the operation in drive cycles. To this end some high performance tests have been conducted.

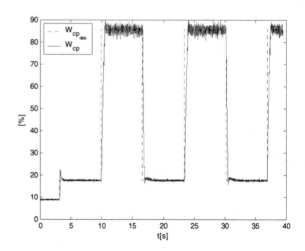

Figure 18 Flow dynamical performance test

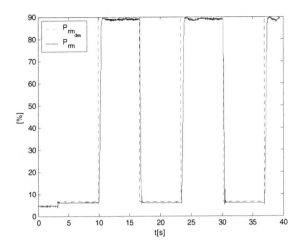

Figure 19 Pressure dynamical performance test

Pressure and flow had been commanded simultaneously. It was required to have the 10% to 90% response performed in less than a second. For both, flow and pressure, the achieved performance is even lower. In Figure 18 the noisy sensor readings of the mass flow sensor are obvious.

CONCLUSIONS AND FUTURE WORK

In this paper a turnkey project for the control of the air (oxygen) path of a fuel cell is presented. A fully nonlinear controller design is utilized as a novel approach. The complete application layer has been implemented using a customized version of the RTW/EmbeddedCoder. The results presented here demonstrate the excellent performance as well as the robustness of the controller design. To achieve this performance even with a noisy MAF sensor, the flow control error for the fast portion of the controller has been computed from estimated values, while concurrently sensor readings have been used for the integral path of the PID. Modeling of the system and using the inverse dynamics simplified the overlaying controller design as well as the calibration. An HIL setup has been used to do basic testing of the control strategy as well as the generated code. With this system it was possible to precisely test the complete communication between fuel cell and ASM controller on one side and between the motor controller and ASM on the other side. The degree of what the plant has been modeled allowed for basic controls testing and was suitable for this type of application. For further projects, a more detailed modeling will be performed.

Some research will be conducted, to investigate the slight opposite pressure transient when performing the step pressure control loop. The next steps will also encompass the implementation of more functionality of the fuel cell stack in the ASM controller to have one system controller for the entire fuel cell.

CONTACT

Johannes Reuter
IAV Automotive Engineering Inc.
4110 Varsity Drive
Ann Arbor, MI 48108
United States of America

Phone: +1 (734) 971-1079 ext. 330
Fax: +1 (734) 971-0570
e-mail: Johannes.Reuter@iavinc.com
web: http://www.iavinc.com

DEFINITIONS, ACRONYMS, ABBREVIATIONS

ASM: **A**ir **S**upply **M**odule

BPV: **B**ack **P**ressure **V**alve

CAN: **C**ontroller **A**rea **N**etwork

HIL: **H**ardware **I**n the **L**oop

MAF: **MA**ssflow

MEA: **M**embrane **E**lectrolyte **A**ssembly

MISO: **M**ultiple **I**nput **M**ultiple **O**utput

OIL: **O**SEK **I**mplementation **L**anguage

OSEK: **Open systems and the corresponding**

interfaces for automotive electronics

RPS: **R**evolution **P**er **S**econd

UCU: **U**niversal **C**ontrol **U**nit

REFERENCES

[1] J.T. Pukrushpan, A.G. Stefanopoulou, H. Peng "Modeling and Control for PEM Fuel Cell Stack System" ACC paper 2002

[2] J.T. Pukrushpan, A.G. Stefanopoulou, H. Peng "Simulation and Analyses of transient fuel cell system performance based on dynamic reactant flow model" ASME paper 2002

[3] J. Nair, et al. "Comparison of 3 different code generators" SAE paper 2004

[4] James Larminie, Andrew Dicks "Fuel cell systems explained"

IV. COMPONENT DEVELOPMENT

Electrical Systems

2002-01-1902

Performance Considerations of a Bi-Directional DC/DC Converter for Fuel Cell Powered Vehicles

Gary R. Flohr
Ford Scientific Research Laboratory

ABSTRACT

Fuel cell powered vehicles require a bi-directional dc/dc power converter in order to provide initial fuel cell startup and then to provide 12 volt battery charging in key run state. An efficient dc/dc converter is required to produce improved vehicle fuel economy. This paper describes the power loss analysis and methods used to develop the key power handling components for the bi-directional dc/dc converter.

INTRODUCTION

A bi-directional dc/dc converter is required in fuel cell powered vehicles to boost the low voltage up to 350 volts to provide power for the compressor and cooling pumps to start up the fuel cell. The converter is required to boost the voltage by up to a 50:1 ratio. This boost function typically lasts for 5 to 20 seconds and may require up to 3 kW of power to start up the fuel cell. Once the fuel cell is operating the dc/dc converter is changed from boost to buck mode, where it now converts power from the fuel cell to recharge the vehicle 12 VDC battery. In buck mode the dc/dc converter is required to produce 2 kW of regulated power over the range of 13.3-14.4 VDC.

The bi-directional dc/dc converter is located between the high voltage bus and the low voltage 12 VDC battery in the fuel cell system. Figure 1 shows the converter within the Fuel Cell Vehicle. The dc/dc converter shares the same cooling system with other components in the system and therefore has an operating temperature of 72 ^{0}C. This high cooling system cold plate temperature requires the power components to be designed for a small temperature rise of 50 ^{0}C from cold plate to case. Converter efficiency must be >80% in boost and between 85% and 88% in buck mode of operation as shown in table 1.

Bi-directional DC/DC Converter in Fuel Cell Vehicle

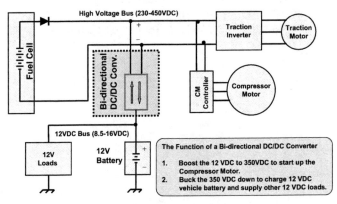

Figure 1

Bi-directional operation at 3/2 kW requires that the converter be designed for a high peak operating current in boost mode, while it must also be very efficient in buck (charging) mode of operation.

Table 1 Converter Efficiency Requirements

Operating Condition	Boost mode	Buck mode
Low Volt DC Input 8-11 vdc	>80%	
Low Volt DC Input 11-16 vdc	>85%	
High Volt DC Input 250-350 vdc		>88%
High Volt DC Input 350-430 vdc		>85%

CONVERTER TOPOLOGY

The converter topology is a current fed full bridge on the 12 VDC side and a voltage fed full bridge on the high voltage side. This topology and its salient features were determined as a part of an earlier study [1] and are shown in Figure 2. The selected converter operates in buck mode with zero voltage switching on the high voltage side and synchronous rectification on the low voltage side. The switching loss is reduced to a small fraction of the conduction loss.

The topology shown in Figure 2 allows bi-directional power flow, and can operate over a wide range of load and input/output voltage if careful design considerations are made up front.

BOOST OPERATION -The boost topology of current fed design provides a startup boost circuit shown with a fly back converter using the inductor and auxiliary winding to charge up the output voltage to approximately one half of Vo. This startup mode provides enough output voltage so that full boost mode can be made via the main transformer without causing excessively high peak currents encountered when there is no output or voltage

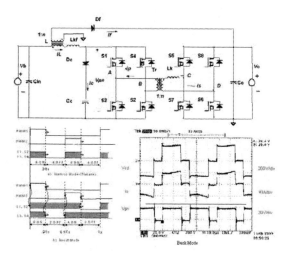

Figure 2 Full Bridge DC/DC Converter

and the primary current is only limited by a small leakage inductance in the main transformer.

Boost operation is accomplished by switching on all 4 of the low voltage mosfet switches to charge up the input inductor. Once the peak input switch current is achieved then S3 and S4 are turned off, thus transferring the inductor current to the transformer primary. At the end of the cycle S3 and S4 are turned on again to increase the inductor current in another half cycle. Here, S1 and

S2 are switched off at the peak current and current is reversed in the transformer primary. The secondary side simply rectifies the output as a full bridge rectifier. In boost mode the switches are turned off at zero voltage.

BUCK OPERATION - The selected converter operates in buck mode with zero voltage switching on the high voltage side and synchronous rectification on the low voltage side. The high voltage devices are controlled in a phase shift mode [2] to allow switching on at zero voltage. This switching mode requires that the high voltage bridge operate in a resonant mode between half bridge conduction cycles. Overall, the switching loss in both bridge sections is reduced to a small fraction of the conduction loss.

CONVERTER LOSS ANALYSIS

LOW VOLTAGE BRIDGE LOSSES - Boost mode operation requires the full bridge switches to be sized for >440 amps rms at low input voltage of 8.5Vdc since 80% efficiency is required. The design goal was to achieve at least 85-90% efficiency therefore the low voltage mosfets and transformer primary could be designed for 392-415 amps rms.

Detailed analysis of the copper cross sectional area was required. The predicted temperature rise indicated that a minimum 12 oz. copper weight was needed to minimize the losses between the input inductor and the transformer primary connection. A heavy copper power board was employed to mount the power mosfets required by the low voltage bridge. This copper board was also used to mount the RCD passive snubber, to provide a connection for transformer primary, secondary and also to mount the high voltage power components.

The losses in the converter are dominated by I^2R losses requiring that a concerted effort be made to keep the R as small as possible. The converter must operate continuously at 72 ^{0}C cold plate temperatures with liquid cooling. To guarantee good performance at this temperature and at 85 ^{0}C ambient the power components were designed with thermal contact area to the cold plate to ensure that the case operating temperature was limited to a maximum of 95 ^{0}C.

This is important since the resistivity of copper increases by .43%/^{0}C in the 25 ^{0}C – 100 ^{0}C range. And a 70 ^{0}C increase will cause a factor of 1.3 for the resistivity of copper and approximately 1.4 for the Rds_on resistance of the mosfets.

Boost Mode

$$Power\__{in} = 3529W_{in} = \frac{3000W_{out}}{.85}$$

$$I_{in\_rms} = 415A = \frac{3529W_{in}}{8.5V_{in\_min}}$$

Buck Mode

$$Power_{\_in} = 2173W_{in} = \frac{2000W_{out}}{.92}$$

$$I_{out\_rms} = 150A = \frac{2000W_{out}}{13.3V_{out\_min}}$$

Loss estimates were made for the converter for both modes of operation. The budget loss for I^2R was determined after the all other I^2R loss values were calculated. This loss was then used to determine the desired R for the converter.

Table 2 - Estimated Losses 3 kW Boost/2 kW Buck (72 C)

Loss Component	Boost Mode 450 Vo, 8.5 Vin	Buck Mode 13.3 Vo, 430 Vin
Control	13W	13W
Low Volt I^2R	378W	90W
LV Switching	41W	16W
Core Losses	15W	10W
HV Switching	49W	15W
High Volt I^2R	33W	29W
Total Losses	529W	173W

Low voltage input R total was determined for boost mode as follows:

$$Boost\_Rlv_{\_in} = .0022\Omega = \frac{378W_{\_loss}}{(415A)^2}$$

During half of the cycle, in boost mode, the current is flowing in both branches (two half bridges in parallel) so the low voltage conduction losses are reduced somewhat.

$$Buck\_Rlv_{\_in} = .004\Omega = \frac{90W_{\_loss}}{(150A)^2}$$

This result indicates that if the low voltage side R could meet the boost condition then the buck mode would easily exceed 90% efficiency over the entire load range.

In order to meet the .0022 ohm boost mode series input resistance the parallel combination of mosfets for each switch would have to be less than .0009 ohms.

Each low voltage switch would require 8 mosfets in parallel, each less than 6 milliohms Rds on.

The mosfet selection criteria were based upon several factors.

- Rds_on resistance must be < 6 milliohms
- Vds of 60v to allow for Vpn spikes of 45vp
- TO-220 package for height requirement
- Ciss < 10,000 pf to allow for 50 kHz operation

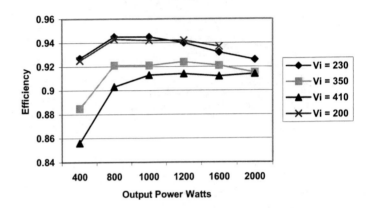

Figure 3 - SN 000019 Buck Effiency at 13.3VDC and 72°C

BUCK MODE LOSS RESULTS - Efficiency test results from the design verification prototype indicate that the required boost and buck mode specification of 80% and 88% efficiency respectively was exceeded by a very good design margin over most of the load range. Figure 3 shows the buck mode efficiency over all operating conditions.

Only two operating points were marginal, namely the 20% load point (400W) at 410 VDC and 350 VDC input. These points were only .5% better than the specification requirement for the converter. The main reason for this is that higher input voltages cause narrow duty cycles in the buck converter (at low load), which in turn lead to greater losses. These added losses are due to the high ratio of peak current to rms current.

The prediction of losses at high input voltage and low load conditions is normally inaccurate early on in the design phase of the product because of the lack of thermal data, and poor accuracy in determining the AC and skin effect losses. Calculation of the losses is usually done at minimal input voltage and full current, and then an approximation of the losses is made at full voltage and low load conditions.

The 230 VDC input condition is the most straight forward since we know that the duty cycle is near maximum. A curve fit can approximate the R total for the converter (R_lv + R_hv) and also estimate the switching losses, duty ratio and fixed losses.

A plot of the losses illustrates these parameters.

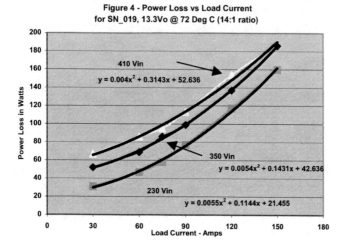

Figure 4 - Power Loss vs Load Current for SN_019, 13.3Vo @ 72 Deg C (14:1 ratio)

Examination of the curve fits in Figure 4 shows that the converter I^2R loss closely matches the design values of the primary resistance R_lv plus the secondary resistance R_hv reflected back into the primary. The secondary resistance is reflected back to the primary by the following relation.

$$R_{\_lv} = .0033\Omega$$

$$R_{\_hv} = .315\Omega$$

$$R_{tot\_lv} = R_{\_lv} + R_{\_hv}\left(\frac{Np}{Ns}\right)^2$$

$$R_{tot\_lv} = .0049\Omega$$

Therefore the equivalent total resistance seen by the load current in buck mode is $R_{tot\_lv} = .0049\Omega$. This value is also apparent in Figure 4 as the total R (.0054) in the coefficient of the x^2 term at both 230 VDC and 350 VDC input.

The remaining loss term in x is due to duty cycle, switching loss and skin effect losses. The constant is merely the total fixed losses such as control and magnetizing losses.

The converter losses are as expected, however the sensitivity to input voltage is quite apparent, especially at the 410 VDC input condition. Since the converter efficiency is affected by input voltage and the

transformer ratio a detailed analysis was performed to prove that the transformer ratio was optimal.

TRANSFORMER RATIO CALCULATION - The early prototype design used a 14:1 ratio transformer and worked well. However, operation over a wide voltage range in both boost and buck with the efficiency specified from 20% load to full load at 72 ^{0}C cold plate required a detailed analysis of each operating condition.

There are four operating conditions for buck and the same for boost. Each condition was evaluated to ensure that the converter could meet output regulation with design safety margin.

Table 3 Worst Case Operating Conditions

Mode	Vin_min Vo_max	Vin_min Vo_min	Vin_max Vo_max	Vin_max Vo_min
Boost	Max Duty Cycle	Max Current	Min Current	Min Duty Cycle & Max Ratio
Buck	Max Duty Cycle & Max Ratio	Max Current	Min Current	Min Duty Cycle

The worst case conditions of max. current, min. and max. duty cycle, and max. transformer ratio are shown in Table 3. The design process requires simultaneous design efforts for both boost and buck modes.

The design process was to first complete the max current analysis and determine the required R_{tot_lv}. Then the duty cycle analysis was completed at a nominal ratio of 14:1. Finally, the transformer ratio was checked again to make sure that we rounded down to the nearest integer since we desired only one primary turn.

The transformer ratio calculation is shown below.

Boost

$$Ratio_{max\_boost} = \frac{(Vo_{\_min}+2)(1-D_{min})}{(Vin_{\_max}-IR_{drop\_max})}$$

$$Ratio_{max\_boost} = 15.7$$

Buck

$$Ratio_{max\_buck} = \frac{(Vin_{\_min} - IR_{drop\_max})D_{max}}{(Vo_{\_max} + IR_{drop\_max})}$$

$$Ratio_{max\_buck} = 14.0$$

The buck ratio of 14.0:1 is used to meet both modes of operation. No rounding down was required and therefore a lot of consideration was given to using a transformer ratio of 12:1 to provide some safety margin for the regulation of the output at Vin=230 VDC and Vo=14.4 VDC. However, the 14:1 ratio was retained because we had included all of the worse case IR drops and thermal effects within the converter.

The decision to use a 14:1 ratio was pivotal in reaching the efficiency requirement at Vin_max and Vo_min. A plot of the Model Efficiency in Figure 5 shows the effect of a 12:1 and 10:1 transformer ratio on the losses.

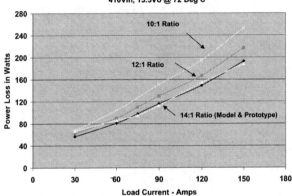

Figure 5 - Model - Power Loss vs Load Current
410Vin, 13.3Vo @ 72 Deg C

The losses increase dramatically with the 12:1 transformer ratio and reduce the efficiency below the specified 85% minimum. At 20% load (30 amps) the loss increases to 78 Watts and results in 83.7% efficiency. At full load (150 amps) the loss increases to 217 Watts and results in 90.2% efficiency. A 13:1 ratio was not used because a single primary turn was required and equal number of turns per layer was required in the secondary. Half turns were not considered due to ratio error effects.

The highest possible ratio transformer 14:1 also produced an added advantage to the converter high voltage bridge phase shift switching at zero voltage. The advantage is that the higher ratio transformer provides a greater reflected primary leakage inductance into the secondary. This higher leakage inductance is required to guarantee that the leakage current does not decay to zero during the transition period.

A 12:1 ratio transformer would reduce the leakage inductance by 30% allowing the current to decay faster which could prevent true zero voltage switching of the high voltage devices.

Zero voltage switching requires that the leakage current charge the leading leg mosfet output capacitance and then maintain the current during the free wheeling interval [3]. Also, the current must be ample enough to charge the trailing leg mosfet Coss fast enough to allow enough dead time margin between the turn off of the opposite device (in this case S5 in Figure 6) and the turn on of the lower device. The current must be maintained positive during the transition period of both leading leg and lagging leg.

The leakage inductance is a common element to all three phases (time periods) which comprise the total time between conduction of one half bridge and the other half bridge. The leakage inductance must store enough energy to charge two bridge device output capacitances Coss and also must keep the current flowing through the free wheeling interval. Therefore the increasing the leakage inductance produces a greater effect than does reducing the device output capacitance Coss.

In order to guarantee zero voltage switching it is recommended that the energy stored in the leakage inductance be much greater than the energy transferred into two times the device output capacitance Coss.

The leakage inductance in this design was heavily dependant upon the primary winding shape and is reflected back by the turn's ratio squared (N^2) into the secondary. Since the shape of the primary could not be altered, (size and shape were optimized to reduce primary leakage inductance and constrained by the mechanical packaging) the zero voltage phase shift switching was quite dependent upon the transformer ratio and winding construction.

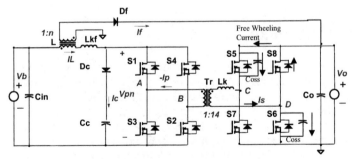

Buck Mode Zero Voltage Switching

Step 1 S5 & S6 turn on to drive transformer

Step 2 S6 turns off → S5 on only

Step 3 Leakage Current →Charges S6 Coss

Step 4 Current Freewheels through → S5 & S8

Step 5 S8 Turns on @ zero voltage via body diode

Step 6 S5 Turns off →Current charges S5 Coss

Step 7 S7 turns on @ zero voltage via body diode

Step 8 S7 & S8 are on to drive transformer

Figure 6 Full Bridge DC/DC Converter Phase Shift Switching Example

The construction of the transformer windings (layered vs. not layered) can alter the leakage inductance substantially. The non-layered design increased the leakage inductance by 40% (primary winding wrapped over the entire secondary winding).

The waveform shown in Figure 7 shows a voltage spike at light converter loads due to not enough leakage inductance reflected into the secondary. This particular case occurred with a 14:1 ratio transformer so we altered the transformer winding construction to gain another 40% of the original total leakage inductance. Note the voltage waveform rises up just prior to switching on of the opposite leg device.

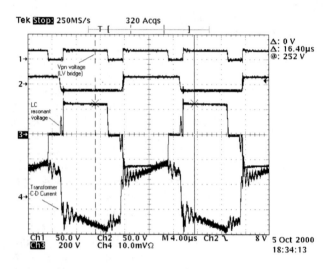

Figure 6 - Full Bridge Phase Shift Switching showing LC resonant voltage during Transition Period

Figure 7 - Full Bridge Phase Shift Switching showing LC resonant voltage spike during Transition Period

CONCLUSION

The performance of bi-directional, soft switching dc/dc converters that must operate over a wide input voltage range and load range is very sensitive to IR drop, input voltage and the transformer turns ratio, which in turn affect the operating duty cycle of the converter. The I^2R losses must be accurately know up front in order to accurately determine the overall losses. The IR drop at maximum temperature must then be included into the transformer ratio calculations in order to allow selection the highest ratio transformer. The higher transformer ratio also provides the added benefit of a higher reflected primary leakage inductance into the secondary. The added leakage inductance is required to help guarantee zero voltage switching in a phase shift-switching converter.

ACKNOWLEDGMENTS

This project originated by a joint contract between Ford Scientific Research Labs, Virginia Tech and Oak Ridge National Labs and was later transferred from Ford Research to Ballard Electric Drive Systems (formerly Ecostar Electric Drive Systems).

REFERENCES

[1] K Wang, C Y. Lin, L. Zhu, D. Qu, F.C. Lee, J.S. Lai, "Bi-directional DC to DC Converter for Fuel Cell Systems", Proc. of IEEE Power Electronics in Transportation 1998, pp 47-51.

[2] K Wang, C Y. Lin, L. Zhu, D. Qu, H. Odendaal, J.S. Lai, F.C. Lee, "Bi-directional Full Bridge DC/DC Converter With Unified Soft-switching Scheme, Part II: Design, Implementation, and Experimental Results," Proc. of VPEC Annual Seminar 1998, pp 151-157.

[3] B. Andreycak, "Phase Shifted, Zero Voltage Transition Design Considerations and the UC3875 Controller", Application Note U-136A Unitrode Applications Handbook, pp 300-313.

CONTACT

Gary Flohr is a Senior Research Engineer with Ford Scientific Research Labs in Dearborn, MI. Email is gflohr@ford.com

2004-01-1006

Electric Power Control System for a Fuel Cell Vehicle Employing Electric Double-Layer Capacitor

Akira Ohkawa
Honda R&D Co., Ltd.

ABSTRACT

A fuel-cell-vehicle has been provided with an electric-double-layer-capacitor system (capacitor) to act as a back-up power source. The fuel cells and the capacitor have different voltages when the system is started, and for this reason the system could not be reconnected by relays.

A VCU (Voltage and current Control Unit) has been positioned in the path of electrical connection between the fuel cells and the capacitor as a method of dealing with this issue. The VCU enables the charging of the capacitor to be controlled in order to equalize the voltage of the two power sources and allow a connection.

INTRODUCTION

The motor in a fuel cell vehicle (FCV) employs the electrical energy produced when hydrogen and oxygen combine to form water.[1] The basic system of the Honda FCV was formulated on the basis of using a capacitor to enable the vehicle to provide a starting and running performance meeting normal automotive standards. This paper discusses the research conducted on methods of enabling the transfer of power from the fuel cells and capacitor, and the application of these methods when the capacitor is used in the vehicle system.

UTILIZATION OF A CAPACITOR

THE NECESSITY OF UTILIZING AN ENERGY BUFFER

The fuel cells in an FCV generate electricity in accordance with the power consumed by the equipment. The current produced by the cells is proportional to the amount of hydrogen and oxygen supplied to them. Therefore, when the equipment requires extra power, increased amounts of hydrogen and oxygen must be supplied to the fuel cells to enable the extra power to be generated. Because it takes time for the supply of hydrogen and oxygen to be increased, the response of the fuel cells to the demand for extra power is delayed.

In addition, fuel cells cannot store energy which is regenerated when the vehicle decelerates. This means that the kinetic energy of the vehicle during deceleration is lost as heat in the working of the mechanical brakes.

Energy is also required to supply hydrogen and oxygen to the fuel cells to enable them to generate electricity. When the vehicle is started, energy is required to enable the fuel cells to commence generating power. However, the fuel cells do not function before the vehicle is started, and therefore cannot supply energy.

These considerations indicate the necessity of employing an energy buffer in an FCV.

SELECTING AN ENERGY BUFFER

The primary requirement for the energy buffer in the system was the ability to compensate for the delay in fuel cell response by supplying the exact amount of energy required by the system. This made it necessary to select a buffer with a large electrical capacity, given that the power required by the FCV is around three times that required by usual hybrid vehicles. To enable a balance to be achieved between the provision of the required power and the achievement of weight savings, it was necessary to utilize an energy buffer with a high power density. Table 1 shows the power density of two types of battery and a capacitor. The Li-ion battery and the capacitor are clearly superior in terms of power density.

	Power density (W/kg)	System voltage (V)	Electromotive force (V)
Ni-MH	Over 500	400 ~210	320
Li-ion	Over 1000	400 ~285	340
Capacitor	Over 1000	400 ~0	-

Table 1 Comparison of power density of batteries and capacitor [2]

The secondary requirement of the energy buffer was controllability, to ensure that it neither over-charged nor over-discharged. Figure 1 shows the voltage-current characteristics of the fuel cells used in the FCV.

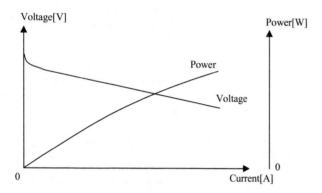

Fig.1 Fuel cell characteristic

Nickel metal-hydride and Li-ion batteries utilize chemical reactions to generate energy and their voltage is therefore stable. However, this fact makes it impossible to control charge and discharge if they are simply connected in parallel with fuel cells. By contrast, the voltage of a capacitor is proportional to its state of charge, and this characteristic makes it possible to directly connect capacitors in parallel with fuel cells.

Figure 2 shows a simple equivalent circuit for a parallel connection between fuel cells and a capacitor.

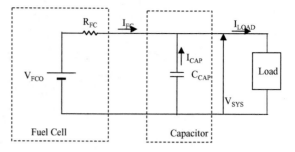

R_{FC}:Impedance of Fuel Cell
V_{FCO}:Electromotive force voltage of Fuel Cell
I_{FC}:Current of Fuel Cell
I_{CAP}:Current of Capacitor
C_{CAP}:Capasitance of Capacitor
V_{SYS}:System Voltage

Fig. 2 Schematic of fuel cell capacitor connection

The transfer function for transient response in this circuit is found using equation (1).

$$\frac{I_{FC}(S)}{I_{LOAD}(S)} = \frac{\frac{1}{R_{FC} \cdot C_{CAP}}}{S + \frac{1}{R_{FC} \cdot C_{CAP}}}$$ Eq. (1)

This function is the primary delay function of I_{FC} against I_{LOAD}. For example, if the current drawn by the load varies in steps, the transient response shown in Fig. 3 is obtained.

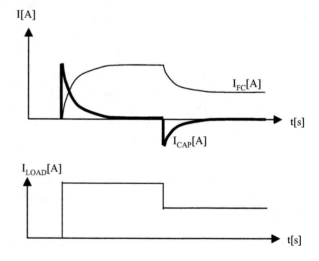

Fig. 3 Transient current characteristic of fuel cells and capacitor

This characteristic enables a capacitor to provide the required power in sudden, transient changes in load, enabling output from the fuel cells to be intentionally delayed. Utilizing such a delay gives the fuel cells time to generate the required extra power.

Another characteristic of capacitors is that the current converges on 0A after a fixed period of time has elapsed after they have been charged with current or have discharged current in response to changes in the load. At the same time, the fuel cell current increases to approach the current being drawn by the equipment, until they reach equilibrium at $I_{FC} = I_{LOAD}$. The use of a capacitor as an energy buffer therefore enables energy management without the requirement of utilizing any special electrical control devices.

The power density of capacitors and the ease with which their energy can be managed determined the selection of a capacitor system as an energy buffer for the FCV.

ISSUES ARISING FROM INSTALLATION OF A CAPACITOR SYSTEM

VOLTAGE DIFFERENCE WHEN THE SYSTEM IS STARTED

If the fuel cells, the capacitor and the equipment are constantly connected, the equipment draws dark current from the power sources when the system is stopped. This can cause the capacitor and fuel cells to over-discharge if the system is left off for an extended period, preventing it from being restarted. It was therefore necessary to design a system structure in which the power sources were isolated from the high-voltage equipment when the system was turned off. In the FCV, relays have been used to electrically separate the capacitor and fuel cells from the equipment.

However, the employment of this system configuration means that there is a voltage difference between the fuel cells and the capacitor when the system is next started. The capacitor voltage is varied by self-discharge. If they were to be connected by relays, the low impedance of both power sources would cause a high transient current flow, interfering with the normal functioning of the system.

CURRENT FROM THE FUEL CELLS DURING REGENERATION

In a system employing the fuel cell-capacitor connection shown in Fig. 2, fuel cell power responds to capacitor voltage (V_{SYS}). Current is supplied to the capacitor from the fuel cells, which have a starting voltage of V_{FC} and an internal resistance of R_{FC}. The equation for the current at this time is: $\dfrac{V_{FC} - V_{SYS}}{R_{FC}}$

Fuel cell power, P_{FC}, is therefore shown by equation (2):

$$P_{FC} = V_{SYS} \cdot \frac{V_{FC} - V_{SYS}}{R_{FC}} \qquad \text{Eq. (2)}$$

This relationship indicates that fuel cell power is determined solely by V_{SYS}, without consideration of the amount of power consumed by the vehicle. For this reason, when the motor is regenerating power during deceleration, power from the fuel cells charges the capacitor, and regenerated energy is lost, without being stored in the capacitor.

EMPLOYMENT OF VCU AS SOLUTION

CONSTRUCTION OF VCU SYSTEM

A VCU has been employed in the path of electrical connection between the fuel cells and the capacitor to provide a solution to the two issues outlined above. This VCU is a DC/DC converter which limits the power output of the fuel cells. The VCU circuit configuration is shown in Fig. 4.

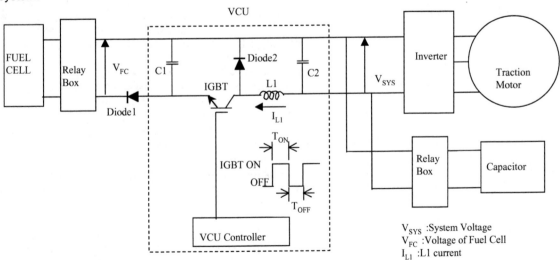

V_{SYS} :System Voltage
V_{FC} :Voltage of Fuel Cell
I_{L1} :L1 current

Fig. 4 VCU circuit

313

The VCU circuit provides control via a signal input to an IGBT. The IGBT has two states: ON and OFF. The VCU employs three modes, depending on the state of the IGBT switch.

	Mode	Status of IGBT
1	VCU OFF	OFF
2	Fuel Cell Power Control	Chopping
3	Connection	ON

Table2 VCU Mode

In VCU OFF mode, the IGBT is OFF, and no current flows from the fuel cells to the capacitor. In Fuel Cell Power Control mode, the IGBT switches ON and OFF at high speed, and the ratio of time between the ON and OFF states is regulated to enable control of fuel cell power. When the potentials of the fuel cells and the capacitor have been equalized by the Fuel Cell Power Control Mode, the IGBT remains ON, and the VCU switches to Connection mode.

In Fuel Cell Power Control mode, the IGBT switches between storing electromagnetic energy by L1 and releasing that energy.

When the IGBT is ON, V_{FC}-V_{SYS} is impressed in L1. The current from L1, I_{L1}, is therefore shown by equation (3):

$$I_{L1\_ON}(t) = \frac{1}{L} \cdot \int_{T_{ON}} (V_{FC} - V_{SYS}) \cdot dt \qquad \text{Eq. (3)}$$

When the IGBT is OFF, the energy stored in L1 when it was ON flows to Diode2. When the system is in this state, -V_{SYS} is impressed in L1.

I_{LI} is determined by the relationship shown in equation (4):

$$I_{L1\_OFF}(t) = \frac{1}{L} \cdot \int_{T_{OFF}} -V_{SYS} \cdot dt \qquad \text{Eq. (4)}$$

$$I_{L1\_OFF}(t) \geq 0$$

The relationships shown in equations (3) and (4) indicate that $\Delta I_{L1\_ON} = \Delta I_{L1\_OFF}$ is the condition at which the power output of the fuel cells becomes steady and constant. A solution on this basis produces the relationship shown in equation 5.

$$\frac{T_{ON}}{T_{ON} + T_{OFF}} = \frac{V_{SYS}}{V_{FC}} \qquad \text{Eq. (5)}$$

When the ratio of time that the IGBT is in its ON and OFF states is in the relationship shown in Eq. 5, I_{L1} is

constant. If T_{ON} increases, I_{LI} also increases, and a reduction in T_{ON} produces a consequent reduction in I_{L1}.

This operating principle enables the VCU to actively control the output of the fuel cells by varying the ratio of time between the ON and OFF states of the IGBT.

RESPONDING TO VOLTAGE DIFFERENTIAL AT STARTUP

Figure 5[3] shows the voltage and current profiles of the fuel cells and capacitor and the operation of the VCU when the FCV is started.

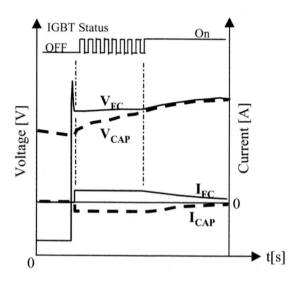

Fig.5 Voltage and current profiles of fuel cell and capacitor at starting up

At startup, there is a difference in potential between the fuel cells and the capacitor. At this time the IGBT is OFF and the VCU is on standby. Next, the capacitor and fuel cell relays are connected. The electrical paths are isolated by the VCU, and no current flows from the fuel cells to the capacitors. The VCU next limits the amount of power supplied from the fuel cells if the amount of power being generated makes it necessary, and gradually provides current to the capacitor. The voltage of the capacitor increases over time and the capacitors are charged, until its potential is equivalent to the potential of the fuel cells. The addition of the VCU to the system has provided an effective solution to the issue of the voltage differential between the fuel cells and enabled a connection to be established between the power sources.

The VCU limits I_{FC} in accordance with the I_{FC} control target (I_{CMD}) established on the basis of the amount of power generated by the fuel cells. If the VCU does not provide sufficient control and IFC exceeds I_{CMD}, the fuel cells will generate excess power. Because this increased power will simply be lost, the VCU must be capable, to the greatest extent feasible, of applying control to ensure that there is no overshoot of I_{CMD}. Figure 6 shows a control diagram of VCU control of I_{FC}. Originally, switching duty was calculated on the basis of the difference between I_{FC} and I_{CMD}. However, when I_{CMD} was given in steps at vehicle startup, I_{FC} would overshoot the command value.

The application of differential control can be expected to ensure that such overshoots do not occur, but this tends to make control unstable. It was difficult to maintain stable control while preventing overshoot. The control layout shown in Fig. 7 was therefore suggested as a solution. This system adds filters for I_{FC} to the original system, enabling the oscillating components of the I_{FC} frequencies to be amplified. The transfer function is as follows:

$$\frac{I_{FC}2(S)}{I_{FC}(S)} = LPF\ 2(S) + \{LPF\ 1(S) - LPF\ 2(S)\} \cdot G(S)$$

Eq. (6)

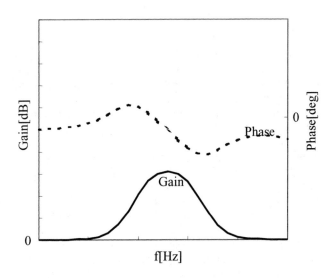

Fig. 8 Frequency response of I_{FC} filter

Figure 8 shows the frequency response of this function.

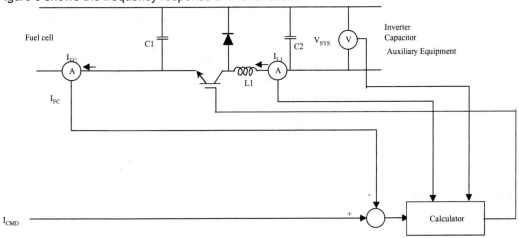

Fig. 6 Control of I_{FC} by VCU

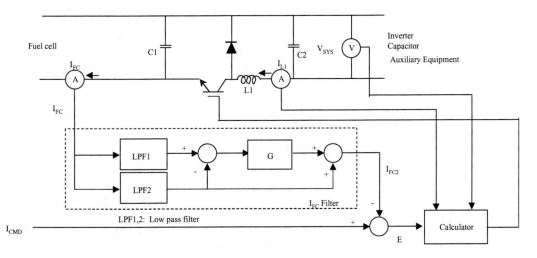

Fig. 7 Control of oscillation of I_{FC}

315

The cutoff frequencies of LPF1 and LPF2 are set to cause the oscillating frequencies (overshoots) to become gain peaks. In addition, even if the oscillating frequencies are fluctuating, the cutoff frequencies for LPF1 and LPF2 can be adjusted to provide gain across a broad range.

This enables the oscillating frequencies causing overshoot to be amplified and added to I_{FC}. This is extremely effective in controlling overshoot, because the oscillating frequencies causing overshoot are highlighted and fed back to the deviation. The original stability of control of frequencies of I_{FC} other than the oscillating component has been maintained, enabling complete control of oscillation in I_{FC}. Figure 9 shows the difference in current waveforms when the I_{CMD} command is given in steps with and without the filter in the system.

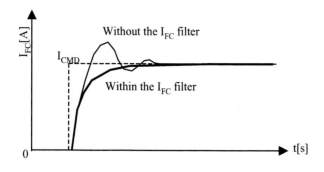

Fig. 9 I_{FC} transient response

LIMITATION OF CURRENT FROM FUEL CELLS DURING REGENERATION

The voltage and current profiles of the fuel cells and capacitor in regeneration mode are shown in Fig. 10[3].

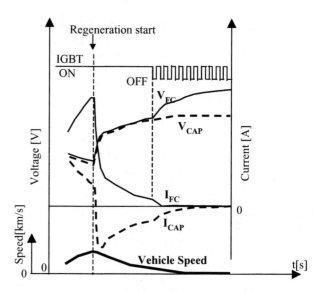

Fig.10 Voltage and current profiles of fuel cell and capacitor at regeneration mode

When the vehicle decelerates and the motor commences regeneration, the capacitor receives both regenerated power and power from the fuel cells, meaning that V_{FC} and V_{CAP} increase in unison. In this mode of operation, the amount of power coming from the fuel cells reduces the regenerated power taken in by the capacitor by the same amount. Control is therefore applied to restrict the power from the fuel cells when V_{CAP} reaches a predetermined level. This causes V_{FC} to increase independently of V_{CAP}, and ensures that the capacitors receive only regenerated power. This enables the capacitor to be charged only with regenerated power from the predetermined voltage by the VCU to its rated voltage.

ADDITIONAL EFFECT OF VCU

The employment of a VCU in the system was not only effective in providing a solution to the two issues involved in utilizing a capacitor in the system, as discussed above, but also contributed to enabling the equipment to be reduced in size.

As Fig. 1 shows, as the output of the fuel cells decreases, their voltage increases. Therefore, the voltage of the fuel cells rises when the equipment is not drawing much power and when the system is regenerating power. At these times, the VCU functions to limit the power from the fuel cells in order to prevent V_{SYS} from increasing. This resulted in a reduction of the rated voltage required from the traction motor inverter, the equipment and the capacitor by more than 100V, enabling the equipment connected to V_{SYS} to be made approximately 20% smaller.

CONCLUSION

A VCU has been employed to control the power from the fuel cells in a fuel cell vehicle utilizing an electric double-layer capacitor, with the following results:

(1) The limitation of power from the fuel cells by the VCU ensures optimum charging of the capacitor, eliminating the voltage differential between the fuel cells and the capacitor and enabling the two power sources to be connected.

(2) Electrical separation of the fuel cells and the capacitor by the VCU when the system is regenerating braking energy limits the amount of energy provided by the fuel cells to the capacitor, enabling a greater quantity of regenerated energy to be used in charging the capacitor.

(3) The step-down function of the VCU has reduced the rated voltage required from the traction motor inverter, the equipment and the capacitor by more than 100V, enabling the equipment to be reduced in size by approximately 20%.

REFERENCES

1. Satoshi KAWASAKI, et al.:Development of the Honda FCX Fuel Cell Vehicle, Honda R&D Technical Review, Vol.15, No.1, p.1-6 (2003)
2. Koji TAMENORI, et al.: Development of Electric Double-layer Capacitor System for Fuel Cell Vehicle, Honda R&D Technical Review, Vol.15, No.1, p.25-30 (2003)
3. Hidekazu TAKENAKA: Research on Electric Power System for Honda FCX Fuel Cell Vehicle, Honda R&D Technical Review, Vol.15, No.2, p.1-6 (2003)

CONTACT

E-mail address: akira_okawa@n.t.rd.honda.co.jp

Design and Testing of a Fuel-Cell Powered Propulsion System Supported by a Hybrid UC-Battery Storage

Di Napoli, F. Crescimbini, A. Lidozzi and L. Solero
University "ROMA TRE", Dept. of Mechanical & Industrial Engineering

M. Pasquali, A. Puccetti and E. Rossi
ENEA Research Center "Casaccia", Electrical & Hybrid Vehicles Testing Laboratories

ABSTRACT

To date hybrid-electric vehicles (HEV) make use of high-power density ac propulsion systems to provide comparable performance with vehicles using internal combustion engine (ICE) technology. Electric motor, inverter, and associated control technology has made substantial progress during the past decade and it is not the limiting factor to either vehicle performance or the large-scale production of hybrid vehicles. The search for a compact, lightweight, and efficient energy storage system that is both affordable and has acceptable life cycle remains the major roadblock to large-scale production of HEVs.

This paper deals with an original HEV propulsion system that includes fuel cell generator and an energy storage system combining ultracapacitor tank and battery. The three on-board power sources supply the vehicle traction drive through a multi-input dc-to-dc power converter which provides the desired management of the power flows. In particular, in the proposed arrangement the ultracapacitor tank is used for leveling the battery load current during transients resulting from either acceleration or braking operation of the vehicle. The paper outlines the features of the dc-to-dc power converter being used in the proposed propulsion system. The control strategy adopted for power flow sharing among the on-board sources is described and main characteristics of a 35 kW prototype of the propulsion system jointly developed by ENEA (Italian National Agency for New Technologies, Energy and Environment) and University ROMA TRE are depicted; both simulation and experimental results are presented.

INTRODUCTION

Proton exchange membrane (PEM) fuel-cells are being increasingly accepted as the most appropriate power sources for future generation vehicles. This acceptance is evident by the formation of a new global alliance for the commercialization of this technology and by the growing participation of major automotive manufacturers in producing sophisticated demonstration vehicles.

With the rapid rate of technical advances in fuel-cells (FCs), and with major resources of the automotive industry being directed to commercialization of FC propulsion systems, it looks more promising than ever that FCs will soon become a viable alternative to internal combustion engine technology.

One issue that has been driving the development of FCs for automotive applications is their potential to offer clean and efficient energy without sacrificing performance or driving range. In the case of a PEM FC powered vehicle realizing this potential means ensuring that the complete FC system operates as efficiently as possible over the range of driving conditions that may be encountered.

A battery storage unit (BSU) can be employed combined with a FC stack in order to achieve the operating voltage-current point of maximum efficiency for the FC system. In such a conventional arrangement, the BSU is sized to deliver the energy amount required to match the traction drive demand, as well as it has to deal with power peaks being on demand during acceleration or overtaking phases resulting from the driving cycle on which the vehicle is expected to operate. Such peak power transients result in a hard constraint for batteries, as in the battery a higher peak than rated power implies an increasing of losses and temperature and so a decreasing of lifetime. Thereby, it is desirable to reduce these power peaks by introducing an additional auxiliary power device; load leveling of the FC–BSU system could be accomplished by means of ultracapacitors (UCs).

UCs, which have high power density, obtain regeneration energy at high efficiency during decelerations and supply the stored energy during accelerations in order to reduce the peak power requirements for the FC–BSU. The capacitors are sized to meet energy storage load leveling requirements resulting from the driving cycle on which the vehicle is expected to operate. The UC tank must supply all the power required in excess of the FC–BSU system rated power, provided that the UC state of charge (SOC) is greater than a specified minimum. Whenever the power required to operate the vehicle is lower than the FC-BSU rated power, the UCs can be charged with the power in excess. Whenever regenerative braking operations occur, energy is put into the UC tank provided this device is not fully charged yet.

Limitations to the use of UC tanks primarily originate from the characteristics of the UCs that to date are being available in the market. In fact, at the to-date stage of development, UCs have a too low value of the cell voltage, as well as a cell leakage current that may change from one cell to the others, which fact may result in significant voltage unbalances in stacked unit. The use of a dc-to-dc power electronic converter in the dc-link of the propulsion drive substantially reduces the drawback resulting from having a cell voltage lower than 2.35 V and allows a suitable regulation of the energy flow coming in and out the UC tank.

This paper deals with the study of an original propulsion system that uses a FC stack together with a combined storage including BSU and UC tank. The study reported in this paper is the main subject of a research project being jointly developed by ENEA and the University "ROMA TRE" to the purpose of designing and testing a FC-powered drive train that is expected to find application in future city-car vehicles.

FUEL-CELL PROPULSION SYSTEM WITH HYBRID STORAGE

The proposed fuel-cell propulsion system is schematically shown in Fig. 1. A Multi-Input Power Electronic Converter (MIPEC) is used as power interface between the vehicle traction drive and the on-board power generating system being including FC stack, BSU and UC tank.

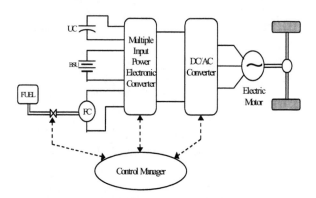

Fig.1. FC powered vehicle supported by hybrid UC-battery storage

Fig. 2 shows the circuit layout of the proposed MIPEC. As such a power converter interface is accomplished by means of connecting in parallel among them the output circuits of three bi-directional step-up/step-down dc-to-dc converters, the joint output terminals of such step-up/step-down converters are used to provide voltage regulated dc power supply to the traction drive. On the other hand, the three couples of MIPEC input terminals are separately fed through FC stack, BSU and UC tank, respectively. In each dc-to-dc converter the step-up mode of operation is used to supply power to the dc input terminals of the traction drive, whereas the step-down mode of operation is used whenever the power flow is required to reverse because of regenerative braking operations being commanded in the traction drive. Each step-up/step-down converter includes two power switches (IGBT) and two power diodes being arranged among them to achieve a single three-terminal assembly, as shown in Fig. 2.

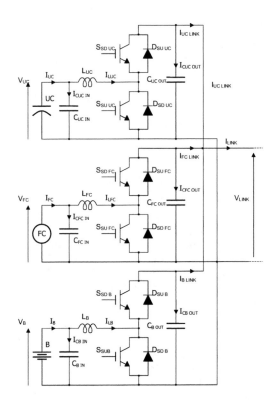

Fig. 2. Circuit layout of the proposed MIPEC

Such an IGBT-diode assembly is commercially available at reasonable cost in the form of the so-called "dual-IGBT power module", and the use of such a quite inexpensive power semiconductor device justifies the useless presence of one switch and one diode in the FC-fed dc-to-dc converter which actually is not required to operate as step-down converter during regenerative braking mode of operation. In addition to a dual-IGBT power module, the power circuit of each step-up/step-down converter includes both an input inductor/capacitor filter and an output capacitor filter. Both such power filters are required in order to limit the current ripple in the output circuit of each power source, as well as to minimize the voltage ripple at the input terminals of the traction drive. Due to the presence of such a quite large capacitor filters in the MIPEC circuit, capacitor pre-charge circuits are included in the complete layout to avoid dangerous over-currents at start-up.

In the MIPEC the IGBT duty cycles are controlled in order to meet the power demand of the traction drive. In doing that, the MIPEC control manager provides sharing among the three power sources of the power flow being on

demand. Such a control strategy is accomplished by taking into account the states of charge (SOC) of UC and BSU, as well as both the maximum admissible power flow variation and the efficiency map for each power source. For both the FC-fed and BSU-fed dc-to-dc converters the IGBT duty cycle is regulated in order to achieve the desired control of the converter input current. As a result, desired working point on FC generator and BSU V-I characteristic is achieved. On the other hand, the MIPEC output voltage is regulated at the desired value by controlling the duty cycle of the UC-fed dc-to-dc converter assuring appropriate dynamic response during either acceleration or braking operation of the vehicle.

VEHICLE CONTROL STRATEGY

Due to the original MIPEC structure, it is possible to suitably control the instantaneous values of the output currents from three power sources. In order to get the best performance, the FC is used to supply the average power required by the vehicle driving cycle whereas the UC tank provides load-leveling of the power peak demand and the BSU supplies the extra power - with respect to the FC maximum output power - required to meet the actual power demand resulting from vehicle operation.

As during the driving cycle the vehicle is continuously either accelerated or braked the system has to either regenerate energy during decelerations or supply the stored energy during accelerations with as high as possible efficiency. At the same time, it should be assured that both battery and UC tank become not discharged during an acceleration transient. Thereby, the system control algorithm has to include suitable regulation of the SOC values of BSU and UC, so that the actual SOC is always kept within preset upper and lower bounds. In addition to that, for each of the three power sources the control algorithm must provide limitation of the output current as well as it has to assure that any of the MIPEC input voltages do not come down under a threshold value.

The control algorithm is based on knowledge of instantaneous traction power request (Pload): control system takes appropriate action as a consequence of the value and sign of

dP_{load}/dt. Fig. 3 shows considered situations for Pload and dP_{load}/dt values corresponding to a generic driving cycle section.

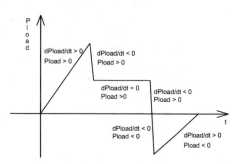

Fig. 3. Generic driving cycle section: variation of traction power request

Vehicle acceleration and deceleration actions are handled by regulating the maximum value of instantaneous power flow that should be related to either FC generator or BSU; gFC and gB are assumed as maximum power per sec. variation for respectively FC and BSU, PFCmax and PBmax and PUCmax as absolute maximum power related to each on-board source. If at time t* on-board sources are controlled at PFC(t*), PB(t*) and PUC(t*) values, at time t*+Δt the required traction power will be Pload(t*+Δt) and therefore

$$
\begin{cases}
P_{FC}\!\left(t^* + \Delta t\right) = \min\!\left[P_{load}\!\left(t^* + \Delta t\right), P_{FC}\!\left(t^*\right) + g_{FC} \cdot \Delta t, P_{FCmax}\right] \\[2mm]
P_{B}\!\left(t^* + \Delta t\right) = \min\!\left[P_{load}\!\left(t^* + \Delta t\right) - P_{FC}\!\left(t^* + \Delta t\right), P_{B}\!\left(t^*\right) + g_{B} \cdot \Delta t, P_{Bmax}\right] \\[2mm]
P_{UC}\!\left(t^* + \Delta t\right) = \min\!\left[P_{load}\!\left(t^* + \Delta t\right) - P_{FC}\!\left(t^* + \Delta t\right) - P_{B}\!\left(t^* + \Delta t\right), P_{UCmax}\right]
\end{cases}
$$

$$(1)$$

When the traction power variation is smaller than FC maximum power per sec. variation, FC generator will supply the entire power gap, otherwise BSU will balance the traction power request up to its maximum power per sec. amount and then UCs will cover the ultimate remaining difference. Same approach is used in case of either negative variation of traction power (i.e. regenerative braking) or constant power request; Figs. 4a and 4b show power flow sharing during respectively vehicle acceleration and vehicle braking, in Fig. 4b FC power is

assumed to be zero (i.e. pure electric vehicle mode of operation).

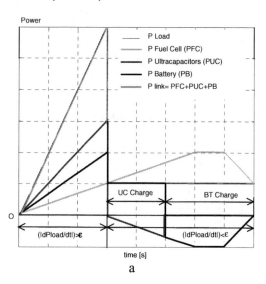

a

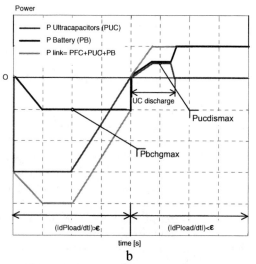

b

Fig. 4. Power flow sharing: a) vehicle acceleration, b) vehicle braking

In order to investigate the control system behavior, a fuel cell powered FIAT 600 vehicle has been simulated in Matlab/Simulink. The vehicle has to meet the speed trace of a real driving cycle experimentally measured in the south part of the city of Rome. This driving cycle is characterized by 9.5kW and 28kW respectively as average and peak power; Figs. 5a and 5b show vehicle speed vs. time and traction power vs. time for the chosen driving cycle.

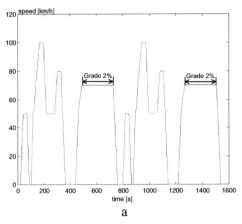

a

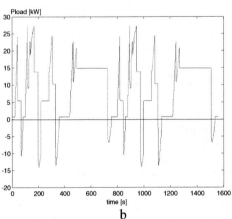

b

Fig. 5. Driving cycle: a) vehicle speed vs. time,
b) traction power vs. time

whereas power flow sharing among on-board sources and SOC for both BSU and UCs are presented in Figs. 6a and 6b.

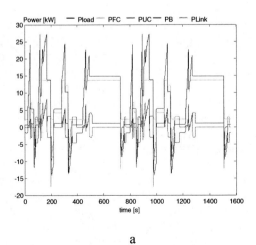

a

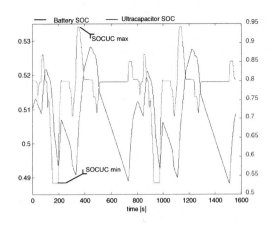

b

Fig. 6. Simulation results: a) power flow sharing, b) SOC performance

Details on power flow sharing during vehicle acceleration and regenerative braking are shown respectively in Fig. 7a and 7b.

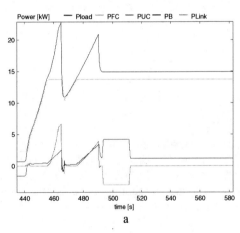

a

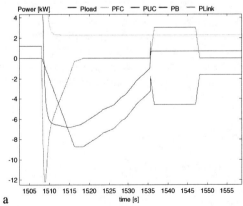

b

Fig. 7. Power flow sharing: a) vehicle acceleration, b) regenerative braking

323

MIPEC CONTROL STRATEGY

As the MIPEC output voltage is being kept constant, at any given operating condition of the traction drive the demanded power is proportional to the dc-link current. Thereby:

$$\frac{dP_{link}}{dt} = V_{link}\frac{dI_{link}}{dt} = V_{link}\left(\frac{dI_{FC}}{dt} + \frac{dI_B}{dt} + \frac{dI_{UC}}{dt}\right) \quad (2)$$

where IFC, IB and IUC, are the current contributions of the three power sources to the MIPEC output current.

Through (2) for each power source it is possible to superimpose the maximum permissible power variation by means of fixing the maximum permissible value for the derivatives. Then, in order to impose the maximum power value it is sufficient to set a maximum admissible current value. After that all the constraints have been fixed it results:

$$\left\|\frac{dI_{FC}}{dt}\right\| \le dI_{FCmax} \; ; \; \left\|\frac{dI_B}{dt}\right\| \le dI_{Bmax} \; ; \; \left\|\frac{dI_{UC}}{dt}\right\| \le dI_{UCmax}$$

$$0 \le I_{FC} \le I_{FCmax} ; \; \|I_B\| \le I_{Bmax} \; ; \; \|I_{UC}\| \le I_{UCmax} \quad (3)$$

In addition to the above constraints, it needs that the SOC values of BSU and UC remain within preset upper and lower bounds. This can be achieved by regulating the currents through a proportional controller.

The reference signals for the control loops are derived from many parameters: the instantaneous load current, the DC link voltage, the BU and UC state of charges, the FC output power, etc. In the following the expressions for reference signals, to be used in MIPEC control, are provided:

$$I_{FC}^* = \frac{I_{Lm} + I_{BUc-d} \cdot (1 - d_{BU})}{1 - d_{FC}}$$

$$I_{BU}^* = \frac{I_{Lm} - I_{FCm} \cdot (1 - d_{FC}) + I_{UCc-d} \cdot (1 - d_{UC})}{1 - d_{BU}}$$

$$V_{link} = cost \quad (4)$$

where I_{Lm} and I_{FCm} are respectively the dc-link and FC measured current, I_{UCc-d} and I_{BUc-d} are the current values of charging and discharging for UC tank and BU whenever storage units' SOC is either lower or higher of the ordinary admitted range, d_{BU} and d_{FC} are duty cycles of respectively BU and FC converters and V_{link} is the dc-link voltage.

The first two expressions (4) give reference currents for FC and BU converters, reference signals' variation is controlled on the basis of generator and storage unit characteristics. UC converter is regulated to keep dc-link voltage either constant or at the most suitable value for traction drive current mode of operation, the third expression.

According to the above described management of the power flows, the three power sources concur to supply the required output power, with the UC tank being active to achieve slow current (power) time variations for both BSU and FC.

EXPERIMENTAL ACTIVITY

In order to evaluate the performance of the proposed propulsion system an experimental campaign has been carried out by using a laboratory prototype. The experimental campaign has been completed at the ENEA Casaccia laboratory facilities, where suitable testing facilities for hybrid vehicles are available. Fig. 8, in the rear, shows the plant of Enea laboratories for testing electric and hybrid vehicles and the experimental fitting in each test room of the components that constitute the drive train: the experimental test rig includes a FC generator, shown in Fig. 9 still in the rear, first developed by ENEA on the base of a PEM FC stack manufactured by De Nora.

Throughout this initial testing phase of the drive train, the FC generator has been replaced by a storage device which generates the requested electrical behavior during power delivery to the input of the MIPEC. To conduct laboratory experiments a MIPEC prototype developed by the University ROMA TRE was arranged to supply a 35kW - 216V electric drive including IGBT inverter and induction motor.

Table 1, in the rear, shows the electric characteristics of the three power sources used, together with the MIPEC prototype, to supply the traction drive. As shown in Fig. 11, always in the rear, the MIPEC power circuit includes three independent dual IGBT power modules (i.e., named "phase-legs" in Fig. 11), ferrite-core inductors and electrolytic capacitors; the MIPEC control algorithm is implemented on a DSP-based board.

Fig. 10 , in the rear, shows the electrical knot and the power connections the laboratory which, taking advantage of the layout of the existing plant, appropriately corrected and upgraded, realize

- the one way connection from primary generator FC to the Electronic Converter MIPEC
- the bi-directional connections (input output) from the Converter towards the stack of UC, towards BSU and towards electrical motor.

The latter is mechanically connected to a computer controlled dynamometer able to reproduce the real behavior of the vehicle in all driving phases (acceleration, constant speed, coast down and braking).

The integration of the Converter and the layout of existing plant has allowed to directly utilize the equipment and the safety systems already working in the test rooms (protections and alarms); the instrumentation of the existing system has been upgraded with measuring and control equipment installed directly on the Converter MIPEC.

The behavior of main components that realize the Drive Train together with experimental results achieved from conducted test campaign are illustrated in Figs. 12, 13, 14 and 15. Figs. 12a and 12b follows respectively show the initial phase of charge of ultracapacitors up to reaching of rated voltage VUC and the phase of voltage growth from the value corresponding to VBSU up to Vlink equal to 216 V.

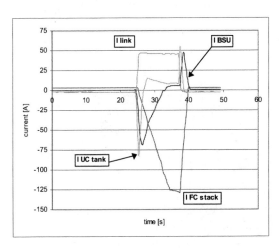

b

Fig. 12. Starting up of the Propulsion System: a) Initial charge of UCs, b) Vlink regulation

Fig. 13 shows the behavior of the currents of the FC generator, of the UC stack and of the BSU corresponding to a step variation of Ilink (traction power).

Fig. 13. Current variations corresponding to a step of Ilink (traction power)

Fig. 14a and 14b depict experimental Vlink oscillations and behavior of current variation in FC generator, UC tank and BSU corresponding to different power traction requests.

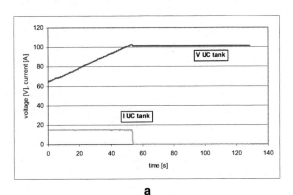

a

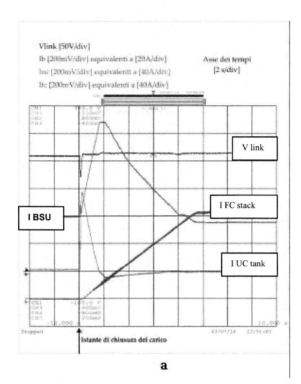

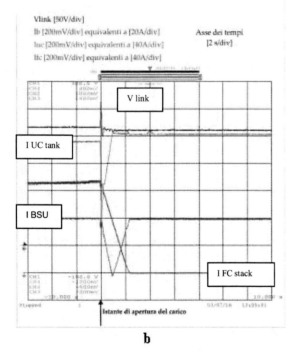

Fig .14. V link oscillations and current variations:
a) dPload/dt >0, Pload>0; b) dPload/dt<0,
Pload>0

Afterwards the behaviour of the MIPEC Converter has been tested in a representative driving cycle derived from the New European Driving Cycle (NEDC) - Urban part (UDC) with speed up to 50 km/h. The driving schedule used is based on the UDC modified in the transient phases, passing from 42 to 28 seconds of acceleration and from 34 to 24 seconds of deceleration, in order to obtain a more severe dynamics for the system.

Fig. 15 shows the power request of the electrical drive (Plink) supplied by FC stack , BSU and UC tank during kinematics cycle. The road load for the dynamometer set up is determined by a coast down test; the vehicle curb weight for testing is equal to 1400 kg.

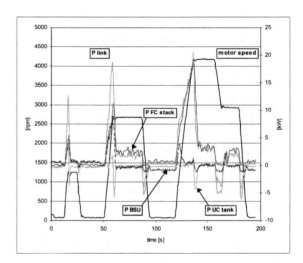

Fig.15. Urban Driving Cycle: Power flow sharing

CONCLUSIONS

An original HEV propulsion system that includes a fuel cell generator and combined UC and battery storage has been presented and discussed. The three power sources are used to supply the vehicle traction drive through a MIPEC which provides the desired management of the power flows. In the proposed arrangement the UC tank is used for providing load-leveling during either acceleration or braking transients. The main features of the MIPEC have been described together with experimental results achieved from a 35 kW prototype, jointly developed by ENEA and University ROMA TRE, in order to conduct laboratory experiments and validate the control strategy.

REFERENCES

[1] "Ultracapacitors Employment in Supply Systems for EV Motor Drives: Theoretical Study and Experimental Results", F. Caricchi, F. Crescimbini, F. Giulii Capponi, L. Solero, Proc. of the 14th International Electric Vehicle Symposium, 1997, cd-rom.

[2] "Employment of Ultra-Capacitors for Power Leveling Requirements in EV: a State of the Art", F. Brucchi, G. Lo Bianco, P. Salvati, F. Giulii Capponi, L. Solero, Proc. of the 32th ISATA, 1999.

[3] "Power Converter Arrangements with Ultracapacitor Tank for Battery Load Leveling in EV Motor Drives", Di Napoli, F. Giulii Capponi, L. Solero, Proc. of the 8th European Conference on Power Electronics and Applications, 1999, cd-rom.

[4] "Ultracapacitor Tests for EV Applications: Introduction of New Equalisation Coefficients", F. Brucchi, M. Conte, F. Giulii Capponi, G. Lo Bianco, P. Salvati, L. Solero, Proc. of the 16th International Electric Vehicle Symposium, 1999, cd-rom.

[5] "On the bench", G. Bernardini, M.Conte, L.De Andreis, G.Pede, E.Rossi & R. Vellone, ElectRic and Hybrid Vehicles Technology 1997

[6] "Vehicle Testing in ENEA Drive-train Test Facility", Giulia LO BIANCO, Giovanni PEDE, Angelo PUCCETTI, Ennio ROSSI, ENEA, Giorgio MANTOVANI, ALTRA, Advanced Hybrid Vehicles Powertrains, SAE SP-1607, Detroit, March 2001

[7] "Ultracapacitors and the Hybrid Electric Vehicle", B. Maher, Applications Engineer, PowerCache

[8] "Evaluating Commercial and Prototype HEVs", Feng An et alii, Argonne National Laboratory, SAE SP-1607, Detroit, March 2001

[9] "Combining Ultra-Capacitors with lead–acid Batteries", B.J.Arnet, L.P.Haines, Solectria Corporation, EVS-17, Montreal, October 2000

[10] "Advanced Batteries for Electric Vehicles: An Assessment of Performance, Cost, and Availability" June 2000, The Year 2000 Battery Technology Advisory Panel, (draft), prepared for State of California Air Resources Board

[11] "Recent Accomplishment of the Electric and Hybrid Vehicle Storage R&D Programs at the U.S. Department of Energy : a Status Report ", R. Sutula et alii, OATT, EVS-17, Montreal, October 2000.

[12] "Ultracapacitor and Battery Storage System Supporting Fuel-Cell Powered Vehicles", A. Di Napoli, F. Crescimbini , L. Solero, G. Pede, G. Lo Bianco, M. Pasquali, EVS 18, Berlin, October 2001

[13] "Hybrid Storage System: an optimization case" A. Di Napoli, University Roma 3, G. Pede , ENEA, SAE Paper 2002-01-1914, Future Car Congress, Arlington June 2002

CONTACT

Di Napoli, F. Crescimbini, A. Lidozzi and L. Solero

University "ROMA TRE", Dept. of Mechanical & Industrial Engineering

Via della Vasca Navale,
79 – 00146 Rome, ITALY

Ph. +39-06-5173277
Fax +39-06-5593732
e-mail solero@uniroma3.it

M. Pasquali, A. Puccetti and E. Rossi

ENEA Research Center "Casaccia", Electrical & Hybrid Vehicles Testing Laboratories

Via Anguillarese, 301 – 00060 S. Maria di Galeria (Rome), ITALY

Fax +39-06-30484327
e-mail angelo.puccetti@casaccia.enea.it

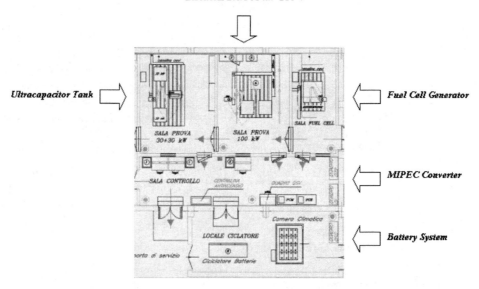

Electrical Drive 35 kW-216 V

Ultracapacitor Tank

Fuel Cell Generator

MIPEC Converter

Battery System

Fig.8. ENEA Casaccia-Laboratories for testing electric and hybrid vehicles

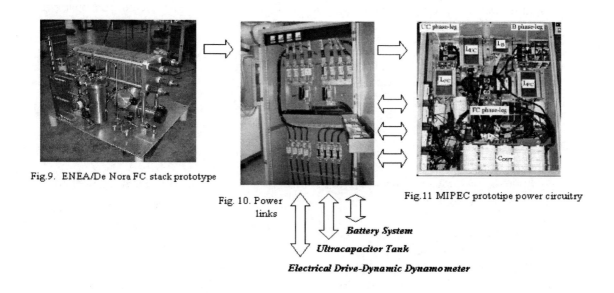

Fig.9. ENEA/De Nora FC stack prototype

Fig. 10. Power links

Fig.11 MIPEC prototipe power circuitry

Battery System

Ultracapacitor Tank

Electrical Drive-Dynamic Dynamometer

Fuel Cell Generator		Ultracapacitor Tank		Battery System	
Open circuit voltage [V]	200	Max voltage (@ SOC=1) [V]	140	Rated voltage [V]	168
Voltage @ max power [V]	120	Min. voltage (@ SOC=0.6) [V]	85	Min. voltage [V]	140
Current @ max power [A]	160	Max current [A]	200	Max current [A]	80

Table 1. Electric characteristic of the propulsion system power sources.

2004-01-1304

Development of Next Generation Fuel-Cell Hybrid System
- Consideration of High Voltage System -

**Tetsuhiro Ishikawa, Shigeki Hamaguchi, Tatsuhiko Shimizu,
Tsuyoshi Yano and Shoichi Sasaki**
Toyota Motor Corporation

Kenji Kato and Masao Ando
AISIN AW Co., Ltd.

Hiroshi Yoshida
Toyota Communication Systems Corporation

ABSTRACT

Toyota Motor Corporation began leasing a new generation fuel cell vehicle the FCHV (Fuel Cell Hybrid Vehicle) in December 2002. That vehicle includes a new variable voltage power electronics system and uses the Nickel Metal Hydride (Ni-MH) battery system from the Prius hybrid gasoline electric vehicle.

This paper describes on-going efforts to model optimum secondary storage systems for future vehicles. Efficiency modeling is presented for the base Ni-MH storage system, an ultra capacitor system and a Lithium ion (Li-ion) battery system. The Li-ion system in combination with a new high efficiency converter shows a 4% improvement in fuel economy relative to the base system. The ultra capacitor system is not as efficient as the base system.

INTRODUCTION

The entire world is clamoring for environmental protection and demanding that the automobile industry make cleaner and more energy-efficient vehicles than it has to date. The fuel cell is a clean, highly efficient energy conversion device that generates electricity from hydrogen and oxygen and produces only water. For that reason, the making of a practical automobile that uses the fuel cell as a power source has long been anticipated.

Ever since Toyota started the development of fuel cell vehicles in 1992, the company has searched for a means of carrying hydrogen fuel on board the vehicle. In 1996, Toyota developed a fuel cell automobile with a hydrogen storage device that used a hydrogen-absorbing alloy, and in 1997, the company announced the world's first fuel cell automobile to employ a methanol reformer. Also, in recent years,

Toyota became the first company in the world to put on limited sale Toyota FCHV that uses a high-pressure tank as the hydrogen storage device and that incorporates the hybrid technology developed for the Prius.

Even as this fuel storage research has continued, Toyota has pursued studies of different types of high-voltage electrical systems for fuel cell automobiles, which are described below.

OVERVIEW OF TOYOTA FCHV SYSTEM

This section provides an overview of the Toyota FCHV system that was the basis for the studies reported here. The hybrid system uses a secondary battery as one of its power sources in order to make use of the energy recovered by regenerative braking. A nickel-metal hydride battery is used for its demonstrated high reliability. Given the fuel economy in cyclic operation, the output ratio of the fuel cell to the secondary battery was set to 90 kilowatts for the fuel cell and 21 kilowatts for the secondary battery. (Refer to Tables 1 and 2.) To handle the range of voltages used by both power sources combined, a full bridge-type converter is used to increase and decrease the voltage, and it uses a three-phase chopper to reduce the ripple current.(Fig. 1)

Type	Polymer Electrolyte
Fuel	Hydrogen
Maximum Power	90kW

Table1　Fuel Cell Specification

Type	Sealed Nickel Metal Hydride
Capacity	6.5Ah
Cooling	Forced Air Cooling

Table2　Battery Specification

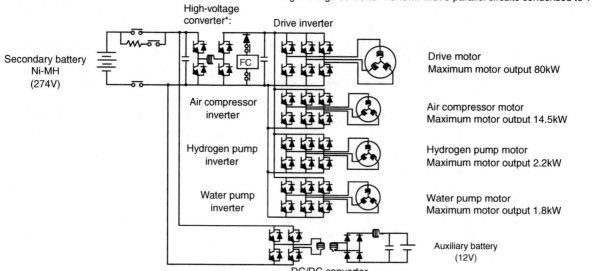

Fig.1 Toyota FCHV High-voltage System Configuration

HYBRID POWER SOURCE CONFIGURATION

This section describes the optimum characteristics for electrical power storage devices which are used together with the fuel cells.

POWER STORAGE DEVICE

The determination of what sort of power storage device is best suited to the characteristics of a fuel cell was approached from the two points of view described below. Note that a comparison between the Toyota FCHV and a gasoline hybrid of the same class is also shown for reference purposes.

Power response

The output response of the fuel cell is shown in Fig. 2. The output response of the fuel cell is faster than that of the engine in the gasoline hybrid, so the amount of energy extracted from the power storage device is small.

net efficiency than the gasoline hybrid throughout the range where the drive output is low, so the region in which the fuel cells operate intermittently is smaller than the region of intermittent engine operation in the gasoline hybrid. For that reason, in cyclic operation, the time during which the fuel cell operates intermittently is shorter than for the gasoline hybrid, so the amounts of power that are stored and discharged are also smaller. Fig. 4 shows the provisional calculations for amount of change in the state of charge (SOC) when the system is operated in LA#4 mode. It can be seen that the range of SOC change for the fuel cell system is smaller than for the gasoline hybrid.

These two observations led to the conclusion that a power type of power storage device is better suited to the FCHV system than is an energy type of device.

The current technologies for power-type power storage devices include the lithium battery and the power capacitor, among others. The results of a study of different power storage devices are shown below (Fig.

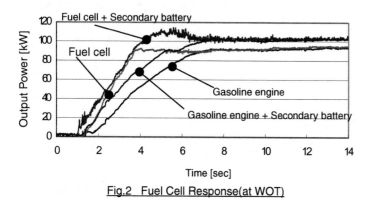

Fig.2 Fuel Cell Response(at WOT)

Net efficiency

As shown in Figure 3, the fuel cell system exhibits better

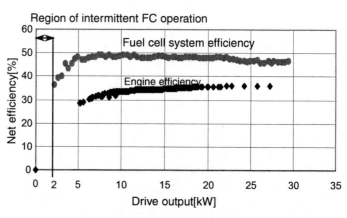

Fig.3 Comparison of Net Efficiency of Fuel Cell and Engine (in Japan 10-15 Mode)

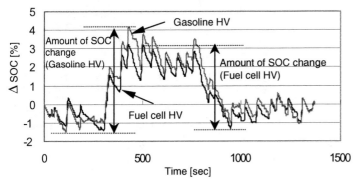

Fig.4 Amount of Battery SOC Change During Operation In LA#4 Mode (Simulation)

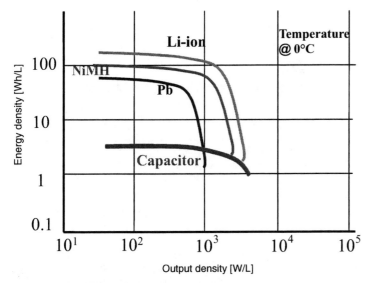

Fig.5 Ragon Plot Characteristics for Different Electrical Power Storage Devices

POWER CAPACITOR

The key point in using a power capacitor is how large a capacitance to specify. For a high-voltage circuit configuration, a converter-less configuration like that shown in Fig. 6 is suitable, because it naturally takes advantage of the broad voltage range over which the capacitor can be used. However, in the FCHV, because the output from the fuel cell is determined only by the load and the voltage, this configuration does not allow the output from the power source to be controlled.

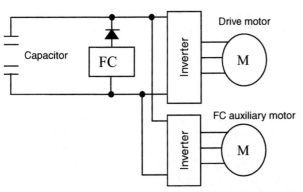

Fig.6 Capacitor System High-voltage Circuit Configuration

In determining such capacitor specifications as the capacitance one must take into consideration such factors as the vehicle's acceleration and deceleration, the repetition of acceleration and deceleration, and the recovery of deceleration energy during cyclic operation. The results from a simulation model study of these factors were compared to results from an actual vehicle study to assess the adequacy of the model. Both the simulation and the actual vehicle evaluation were carried out on a scaled-down equivalent of the Toyota FCHV. Table3 shows the specification of system for verification. Fig. 7 shows data collected during cyclic operation, and Fig. 8 shows the results from an actual vehicle during repeated acceleration and deceleration and calculated values. The results show an error of only 3% for the model, making it nearly identical to the actual vehicle studies.

Maximum Motor output [kW]	3.5
Maximum FC output [kW]	4.28
FC voltage area [V]	96 - 50
Capacitor capacitance [F]	39.2
Capacitor resistance [ohm F]	3.3

Table.3 Specification of the System for Verification

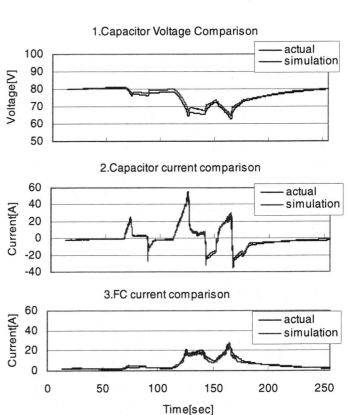

Fig.7 Congruence of Simulated and Actual Results in Mode Operation

331

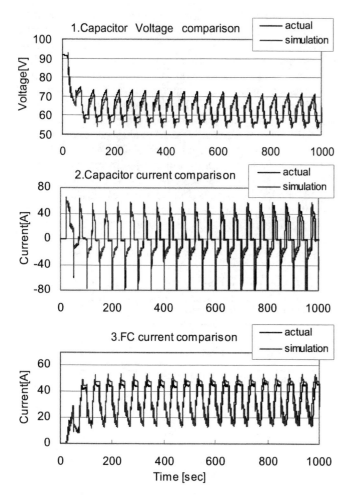

Fig.8 Congruence of Simulated and Actual Results In Acceleration and Deceleration

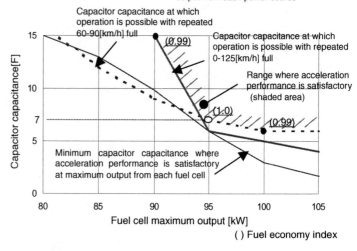

Fig.9 Fuel cell Maximum Output and Capacitor Output

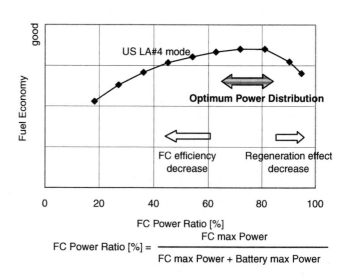

$$\text{FC Power Ratio [\%]} = \frac{\text{FC max Power}}{\text{FC max Power} + \text{Battery max Power}}$$

Fig.10 Fuel Cell Maximum Output and Li-ion Battery Output

Fig. 9 shows the results of a parameter study of vehicle acceleration and deceleration and the like in which the model was used. It shows that while the vehicle meets the acceleration performance requirements, when acceleration is repeated on a flat road, there is a lower limit that does not fall below the designed voltage of the motor. In the graph, the shaded area represents the region where the prescribed requirements are met. The point where the fuel economy is best is where the capacitor capacitance is 7 farads and the maximum output from the fuel cell is 95 kilowatts.

LITHIUM BATTERY

The fuel economy characteristics of a system that uses a lithium battery when the output ratio of the lithium battery to the fuel cell is the parameter that is varied are shown in Fig. 10. The fuel economy is best when the ratio of fuel cell output to total system output is in the range from 65% - 85%. The fuel cell output was therefore set to 80 kilowatts and the lithium battery output to 31 kilowatts.

NEXT-GENERATION FCHV HIGH-VOLTAGE SYSTEM

This section describes the different circuit configurations of the high-voltage electrical system used for the different hybrid power sources that are suitable for the fuel cell described in the previous section.

HIGH-VOLTAGE SYSTEM CONFIGURATIONS

Power capacitor system

As stated earlier, where a power capacitor is used, the high-voltage electrical system is configured so that the power capacitor is connected in parallel with the fuel cell to take advantage of the wide voltage range over which the capacitor can be used.

Lithium battery system

Where a lithium battery is used as the power storage device, a converter becomes necessary, because the usable voltage range of the battery is limited. However, because the lithium battery has high power density, the configuration of the converter circuit can be changed from a full bridge type to a more efficient half bridge type. This is the situation shown in Fig. 11. The change to a half bridge converter requires that the usable voltage range of the fuel cell be separated from the usable voltage range of the lithium battery, but this is feasible as long as the number of battery cells is sixty or less. It can also be seen that even when one takes into account the increase in converter loss caused by the increase in the transformer ratio, the half bridge converter is better than the full bridge type.

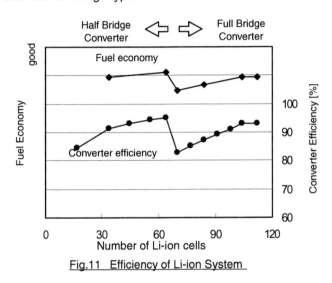

Fig.11   Efficiency of Li-ion System

FUEL ECONOMY CHARACTERISTICS

Table 4 shows a comparison of the fuel economy characteristics of the high-voltage system configurations described above. The lithium battery system shows better fuel economy than the current nickel-metal hydride battery system, indicating that a 4% improvement is possible. Because the power capacitor system can not operate the fuel cell intermittently, the auxiliary motor loss increases, as shown in Fig. 12, so that the fuel economy is worse than that of the current system.

	Capacitor	Ni-MH	Li-ion
Volume[L] (with converter & FC stack)	235	220	196
Weight[kg] (with converter & FC stack)	259	252	206
Fuel economy index (Ni-MH=100)	93.7	100	104.2
Fuel economy Improvement [%]	11.3	4.2 ⟶	⟶ UP

Table.4   Comparison of Fuel Economy in LA#4 Mode

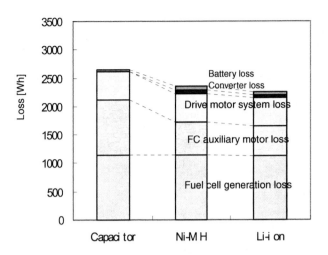

Fig.12   Comparison of Loss in Different Systems

Fig. 13 and Fig. 14 shows the loss transitions and voltages of each component, as well as the state of SOC changes, when each system is used in cyclic operation. With the lithium battery systems, which use converters, the fuel economy is good, because the fuel cell can be operated intermittently when the vehicle is stopped or when energy is being recovered during deceleration.

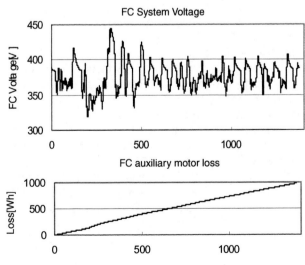

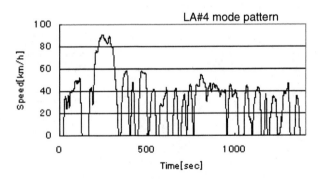

Fig.13   Voltage and Power Transitions in Capacitor System

333

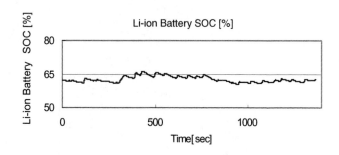

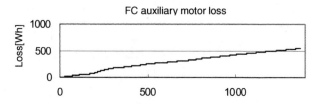

Fig.14 SOC and Loss Transitions in Li-ion Battery System

CONTACT

E-mail address of Tetsuhiro Ishikawa, main author , is the following: tetsu@Ishikawa.tec.toyota.co.jp

CONCLUSIONS

(1) A high-voltage system was developed for the highly efficient next-generation FCHV. The system uses a lithium battery, the optimum power storage device for the characteristics of the fuel cell system, and a highly efficient half bridge converter circuit.

(2) The change to a lighter secondary battery and the reduction of converter losses make it possible to improve the fuel economy by 4% over the current Toyota FCHV.

(3) Because a system that uses a power capacitor does not allow the fuel cell to be operated intermittently, its fuel economy is 11% worse than that of the Li-ion system.

REFERENCES

1. T. Matsumoto, N. Watanabe, H. Sugiura, T. Ishikawa. "Development of Fuel-cell Hybrid Vehicle", The 18th International Electric Vehicle Symposium, (2001).
2. S. Sasaki, T. Takaoka, H. Matsui, T. Kotani. "Toyota's Newly Developed Electric-Gasoline Engine Hybrid Power train System", The 14th International Electric Vehicle Symposium, (1997).
3. T. Matsumoto, N. Watanabe, H. Sugiura, T. Ishikawa. "Development of Fuel-cell Hybrid Vehicle", the 2002 SAE World Congress, (2002).
4. Y. Hori, T. Teratani, R. Masaki. "Motor Technology for Automobile", The Nikkan Kogyo Shimbun, LTD (The Business & Technology Daily News), (2003).

IV. COMPONENT DEVELOPMENT

Fuel Cell Stacks

2002-01-0407

Advanced MEA Technology for Mobile PEMFC Applications

K. A. Starz, J. Koehler, K. Ruth and M. Vogt
OMG dmc² Division

ABSTRACT

PEMFC systems are gaining increased importance as clean and efficient energy sources for electric vehicles. Research and development at OMG dmc² division is focused on catalyst systems for the growing PEMFC market, particularly on Membrane Electrode Assemblies (MEAs) for PEMFC stacks [1-4]. The paper describes the current technology for MEAs operating on hydrogen as well as on CO containing reformate gases. Characteristics and technical features of the MEA products are described with emphasis on steady state and transient testing procedures.

INTRODUCTION

Membrane Electrode Assemblies (MEAs) are key components of the PEM fuel cell. Important characteristics of a PEMFC stack such as power density, dynamic behaviour, durability and cost are determined by the type, quality and performance of the MEA incorporated into the stack.

As a consequence, major advances and breakthroughs achieved on the MEA level will immediately influence the state-of-the-art of PEMFC stacks and thus facilitate the widespread commercial introduction of PEMFC technology. Key for a rapid commercialization of mobile PEMFC technology is the availability of MEA products tailored for this application.

PRODUCT CHARACTERISTICS

MEA ASSEMBLY - The development of a proprietary MEA technology over the past five years resulted in a leading position as a supplier of these products to the fuel cell industry [5]. In parallel to these activities, a range of Pt-based electrocatalysts was developed, which are used in the MEA products. The manufacturing process for MEAs is based on the continuous production of a catalyst coated membrane (3-layer CCM) consisting of an anode catalyst layer, the ionomer membrane and the cathode catalyst layer. The CCM is subsequently sandwiched between gas diffusion layers (GDLs) to generate a 5-layer MEA product. With appropriate sealings, it is built into a PEM single cell or stack. Along with the CCM, tailor-made GDLs (or "backing" materials) are supplied to ensure optimum performance of the MEA. Figure 1 depicts a schematic drawing of the assembly of a 5-layer MEA into a PEM single cell.

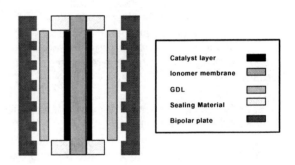

Figure 1: Assembly of a typical 5-layer MEA into a PEM single cell (schematic drawing).

A key feature of the continuous MEA manufacturing process is its wide processing window. A variety of ionomer membrane materials can be processed, including solid ionomer membranes (which are commercially available from various suppliers [6]) as well as composite membranes. The membrane thickness can vary between 0.5 mil (12.5 microns) and 10 mil (250 microns). Furthermore, the MEAs are available in various formats and sizes, ranging from 50 cm² to 1,000 cm² of active area. This is in accordance with the different application fields of PEMFC technology (i.e. residential, portable, automotive or premium power), which in turn have different requirements for power density, cost, reliability and safety. In figure 2, a large format 3-layer MEA (CCM) for mobile applications is shown.

Figure 2: Large format 3-layer MEA (CCM) for mobile applications.

HIGH POROSITY MEAs - As a common feature of all MEAs produced by the company, the electrode layers are characterized by a very porous structure. The introduction of additional porosity, i.e. macropores in the range of 0.5 to 5 microns, improves the MEA performance, particularly when operating with highly diluted reformate gases and air at ambient pressure and low stoichiometries. Details are given later in this article. The benefits of this structure are twofold: The higher porosity causes

- a better access of reactant gases to the active sites inside the electrode and

- a better removal of product water from the cathode.

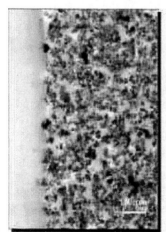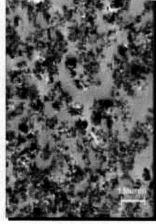

Figure 3: Comparison of MEAs with low porosity (type A, left) and high porosity (type B, right) electrode structure (Microsection, TEM).

Figure 3 presents a comparative TEM analysis of a low porosity and a high porosity electrode structure. There is a distinct influence of the increased porosity on MEA performance. In standard hydrogen/air operation, a markedly higher cell voltage is obtained for a MEA cathode structure type B (~ 80% porosity) compared to type A (~ 40% porosity), particularly at high current densities above 500 mA/cm². In the low current density area (i.e. < 200 mA/cm²), the performance of both MEA structures are quite similar.

IONOMER MEMBRANE MATERIALS - As already outlined, a variety of different ionomer membrane materials can be used for MEA manufacturing. As an example, the H_2/air-performance of two MEAs containing perfluorinated ionomer membranes with identical EW but different thicknesses are shown in figure 4. For ambient (1 bar) and pressurized (3 bar) conditions, the performance of the thin 1 mil (25 microns) membrane is slightly better than the 2 mil (50 microns) membrane, with the 1 mil type showing an 30-50 mV higher cell voltage at a current density of 1 A/cm². Humidification parameters were kept constant in this experiment.

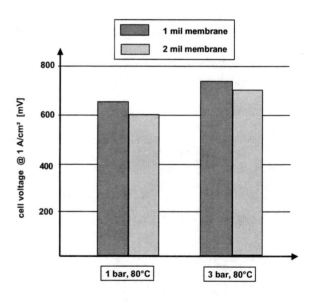

Figure 4: Performance of MEAs based on 2 different ionomer membranes (H_2/air operation at 1 and 3 bar, cell temperature 80°C); humidification anode: 80°C; humidification cathode: 60°C; stoichiometry: H_2: 1.3; air: 2.0.

GAS DISTRIBUTION LAYERS (GDLs) AND SEALING MATERIALS - Generally, GDLs are electrically conductive and serve for an adequate transport of the humidified reactant gases into the electrode layers, as well as for a rapid product water removal from the cathode. The selection of the appropriate GDL types depends on the operating conditions and geometry of the PEMFC stack and is critical for optimum MEA performance. For the three layer MEAs (CCMs), taylor-made GDLs were developed and are supplied to the user. These GDLs consist of a highly porous carbon substrate (either carbon fiber paper or carbon cloth) and are treated to obtain the appropriate degree of hydrophobicity for anode and cathode side.

Typically, the GDLs are further coated with a porous microlayer on the side which is in contact with the catalyzed electrode layer of the CCM. Since the catalyst layer is already applied to the CCM, no further hot-pressing or laminating steps are necessary. This improves the overall reliability of the MEAs made from the CCMs. Mechanical failures, such as pinhole formation or perforation of the membrane, which can occur at the elevated temperatures and high pressures commonly applied in the lamination processes of catalyst coated GDLs with ionomer membranes, do not occur. The optimum contact between CCM and GDLs is achieved by the compressive load applied to the end plates in the PEMFC single cell or stack assembly. Therefore, the sealing materials play an important role for proper CCM and GDL assembly. They have to be carefully selected in terms of thickness, materials compressability, dimensional stability and endurance. A range of suitable sealing materials are available on the market for use in MEA and stack assembly.

PERFORMANCE IN HYDROGEN / AIR OPERATION

PLATINUM LOADING - Standard MEAs designed for hydrogen/air operation contain supported Pt/C catalysts only. The typical total loading is 0.6 mg Pt/cm² with 0.2 mg Pt/cm² on the anode and 0.4 mg Pt/cm² on the cathode side. Ionomer membranes with 50 micron thickness are used in the standard materials.

INFLUENCE OF PRESSURE CONDITIONS - Operating parameters such as pressure, temperature, humidi-fication and stoichiometry play an important role for the MEA performance in single cells and stacks. To illustrate this, the MEA performance for hydrogen/air operation at three different pressures [1, 2 bar and 3 bar (abs)] is shown in figure 5. For the pressurized operation at 3 bar, the peak power density at 1 A/cm² is markedly improved from 0.6 W/cm² to 0.7 W/cm².

CELL TEMPERATURE AND HUMIDIFICATION - Generally, to obtain best performance of the MEAs based on 50 microns solid perfluorinated membranes, the anode should run under fully humidified (saturated) conditions, whereas the cathode can be slightly lower in humidification temperature. For a given cell temperature of 80°C, the recommended anode humidification should be in the range of 75 - 80°C, the cathode humidification at about 60 - 70°C. In most cases, performance losses are detected when the humidification is too low at the

anode side. High cell temperatures (e.g. above 90°C) are not recommended on the long run due to problems with membrane dry-out and water management.

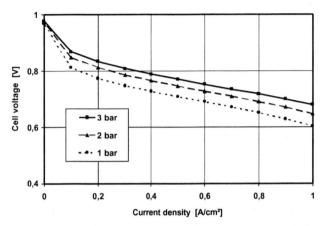

Figure 5: Performance of MEAs operating at different pressures. Testing conditions: H_2/air, cell temperature 80°C, pressure 1, 2 and 3 bar(abs); humidification anode: 80°C; humidification cathode: 60°C; stoichiometry: H_2: 1.3; air: 2.0.

LIFETIME EVALUATION - In figure 6, a MEA lifetime test for 1,400 hours of continuous operation at a constant current density of 500 mA/cm² is depicted. The feed gases are H_2/air and the humidification conditions are fully saturated. As shown, the cell voltage remains at a very constant level and the degradation rate is less than 5 mV/1,000 hours. This lifetime test documents the durablity and endurance for the MEAs under H_2/air operating conditions.

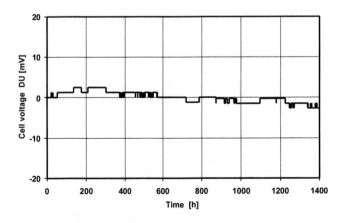

Figure 6: Lifetime test of MEA in operation for 1,400 hours under continuous, non-interrupted testing. Testing conditions: H_2/air, cell temperature 75°C, pressure 1 bar(abs), constant current density 500 mA/cm².

PERFORMANCE IN REFORMATE/AIR OPERATION

PRECIOUS METAL LOADING - Standard MEAs designed for reformate/air operation contain advanced PtRu/C electrocatalysts on the anode and pure Pt catalysts on the cathode side of the MEA. Typical total precious metal loadings are 0.85 mg (Pt+Ru)/cm² with 0.4 mg Pt/cm² at the cathode. Mechanisms and theories for CO tolerance improvement by PtRu/C anode catalysts are reported in the literature [7-9]. In the following, the typical MEA characteristics for reformate/air operation are described.

INFLUENCE OF REFORMATE GAS COMPONENTS - MEA performance for operation with simulated MeOH reformate at 3 bar(abs) pressure is shown in figure 7. The reformate composition is based on simulated methanol reformate containing 60 vol.% hydrogen, 25 vol.% carbon dioxide and 15 vol.% nitrogen. As a reference, the performance for pure hydrogen is shown. It is visible, that the influence of CO_2 and N_2 is predominantly a Nernst type dilution effect (loss of 20-30 mV), a significant poisoning effect by CO_2 cannot be detected.

stage) are required. Thus it would be beneficial to have electrocatalysts tolerating high CO concentrations (up to 1,000 ppm CO) since the complete PrOx unit could then be omitted. This would lead to a size and cost reduction of the PEMFC system.

Furthermore, the amount of air bleed added to the feed gas should be limited due to system efficiency (i.e. reduction of hydrogen yield) and safety reasons. The advanced anode catalysts employed in the MEAs are offering a compromise to this situation. As outlined in figure 8, steady state concentrations of up to 300 ppm CO in the reformate feed can be tolerated in combination with an air bleed level of 3 vol.%. This level is needed to maintain the cell voltage above 0.6 V under steady state operating conditions. It is understood, that lower CO concentrations (such as 100 ppm CO) in turn need lower air bleed levels (typically 0.5 to 1 vol.%) to obtain the identical performance.

Additionally, as shown in the next chapter, transient concentrations of 300 ppm CO are tolerated by the MEAs at much lower air bleed levels.

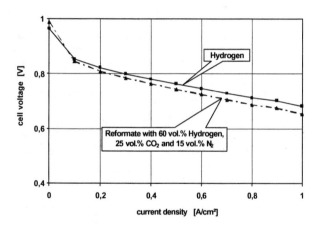

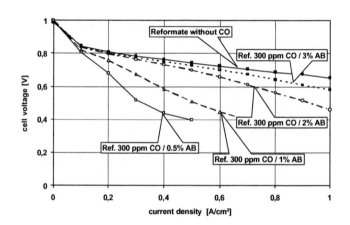

Figure 7: MEA performance running on hydrogen and simulated methanol reformate (60 vol.% H_2, 25 vol.% CO_2, 15 vol.% N_2). Testing conditions: single cell 50 cm²; T: 80°C; pressure: 3 bar(abs); humidification anode: 80°C; humidification cathode: 60°C; stoichiometry: H_2: 1.3; air: 2.0.

Figure 8: MEA performance in reformate/air operation with 300 ppm CO at various air bleed levels. Testing conditions: ref. to figure 7; reformate composition: 60 vol.% H_2, 25 vol.% CO_2, 15 vol.% N_2.

INFLUENCE OF CO-POISONING (STEADY STATE CONDITIONS) - As well described in the literature [7], carbon monoxide (CO) is poisoning the anode catalyst in a reversible process. Typically, 40-50 ppm CO in the reformate feed gas can be tolerated by standard PtRu electrocatalysts with a level of 0.5-5 vol.% air injected into the anode ("air bleed").

In order to achieve this low CO concentrations in the feed gas, bulky PrOx reactors, sometimes containing two stages (i.e. low temperature and high temperature

INFLUENCE OF TRANSIENT CO-POISONING (DYNAMIC CONDITIONS) - In addition to steady state testing, initial studies on transient performance decay and recovery in MEAs were published by the authors in a previous SAE paper [3]. Aim of the tests described there was to simulate conditions that occur upon driving of a fuel cell vehicle system, i.e. during practical use. It was shown that sudden load changes in the PrOx reactor of the fuel processor system can lead to high CO concentrations in the feed gas. In the experiments, 200 pulses of 300 ppm CO were added to a "base load" reformate in constant intervals of 10 mins to simulate CO peak concentrations. The PEM single cell was held at a constant current density of 600 mA/cm² and was running on a continuous simulated MeOH reformate gas flow with 40 ppm CO.

In our recent work, the dynamic testing procedures were optimized to become a tool for the development of CO tolerant anode electrocatalysts. Furthermore, the method allows to study the influence of various air bleed conditions on dynamic performance and stability of the anode electrocatalysts. For this purpose, the dynamic testing was extended to 1,000 CO pulses and the CO induced cell voltage drop (ΔU_i) was monitored over the complete testing cycle. In figure 9, a typical test sequence is shown after 80 hours of CO pulsing (corresponding to 960 pulses) for an air bleed level of 0.5 vol.%. As outlined, the cell voltage drop ΔU_i is in the range of 30-40 mV and the overall cell voltage U_o remains very constant at about 0.7 V during the complete experiment.

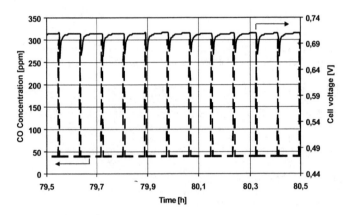

Figure 9: Dynamic testing of MEAs: Testing conditions: single cell 50 cm²; T: 80°C; pressure: 3 bars (abs); humidification anode: 80°C; humidification cathode: 60°C; stoichiometry: H₂: 1.3; air: 2.0; reformate composition: 60% H₂ 25% CO₂ 15% N₂, 40 ppm CO; air bleed 0.5 vol.%; CO peaks: 40 to 200 ppm for 10 secs, 5 peaks with intervals of 10 mins.

In order to evaluate the influence of different air bleed levels on the dynamic CO tolerance of MEAs, the same testing procedure was performed employing 3 air bleed conditions (0, 0.5 and 1 vol.% ; Note: air bleed concentrations are related to total gas composition of 100 vol.%). The CO-induced cell voltage drop (ΔU_i) was examined after 50 and 650 pulses.

As depicted in figure 10 (after 50 pulses) and figure 11 (after 650 pulses), the air bleed levels of 0.5 and 1 vol.% cause only a small cell voltage drop ($\Delta U_i \sim 35$ mV) and the recovery time after the reversible CO poisoning process is significantly reduced. It is interesting to note that in the case of 0 vol.% air bleed, the cell voltage drop is gradually increasing during the experiment (ca.

120 mV after 50 pulses, 210 mV after 650 pulses). Since in parallel the cell voltage U_o was also slightly lower at the end of the test (by about 30 mV), it is obvious that the CO tolerance of the anode catalyst is reduced.

An explanation for this finding could be that an irreversible CO poisoning of some of the reactive sites on the PtRu/C catalyst occurs. Furthermore, the CO-oxidation by adsorbed OH molecules on the Ru (as proposed in the bifunctional spill-over mechanism [8]) is slower than "gas phase type" oxidation of CO occuring in the air bleed process, thus, without air bleed, the recovery time for the cell voltage is lenghtened as well (up to 3 mins at the end of the pulsing experiments). As a consequence from these results, a small level of air bleed is recommended for optimum MEA performance in the dynamic operating conditions of mobile PEMFC applications.

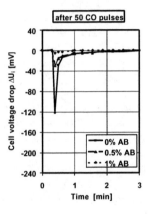

Figure 10: Dynamic testing of MEAs: CO-induced cell voltage drop (ΔU_i) for three different air bleed levels after 50 CO pulses. Testing conditions: ref. to figure 9.

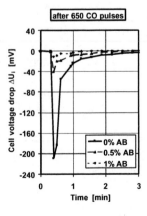

Figure 11: Dynamic testing of MEAs: CO-induced cell voltage drop (ΔU_i) for three different air bleed levels after 650 CO pulses. Testing conditions: ref. to figure 9.

CONCLUSION

Membrane Electrode Assemblies (MEAs) are key components of PEM fuel cells. Typical characteristics of the advanced MEA products were described. The standard materials consist of catalyst coated membranes (CCMs) and suitable gas diffusion layers (GDLs) which are assembled together in a single cell or PEMFC stack. Because of the broad processing window of the proprietary manufacturing processes, MEAs based on various ionomer membrane materials can be manufactured in different sizes. Due to the high porosity structure of the electrode layers, a significant performance improvement is obtained, particularly at high current densities and ambient pressure conditions.

Two different MEA product types are available for hydrogen/air as well as for reformate/air operation. For hydrogen/air operation, the standard MEAs contain Pt catalysts at a total loading of 0.6 mg/cm². The influence of the key operating parameters such as pressure, humidification and temperature on the MEA performance are descibed. A MEA lifetime test was run for 1,400 hours in non-interrupted operation showing a very stable cell voltage performance.

In steady state reformate operation, the MEAs tolerate various concentrations of carbon monoxide in the anode feed gas depending on the air bleed level applied (300 ppm CO with 3 vol.% air and 100 ppm CO with 0.5 to 1 vol.% air). A simulated methanol reformate containing 60 vol.% H_2 was used in these experiments. A major poisoning effect by CO_2 could not be detected.

As the CO pulsing tests for transient poisoning indicate, up to 1,000 pulses of 300 ppm CO with air bleed levels of 0.5 vol.% can be tolerated in the reformate feed gas without cell voltage decay, thus demonstrating a stable performance of the MEAs under the dynamic operating conditions required for mobile use.

A low level of air bleed (0.5 to 1 vol.%) is recommended to compensate for the repeated CO pulses which can occur in the feed gas due to dynamic load changes of the PrOx reactor. When omitting the air bleed completely, the CO tolerance of the anode electrocatalyst is gradually degrading due to irreversible poisoning of some PtRu sites of the catalyst.

Additional work is under way to develop further improved MEAs and electrocatalysts to fulfill the requirements for mobile PEMFC application.

ACKNOWLEDGMENTS

The authors wish to thank the Fuel Cell Team of OMG dmc² division for their contributions and continuous support of the work. In particular, we want to thank Mr. G. Heinz for developing the dynamic testing program, Mrs. R. Kleisinger and Mr. U. Stenke for performing the steady state and dynamic MEA measurements.

CONTACT

Dr. Karl Anton Starz, OMG dmc² division, P.O.Box 1351, D-63403 Hanau/Germany

Dr. Markus Vogt, OMG dmc² division, P.O. Box 1351, D-63403 Hanau/Germany

REFERENCES

[1] K.A. Starz, E. Auer, Th. Lehmann and R. Zuber, Proc. 3rd International Fuel Cell Conference, Nagoya, Japan, 93-98 (1999).
[2] K. Ruth, E. Auer, Th. Lehmann, K.A. Starz and R. Zuber, Abstracts Fuel Cell Seminar, Portland, Oregon, 40-43 (2000)
[3] K.A. Starz, E. Auer, F. Baumann, Th. Lehmann, S. Wieland and R. Zuber, SAE Technical Paper Series 2000-01-0013.
[4] K.A. Starz, K. Ruth, M. Vogt and R. Zuber, Proc. 3rd Internat. Fuel Cell Conference, Nagoya, Japan, 210-215 (2000)
[5] US 6,309,722, US 5,861,222, US 6,007,934, US 6,006,410, EP 1,037,295 and others (OMG dmc² division)
[6] O. Savadogo, J. of New Materials for Electrochemical Systems 1, 47-66 (1998)
[7] L. Gubler, G. Scherer und A. Wokaun, Chem. Eng. Technol. 24, 59-67 (2001)
[8] H.F.Oetjen, V.M. Schmidt, U. Stimming, F. Trila, J. Electrochem. Soc. 143, 12 (1996) 3838.
[9] T.A. Zawodzinski Jr., J. Bauman, S. Savett, T. Springer, F. Uribe and S. Gottesfeld, Proc. of International FC Seminar, Palm Springs, USA (1998), S. Gottesfeld and J. Pafford, J. Electrochem. Soc. 135 (1988) 2651-265.

2002-01-0410

Solid Oxide Fuel Cells for Direct Oxidation of Liquid Hydrocarbon Fuels in Automotive Auxiliary Power Units: Sulfur Tolerance and Operation on Gasoline

Gary M. Crosbie, Erica Perry Murray and David R. Bauer
Research Laboratory, Ford Motor Co.

Hyuk Kim, Seungdoo Park, John M. Vohs and Raymond J. Gorte
University of Pennsylvania

ABSTRACT

To be practical, auxiliary power units (APUs) should operate on the same fuels that the internal combustion engine (ICE) uses for vehicle propulsion. Solid oxide fuel cells (SOFCs) have previously been shown to be able to convert the chemical energy of certain room-temperature-liquid hydrocarbon fuels (toluene and synthetic diesel fuel) to electricity by direct oxidation. Because such SOFCs operate without reformers, the systems based on these SOFCs are expected to be compact. To work with existing infrastructure fuels, the cells must be able to tolerate typical contaminants such as sulfur that are found in the everyday fuels. In this paper, we report on recent laboratory results that show direct oxidation SOFCs with ceria-copper anodes can provide at least 2 hours operation in the presence of 200 ppm sulfur in the fuel. Also, a laboratory cell has been run for 12 hours on regular unleaded gasoline.

INTRODUCTION

On-vehicle electricity requirements continue to grow due to demands for more internal-combustion-engine control systems and for more customer features. For high chemical-to-electrical conversion efficiency,[1] fuel-cell based APUs are of interest for low noise and low emissions. With only the $\Delta G = -nFE$ Nernst limit [discussed herein], fuel cells promise higher efficiency for the chemical-to-electric conversion than can be obtained with the indirect chemical-to-mechanical-to-electric conversions, which are constrained by the Carnot P-V cycle limit.

Several types of fuel cells exist for consideration for use as APUs.[2,3] Since, as a practical matter, the APU fuel must be the same as that powering the internal combustion engine used for propulsion, the existing fuel delivery infrastructure establishes a need for an APU that can operate on liquid hydrocarbon (HC) fuels, such as gasoline or diesel fuels, at least for the nearer-term. Although reforming of such HCs to H_2 is feasible (and necessary) for proton exchange membrane (PEM) fuel cells, such an added indirect conversion step requires reformer and purification subsystems, which bring efficiency losses, complex-to-control transient responses, and take up added volume that may diminish systems interest in any APU systems.

With solid oxide fuel cells for APUs,[1,4,5] several potential benefits have been recognized: no need to purify after reforming, a low cost due to absence of precious metals, and low maintenance. In a previous paper,[6] we examined the implications on the design of SOFC APUs that followed from earlier demonstrations of direct oxidation of toluene and synthetic diesel fuels. The lab results had indicated that solid oxide fuel cells can provide a means for direct oxidation of room-temperature-liquid hydrocarbon fuels[7] without the use of a reformer, if the proper materials are present in the anode. Analogies were drawn[6] to the chemical similarities between the direct-oxidation SOFCs and present ICE systems. The systems-level implications presented as simplicity and compactness that can follow from a system that has no reformer at all.

In this paper, we report on the implications of more recent laboratory results that are also likely to be of particular interest to an ICE-oriented readership: demonstration of stable operation of the solid oxide fuel cells (with special anode compositions) on sulfur-containing fuels and on ordinary unleaded gasoline.

AUXILIARY POWER UNIT GOALS

A FC-APU is intended to service not only the present alternator loads such as lights and engine controls, but also electrical loads for items that are now provided mechanically by the front end accessory drive (FEAD). A vehicle propulsion future analysis[8] illustrates increases of auxiliary electrical loads (for the engine), since even after allowing for more efficient power electronics, the 2020 vehicles are expected to have more on-board electrically driven systems, drawing more power.

The list of the anticipated supplementary electrical power needs in Table I is based on discussions within the Ford Motor Company. This list (abridged from Ref. 6) reflects one viewpoint. It is intended to be unbiased by the present capabilities of any particular APU technology.

Table I. Fuel Cell - Auxiliary Power Unit (FC-APU) System Performance Goals (Abridged from Ref. 6)

Fuel	gasoline or diesel	
Sulfur tolerance	50	ppm
Power	5	kW
Efficiency	>35	% LHV
Emissions	(including start-up)	
NO$_x$	<0.01	g/kWh
HC	<0.005	g/kWh
CO	<0.1	g/kWh
Mass	<50	kg
Volume	<50	L
Start-up time	<3	min

THERMAL CYCLING AND MECHANICAL ISSUES

Two critical issues for most transportation-related SOFC-APU applications are fast start-up (<3 min, in Table I) and thermal cycling. Since high-volume automotive ceramics survive in harsh thermal cycling and vibration environments (in spark plugs, catalysts, exhaust sensors and particulate filters), it is expected by analogy that fast-heat-up, long-life, high-value automotive SOFCs can be designed and manufactured. The geometry-dependent mechanical aspects were discussed in a recent paper.[6]

The focus here is on the chemical aspects of the anode that allow at least laboratory-scale direct oxidation of room-temperature liquid hydrocarbons with sulfur impurities.

SOFC CHEMISTRY -- COMPARISONS WITH PEM FUEL CELLS AND IC ENGINES

The SOFC is closer in chemistry to an ICE (than a PEM fuel cell is) in that any fuel that reacts with oxygen can, in concept, be used within the microscopic reaction zone. (In contrast, PEMFC systems rely upon hydrogen as the single, preferred fuel to be present in its anode.)

Before combustion in an internal combustion engine cylinder, fuel and oxidant are mixed together. Mechanical energy is produced as electrons are transferred from molecule to molecule, with a volume expansion due primarily to the heating of the gases present and secondarily to the creation of additional molecules. Power follows from the displacement volume.

In contrast, in electrochemical systems (batteries and fuel cells), the fuel and oxidant are physically separated by a medium such as a membrane that conducts only ions, instead of being mixed. Because the electrons have no choice but to move in the external electrical circuit, the motion of electrons can produce electrical work. Power follows from the membrane surface area, to a first approximation.

In contrast to batteries, fuel and oxidant are continuously added to fuel cells. Since air is drawn as needed to be the oxidant (at least for near-surface transportation), only the fuel needs to be carried for the fuel cell to operate. This aspect gives fuel cell compactness and low mass density advantages compared to batteries, as has been noted.[9] Fuel cells and ICEs share this system characteristic of only needing to carry a single reactant, the fuel.

When the fuel cell fuel is a room-temperature liquid hydrocarbon, then the chemical energy for the APU can be stored simply and with weight and volume advantages, as is the case with the established fuels for ICE-powered vehicles, today. Since some fuel supply system components (such as the fuel tank and pump) may be shared between the ICE and the APU, handling standards and an experience base for such fuels are already in place.

The type of fuel cell is usually identified by the ionic separation membrane (or liquid) that allows only the passage of ions, not electrons, within the cell. In liquid-based fuel cells such as phosphoric acid, alkaline KOH, or molten carbonate, the ionic conducting medium is a concentrated ionic solution. In PEM fuel cells, a humidified Nafion® membrane conducts only protons. Consequently, to operate PEM fuel cells on hydrocarbon fuels, it is necessary to reform (at a temperature above 750°C) to reach the highest H$_2$ concentration levels and then to take subsequent steps to purify the reformate to remove membrane-catalyst poisons, such as carbon monoxide.

In a SOFC, the ion-conducting separation is a polycrystalline ceramic membrane that is stabilized in a cubic fluorite crystal structure with many oxygen sites vacant. In the typical yttria-stabilized zirconia (YSZ) membranes, no purification with respect to CO is needed, since the SOFC can use any CO that is present as an additional fuel. (If we reach a time when the infrastructure is in place to have H_2 be widely available as the vehicular propulsion fuel, the SOFC APU can operate well on that, too.)

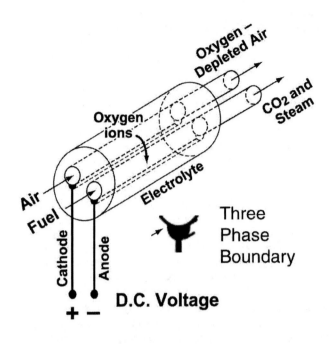

The chemical reaction of a hydrocarbon fuel with oxygen is similar to the combustion reaction in an ICE, except that O enters the reaction as an ion. The contacting phases are indicated below in brackets. (For the ICE all the constituents are gas phase.)

$$\text{fuel} + n\,O^{2-} \longrightarrow \text{reaction products} + 2n\,e^-$$

$$\| \qquad \| \qquad\qquad \| \qquad\qquad \|$$

[gas] [electrolyte]   [gas]   [current collector]

In the schematic drawing of Fig. 1, the flows of fuel and oxidant (air) are separated by a solid electrolyte. It is noted that at certain particular lines, called three phase boundaries (TPBs), three phases come together. It is at the TPBs that the electrochemical reactions occur, with mutual contact of electronic conductor (metallic), the electrolyte, and the gas phase. For high power output, the TPB length is extended by use of a fine porous intermixture of one ionically- and one electronically-conducting phases.

At the cathode (positive terminal for fuel cells), molecular O_2 is (chemically) reduced to two O^{2-} ions. The ions then diffuse through the electrolyte membrane to the anode.

At the anode (negative terminal for fuel cells), the chemical combustion reaction takes place. At the anode interface, oxygen ions that have been carried through the electrolyte react with the fuel. Because the anode contacts the fuel, the anode is the focus of the developments related to direct oxidation of hydrocarbons.

Figure 1. Schematic diagram of a fuel cell. As oxygen ions move from the cathode to the anode through the solid electrolyte, electrons perform work in an external circuit, where the d.c. voltage is indicated. The electrochemical reactions occur at the three-phase boundaries, where there is a meeting of gas, solid electrolyte, and electronic (metallic) conductor.

The limiting cell voltage (at constant pressure) is from the Nernst equation,[10]

$$\Delta G = -n\,F\,E,$$

wherein ΔG is the free energy of reaction (for a reversible process), n is the number of electrons transferred, F is the Faraday constant and E is the open circuit voltage. For a wide range of hydrocarbon fuels of interest, the calculated Nernstian voltages are between 1.1 and 1.2 V, with only a weak dependence upon temperature. Of course, for actual fuel cells, there are loss mechanisms, such as electrode polarization and current collector resistance, to consider. The ICE analogy is that a calculated Carnot limit cannot be reached in a practical IC engine.

GASEOUS HYDROCARBON-FUELED STATIONARY SOFCS

REFORMED METHANE FUELED SOFCS —Designed for stationary co-generation of electricity and heat, some SOFCs are presently nearing commercial production. These include the fraction-megawatt central generation units of Siemens-Westinghouse[11] and much smaller ones for residential use.[12] Regardless of the size, the typical fuel is natural gas, which is mostly methane.

By the addition (within the anode chamber) of steam and recycled anode exhaust, these stationary units typically operate with internal reforming. The HC is converted [reformed] to hydrogen and CO, without deposition of carbon even at the operating temperatures near 1000°C, so long as the H_2O:HC ratio is large. The reformer can be as simple as a controlled recycle loop or as complex as several separate modules. From the vehicle systems point of view, the reformer takes up more space, provides more to go wrong, and adds more weight — factors that are less important in large, stationary applications. The control of the added process stream (and possibly a recycle stream) for water vapor remains relatively simple for steady-load utility operation, but is complicated for the many transients expected for vehicular APU electrical loads.

A gas volume expansion occurs during the oxidation reaction at the anode, as each HC molecule in the fuel becomes multiple exhaust molecules. In stationary systems there is the opportunity to add recovery cycles (for example, turbines) to capture this energy and thereby to produce even higher overall efficiencies, but such bottoming cycles would add complexity to automotive system.

Unlike the high flame temperatures associated with the rapid flame propagation in the ICE, the microscopic temperature of reaction in a fuel cell is expected to be near the steady-state, macroscopic one. Nitrogen is present only at the cathode. Furthermore, NOx is produced by reaction of O_2 and N_2 only above 1300°C. Therefore, there can be little NOx produced, even at the near-1000°C temperature of operation of stationary SOFCs. Residual amounts of unreacted fuel may be catalytically reacted at low temperatures, before exhausting from the system.

DIRECT OXIDATION OF METHANE AND OTHER GASEOUS HYDROCARBONS — If the above (internal reforming) fuel cells were run without steam addition and reforming, a carbon char would rapidly form in the anode pores in operation at 700°C and above, disrupting (stopping) the useful-output-power operation of the cell. An ICE analog of the pyrolysis is found in operation with overly rich fuel mixtures, leading to spark plug fouling. (Note: This analogy is not particularly strong, because it does not encompass the catalytic action of the nickel in HC pyrolysis.)

Laboratory reports[13] describe the use of dry methane (without reforming outside of the anode chamber) in a temperature window below 700°C, by use of anode compositions with ceria and doped ceria. Ceria-zirconia plays a key role in oxidation catalysts for ICE exhaust gas treatment. Ceria (and ceria-zirconia) appear to aid in the oxidation of the methane. The operation of the fuel cell is simplified without water addition. The low temperature of operation (needed to reach the pyrolysis-

free window for methane) requires exceptionally thin ceramic membranes and improved cathode exchange currents. One approach to reduce the interfacial resistance of cathodes for use at such low temperatures is to provide a fine mixture of electronic and ionic conductors[14] to increase the TPB length.

Water appears as a reaction product at the anode, although it is not added directly. Nevertheless, the product water may also help kinetics of the reactions. Nickel is known to be a steam reforming catalyst. [Note: A SOFC differs from a PEMFC in that the exhaust products appear on the fuel side, as the reactions proceed.]

The operation of a Ni-ceria-YSZ-anode SOFC in the temperature window (below 700°C) relies on a substantial purity of the methane with respect to higher hydrocarbons. The storage of natural gas is rather too voluminous for ICE propulsion to meet all customer expectations for range and cargo space for light trucks and passenger automotive use. Of course, some market for CNG vehicles is to be mentioned, for taxis and truck fleets. Again, having a separate supply of compressed natural gas fuel for an APU is seen as impractical. For storage compactness reasons, one would prefer higher molecular weight hydrocarbons.

Higher hydrocarbons do not appear to have any temperature range that is both thermodynamically free of pyrolysis and practical for operation of a SOFC. These are minor components of natural gas. For natural gas, one would need to return to a requirement for purification of the fuel of these higher hydrocarbons above that needed for combustion operation of a CNG-fueled ICE, which seems, again, to be impractical.

In these reports, the metal in the anode is still the nickel of the stationary reformer SOFC systems. But here, the nickel appears to act to catalyze the pyrolysis of the hydrocarbons (forming an infinitesimal charring). The ceria aids the simultaneous oxidation of the carbon so produced, as long as the fuel cell is running under load (and delivering oxygen ions to the anode). What would otherwise become a char deposit is either oxidized with oxygen ions appearing at the solid electrolyte surface or steam reformed with steam produced from the reaction.

There are indications that gaseous HC (including n-butane, a liquid under mild pressure) could be operated with the CeO_2 and copper-cermet anode system.[15] Perhaps not surprisingly, the innovation has come from experience in catalysis for internal combustion engine exhaust treatment. Materials and chemistry aspects for the direct oxidation of these gaseous HC have been published.[16] The strongest supporting evidence that the process is indeed direct oxidation, rather than internal reforming in the vicinity of the anode, is the agreement between the rate of CO_2 production and the electrical current, what are related by Faraday's equation. In Fig. 2, such a plot is shown for butane and methane. From gas chromatography analysis of the products, it was demonstrated that there is no internal reforming.

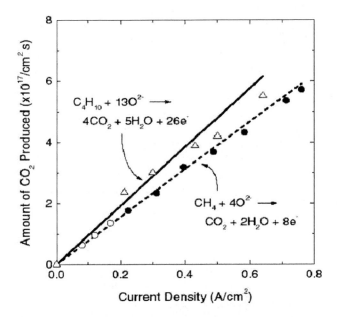

Figure 2. The amount of CO₂ formed corresponds directly to the number of electrons produced. The lines are not a best fit to the data, but simply a stoichiometric prediction based on Faraday's equation.

ROOM-TEMPERATURE-LIQUID HYDROCARBON FUELS

DILUTE REFORMING SYSTEMS — With sufficient dilution with CO_2, other higher HC systems can be made to function. There is a recent report[17] of the direct oxidation of iso-octane in a solid oxide fuel cell, if diluted sufficiently. By recirculating part of the CO_2-rich anode exhaust, carbon deposition can be suppressed for at least 3 hr of operation. The presence of the CO_2 encourages reforming to occur within the anode chamber. Although this dilute system does represent an example of a direct supply of a room temperature liquid HC to an SOFC, the approach appears of limited practical benefit: The extra heating of the non-reacting diluent gas will necessitate larger heat exchangers and the recycle line will add various transient control complications.

CONCENTRATED LIQUID HYDROCARBON SYSTEMS — The breakthrough (cited in the previous SAE paper[6]) is in operation of SOFCs with HC(liq) fuels without the necessity of high dilution by the use of copper-ceria-stabilized zirconia anodes.[7] Previous attempts with similar higher hydrocarbons (without reforming) extinguish the cell performance in short times as carbon deposits mask nickel catalyst sites.

The direct oxidation impact on the APU system is that by removing the reformer, one helps obtain compactness of reformerless APU system. As a compactly storable liquid, the same fuel can then be used for both the ICE and the APU.

The first published examples[7] for concentrated room-temperature liquid hydrocarbon fuels also include anode compositions with CeO_2 (of ICE exhaust catalysts), as in the earlier cases for direct oxidation of gaseous HCs. However, the difference is the substitution of a non-char-catalyzing metal, namely copper, for the conventional nickel. Copper has the additional advantage of a higher electrical conductivity, which may reduce the resistance losses in the anode current collection, for a given design.

A sampling of the first experimental results with liquid HCs is given in two figures, which were adapted from those in Ref. 7.

Fig. 3 shows stable chemical-to-electrical conversion separately with both toluene and synthetic diesel, as examples of higher hydrocarbons fuels. At a constant load current, the cell voltages are stable, over many hours, even with 40% concentrated levels of toluene, a gasoline additive. Although this open-ended dataset is far short of the 40,000 hour operation of central generation SOFC stationary demonstrations[11] and the >8000 hour (with <10% loss) APU durability goal,[6] the observed stability result is nonetheless promising. Although it might be considered that, as in the case of dry-methane direct oxidation, the exhaust products help keep the anode sufficiently oxidized to prevent char deposition, the conversion in these cells is less than 1%, so any product H_2O is only at a very low level.

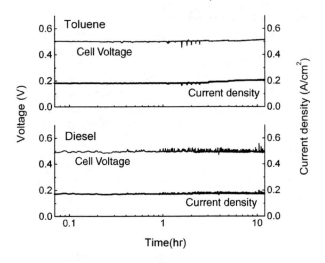

Figure 3. Plots of cell potential and current density as a function of time for toluene and the diesel fuel. Each of the fuels was fed to the cell with N₂ at a concentration of 40 wt.% hydrocarbon.

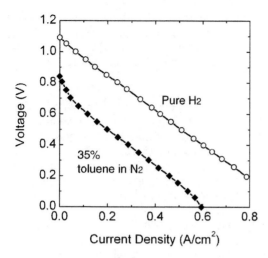

Figure 4. A plot of the cell potential as a function of current density for the SOFC at 973 K. The data are shown for the following fuels: (○) pure H₂ and (◆) 35 vol.% toluene in N₂.

Fig. 4 illustrates that there are, in these first examples, lower voltages with the HC (liq) fuel than with H_2 at equivalent current densities. For H_2 at 700°C, the observed 1.1 V open circuit voltage (OCV) matches the Nernst calculation. For toluene, the observed OCV is 0.85 V, not the 1.14 V from the Nernst equation.

Such a lower voltage bears directly on the efficiency. The working interpretation is that the lower voltage results from equilibrium with incomplete reaction products. If this is so, then other catalytic anode additives will allow higher voltages to be obtained. For additional details about the chemical interpretation, the reader is referred to original journal source.[7] In support of this expectation is earlier work[18] in which a no-load (open circuit voltage) of 1.2 V open circuit voltage observed in fuel cell operation using methane with 1%Rh-CeO₂ in the SOFC anode. Although breaking the earlier "no-precious metals" SOFC promise, the precious metal needs only to be present in the porous three-phase boundary region extending 10 to 20 micron from the electrolyte surface.[19] Because there are substantial efficiency losses in reforming/purifying HCs to H_2, it is worth noting that the efficiency of a SOFC stack/bundle operating with direct oxidation of HCs does not have to be as high as the efficiency of a fuel cell stack/bundle operating with H_2 from a reformer.

The tests are run with high feed concentrations, since the conversions are small in laboratory tests. Expectation is that current flow helps work against carbon deposition, at least in the TPB region. Steam and CO_2 will be responsible for removing carbon farther from the electrolyte membrane. The ICE analogy here is to have lean or stoichiometric conditions at the anode-electrolyte interface.

In this way, there is strong evidence that reformerless SOFCs can operate with direct oxidation of room-temperature-liquid HCs. Therefore, there is expectation of more compact systems which share fuel handling with the ICE, based on laboratory-level anode performance. Development of catalysts for higher voltages is desired for efficiency. Otherwise, one will continue to need some reforming, although the associated purification will be simpler than for PEM.

One aspect of sharing the same fuel with the ICE is the need for tolerance of some sulfur. In particular, in Table I, the target level of sulfur for the APU to tolerate is listed as 50 ppm. Recent laboratory results[20] from tests at the University of Pennsylvania show that the copper-cerium oxide anodes are stable (Fig. 5) in fuels with sulfur concentrations that are typical of today's gasoline and natural gas.

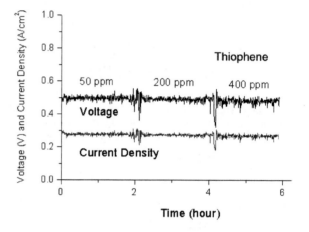

Figure 5. A plot illustrating the sulfur tolerance of the copper-cerium oxide anodes. The cell potential and current density are shown as functions of time at levels of thiophene in n-decane of 50, 200, and 400 ppm. For these levels of sulfur in the fuel, the voltages and current densities are stable.

A significant related result is the demonstration of operation of the laboratory solid oxide fuel cell on an ordinary example of gasoline. The gasoline was purchased from a "regular unleaded" fuel pump at a commercial service station in Philadelphia. As seen in Fig. 6, the voltage and current densities are essentially stable over 12 hours of operation with this gasoline being the sole input (100%) to the anode chamber.

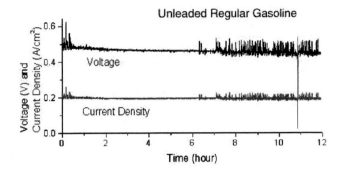

Unleaded Regular Gasoline

Figure 6. Operation of the solid oxide fuel cell laboratory cell with unleaded regular gasoline purchased from a commercial service station. With the copper-cerium oxide anode, the cell potential and current density are essentially stable for at least 12 hours of operation.

CONCLUSIONS

SOFCs are shown to be able to tolerate 200 ppm sulfur (Fig. 5) in operation at moderate current densities for periods of time greater than 2 hours.

In much the same way that the tolerance of carbon monoxide by SOFCs simplifies the fuel preparation processing designs for SOFC systems, the sulfur tolerance suggests that compact auxiliary power units can operate with existing infrastructure fuels without a stage of sulfur removal. Indeed, it is demonstrated (Fig. 6) that the SOFCs can operate with retail gasoline as the fuel, given the appropriate anode chemistry.

The points made in a previous report[6] about the suitability of SOFCs for auxiliary power units are restated in an abridged manner. In the present work the chemical description is amplified. Experimental evidence is given (Fig. 2) that strongly supports the interpretation of direct oxidation (not internal reforming) on the ceria-copper-containing anodes.

Together with the high chemical-to-electric efficiencies of fuel cells (even at part-load), the new results of tolerance of sulfur impurities and of operation on gasoline increasingly make SOFCs an attractive system match in a vehicle configuration consisting of an ICE propulsion unit (for mechanical power) plus a fuel cell auxiliary power unit (for electrical power).

REFERENCES

[1] Tachtler, J., Dietsch, T., Goetz, G., "Fuel Cell Auxiliary Power Unit — Innovation for the Electric Supply of Passenger Cars," SAE Paper No. 2000-01-0374, Society of Automotive Engineers, Intl., March 2000.

[2] Jost, K., "Fuel Cell Concepts and Technology," *Automotive Engineering Intl.*, Vol. 107, No. 3, p. 173-85, March 2000.

[3] Hirschenhofer, J. H., Stauffer, D. B., Engleman, R. R., and Klett, M. G., *Fuel Cell Handbook, Fourth Edition*, DOE/FETC-99/1076, November 1998, 268 pp.

[4] Pearsall, R. D., and Hirschenhofer, J. H., "Preliminary Assessment of Planar Solid Oxide Fuel Cells for Transportation Power Applications," U.S. DOE, Contract No. 992352402, Technical Report, November 2000, 44 pp.

[5] Read, C. J., Thijssen, J. H. J., and Carlson, E. J., "Fuel Cell Auxiliary Power Systems: Design and Cost Implications," SAE Paper No. 2001-01-0536, Society of Automotive Engineers, Intl., March 2001.

[6] Crosbie, G. M., Perry Murray, E., Bauer, D. R., Kim, H., Park, S., Vohs, J. M., and Gorte, R. J., "Application of Direct Oxidation of Liquid Hydrocarbon Fuels in Solid Oxide Fuel Cells to Automotive Auxiliary Power Units," SAE Paper No. 2001-01-2545, Society of Automotive Engineers, Intl., Sept. 2001.

[7] Kim, H., Park, S., Vohs, J. M., and Gorte, R. J., "Direct Oxidation of Liquid Fuels in a Solid Oxide Fuel Cell," *J. Electrochem. Soc.*, Vol. 148, No. 7, p. A693-A695 (2001).

[8] Weiss, M. A., Heywood, J. B., Drake, E. M., Schafer, A., and AuYeung, F. F., "On the Road in 2020: Life Cycle Analysis of New Automobile Technologies," MIT Energy Lab Report, MIT-EL00-003, October 2000. (Published on-line at http://web.mit.edu/energylab/www.)

[9] Patil, A. S., and Jacobs, R., "US Army Small Fuel Cell Development Program," SAE Paper No. 1999-01-2726, Society of Automotive Engineers, Intl., 1999.

[10] Hammou, A. and Guindet, J., "Solid Oxide Fuel Cells," Chap. 12, p. 407-43 of *CRC Handbook of Solid State Electrochemistry*. Gellings, P. J., and Bouwmeester, H. J. M., eds., CRC Press, Boca Raton, Florida, 1997.

[11] Singhal, S. C., "Science and Technology of Solid-Oxide Fuel Cells," *Mat. Res. Soc., Bull.*, Vol. 25, No. 3, p. 16-21, 2000.

[12] Costamagna, P. and Honegger, K., "Modeling of Solid Oxide Heat Exchanger Integrated Stacks and Simulation at High Fuel Utilization," *J. Electrochem Soc.*, Vol. 145, No. 11, p. 3995 (1998).

[13] Perry Murray, E., Tsai, T., and Barnett, S. A., "Direct-Methane Fuel Cell with a Ceria-Based Anode," *Nature*, Vol. 400, p. 649-51, August 19, 1999.

[14] Perry Murray, E. and Barnett, S. A., "(La,Sr)MnO3-(Ce,Gd)O2-x Composite Cathodes for Solid State Fuel Cells, Solid State Ionics, Vol. 143, p. 265-273 (2001)>

[15] Park, S., Vohs, J. M., and Gorte, R. J., "Direct Oxidation of Hydrocarbons in a Solid Oxide Fuel Cell," *Nature*, Vol. 404, p. 265-67, 2000.

[16] Gorte, R. J., Park, S., Vohs, J. M., and Wang, C., "Anodes for Direct Oxidation of Dry Hydrocarbons in a Solid-Oxide Fuel Cell," *Adv. Mater.*, Vol. 12, No. 19, p. 1465-69, October 2000.

[17] Kendall, K. and Saunders, G., "Performance of Several Fuels in Small Tubular SOFCs," to appear in *J. Power Sources*, Feb. 2002.

[18] Putna, E. S., Stubenrauch, J., Vohs, J. M., and Gorte, R. J., "Ceria-Based Anodes for the Direct Oxidation of Methane in Solid Oxide Fuel Cells," *Langmuir*, Vol. 11, p. 4832-37, 1995.

[19] Tsai, T. and Barnett, S. A., "Increased Solid-Oxide Fuel Cell Power Density Using Interfacial Ceria Layers," *Solid State Ionics*, Vol. 98, p. 191-96, 1997.

[20] Kim, H., Vohs, J. M., and Gorte, R. J., "Direct Oxidation of Sulfur-Containing Fuels in a Solid Oxide Fuel Cell," *J. Chem. Soc., Chemical Communications*, p.2334-35 (2001).

CONTACTS

Gary M. Crosbie, Ph: 313-337-1208;   Fax: 323-1129; MD3182 SRL Bldg.,Ford Motor Co., 20000 Rotunda Dr., Dearborn, MI 48121-2053; e-mail: gcrosbie@ford.com.

Raymond J. Gorte, Ph: 215-898-4439;   Fax: 573-2093; 311 Towne, University of Pennsylvania, 220 South 33rd St., Philadelphia, PA 19104;   e-mail: gorte@seas.upenn.edu.

Available online at www.sciencedirect.com

Journal of Power Sources 118 (2003) 424–429

www.elsevier.com/locate/jpowsour

PEMFC systems: the need for high temperature polymers as a consequence of PEMFC water and heat management

Ronald K.A.M. Mallant

Energy Research Centre of the Netherlands (ECN), P.O. Box 1, 1755 ZG Petten, The Netherlands

Abstract

The proton exchange membrane fuel cell (PEMFC) is usually operated at elevated pressure, requiring the use of a compressor. Also the operating temperature is low, generally below 80 °C. For many applications, a somewhat higher operating temperature would be preferable. This paper describes the reasons for pressurised operation at, or close to, 80 °C in terms of water and heat management issues. It is concluded that a new type of proton conducting material is highly desirable, having many of the good Nafion properties, but based on a proton conduction mechanism that does not require the presence of large amounts of water in the electrolyte.
© 2003 Elsevier Science B.V. All rights reserved.

Keywords: PEMFC; Water management; Heat management; Compressor

1. Introduction

It is quite common to operate proton exchange membrane fuel cell (PEMFC) systems at pressures of 2–3 bar (a) and at or below 80 °C. Pressurised operation requires that the system includes a compressor, possibly an intercooler and an expander to reclaim part of the energy that was required for compression. A typical air management system is given in Fig. 1. The relevant components are

- a compressor or blower;
- the humidifier, which may be integrated with the cooling circuit;
- the stack (cathode);
- a water separator;
- a pressure release device (backpressure controller or expander);
- a cooling circuit: pump, buffer and heat exchanger.

In case of pressurisation above about 1.6 bar an air cooler is required. Cooling of the inlet air can also be done by injection of water into the compressor, this option is however not considered here.

For many systems, the use of a compressor is not a viable option. In particular this is the case for small systems. These systems require low airflows, for which efficient compressors are not available. Also noise and wear may not be acceptable. Such systems are operated close to atmospheric pressure. In this case, the operating temperature will be limited to 60–65 °C. This may or may not be acceptable,

depending on the application. Some applications however, would benefit from a higher operating temperature in order to make use of the waste heat or to reduce heat rejection problems. This paper addresses why PEMFC systems have difficulty in being operated at both a low pressure and high (>70 °C) temperature.

2. Theory

Application of Faraday's law yields the relation between the current I and the dry air feed (g/s) [1].

$$q_{air} = \frac{M_{air} I \lambda_{air}}{nNex} \tag{1}$$

where I is the current produced (A), n the number of electrons involved in the reaction (in this case, $n = 2$), e is elementary charge (1.6022E−19 Coulomb), N the Avogadro's constant (6.022045E ± 23 mol⁻¹), λ_{air} is the air stochiometry (the inverse of the oxygen utilisation), x the fraction of oxygen in air (0.20946), M_{air} the molar mass of air (28.964 g/mol)

For the remainder, all calculations are based on the general gas law.

$$\frac{PV}{T} = R$$

and the associated equations

$$p_i = \frac{m_i}{V} \frac{R}{M_i} T$$

0378-7753/03/$ – see front matter © 2003 Elsevier Science B.V. All rights reserved.
doi:10.1016/S0378-7753(03)00108-3

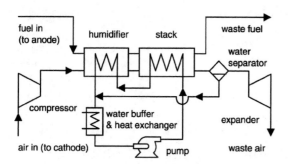

Fig. 1. Simplified lay-out (intercooler not shown).

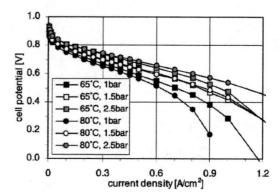

Fig. 2. PEMFC polarisation curves.

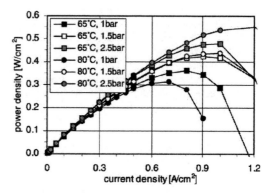

Fig. 3. Power density curves.

to calculate the partial pressure for each component in a gas mixture and

$$p = \rho \frac{R}{\bar{M}} T$$

for calculations on gas mixtures, in which the weighted harmonic mean $\bar{M}$ of the individual molecular weights M_i of the gases present in masses m_i in the gas mixture is calculated using

$$\bar{M} = \frac{m_1 + m_2 + \dots}{m_1/M_1 + m_2/M_2 + \dots}$$

The calculation of $p_{\max,\text{H}_2\text{O}}(T)$, the maximal water vapour pressure as function of temperature, was done using a polynomial approximation of water vapour pressure data taken from [1].

3. Experimental

To assess the effect of pressure and temperature on cell performance, a 7 cm^2 cell was operated at 1, 1.5 and 2.5 bar (a), and at temperatures of 65 and 80 °C. An air stoichiometry of 2 was maintained by continuously adjusting the cathode flow rate as function of current density. The MEA was made using E-Tek electrodes and a membrane prepared by filling a 50 µm 85% porous substrate (Solupor) with Nafion ionomer [2]. Both anode and cathode gases were fully humidified. Polarisation curves obtained with this cell under various P, T conditions are given in Fig. 2.

Although the PEMFC can be operated over a wide range of the displayed curves, it should be noted that cell efficiency is proportional to cell potential, and is in first approximation given by $\eta_{\text{cell}} = 0.8 V_{\text{cell}}$ (V_{cell} in volts, η_{cell} is with respect to LHV, at 100% H_2 utilisation). For that reason, operation at a potential lower than 0.6–0.7 V hardly is attractive. Power density curves are given in Fig. 3. Opting for the highest power density is tempting in order to reduce the power specific size, weight and cost of the stack.

However, compression of air takes energy, and these losses should be accounted for. The polarisation curves given in Fig. 2 can be corrected to account for the losses for compression. A straightforward way to do this is to

reduce the cell potential at given current I with a voltage ΔV, such that $I\Delta V$ equals the power required for compression of the air required at that current. The amount of air depends on the operating conditions. At an individual cell potential of 0.675 V and an air stoichiometry of 2, the air consumption of a 50 kW system is 150 Nm^3/h. The calculated compressor power as function of pressure ratio is given in Fig. 4 ([3], based on 20 °C inlet temperature and 70% compressor efficiency).

Using the calculated energy consumption for air compression, the curves of Fig. 2 were corrected, see Fig. 5. For the 50 kW (gross power) system referred to in this paper the compressor uses 8.6 kW, equivalent to 17% of the power

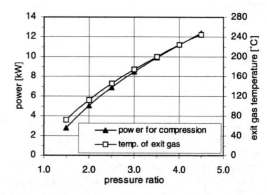

Fig. 4. Calculated shaft power and gas exit temperature for compression of 150 Nm^3/h air.

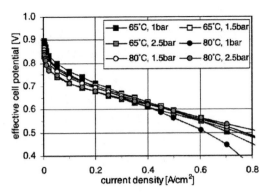

Fig. 5. $V_{effective}$, based on Fig. 2, corrected for energy consumption for compression.

produced by the fuel cell. The actual power losses will be slightly higher due to losses in the drive motor and power electronics. On the other hand, for large systems for which a compressor expander can be used the losses can be reduced. For very small systems (less than several kW), more severe losses must be accounted for, since small air compressors are not likely to attain 70% efficiency.

Ultimately, the two parameters that are of relevance for the system developer and user are the net power density and the system efficiency. For the system dealt with here, the efficiency is mainly determined by cell efficiency and losses for compression. Using the approximate relation between cell potential and cell efficiency, $\eta_{cell} = 0.8V_{cell}$, the system efficiency is given by $\eta_{system} = 0.8V_{effective}$. The effective cell potential can also be used to calculate the net power density. The resulting relation between net cell power density and system efficiency is given in Fig. 6. From this, it is concluded that the net effect of pressurisation is negligible at preferred operating conditions, that is, at current densities of less than about 0.6 A/cm². So, if for reasons of efficiency, high cell potential/low cell current density operation is preferred, and if no net system power gain is achieved under such conditions, then why would one prefer pressurised operation? The reason is that at pressurised conditions, the water vapour pressure has less effect on water balance, gas flow rates and gas composition. In the following, this will be shown for the air system only.

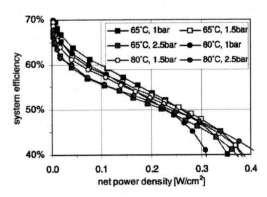

Fig. 6. Relation between system efficiency and net power density.

4. Consequences of high relative humidity

The performance of the PEMFC strongly depends on the conductivity of the polymer [4], which is strongly related to the relative humidity of the gas the cell is in contact with. The requirement of working at saturated gas conditions may perhaps not make it impossible to operate the PEMFC at low P and relatively high T, but such conditions require an adapted design of stack and system. High water vapour pressures lead to strong dilution of reactants by water vapour and large heat exchange surfaces.

In terms of water management, the following occurs in the cathode line.

4.1. Humidifier

The air is humidified and heats up to stack temperature. Using saturation vapour pressure data found in literature [1], the consequences of working at 100% RH are analysed. First of all, the addition of water vapour increases the volume of gas flowing through the system (when pressure is kept constant). Especially at high temperature and low pressure, the increase in gas volume is substantial (Fig. 7). The hydraulic implications of the increase in gas volume have to be taken into consideration in the design of the fuel cell stack and the system. The pressure drop in a tube is given by the equation

$$\frac{dp}{dl} = -\frac{\lambda^|}{d}\frac{1}{2}\rho v^2 \tag{2}$$

where dp is the pressure drop over length dl, $\lambda^|$ the dimensionless friction factor, d the inside pipe diameter, ρ the average gas density over length dl, and v the average gas velocity. The pressure drop is proportional to the product of the gas density and the square of the gas velocity, although $\lambda^|$ is dependent on the Reynolds number, and therefore gas velocity as well. The effect of humidification is that due to the addition of water vapour, the gas density decreases whereas the gas volume increases. The lower gas density results in a lower pressure drop in pipes and bends. The

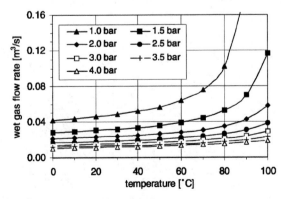

Fig. 7. The calculated wet gas volume at various temperatures and pressures. The original dry gas flow rate used in this calculation is 150 Nm³/h, or 0.042 Nm³/s.

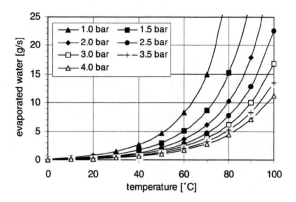

Fig. 8. The amount of water required to humidify 150 Nm³/h of initially dry air, as function of temperature and pressure.

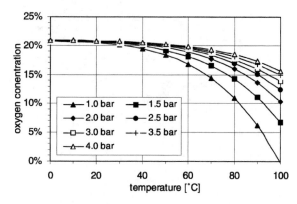

Fig. 10. Calculated oxygen concentration in fully humidified air, in dependence of pressure and temperature.

increased volume however will have a much stronger effect, resulting in an increased pressure drop. This can be compensated for by increasing the diameters of flow channels in the fuel cell stack and system piping. This will however, result in increased system volume and weight.

A trivial consequence of humidification is that water is consumed (Fig. 8). This will particularly becomes relevant if the water that is produced in the stack is not sufficient to compensate for the water used in the evaporator. The evaporation of water takes energy (Fig. 9). This energy can be taken from the cooling circuit, in which case no extra energy is required. The waste heat of a 50 kW fuel cell stack operating at 0.675 V individual cell potential (the basis for these calculations) under condensing conditions is 61 kW. Comparing this number with the values given in Fig. 9, it can be concluded that there should generally be no problem to fulfil this heat demand.

4.2. Cathode

The addition of water vapour results in a dilution of the oxygen present in air (Fig. 10). The concentration shown here is that at the cathode inlet. It can be concluded that working at low P, high T conditions leads to high oxygen dilution, which decreases cell performance. Since oxygen is

used in the fuel cell reaction, the downstream part of each cell is exposed to substantially lower concentrations.

Under the conditions assumed here (50% oxygen utilisation), half of the oxygen is used by the electrochemical reaction taking place in the fuel cell. Since this reduces the volume of gas, some of the water originally present in the vapour phase will condense out to form liquid water. This produces heat, which has to be removed by the stack coolant circuit. Also, some water is produced, in addition of the water produced by the electrochemical reaction. So, in the cathode, a two-phase flow is established.

4.3. Water separator

The liquid water is separated from the cathode exit gas by a water separator. The amount of water that is collected should match the amount needed for humidification. Fig. 11 shows the water surplus of a system as function of P and T. Negative numbers indicate a net water loss from the system. One way to prevent a system from turning into a net water consumer is to cool the water separator, such that it now becomes a condenser. This may seem unattractive since it apparently increases the amount of heat that has to be discarded to the surroundings, which is very undesirable for automotive applications. Theoretically, this is not the

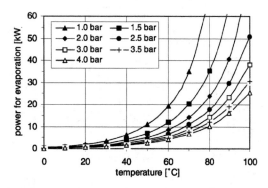

Fig. 9. The power required to fully humidify 150 Nm³/h of initially dry air, as function of temperature and pressure.

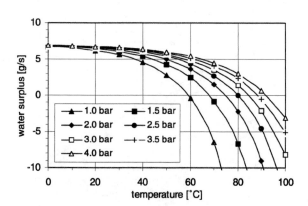

Fig. 11. Water surplus for the system shown in Fig. 1.

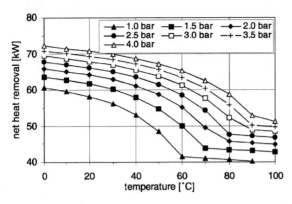

Fig. 12. The calculated amount of sensible heat that has to be discarded to the surroundings.

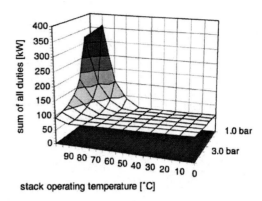

Fig. 13. Sum of duties of air cooler, humidifier, condenser and external heat exchanger as function of stack operating temperature and pressure.

case however since under these conditions a large amount of heat is taken from the stack coolant flow, to provide the heat required for evaporation of the large amount of water for humidification. So, there is merely a shift of the heat removal problem.

4.4. Overall heat management

One can try to match heat sources and sinks, to avoid the need for external heating and to reduce the amount of heat that has to be discarded by a radiator. Even if such matching could be done perfectly, a net cooling duty is still necessary. This is shown in Fig. 12. At low temperatures, the water content of the cathode exit gas is low. Little heat is removed in the form of latent heat, but as sensible heat instead. As temperature increases, cathode exit gas humidity increases, and the net sensible heat removal from the system is reduced. At even higher temperatures the water separator has to be cooled in order to become a condenser, e.g. at 60 °C for the 1 bar situation. For high operating pressures, the heat removal by latent heat is relatively low, since most water is in the liquid form. In addition, a significant cooling duty is required for the inter cooler.

5. Discussion

From Fig. 12 it seems logical to conclude that working at low P and high T is most attractive. A compressor is not required, and heat rejection to the surroundings is minimised. However, to make such a system work, huge amounts of water have to be evaporated in the humidifier, requiring large amounts of heat. The water condenser has to make up for the water used in the humidifier and therefore has to withdraw large quantities of heat from the humid cathode exit flow. In the calculations used to produce Fig. 12, the heat sources and sinks are coupled. This is however, not realistic at high T, low P combinations, as can be deduced from Fig. 13. Here, the sum of the duties of the air cooler after the compressor, the humidifier, the condenser and the external

heat exchanger (e.g. car radiator) is presented. From Fig. 13, it can be concluded that working at high T and low P will lead to huge heat duties (hence sizes) for the heat exchanging components. In combination with the large wet gas volumes (Fig. 7), there will be a large effect on system size.

To resume the essentials of all the above, an analysis is made of six system cases, each in the range of 40–50 kW (gross) power (Table 1). The systems are operating at 65, 80 °C and 1, 1.5 and 2.5 bar based on single cell performances given in Table 1. It is assumed that the systems all operate at a current density of 0.6 A/cm^2. This assumption implies that the systems use the same amount of reactants (H$_2$: 30 Nm^3/h, air: 145 Nm^3/h). In the table, a new term is introduced, the "relative pressure drop factor" derived from Eq. (2): $\rho_{T,p}(V_{T,p})^2 / \{\rho_{80\,°C,2.5\,bar}(V_{80\,°C,2.5\,bar})^2\}$. Here, $\rho_{x\,°C,y\,bar}$ is the humid gas density at x °C/y bar, whereas $V_{x\,°C,y\,bar}$ is the wet gas volume at x °C/y bar. The reference is the 80 °C/2.5 bar case. Assuming equal stack design and system piping in all 6 sample cases, and neglecting the effect of the Reynolds number, the "relative pressure drop factor" is presented as an indicator for the effects of operating conditions on pressure drops in the system.

From Table 1, it can be concluded that the most pronounced differences in the sample systems are in the amounts of heat that have to be rearranged in the systems, and in the relative pressure drop factors. It is also concluded that working at atmospheric pressure and temperatures of 65 °C and higher is rather complicated. The 80 °C/1 bar case is the most extreme. The high wet gas flow has significant consequences for the pressure drop in the system. The heat required for humidification can be equal to the waste heat of the stack, making it difficult to humidify the gases with the stack coolant water. The oxygen concentration at the cell inlet drops to just 11%. Also, the sum of all heat exchanger duties is increasing significantly, leading to large system dimensions.

It seems that the use of a compressor is inevitable if heat rejection to the surroundings demands for operating temperatures of close to 80 °C. For many applications a different parameter setting would be much more attractive though.

Table 1
Overview of essential numbers for six sample systems

	65 °C			80 °C		
	1 bar	1.5 bar	2.5 bar	1 bar	1.5 bar	2.5 bar
Power at the rate of 600 mA/cm^2 (kW)	39.6	43.5	46.3	37.0	43.0	47.4
Compression losses (kW)	0	2.8	6.7	0	2.8	6.7
Net power (kW)	39.6	40.7	39.6	37	40.2	40.7
Wet gas flow rate (m^3/h)	241	145	80	358	184	93
Oxygen concentration[a] (%)	15.7	17.5	18.9	11.0	14.3	17.0
Duty of air cooler (kW)	Not required	0.3	4.3	Not required	Not required	3.5
Heat requirement humidifier (kW)	25.1	15.1	8.4	66.7	34.2	17.3
Heat production stack[b] (kW)	69.8	64.9	61.4	76.8	67.4	61.2
Duty condenser (kW)	6.7	Not required	Not required	44.2	15.1	Not required
Net system cooling duty (kW)	51.4	50.1	57.3	54.3	48.3	47.4
Sum of all duties[c] (kW)	101.6	80.3	74.1	187.7	116.7	82
Relative pressure drop factor[d]	16.4	3.8	0.7	41.9	6.9	1

[a] At cathode inlet.
[b] Includes heat released due to condensation of water as a result of gas volume reduction.
[c] Sum of absolute value of duties, relevant for overall system dimensions.
[d] $\rho_{T,p}(V_{T,p})^2/\{\rho_{80\,°C,2.5\,bar}(V_{80\,°C,2.5\,bar})^2\}$, Relevant for the pressure drop in the system.

Preferably, the temperature should be even higher. That is, close to or above 100 °C. The pressure should be close to atmospheric to avoid compressors. For cars, the higher operating temperature helps to alleviate the problem of heat rejection to the surroundings[1]. For stationary applications, the benefit of a higher operating temperature would be in the more versatile use of the waste heat for space heating and hot tap water.

This paper demonstrated the effect of maintaining 100% relative humidity conditions on operating conditions. The need for operation at such high relative humidity is linked to the characteristics of Nafion and similar proton conductors. It will be clear that as long as such polymers are used in the PEMFC, the problems of pressurised operation and heat and water management will not be solved. This will only be the case if a new type of proton conductor is found, allowing for operation at higher temperatures and low relative humidity. Preferably, such a polymer should still have the good properties of Nafion type polymers: cold start capabilities, toughness and good proton conductivity.

References

[1] CRC Handbook of Chemistry and Physics, 64th ed., CRC Press, Boca Raton, Fl, pp. 83–84.
[2] R.K.A.M. Mallant, et al., Electrolytic membrane, method of manufacturing it and use, Patent WO98/20063.
[3] J.M Smith, et al., Introduction to Chemical Engineering Thermodynamics, fourth ed., McGraw Hill, Singapore, 1987, ISBN 0-07-100303-7.
[4] A. V Anantaraman et al., Studies on ion-exchange membranes. Part 1. Effect of humidity on the conductivity of Nafion®, J. Electroanal. Chem. 414 (1996) 115–120.

---

[1] The problem of heat rejection is evident when examining current experimental fuel cell vehicles, such as the Daimler Chrysler Necar 4, which have radiators that are substantially larger than those found in the standard ICE powered cars.

IV. COMPONENT DEVELOPMENT

Water and Thermal Management

2003-01-0802

Leaching of Ions from Fuel Cell Vehicle Cooling System and Their Removal to Maintain Low Conductivity

Sherry A. Mueller, Byung R. Kim, James E. Anderson, Mukesh Kumar and Chendong Huang

Ford Motor Company

ABSTRACT

The deionized water/ethylene glycol coolant used in the Ford Focus Fuel Cell Vehicle (FCV) requires very low conductivity (< 5 μS/cm) to avoid current leakage and short circuiting, presenting a unique water chemistry issue. The coolant's initially low conductivity increases as: 1) ions are released from system materials through leaching, degradation and/or corrosion, and 2) organic acids are produced by ethylene glycol degradation. Estimating the leaching potential of these ions is necessary for design and operation of fuel cell vehicles. An on-board mixed-bed, ion exchange resin filter is used to maintain low conductivity by removing leached or produced ions.

Various candidate materials were evaluated for leaching potential by exposing them to coolant at the design operating temperature for several months and periodically analyzing the coolant for ions. There was a wide range of leaching potential among the materials evaluated, which led to careful selection of materials for the cooling system. Ethylene glycol was also found to slowly degrade to organic acids, which are dissociated to organic ions, and significantly affect the conductivity of the coolant.

The ion exchange capacity of several anion and cation exchange resins were also assessed under typical system operating conditions to determine their ion exchange capacity.

Recommendations were made for cooling system manufacture, including a system wash prior to installation of the ion exchange resin. In addition, the size and frequency of replacement of the on-board ion exchange resin filter was determined. A conservative estimate of the expected useful life of the on-board ion exchange resin filter is 60 days. The expected life can be extended to as much as 180 days by an initial rinse of the system.

INTRODUCTION

The Ford Focus Fuel Cell Vehicle (Figure 1), introduced at the 2002 New York International Auto Show, is a hybrid proton-exchange membrane (PEM) fuel cell vehicle. Several of these vehicles are scheduled to be built in 2002 and will be delivered to the California Fuel Cell Partnership where they will be part of a three-year demonstration project to assess long-term viability of this technology.

Figure 1. Ford Focus Hybrid Electric - Fuel Cell Vehicle

The FCV coolant must meet strict conductivity requirements (targeted at 5 μS/cm) to avoid current leakage and short-circuiting because the coolant is not electrically insulated from the fuel cell stack. Therefore, conductivity management is of utmost concern for the continued operation of the fuel cell.

Traditional internal combustion engine coolant formulations primarily consist of water and ethylene glycol. Although water and ethylene glycol themselves do not conduct an electrical current very well, these formulations will not meet this conductivity requirement. Additives, such as corrosion inhibitors, pH adjustors, and dyes, are ionic in nature, thereby increasing coolant conductivity. Therefore, the coolant selected for the Focus FCV is a mixture of deionized water and ethylene glycol with no additives. However, this only ensures initially low coolant conductivity. Ethylene glycol decomposes in the presence of oxygen to form acidic by-products, such as glycolic, glyoxylic, and formic acids[1]. The presence of these by-products not only increases the conductivity, but also can significantly accelerate corrosion processes in the coolant loop due to the lack of corrosion inhibitors and pH buffers.

Further, the rate of ethylene glycol decomposition increases with increasing temperature.

In addition, plastic, aluminum, and rubber components used in the cooling system add ions (organic and inorganic) into the coolant through leaching, degradation and/or corrosion processes, increasing conductivity.

An on-board ion exchange resin filter, containing both anion and cation exchange resins, has been used to remove unwanted ions by exchanging them with H^+ or OH^-.

In this paper, the results of the ion leaching and ion exchange experiments are described in terms of leaching potential of candidate materials, an initial washing procedure to remove easily leachable ions from the cooling system is recommended, and the size and replacement frequency for the on-board ion exchange resin filter are determined.

EXPERIMENTAL PROCEDURE

LEACHING OF CANDIDATE MATERIALS

Materials Evaluated.

Candidate rubber, plastic, and aluminum materials were evaluated for ion leaching potential in two-phase laboratory experiments (Table 1). Some candidate materials were selected due to use in other automotive applications, while others were chosen for their chemical resistance and desired physical properties.

Material Class	Candidate Materials Evaluated
Plastic	Polypropylene
	Fluorosilicone
	Glass-Reinforced Nylon
	Polyurethane
Metal	Aluminum 3003, 4043, and 7072
Rubber	Ethylene Propylene Diene Monomer (EPDM), Sulfur and Peroxide Cured
	Silicone

Table 1. Candidate Materials Evaluated for Ion Leaching Potential

Experimental Procedure:

Phase 1 (Initial Materials Rinse): Rubber and plastic samples were cryogenically ground into pieces <1 mm in diameter and aluminum samples (approximately 1 mm thick) were cut into approximately 1-cm^2 coupons. Weighed samples of materials of this size fraction were placed in glass vials containing a 60/40 mixture of water and ethylene glycol (WEG), and mixed continuously at system operating temperature (80 ^{o}C) for 24 hours to remove easily leachable ions. The initial rinse solution was removed (without removing sample materials) and the conductivity was measured. The initial conductivity of the 60/40 WEG is approximately 0.8 μS/cm.

Phase 2 (Continued Ion Release): Fresh WEG was added to the vials and mixing continued for up to six months at the above temperature. The increased conductivity of the aqueous solution was periodically measured. The conductivity was also expressed as an equivalent concentration of Na^+ and Cl^-. This allowed a direct comparison of results between leaching and ion exchange experiments.

DEVELOPMENT OF SYSTEM RINSE PROCEDURE

Weighed samples of a selected Ethylene Propylene Diene Monomer (EPDM) rubber (the candidate material with the highest leaching potential) were added to vials along with either deionized water (DIW) or WEG. Vials were mixed continuously at 25 ^{o}C or 80 ^{o}C and the conductivity of the solution was periodically monitored. After 24 hours, the increase in conductivity slowed considerably and the experiment was terminated. This data was used to develop a recommended rinse procedure for the system prior to installation of the ion exchange resin filter.

DEGRADATION OF ETHYLENE GLYCOL

60/40 WEG was continuously mixed at 80 ^{o}C for sixteen months. The solution was periodically sampled for conductivity and pH analysis. The formation of organic acid by-products due to degradation was quantified by ion chromatography[1].

ION EXCHANGE CAPACITY EXPERIMENTS

Several anion and several cation exchange resins were selected and evaluated for ion exchange capacity in either DIW or WEG using a semi-batch technique[2]. Resins were pre-conditioned through alternating washings with acidic and alkaline solutions and stored in deionized water prior to use[2].

First, weighed resins were placed in glass vials containing DIW or WEG with varying concentrations of NaCl and equilibrated for two hours on a shaker (determined to be an adequate length of time for ion exchange reactions to occur). After equilibration, the loss of Na^+ (in the case of cation exchange resins) or Cl^- (in the case of anion exchange resins) from solution was quantified by ion chromatography and the loss from solution was attributed to ion exchange onto the resin. Then, fresh NaCl solution was added and the process was repeated until no additional ion exchange was observed. From this data, the ion exchange capacity in milliequivalents per gram of resin (meq/g) was calculated.

RESULTS AND DISCUSSION

LEACHING

Table 2 shows the observed leaching rate, defined as meq of NaCl/g-day, for the candidate with the highest leaching potential in each material class. Due to the nature of the leaching experiments, in which the entire material sample was continuously exposed to the coolant, leaching rates were expressed on a mass, rather than surface area, basis. These rates were used in calculating the estimated useful life of the on-board ion exchange filter, providing a conservative estimate of its replacement frequency. The ion leaching potential from the fuel cell stack was not considered in this study. The possibility of assessing this potential ion source is under investigation. However, we do not anticipate significant leaching of ions from the graphite flow field or its chemically inert seals. Coolant does not come in contact with the proton exchange membrane.

Material Class	Candidate with Highest Leaching Potential	Observed Leaching Rates (meq NaCl/g-day)
Plastic	Glass Reinforced Nylon	$(1.8 \pm 0.21) \times 10^{-4}$
Metal	Aluminum	$(1.4 \pm 0.1) \times 10^{-4}$
Rubber	EPDM	$(4.3 \pm 1.4) \times 10^{-4}$

Table 2. Maximum Rate of Ion Leaching from Candidate Materials

During the course of carrying out materials leaching experiments, we observed that a significant portion of the easily leachable ions were released within the first 24 hours (Figure 2), followed by a slower, continuous release with time (either from leaching, degradation, or corrosion processes, depending on the material class). This provided an opportunity to increase the life of the filter by removal of these initially released ions with an initial materials rinse.

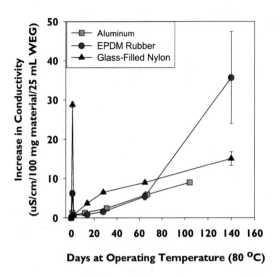

Figure 2. Maximum Observed Leaching/Production of Ions from System Materials

DEGRADATION OF ETHYLENE GLYCOL

Ethylene glycol degraded to organic acids at 80 °C (Figure 3). The production of glycolic, glyoxylic, and formic acids will likely decrease the expected life of the ion exchange filter due to exchange of these organic ions onto the resin. This reduction in capacity was not included in our calculations, but is expected to be significant after approximately 250 days at 80 °C.

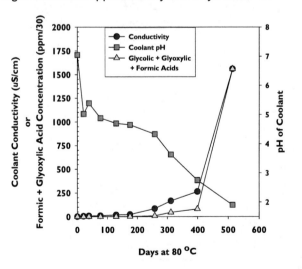

Figure 3. Ethylene Glycol Degradation Produces Organic Acids

ION EXCHANGE CAPACITY

Ion exchange capacity experiments were conducted to determine the effect of low ionic strength, temperature, and ethylene glycol on the ion exchange capacity of several resins.

Preliminary results indicate that there is little effect of low ionic strength on the ion exchange capacity. The addition of ethylene glycol to the systems also did not significantly affect the ion exchange capacity.

The life of the resin filter, which contains 500 mL (total) of cation and anion exchange resins, was estimated to be approximately 60 to 180 days using the data experimentally obtained on ion leaching potential and ion exchange capacities (2 meq/g of wet resin for both anionic and cationic resins) as well as the conservative assumptions about materials used and leaching rate. The ethylene glycol/deionized water mixture will have to be replaced along with the filter. The effect of degradation products of ethylene glycol on the life of the filter may not be important because appreciable amounts of degradation were not found to occur before the filter and the coolant are to be replaced.

DEVELOPMENT OF A SYSTEM PRE-RINSE PROCEDURE

As mentioned earlier, a significant amount of ions leaches from materials in the first 24 hours of contact with coolant. Removal of these easily leachable ions by a cooling system flush, prior to installation of the on-board ion exchange filter, would significantly extend the useful life of the ion exchange filter.

Figure 4 shows the effectiveness of DIW or WEG at two temperatures on the removal of easily leachable ions and the life expectancy of the resin filter with and without initial rinsing.

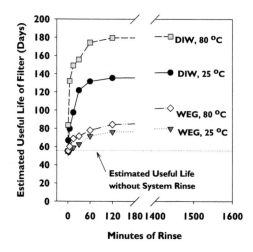

Figure 4. Initial System Rinse Increases Life of Ion Exchange Filter

The following observations can be made from Figure 4.

1. The estimated useful life of the resin filter is conservatively estimated at 60 days without an initial system rinse. This can be increased to up to 180 days with an initial wash of the system with hot deionized water.

2. Pure deionized water removes about three times as many ions as WEG.

3. An 80 °C rinse removes about two times as many ions as a 25 °C rinse.

4. A 60-minute rinse removes most of the easily leachable ions. A 15-minute rinse removes about half as many ions as the 60-minute rinse.

Based on the above findings, a rinse of the cooling system, preferably with hot (at least 80 °C) deionized water, was recommended before installing the resin filter and filling the system with the WEG coolant.

CONCLUSIONS

The ion leaching potential of various candidate materials, e.g., plastic, aluminum, and rubber, were evaluated under varying operating conditions and appropriate materials were selected for use in the cooling system. Recommendations were made for cooling system manufacture, including a system wash prior to installation of the ion exchange resin. In addition, the size and frequency of replacement of the on-board ion exchange resin filter was determined. A conservative estimate of the expected useful life of the on-board ion exchange resin filter is 60 days. The expected life can be extended to as much as 180 days by an initial rinse of the system.

REFERENCES

1. Rossiter, Jr., W., Brown, P. and Godette, M. "The Determination of Acidic Degradation Products in Aqueous Ethylene Glycol and Propylene Glycol Solutions Using Ion Chromatography", Solar Energy Materials, 9:267-279, 1983.

2. Helfferich, F. Ion Exchange. McGraw-Hill Book Company, Inc. 1962.

CONTACT

Sherry A. Mueller, Research Scientist, Ford Research Laboratory, Physical and Environmental Sciences Department, PO Box 2053, MD 3083/SRL, Dearborn, MI 48121, email: smuell12@ford.com

DEFINITIONS, ACRONYMS, ABBREVIATIONS

DIW: Deionized Water

EPDM: Ethylene Propylene Diene Monomer

FCV: Fuel Cell Vehicle

PEM: Proton Exchange Membrane

WEG: 60/40 Water/Ethylene Glycol mixture

2001-01-0537

Fuel Cell Stack Water and Thermal Management: Impact of Variable System Power Operation

P. Badrinarayanan, S. Ramaswamy, A. Eggert and R. M. Moore
Institute of Transportation Studies, UC Davis

ABSTRACT

This paper deals with the analysis of a hydrogen-air fuel cell system based on a Proton Exchange Membrane (PEM) fuel cell stack. The goal of the analysis is to understand the impact of stack water and thermal management on the system both during steady state and dynamic operations. The stack level study is done in terms of liquid and water vapor flows and distribution via a detailed stack water management model. An analysis of the stack and the system level implications of varying the anode saturation temperature is performed. It is shown that increasing the anode saturation temperature potentially enhances stack performance but need not improve system performance.

INTRODUCTION

Proper water and thermal management of a fuel cell system is critical in achieving high overall efficiency and maintaining water self-sufficiency. For fuel cell automotive applications, it becomes important to understand water management especially in a dynamic operating scenario.

When trying to optimize a fuel cell system, we have to take into account the operation of all its components. It so happens that operating conditions that favor one component need not favor the others. For example, high-pressure operation on the cathode side might favor stack performance and overall system water management (Eggert [1]) but might be a substantial load on the compressor (Friedman [2]). On the other hand, high stoichiometry on the cathode side might benefit stack performance but increase the parasitic loads of the compressor and system water management (Friedman [2] and Badrinarayanan [3]). If the fuel cell system is going to be operated at a single power level, it might not be all that tricky to arrive at an operating strategy that takes into account the various components. However, for a case such as an automobile that encounters varying power levels, it becomes increasingly difficult to work out an optimal operating strategy in terms of pressure, stoichiometry, humidification levels etc. One of the several factors that add to this increasing difficulty is the water transport inside the stack. Proper water transport is required to keep the membrane humidified and to keep the resistance under reasonable levels. The water transport properties vary with varying power levels even for a given set of anode and cathode operating conditions. This makes it imperative to have a comprehensive understanding of the water transport processes inside the fuel cell before designing an optimal operating strategy.

A number of factors influence the water transport inside the fuel cell. Many of them (e.g., flow rates, and pressure) cannot be adjusted without affecting the performance of other components. One of the parameters on the anode side that plays an important role in the water transport process is the anode saturation temperature. This is the temperature at which the anode stream is saturated with water prior to its entry into the stack. This temperature can be adjusted relatively independently. Our preliminary analysis shows that an increase in the anode saturation temperature might enhance the stack performance but can potentially increase the parasitic load on the fuel cell system.

This study has been done for a <u>load following direct hydrogen fuel cell system</u>. Similar studies are underway for indirect hydrocarbon fuel cell systems at UC-Davis. A qualitative description of the water management characteristics for variable power operation of indirect hydrocarbons fuel cell systems is done towards the end of this paper.

This analysis has been performed by modeling the water transport in the fuel cell along the same lines as the fuel cell model developed by Springer et al. [4].

The first section includes a brief description of the modeling philosophy and the capabilities of the model. The second section does an analysis of water transport inside the fuel cell. The analysis is done primarily to study the effects of varying the anode saturation temperature. The third section reviews the system level implications of varying the anode saturation temperature. Finally, stack level water requirements are analyzed for the FUDS and the US06 cycles.

THE MODEL

This model is being developed as part of the overall water and thermal management (WTM) model for the UC Davis Fuel Cell Vehicle Modeling Program. The primary objective of this modeling effort is to determine the water and thermal management challenges facing the direct-hydrogen fuel cell system. A schematic of the fuel cell system on which <u>our</u> model is based is illustrated in Figure 1.

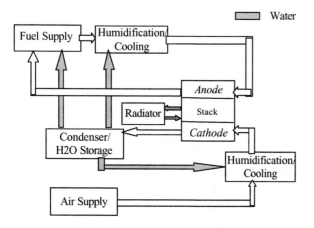

Figure 1: General Schematic of a Direct Hydrogen Fuel Cell System

The model description of the components pertinent to this analysis is given below.

STACK MODEL: The analysis done in this paper is based on an isothermal, steady-state, <u>one-dimensional</u> water transport model that is built along the same lines as that developed by Springer et al. [4]. This model will be referred to as the water transport (WT) model in the text. The important points about the model are mentioned below:

- The modeling here includes the transport of water and reactant species across the gas diffusion layer and the membrane. (The catalyst is assumed to be a thin plane and the transport across the catalyst is not done for this analysis).

- The only voltage losses taken into account in this analysis are the resistance losses in the membrane and losses due to cathode overpotential.

- The impacts of cathode flooding are not considered. (Cathode flooding is a concern and has the potential to constrain the operating characteristics of the fuel cell system. The

authors are aware of this and will draw caveats wherever appropriate)

- The effects of the anode – cathode pressure differences on the water transport have not been modeled.

The primary inputs into the stack model include the operating current density, the anode and cathode operating conditions and the cell temperature. The model outputs the water profile across the fuel cell in terms of the mole fraction of water, the net water dragged across the membrane, the membrane resistance and the cell voltage.

RADIATOR AND CONDENSER: A radiator has been modeled to maintain the stack temperature at around 80 centigrade. The radiator with an area of 0.5 m^2 employs a variable speed fan. The radiator model is based on a lookup table that uses empirical data generated from a standard brazed aluminum, single-pass radiator with 33 tubes. The air-air condenser that aids in water recovery is based on a cross flow heat exchanger and has an area of .33 m^2. The condenser heat transfer coefficient on the cathode side is calculated using an equation for filmwise condensation in vertical tubes. (Chato [5]) The condenser heat transfer coefficient on the air side is calculated from Kroger and is strongly affected by the air mass flow rate.(Kroger [6])

The important parameters used in this analysis are shown in Table 1.

Parameter	Value
Number of Cells	440
Cell Area	750 sq.cm
Anode and Cathode Thickness	.0365cm
Membrane Thickness	.0175cm
Cell Temperature	80 C
Anode Stoic	3
Anode pressure	3atm
System Type	Load Following
Compressor Type	Twin Screw (Variable Pressure)
Anode and Cathode Inlet Relative Humidity	100%

Table1: Modeling Parameters

ANALYSIS OF WATER TRANSPORT

The analysis detailed in this section has been based on the water transport model explained in the previous section. The primary parameter whose impact is studied in this report is the anode saturation temperature. This is the temperature to which the anode inlet stream is humidified (to saturation). It must be understood here that a higher anode saturation temperature means that more water can be carried by the anode stream into the stack.

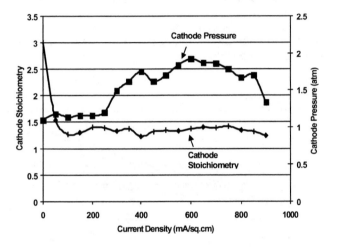

Figure 2: An Optimized Stack-Compressor Operating Control Scheme

The operating conditions, i.e. the flow rate and the operating pressures (for the cathode side) have been derived from work done earlier (Friedman [2]). These inputs arise from an optimization of the stack and air supply system and represent the operating strategy for a well-humidified stack. The water and thermal management parasitic loads on the system have not been taken into account in the above-mentioned optimization. A typical optimized stack compressor operating control scheme (for a twin screw compressor) is shown in Figure 2$^{\psi}$. These characteristic curves will change with different air supply technologies. (See Cunningham et al. [7])

As mentioned before, our existing water management strategy relies on upstream humidification of the anode flow stream. The anode pressure, flow rate and the humidification temperature directly influence the water profile in the fuel cell. In this preliminary study, only the influence of the anode saturation temperature is considered in detail. This is because our initial analyses indicate that this parameter substantially influences the

---

$^{\psi}$ It can be noticed that the curves are not exactly monotonic. This is a result of an optimization process that involves non-linear components and their interactions. Explaining these curves is beyond the scope of this paper.

performance of the fuel cell. However, it must be mentioned that the other anode side parameters such as anode pressure and stoichiometry also affect the performance of the fuel cell. An increase in anode stoichiometry can potentially increase the performance of the fuel cell (Springer [4]). It is important to take all these anode parameters into account while trying to devise optimum anode operating conditions.

WATER DRAG: It is important to understand water drag in a fuel cell and its significance before this analysis is detailed. There are basically two components to water drag - "electro-osmotic" drag and "back diffusion". In simple words, when the hydrogen ions travel across the membrane from the anode to the cathode, they tend to drag some water molecules with them. This phenomenon is known as "electro-osmotic" drag. Water is produced at the cathode and some water tends to accumulate at the cathode from the electro-osmotic drag. As the concentration of water becomes higher on the cathode side than on the anode side, there tends to be some "back-diffusion" of water towards the anode side. The result of the above-mentioned competing effects is the net water dragged. In general, water drag refers to the "net water drag". The mathematics behind modeling water drag can be found in Springer et al. [4]

In this analysis, water drag will be represented by the ratio "alpha".

Alpha = Moles of water dragged per mole of hydrogen utilized

Therefore, the net amount of water dragged across the membrane will be a product of the hydrogen utilized and alpha.

$$nh2O\_drag = nH2\_utilized * alpha$$

Where nh2O_drag is the net water dragged and nH2_utilized is the hydrogen utilized at the cell.

What is the significance of water drag? The amount of water dragged can play an important role in the performance of the fuel cell. The water that is dragged across the membrane helps in keeping the membrane humidified. A membrane that is not well humidified can have high resistance leading to substantial voltage losses in the fuel cell. The total amount of water dragged adds to the total amount of water produced at the cathode layer. So, a large water drag brings up the concern of "flooding" which can substantially bring down the performance of the fuel cell.

ANALYSIS: The WT model was run to simulate the performance of the fuel cell for varying anode saturation temperatures. One of points that become evident from the analysis is that the net water drag across the membrane increases with an increase in anode saturation temperature.

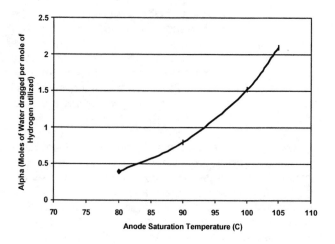

Figure 3: Variation of Water Drag with Anode Saturation Temperature (For Current density =0.5A/sq.cm)

In Figure 3, "alpha" (the number of moles of water dragged per mole of hydrogen utilized) has been plotted for different anode saturation temperatures and a given current density of .5 A/sq.cm. It can be seen that "alpha" increases from around 0.4 for 80 C to almost 2.3 for 105 C.

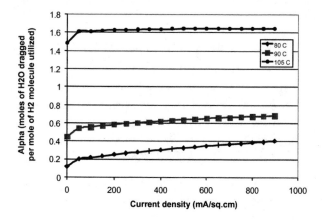

Figure 4: Variation of Water Drag with Current Density for Different Anode Saturation Temperatures

There are several system and stack level implications of this change in net water drag. Figure 4 highlights the variation in water drag with current density for different anode saturation temperatures. The anode stoichiometry and anode pressure have been maintained constant at 3 and 3 atm respectively. The characteristics of Figure 4 show that for a given anode saturation temperature alpha increases with an increase in current density. It can also be noticed that alpha increases with an increase in anode saturation temperature. When the membrane resistance is plotted versus current density (see Figure 5) for different anode saturation temperatures it can be observed that the membrane resistance increases with an increase in current. The resistance is lowest for the case of high anode saturation temperature. This can primarily be attributed to the fact

that the membrane is well humidified due to the high alphas at higher saturation temperatures, which in turn reduce the membrane resistance. (Generally, membrane resistance decreases with humidification)

A polarization plot that takes into account the resistance losses and the cathode overpotential losses is shown in Figure 6. It can be seen that the change in voltage at low current densities is almost negligible. However, at higher current densities higher anode saturation temperatures yield higher voltages leading to enhanced stack performance. Another useful way of viewing the polarization plot in Figure 6 is illustrated in Figure 7 where the stack power has been plotted against the current density. One caveat here is that at high anode saturation temperatures (such as 105 C) where alpha is very high, cathode flooding is a potential concern.

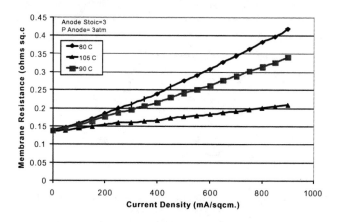

Figure 5: Variation of Membrane Resistance with Current Density for different Anode Saturation Temperatures

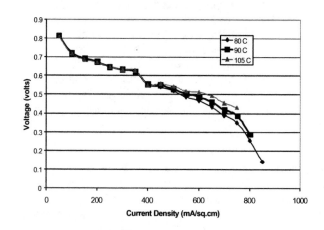

Figure 6: Polarization Curves for Different Anode Saturation temperatures[¥]

[¥] It can be noticed that these curves are not smooth. This is primarily because of the nature of the optimized stack-compressor operating control scheme detailed earlier. A similar effect can also be seen in the plots that follow.

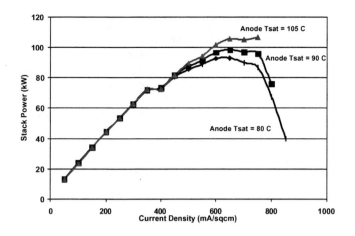

Figure 7: Power- Current Density Curve for the Stack

SYSTEM LEVEL IMPLICATIONS OF INCREASING ANODE SATURATION TEMPERATURE

It is important to bear in mind that the enhancement in stack performance mentioned in the previous section comes at a price. Both the anode and cathode streams are humidified to saturation prior to their entry into the stack. Higher the anode saturation temperature, more amount of water vapor is required to achieve complete saturation. As water vapor is not readily available, liquid water (recovered from the cathode exhaust) is vaporized to meet this water requirement. In this modeling exercise, it is assumed that there is an electric heater on board that supplies the heat to vaporize the water. Another possible method of supplying the heat energy required for humidification would be to employ a hydrogen burner. This method offers the potential of being more efficient as the heat energy from the hydrogen can be used directly and is spared of the "inefficiency" of the fuel cell. Our preliminary analysis indicates that such a burner might bring down the humidification loads but does not change the trends that are being illustrated in this section.

The energy required by the electric heater is a parasitic load on this direct-hydrogen fuel cell system. When we compare an anode saturation temperature of 80 C as opposed to 105 C, the parasitic load due to humidification is higher for 105 C as the amount of water that needs to be vaporized is higher.

To get an understanding of the system level impact of humidifying the anode stream to high temperatures, the stack power-current density curve is corrected for this parasitic load and is shown in Figure 8. The parasitic load is assumed to be the energy needed to vaporize the water (The efficiency of the heater is assumed be 100%). It is important to add that the power-current density curve has been corrected for the heater parasitic load only and not the other loads such as the compressor, radiator and condenser. A hydrogen

recirculation scheme is assumed on the anode side. As a result of that, the water needed for humidification at any operating point is assumed to be the water dragged for that operating point. The purpose of this exercise is to find out the "best" anode saturation temperature, if any, from a sub-system perspective.

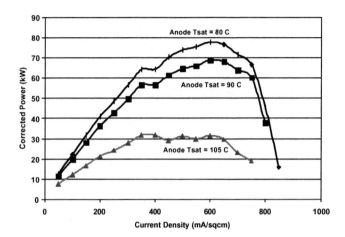

Figure 8: Power- Current Density Curve Corrected for Humidification Loads Alone [*] [¶]

It can be seen here that in Figure 8, the curve that corresponds to a higher anode saturation temperature of 105 C gives the lowest power for a given current density. One might be tempted to conclude that higher anode saturation temperatures are not good from a system efficiency point of view. This, however, would only be part of the overall effect on the fuel cell system. There are three more important effects to understand here

- The first effect is that higher anode saturation temperatures result in higher stack voltages at a given current and hence, the amount of heat that has to be dissipated due to the inefficiency of the stack is lower. This potentially reduces the radiator load.

- Where is the water needed for humidification obtained? The water required by the system is condensed at the cathode exhaust that is rich in water. Some water gets condensed inside the stack itself and the remaining amount of water needed is condensed at the condenser. The condensation that occurs at the stack increases the radiator load, as the heat of condensation of the water has to be rejected by the stack. This water recovery imposes an additional load on the system.

---

[*] The air supply loads are not taken into account in these plots and hence "ups-and-downs" in the curves can be observed.

[¶] The curves shown in the plot will potentially come "closer" if a hydrogen burner is employed instead of an electric heater.

- The radiator loads can possibly increase because of the changes in the gas phase enthalpies when anode inlet streams with temperatures higher than the stack temperature enter the stack. However, our studies indicate that this effect is not significant compared to the other two effects.

Figure 9 illustrates the power - current density curves corrected for humidification, radiator and condenser loads. The characteristics of Figure 9 suggest that at higher anode saturation temperatures, the gain due to the improved stack performance is not able to compensate the increased humidification and radiator loads. So, for the given set of system parameters and for the assumed configurations of the radiator and condenser, increasing the anode humidification temperature does not enhance the system performance.

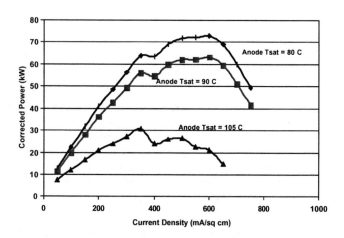

Figure 9: Power - Current Density Curve Corrected for Humidification, Radiator and Condenser loads

REFORMATE FUEL CELL SYSTEMS: The above analysis has been performed for a direct hydrogen fuel cell system. The case for a reformate fuel cell system can be different. A reformate fuel cell system employs a fuel processor that processes a hydrocarbon fuel (such as methanol or gasoline) on-board the vehicle to a hydrogen rich stream. The reformate usually exits the fuel processor at a temperature much higher than the stack operating temperature. This heat of the reformate can be used to humidify the stream. (Eggert et al [1]) The extent to which the reformate can be humidified by using its own heat will be dictated by the nature of the fuel processor.

If the fuel cell system can do away with the humidification load, then a higher anode saturation temperature may potentially lead to enhanced system performance. This prediction will have to be verified by modeling or experimentation in order to fully understand the various interactions that take place in the system.

WATER TRANSPORT OVER A DRIVE CYCLE

To obtain an understanding of water drag and humidification requirements in a realistic driving scenario, the water drag profile is drawn for the FUDS and the US06 cycle. Figure 10 (b) and Figure 11 (b) show the load profiles in terms of current density for a FUDS and a US06 cycle respectively The UC Davis Fuel Cell Vehicle Simulation tool was used to generate these load profiles. These profiles have been generated for a mid-size car and an 80 kW motor. The US06 cycle, being a more demanding cycle, makes the stack operate at higher current densities when compared to the FUDS cycle.

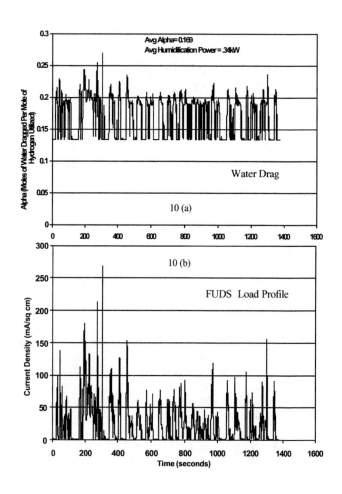

Figure 10: Load Profile and Water Drag Profile for a FUDS cycle

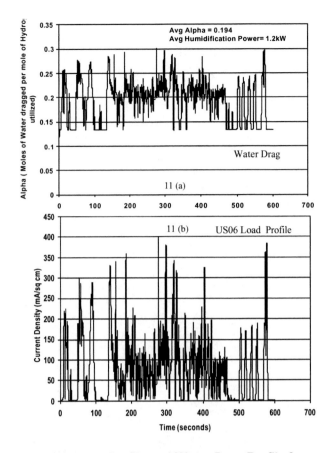

Figure 11: Load Profile and Water Drag Profile for a US06 Cycle

Alpha has been plotted for both the FUDS and the US06 cycles and is shown in Figure 10 (a) and Figure 11 (a) respectively. From the analysis detailed in the earlier sections, (see Figure 4) alpha was found to be higher for higher current operation. As the US06 cycle is relatively a higher power cycle when compared to the FUDS cycle, we can expect to see higher average alpha for the US06 cycle. The average alpha for the FUDS and the US06 cycle was found to be 0.169 and 0.194 respectively. The average alpha will also depend on the nature of the drive cycle. It might be possible that the same average current density might result in different average alphas.

The total amount of water dragged is a function of alpha and the current. As alpha increases with increasing current densities, we can expect a non-linear increase in net water dragged at higher current densities. As a result, we can also expect to see a non-linear increase in the humidification loads when the stack goes towards higher operating current densities, as the humidification load is directly proportional to the amount of water dragged across the cell. The average power required for humidifying the anode stream to 80 C was found to be 0.334kW and 1.2 kW for the FUDS and US06 cycles respectively. As discussed before, this power will increase if we increase the anode saturation temperatures.

CONCLUSIONS AND FUTURE WORK

A water transport model was developed to analyze the effect of various water and thermal management parameters. The primary loss mechanisms in the fuel cell were studied for different operating conditions. One of factors that play a big role in stack level water management is the anode saturation temperature. It is shown that increasing the anode saturation temperature has many stack and system level implications. An increase in the anode saturation temperature increases the net water drag across the membrane. This decreases the membrane resistance and hence enhances the stack performance. However, this enhancement in stack performance comes at a cost in terms of parasitic loads to achieve these high levels of humidification. It is shown that, for a direct hydrogen fuel cell system, it might not be beneficial to go towards higher anode saturation temperatures as the gain in stack performance is more than offset by the increase in the humidification, radiator and condenser loads.

It is illustrated that alpha (moles of water dragged per mole of hydrogen utilized) tends to increase for higher current densities. The important implication of this is that as current density increases, there is a non-linear increase in the power required for humidification. This effect has been highlighted by considering realistic operating scenarios, the FUDS and the US06 cycles.

The next step in our modeling exercise will be to create a "segmented two-dimensional" stack model that will help us understand the impacts of the varying concentration of reactants along the gas channels.

In the analysis outlined in this paper, an optimized stack-compressor operating control scheme was employed. While trying to develop optimum operating conditions, all the components of the fuel cell system and their interactions must be taken into consideration. Work in this regard is currently underway. (Friedman et. al. [8])

ACKNOWLEDGMENTS

The authors would like to acknowledge the work and helpful input of the members of the Fuel Cell Vehicle Modeling Program at UC-Davis including Dr. M. A. Hoffman, David Friedman, Karl Hauer, Joshua Cunningham, Meena Sundaresan, Claudia Diniz and Fernando Contadini.

CONTACT

P Badrinarayanan
Institute of Transportation Studies
University of California at Davis
One Shields Ave, Davis, CA 95616
e-mail: pbadri@ucdavis.edu
Tel: 530-754-6770

371

REFERENCES

1. Eggert et al. "Water and Thermal Management of an Indirect Methanol Fuel Cell System for Automotive Applications" Proceedings of the 2000 International Mechanical Engineering Congress and Exposition, American Society of Mechanical Engineers, Orlando, Florida

2. Friedman, D.J., et al., "PEM Fuel Cell System Optimization", Proceedings of the Second International Symposium on Proton Conducting Membrane Fuel Cells II, Edited by S. Gottesfeld et al., Electrochemical Society, Pennington, NJ, 1999

3. Badrinarayanan, P. et al, "Implications of Water and Thermal Management Parameters in the Optimization of an Indirect Methanol Fuel Cell System", 35th Intersociety Energy Conversion Engineering Conference, American Institute of Aeronautics and Astronautics, July 2000, Paper # 2000-3046

4. Springer et al. "Polymer Electrolyte Fuel Cell Model", Journal of Electrochemical Society, Vol. 138, No. 8, pp. 2334-2342, 1991

5. Chato, J.C., "Laminar Condensation Inside Horizontal and Inclined Tubes", J. ASHRAE, 4, 2, 1962

6. Kroger, D. G., "Radiator Characterization and Optimization", # 840380, Vol. 93, SAE Transactions, 1984, pp2.984 – 2.990

7. Cunningham, J. M., et al., "A Comparison of High Pressure and Low-Pressure Operation of PEM Fuel Cell Systems", Presented at the SAE 2001 World Congress, March 5-8, 2001, SAE# 2001-01-0538, SAE International, Warrendale, PA

8. Friedman, D. J., et al., "Balancing Stack Output, Air Supply, and Water and Thermal Management Demands for an Indirect Methanol Fuel Cell System", Presented at the SAE 2001 World Congress, March 5-8, 2001, SAE# 2001-01-0535, SAE International, Warrendale, PA

IV. COMPONENT DEVELOPMENT

Fuel Processors

Reprinted with Permission

ELSEVIER

Journal of Power Sources 106 (2002) 35–41

JOURNAL OF
POWER SOURCES

www.elsevier.com/locate/jpowsour

Fuel processors for fuel cell vehicles

Detlef zur Megede

XCELLSiS AG, Neue Strasse 95, D-73230 Kirchheim/Teck, Germany

Abstract

Hydrogen is the fuel of choice for today's fuel cells with polymer electrolyte. However, for its successful application in vehicles there are some open issues, for example the missing hydrogen infrastructure and difficulties with regard to vehicle storage tanks. This makes on-board fuel processors together with liquid fuels both attractive and necessary. Different types of fuel processors for fuel cell vehicles are described. Some examples together with developmental challenges are presented. © 2002 Elsevier Science B.V. All rights reserved.

Keywords: Fuel cell vehicles; Fuel processors; Fuel infrastructure

1. Introduction

Today fuel cells with a polymer electrolyte are mostly regarded as the best solution for a vehicle with a fuel cell engine. For this type of fuel cells hydrogen is the fuel of choice. Of course there are other and very promising developments like direct methanol fuel cells or fuel cells with different electrolytes, but up to now they do not play a major role in the transportation sector. The two fundamentally different ways to supply the fuel cell with hydrogen are either to store hydrogen in a high-tech tank in the vehicle or to produce hydrogen on-board by processing a liquid fuel, which is stored in a much less sophisticated tank. The article will discuss some pros and cons of the hydrogen solution first and then describe several different types of fuel processors for fuel cell vehicles by presenting some examples. An outlook on some of the technical challenges, which still need to be addressed, is given at the end of the article.

2. Some remarks about the need of fuel processors in a hydrogen world

Why should fuel processors deserve consideration? It seems to be obvious to supply fuel cells directly with the fuel they are designed for. By doing this many successful demonstration vehicles have been built since a couple of years. Fuel processors are often described as bulky and heavy, requiring long start-up times and providing no dynamics. Also, the general consensus seems to be that fuel processors reduce vehicle efficiencies, all while increasing fuel production cost. The infrastructure requirements that accompany fuel processors seem to be much more critical

than the requirements for using fuel cells and hydrogen, especially because the fuels for fuel processors are often regarded as transition fuels being on the market only for a limited period of time. Many people believe, that in some 20 years from now the hydrogen economy will be on the way and it will last for many decades.

2.1. Weight and volume

The issues around weight and volume can be discussed by looking at two well-known fuel cell vehicles, DaimlerChrysler's Necar 2 and 3 (Fig. 1). Both vehicles have a 50 kW fuel cell engine inside, but the first one uses gaseous hydrogen as fuel while the other one uses liquid methanol in conjunction with a fuel processor. Necar 2 appears to be a fully usable minivan with enough room for six passengers as well as some luggage. What has been left of the Mercedes A class in the Necar 3 is a two-seater with no trunk at all.

Taking a closer look at both vehicles, it becomes clear that Necar 2 is considerably larger than Necar 3. There is a difference in the basic design. In addition to this, Necar 2 has two hydrogen pressure vessels on its roof in a dedicated compartment, sporting a volume of more than 750 l. The fuel cell systems, disregarding the fuel storage and the fuel processor, are more or less the same for both vehicles. They are located either below the floor or under some of the seats. A direct comparison of the storage systems including the fuel processor shows that Necar 3 offers a slightly greater range than Necar 2 at a comparable volume!

Of course a storage tank for liquid hydrogen is much smaller, but it is possible to make a similar comparison of DaimlerChrysler's Necar 4 (liquid hydrogen) and Necar 5 (methanol fuel processor). The question of how to store

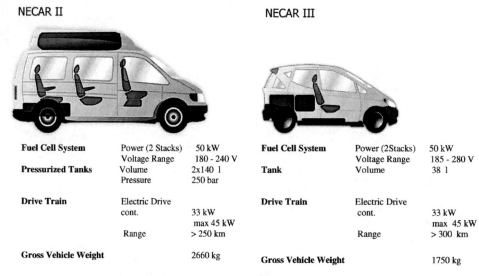

Fig. 1. DaimlerChrysler's fuel cell vehicles Necar 2 and 3 [1,2].

NECAR II

Fuel Cell System	Power (2 Stacks)	50 kW
	Voltage Range	180 - 240 V
Pressurized Tanks	Volume	2x140 l
	Pressure	250 bar
Drive Train	Electric Drive	
	cont.	33 kW
		max 45 kW
	Range	> 250 km
Gross Vehicle Weight		2660 kg

NECAR III

Fuel Cell System	Power (2Stacks)	50 kW
	Voltage Range	185 - 280 V
Tank	Volume	38 l
Drive Train	Electric Drive	
	cont.	33 kW
		max 45 kW
	Range	> 300 km
Gross Vehicle Weight		1750 kg

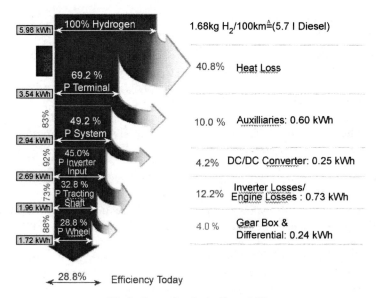

Fig. 2. Energy flow in the Necar 2 [3].

hydrogen in a vehicle still remains unanswered, and the demand for an acceptable range for fuel cell vehicles requires liquid fuels together with fuel processors. Only a new breakthrough technology would be a game changer in this situation.

2.2. Emissions

The hydrogen-fuelled vehicle is regarded as a "zero emission vehicle (ZEV)," while the methanol-fuelled vehicle "only" complies with the Californian "SULEV" standard.[1] The reason for this are very low hydrocarbon emissions of the fuel processor. CO_2 is not an issue in California. In a well-to-wheel scenario the result of this comparison might look different, especially with both fuels being produced from renewable resources (including CO_2!). In this case both fuel cell vehicles should perform very similar. Methanol is here only used as one example and the other liquid fuels can be discussed with equivalent results in a similar way.

2.3. Infrastructure cost

A comparison of the infrastructure cost shows that hydrogen fuel stations will cost at least 10 times as much as a fuel station for a new liquid fuel (e.g. methanol or ethanol). Naturally, in the case of today's already wide-spread used gasoline or diesel there would be no cost at all, which is almost also the case for a special new "fuel cell grade" gasoline.

---

[1] SULEV: 1.0 g/mile CO, 0.11 g/mile NMHC, 0.02 g/mile No_x, 0.01 g/mile PM over the FTP 75 cycle (ZEV: all 0.0 g/mile).

Its cost would be in the same order of magnitude of those for a completely new fuel, maybe lower. The cost for introducing a completely new liquid fuel can also vary greatly depending on the comfort that the fuel station is offering to its customers, as can be seen at the broad variety of gasoline fuel stations today and in the past. If opting for hydrogen fuel stations, however, immense expenses simply cannot be avoided since these fuel stations require dedicated tanks and dispensers to handle the fuel safely while operating at high pressures or very low temperatures. It would only make sense to build up a hydrogen infrastructure if it was perfectly clear that it would last for many decades. This, however, is not clear today because of safety reasons and because no suitable hydrogen storage system for vehicles exists.

2.4. Availability

Taking a look at today's availability of the above mentioned fuels shows also that liquid fuels are closer to reality than hydrogen. Besides today's gasoline and diesel large production capacities especially for methanol and in increasing amounts for ethanol are already available that are sufficient for hundreds of thousands of fuel cells vehicles. This is not the case for hydrogen, where almost no overcapacities exist and the necessary production plants would have to be built up. On the other hand, new capacities for clean synthetic gasoline and diesel are already planned and will be installed in the next years. These fuels may be used in fuel cell vehicles, too.

Of course there are many more reasons to why fuel processors are beneficial and necessary for fuel cell vehicles but that would be outside the scope of this article. Instead, we will continue by taking a closer look at some of the technical issues surrounding using fuel processors in a vehicle, starting with considering the aspect of efficiency.

3. Thermal efficiencies of fuel processors in fuel cell vehicles

The efficiency of a fuel cell vehicle is determined by many parameters. Many of them have nothing to do with the fuel cell engine itself or even with a fuel processor. An analysis of the energy flow in Necar 2 shows that the efficiency of the fuel cell system of close to 50% is lowered to 28.8% due to the drive train and its individual components (Fig. 2). The inverter and the engine cause an efficiency loss of 27%. This is typical for all fuel cell-driven vehicles and of course in a similar way also valid for conventional ICE-driven vehicles. On the other hand, this offers a quite reliable way to define efficiency by dividing the energy brought to the road by the energy leaving the storage tank in the vehicle. One should not forget, however, the big influence of vehicle components such as inverters, gear boxes, or even tires.

A realistic discussion about the efficiency of a fuel cell engine needs to start with the interfaces to the vehicle.

The definition that is used at Xcellsis is based upon interfaces to the fuel storage, the vehicle control unit, a 12 V battery, a DC delivery port for the traction motor inverter, the connections to the car radiator and to the outside air with an inlet, and an outlet to the exhaust pipe. Everything within these limits is defined as a fuel cell engine (Fig. 3). The Xcellsis efficiency values are based upon this.

Hence, a system with a fuel processor is much more complex, in particular when considering its many internal interfaces in addition. When discussing fuel processors, those interfaces become very important (Fig. 4). Because it is very easy to calculate a good efficiency for the fuel processor resulting in a bad efficiency of the fuel cell stack or vice versa. This can be continued for the fuel cell engine and its relation to the vehicle. However, this is more or less useless because the task is to optimize the vehicle efficiency as a whole and not just the efficiency of a single component by lowering the efficiency of another component. So, it does not make a lot of sense, in regards to system aspects, to talk about the efficiency of "the reformer" alone as long as this word is not used for the complete fuel processor within its well-defined limits.

Fuel processors have two important interfaces to the fuel cell system:

1. An interface to the air supply of the fuel cell to import the air for the air-consuming components such as partial oxidation or preferential oxidation reactors. The air compressor is usually not considered in the efficiency calculations for the fuel processor, but the hydrogen consumed by the air is considered since it leads to efficiency losses.
2. The second important interface is to the fuel cell that is embodied in the fuel processor. The fuel cell can be considered a hydrogen consumer with an efficiency of less than 100%, which means that some hydrogen is leaving the fuel cell again and flowing back into the fuel processor.

The second interface is now considered in more detail. The system schematic in Fig. 4 shows a fuel processor that needs thermal energy for the vaporizer and the reformer, wherefore a catalytic burner is included. This burner can be used to burn the excess hydrogen that is leaving the fuel cell. The fuel cell will always deliver some hydrogen because it cannot be operated as a dead-end unit with a reformate containing more than pure hydrogen. How much hydrogen a fuel cell can utilize depends on its internal structure (e.g. non-uniform flow distribution will result in lower utilization) and on the performance of the MEA's. Usually the utilization can be lowered but not easily increased. As long as the fuel processor needs all the hydrogen leaving the fuel cell there are no efficiency losses. But if the fuel processor is already exothermic, as it is in the case of a partial oxidation process the excess hydrogen causes efficiency losses based on the hydrogen utilization of the fuel cell stack. The fuel processor efficiency including heat losses can vary between

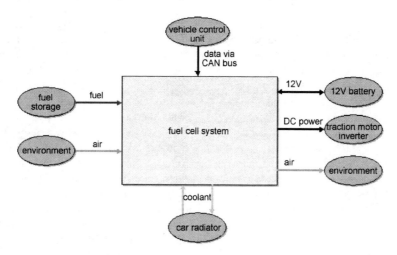

Fig. 3. Schematic view of a fuel cell engine.

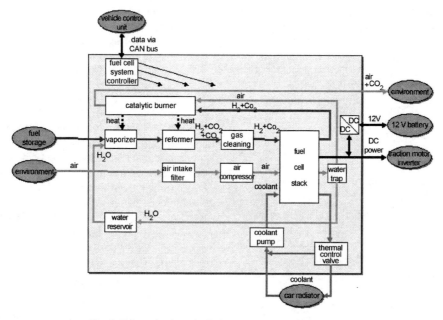

Fig. 4. Schematic view of a fuel cell system with fuel processor.

63 and 85% with the highest efficiency at 100% utilization of the fuel cell (Fig. 5).

The possible need of hydrogen by the fuel processor depends of course on the fundamental process of the hydrogen production. Depending on the amount of oxygen present in the process the fuel processor can be operated either exothermically or endothermically, varying from partial oxidation via autothermal reforming to pure steam reforming. Therefore, a fuel processor that can be operated under varying conditions is very helpful in optimizing the fuel cell engine's efficiency during the varying conditions of a drive cycle.

This situation can partly be avoided by using a hydrogen separation membrane in the fuel processor. Assuming that such a membrane is 100% selective for hydrogen as

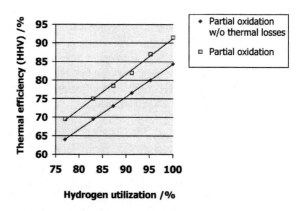

Fig. 5. Impact of hydrogen utilization on fuel processor efficiency (partial oxidation of methanol) [4].

in the case of a non-porous palladium membrane we have a pure hydrogen system on the fuel cell side. This eliminates the utilization problem with a single stroke but does not remove the need for making use of the unused reformate leaving the membrane unit on the fuel processor side. Hence, the previous discussion regarding exothermic and endothermic fuel processors comes up again.

There are other possibilities to use the excess hydrogen, including turbines, expanders, or even additional fuel cells, but they only add complexity to a system that is already regarded as very complex by many people.

As a general conclusion of the discussion about fuel processor efficiencies it can be stated that

1. fuel processors with steam reformers are the most efficient followed by those with autothermal reformers;
2. fuel processors based on pure partial oxidation offer the lowest efficiency.

4. Operating temperatures

Another aspect is the choice of the operating temperature of the fuel processor, and this is where the chemical structure of fuel plays an important role. Basically, there are two alternatives for a fuel processor:

1. it can be operated at a low temperature (250–300 °C) with all components (vaporizer, reformer, gas purification and perhaps catalytic burner) operating more or less at this temperature; or
2. the individual components operate at different temperature levels (250–800 °C) depending on their individual chemical reactions.

The first version can only be realized with fuels such as methanol or DME, which are easily processed. The second alternative is valid for all so-called "multi-fuel" systems with a high-temperature step for cracking fuels such as methane or hydrocarbons, followed by at least one shift unit, and a further gas cleaning unit, for example, preferential oxidation. Today, most high temperature fuel processors for vehicles use autothermal reforming or partial oxidation in the high temperature step.

Systems based on first alternative can be integrated to a higher degree than those based on second alternative. The Xcellsis ME 50-3 and ME 75-5 fuel processors may be regarded as examples showcasing the benefits of a high degree of integration.

5. Examples for fuel processors for vehicles

5.1. Low temperature fuel processors

The ME 50-3 fuel processor that is built into Necar 3 is a system with a low degree of integration. It has one component per function and makes you think of a small chemical plant with two boxes—one for the reformate production, containing catalytic burner, vaporizer, and steam reformer connected by a loop of thermal oil—as well as a second box with the water-cooled selective oxidation unit. It is perfectly clear why Necar 3 became a two-seater with these boxes in the trunk. Nonetheless, it was hereby possible to demonstrate that a vehicle with a fuel processor can be operated dynamically without an auxiliary battery or hydrogen storage. The dynamics were possible by means of a special vaporizer as a heat exchanger using µ-technology.

The ME 75-5 fuel processor is much more integrated, featuring components with several functions (Fig. 6):

1. vaporizer and superheater in a combined catalytic burner/vaporizer,
2. reformate production in a combined reformer and preferential oxidation reactor,
3. start-up and dynamics in a novel reactor allowing exothermic, autothermal, and endothermic operation.

This technology allows us to design a very compact fuel processor as one part of a fuel cell engine with in total four boxes that all fit together under the floor of a vehicle, as demonstrated in the Necar 5 as well as other vehicles.

5.2. High temperature fuel processors

High temperature fuel processors are necessary for fuels that are based on hydrocarbons with very stable chemical bonds. This is the case for methane with its very high symmetry and for molecules containing carbon–carbon bonds as in the case of LPG, gasoline, and diesel-type fuels. It is well known, in the chemical and petrochemical industry, that those compounds can be handled in very efficient processes that are more or less state of the art. All varieties from steam reforming to partial oxidation are possible and are also used within the corresponding surroundings of a large plant. In principle, there is always a step needed to produce a syngas consisting of hydrogen and carbon monoxide, which is then enriched in hydrogen in one or more shift stages. The hydrogen-rich reformate is then cleaned in additional steps such as preferential oxidation, membranes, or pressure swing adsorption.

The challenge enters when wanting to transfer these processes into a car. All of the above mentioned processes operate at different temperatures and sometimes also at different pressure levels. This means that there is a need for heat exchangers and maybe also for additional compressors in our system. Then, one has to deal with the possibility of soot formation, which will always happen if something gets out of control. The important parameters for a successful system operation are:

1. temperature control during start up, shut down and operation, especially during load transients;
2. control of S/C;
3. control of O/C (if necessary).

Fig. 6. ME 75-5 fuel processor.

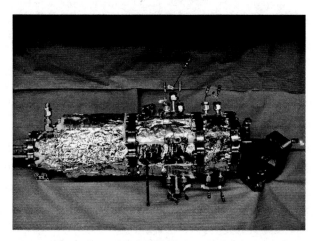

Fig. 7. Example of a gasoline reformer.

The necessity of large amounts of process water adds further complexity to the system as there are additional means needed for recovering the water and cleaning it before it can be used again in the fuel processor. Of course this is also the case for a low temperature system, but in the latter case we need less water.

Compact gasoline fuel processors can be constructed based on autothermal high temperature technology with integrated heat exchangers and shift stages (Fig. 7). Xcellsis demonstrated this technology both in 5 and 50 kW fuel cell lab systems (MF 5-1 and MF 50-1). A packaging study shows that, with our current level of knowledge, fuel processors can only be designed with an integration level similar to Necar 3. Until now we have yet to integrate such a fuel processor into a vehicle. Our simulations show that such a vehicle would need a large battery to ensure drivability. Further, it would have difficulties to compete with a very good diesel ICE (as hybrid or not) in terms of efficiency.

So, maybe high temperature fuel processors are better suited for non-mobile applications including fuel stations or

in combination with large electric batteries to offer vehicles with better efficiencies than today's gasoline ICE and lower emissions than today's diesel ICE.

6. Development challenges

Nevertheless, fuel cell vehicles featuring any kind of fuel processor still present us with challenges, which need to be met until commercial fuel cell vehicles with fuel processors will be on the streets.

1. The interfaces between the fuel processor and the peripheral components:
   - pump and flow controller-dosing accuracies (stoichiometries);
   - water trap efficiencies (contaminants, thermal shocks);
   - contaminants (e.g. lubricants, corrosion products);
   - condensates;
   - catalyst particle emissions (valve operation).
2. The interfaces between the fuel processor and the fuel cell stack:
   - CO content during cold start and mode transition;
   - temperature and humidity control of reformate;
   - high-anode stoichiometries;
   - load-variable anode and cathode stoichiometries.
3. Control issues:
   - H_2-stoichiometry control (e.g. balancing of H_2-generation and consumption);
   - optimisation of dynamic response time;
   - reduction of dynamic losses;
   - control of mode transitions.

7. Conclusions

Hence, the overall and possibly thought-provoking conclusions are as follows:

1. Although the use of hydrogen seems very attractive at first glance both in terms of the fuel cell vehicle and of infrastructure, there are good reasons for the use of a liquid fuel together with an on-board fuel processor. Today and in the foreseeable future it is an open issue when a hydrogen economy will be installed for commercial use of fuel cell vehicles.
2. A discussion about fuel processor efficiencies has to be based upon a clear definition of the interfaces to the surrounding system. The chemical technology used for the hydrogen production (steam reforming, partial oxidation or autothermal reforming) has a strong impact on the efficiency of a fuel cell engine.
3. Low temperature processes are advantageous from the technical point of view for vehicular applications. High temperature processes are also possible, but they require additional efforts in the vehicle. From today's point of view, however, they may be better suited for stationary

applications than for vehicle drive trains, but there is no clear decision.

Although the technology has been successfully demonstrated since some years, there are still several challenges to be met for all types of fuel processors in vehicle systems, but there are no insurmountable hurdles any more. Fuel processors are more than just an option for fuel cell vehicles. They seem to be essential for a successful commercialization in a mass consumer market.

References

[1] Necar II—Driving without Emissions, Daimler-Benz AG, Communication, Stuttgart, Germany, May 1996.
[2] Necar 3—A Methanol Car Hits the Road, Daimler-Benz AG, Communication, KOM 5746-0204.02/1097, Stuttgart, Germany.
[3] R. Krauss, J. Friedrich, K. E. Noreikat, in: Proceedings of the 30th ISATA, Florenz, Italy, 16–19 June 1997, pp. 195–207.
[4] M. Schüßler, Fortschr.-Ber. VDI Reihe 6 Nr.401, VDI Verlag Düsseldorf, ISBN 3-18-340106-1, 1998, p. 34.

2003-01-1366

Performance of Microlith Based Catalytic Reactors for an Isooctane Reforming System

Marco Castaldi, Maxim Lyubovsky, Rene LaPierre, William C. Pfefferle and Subir Roychoudhury

Precision Combustion, Inc.

ABSTRACT

Use of catalytically coated short contact time (SCT) design approaches for application in mass transfer controlled reactors such as Auto Thermal Reformers (ATR's) is an area of much recent interest. Precision Combustion, Inc. (PCI) has developed an efficient and compact ATR using ultra-short channel length, high cell density SCT substrates (Microlith®). PCI has also extended this Microlith technology to other fuel processor reactors that operate at lower temperatures and are not mass transfer limited. Namely, reactors for the Water Gas Shift (WGS) and Preferential Oxidation (PROX) of CO have been developed. Due to the higher surface area per unit volume of the Microlith substrate compared to conventional monoliths, size advantages have been observed for these reactions, which are more kinetically controlled. This results not only in shortened startup times and quick load following capability but also allows much smaller and lighter reactors – required attributes for automotive fuel cell applications. In this paper, experimental data on the performance of Microlith based ATR, WGSR and PROX reactors for reforming isooctane is presented. Transient and durability characteristics have also been included and compared to Department of Energy (DOE) targets.

INTRODUCTION

One of the major obstacles in widespread use of fuel cells is the lack of a hydrogen supply infrastructure, which is constrained by significant technical and economic hurdles. On-board or localized reforming of liquid fuels, such as gasoline, which have extensive supply networks, is a sensible approach until a hydrogen supply infrastructure is available. To realize this goal there have been a number of attempts to develop on-board fuel processors. Although significant progress continues to be made, major remaining hurdles need to be overcome before on-board fuel processing technology can be successfully implemented. Current fuel processors are limited by low power density, sluggish transient response and slow startup time mostly as a consequence of large size and weight. The required size and weight goals for a practical system have been outlined in the DOE PNGV targets.

Short Contact Time (SCT) reactor design approaches offer the potential for development of advanced fuel processors with a high likelihood of overcoming these barriers. Precision Combustion, Inc. (PCI), using a SCT Microlith technology has developed extremely compact, lightweight and efficient fuel processor reactors – Auto Thermal Reformers (ATR), Water Gas Shift Reactors (WGSR) and Preferential Oxidation Reactors (PROX) – with very fast transient response capability. While These reactors have been demonstrated over a range of conditions and fuels. In this paper we report the performance of these reactors under conditions suggesting consecutive operation, i.e. same H_2O/C and O_2/C ratio with the objective of demonstrating feasibility of a compact and fast transient response integrated reformer system.

SHORT CONTACT TIME APPROACH

Short contact time approach to chemical reactor design essentially consists of passing a reactant mixture over a catalyst at very high flow velocities, such that the residence time of the gas mixture inside the catalyst bed is on the order of milliseconds). Such SCT processes have commercially been used for a long time, for example in ammonia oxidation reaction in nitric acid production, where a mixture of ammonia and air is passed over precious metal gauzes. Near 100% conversion of ammonia with very high selectivity to the desired product is achieved. PCI has been developing application of similar catalytic systems based upon wire mesh coated with precious metal catalysts for many applications [1]. In more recent years Prof. Lanny Schmidt has proposed application of this system to a range of partial oxidation reactions by using either monolith supported catalysts [2] or a single gauze of bulk precious metal catalyst [3]. A research group at Shell performed similar partial oxidation over catalyst coated on a foam substrate with high porosity and tortuosity [4].

Using coated metal screen catalytic systems (Microlith) we have designed reactors operating at very high gas hourly space velocities for both mass transfer and kinetically controlled reactions. This provides many advantages over traditional packed bed or monolith bed approaches e.g. smaller reactor size and improved selectivity over conventional substrates. For example in mass transfer limited reactions (e.g. ATR) SCT Microlith catalyst and substrate designs show higher selectivity to partial oxidation reactions, allowing operation within the material limits of commonly available materials. In kinetically controlled reactions (e.g. WGSR) near equilibrium operation at high space velocities (i.e. small reactor sizes) with lower selectivity to methanation has been observed. This is due to the higher Geometric Surface Area (GSA) per unit volume of the reactor combined with a high Specific Surface Area of the catalyst support/washcoat.

The Microlith technology consists of a series of discrete, ultra short channel length, low thermal mass metal monoliths [1]. Figure 1 shows magnified pictures of a conventional monolith and the Microlith substrates to give a conceptual understanding of its physical attributes. The effectiveness of the Microlith technology and long-term durability of PCI's proprietary catalyst coatings have been demonstrated in many applications e.g. exhaust aftertreatment [5], trace contaminant control [6, 7], catalytic combustion [8], partial oxidation of methane [9], and hydrogen peroxide decomposition. The ultra short channel length avoids the boundary layer buildup observed in conventional long channel monoliths and is justified in the following paragraph.

The breakup of the boundary layer in the Microlith substrate (due to the very small L/D ratio) enables higher heat and mass transfer coefficients than that for long channel monoliths or foam substrates, as shown in Figure 2. This Figure shows the results of a prediction of mass transfer coefficients as function of channel length to channel diameter (L/D) ratio and flow velocity for monoliths with a channel diameter of 3 mm. The mass transfer coefficients are determined by using Reynold's analogy between heat and mass transfer coefficients and correlation given for heat transfer coefficients in Shah and London [10]. Conventional monoliths have L/D ratio much greater than 10, whereas for Microlith substrate the L/D ratio is in the range of 0.1 to 0.5. The 3-D plot suggests that while increasing the flow velocity over conventional monolith substrates does provide some increase in mass transfer rate, a similar increase can be obtained over the Microlith substrate at

much lower flow velocities due to the lower L/D ratio. As an example, consider a Microlith substrate with L/D of 0.1 and monolith with L/D of 10. In order to achieve a mass transfer coefficient, K_g, of 100 cm/sec, the flow velocity would have to be much greater than 200 ft/sec. However the Microlith could operate at a K_g of 100 cm/sec at only 75 ft/sec. The data also suggests that the higher the velocity, the greater the difference between the mass transfer coefficients Microlith and Monolith substrates. Higher mass transfer coefficients also imply higher heat transfer coefficients. At a velocity of 150 ft/sec, the correlation predicts mass transfer coefficient of 228 k_g (cm/sec) for Microlith substrate (L/D of 0.1) versus 28 for monolith substrate (L/D of 10). The implication therefore is that short contact time coupled with high mass transfer coefficients are available only for low L/D substrates.

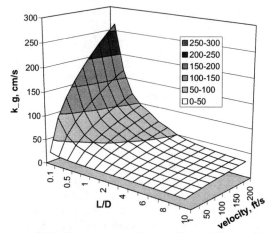

Figure 2: Plot of mass transfer coefficient (k_g) vs. length to diameter ratio (L/D) vs. velocity.[9]

Convective heat exchange with the gas phase is also strongly dependent on the boundary layer buildup. A Lumped Sum Capacitance analysis yielded time constants of 0.12 and 3.4 sec for the Microlith and ceramic monolithic substrate, respectively; which is a 30-fold improvement in thermal response [11]. A theoretical prediction of the heat up of the substrate in a non-reacting gas stream shows excellent convective heat transfer. Additionally the low thermal mass of the Microlith substrate results in rapid heat exchange with the gas, allowing equilibrium conditions to be quickly achieved. The fast startup potential of the Microlith substrate has also been seen in automotive tests [12].

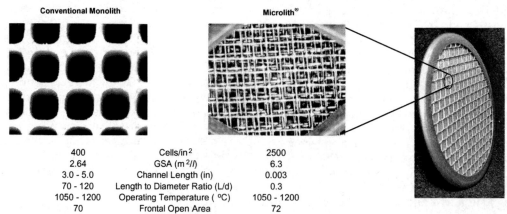

	Conventional Monolith	Microlith®
Cells/in²	400	2500
GSA (m²/l)	2.64	6.3
Channel Length (in)	3.0 - 5.0	0.003
Length to Diameter Ratio (L/d)	70 - 120	0.3
Operating Temperature (°C)	1050 - 1200	1050 - 1200
Frontal Open Area	70	72

Figure 1: Physical characteristics of conventional monolith and Microlith substrate.

384

In exothermic reactions control of reactor temperature is a primary concern. The Microlith substrate is well suited for applications where control of reactor temperature is critical. One of the issues with on-board hydrogen generation for automotive applications is the presence of transients during startup, sudden accelerations and decelerations that can lead to abrupt changes in load demands (throughput) on the reactor. The enhanced transport properties of the Microlith combined with the lower overall mass offers the potential for improved transient response. Tests have shown that Microlith bed temperatures track inlet temperatures to within 2- 3 sec.

In this paper we report the results of separate testing of the three Microlith based reactors, comprising the principal components of a fuel processor system, namely, an ATR reactor, a WGS reactor and a PROX reactor. All reactors were tested using synthesized gas mixtures in a rig setup with mass flow controllers and heaters providing specified inlet conditions. These conditions were picked for each reformer component test such that the effluent of a previous reactor corresponds to the inlet to the following reactor. We realize that some additional conditioning of the stream between the consecutive reactors may be required, such as cooling and addition of extra steam between the ATR and WGS and cooling and air addition between the WGS and PROX. Integration with balance of plant components such as pumps, heat exchangers, anode gas burners, etc. are not reviewed in this study since these are often system-specific and is an outcome of system optimization analysis. The objective of this paper is to show the advantages of using Microlith based reactors for each of the reformer components and to demonstrate feasibility of a complete reformer, which exceeds current DOE size and weight targets.

THE AUTO THERMAL REFORMER

Hydrocarbons may be converted to H_2 either through steam reforming (reaction 1) or partial oxidation (reaction 2) usually followed by a water gas shift reaction (3), which converts CO formed on the first step to additional hydrogen.

$CH_x + H_2O \rightarrow CO + (x/2+1)H_2$ $\Delta H > 0$ (1) steam reforming
$CH_x + 1/2O_2 \rightarrow CO + x/2H_2$ $\Delta H = < 0$ (2) partial oxidation
$CO + H_2O \rightarrow CO2 + H_2$ (3) water gas shift

The steam reforming reaction is endothermic and requires external heat, while traditional non-catalytic partial oxidation of liquid hydrocarbons proceeds through a complete oxidation step, such that very high temperatures are reached on the front of the reactor. For improved efficiency and control of the process it is desirable to operate a primary reformer in the autothermal regime through combined steam reforming and partial oxidation in one reactor.

Typically if an autothermal reformer is operated at a stoichiometry close to that of the ideal reforming process, with good thermal integration, high reactor efficiency can be achieved e.g. PCI's experience with autothermal reforming of methanol suggests that efficiency of up to 87% can be obtained in these processes. In addition to high efficiency the autothermal reaction can be very fast, carried out in a small volume under adiabatic conditions (without external heat) and at high space velocities (~500,000 hr^{-1}).

An autothermal reactor (ATR) based on the Microlith catalyst substrate technology was tested for reforming isooctane into hydrogen. The reactor was made of stacked Microlith screens coated with La-stabilized alumina washcoat and a precious metal based catalyst. A schematic diagram of the reactor is shown in Figure 3. The reactor diameter was ~ 4 cm and length ~1.2 cm, which corresponds to a volume of about 15 cm^3 (~1 in^3) and a mass of ~12 gms for the catalyst and substrate. The isooctane feed rate corresponded to about 3.4 kW$_t$ of thermal energy input and water and airflow were regulated to provide the specified steam to carbon and oxygen to carbon ratios. This implies a power density of 160 kW/L and 200 kW/kg. Several thermocouples and gas sampling probes were placed between the screens along the axis of the reactor to elucidate the reaction sequence. This allowed measuring axial temperature and gas composition profiles along the reactor. In this test the reactor feed comprised of isooctane (2,2,4-trimethylpentane), steam and air with H_2O/C ratio of 2.0 and O_2/C ratio of 0.5. Upstream of the reactor water was vaporized into superheated steam by an electrically heated vaporizer and mixed with pre-heated air. Fuel was injected into the stream as liquid upstream of a static mixer, where it was vaporized and mixed with the stream before entering the catalyst bed. The heat required to vaporize the water in the test would be expected to come from a thermally integrated system. Calculations of the amount of heat needed to vaporize the water and fuel indicate no external energy input would be needed for a complete processor system.

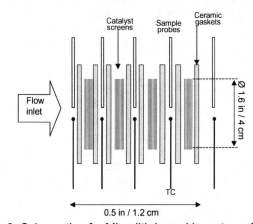

Figure 3. Schematic of a Microlith based isooctane ATR.

The catalyst was initially preheated to about 200 °C by flowing hot air through the reactor. The water vaporizer was preheated to ~200°C without water flow. Water and fuel flows were simultaneously started leading to lightoff of the catalyst bed. Thermocouple readings of the catalyst temperature during the start up process are shown in figure 4. Fuel and water flows were started at t = 0, after which the catalyst bed temperature rapidly increased to a steady state value. Lightoff of the reactor started at the front of the bed and rapidly propagated towards the back. From the figure 4 it can be seen that the start up process, from lightoff to steady state, takes less than 30 seconds, with the front of the reactor being fully operational after less than 10 seconds. Finally, based on previous data from similar reactors and the transient response of the WGS and PROX reactors, the ATR is anticipated to meet DOE targets of 5-second response time for 25% to 100% load transient.

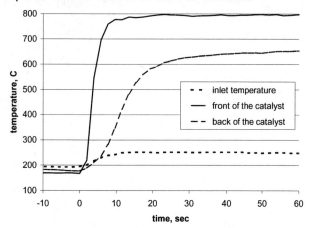

Figure 4. ATR reactor lightoff characteristics.

Results of the test of the Microlith based ATR reactor under the specified condition is shown in figure 5. The figure shows the catalyst temperature profile, hydrogen and oxygen concentrations in the reformate mixture on a dry basis and fuel conversion. These data show that complete conversion of fuel to primarily H_2 and CO was achieved at a peak catalyst temperature of ~ 850°C. In addition, the reactor achieved nearly 90% of the final, or equilibrium, hydrogen concentration within 2 mm into the reactor. Exit mixture composition on a dry basis was 34 % H_2, 7.7 % CO, 14 % CO_2, 0.5 % CH_4, 44 % N_2 and trace amounts of higher hydrocarbons. Note the low amount of methane produced in the ATR reactor under these conditions. This is an important advantage of the Microlith based catalyst, as the methanation reaction irreversibly consumes hydrogen, such that it cannot be extracted in the downstream reactors. There was no evidence of coke formation in the reactor after about 5 hours on stream.

Assuming a 1 to 1 conversion of CO into H_2 in a downstream water gas shift reactor, the thermal efficiency of the tested reactor (based on the ratio of lower heating values of product hydrogen to input fuel, as defined by the DOE fuel cell handbook [13]) was ~70 %. Note, that using this definition the maximum achievable efficiency for isooctane reformation is 90 %.

While long-term ATR durability data for the reactor described above is currently unavailable, the durability of a nearly identical reactor was tested for the partial oxidation of methane for 500 hours. No performance degradation was observed during that test which was aged at much higher temperatures, which can be considered a severe aging condition. More importantly the washcoat was adherent even in the hottest portions of the reactor [9].

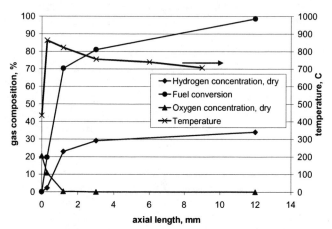

Figure 5: Temperature and Gas composition profile along the reactor.

THE WATER GAS SHIFT REACTOR

A compact and lightweight water-gas shift reactor has been developed that can be operated in both single and dual stage configurations as per system requirements. These attributes arise from the high operating space-velocities ranging from 11,000 hr^{-1} in a two-stage configuration to 40,000 hr^{-1} in a single stage. These two configurations achieved outlet CO concentrations of 1.0% to 0.5%, which then are processed in a PROX reactor for further reduction to permissible levels for PEM fuel cells, usually between 50 and 10 ppm. The formulations used for operating in these configurations, produced a WGS rate constant of 684 mol/(min-l catalyst). This allows operation at even higher space velocities of 30,000 hr^{-1} to 70,000 hr^{-1} at 280°C (subject to inlet concentrations, especially the water to CO ratio). In comparison, other reactor/catalyst combinations have shown WGS rate constants of about 17.5 mol/(min-l catalyst) at the same temperature [14].

Currently, the PCI WGSR exceeds DOE 2004 primary performance targets. Table 2 compares the goals achieved by PCI's WGSR with DOE's targets for

Table 1: Performance comparison of Microlith-based ATR vs. DOE targets.

Parameter	ATR	DOE ATR Targets
[1] Power Density (catalyst only), (kW/L)	160	>77
Weight (catalyst only), (kW/kg)	200	>67
Durability, (hrs)	underway	5000
Fuel conversion, (%)	> 99	> 99.9
Startup, (sec)	<30	<60

1: Based on thermal energy input, reactor volume/weight and efficiency.

the shift reactor. All calculations were done for simulated conditions of a 75 kW PEM fuel cell automotive application. Two different concepts were explored, a single stage unit that would operate adiabatically and a dual stage unit that would employ an inter-stage heat exchanger. This translates into a specific power density of more than 25 kW/L for a representative WGSR. Note that these configurations are considered un-optimized, yet demonstrate the potential of the Microlith substrate to be used for lightweight compact applications.

The operating condition selected for showcasing the WGSR in this paper is representative of the composition that would be delivered from many types of reformers operating on various fuels. The condition consists of about 6% CO, 30% H_2, 34% H_2O and 10% CO_2 with N_2 making up the balance with an operating temperature of about 300°C. This is usually considered a high temperature shift stage and would normally be followed by a low temperature shift stage to further reduce the CO. Yet, the Microlith-based design requires only the high temperature stage reactor since the effluent contains less than 1% CO and can safely be introduced into the PROX reactor. Two additional important constraints for reactor design are the transient response capability and durability. The transient capability manifests itself in the ability to rapidly respond to changes in load, which is particularly critical in automotive applications. In addition, durability is essential to the long term, low maintenance operation of the reactor within the system. To address these concerns and determine where improvements are needed, a promising water gas shift formulation, a precious metal on a stabilized alumina support, in terms of performance was chosen and tested. Observations and calculations indicate that the durability of the water gas shift reactor is mostly dependent on the catalyst formulation whereas the transient behavior is a combination of reactor substrate design and formulation.

A durability test, alternated between low and high temperature shift conditions, was done on the same formulation. Equilibrium CO concentrations for the high and low temperature conditions are 2.0% and 1.0% respectively, from inlet concentrations of 5.65%. This testing produced an aggregate of 215 hours of operation, 123 hours at low temperature conditions and 92 hours at high temperature conditions. In addition, these tests cycled the reactor about 5 times which can be thought of as a simulation of start-stop cycles. Each time the reactor was started for a new durability run, it immediately arrived at its steady-state operational level. The activity was very stable for the high temperature conditions, but a slight decrease in activity was measured for the low temperature conditions.

Transient response tests were done to determine the ability of the Microlith WGS reactor to follow load conditions. The concentrations and flow rates chosen simulated a nominal 25% and 100% load for a 75 kW automotive fuel cell application (Figure 6). Figure 7 is an exploded view of one cycle from Figure 6 to allow a clear idea of the time response for steady state to be achieved after a load switch. It can be seen that the system reacts immediately to the changing influent

and takes about 10 seconds to settle out and achieve steady state operation (this includes rig response time and has not been separately called out). If the transient condition is considered complete when the reactor achieves 90% of steady state, the response time is 25% faster.

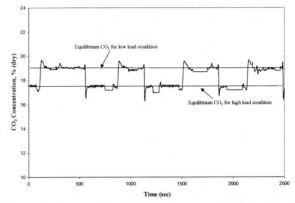

Figure 6: Composite plot showing reproducible behavior of Microlith WGSR for load following transients between 25% and 100% load simulations.

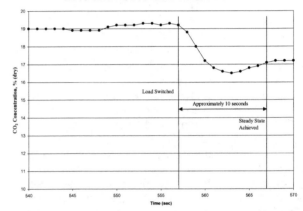

Figure 7: Expanded view of a 25% to 100% load transient. Response time to steady state is ~10 sec.

In this test, an NDIR CO_2 analyzer was coupled to the reactor effluent to measure the transient change in CO_2 concentrations when a solenoid valve switched conditions. The NDIR was used because the response time of the unit enabled multiple data points to be taken during the transients. The available CO analyzer had too long of a response time to track performance. Once the reactor achieved steady state, GC measurements were taken to confirm the CO_2 NDIR data. The plot in Figure 6 below shows CO_2 concentrations on a dry basis for a set of switches done to confirm reproducibility of the reactor system. Equilibrium CO_2 concentrations for both load conditions are shown on the plot as well. This experiment clearly demonstrates the potential for the Microlith WGSR to continually convert CO at constant levels for a given influent while responding to rapid changes that could be experienced in an automotive application.

PREFERENTIAL OXIDATION OF CO REACTOR

Following the WGS reactor is a preferential oxidation reactor (PROX) to reduce the CO further. In order to achieve the sub-10 ppm CO levels needed for the PEM fuel cell application, CO conversions of 90% or more are needed in the PROX unit. Representative values of shifted reformate were taken from equilibrium calculations to allow determination of the constraints imposed on the PROX reactor. CO conversions, in the PROX reactor, computed for a range of inlet temperatures for the shifted reformate show that thermodynamics do not favor 100% selectivity of O_2 to convert CO and reverse water gas shift reactions need to be avoided. Even at temperatures of $25^{\circ}C$, the calculations show that all the O_2 is converted mostly reacting with H_2 to form water. At higher temperatures, the reverse water-gas shift reaction occurs which increases CO concentration in the effluent. The most dominant reactions that can occur within the reactor for this type of composition are:

$$CO + \tfrac{1}{2} O_2 = CO_2 \quad (1)$$
$$CO + H_2O = CO_2 + H_2 \quad (2)$$
$$CO + 3H_2 = CH_4 + H_2O \quad (3)$$
$$H_2 + \tfrac{1}{2}O_2 = H_2O \quad (4)$$

The catalyst formulation chosen has to be selective toward CO oxidation (reaction 1) over methane and water production reactions (reactions 3 and 4), as well as be active at low temperatures where the reverse rate of reaction 2 is slow. The formulation also needs to be highly active for reaction 1, thus allowing the use of near stoichiometric amounts of O_2. Stoichiometric amounts of O_2 are desirable to keep the selectivity toward CO conversion high. Any O_2 remaining after conversion of CO will react with H_2, thus decreasing the effectiveness of the reactor. Typical formulations that can operate within these constraints are precious group metal formulations on stabilized alumina supports of which PCI has developed a proprietary formulation that achieves nearly 50% selectivity toward CO conversion.

Since the oxidation of CO and H_2 are highly exothermic (ΔH = -67 kcal/mol and -58 kcal/mol, respectively), control of reactor temperature becomes a critical issue. A calculation of the adiabatic temperature rise for a typical reformate stream exiting a water gas shift reactor with 5000 ppm CO and with 2500 ppm added O_2 was ~ 45 °C (assuming 100% selectivity to CO oxidation). The Microlith catalyst substrate is especially suited for applications where control of reactor temperature is critical. This enables the reactor to run closer to the gas phase temperature than conventional monolithic or pellet bed reactors before the onset of bulk mass transfer controlled reaction.

The durability of the PROX catalyst has been tested for 500 hours to allow reasonable extrapolation of long term operation (5000 hr) and to investigate possible deactivation mechanisms. As evidenced by Figure 8, there was no degradation observed over 500 hours. Temperature scans, shown on Figure 8 (a), were done to evaluate the performance of the catalyst with time over an operational window for catalyst surface temperature. The data falling on the same general curve indicated that the performance is not affected over the entire operating window by aging. The graph in Figure 8 (b) shows the composite 500 hours of aging at a constant surface temperature, again with no evidence of decay. These two tests were done to ensure the ability of the catalyst to perform well over a range of conditions.

To begin the elucidation of a decay mechanism, BET measurements and XRD analysis were done for the fresh catalysts and compared to the aged catalysts from the durability study, which were exposed to PROX reaction conditions for 510 hours. The results from N_2 adsorption BET measurements indicate that the specific surface area decreased by ~ 28% from the fresh catalyst measurements when compared to the final measurement after 510 hours of operation. The catalyst surface area was still considered to be relatively high. This is an indication that the γ-Al_2O_3 structure did not collapse under PROX reaction condition, remaining fairly stable during reactor operation. Both techniques BET and XRD, indicated that precious metal agglomerates and forms larger crystallites with time on stream. Despite the agglomeration process no decrease in the catalyst activity in terms of CO/CO_2 conversion and selectivity was observed after 510 hours of testing.

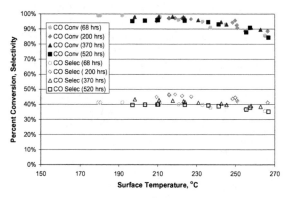

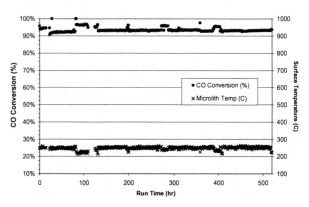

Figure 8 (a), (b): Results of 500-hour PROX test. 8 (a) on the left shows the performance versus catalyst surface temperature vs. time. 8 (b) on the right shows composite aging at a constant surface temperature.

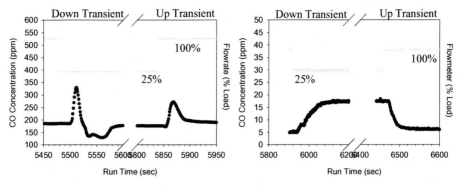

Figure 9: Transient response of PROX 1 (a) and PROX 2 (b) reactors. While there is a CO breakthrough in PROX 1 it is not seen in PROX 2.

The PROX reactor was tested to determine its performance under realistic transient operation via a simulated load change from 25% to 100%. Figure 9 shows the transient response results of the PROX reactor stages for a 4 to 1 turndown in load. The inlet and outlet conditions are given in Table 3. The outlet CO concentrations from PROX 1, Figure 9 (a), is the input concentration to PROX 2, Figure 9 (b).

It is seen that the PROX 1 reactor responds very rapidly to the change in flow conditions. The lag time between the load change and CO concentration change is test-stand limited and is attributed to the time required for the new flow conditions to reach the reactor. Once the new conditions reach the reactor, there is a small CO concentration spike of ~150 ppm, however it quickly returns to the previous steady state CO concentration, indicating that the activity of the reactor is unaffected. This behavior is indicative of O_2 limited conversion. In addition, a change in selectivity occurs with the change in space velocity. That is, the lower space velocity has a lower selectivity toward CO conversion, and thus the H_2 consumption increases slightly. The combination of the two types of behavior keep the CO conversion constant for a 4 to 1 change in space velocity.

For the PROX 2 reactor, there is a steady approach to the step changes in load. The approach to steady state occurs when the new flow conditions enter the reactor and achieve steady state operation in about 40 seconds. In addition, there is no evidence of a CO spike during this change. It can be concluded that any CO spike that occurs in PROX 1 is greatly damped, or eliminated in PROX 2. A simple calculation of conversion efficiency demonstrates this damping effect. The CO

spike in PROX 1 amounts to a maximum of 150 ppm. PROX 2 conversion efficiency being near 99%, the 150 ppm spike entering PROX 2 will be reduced to a 1.5 ppm spike, if it is not totally eliminated.

Startup performance of the WGS and PROX reactors has not been explicitly tested thus far. However, based on lightoff data from similar reactor configurations and the transient test results, it is anticipated that these Microlith reactors will achieve DOE startup targets of less than 60 seconds.

INTEGRATED ATR, WGSR, PROX REACTORS:

A conceptual fuel processor train is depicted in Figure 10, with placement of heat exchangers. The experimentally obtained data from the separate reactors presented above were combined in Table 3 for examination of an integrated prototype fuel processor arrangement. Since the reactors were tested separately, the effluent from each reactor does not exactly match the influent for the subsequent reactor. However, the critical species, such as carbon monoxide and hydrogen, were closely matched. It is recognized that balance of plant and ancillary equipment needs to be incorporated to produce a complete fuel processor and the integration of those pieces are not a trivial task. However, the technology that is needed for the balance of plant components currently exists and is assumed to be available, once defined. From the patchwork of data, the overall fuel processor size for a 75 kW system, without heat exchangers and balance of plant components is estimated to be 6 liters (~0.2 ft^3). This small size arises from the ability of the Microlith reactors to operate at very high space velocities (40,000 – 220,000 hr^{-1}).

Table 2: Performance comparison of Microlith-based WGSR and PROX vs. DOE targets.

Parameter	WGSR	PROX	2004 DOE Targets
Power Density, (kW/L)	> 25	> 30	2.25
Specific Power, (kW/Kg)	> 28	> 35	2.25
CO conversion (%)	>92 [6% to 0.5%]	99.8 [0.5% to 10ppm]	> 90
Selectivity [H_2 in product/H_2 extractable from feed]	> 99%	<3% H_2 lost	>99% & <20% H_2 lost
Durability, (hrs)	200 proven	500 proven	4000
Transient: 10-90% Power (sec)	3	3	5

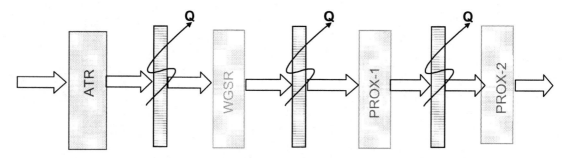

Figure 10: Schematic of integrated reactors.

Table 3: Reformate concentrations with integrated fuel processor reactors.

	ATR in	ATR out	WGS in	WGS out	PROX 1in	PROX 1out	PROX 2in	PROX 2out
O/C	1.06				(lambda=2.4)		(Lambda=8)	
H2O/C	2.01							
Sp vel (hr)-1	175,000		40,000		150,000		220,000	
N2		32.6	21.5	21.5	20.1	20.2	20.5	20.5
H2O		25.9	34.1	29.3	32.9	33.8	33.6	33.6
H2		25.1	29.1	33.9	32	31.5	31.3	31.3
CO		5.7	5.7	0.85	0.55	0.019	0.018	0.0001
CO2		10.3	9.7	14.5	13.9	14.5	14.4	14.4
O2		~0			0.66	0.05	0.16	0.13
Tin (C)	237		320		170		132	
Tout (C)		~700		340		250		150

CONCLUSIONS

The results of this paper demonstrate that the performance of the short contact time based Microlith catalytic reactors is beneficial for ATR, WGS and PROX stages of an isooctane reforming system and exceeds DOE size and transient response targets with the potential of meeting startup targets. While each reactor was tested separately, the conditions allow conceptual integration with a high likelihood of success. This provides a strong basis for achieving the DOE performance targets for an integrated fuel-reforming system capable of efficient on-board reforming of gasoline into PEM cell quality hydrogen. Challenges expected during integration of these reactors will include heat exchange for flow conditioning between the reactors, flow distribution, system startup, transient response, tolerance to poisoning, minimization of parasitic losses e.g. pressure drop, and control issues.

ACKNOWLEDGMENTS

We gratefully acknowledge the National Science Foundation's support in conducting a significant portion of this research. Any opinions, findings, and conclusions or recommendations expressed in this publication are those of the authors and do not necessarily reflect the views of the National Science Foundation.

REFERENCES

1. U.S. Patent # 5,051,241, September 24, 1991.
2. U.S. Patent # 5,648,582, July 15, 1997.
3. U.S. Patent #5,654,491, August 5, 1997.
4. U.S. Patent # 5,510,056, April 23, 1996.
5. S. Roychoudhury, J. Bianchi, G. Muench, W.C. Pfefferle SAE 971023, SAE Intl, 1997.
6. R. Carter, J. Bianchi, W. Pfefferle, S. Roychoudhury, J.L. Perry, SAE 972432, SAE Intl, 1997.
7. J.L. Perry, R.N. Carter and S. Roychoudhury, SAE 1999-01-2112, SAE International, (1999).
8. G Kraemer, T. Strickland, W.C. Pfefferle and J. Ritter, Proc., International Joint Power Generation Conference, ASME International 1997.
9. M. Lyubovsky, et al., *Catalysis Today*, Proc. of "The 5th Intl. Workshop on Cat. Comb", April 2002, *in print*.
10. Shah, R. K. and London, A. L., Laminar flow forced convection in ducts, Advances in Heat Transfer, Academic Press, New York, (1978).
11. R.N. Carter, S. Roychoudhury, W. Pfefferle, G. Muench, H. Karim. *MRS Symposium*. Proc. Vol. 454 1997 Materials Research Society.
12. R. N. Carter, P. Menacherry, W. C. Pfefferle, G. Muench, S. Roychoudhury, SAE 980672, SAE Intl, Warrendale, PA, 1998.
13. Fuel Cell Hand Book 5th ed., by EG&G Services Parsons, Inc. Contract No. DE-AM26-99FT40575 (2000).
14. Alternative Water-Gas Shift Catalysts; D. Myers, J. Krebs, T. Krause, and M. Krumpelt; Argonne National Laboratory; Annual National Laboratory R&D Meeting, DOE Fuel Cells for Transportation Program, June 2000.

CONTACT

Author for correspondence: Subir Roychoudhury;
Fax: 203 287 3700.
Email: sroychoudhury@precision-combustion.com

Efficiency, Dynamic Performance and System Interactions for a Compact Fuel Processor for Indirect Methanol Fuel Cell Vehicle

Sitaram Ramaswamy, Meena Sundaresan, Karl-Heinz Hauer, David Freidman, Robert M. Moore
Fuel Cell Vehicle Modeling Program, University of California, Davis

ABSTRACT

Fuel cell vehicles powered using Hydrogen/air fuel cells have received a lot of attention recently as possible alternatives to internal combustion engine. However, the combined problems of on-board Hydrogen storage and the lack of Hydrogen infrastructure represent major impediments to their wide scale adoption as replacements for IC engine vehicles. On board fuel processors that generate hydrogen from on-board liquid methanol (and other hydrocarbons) have been proposed as possible alternative sources of Hydrogen needed by the fuel cell.

This paper focuses on a methanol fuel processor using steam reformation of methanol to generate the Hydrogen required for the fuel cell stack. Since the steam reformation is an endothermic process the thermal energy required is supplied by a catalytic burner. Though a number of different designs for this type of fuel processor have been investigated, the general class of compact fuel processors characterized by close thermal integration of the burner and the reformer show the greatest promise for fuel cell vehicle applications.

Through the use of a model for the fuel processor, comprising of detailed models for the reformer, burner and CO cleanup units, we investigate the performance of the fuel processor from the viewpoints of efficiency and transient performance. In particular, the model can generate results in terms of a) steady state efficiencies b) response to transient step inputs c) Efficiency and dynamics in the context of its use in a fuel cell vehicle under standard driving cycles.

Our results illustrate the complex nature of the interaction between the fuel processor, the fuel cell stack and auxiliaries and the vehicle subsystem. In particular, they highlight a) the impact of fuel processor response times, b) the impact of control strategies, c) fuel processor-stack interactions, all, in the context of the ability of the vehicle to meet both the driving cycle and achieve high efficiencies. The results also suggest that steady state fuel processor efficiencies would not be very accurate predictors of fuel processor performance in an overall fuel cell vehicle.

FUEL PROCESSOR MODEL

Steam reformation for the purposes of hydrogen generation is widely used in industrial applications. The primary fuel in most cases tends to be natural gas. For fuel cell vehicle applications, methanol, being a liquid fuel has been considered as a possible candidate for use in onboard fuel processors by a number of different companies (Ref 1). Some of the research into onboard fuel processors have also focused on the use of partial oxidation for the generation of hydrogen reformate. We will not discuss the relative merits of either process in this paper but would like to highlight the main differences between steam reformation and partial oxidation. The steam reformation process is endothermic and therefore requires a heat source and the rate of heat transfer to the reaction ends up determining the transient response of the fuel processor (Ref 2). The POX reformation process however is inherently exothermic and hence offers the possibility of faster transient response. In order to address the limitation due to heat transfer, which is often cited as a reason why steam reformation may be unable to meet transient vehicle demands, a number of novel designs emphasizing close thermal integration between the reformer and the burner have been proposed. Our current study models such a design.

Figure 1 shows the schematic layout of the fuel processor system using steam reformation. Two major components of the fuel processor are the steam reformer and catalytic burner. The other components are the fuel preparation unit (preheater./vaporizer /superheater) and the CO-cleanup units. The steam reformation of methanol is an endothermic process that needs an energy source for providing the necessary energy for the process.

$$CH_3OH + H_2O => CO_2 + 3H_2 - \Delta H$$

Typically, for liquid reactants, the endothermic energy requirement is around 131 kJ/mole at 298 K.

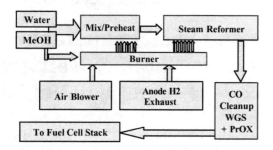

Figure 1: Schematic representation of the fuel processor for the fuel cell vehicle using the steam reformation of methanol to generate hydrogen

Designing compact steam reformers with quick transient response requires that we address the issue of thermal lag from the heat source (burner) to the heat sink (reformer). This essentially boils down to an issue of "parallelizing" mass and heat transfer. This can be accomplished by using what we term as a "bi-catalyst" plate type reformer configuration (Fig. 2). In this configuration, one side of a metal plate is coated with the reformer catalyst and the other side with the burner catalyst. The heat transfer occurs across the narrow width of the plate (and the catalyst substrate). This approach has been explored in practice by a number of researchers (Ref 3).

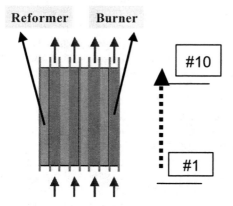

Figure 2: Representation of a "bi-catalyst" plate type configuration of showing the close reformer/burner thermal integration. One side of each plate is coated with the reformer catalyst and the other side with the burner catalyst.

The catalytic burner provides the energy for the endothermic steam reforming reactions to occur. The catalytic burner in the Indirect Methanol Fuel Cell Vehicle (IMFCV) fuel processor is assumed to be a heterogeneous, platinum-loaded gamma-alumina monolith reactor for the oxidation of methanol in a one-step reaction. The burner is assumed to operate at atmospheric pressure. Kinetic rate constants are included for methanol oxidation and complete conversion of methanol to carbon dioxide and water is assumed. Equations are modeled for mass and energy

for the solid and gas phases as well as for pollutant (NOx) formation in the burner (Ref 4).

The reformer subsystem models a steam reformation process using the kinetics reported by Amphlett (Ref 6). ($CuO/ZnO/Al_2O_3$ catalyst). The steam reformer is modeled as a PFR (plug flow reactor) represented by a series of Continuous Stirred Tank Reactors (CSTRs). The choice of the number of CSTRs is dictated by a compromise between the desired accuracy and computational times. Based on testing and literature survey, a set of ten CSTRs were used to model the steam reformation process.

The catalytic burner is similarly modeled as a plug flow reactor represented by the CSTR-in-series method for the simulation. Since the burner and reformer are thermally integrated within the fuel processor, ten CSTRs are also used in the burner model. The catalyst burner model is discussed in detail elsewhere (Ref 5).

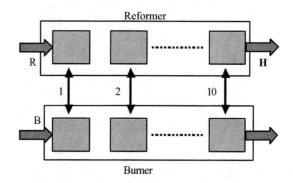

Figure 3: Illustration of the Burner Reformer integration as modeled in Simulink. "R" represents the methanol flow to the reformer; "B" represents the flow to the burner. "H" represents the hydrogen flow rate at the reformer exit. The wall is included as a part of the reformer subsystem block.

The fuel preparation units are represented by a lumped parameter thermodynamic model that accounts for the hydrogen present in the anode exhaust, the energy content of the exhaust from the burner and the additional energy needed by the burner for fuel/water vaporization and superheating.

The CO cleanup unit is modeled as a well functioning low temperature Water Gas Shift unit followed by a Preferential Oxidation unit. The CO cleanup units have an overall CO selectivity of 30 % and fast transient response.

FUEL PROCESSOR EFFICIENCY

The efficiency of the fuel processor can be defined in a number of different ways and it is important to keep in mind that the exact definition adopted becomes quite important when comparing efficiency numbers from different sources or when predicting the overall system efficiencies. In the following discussion, we will describe our approaches to defining fuel processor efficiencies both in the context of the fuel processor as a stand-alone unit and as a part of an overall fuel cell system on board an indirect methanol fuel cell vehicle. For stand-alone fuel processors, the efficiency is typically defined in terms of the hydrogen produced and the total methanol (burner + reformer) consumed.

$$\eta = \frac{H_2\_produced \times LHV_{Hydrogen}}{Total\_MeOH\_in \times LHV_{Methanol}}$$

With this definition, the maximum possible efficiency with liquid methanol and water reactants at 298 K is 94% assuming full conversion. Due to the energy lost just in the burner exhaust under real world conditions, the efficiency will definitely be lower than this figure.

Since stack efficiencies are typically stated in terms of the hydrogen consumed in the stack, it could be argued that implicit in this definition is the assumption of full hydrogen utilization in the fuel cell stack i.e. all the hydrogen produced can be used stack and converted into electric power. However, in a fuel cell system, typically, not all the hydrogen generated by the fuel processor is used up in the fuel cell stack. Therefore it makes more sense to define the efficiency of the fuel processor (when part of a fuel cell system) in terms of the actual hydrogen utilized in the fuel cell stack as shown in Figure 4.

$$\eta = \frac{H_2\_produced \times Util \times LHV_{Hydrogen}}{Total\_MeOH\_in \times LHV_{Methanol}}$$

$$Util = \frac{H_2\_produced - H_2\_anode\_exhaust}{H_2\_produced}$$

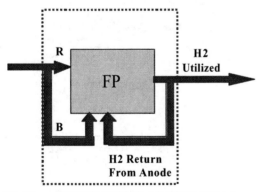

FP_Eta = (LHV H2 Utilized) / (LHV MeOH in)

Figure 4: Fuel processor efficiency defined in terms of hydrogen utilized in stack; 'R' – Reformer Methanol; 'B' – Burner Methanol feed. Also shown is Hydrogen from anode exhaust.

When fuel processor efficiency is defined in the manner shown in Figure 4, the extent of hydrogen utilization in

the stack can have a strong influence on the efficiency numbers obtained from any analysis of the fuel processor. In addition the amount of burner methanol flow 'B' needed for a given reformer methanol flow 'R' will also strongly influence the efficiency. [NOTE: In subsequent discussions, "efficiency" refers to fuel processor efficiency unless otherwise stated]

Optimization

For a given catalyst mass and reformer methanol flow, the conversion of the methanol in the reformer depends both on the operating temperature (which influences kinetics) and the availability of endothermic heat. The burner methanol flow rate determines the endothermic energy available and consequently the temperature for the reformation reaction. Since the endothermic energy requirement increases with increasing value of 'R' (Fig.4), the burner methanol flow 'B' is a strong function of 'R'. In order to investigate the exact dependence, we define a burner flow controller parameter 'B/R' and investigate its impact on the overall performance of the fuel processor.

From a purely thermodynamic point of view, one would expect the ratio 'B/R' to be nearly a constant. However, since we have a finite catalyst mass, the conversion is determined both by the amount of endothermic energy supplied and the temperature of the catalyst. This is due to the fact that catalyst kinetic activity and consequently the conversion increases as we increase the temperature of the catalyst. Since the heat flux (endothermic energy) has to travel from the burner to the reformer, a temperature gradient has to be established between the burner and the reformer and the magnitude of the temperature difference will be a function of the heat flux which itself would be a function of the operating level of the fuel processor. Higher catalyst temperatures can be achieved by having a value of 'B/R' larger than what would be dictated by the endothermic energy requirements. Figure 5 shows the effect of increasing B/R for a fixed value of input methanol flow (corresponding to a normalized power of one).

The three curves shown in the figure show i) methanol consumed in the steam reformation process, ii) CO produced and iii) the exit methanol as a function of 'B/R'. All values are normalized with respect to methanol flow rate 'R' supplied to the reformer. As expected, the conversion of MeOH to Hydrogen represented by *' in Fig. 5, initially increases with increasing 'B/R'. However, after reaching a maximum at around 'B/R' = 0.1, it starts to decrease again. The reason for this is directly tied to the problem of CO production at the higher temperatures. At higher temperatures, both the reformation reaction rate and the CO production due to methanol dissociation increase.

This causes the CO production to shoot up dramatically if the temperatures are not "optimal". Increase in CO production directly implies that smaller amounts of the incoming methanol flow actually undergoes steam reformation and therefore directly affects the rate of hydrogen production. Furthermore, (though not illustrated in Fig. 5), there is the energy penalty to be paid for CO (and unconverted Methanol) cleanup in the CO cleanup stages. It should be noted that Fig. 5 refers to the outputs at the exit of the reformer and not the exit of the fuel processor (i.e. after the CO cleanup units).

methanol flow controller that will optimize steady state efficiencies over the range of operating power.

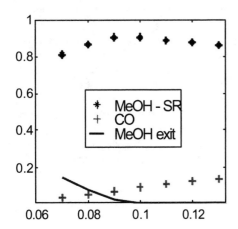

Figure 5: Comparison of methanol consumption for reformation(*) and CO pr B/R ı(+) as a function of burner controller parameter 'B/R' for a normalized power level of one. Also shown is the amount of unconverted methanol (-) at the exit of the reformer.

The discussion above points the way to optimizing the performance of the fuel processor from the viewpoint of steady state efficiency by estimating the appropriate value of **'B/R'** for different values of **'R'**. Results from such an analysis are shown in Fig. 6. The model predictions, in terms of efficiency as a function of the normalized power (ranging from 0 to 1) and **'B/R'** (ranging from 0 to 0.2) indicate that there is a complex dependence between efficiency, **'R'** and **'B/R'**.

Figure 7 presents another view of this complex dependence where each curve corresponds to a fixed value of **'B/R'**. From the viewpoint of estimating maximum efficiency, the envelop of these curves should give us the maximum possible efficiency at each operating power level. The resulting "maximum-efficiency" curve is shown in Fig. 8.

Though not indicated explicitly in Fig. 8, the above curves also reveal the dependence of **'B/R'** on the operating power from the viewpoint of optimizing efficiency under steady state conditions. Each point in the curve in Fig. 8 corresponds to a different value of **'B/R'**. By incorporating this dependence into our burner controller parameter, we can come up with a burner

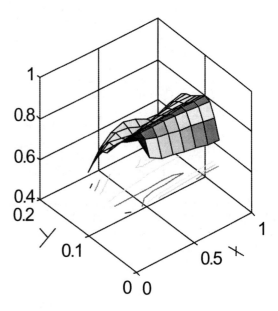

Figure 6: Fuel processor efficiency as a function of normalized operating power level (x-axis : 0-1) and controller parameter (y-axis : 'B/R'). This shows the results for the case when the hydrogen is fully utilized in the stack.

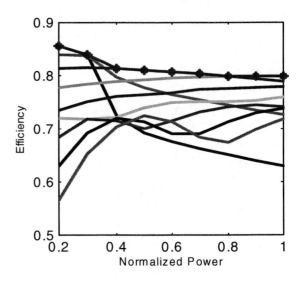

Figure 7: Efficiency vs. normalized power ; each curve corresponds to a single value of 'B/R' controller parameter. Impact of controller parameter non-linear. Also shown is a (-*-) curve representing maximum efficiency possible at each power level.

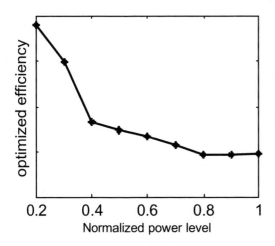

Figure 8: : **Optimized fuel processor efficiency curve vs. normalized power level obtained from the results presented in Fig 4. Efficiency range on y-axis 0.7 to 0.9**

Utilization

The results shown in the Fig. 6, 7 & 8 assume that the hydrogen utilization in the stack is 100 %. However, the actual utilization of hydrogen in the stack can vary over a wide range depending on the operating conditions. This is because for load following systems (without hybridization), the stack subsystem ends up determining how much energy it should generate to meet demand and so if there is an oversupply of hydrogen at any operating point, the hydrogen utilization value drops. This hydrogen is then available to the fuel processor burner and can be used to supply some of the endothermic heat requirements and reducing the amount of methanol supplied to the burner.

It should be noted that the highest efficiencies occur when there is full utilization of hydrogen in the fuel cell stack. At lower utilizations, even though the amount of burner methanol flow needed is reduced, the utilization of hydrogen in the burner reduces efficiency. The reason for this is directly related to our definition of fuel processor efficiency as illustrated in Fig. 3 where the efficiency is defined not in terms of hydrogen produced by the fuel processor but in terms of hydrogen utilized by the fuel cell stack, an appropriate definition since the only place the anode exhaust hydrogen is consumed is in the fuel processor (i.e. burner).

For each utilization level, it is possible to estimate the operating strategy for maximum efficiency using the approach illustrated earlier in Fig. 6, 7 & 8. Figure 9 shows the results for optimized efficiency vs. power level for stack hydrogen utilization values from 0.6 to 1.0 (in steps of 0.05). As expected, Fig. 9 reveals that the efficiency will decrease monotonically with decreasing hydrogen utilization in the stack for all power levels.

However, Fig 9. also reveals that the relationship between efficiency and utilization is not linear.

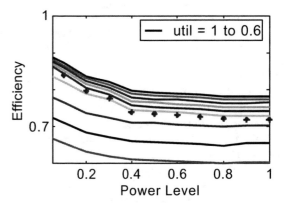

Figure 9: **Optimized fuel processor efficiency vs. normalized power level for different utilization of hydrogen in the stack. Efficiency decreases monotonically with utilization. Utilization descretized in steps of 0.05.**

The pseudo-curve defined by the set of ' * 's in Fig. 9, defines the border between the regions where the hydrogen in the anode exhaust can be fully utilized in the burner for the endothermic heat requirements of the reformer and the region where there is actually an *excess* of hydrogen in the anode exhaust stream.

As explained in the section leading up to Fig. 6, 7 and 8, there is actually a negative effect on the fuel processor hydrogen output when *excess* fuel is supplied to the burner due to CO production. In actual practice, excess fuel supply to the burner can also lead to reformer catalyst deactivation, illustrating why limiting burner fuel supply would be a reasonable strategy to follow for reasons other than simple efficiency considerations. In our fuel processor model, we effectively shunt this excess hydrogen exhaust away from the burner.

Therefore, at low utilizations, in our current model, the excess hydrogen represents a energy loss mechanism that has a significant impact on the fuel processor efficiency. It also illustrates the fact that though steam reformation based fuel processors can use the energy in the anode exhaust, the actual use of this energy stream and its contribution to the overall energy efficiency is not straightforward.

Though it could be argued that this excess hydrogen could be used elsewhere (i.e. in a turbo-expander) and thus increase the overall system efficiency, this would also increase the system complexity

Stack considerations

The results that we have seen above are primarily steady state results. Even though we have looked at the impact of anode hydrogen utilization from full (u=1) to partial (u=0.6), in actuality, it is difficult to operate the stack at high utilizations. High overall utilizations imply that portions of the stack flow channel see very low concentrations of hydrogen and this has an impact on the overall voltage produced by the stack. In addition the stack also is more vulnerable to contaminants and CO poisoning under these considerations. Typically, the stack utilizations max out roughly at around 90% utilization.

DYNAMIC OPERATION

With the help of the IMFCV model developed at UC Davis (Ref 7), which was updated to include our latest fuel processor model, we were able to test the performance of the fuel processor over different drive cycles and possible operating strategies. We shall present some of the results from our simulations below that illustrate the issues related to running the fuel processor as a part of the full fuel cell system.

The typical response of the fuel processor to a step change in the input from idle to full power is under four seconds. However, the power demand placed on the vehicle subsystem under dynamic conditions can be quite significant and require appropriate control strategies to ensure quick ramp-up of the fuel processor output. In the event that the fuel processor output cannot ramp up in time, either due to its intrinsic limitations or due to non-optimized controllers, the utilization in the stack can reach very high levels (and lead to a drop in the stack voltage to low levels).

This effect of utilization on the voltage is illustrated in Fig. 10 and it is seen that the voltage of the fuel cell stack subsystem dips to low levels at around 409 seconds. The reason for this is that there is an acceleration period in the drive-cycle where the vehicle velocity jumps from 20km/hr at t=406sec to 40km/hr at t=412sec. During this period, there is a large power demand placed on the fuel cell subsystem and has an impact on the stack hydrogen utilization values.

Since the stack is capable of responding instantly, the limiting factor becomes the fuel supply. Figure 11 shows the hydrogen utilization in the stack as a function of time for this same time period. It can be seen that the stack anode utilization reaches its highest value at around 409 seconds, corresponding to a high current draw from the fuel cell system. This corresponds to roughly the same point at which we see a dip in the stack voltage.

Since such high utilizations can have detrimental effect on the stack performance, it is important to ensure proper supply of the hydrogen reformate from the fuel processor to the fuel cell stack under these transient conditions. It should be noted that the results shown above are for the case where the fuel processor controller is fairly simple. In the example shown above, the methanol fuel flow to the fuel processor is driven (during the period shown) primarily by the stack voltage. Thus the fuel processor is unable to anticipate or

respond quickly to the stack voltage drop which, in any case, lasts for a short period of time since the motor controller immediately backs off on the current demand from the stack. This decreased current demand then causes the utilization to fall (since, as shown below, the fuel processor is just beginning to start supplying hydrogen) and the voltage to rise up again.

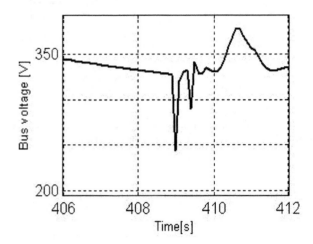

Figure 10: Bus Voltage as a function of time: snapshot from a FUDS cycle simulation.

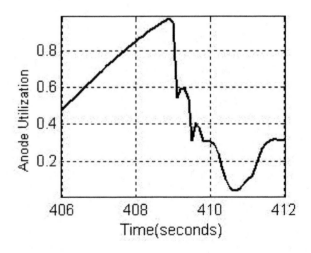

Figure 11: Stack anode utilization as a function of time. Snapshot from a FUDS driving cycle.

The corresponding hydrogen outflow and the controller driven methanol inflow associated with the fuel processor are shown in Fig. 12. It is clear that in this particular case, the lack of sufficient hydrogen supply to the stack is mostly due to the controller limitations in terms of failing to track the stack hydrogen utilization and to a lesser extent, due the transient response limitations of the processor.

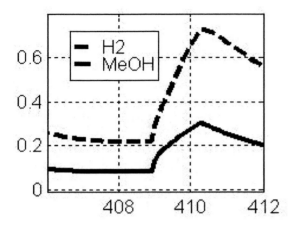

Figure 12: Normalized Hydrogen output (dotted line) and Methanol input to the fuel Processor over time (seconds)

It should be noted however, that the controller parameters used here were tailored for this specific fuel processor. If we had a fuel processor that could respond in fractions of a second, the controller methanol flow would have risen at faster rates. Alternatively, if the controller had a sense of the utilization in the stack, it's response could have be better tuned to meeting the stack fuel supply requirements. One approach to avoiding stack starvation would be to operate the fuel processor in a manner such that there would always be some excess hydrogen flow in the stack (i.e. low utilization). One extreme would be to operate the fuel processor at full power. The stack would then never be starved of fuel but, as shown in Fig. 9, this would have a drastic impact on the overall efficiency.

It might however be possible to operate the stack at utilization levels corresponding the "self-sufficiency curve" shown in Fig. 9 without significantly affecting the overall efficiency. If we do so, then at each operating power level, we will have some excess amount of hydrogen available in the anode flow to meet transient demands instantaneously (the shortfall in hydrogen available to the burner would be made up by feeding methanol to the burner). However, our attempts to fashion a control scheme by relying on this factor proved to be sub-optimal when tested for even simple drive-cycles. The reason for this is hinted at by the Fig. 13 below.

Fig. 13 shows that the amount of excess power (or hydrogen flow rate) available is proportional to the operating power level of the fuel processor. If we choose

to operate the fuel processor continuously at a power level of 75%, we would be in a position to meet all transient requirements without any lag. If we were to operate at power levels lower than 75%, say 30%, we would only be able to handle transient power spikes of around 10%.

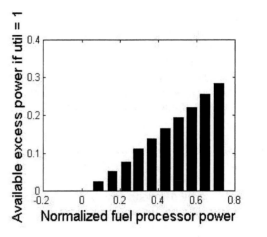

Figure 13: Impact of operating in an self-sufficient mode on the ability to meet transient requirements. Excess power available versus normalized fuel processor power.

Figure 14 illustrates the impact of adopting a variant of our initial control scheme in an attempt to explore whether it might be possible to improve on our initial attempt.

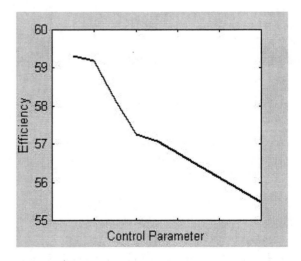

Figure 14: Drive cycle efficiency decreases when the control scheme tries harder to meet transient power demands.

Our previous control scheme was modified to use the derivative of the stack voltage as an additional input.

397

When operating the vehicle over drive-cycles, the sudden down-spikes in the voltage typically signal regions of high transients. Simply put, by quickly ramping up the fuel processor operating power level at the first sign of transient region and only slowly bringing ramping down (using the ramp down time constant as a control parameter), it was hoped that the danger of stack starvation would be somewhat minimized. Figure 14 illustrates the impact on efficiency (FUDS cycle) when adopting this control strategy. The x-axis represents the time constant associated with the ramp-down rate. As this increases, the efficiency drops monotonically since more and more of the hydrogen generated is being wasted. Though not shown here, there is some limited impact on the number of voltage spikes seen in the results but this limited improvement is offset by the large drop in efficiency.

To summarize, given that it is difficult to predict the transient power requirements of a generic drive cycle, attempts to explore the feasibility of using excess hydrogen flows in the anode as a way of meeting transient requirements proved unsuccessful.

Utilization-control

We therefore have adopted what we call an "utilization-control strategy". Instead of operating the fuel processor under constant pressure conditions, we effectively permit the fuel processor operating pressure to vary slightly about the nominal pressure via a backpressure valve. Figure 15 shows a schematic representation of the utilization based control strategy.

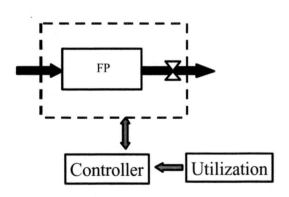

Figure 15: Schematic of utilization control implemented using fuel processor backpressure control

This makes use of the hydrogen rich reformate in the latter half of the fuel processor volume to meet some of the transient demands and this volume effectively acts as an *internal* hydrogen buffer. The controller is now modified to be driven by variation of utilization from a preset value(~85%). The control scheme also ensures that the pressure fluctuations are bounded to within 5% of fuel processor operating pressure (3atm) so as to not

unduly impact the performance of the fuel-processor. The controller actuates both the backpressure valve and determines the operating power level of the fuel processor. The variation of backpressure is shown in Fig. 16 and the corresponding variation in stack hydrogen utilization is shown in Fig. 17.

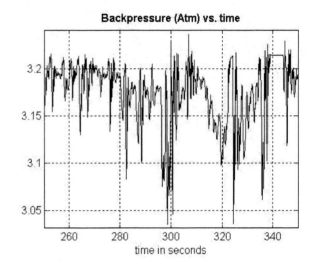

Figure 16: Fuel processor pressure variation during FUDs cycle when using backpressure control

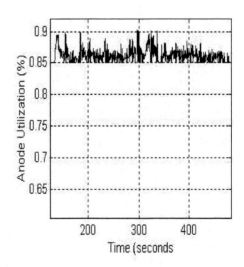

Figure 17: Anode hydrogen utilization variation over a FUDs cycle using backpressure control

The hydrogen rich reformate in the fuel processor is able to meet the immediate transient fuel requirements of the stack during the period when the fuel processor slowly ramps-up to the required power level. The pressure drop during this process is made up by operating the fuel processor for incrementally longer duration at the higher power level. This effectively replenishes the hydrogen in the fuel processor volume depleted during previous power transients. Another advantage of this approach is that some of the hydrogen that might have otherwise been wasted by the burner during down-transients is stored within the fuel processor volume provided the resulting rise in pressure does not exceed specified limits. When using this approach, the stack anode utilization hovers around 85% and never exceeds 90%. We do not explicitly address the question of the proper method to estimate the stack utilization here but assume that it is feasible to do so. It should be noted that this is one possible implementation using utilization as an input to the controller and other approaches to achieving the goal of good transient response and high efficiencies are possible and are currently under investigation.

DRIVE CYCLE RESULTS

Using the utilization based controller for the fuel processor, we were able to run the fuel cell vehicle over a number of different cycles with varying overall efficiencies. In all the simulations, the vehicle was able to meet the drive-cycle requirements to within the tolerances stated in the specifications. We shall now present the some of the results from our simulations.

The effect of transient operation on the overall efficiency of the fuel processor can be seen in Fig. 18. When compared to the steady state curves shown, both the FUDS and the USO6 cycles indicate that there is a fair amount of loss in efficiency associated with these cycles. We also see that the efficiency of USO6 cycle is worse than that of the FUDS cycle. The average power of the hard USO6 cycle is higher than that of the FUDS and the transient demands on the fuel processor are also higher.

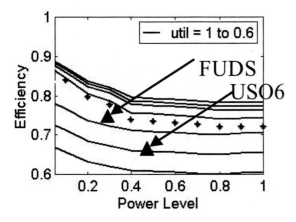

Figure 18: Comparison of the results from drive cycle tests for the FUDs and USO6 cycles. Shows impact of transient operation on efficiency.

This efficiency impact due to transient operation is further illustrated in Fig. 19 which shows the time averaged (10s) dynamic efficiency of the fuel processor over the FUDS cycle. In this case, if we use the steady state efficiency values as a function of power level to estimate the efficiency of the fuel processor, it appears that the estimated efficiency over the drive cycle would be 89% of maximum fuel processor efficiency. If however, we calculate the actual efficiency over the drive cycle and account for the transients, the efficiency value of the fuel processor drops to 82% of maximum. This illustrates the point that neglecting dynamics in the evaluation of fuel processor efficiency over drive cycles can lead to errors in estimates of fuel processor and subsequently system and vehicle efficiencies.

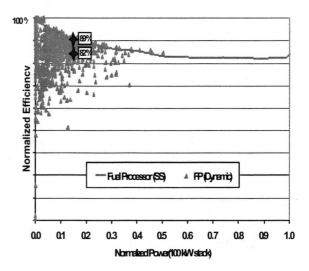

Figure 19: Impact of a FUDS drive-cycle on the fuel processor dynamic efficiency.

CONCLUSIONS

We have presented the results from our study of an Indirect Methanol Fuel Processor for a Fuel Cell Vehicle. The fuel processor efficiency and dynamic response can be optimized by selecting proper fuel processor designs emphasizing close integration between the burner and the reformer and determining the optimal operating strategy from the viewpoint of maximizing the hydrogen output. The dynamic control and operation of the reformer is closely coupled to the characteristics of the rest of the fuel cell system and in particular to characteristics of the fuel cell stack. The choice of appropriate control strategies is very crucial to ensuring that the fuel processor has the ability to meet both the demands of a driving cycle and the need for high efficiencies.

ACKNOWLEDGMENTS

This work is supported by the Fuel Cell Vehicle Modeling Project at the University of California. We would like to acknowledge the contribution of the other members of the program namely, P. Badrinarayanan, Joshua Cunningham and Anthony Eggert.

REFERENCES

1) Geyer, H.K. et. al., "Dynamic Response of Steam-Reformed, Methanol-fueled, Polymer Electrolyte Fuel Cell Systems", Proceedings of the 31st IECEC v2, p1101, 1996

2) "Fuel Processing Options for PEM Fuel Cell Systems for Mobile Applications". Dams, R.A.J. et al., 218th ACS Meeting: New Orleans - August 22-26 1999

3) "A Novel Compact Steam Reformer for Fuel Cells, With Heat Generation by Catalytic Combustion Augmented by Induction Heating", Marinus van Driel, Marianne Meijer, 1998 Fuel Cell Seminar at Palm Springs – Nov 16-19 1998

4) "Structured Catalysts and Reactors" Cybulski, Andrzej and Moulijn, Jacob A., editors, 1998, , New York: Marcel Dekker, Inc., pp. 151-2.

5) "Modeling a Catalytic Combustor for a Steam Reformer in a Methanol Fuel Cell Vehicle" Meena, S. et. al. IMECE 2000

6) "Hydrogen Production by Steam Reforming of Methanol for Polymer Electrolyte Fuel Cells", Amphlett, J.C. et. al., Journal of Hydrogen Energy, Vol.19, No.2, 131-137 (1994)

7) "Simulation Model for an Indirect Methanol Fuel Cell Vehicle" , Karl-Heinz Hauer, SAE FTT 2000, Costa Mesa.

2003-01-0806

Enhancing Hydrogen Production for Fuel Cell Vehicles by Superposition of Acoustic Fields on the Reformer: A Preliminary Study

Paul Anders Erickson
Mechanical and Aeronautical Engineering Department, University of California Davis

Vernon Roan
Mechanical and Aerospace Engineering, University of Florida

ABSTRACT

Because of recent interest in energy independence, efficiency, and environmental issues, fuel cell vehicles are seen by many to be the way of the future. As near term fuel cell vehicles will likely use the existing liquid fuel infrastructure, the efficient reformation of hydrocarbon fuels is one technological hurdle that must be addressed.

An investigation has been made into the possibility of enhancing reformation processes through superposition of an acoustic field on the catalyst bed of a methanol-steam reformer. As part of this study, background is given outlining the difficulties and liabilities of steam-reformation for transportation applications. The facility studied includes a steam-reforming reactor that has been modified to accept an acoustic field. The effect of the acoustic field was experimentally investigated with relation to fuel conversion and temperature profile.

Although the facility used has not been optimized for utilizing acoustic waves, significant acoustic enhancement of the steam-reformation process is demonstrated. It is expected that for different fuels and/or reforming methods, similar results would be obtained for comparable process constraints.

INTRODUCTION

Because of high energy conversion efficiencies and low emissions, fuel cell power plants (or "engines") are currently receiving a high degree of interest as potential replacements for the internal combustion engines in transportation applications. Because of the infrastructure and energy density issues, near-term true consumer (as opposed to limited fleet operation) fuel cell vehicles will likely utilize a liquid hydrocarbon fuel. Although smaller environmental benefits are projected for using these liquid fuels, rather than using hydrogen from a renewable resource, the use of fossil-derived liquid hydrocarbon fuels can serve as a precursor to enable the move toward a renewable energy economy.

Liquid hydrocarbon fuels need to be decomposed or "reformed" into a hydrogen-rich stream before they can be used by a Polymer Electrolyte Membrane (PEM)[1] or a Phosphoric Acid Fuel Cell (PAFC). This can be accomplished by a steam reforming, partial oxidation, or an autothermal reaction. For comparisons of the reforming methods, see Ahmed and Krumpelt [1], Brown [2], and Docter and Lamm [3]. The on-board reforming process further complicates the fuel cell vehicle and increases the cost, weight, and under-hood volume requirements of the fuel cell engine. The reforming process also penalizes the overall on-board energy conversion efficiency because of an associated increase in entropy. Furthermore, the liquid-fueled vehicle may require clean-up units to purify the reformate stream before hydrogen-rich mixture can be utilized in the fuel cell. The reforming process also may further complicate the overall vehicle system by requiring long start-up [4] and shutdown [5] periods. Steam reforming, where fuel and steam are externally heated and catalytically reacted is currently the most widespread method of reforming a hydrocarbon fuel into a hydrogen-rich gas. Although steam reforming is widely used, typical steam reformers are not designed for transportation application and thus they have large volume with corresponding high mass, slow transient response (especially at start-up), and are often characterized by having hot and cold spots throughout [6]. While it has been demonstrated that the endothermic steam-reforming process can be accomplished in a catalytic converter [7,8,9], the process has liabilities of weight, size, and complexity, in addition to long start-up and transient response times [10]. Although liquid-fueled fuel cell vehicles are

---

[1] There are some exceptions such as the direct methanol PEM fuel cell but it is still in the laboratory research phase.

demonstrated, the efficient reforming of the liquid fuel, utilizing practical on-board reformers is considered one of the major technological hurdles delaying widespread introduction of fuel cell vehicles [11,12]. Because of these liabilities any enhancement scheme for the steam-reforming process that could lead to reducing size, weight, or response time of the reformer should be investigated.

A steam-reformation process that utilizes acoustic fields in critical fluid paths is proposed. This study builds on a theoretical approach for the use of acoustic fields in conjunction with steam reforming of methanol, and demonstrates experimentally that the resulting acoustic enhancement is one approach that can reduce the above liabilities of the steam-reforming process.

LIMITING STEPS IN THE REFORMATION PROCESS

In order to design an optimal reformer for fuel cell vehicles one would desire the smallest possible, lightweight reformer that would have enough catalyst surface area to react the fuel into a hydrogen-rich stream at the maximum design flow at the end of the useful lifetime. Steam reforming of hydrocarbons can be limited by heat transfer rates, diffusion or mass transport, chemical kinetics and/or by degradation of the catalyst with time. These limitations are evidenced by restrictions in reformer capacity, slow transient response and/or limited lifetime.

Because of its endothermic nature, the reformation of liquid hydrocarbon fuels requires heat transfer to the reactants and the catalyst. In a steam reformer, heat comes from an external source and is transferred through the catalyst bed-housing wall and into the catalyst bed as shown in Figure 1. Because of this endothermic nature and irregular flow patterns associated with packed bed catalysts, the typical steam reformer is plagued with temperature gradients. Typically high temperatures are found nearest to the heating surface, and low temperatures are found toward the midpoint of the catalyst bed. Furthermore, if the heat transfer rate to the midpoint of the catalyst bed is not high enough, unreacted species may pass through the reactor at this location resulting in a limited reformer capacity. Additionally, species condensing where heat transfer is insufficient may damage the catalyst. Heat transfer can take place as convection, conduction, or radiation. Radiation is typically negligible, as the local temperature difference is relatively small. Heat conducts through the catalyst bed-housing wall and possibly into the first layer of catalyst particles coming into contact with the wall. Because of point-to-point contact between the catalyst pellets, conduction from catalyst pellet to catalyst pellet is minimal. The bulk of the heat transfer to the center of the packed bed is via convection through the fluid inside the reactor. As the explained

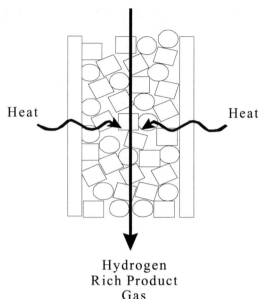

Figure 1: Simplified Packed Bed Catalyst Methanol Steam Reformer

temperature gradient manifests, heat transfer via convection is typically the limiting factor. Heat transfer limitations can be compensated for by an increased size of the reformer to increase the heat transfer area [10]. Such design results in undesirable large processors and corresponding high mass. The large size and thermal mass then penalize the process for transportation applications.

Diffusion can also limit the reformation process. The processes of transport within the catalyst bed and conversion to a hydrogen-rich gas are represented in Figure 2. As this figure shows, the hydrocarbon and steam mixture must first be transported to the surface of

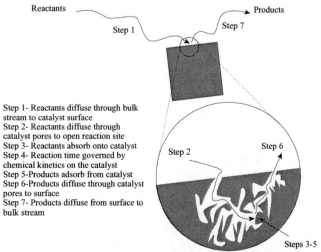

Step 1- Reactants diffuse through bulk stream to catalyst surface
Step 2- Reactants diffuse through catalyst pores to open reaction site
Step 3- Reactants absorb onto catalyst
Step 4- Reaction time governed by chemical kinetics on the catalyst
Step 5-Products adsorb from catalyst
Step 6-Products diffuse through catalyst pores to surface
Step 7- Products diffuse from surface to bulk stream

Figure 2: Steps required within the catalytic steam-reformation process.

the catalyst. Next the mixture must diffuse through the pores to an open catalyst site to react. The reaction rate on the catalyst is then governed by chemical kinetics. Products of the reaction must then diffuse back out of the catalyst pores and into the bulk flow stream. Increasing the amount of catalyst sites available can compensate for limitations in diffusion. Normally this means increasing the amount of catalyst to provide sufficient reaction area and open reaction sites closer to the surface of the catalyst, allowing for a decreased diffusion distance. This typically requires an increase in the reactor volume or conversely, a decrease of mass flow rate of the reactants. This is a decrease in what is known as "space velocity" as shown in Equation 1. This decrease of space velocity once again results in larger reforming processors or a more limited processing capacity for a set reformer volume.

$$Space\ Velocity = \frac{Volumetric\ Flow\ Rate}{\mathrm{Re}\,actor\ Volume} \quad \textbf{(Eq. 1)}$$

Considering Arrhenius type rate equations, reactions will proceed more quickly for higher temperatures. Thus, increasing the temperature can usually alleviate slow chemical kinetics. Because temperature gradients exist within the catalyst bed, temperature cannot be increased at the midpoint of the catalyst bed without possibly damaging the catalyst near the wall surfaces by overheating. Thus, restriction by chemical kinetics is normally a manifestation of convective heat transfer limitations to the center portion of the catalyst bed.

USE OF AN OSCILLATING FLOW FIELD IN CONJUNCTION WITH A STEADY FLOW COMPONENT

In several past studies (see overview by White [13]) an acoustic wave superimposed on a flow field has been shown to change the characteristics of the flow field. Periodic plane waves in a wave guide can be represented by their pressure (p) and velocity (u) components as shown in Equations 2 and 3 [14].

$$p = Ae^{j\left(2\pi ft - \frac{2\pi f}{c}x\right)} + Be^{j\left(2\pi ft + \frac{2\pi f}{c}x\right)} \quad \text{(Pa)} \quad \textbf{(Eq. 2)}$$

$$u = \frac{A}{\rho_o c}e^{j\left(2\pi ft - \frac{2\pi f}{c}x\right)} + \frac{B}{\rho_o c}e^{j\left(2\pi ft + \frac{2\pi f}{c}x\right)} \quad \text{(m/s)} \quad \textbf{(Eq. 3)}$$

In these equations x (m) is position, t (s) is time, f (Hz) is frequency, c is the speed of sound (m/s), j is the imaginary constant, ρ_o (kg/m3) is the non-oscillating density, and A and B are found from the initial and boundary conditions. When the acoustic velocity component is larger than a steady flow component, the motion of the fluid periodically changes direction depending on space and time. The use of resonance can allow for high acoustic pressures and velocities to build with the resulting standing waves, yielding a periodic change of motion although dependant on the spatial position associated with the specific mode of the standing wave.

In the development of fluid mechanics it has been noted that acoustic waves can influence several parameters of the flow field. These parameters include the boundary layer and flow characterization. The acoustically thinned boundary layer in laminar flows as well as a jump to the turbulent flow regime, due to the employment of an acoustic field has direct positive implications on heat transfer [15]. In combustion studies acoustic wave amplitude has been correlated to heat transfer [16], temperature [17], emissions [18], particle dispersion [19], mixing of species [20], and flame structure [21]. An example of the effect of acoustics on flame structure is shown in Figure 3 (Photographs by author). Frame (A) shows the flame front without flow oscillation and Frame (B) is the flame front with acoustic oscillation.

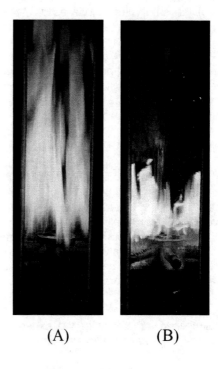

(A)   (B)

Figure 3: Propane diffusion flames showing the difference between non-oscillating (A) and oscillating (B) flow fields.

These photographs of propane diffusion flames show marked changes in the flow field due to acoustic oscillation. The compacted flame zone for the case with the oscillating flow field demonstrates enhanced mixing. The flame zone being moved downward below the burner arms manifests a change of flow field direction. These flame structures visually show the dramatic effect of an acoustic field on a chemical reaction. Some of these changes seen in past processes can be expected with implementation of an acoustic field in the steam-reforming process.

STEAM-REFORMING IN COMBINATION WITH ACOUSTICS

Superimposing an oscillating flow in the axial direction of the reformer (also referred to as organ pipe oscillations) with the steady flow component is proposed [22]. High acoustic pressures and corresponding local particle velocity perturbations are achieved by establishing resonance in the reactor. If the acoustic velocity component is high enough, the steady flow component carries the overall flow forward and the oscillating flow portion incrementally steps the fluid forward and backward across portions of the catalyst bed. This allows for a "mixing" of the boundary layer along the reactor walls and past the catalyst particles. Under the proper conditions, increased heat and mass transfer rates occur as a result of the thinned boundary layers and acoustic-induced mixing in the reactor. The heat and mass transfer in the acoustic path is thus affected. However, it is not expected that the mass transfer within the interior of the catalyst will be directly affected by the acoustic wave because of the small dimensions associated with the pore structure. With increased heat and mass transfer in the acoustic path, temperature gradients are also reduced throughout the acoustically enhanced reactor. The projected effects of acoustics on the limiting steps of heat transfer, mass transfer and chemical kinetics of the overall process are discussed as potential theory-based models by Erickson [23].

EXPERIMENTAL APPROACH AND FACILITY

The experimental approach was designed to quantify the effect of the acoustic field on the reformation process. This included adaptation of a reformer facility to accept acoustic waves, development of procedures for operation of the facility, establishment of temperature and acoustic control in the catalyst bed, the acoustical analysis of that facility, and the design of an experiment from which, a statistically valid empirical model could be developed.

Figure 4 is a simplified schematic of the experimental facility. In this system methanol:water premix is pumped through an electrically heated vaporizer. The vaporized gas is routed through an electrically heated liquid trap or secondary vaporizer. This liquid trap insures complete vaporization of the premix entering the reformer at high flow rates. The gaseous premix then may be routed to the exhaust or directed to the superheater by manipulation of the valves shown. If routed to the electrically heated superheater the premix follows the path shown to the catalyst bed. An oscillating pressure is directed through the catalyst bed from a nitrogen-purged and water-cooled acoustic driver attached at the lower end of the catalyst bed. After traveling through the electrically heated catalyst bed the reacted mixture can be routed through the condensing unit or to the exhaust by manipulation of the valves shown. After passing through the condensing unit the dry gas may be sampled through a gas sampling port as it travels to the exhaust. A carbon dioxide purge unit attaches to the superheater assembly to keep the superheater and catalyst bed free from oxygen during non-operating periods. A scale and a timing device were used to calculate the actual mass flow rate of premix fed into the reactor. Changing the volumetric flow setting on a peristaltic pump enabled rough control of premix mass flow rate. The volumetric flow setting could be set at increments of 0.1 ml/min. The actual mass flow rate was found by measuring the mass change of the reservoir per unit time.

The catalyst bed was housed in a 2.68 cm O.D. (nominal 3/4 in. schedule 40) stainless steel pipe cut to a 63.5 cm (25 in.) length. Receptacles for fittings were made at both ends of the pipe. The bed was filled with a pelletized low-temperature water-gas shift catalyst. The catalyst was designed for a temperature range of 200-300°C. This catalyst is a proprietary combination of copper oxide, zinc oxide, and aluminum oxide and had to be reduced before use. The pelletized catalyst was dimensionally 0.52 cm in length and 0.54 cm in diameter and had a bulk density of 1.2-1.3 kg/liter (the catalyst dimensions did not change throughout the experiments performed). This catalyst has been shown to perform adequately for methanol-steam reforming studies in the past [24] and was shown to allow acoustic pressure transmittance through the reactor. The catalyst bed was divided into 6 zones. At the end of each zone and before the first zone a 0.93 cm diameter (1/8 in. NPT) port was machined into one or both sides of the stainless steel pipe. Fittings could be placed into the machined holes to extract temperature or pressure information at each respective location in the bed. The fittings were sealed by a high temperature silicone sealant. Above the first zone, two ports are used for monitoring dynamic pressure and the exit temperature of the superheater. A 250-Watt band heater individually heated each zone. The temperature at the respective zone exit determined control for these heaters. Temperature was monitored using k-type stainless steel sheathed thermocouples. Each fitting was machined to be flush with the interior surface of the pipe at the zone exits. The control thermocouples were aligned as to be near the interior wall location in the fitting. These temperatures represented the interior wall temperature at the specified axial location.

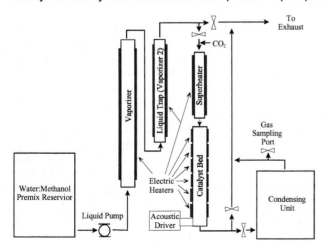

Figure 4: Simplified schematic of the experimental facility.

At the bottom of the catalyst bed an adapter was threaded into the machined receptacle to allow an acoustic driver to modulate the catalyst bed pressure. Transmission of acoustic pressure and associated velocity fluctuations are a product of geometric conditions, and also the fluid properties. Additionally, transmission is typically inhibited by obstruction in the acoustic path. In a packed catalyst bed, the catalyst itself partially obstructs the propagation of acoustic waves. Notwithstanding this difficulty, standing acoustic waves can build up within the bed as a result of wave reflection at the ends of the reactor.

An acoustic field was generated by pulsating an acoustic driver attached to an adapter at the bottom of the reactor as shown in Figure 4. The acoustic driver used was a standard horn driver typically used in amphitheater applications. The method of using loudspeakers to modulate pressure in a chemical reaction has been proven by several researchers including McManus et al. [21], Poinsot et al. [25], Heckl [26], Lang et al. [27], Zikikiout et al. [28], and many others. The horn driver was attached to mounting devices and directed into the catalyst bed as shown. As the horn driver was not designed for use in relatively high temperature chemical applications, a water-cooling jacket was designed around a fitting to eliminate conduction from the catalyst bed walls. This water-cooled section was also a heat sink for the reactor as it pulled heat away from the bottom zone. A nitrogen purge was directed to both the front of the driver diaphragm and the back of the driver housing. The nitrogen at the front of the diaphragm was used to inhibit hot reacted species from flowing into the driver. The nitrogen purge at the back of the housing was a safety precaution as there was possibility of electrical ignition in the voice coils in the presence of hydrogen. A piezoelectric pressure transducer monitored the acoustic wave. This piezoelectric transducer was factory calibrated and was mounted at the port above the first zone in the catalyst bed for the experiments performed. This position was selected so that the monitored acoustic wave would have to travel through the entire length of the catalyst bed. The mounting took place through a 1.0 cm O.D. (1/8 in. schedule 40) stainless steel pipe cut to a 12.7 cm (5 in.) length and an adapter to connect the pressure transducer.

Conversion of the methanol is defined as shown in Equation 4 where X is the conversion of the methanol fuel, and methanol is determined on a mass basis. The calculated conversion is then independent of the sample time period. From the methanol mass fraction of the condensate, the mass of unused methanol (output) is found, and the conversion can then be determined.

$$X_{CH_3OH} = \frac{CH_3OH_{Input} - CH_3OH_{Output}}{CH_3OH_{Input}} \quad \textbf{(Eq. 4)}$$

Space velocity with regards to liquid methanol is used. Liquid Hourly Space Velocity of Methanol (LHSV-M), as defined in Equation 5 is used throughout the presentation of the results.

$$LHSV - M = \left(\frac{m^3/hr \ liquid \ methanol \ input}{m^3 \ reactor \ volume} \right) \quad \textbf{(Eq. 5)}$$

The magnitude of the acoustic field was dependant on the driver power and the resonant properties of the catalyst bed. The driver was used to excite a sound pressure level of 165dB (re20μPa) at 650 Hz for the results presented.

RESULTS AND DISCUSSION

In Figure 5, the conversion of methanol is shown as a function of methanol liquid hourly space velocity (LHSV-M) for conditions with and without an acoustic field. In this figure, adjusting the flow rate in a 0.576 m length

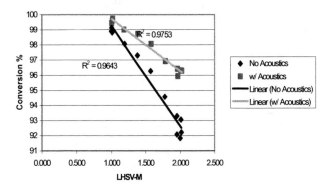

Figure 5: Percent conversion as a function of space velocity with and without the 165dB (re20μPa) acoustic field.

catalyst bed changed space velocity. Note that there is a significant increase in the conversion for those cases utilizing the acoustic field. This figure shows that one can increase the average conversion by up to 4% over the non-oscillating cases for the flow rates shown. This difference in conversion increases for higher space velocities.

Notice that in Figure 5 this increase in conversion by using an acoustic field reduces the amount of unconverted methanol by approximately 50% at 2.0 LHSV-M. At a given conversion utilizing the acoustic field can increase the allowable space velocity. For example, in Figure 5 at a 96% conversion the allowable space velocity increases from approximately 1.5 hr-1 to 2.0 hr-1 using the linear regressions. This is an increase of 33% in generating capacity or reduction in reactor volume over the non-oscillating case. While 96% fuel conversion still may be an intolerable amount of fuel loss, this demonstrates that the acoustic field can increase the capacity of the reformer. At the lower end of space velocities employing an acoustic wave actualizes average conversion increases of 0.6%. While this does not seem overly significant as a conversion increase, the amount of wasted fuel is reduced by a factor of 2.5. This increase in conversion at the lower end of space velocities becomes more important considering that certain types of fuel cells can be poisoned by unconverted methanol [29]. While the conversion increase is the perhaps the most significant output parameter found using acoustics, other benefits are also demonstrated.

Temperature profile is flattened in the catalyst bed when using acoustics as shown in Figure 6. This figure shows the average temperature at the exit of the second zone at specified radial locations and a best linear fit of that data. Note that the profile is more flat for the case with acoustics than without the enhancement method and the centerline temperature is higher. The flattened profile for acoustics is considerable when observing data points

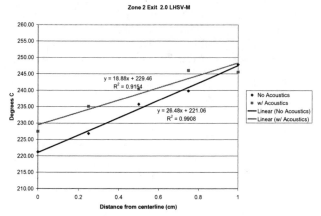

Figure 6: Average radial temperature profile at the exit of the second heating zone at a space velocity of 2.0 LHSV-M with and without the acoustic field.

collected near the wall. Increasing centerline temperature when using the acoustic enhancement method demonstrates that heat is more effectively transferred into this region. Similar results were found for other zone exits. This shows that the limiting steps of heat transfer and the associated chemical kinetics are alleviated by proper application of the acoustic field.

The acoustic field represents energy dissipated within the reactor and comes at an added cost of complexity and electrical power. Preliminary data collected at 2.0 LHSV-M show that average hydrogen flow rate increased by 82 Watts (LHV) over the non-oscillating reactor at the same space velocity. The acoustic oscillation required 9.3 Watts (VI) to the acoustic driver and increased band heater power draw by 9 Watts. This corresponds to a power benefit-cost ratio of 4.4. The power benefit-cost ratio is lessened at the lower space velocity of 1.0 LHSV-M as the conversion difference decreased, but it still remains over 1.4.

It is important to note that the acoustic driver and the reformer were not optimized for the application and thus the size and overall power required for the acoustic subsystem (including signal generation and amplifier standby power) is prohibitive for practical use in a reformer. However it has been proven that acoustic fields are a viable method of enhancing the reformer performance at a number of space velocities. Comparisons of this enhancement method to more passive methods are also warranted. Further study and optimization is necessary before fuel cell vehicles may

be able to decrease the overall fuel processing weight and power requirement using this acoustic enhancement method.

CONCLUSION

A preliminary study investigating the possibility of acoustic enhancement of steam reformation has taken place. Background to this study is given as a review of the limiting steps of the steam-reformation process, the use of acoustic fields with related processes and the experimental approach and facility. The superimposed acoustic field changes the main mechanisms of heat and mass transfer significantly. This study demonstrates that superposition of an acoustic field on the reformer increases conversion levels of methanol to hydrogen. Although further study and optimization of the reformer to work with acoustic fields is warranted, validation of the enhancement method has taken place in non-ideal conditions using a simple methanol-steam reformer and a standard acoustic driver.

ACKNOWLEDGMENTS

The authors would like to acknowledge the Ford Motor Company and Air Products and Chemicals Inc. who funded this research project.

REFERENCES

1. Ahmed, S. and M. Krumpelt, "Hydrogen from Hydrocarbon Fuels for Fuel Cells," International Journal of Hydrogen Energy, v 26 n 4, (2001)
2. Brown, L.F., "A Comparative Study of Fuels for On-board Hydrogen Production for Fuel-cell-powered Automobiles," International Journal of Hydrogen Energy, v 26 n 4, (2001)
3. Docter, A. and A. Lamm, "Gasoline Fuel Cell Systems," Journal of Power Sources, v 84 n 2, (1999)
4. Erickson, P. A., D. A. Betts, T. C. Simmons and V. P. Roan, "An Analysis of Start-up for an Operational Fuel Cell Transit Bus," SAE Transactions Journal of Commercial Vehicles, Vol 109 S2, (2001)
5. Heckwolf, M. J., P. A. Erickson, T. C. Simmons and V. P. Roan, "An Analysis of Shutdown for an Operational Fuel Cell Transit Bus," SAE Transactions Journal of Commercial Vehicles, Vol 110 S2, (2002)
6. deWild P.J. and M. J. F. M. Verhaak, "Catalytic Production of Hydrogen from Methanol," Catalysis Today, v 60 n 1, (2000)
7. Idem, R. O. and N. N. Bakhshi, "Production of Hydrogen from Methanol. 2. Experimental Studies," Industrial Engineering Chemical Research, 33, (1994)
8. Ledjeff-Hey K., V. Formanski, T. Kalk, and J. Roes, "Compact Hydrogen Production Systems for Solid Polymer Fuel Cells," Journal of Power Sources, 71, (1998)

9. Peppley, B. A., J. C. Amphlett, L. M. Kearns, R.F. Mann, and P. R. Roberge, "Hydrogen Generation for Fuel Cell Power Systems by High-Pressure Catalytic Methanol-Steam Reforming," in Proceedings of the Intersociety Energy Conversion Engineering Conference, IECEC 97093, (1997).

10. Ohl, G. L., J. L. Stein, and G. E. Smith, "Fundamental Factors in the Design of a Fast-Responding Methanol-to-Hydrogen Steam Reformer for Transportation Applications," ASME Transactions, 112 Vol 118, (1996).

11. Erickson, P. A., and V. P. Roan, "Fuel Cells for Transportation: A Review of Principles and Current Technology Status," in Proceedings of the Renewable and Advanced Energy Systems Conference, RAES99-7610, (1999).

12. Romney, G., "Fueling the Fuel Cell: Fuel Processing Sub-systems for the Real World with Texaco Fuel Processing Technology," Seventh Grove Fuel Cell Symposium, (2001).

13. White, F. M., Viscous Fluid Flow 2nd ed., McGraw-Hill: New York, (1991).

14. Kinsler, L. E., A. R. Frey, A. B. Coppens and J. V. Sanders, Fundamentals of Acoustics 3rd. Edition, John Wiley: New York, (1982).

15. Fraenkel, S. L., L. A. H. Nogueira, J. A. Carvalho-Jr., and F. S. Costa, "Heat Transfer Coefficients for Drying in Pulsating Flows," International Communications in Heat and Mass Transfer, v 25 n 4, (1998).

16. Kwon, Y., and B. Lee, "Stability of the Rijke Thermoacoustic Oscillation," Journal of the Acoustical Society of America, 78 n 4, (1985).

17. Crocco, L. "Research on Combustion Instability in Liquid Propellant Rockets," In Proceedings of the Twelfth Symposium (International) on Combustion, (1969).

18. Keller, J., and I. Hongo, "Pulse Combustion: The Mechanisms of NOX Production," Combustion and Flame, 80, (1990).

19. Barrere, M. and F. A. Williams, "Comparison of Combustion Instabilities Found in Various Types of Combustion Chambers," In Proceedings of the Twelfth Symposium (International) on Combustion, (1969).

20. Dubey, R. K., P. A. Erickson and M. Q. McQuay, "Combustion Characteristics of an Ethanol Spray-Fired Rijke-Tube Combustor in an Actively Controlled Acoustic Field," 1996 National Heat Transfer Conference, American Society of Mechanical Engineers, Heat Transfer Division, HTD v 328 n 6, (1996).

21. McManus, K., T. Poinsot, and S. Candel, "A Review of Active Control of Combustion Instabilities," Progress in Energy Combustion Science, 19, (1993).

22. Erickson, P. A. and V. P. Roan, "Enhancing the Steam Reforming Process with Acoustics: Potential Benefits for Fuel Cell Vehicles," in Proceedings of the ASME Advanced Energy Systems Division, AES-Vol. 40, HO1194, (2000).

23. Erickson, P. A., "Enhancing the Steam-Reforming Process with Acoustics: An Investigation for Fuel Cell Vehicle Applications," Ph. D. Dissertation, University of Florida, Gainesville (2002)

24. Sterchi, J. P., "The Effect of Hydrocarbon Impurities on the Methanol Steam-Reforming Process for Fuel Cell Applications," Ph. D. Dissertation University of Florida, Gainesville (2001)

25. Poinsot, T., F. Bourienne, S. Candel, E. Esposito, and W. Lang, "Suppression of Combustion Instabilities by Active Control," Journal of Propulsion, v 5 n 1, (1989).

26. Heckl, M., "Active Control of the Noise from a Rijke Tube," Journal of Sound and Vibration, v 124, n 1, (1988).

27. Lang, W., T. Poinsot, and S. Candel, "Active Control of Combustion Instabilities," Combustion and Flame, 70, (1987).

28. Zikikout, S. S. Candel, T. Poinsot, A. Trouve, and E. Esposito. "High-Frequency Combustion Oscillations Produced by Mode Selective Acoustic Excitation," In Proceedings of the 21st Symposium (International) on Combustion, (1986).

29. Amphlett, J. C., R. F. Mann, and B. A. Peppley, "On-board Hydrogen Purification for Steam Reformation/PEM Fuel Cell Vehicle Power Plants," International Journal of Hydrogen Energy, v 21 n 8, (1996).

CONTACT

Paul Erickson is an assistant professor at the University of California, Davis. The study presented is a portion of work undertaken at the University of Florida under the supervision of Dr. Vernon Roan. Dr. Erickson can be reached at paerickson@ucdavis.edu or at 530-752-5360.

Development of an Onboard Fuel Processor for PEM Fuel Cell Vehicles

Brian J. Bowers, Jian L. Zhao, Druva Dattatraya and Vincent Rizzo
Nuvera Fuel Cells Inc.

Fabien Boudjemaa
Renault

ABSTRACT

Reduction of pollutants and greenhouse gas emissions is one of the main objectives of car manufacturers and innovative solutions have to be considered to achieve this goal. Electric vehicles, and in particular Fuel Cell Electric Vehicles, appear to be a promising alternative. Renault is therefore investigating the technical and economic viability of a Fuel Cell Electric Vehicle (FCEV). A basic question of this study is the choice of the fuel that will be used for this kind of vehicle. Liquid fuels such as gasoline, diesel, naphtha, and gas-to-liquid can be a bridge for the introduction of fuel cell technologies while hydrogen infrastructure and storage are investigated. Therefore, multi-fuel Fuel Processor Systems that can convert liquid fuels to hydrogen while meeting automotive constraints are desired. Renault and Nuvera have joined forces to tackle this issue in a 3-year program where the objective is to develop and to integrate a Fuel Processor System (FPS) on a vehicle.

This paper presents progress in the FPS development and its integration in a vehicle power plant. This PEM fuel cell power plant is being developed by Renault and will contain a FPS developed by Nuvera. The program goal is to meet vehicle requirements including high efficiency, low emissions, low cost, and long driving range. To date, a 75 liter fuel processor has shown operation up to 200 kWth on gasoline with 78% hydrogen efficiency and CO under 100 ppm.

INTRODUCTION

The 1997 Kyoto Protocol[1] requires that nearly all of Europe reduce quantified greenhouse gas emissions to a level 8% below 1990 levels in the commitment period of 2008 to 2012. Following this important decision, European, Japanese, and Korean carmakers agreed to reduce emissions of their vehicles[2]. In 1998, The European Automobile Manufactures Association (ACEA) made a voluntary commitment to reduce CO_2 emissions by 25% from the 1995 levels of 186 g CO_2/km to 140 g

CO_2/km in 2008[3]. Further plans could put a target of 120 g CO_2/km in 2012[3,4].

Following these guidelines, carmakers are making significant efforts in the development of innovative propulsion system. Fuel Cell Vehicles (FCVs) are one of the most attractive solutions to have clean, quiet, high-efficiency systems[5]. Today, carmakers are carrying out numerous demonstration programs to prove feasibility of such a power train[6].

The choice of fuel for the future fuel cell systems is a key point and it generates numerous discussions between carmakers, fuel suppliers, governments, and other actors in transportation[7]. Two solutions are available : onboard reforming or hydrogen storage. Trade-offs between technologies including performance, system cost, operating cost, flexibility, customer acceptance, security, and reliability must be considered[8].

Considering that hydrogen storage and infrastructure issues are not yet resolved, Renault has decided to pursue onboard fuel processing. Renault is focusing its research on a fuel-cell vehicle with a gasoline reforming system to be marketed after 2010. This option will permit a smooth introduction of fuel cell stack technology on vehicles, and it will allow FCVs to use the current fuel distribution network. With this onboard fuel processing FCV experience, it would be relatively easy to move to direct hydrogen FCVs when storage technology improves and a hydrogen network becomes available.

Renault is pursuing fuel-cell research through a partnership agreement with Nuvera Fuel Cells. The joint program targets development of a fuel processing system specifically designed for onboard hydrogen production. Features of the Nuvera system include multi-fuel capability as the technology could reform fuels such as petrol, natural gas, LPG, ethanol, and potentially diesel. Design flexibility is built-in to ensure the system can readily integrate new technologies and keep pace with fast-moving developments. The program starts with two phases. Phase 1 focuses on a laboratory demonstration, while Phase 2 seeks a 2004 delivery of a fuel processor closely adapted to onboard requirements[9].

FUEL CELL VEHICLE POWER PLANT

To be competitive, fuel cell vehicles have to show better performance than current and emerging power trains. Renault's target for a commercial onboard fuel processor / fuel cell vehicle is a well-to-wheels CO_2 emissions of 100 g of CO_2/km with 80 g of CO_2/km for the onboard tank-to-wheels portion. Correlating these emissions with the properties of the primary design fuel (gasoline), the fuel consumption should be around 3.2 liters/100 km. The second important objective of this vehicle is to require no water addition with an external temperature of 45°C (conventional Renault vehicle specification).

These major vehicle targets have been translated to power plant targets and then to sub-component targets. The power plant is composed of a fuel processor (reformer), fuel cell stack, and management systems for water, gasoline, and air as described in Figure 1.

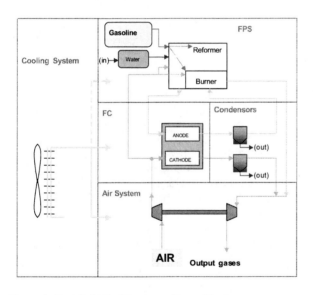

Figure 1 : Fuel Cell / Fuel Processor Power Plant

To reach the 3.2 liters/100 km, a high efficiency power plant is needed. An initial target of 40% efficiency has been defined for the 2004 power plant prototype, although higher efficiencies may eventually be needed. Of the 200 kWth fuel input, more than half of the energy is rejected via thermal power, creating a challenge for radiator packaging. The parasitic power, which is dominated by the air compressor system, should not be above 8% to reach our objective of 40% efficiency for the power plant. This value will define the maximum allowable power of our air compressor system.

To be integrated in the vehicle without modification of the passenger space, the power plant's volumetric density should be around 0.25 kWe/l with a mass density around 0.3 kWe/kg. For the targeted SCENIC II vehicle, approximately 200 mm of height will allow integration under the passenger floor. A first layout is shown in Figure 2. This integration shows lack of space for a large water tank. As a consequence, the power plant should internally recover and recycle enough water for hydrogen production. In theory, this water balance can be positive since the fuel cell stack produces water. Operation parameters just need to be defined to condense enough water. The key parameter is the operating pressure of the power plant. With an external temperature of 45° and a PEM fuel cell stack operating at 80°C, a pressure of around 3 bars at the condenser is needed to recover enough water for the fuel processor at full power (see Figure 3).

Figure 2 : "2004 gasoline power plant" layout in SCENIC II vehicle.

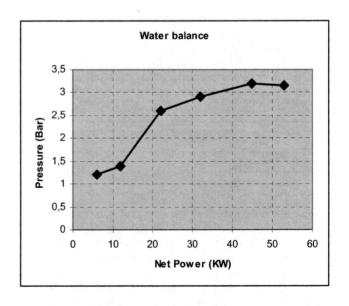

Figure 3 : System pressure required for positive water balance versus electrical output of the fuel cell

410

To be competitive with emerging and future power trains, the onboard fuel cell vehicle has to allow similar speeds, response, and startup time. Our objective is for the power plant to produce electricity within 2 minutes to minimize the battery size. Battery technology developed for electric vehicles can provide energy to move the vehicle during the power plant startup. The electric architecture is still being developed but may be similar to the one presented in Figure 4.

Using the constraints of vehicle integration and performance, specifications of the overall power plant were defined and are summarized in Table 1.

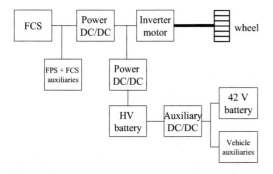

Figure 4 : Electric architecture of "2004 power plant"

Table 1. Renault Power plant objectives for 2004

Characteristic	Objective
Fuel Type	Sulfur Free Gasoline
Electric Power	70 kWe
Volume	280 liters
Mass	235 kg
Full power Efficiency	40 %
Water requirements	No water addition to vehicle
Starting time	2 min

FUEL PROCESSOR SYSTEM

FUEL PROCESSOR GOALS

The overall power plant objectives from Table 1 were used to set the design goals for the Phase 1 fuel processing system that are shown in Table 2.

The highest priority goal was the fuel processor volume since it is essential for vehicle installation. Preliminary packaging feedback suggested a volume of 80 liters and a height of 229 mm (9 inches) would be acceptable in Phase 1. Based on the volume goal, an autothermal reforming (ATR) process followed by a CO clean-up process of water-gas-shift (WGS) and preferential oxidation (PrOx) was chosen to efficiently convert gasoline or other liquid fuels into fuel-cell-quality hydrogen. To burn hydrogen exiting the fuel cell, a tail gas combustor (TGC) is included. To meet the volume goal, only substrate-based catalysts were to be used and pellet-type catalysts were not allowed. The 80 liters includes everything needed to take ambient temperature fuel, air, and water and convert it to fuel-cell-quality reformate at approximately 100 C. Thus, the ATR, WGS, PrOx, TGC, and heat exchangers needed for this are all inside the fuel processor package.

The primary fuel for the design is gasoline. Since gasoline is a relatively challenging fuel, past experience suggests a successful gasoline design will allow conversion of several other fuels. For laboratory testing, sulfur-free gasoline is used due to the assumption that sulfur would be removed either in the fuel refinery or in an onboard desulfurizer that is external to the fuel processor.

The goals of 1.3 g/s of hydrogen production and 78% hydrogen efficiency lead to a fuel input of about 200 kWth (based on lower heating value). As another measure of processing efficiency, the residual (unconverted) fuel was set as less than 1% methane. The TGC was designed to burn anode exhaust assuming 80% hydrogen utilization in the fuel cell. Thus, the electrical power from a matching fuel cell can be approximately 65-70 kWe with current PEM technology. This technology also indicates a need for less than 100 ppm CO in the reformate for steady state operation.

To help with the water balance and size requirements, a system operation pressure up to 3 bar was chosen. To minimize the parasitic air compressor load, the Phase 1 goal for pressure drop was set at 1 bar.

Table 2. Phase 1 Fuel Processor Design Goals

Characteristic	Design Goal	Comments
Fuel Processor Volume* *without balance of plant components or plumbing	≤ 80 Liters	Includes everything between cold feed streams and 100 °C fuel-cell-quality reformate outlet stream.
Height	<229 mm	"Flat" aspect ratio for vehicle installation
Catalysts	Non-pellet-catalyst	Minimize volume
Fuel Type	Sulfur Free Gasoline	Onboard or refinery desulfurizer assumed
Maximum hydrogen in reformate	1.3 gram/sec	~157 kWth based on LHV
Full power hydrogen Efficiency	≥ 78 %	LHV H2 / LHV ATR fuel
Residual fuel (as CH4)	< 1 % (dry)	At PrOx exit
Assumed FC H2 utilization	80%	At peak power
CO Concentration -steady state - transient	 ≤100 ppmv (dry) < 1000 ppmv (dry)	At PrOx exit
Reformate pressure	3 bar	At PrOx exit
Pressure loss	≤ 1 bar	ATR air inlet – PrOx exit

411

FUEL PROCESSOR DEVELOPMENT AND DESIGN

Catalyst Development

To meet the strict volume requirements, Nuvera expended a significant effort over several years to investigate substrate-based catalysts for the five reaction zones:

1. Autothermal reforming (ATR) catalyst
2. High temperature water gas shift (WGS) catalyst
3. Low temperature water gas shift (LTS) catalyst
4. Preferential oxidation (PrOx) of CO catalyst
5. Tail Gas Combustion (TGC) catalyst to burn hydrogen from the anode exit

The catalyst testing included investigation at both micro-reactor scale and full scale. Nuvera worked with several suppliers to screen catalysts for activity and selectivity in sub-kilowatt micro reactors. This testing narrowed down the choices and allowed the development of catalytic rate expressions for later use in the final fuel processor design. Once promising candidate catalysts were identified, full scale testing was done in Nuvera's Modular Pressurized Reactor (MPR) facility. The MPR tests allowed verification of the micro reactor results and exploration of full-scale phenomena such as flow distribution and temperature distribution throughout the catalyst zone. In addition, it allowed the development of relationships between catalytic conversion and space velocity at full-scale that also fed into the final fuel processor design. Once the small and large scale testing was completed, Nuvera was able to choose catalysts most appropriate for the 5 reaction zones. The result of this investigation was a high-power-density substrate-based catalyst suite suitable for the Phase 1 fuel processor.

Heat Exchanger Development

In addition to the work on catalysts, Nuvera developed custom heat exchangers for every part of the fuel processor. These heat exchangers allow for the proper control of the temperatures in the catalyst zones and they also integrate the overall heat flow of the system to achieve the maximum thermal efficiency. To meet the pressure drop, size, and heat duty requirements, Nuvera explored various heat exchangers including custom tube-in-tube, shell-in-tube, finned-tube, and direct and indirect heat exchange between water, steam, fuel, air, and reformate. The result is a unique combination of multiple heat exchangers inside the small package.

Fuel Processor Design

Using the selected catalysts and heat exchangers, the fuel processor design integrated numerous features to meet the goals of Table 2. The result is a compact fuel processor that is small enough to fit in a vehicle. Figure 5 shows the actual Phase 1 fuel processor. The dimensions are approximately 220 mm x 440 mm x 850 mm. Taking into account the rounded sides, the volume is approximately 75 liters. This volume contains all of the catalysts and heat exchangers necessary to take room-temperature fuel, air, and water and convert it into a fuel-cell-quality reformate stream at approximately 100 C. If lower temperature reformate is desired, an optional cooling system can be installed.

This design represents a huge leap in fuel processing technology that in only 3 years reduced the volume more than 10 times over the previous fuel processors that used pellet-based catalyst. To accommodate the large energy releases and energy transfers in such a small volume, the design process gave special attention to thermal expansion issues. In addition, the design incorporates separate modules inside the package, which allows easy servicing during the prototype stage and would be useful for servicing a consumer vehicle.

Figure 6 shows that vehicle packaging work has already started and that the fuel processor and its major controls components can fit into a volume and shape appropriate for vehicle installation.

Figure 5 : Phase 1 Fuel Processor on test stand

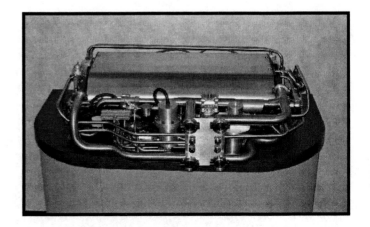

Figure 6 : Fuel Processor in vehicle packaging study

FUEL PROCESSOR TESTING

To test the advances of the Phase 1 fuel processor, Nuvera created a flexible laboratory test stand. The test stand controls the flows of air, fuel, water, and reformate along with the system pressure in a manner that simulates the final fuel cell / fuel processor system. Figure 7 shows a simplified process flow diagram of the fuel processor system.

Since the initial focus was on the fuel processor performance, the test stand simulates a fuel cell hydrogen consumption by diverting most of the reformate to an external afterburner. The Phase 1 test stand is intended to be a flexible way to test the fuel processor and is not indented to be installed in a vehicle, so it uses a combination of industrial and automotive balance of plant (BOP) components. In parallel to the fuel processor testing, a BOP development effort is producing automotive components that are systematically replacing the test stand BOP to allow for eventual vehicle installation. For example, the water system has already been replaced with a compact automotive-type system that significantly reduces volume and increases the turndown ratio of high to low power. Similarly, a compact automotive-type fuel system was installed after the successful testing with the water system.

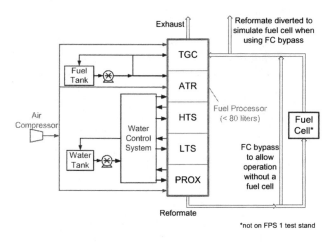

Figure 7 : Fuel Processor System Process Flow Diagram

The test process is following a rigorous procedure:

1. Validation testing to confirm steady state operation
2. System identification testing to confirm transfer functions of the system model
3. Control system design using the validated system model
4. Controls tests to optimize transient performance
5. Ongoing iteration between testing, modeling, control design, and FPS design

At the time of the writing of this paper, the system has completed the step 1 validation testing with the system model verified against steady state results; the most important transfer functions of the step 2 system identification have been found; the step 3 control system design has begun using the theoretical model; and the first controls tests of step 4 have begun.

PHASE 1 FUEL PROCESSOR TEST RESULTS

To date, the fuel processor tests have verified all of the specifications of Table 2. These achievements are the results of years of work with the most significant breakthrough being the high power density that finally allows a fuel processor small enough to fit in a car and powerful enough to move it. This is shown by the conversion of gasoline into 1.3 g/s of hydrogen in only 75 liters while maintaining high efficiency and low CO.

Figure 8 shows the system operating at 60, 100, 130, and 195 kWth of gasoline input during the validation testing. Over this entire power range the fuel processor achieves a hydrogen efficiency (defined in Table 2) of 77 to 80% with a dry hydrogen concentration of about 40%. In addition, the CO is kept right at the 100 ppm target for most of the time while the entire time it is well under the 1000 ppm limit for short transients. The hydrogen efficiency and hydrogen concentration are obtained using a gas chromatograph (GC) every 4 minutes while the CO is measured with a continuously online IR analyzer. All samples are from the exit of the PrOx. Although the validation tests focused on steady state operation, this data already demonstrated that the transient goals are achievable under moderate power changes.

Figure 9 shows the fuel processor system operating at high power with the compact automotive-type water control system installed. The gasoline input is between 180 and 200 kWth. Once again, the hydrogen efficiency is high and steady at between 79 and 81% while the CO meets the 100 ppm target. The dry hydrogen concentration ranges from 38 to 39%. This test confirms that the excellent high power performance is preserved while using the automotive water system.

Figure 10 shows the fuel processor system operating at low power with the automotive-type water control system. The gasoline input to the ATR 33 kWth. This low value was not possible using the standard laboratory water control system. Following a typical fuel processor efficiency curve, the low-power hydrogen efficiency of 75.3% is slightly less than at high power. The CO is in line with the target with a value of 110 ppm. The dry hydrogen concentration ranges from 38 to 39%. This test shows another step in the development process as the improved water system has allowed system operation to be proven over a 6 to 1 turndown (200 to 33 kWth).

Figure 11 shows the fuel processor operating with both the automotive-type fuel and water systems during the initial control tests. These systems and the preliminary controller designs allow excellent performance over a range of power transients. Once again the hydrogen efficiency is at or above the targets while the CO is as low at 27 ppm. The CO is maintained well under both the steady state goals of 100 ppm and transient goal of 1000 ppm the entire time.

This test data shows significant progress in the development of an onboard fuel processor for use in a PEM fuel cell vehicle. The testing covers an ATR gasoline input power range of 33 to 200 kWth with hydrogen efficiencies ranging from 75% to 81% and CO levels controllable within the steady state and transient goals.

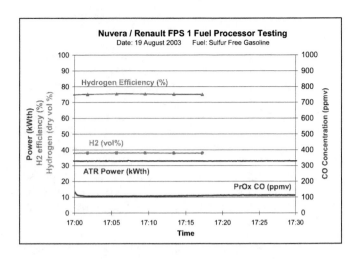

Figure 10 : Phase 1 Fuel Processor - 33 kWth with automotive water system

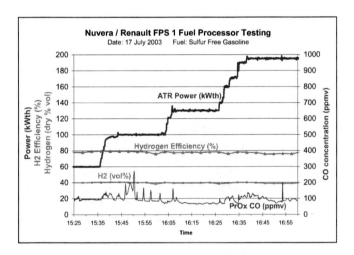

Figure 8 :   Phase 1 Fuel Processor - 60 to 195 kWth operation with laboratory control system

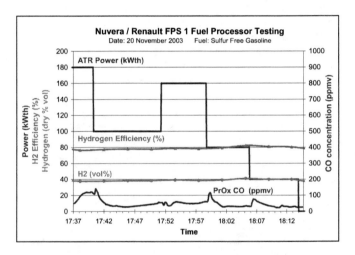

Figure 11 : Phase 1 Fuel Processor - Power transients with automotive fuel and water system

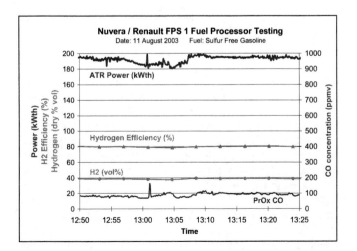

Figure 9 :   Phase 1 Fuel Processor - 195 kWth with automotive water system

CONCLUSION

The commitment by European automakers to reduce vehicle CO2 emissions has spurred research into many areas including fuel cell vehicles. While technical barriers to hydrogen storage are being investigated, onboard reforming of fuels such as gasoline can provide a source of hydrogen for the fuel cell. Renault and Nuvera are actively pursuing a fuel cell power plant using an onboard fuel processor. The program starts with a Phase 1 laboratory demonstration and seeks a Phase 2 delivery of a fuel processor closely adapted to onboard requirements.

Renault has already led work on the power plant design, which sets targets of 70 kWe output at 40% efficiency in a 280 liter volume. In addition, the specifications call for no water addition to the vehicle.

Nuvera's Phase 1 fuel processor has already shown that 200 kWth of gasoline can be converted into fuel-cell-quality reformate in the compact 75 liter volume while maintaining hydrogen efficiencies of 78% and CO levels of 100 ppm. This gives a world-class power density of 2.1 kW-H2 per liter of fuel processor and proves that onboard fuel processors can be small enough and powerful enough to be integrated into the power plant of a fuel cell vehicle. In addition, preliminary work on vehicle controls already shows successful transient performance over a wide range of power.

This technology represents a dramatic reduction of fuel processor volume and a significant step toward vehicles with onboard fuel processors. Ongoing research will continue to optimize the fuel processor, control system, and balance of plant components to allow packaging in a vehicle.

REFERENCES

1. *Kyoto Protocol to the United Nations Framework Convention on Climate Change.* Adopted at the third session of the Conference of the Parties to the UNFCCC in Kyoto, Japan, on 11 December 1997.
2. *Detailed Description of Best Practices – European Union No. 2.* G8 Environmental Futures Forum 200 on "Domestic Best Practices Addressing Climate Change". http://www.env.go.jp/earth/g8_2000/forum/g8bp/detail/eu/eu02.html
3. European Automobile Manufactures Association (ACEA). *The European Automobile Manufacturers commit to substantial CO2 emission reductions from new passenger cars.* http://www.acea.be/ACEA/290798.html. Press Release. 29 July 1998.
4. European Automobile Manufactures Association (ACEA). *ACEA's CO2 commitment.* http://www.acea.be/ACEA/brochure_co2.pdf. 5 December 2002
5. F. Stadolsky, L Gaines, C. L. Marshall, J. Eberhard, SAE 1999-01-0322
6. Fuel Cell Vehicles (from auto manufactureres), http://www.fuelcells.org/fct/carchart.pdf
7. S. Carter, P. Teagan, R Stobart, SAE 2000-01-0001
8. S. G. Deshanais, J. Y. Routex, M Holtzapple, M. Ehsani, SAE 2002-01-0097
9. Renault/Nuvera press release, June 2002, www.renault.com

CONTACT

Brian Bowers is the project manager for the Nuvera automotive fuel processor project and holds a Master of Science degree from the Massachusetts Institute of Technology. He has worked in fuel processing for 5 years with publications on system emissions and performance.

Dr. Fabien Boudjemaa is the project manager for the Renault fuel processor development project and holds a Master of Science degree in chemistry and a Doctor of Philosophy degree in catalysis from University of Lyon (France). He joined the Renault Research Division in the alternative fuel department in 1999 to work on fuel cell systems. Since 2000 he has managed the Fuel Processor System activity for Renault within the ALLIANCE Renault-NISSAN Fuel Cell Program.

Dr. James Zhao is the senior engineer for the Nuvera automotive fuel processor design and holds a Doctor of Philosophy degree from Northeastern University. He has worked in the chemical engineering field for more than 10 years and has publications on fuel processor and fuel cell systems.

Vincent Rizzo is the senior control system engineer for Nuvera and holds a Master of Science degree from Tufts University. He has 5 years experience in fuel processing with publications on system controls.

Druva Dattatraya is the lead system engineer for the Nuvera automotive fuel processor and holds a Master of Science degree from Rutgers University. He has worked on fuel processing for 3 years and has published on system performance.

V. DEVELOPMENT, TESTING AND LIFE CYCLE ISSUES

V. Development, Testing and Life Cycle Issues

In recent years the number of papers published on the subject of development issues in fuel cell technology has increased significantly. At the end of the 1990s the major pre-occupation in the fuel cell community was an understanding of the fundamentals of fuel cell operation. The complex questions of materials developments to meet new and demanding requirements are beginning to be resolved and there is a general agreement about where effort should be invested. With a clearer perspective on the research issues, there has been a growth of interest in design processes, design support– particularly in end-of-life requirements for fuel cell systems.

Questions of test methods, test planning techniques, measurement methods and data analysis although established in the wider industry are raising new issues. The methods needed to measure fuel consumption in a hydrogen-powered car cannot be developed from existing methods used with conventional fuels. Innovations are needed and some new ideas are presented in one of the papers included in this section.

Life Cycle

There are two distinct aspects to the environmental impact of fuel cells. The first, covered comprehensively in this collection of papers is the impact on air quality that will be made by a progressive substitution of conventional IC engined vehicles with fuel cell powered equivalents. The second is the impact of manufacturing on environment through component manufacture and end of life recycling and management of waste products.

An SAE Committee for Fuel Cell Standards was started in 1999. Paper 2003-01-1141 is a report on the preferred practice standard, J2954 that gathers together some existing material on recyclability. Its scope is restricted to major components including fuel supply and storage. It does not cover material preparation, component manufacture and as yet offers no guidelines on how the re-cyclability of a fuel cell stack can be assessed.

The paper *Impact of the European Union vehicle waste directive on end-of-life options for polymer electrolyte fuel cells* was presented at the Grove Fuel Cell Symposium in 2001. Because the directive is based on weight, it focuses interest on the heavy parts of the fuel cell stack, notably the bipolar plates. The authors analyse the materials usage in a 70kW fuel cell stack and propose recycling processes for the major components.

Test Methods

In paper 2004-01-1008, the authors present results from the application of a number of methods to determine the fuel consumption of a hydrogen fuelled vehicle.

The authors of paper 2004-01-1013 investigate the behaviour of a vehicle hydrogen storage tank in a fire. The conclusions from the test concern the positioning and design of a pressure valve that will prevent the storage vessel from rupturing.

In another development, the authors of paper 2004-01-1305 measure the fuel economy of the Honda FCX using the weight method in which a separate fuel tank is weighed before and after the test. The authors describe how they handle the various sources of measurement error to apply the method to the required degree of accuracy.

V. DEVELOPMENT, TESTING AND LIFE CYCLE ISSUES

Life Cycle

2003-01-1141

Development of Recycling Guidelines for PEM Fuel Cell Systems

Stella Papasavva, Angie Coyle, Stefanie Goldman, Robert Privette, Renato Legati, Connie Huff, Larry Frisch and Richard Paul

General Motors Corp., Delphi, Millennium Cell, OMG, Ballard, Ford Motor Company, Dow Corning, American Honda Consultant

ABSTRACT

In 1999, the Society of Automotive Engineers established a Committee for Fuel Cell Standards. The Committee is organized in subcommittees that address issues such as Emissions and Fuel Consumption, Safety, Performance, Terminology and Recycling. The mission of the Recycling Subcommittee was to develop a preferred practice document, SAE J2594, that incorporates existing information on recycling practices and infrastructure, and identifies technical, economic, and environmental sustainability issues and applies them to proton exchange membrane (PEM) fuel cell (FC) systems. Recyclability should be considered early in the product engineering design/development process in order to enhance its potential for reuse or recycling at the end of the vehicle's useful life.

The purpose of this technical report is to provide an overview of the recommendations of the SAE J2594 recommended practice document. The boundary of the system studied includes: Fuel Supply and Storage, Fuel Processor, the PEM Fuel Cell Stack, and Balance of Plant. J2594 does not address material preparation, component fabrication or other in-use applications. Another important limitation is that the recommended practice does not provide a methodology for calculating the recyclability of a FC, an entire vehicle or its other components.

INTRODUCTION

Environmental management is a systematic approach to the review, assessment and improvement of a company's products and operations to enhance environmental performance [1]. One important tool to improve environmental performance of products is design for recyclability (DFR) or design for environment, beginning at the product engineering stage and includes the selection of materials that can be readily recycled, the prediction of reuse potential, disassembly design selections and other component fastening aspects. Case studies and other examples of using DFR and DFE in the early stages of product planning are abundant in the literature [2,3].

As fuel cell systems are developed to power the automobiles of the future, other reasons to include recyclability considerations are the increased emphasis on product stewardship by manufacturers, heightened consumer focus on environmental and recyclability aspects of products, concerns on natural resource depletion and use of substances of concern.

Broader incentives to address preferred practices for recyclability of PEM FC include the passage and implementation of laws adopted by governments in Europe [4] and Japan [5] to increase the reuse and recycling of end-of-life automobiles in their regions. These laws focus on the handling and treatment of End of Life Vehicles (ELV's), diversion of hazardous materials from landfill, minimum levels of parts reuse and increased material recycling. In N. America, similar legislation does not currently exist, but myriad state and federal laws are approximately equivalent to many of the European and Asian requirements and apply directly or indirectly to ELV treatment [6].

In this context, the members of the SAE subcommittee developed preferred design practices to improve the recyclability of PEM fuel cells, SAE J2594. Many of the recommendations in the document are based on the technology, business practices, markets and processes of the existing infrastructure for recycling End of Life Vehicles (ELVs) automobiles in N. America. Automobile recycling has been practiced since the early 1900s, shortly after the first automobiles were built, specifically in the area of metals recovery. Generally, over 94% of all ELV's are recycled with an average rate of recycling of 75% by weight per vehicle [7]. The remaining 25%, made up primarily of foam, glass, plastic, and rubber, is known as Automotive Shredder Residue (ASR) and is usually disposed of in landfills [4]. Worldwide, approximately 30 million end-of-life vehicles and 8 million tons of ASR are generated annually.

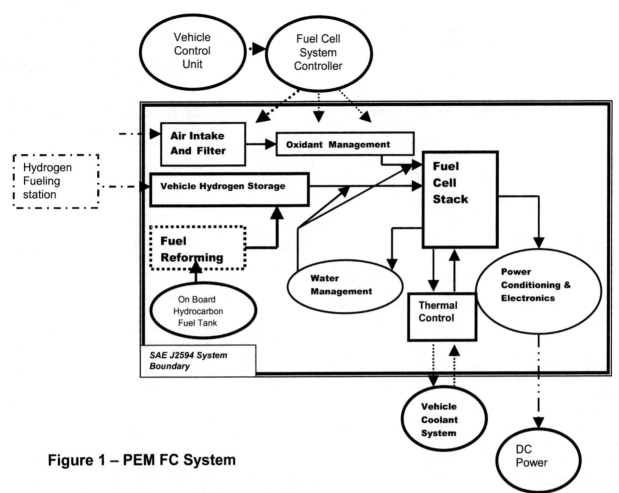

Figure 1 – PEM FC System

The handling of an entire end-of-life vehicle is a four-step process: pre-treatment (depollution), dismantling (also known as disassembly), metal separation (shredding), and residue treatment. Within each step of the ELV process, a variety of vehicle components and/or materials may be recovered for reuse, recycling, remanufacture, or energy recovery. Impacts of reusing or recycling PEM FC components, and having decreasing the landfill portion, have the potential to be quite significant. Although there is a range of external pressures (in different global regions) that influence the selection of suitable end-of-life management strategies and opportunities for reuse may be limited in some regions, all components of the PEM FC stack could be recycled in principle. [8]

The purpose of this recommended practice document is to provide a tool for the FC system designers and engineers to incorporate recyclability into the PEM FC design process. PEM FC recyclability, as used in this document, refers to the reuse, remanufacturing, recovery, and/or recycling of materials used in the components of the PEM FC system. Fuel cell system and subsystem recyclability depends on two factors: the ease of disassembly (part removal) and recovery from a vehicle, and the inherent recyclability of the components/materials. For the entire fuel cell system, a variety of recycling methods may be applied to the various subsystems and components. These include

manual separation (where a dismantler removes the sub-components by hand), chemical recovery (thermal, chemical or electro-chemical separation to recover materials such as precious metals), and mechanical treatment (where materials are separated and/or shredded together then separated by material properties e.g. magnetic, density, etc.). It may be possible to selectively remove sub-components containing a number of mixed materials so that the remainder of the fuel cell system can be recycled more easily. High value materials may be manually separated, with the balance of the system being recycled along with the bulk of the vehicle during the shredding process.

Reuse of parts is the highest form of recycling (parts may be used directly without any additional processing, e.g. fender, trim, tail light lenses) and depends on the ease of disassembly, market demand for use in repair or replacement, part durability, cost of a new part, and the existing infrastructure for collection, distribution and service for part replacement. Other parts may be remanufactured by restoring original quality and performance levels through the replacement of all worn or deteriorated components and re-tested to OE specifications. This saves energy, natural resources, landfill space, and cost; e.g., alternators, calipers, electronic control modules, etc.

Factors influencing remanufacture are similar to reuse. Material recycling is based on technology available to process the material, the existence of a supporting infrastructure, cost of virgin material, ability to separate the materials into pure streams, hazardous materials content and the inherent economic value of the material. These points are discussed in more detail throughout the guideline.

Recyclability should be considered early in the product engineering design/development process. The design engineer must be concerned with the product after its useful life by adapting to a new mindset of designing for disassembly and recycling. The derived (disassembly/recyclability) ratings, discussed below, are used as a PEM FC component design tool for continual improvement opportunities and not for purposes of calculating recyclability of the entire vehicle.

The purpose of this recommended practice document is to provide a tool for the FC system designers and engineers to incorporate recyclability into the PEM FC design process. This recommended practice document was derived by considering existing recycling practices [5] then applying them to assess and evaluate the recyclability of the PEM FC system. This recommended practice should be applied to assist a process for continually assessing the recyclability of component and assembly designs during the early design phase, in order to reach an optimized recyclability, recycled content and minimized environmental impact associated with those designs. This recommended practice defines a PEM FC rating system that assesses the ease of removal of the PEM-FC system and/or components from a vehicle; then upon removal from the vehicle, the ease of recycling. While other trade-offs such as mass, piece-cost, volume, etc. must also be considered when designing these systems they are not discussed in this document.

The PEM FC system boundary considered for recyclability issues is shown in Fig. 1. It does not address material preparation, component fabrication or other in-use applications. Another important limitation is that the recommended practice does not provide a methodology for calculating the recyclability of a FC, an entire vehicle or its other components. Such a method is detailed in the International Standards Organization (ISO) calculation methods for road vehicle recyclability [9].

FUEL CELL RECYCLABILITY CHART

A FC recyclability chart was created for the guideline as a tool to identify major recyclability and environmental sustainability issues associated with PEM FC systems and sub-systems. The chart is organized by subsystem and includes all of the principal components within an automotive FC system.

Column headings on the chart include material types, recyclability (disassembly issues, reuse or alternative use, technical, infrastructure), and sustainability, (resource depletion and end-of-life environmental issues) and provide information in each category on the major FC systems and system components.

The purpose of the chart is to provide specific recyclability data on major PEM Fuel Cell subsystem components per currently known technology. Three major subsystems are included in the chart:
- Fuel Supply
- Balance of Plant
- PEM Fuel Cell Stack
  A sample of the chart layout is shown in Fig. 2.

The main chart headings address system and system components, material types recyclability issues, disassembly, reuse or alternative use, technical issues, infrastructure issues, sustainability issues *(environmental only)* resource depletion and end-of-life environmental issues.

This chart serves as a quick reference, supported by the text of the standard, to evaluate specific choices by design engineers and determine the implications of those choices for reuse, recyclability and landfill potential.

DISASSEMBLY AND RECYCLABILITY RATINGS –

Disassembly ratings (Fig. 3) refer to the ease of removing the component/assembly from the vehicle.

In general, the disassembly of the component/assembly from the vehicle is dependent on the part location/accessibility on the vehicle, its value, and the treatment method used. Recyclability ratings (Fig. 4) refer to the ability to re-use, remanufacture, recycle, or recover the materials in the component/assembly, assuming removal/disassembly from the vehicle. J2594 provides easy to use tables, which will guide the FC engineer and designer in evaluating the disassembly and recycling potential of the materials and fastening systems selected.

The lower the rating-score (with respect to disassembly and recycling) of the system the better the design.

FIGURE 2.   PEM FUEL CELL RECYCLABILITY

System	System Components	Material Types	Recyclability Issues				Sustainability Issues	
			Disassembly	Reuse or Alternative Use	Technical	Infrastructure	Resource Depletion	End of Life Environmental
Fuel Supply	**On Board H₂ Storage**							
	Compressed H₂							
	- Type 1 Tank	Non-composite metallic	Tank must be purged prior to disassembly	Dependent on tank condition	No issue	No issue	Depends on metal	No issues
	- Type 2 Tank	Composite metallic hoop wrapped	Tank must be purged prior to disassembly	Dependent on tank condition	Material separation	Infrastructure for composite recycling	Fossil fuel depletion for composites	Composites waste generation
	- Type 3 Tank	Composite metallic full wrapped	Tank must be purged prior to disassembly	Dependent on tank condition	Material separation	Infrastructure for composite recycling	Fossil fuel depletion for composites	Composites waste generation
	- Type 4 Tank	Composite non-metallic full wrapped	Tank must be purged prior to disassembly	Dependent on tank condition	Material separation plastic compatibility	Infrastructure for composite recycling	Fossil fuel depletion for composites	Composites waste generation
	Liquid H₂							
	- Hardened Aluminum	w/composite wrap	Tank must be purged prior to disassembly	Dependent on tank condition	Material separation	Infrastructure for composite recycling	Fossil fuel depletion for composites	Composites waste generation
	- Stainless Steel Tank	Stainless Steel, Insulation (mineral wool or fibreglass)	Tank must be purged prior to disassembly	Dependent on tank condition	Material separation Stainless steel is 100% recyclable	Infrastructure for insulation recycling	Ni	Solid wastes (insulation)
	Metal Hydrides	Titanium, Magnesium, Nickel Alloys doped w/ rare earths (e.g. Lanthanum nickel, iron - titanium)	Potentially pyrophoric when exposed to air	Lifetime restrictions for recharging	No issue	Infrastructure needed	Ni and La alloys may have issues	Hydrides cannot be landfilled
	- Hydride Tank	Composites or Metals	Remove from vehicle prior to shredding; special procedures required to remove hydrides	Dependent on condition of tank. Lifetime restrictions for recharging.	No issue	Infrastructure for composite recycling	Depends on metal	Potential hydride contamination
	- Hydride Vessel	Composites or Metals	Purge fluids	None	No issue	No issue	Depends on metal	No issue
	- Burner	Metals	No issue	Potential for reuse	No issue	No issue	Depends on metal	No issue
	- Heat Exchanger	Metals	Purge fluids	Potential for reuse	No issue	No issue	Depends on metal	
	Chemical Hydrides	Lithium Hydride and Mineral oil and polymeric dispersant	Tank must be purged prior to disassembly; chemical is caustic	Potential for reuse or recycling.	No issue	Spent hydride infrastructure needed	Lithium, fossil fuel depletion for dispersant	Neutralize prior to disposal; lithium
		Sodium Hydride and polyethylene or other inert plastic coating	Tank must be purged prior to disassembly; chemical is caustic	Potential for reuse or recycling.	No issue	Spent hydride infrastructure needed	Fossil fuel depletion for plastic	Neutralize prior to disposal
		Sodium Borohydride	Tank must be purged prior to disassembly	Potential for reuse or recycling.	No issue	Spent hydride infrastructure needed	No issue	Neutralize prior to disposal

FIGURE 3 – DISASSEMBLY RATING CHART

Fuel Cell Disassembly Rating

Joining Aspects	Joining Configuration	Joining Score		
1. Location	Visible	1		
	Covered		2	
	Hidden			3
2. Disconnectability	Disconnected nondestructively	1		
	Partial destruction		2	
	Disconnected only by part destruction			3
3. Accessibility	Axial dismantling direction	1		
	Axial accessible direction		2	
	Radial or difficult direction			3
4. Number of joinings	One or few joining	1		
	Low number of joining		2	
	High number of joining			3
5. Joining Tools	Joining elements standardized	1		
	Standardized within type of joining		2	
	Not (or almost not) standardized			3
6. Joining Types	No fastner, pressure fit, snap fits	1		
	Clips, screws, bolts, etc.		2	
	Rivets, welding, soldering, adhesives			3
Total Score (Lower score is better)		+	+	=

FIGURE 4 – RECYCLABILITY RATING CHART

Recyclability Rating	Category	Description/Examples
1	Re-Use	A radio or bumper/fascia can be removed from an ELV and be re-used as a replacement part in an appropriate make/model vehicle.
2	Remanufacture	Products such as brake cylinders, alternators, pumps, and motors can be removed from vehicle and reconditioned for use as after-market parts.
3	Recycled	Part removed for materials recycling or materials separated and recycled, e.g. metals, batteries, catalytic converters, fluids, etc.
4	Technically Feasible, but not recycled	Parts made out of pure materials, compatible materials, or materials that can be separated into recyclable streams using technologies that have no infrastructure and/or are not economic to recycle, e.g. polypropylene, glass, elastomers.
5	Energy Recovery	Part is made out of mixed or contaminated materials that cannot be readily recycled, but contains materials that can be incinerated for generation of energy, e.g. automotive shredder residue, tires.
6	None of the above	Part has no calorific value, e.g. ceramics, mineral fibers.

427

DESIGN FOR DISASSEMBLY AND RECYCLABILITY RECOMMENDATIONS

General recommendations are provided which, when used in conjunction with the tables described above, will accomplish changes that will increase the disassembly and recycling potential of the FC system and it's components. The recommendations address design for effective removal of components and assemblies, part durability for reuse, use of hazardous materials, plastic parts selection, use of recyclable materials, use of recycled materials, reduction or elimination of coatings/finishes, minimal use of adhesives and parts marking.

It is important to understand clearly the philosophy used by the authors in assessing the end-of-life environmental issues of materials. For the purpose of the J2594 document, materials of concern have been considered with a view toward the following items:

- Probability of escape/release to the environment (how the substance is bound or contained, physical state of substance -gas/liquid/solid)
- Concentration (mass/volume)
- Inherent toxicity
- Probability/opportunity of reuse/recycling using existing collection and recycling system
- Probability/opportunity of interaction with humans
- Generation of non-recyclable waste

In cases where the element or material can be considered to be permanently bound within a substance and in this bound condition is not an environmental threat considering its expected end of life treatment, and further at end-of-life has a high probability of reuse/recycling, then the element or material is not identified as an end-of-life environmental issue. For example, while stainless steel contains nickel and chromium, these elements are considered to be permanently bound within the steel and if reused or recycled will not escape into the environment and therefore the nickel and chromium within stainless steel are not identified as end-of-life issues. As another example, glycol ether is expected to be processed using existing processes, however as a liquid it has a higher chance of release into the environment during draining, leading to potential release or human exposure and therefore it is identified as an end-of-life environmental issue.

SUBSYSTEMS, MATERIALS AND RELATED ENVIRONMENTAL ISSUES

In this section of the document, information is presented on the various subsystems of the PEM FC.

The systems are divided as follows:
- Fuel Supply
  - Gaseous H2 Storage
  - Liquid H2 Storage
  - Metal Hydrides
  - Chemical Hydrides
  - On-board fueling processing
- Balance of Plant
- Fuel Cell Stacks

For each subsystem, component and material types are discussed, which includes basic design information and likely materials used in the components. The most important recycling engineering issues and recycling environmental issues are also presented. Recycling engineering issues refer to the ability to disassemble the parts and recycling environmental issues refer to the anticipated air, water or solid waste impact of part disposal at end-of-life.

CONCLUSION

The Recommended Practices for Recycling PEM Fuel Cell Systems, J2594, was created as a design tool and is not intended to be a prescriptive process for fuel cell systems. . It is acknowledged that there are numerous competing aspects to fuel cell design. Recyclability of any product, including fuel cells, is dependent upon numerous factors, many of which are beyond the control of the design engineer. As a result J2594 does not address material preparation, component fabrication or other in-use applications. Another important limitation is that the recommended practice does not provide a methodology for calculating the recyclability of a FC, an entire vehicle or its other components. Trade-offs must be evaluated and final design based on program imperatives (safety, performance, durability), which can only be decided by the design engineer. PEM fuel cell technology is evolving and the recycling infrastructure, recycling technology and market demands are also improving and changing. Therefore, the chart and supporting text will need to be updated for evaluation of new materials and designs.

ACKNOWLEDGMENTS

The authors acknowledge the valuable comments received from the members of SAE's Fuel Cell Standards Committee and those from the reviewers.

REFERENCES

1. Practical Guide to Environmental Management, 8th Edition, May 2000, Environmental Law Institute.
2. Integrating LCA and DfE in the Design of Electrical and Electronic Products in the Automotive Sector, J.C. Alonso, et al. Proceedings of the 2002

Environmental Sustainability Conference and Exhibition, Society of Automotive Engineers. 2002

3. Analysis and Assessment of Automobiles with Regard to the Requirements of Design for Recycling, J. Boes, et al., Proceedings of the 2002 Environmental Sustainability Conference and Exhibition, Society of Automotive Engineers. 2002
4. Japan Automotive Digest, Volume VII, Number 4, February 5, 2001
5. Directive 2000/53/EC of the European Parliament and of the Council of 18 September 2000 on End-of-Life Vehicles, Official Journal of the European Communities, 21.10.2000.
6. How prepared are US dismantlers to Meet the EU Directive for ELV Recycling, R. T. Paul, SAE International Sustainability Conference, Graz, Austria, 2002.
7. Automotive Facts and Figures, Alliance of Automobile Manufacturers, 2001
8. Impact of the European Union Vehicle Waste Directive on End-of-Life Options for Polymer Electrolyte Fuel Cells, C. Handley, et al., Journal of Power Sources, 2002.
9. Design for Recycling Guidelines, Vehicle Recycling Partnership/USCAR, 1996
10. Road Vehicles-Recyclability and Recoverability-Calculation Method: 22628, International Standards Organization, 2002.
11. SAE J1344, Marking of Plastic Parts, 1993
12. Worldwide Design Requirements, Complete Vehicle – Recycling, Ford Motor Company, February 1996.

CONTACT

Corresponding author. E-mail address: stella.papasavva@gm.com

DEFINITIONS, ACRONYMS, ABBREVIATIONS

Automotive Shredder Residue (ASR): The material in a vehicle that is not sold for reuse after dismantling, or separated for recycling after shredding and is typically land filled. This material contains about 34% plastic, 17% fluids, 12% rubber, 16% glass, 21% other contaminants (e.g. foam, fibers, dirt, and contaminants).

Dismantling: process of removing component parts from the vehicle.

End-of-Life Vehicle (ELV): A vehicle that has completed its useful life and is taken out of service for disposal.

Post-Industrial Recycled Content: The portion of a material's or product's mass that is composed of materials that have been recovered from or otherwise diverted from the solid waste stream during the manufacturing process (pre-consumer).

Post-Consumer Recycled Content: The portion of a material's or product's mass that is composed of materials that have been recovered from or otherwise diverted from the solid waste stream after consumer use.

Reuse or Alternative Use: Any operation by which component parts of end-of-life vehicles are used for the same purpose for which they were initially intended.

Remanufacture: The process of restoring used durable product to a "like new" condition

Recovery: Reprocessing a production process of waste materials for the original purpose or for other purposes, and the processing as a means of generating energy.

Recyclability: The potential to recover or otherwise divert products or materials from the waste stream for purposes of recycling.

Recyclability Rate: Percentage by mass of the vehicle that can potentially be recycled, reused, or both.

Recyclability Rating: The numeric value (as defined in this document) given to assess the potential to re-use, remanufacture, recycle, or recover the parts or materials from a vehicle.

Recycling: Reprocessing a production process of waste materials for the original purpose or for other purposes, excluding the processing as a means of generating energy.

Reprinted with Permission

Journal of Power Sources 106 (2002) 344–352

JOURNAL OF
POWER
SOURCES

www.elsevier.com/locate/jpowsour

ELSEVIER

Impact of the European Union vehicle waste directive on end-of-life options for polymer electrolyte fuel cells

C. Handley[a,1], N.P. Brandon[b,*], R. van der Vorst[a]

[a]Department of Environmental Science and Technology, Imperial College of Science, Technology and Medicine, London SW7 2BP, UK
[b]Department of Chemical Engineering and Chemical Technology, Imperial College of Science, Technology and Medicine, London SW7 2BY, UK

Abstract

Polymer electrolyte membrane fuel cells (PEMFCs) may well be powering millions of cars by 2020. At its end-of-life, each car will have a redundant PEMFC stack. The EU vehicle waste directive sets tough recycling and re-use requirements for the cars of the future. The criteria for assessing the end-of-life options are based on technical, economic and environmental feasibility. The optimum strategy will require stack dismantling and separation of the major components. Steel and aluminium parts can enter the general recycling stream, but the membrane electrode assembly and bipolar plates will require a specialised recycling process. One option is to shred the MEA, dissolve and recover the membrane, burn off the carbon, and recycle the platinum and ruthenium catalysts using solvent extraction. The heaviest part of the PEMFC stack is the bipolar plates. If carbon fibre based, the bipolar plates could enter a fluidised bed recovery process where the constituent materials are recovered for re-use. The EU vehicle waste directive sets high recycling targets based on weight, and thus it is strongly advisable for the relatively heavy bipolar plates to be recycled, even though energy recovery by incineration may be a cheaper and possible more environmentally benign option. The EU vehicle directive will put pressure on the end-of-life options for the PEMFC stack to be weighted towards recycling and re-use; it will have a significant impact on the design and end-of-life options for the PEMFC. The overall effect of this pressure on the end-of-life treatment of the PEMFC and the consequential contribution to environmental life cycle impacts is discussed. It is concluded that a range of external pressures influence the selection of a suitable end-of-life management strategy, and while opportunities for re-use of components are limited, all components of the PEMFC stack could in principle be recycled. © 2002 Elsevier Science B.V. All rights reserved.

Keywords: Polymer electrolyte fuel cells; End-of-life; Life cycle analysis; Recycling; EU vehicle waste directive

1. Introduction

Polymer electrolyte membrane fuel cells (PEMFCs) are an emerging technology, which offer many advantages over conventional methods of electricity generation. They are under development for both transportation and stationary power applications. Research efforts are presently focused on issues such as stack performance, durability and cost. Information on the present status of PEMFC development can be found in e.g. [1,2].

The increasing emphasis on fuel cells as a candidate power generation system of the future [3] means that there is a growing need to look at the environmental impact of the whole life cycle of the system. This includes the manufacturing, in-use, and end-of-life stages. Such an approach is

termed a life cycle assessment (LCA). LCA is an analytical environmental management tool (see for example [4]) used to inform decision making within environmental product, process and systems design, as well as a recommended step in the implementation of environmental management systems [5]. It is used to assess the environmental burden of a product, process or activity over its entire life cycle starting with raw materials extraction and ending with the final waste disposal. The full process of making this assessment involves: a goal definition and scoping phase, an inventory analysis, an impact assessment and finally an improvement assessment [4].

While the potential environmental impacts of PEMFCs in-use are well documented [6] there is a great deal of uncertainty concerning the environmental impact of the manufacturing and end-of-life stages. A first step towards exploring the potential environmental impacts of the manufacturing stage has been reported by Karakoussis et al. [7] who analysed the materials and energy flows associated with the manufacture of a PEMFC system using an LCA approach. A similar approach has been reported for assessing

* Corresponding author. Tel.: +44-20-7594-5704;
fax:+44-20-7594-5604.
E-mail address: n.brandon@ic.ac.uk (N.P. Brandon).
[1] Present address: AEA Technology, Policy Group, Culham, Abingdon, Oxfordshire OX14 3ED, UK.

Table 1
Materials inventory of a representative 70 kW PEMFC stack [7]

Component	Material	Weight (kg)	Wt.%
Electrode	Platinum	0.06	0.1
	Ruthenium	0.01	0.0
	Carbon paper	4.37	6.3
Membrane	Nafion membrane	5.64	8.1
Bipolar plate	Polypropylene	16.14	23.1
	Carbon fibres	16.14	23.1
	Carbon powder	21.52	30.8
End-plate	Aluminium alloy	2.80	4.0
Current collectors	Aluminium alloy	1.14	1.6
Tie-rod	Steel	2.05	2.9
Total		69.87	100.0

the environmental impact of manufacturing a solid oxide fuel cell system [8–10]. The study reported in this paper extends this earlier work, and then seeks to set it in the context of the likely impact of the European Union vehicle waste directive.

The PEMFC considered was one suitable for use in light duty vehicles, with a power output of 70 kW. The stack construction was based on that detailed in [11]. Table 1 summarises the materials inventory used for this study. End-of-life management options were considered for the individual components of the stack; the Nafion-type membrane, the platinised electrodes, the bipolar plates and the ancillary components (tie-rods, end plates and casing). For each component a range of options was examined, following the waste hierarchy, which provides a framework for the discussion of end-of-life options and their environmental impact. Using the waste hierarchy, which suggests that re-use is (normally) environmentally better than recycling, recycling better than incineration with energy recovery, and incineration better than disposal, an assessment of the options identified was made and an end-of-life strategy suggested. Data for the study was gathered via a series of interviews with key players in the field, together with a review of relevant literature.

2. Results and discussion

2.1. Electrolyte

The fluorinated Nafion-type membrane is the most frequently used membrane in the PEMFC. This type of membrane has a high production cost, with projected Nafion costs of US$ 50 m^{-2} [12]. Hence, both economic and environmental factors drive membrane re-use and recycling.

Re-use of the membrane would be preferred, both in environmental and economic terms. However, there are many barriers to re-use. In particular, membrane dehydration and pin-holing [13] are a common cause of failure.

Dehydrated membranes are unlikely to be re-usable as they are weak, and as they would probably be damaged when the membrane electrode assembly was removed from the stack. Furthermore, contaminants will accumulate on the membrane during use and the structure will be degraded, thus reducing the efficiency of the membrane. A process to mend pin-holes in Nafion exists [14], but could be difficult to apply to a membrane electrode assembly as the resin injected could well damage the gas diffusion layer.

Recycling the membrane is therefore likely to be more feasible than re-using it. To recycle, the membrane must first be removed from the MEA. The membrane is sandwiched between two electrodes so mechanical separation from the carbon structure would be difficult. Chemical extraction of the membrane for recycling is a more viable option. To remove the membrane, it must be dissolved and then returned to its ionomer form.

Dissolution of the membrane and re-casting as a polymer film [15] is one possible method of membrane recycling. The MEA would need to be shredded, the ionic groups on the membrane converted to the Li$^+$ salt, and the membrane dissolved in a water:ethanol mixture at 250 °C under pressure. The soluble membrane could then be separated from the other MEA components, and the solvent evaporated to produce powder. This could then be converted to the Nafion ionomer powder using nitric acid. To re-form a Nafion membrane the Nafion powder would be dissolved in ethanol and dried at ambient temperature. While the resultant film has been described as brittle [15], this is nonetheless a potential method of reproducing the Nafion membrane from an MEA at its end-of-life. The recycled Nafion powder could be re-entered into the Nafion membrane production process.

Major uncertainties in analysing the recycling of Nafion lie in the amount of energy consumed, the cost of the process, and the purity of the recycled Nafion. However, it has been shown that the energy requirements for recycling should be below the 14 kJ kg^{-1} used in the production of Nafion powder from its raw materials [7].

A possible problem when trying to recycle the membrane is that the degraded parts of the polymer may contaminate the recyclate. It may then be difficult to recover the monomer after polymer degradation. Hence, the recyclate may need to be purified during the process, or the membrane boiled in nitric acid prior to recycling to remove the degraded areas.

If the above method can efficiently and cost effectively recycle the membrane, then this appears to be an attractive end-of-life management route. In comparison, incineration is not a favourable option, as highly toxic hydrogen fluoride would be emitted, and a costly HF recovery plant would need to be built alongside the incinerator. Dissolving the membrane out of the MEA is therefore the alternative preferred to incineration or re-use.

Other options to perfluorinated membranes are under development, and are briefly considered here from an end-of-life perspective. Hydrocarbon based membranes offer lower cost than the fluorinated forms, but are less resistant to oxidation. However, absence of fluorine in the membrane means that they can be readily incinerated without hydrogen fluoride emission. Recycling however is likely to be more difficult, due to high levels of chemical degradation within the membrane likely at its end-of-life. One developer has reported a membrane that is both non-fluorinated and offers high performance [26]. In this case recycling may be possible. However, few details of the membrane are available, precluding discussion of end-of-life options. Fig. 1 summarises the end-of-life options for all three membrane types.

2.2. Platinum/ruthenium electrocatalysts

There are fundamental economic and environmental reasons for recycling the platinum and, if present, ruthenium, in fuel cells. The value of platinum and ruthenium in a 70 kW stack is around US$ 1000 at precious metal prices ≈US$ 16 g^{-1}.

The Department of Energy and the USGS minerals information team in USA agree that it will be essential to recycle the platinum in the fuel cell for the product to be sustainable in the long term [16]. The environmental argument for recycling platinum is strong. Emissions of sulphur dioxide (SO_2) are decreased by a factor of 100, and the primary energy demand is reduced by a factor of 20, when the platinum is recycled in comparison to its production from primary sources [17].

Both platinum and ruthenium can be recovered with high yield using a chemical recovery process. The use of solvent extraction to recycle both platinum and ruthenium is well established [18]. The electrolyte membrane must be removed before solvent extraction, as every catalyst particle is in contact with the membrane in the MEA. After electrolyte removal, residual organics must be removed, which would otherwise cause problems in the extraction process, such as crud formation and poor phase separation. Given that the fluorinated polymer has already been removed, it would then be possible to burn-off residual organic material to leave an incinerator ash with high platinum and ruthenium content suitable to enter the extraction process.

A detailed review of the solvent extraction process for the recovery of platinum and ruthenium has been given by Barnes and Edwards [18]. Ruthenium would be extracted before platinum by distillation of the tetraoxide. Platinum is extracted into tri-n-butyl phosphate in 5 M HCl, using a counter-current process of extraction, scrubbing, and stripping of platinum. Residual scrub liquor is recycled back to the extraction stage, or piped away for waste liquor treatment depending on its composition, thus reducing waste emissions. The final stage uses water to strip the platinum from the organic phase and the resultant strip liquor contains pure H_2PtCl_6. Ammonium chloride is used to precipitate out the platinum as 99.95% pure $(NH_4)_2PtCl_6$. Emissions to air from the process include ammonia, chlorine, nitrogen dioxide and hydrogen chloride. Any base metals released during

platinum recycling are precipitated out using lime and then landfilled.

Platinum recycling is crucial to the sustainable future of PEMFCs due to limited platinum reserves, coupled with the saving in energy by a factor of 20 in comparison to extraction from the ore. The platinum industry will see a large increase in recycling if the prediction that the world fuel cell market will exceed 2.1 million fuel cell passenger cars by 2010 is realised. At the current level of platinum loading of 60 g per stack [7], this would result in 126 tonne of platinum becoming available for recycling by 2020 (assuming a 10 year car/fuel cell life). At US$ 16 g^{-1} this would be a US$ 2×10^9 market for platinum and ruthenium recycling.

2.3. Bipolar plates

There are several different materials under development for the fuel cell stack's bipolar plate, including stainless steel, graphitic carbon, and carbon composites.

Graphite-based bipolar plates are used by a number of manufacturers. To reduce cost, volume and weight, resin impregnated graphite plates are also being developed [19]. Graphite is resistant to corrosion, so the bipolar plate should have a lifetime well beyond that of other components. However, even if the bipolar plate was undamaged after use, improvement in fuel cell design will mean that the plate design will be obsolete, such that re-use would not be possible. The only practical end-of-life option for a graphite plate would therefore be to burn it for energy recovery.

Other fuel cell developers are pursuing the use of steel bipolar plates. This is a fairly low cost option. The drawbacks are the potential for corrosion and subsequent damage to the cell. Makkus et al. [20] has suggested that steel bipolar plates are re-usable once cleaned and the outer oxide layer removed. This work also confirms that the steel could be recycled. The recycling option is the most likely, as it is probable that the design of the bipolar plate will be obsolete at cell's end-of-life, such that re-use would not be possible.

Given that the end-of-life options for these two materials appear straightforward, this study focussed on carbon composite bipolar plates. It has been suggested that these have a potential for low cost, large scale manufacture [11]. However, an end-of-life management route is not straightforward. Recycling, and incineration to generate energy, are again the likely two options for the end-of-life management of the carbon composite bipolar plates.

Given that the bipolar plates comprise 70–80 wt.% of the stack, the European Union vehicle waste directive [21] will push the manufacturer to recycle rather than incinerate. This is discussed further in Section 4. Table 1 shows that the carbon composite material comprises carbon fibres, carbon powder and polypropylene polymer. Carbon fibres are the most costly material in the plates, and the carbon powder contributes the largest mass to the plates. Three processes have been identified which could be used for recycling:

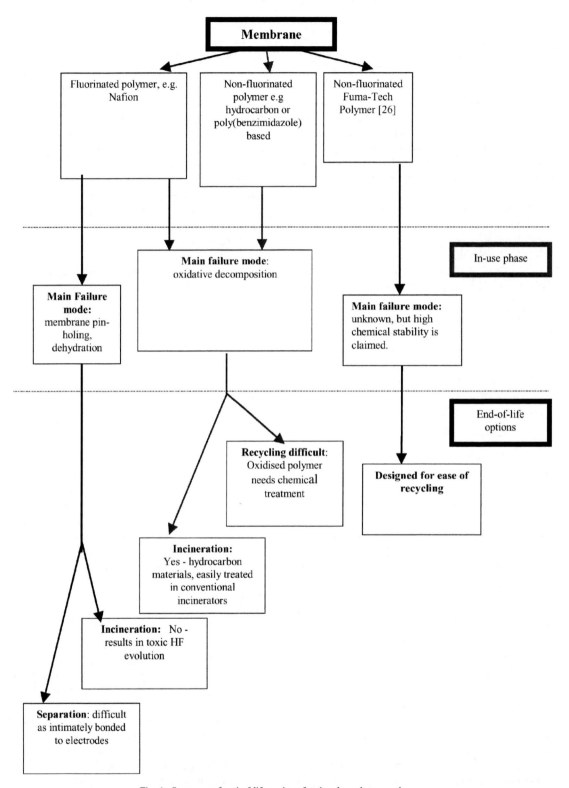

Fig. 1. Summary of end-of-life options for the electrolyte membrane.

catalysed low temperature pyrolysis [22], reverse gasification [22] and fluidised bed fibre recovery [23].

Comparison of these three methods suggests that the fibre quality from the reverse gasification and catalysed low temperature pyrolysis routes may be impaired, which could effect their potential for re-use. Reverse gasification produces high carbon monoxide levels (a major process by-product), which will need to be treated before release. Low temperature pyrolysis operates in a closed system, which minimises environmental emissions, and generates sufficient heat to drive the process without the need for additional fuel.

Fluidised bed fibre recovery appears to be an attractive option. The process is the only one of the three to recover both the carbon fibres and the carbon filler. It has been reported that the quality of the recovered fibres is high [23], which could enable re-use. Avoiding combustion of the carbon filler also reduces carbon dioxide emissions from the process. Given these potential advantages, the use of fluidised bed fibre recovery to recycle carbon composite bipolar plates is considered in further detail in the following paragraph.

The carbon composite bipolar plates are first broken down using a hammer mill. After sizing, the material is fed into a fluidised bed at 450 °C, which breaks down the composite. The filler and the fibres are recovered and the exhaust gases are burnt (with heat recovery). Acid gas scrubbing is used to remove any acid or halogen in the off-gas. It has been estimated that such a processing plant would need to take in around 9000 tonne of composite per year from scrap collected within an 80 km radius [23]. This high tonnage would initially cause a problem if the plant were used solely for bipolar plates—to break even it would require over 160,000 end-of-life stacks per year. Inputting a mixture of carbon composites is an option, although this would mean the carbon fibres could not be recycled directly back into bipolar plates. However, with increasing pressures on car companies to recycle a large proportion of their car, and incorporate recyclates into new cars, there is a potential growth market for recovered carbon fibres.

As stated previously, all three components in the carbon composite bipolar plates are combustible, and thus it would also be possible to burn the plates for energy recovery. The energy value for the composite is 35.6 MJ kg^{-1}, as shown in Table 2, higher than that of coal (26–30 MJ kg^{-1} [22]). As the plates are made of such a high-energy material, the value of energy from waste must not be ruled out.

Certainly, the option of recovering energy from the composite waste would be a straightforward, low cost, flexible option with minimal incinerator ash generated. However, the drawbacks are many, from public opinion to environmental impact. Another problem with energy recovery is the residues of metal and, more importantly, Nafion that may be left on the plates. Emissions of HF in incinerator plumes would be serious and may well stop the plates from being incinerated, as costs of installing a HF recovery plant may be prohibitive.

In the short term, when only small numbers of fuel cell vehicles are reaching their end-of-life every year, it is likely that carbon composite bipolar plates will be incinerated as this capacity already exists. They could also enter into an existing general carbon fibre recovery process, although the grade of fibre produced by a wide variety of scrap would make the re-sale value low and the fibres could not be used in new bipolar plates. The option of transporting low value bipolar plates across the country to a single fibre recovery plant could possibly produce more CO_2 than burning the plate locally in a combined heat and power plant.

Fig. 2 presents a summary of the end-of-life options for the three types of bipolar plate materials discussed in this section.

2.4. Ancillary components

The ancillary components of the fuel cell stack are the non-repeat items. They are significant when considering recycling as they contribute to nearly 16% of the weight of the stack. They consist of the polypropylene housing, the PTFE insulators, the steel tie-rods, the aluminum alloy endplates and the soft aluminum alloy current collectors.

The optimal end-of-life management option for both the steel and aluminium parts of the fuel cell stack is recycling. There are large energy savings if the metals are recycled in comparison to extraction. The polypropylene housing could also potentially be recycled, the energy saving would need to be investigated to assess the feasibility and effectiveness.

The first stage in recovering the steel tie-rods involves removing them from the stack. The rods would be unbolted and, based on present designs, the rods, MEA and bipolar plates should be relatively straightforward to separate. The rods would then be transported to a recycling plant—it is unlikely they could be re-used as after 10 years the dimensions of the stacks will change.

The Boustead Model [24] gives an energy value of 22.4 MJ kg^{-1} to produce steel from iron ore. The International Council on Metals and the Environment, states that recycling steel requires 30–35% of the energy required to manufacture steel from the raw materials [25]. Taking the

Table 2
Energy content of a typical carbon composite bipolar plate[a]

Composition	30 (wt.%) Carbon fibres	40 (wt.%) Carbon powder	30 (wt.%) Polypropylene polymer	Total
Mass per 70 kW stack (kg)	16.14	21.52	16.14	53.8
Mass per kW (kg of fuel)	0.23	0.31	0.23	0.77
Energy recoverable by combustion (MJ kg^{-1})	32 (31)	32 (31)	44 (32)	35.6
Energy recoverable per FU (MJ per FU)	7.36	9.92	10.12	8.19

[a] FU: functional unit (material required to manufacture each kW capacity of PEMFC stack).

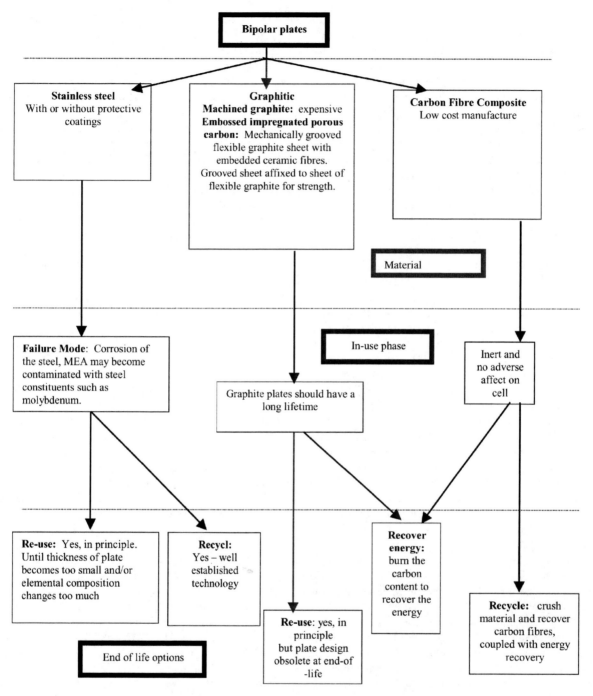

Fig. 2. Summary of end-of-life options for bipolar plates.

mid-value, this gives an energy requirement of 7.3 MJ kg^{-1} for recycling the steel rods. The additional energy to manufacture the steel is 0.03 MJ kg^{-1}, so the entire scrap to new steel rod in the stack process should take around 7.33 MJ kg^{-1}. The infrastructure for recycling scrap car steel already exists along with an excellent market for the recycled product, so these factors will facilitate steel rod recycling. The energy saved also offers a clear advantage from an economic and environmental resource perspective.

The steel rods should not be contaminated with the fluorinated polymer, minimising problems during smelting.

The aluminium end-plates are well positioned for fast removal; they need unscrewing from the stack before the other components can be removed. It has been estimated that recycling requires 5% of the energy required to produce primary aluminium from the ore [17]. This results in the end-plate alloy requiring 28.2 MJ kg^{-1} for recycling and the soft alloy for the current collectors 14.1 MJ kg^{-1}.

Polypropylene is commonly used for the stack casing. This material is already recycled within the automotive industry. This system of cascade recycling, as compared to closed loop recycling—where the recyclate will have the same quality, function and thus value as the primary material—could be used for the fuel cell polypropylene casing, and would allow the integration of recyclate into a new vehicle. The casing would be removed, granulated, compressed and then used as splash shields, wheel guards etc. in new vehicles. To facilitate removal of the casing, design for easy removal should be considered.

3. An end-of-life management route for the PEMFC stack

A proposed end-of-life management strategy for the PEMFC stack is shown in Fig. 3. This assumes a Nafion-type electrolyte and composite carbon fibre bipolar plates. Bipolar plates, platinum electrocatalysts, membrane and ancillary components could all be recycled. The flow chart details the order and the links in the chain for the PEMFC stack at end-of-life, from dismantling to recycling the platinum.

The first step in the dismantling procedure is to remove the fuel cell stack. The next step is to dismantle the stack by removing the casing and unscrewing the steel tie-rods. The rods would go to a steel recycling plant with other steel car scrap, and an equivalent route would be followed for the aluminium end plates. The bipolar plates would enter a fluidised bed treatment process to extract the carbon filler and carbon fibres, potentially for re-use in new bipolar plates.

The MEA would enter a recycling stream different to that of the other components as it contains the fluorine in the polymer, which would emit toxic fumes if heated. It would be advisable to have a specialised treatment plant for the comparatively light but very valuable MEA. Firstly, the MEA needs shredding into small pieces to aid its dissolution. Dissolving the membrane would require ethanol at high

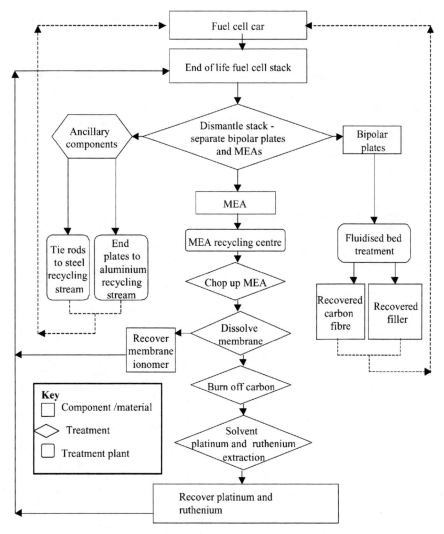

Fig. 3. Proposed flowsheet for the end-of-life treatment of PEMFCs.

C. Handley et al./Journal of Power Sources 106 (2002) 344–352

temperature and pressure, so it would have to be carefully controlled and is likely to be an energy intensive process.

Once the membrane is removed, residual carbon would be burnt off to produce a precious metal bearing ash. This would then enter the solvent extraction process, which in principal is capable of recovering more than 99% of the platinum and ruthenium as a high purity product. This high recovery is essential due to the high demand that fuel cells will put on both platinum and ruthenium resources. It is likely that all of the recovered metal will be required for the manufacture of new fuel cell MEAs.

4. Impact of European Union vehicle waste directive

The end-of-life vehicles directive imposes recycling and re-use regulations on vehicles that will be sold in the future, and also on the vehicle at its end-of-life. The aim of the directive is to harmonise measures concerning end-of-life vehicles, in order to minimise impact on the environment, and to avoid any impact on competition between member states [21]. The directive sets out strict targets, such that vehicles type-approved and put on the market after 1 January 2005 must be re-usable and/or recyclable to a minimum of 85 wt.% per vehicle.

The responsibility for ensuring vehicle take-back and covering the costs of end-of-life process lies with the vehicle manufacturer (OEM). The Society of Motor Manufacturers and Traders estimate the total liability from the take-back scheme at a cost of UK£ 4 billion for the UK Motor industry and UK£ 30 billion for Europe [27]. These high take back costs are largely because the scrap value of recycled IC engine cars is only ~UK£ 20 and the cost of returning the cars, dismantling them and recycling them will be higher. The value of the fuel cell car may well be increased because of the valuable elements, such as platinum, in the fuel cell. The authors suggest that a future scenario is that a specialist tier 1 or 2 company will be contracted by the OEM to recycle the fuel cell stack.

In conclusion, the directive re-enforces the need to recycle and re-use components of the fuel cell stack. With such high recycle and re-use targets, it will put pressure on the car manufacturer to ensure the smaller components, such as the end plates, are recycled even though they make a small contribution to the total mass. The recyclability of every part of the car must be considered in order to reach the target. Issues such as design for recyclability and component labelling will also be important to minimise cost.

5. Conclusions

End-of-life options for the electrolyte, electrocatalysts, bipolar plates and ancillary components of a polymer electrolyte membrane fuel cell (PEMFC) have been considered, and an end-of-life management strategy proposed, in the context of the European Union vehicle waste directive. The optimum strategy will require stack dismantling and separation of the major components. Steel and aluminium parts can enter the general recycling stream, but the membrane electrode assembly and bipolar plates will require a specialised recycling process. One option is to shred the MEA, dissolve and recover the membrane, burn off the carbon, and recycle the platinum and ruthenium catalysts using solvent extraction. The heaviest part of the PEMFC stack is the bipolar plates. If carbon fibre based, the bipolar plates could enter a fluidised bed recovery process where the constituent materials are recovered for re-use. The EU vehicle waste directive sets high recycling targets based on weight, and thus it is strongly advisable for the relatively heavy bipolar plates to be recycled, even though energy recovery by incineration may be a cheaper and possible more environmentally benign option.

Acknowledgements

The authors would like to thank all those who provided data and comments to this paper, particularly A. Kucernak (Imperial College), P. Adcock (Loughborough Univresity), S. Sautley (Ford, UK), P. Davis (US DOE), G. Hards and T. Ralph (Johnson Matthey), R. Lea and S. Sheppard (Inco Europe), I. Kleinwaechter (Degussa Huels Fuel Cells), R. Makkus (ECN), J. McKinley (US Geological Survey), M. Rikukawa (Sophia University, Tokyo), S. Pickering (Nottingham University), and M. Matsukawa (Subaru).

References

[1] D.G. Lovering (Ed.), Proceedings of the 6th Grove Fuel Cell Symposium, J. Power Sources 86 (1/2) (2000).

[2] F.N. Buchi, G.G. Scherer, A Wokaun (Eds.), Proceedings of the 1st European PEFC Forum, Lucerne, Switzerland, 2–6 July 2001.

[3] G. Edge, Markets wake up to fuel cell revolution, FT Energy Econ. 220 (2000) 15–16.

[4] SETAC, 1993, 1994 and 1997.

[5] R. van der Vorst, A. Grafe-Buckens, W. Sheate, JEAPM 1 (1999) 1–26.

[6] Bauen, et al., Fuel cells: clean power, clean transport, clean future, in: D Hart, A Bauen (Eds.), Financial Times Energy Report, 1998, ISBN 1 84083050 6.

[7] V. Karakoussis, M. Leach, R. van der Vorst, D. Hart, J. Lane, P. Pearson, J Kilner, Environmental Emissions of SOFC and PEMFC System Manufacture and Disposal, ETSU Report for the DTI, F/01/00164/REP, (2000).

[8] N. Hart, N.P. Brandon, M. Day, J. Shemilt, in: Proceedings of the Fuel Cell 2000, Lucerne, Switzerland, 10–14 July 2000, pp. 389–397.

[9] V. Karakoussis, N.P. Brandon, M. Leach, R. van der Vorst, J. Power Sources 101 (2001) 10–26.

[10] N. Hart, N.P. Brandon, J.E. Shemilt, Mater. Manufact. Processes 15 (2000) 47–64.

[11] F.D. Lomax Jr., B.D. James, G.N. Baum, C.E. Thomas, Detailed manufacturing cost estimates for polymer electrolyte membrane (PEM) fuel cells for light duty vehicles, August 1998, Prime Contract DE-AC02-94CE50389 to the US Department of Energy Office of Transportation Technologies.

[12] F.R. Kalhammer, P.R. Prokopius, V.P. Roan, G.E. Voecks, Status and prospects of fuel cells as automobile engines, Fuel Cell Technical Advisory Panel (FCTAP), State of California Air Resources Board, July 1998.

[13] J. Matthey, Fuel cell development research group, Interviewed 9 May 2000, personal communication.

[14] W.G. Grot, Production of a liquid composition containing a perfluorinated ion exchange polymer, and the use of this polymer, EP 00066369A1, 8 December 1982, Du Pont de Nemours and Co.

[15] A.T. Tsatsas, W.M. Risen Jr., J. Polym. Sci. Part B: Polym. Phys. 31 (1993) 1223–1227.

[16] J.M. McKinley, US Geological Survey, Minerals Information Team, 2000, personal communication.

[17] M. Pehnt, Proceedings of the Fuel Cell, Lucerne, Switzerland, 10–14 July 2000, pp. 367–378.

[18] J.E. Barnes, J.D. Edwards, Chemistry and Industry, 6 March 1982, pp. 151–155.

[19] R.A. Mercuri, J.J. Gough, Flexible graphite for use in the form of a fuel cell flow field plate, US Patent 6,037,074, UCAR Carbon Technology Corporation, 14 March 2000.

[20] R.C. Makkus, A.H. Janssen, F.A. de Brujin, R.K.M. Mallant, J. Power Sources 86 (1/2) (2000) 274–282.

[21] 300 L0053 Directive 2000/53/EC of the European Parliament and of the Council of 18 September 2000 on end-of life vehicles, Commission Statements, OJ L 269, 21 October 2000, p. 34.

[22] J. Scheirs, Polymer rRecycling, Science Technology and Applications, Wiley, New York, 1998, ISBN 0-471-97054 9.

[23] S.J. Pickering, R.M. Kelly, J.R. Kennerly, C.D. Rudd, N.J. Fenwick, Composites Sci. Technol. 60 (2000) 509–523.

[24] Boustead Model 4.0 and 4.1, Boustead Consulting Ltd., 1999.

[25] M.E. Henstock, The recycling of non-ferrous metals, International Council on Metals and the Environment, ISBN 1-895720-11-7 (1996).

[26] FuMA-Tech GmbH, 2000, personal communication.

[27] SMMT Press release, 24 May 2000, New Euro scrap car rules threaten UK manufacturing.

V. DEVELOPMENT, TESTING AND LIFE CYCLE ISSUES

Test Methods

2004-01-1008

Hydrogen Consumption Measurement for Fuel Cell Vehicles

Yi Ding, John Bradley, Kevin Gady, Mitch Bussineau, Tom Kochis, Ed Kulik
and Virgo Edwards

Ford Motor Company

ABSTRACT

Fuel cell vehicle fuel consumption measurement is considerably different from internal combustion engine vehicle fuel consumption measurement. Conventional Carbon Balance Method and Flow Measurement methods for gas consumption within combustion engines are not suitable for fuel cell vehicles. The small quantities of fuel consumed and the characteristics of hydrogen itself impose a challenge for the hydrogen measurement. This paper addresses fuel consumption measurement for fuel cell vehicles using various methods such as mass flow measurement, pressure/temperature/volume method, weigh method as well as other methods. The advantages and disadvantages of these methods are discussed.

INTRODUCTION

Fuel economy (FE) measurement is an important concern to both vehicle manufacturers and oil formulators, primarily because of environmental and economic concerns. Many companies including Ford, Honda, Toyota, and Daimler-Chrysler have announced plans to put production fuel cell vehicles on the market around 2004. Many auto companies have built and tested the prototype fuel cell vehicle already. Fuel cell vehicles are claimed to be significantly more fuel-efficient and emit fewer emissions than conventional internal combustion engine vehicles or no emissions whatsoever. Vehicle certification becomes an immediate concern. Hydrogen consumption measurement is essential to vehicle certification. Typical fuel consumption measurement methods for combustion vehicles include: the carbon balance and the direct flow measurement methods. The first method is based upon the concept that the carbon content in the exhaust gas should be the same as that in the fuel. The second method is to measure fuel consumption directly with a flow meter. For fuel cell vehicles, the carbon balance method is obviously not appropriate. Fuel cell vehicles are highly efficient and hydrogen consumption is very small as compared to gasoline consumption in a corresponding combustion vehicle. Direct flow measurement become a problem because the 1% measurement accuracy requirement imposes a big restriction to flow meter

selection for a relatively small hydrogen consumption of fuel cell vehicles during standard drive cycles.

Fuel cell vehicles that use pure compressed hydrogen gas as a fuel are considered zero-emission vehicles. Hydrogen measurement is thus the key to determine vehicle fuel economy for EPA certification. Several methods were considered for hydrogen consumption measurement: pressure/temperature measurement, direct mass flow measurement, scale weight method, water content, electrical current measurement etc.

The water measurement method yields a big error in the Hydrogen consumption measurement. Fuel cells utilize a catalyst to perform a chemical process that combines the hydrogen fuel and oxygen from the air to generate electricity and water. The water emitted in the vehicle pipeline is measured and thus allows one to estimate the hydrogen consumption. However, un-reacted hydrogen will not be counted with this method. Hydrogen and oxygen humidification will also affect the precision of this method.

Electrical current method for hydrogen consumption measurement also provides certain errors. For this method, the hydrogen consumption is calculated by measuring the output current of the fuel cell stack. Since electric current from a fuel cell is generated by a flow of hydrogen ions, the hydrogen consumption can be determined by integration of electric current values.

In the Weight measurement method, the hydrogen consumption is by measuring the weight of the high-pressure fuel tank before and after the test. The tank used for the test should be suitable for measuring weight. This method is simple and relatively easy. However, it has the following problems:

The hydrogen tank is usually relatively heavy and hydrogen consumption for fuel cell vehicle is relatively small during a drive cycle. Thus, it is not easy to find a scale for hydrogen tank weight measurement but still provide reasonable measurement accuracy. It cannot be used for on-board measurement.

The pressure changes within tube (which connects high-pressure hydrogen tank and vehicle) will introduce measurement errors if not accounted for.

In the **Pressure/Temperature method**, the hydrogen consumption is calculated from measurements of the pressure and temperature of gas in the high-pressure fuel tank before and after the test. The hydrogen consumption is determined from the change in the number of moles of gas in the storage tank by applying the measured values of pressure and temperature to the state equation. The hydrogen density data can be obtained from the specific pressure and temperature data from the NIST website below:

http://webbook.nist.gov/chemistry/fluid/. A tank with known internal volume that allows the measurement of gas pressure and temperature should be used for this test. Pressure and temperature sensor accuracy as well as tank size is the key to determine the hydrogen consumption measurement accuracy for this method.

For Direct Hydrogen Flow Measurement, the flow meter should be installed in the fuel pipe between the fuel supply source and the fuel cell vehicle. This is the most convenient way to measure hydrogen consumption during dynamometer testing or even road testing. With most liquid or gas flow measurement instruments, the flow rate is determined inferentially by measuring the liquid's velocity or the change in kinetic energy. Velocity depends on the pressure differential that is forcing the liquid through a pipe or conduit. Because the pipe's cross-sectional area is known and remains constant, the average velocity is an indication of the flow rate. Volumetric technologies-such as differential pressure (dP), turbine, vortex or ultrasonic flow meters-follow a complicated path to deliver these measurements. This is because the technologies require compensation for gas density fluctuations, due to the compressible nature gases. Other factors that affect liquid flow rate include the liquid or gas' viscosity. Temperature will be one of the factors that affect this measurement accuracy. The hydrogen flow and its large flow range will likely cause the temperature change. The pressure measured at certain temperatures need to be converted into standard conditions. This requires an accurate temperature sensor with quick dynamic response. This is typically not practical.

There are other methods being used for flow measurement such as the Coriolis meter, which measure mass flow directly. The Coriolis meter method can calculate SCF and Nm^3 directly without the need to compensate for operating pressure, temperature and composition.

Experimental

Based upon the above analysis, the Coriolis mass flow measurement, the pressure/temperature method, and weigh method are the three most promising methods for accurate hydrogen consumption measurements. We concentrated on these three methods in our experiments. The mass flow meter measurement was compared to that of a precision scale measurement for net weight change in the fluid supply tank, and the NIST pressure-temperate state change of the fluid in the supply tank. Theoretically, all three measurements should provide the same fuel consumption result given no experimental error.

Equipment List:

Name	Specification
GFI Regulator	3600 psi max inlet
Dynetek Gas Tank	33L, weight grams
GFI Tank Regulator/Valve	Opens with 12V, 1Amp input
Rosemount Tank Pressure Sensor	Better then 1% accuracy
Tank Temperature Sensor from Analog devices	Better than 0.5°C accuracy
Micromotion Coriolis Mass Flow Meter	Better then 0.35% accuracy
A&D Precision Scale	Accurate to 0.1grams
Fuel Cell Emulator	Max sustained flow rate: 800SLPM
	Transient response: 400 ms
Parker Valves and Fittings	All stainless steel
6033A Power Source	

Test Procedure and Equipment Specifications

Figure 1 illustrates the hydrogen consumption measurement comparison experimental setup.

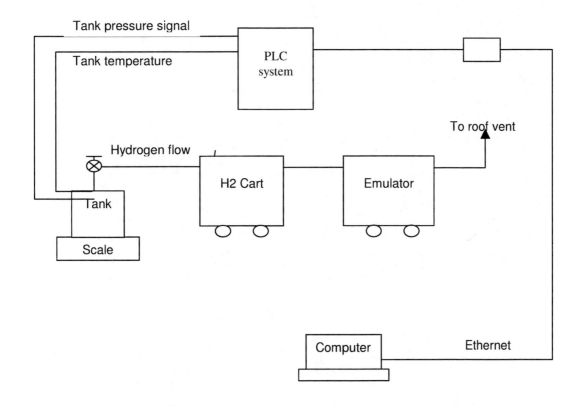

Figure 1. Hydrogen consumption measurement test configuration

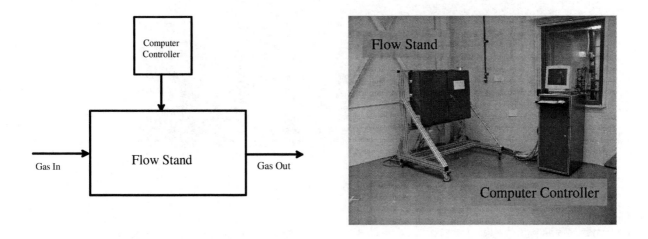

Figure 2. Block Diagram and Photograph of the Fuel Cell Emulator.

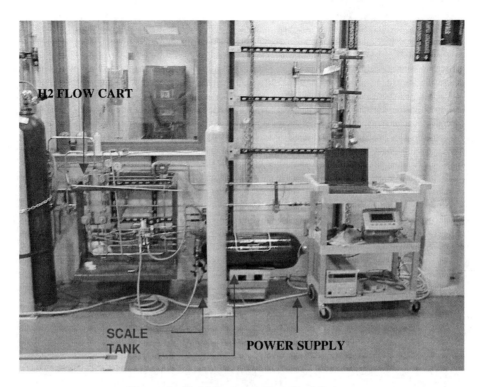

Figure 3. Fluid Consumption Test Set-up

Figure 4. Zoom-in on Regulator, Tank Pressure Sensor, & Mass Flow Sensor

Generic Test Procedure:

1. Fill the Dynetek tank with the test fluid (tank solenoid will open during fill with 30psi differential)
2. Place the tank on the scale, tape down the pressure senor and temperature sensor wires so then hang loosely from the tank, and record the weight.
3. Close manual vent valve in Cart2
4. Connect the regulator input flex line to the tank output fitting.
5. Connect the pressure relief flex line to the tank relief fitting.
6. Open the tank solenoid with 12V and 1Amp from the power supply.

446

7. Check for fitting leaks around tank fittings with Snoop.
8. Open both manual valves in the regulator input line. (Now Cart2 is pressurized up to the first servo valve).
9. Person in control room will record tank pressure and temperature, and totalize the mass flow from mass flow meter.
10. Person in control room will open the Cart2 servo-valves
11. Person in the lab will open the manual vent valve and control its position. This allows fluid to flow through the cart and vent to the atmosphere.
12. Record the mass flow data.
13. After finished collecting data, shut manual vent valve.
14. Close regulator-to-tank manual valves and the relief line manual valves, and disconnect both lines.
15. Person in lab will record tank weight.
16. Person in control room record the tank pressure-temperature, and mass flow reading.
17. Compare the three measurements.

Results and discussions

Tables 1, 2, & 3 list the results for the fluid helium, nitrogen, and hydrogen consumption testing with respect to the test fluid. The testing fluid flow was controlled by Fuel Cell Emulator, which simulates different vehicle drive cycles, including FUDS, HWFET, FTP-75 and hot 505 etc. The table lists the mean for each measurement.

Table 1: Cart2 Fluid Consumption Results for Helium

Test Date	$W_{P, T}$ (g)	W_{Scale} (g)	W_{MF}(g)	Mean (g)
7/31/02	23.97	23.9	23.9	23.92
7/31/02	24.01	22.7	24.2	23.64
7/31/02	47.16	48.8	47.2	47.72
8/1/02	112.94	114	112.4	113.11
8/1/02	133.34	135.9	134.1	134.45

Table 2: Cart2 Fluid Consumption Results for Nitrogen

Test Date	$W_{P, T}$ (g)	W_{Scale} (g)	W_{MF}(g)	Mean (g)
8/6/02	209.1	210	207.5	208.87
8/6/02	310	316.1	307.8	311.30
8/6/02	403.29	400	404.8	402.70
8/7/02	61	46.8	35.4	47.73
8/7/02	306.5	309.3	304.3	306.70
8/7/02	302.1	305	300	302.37
8/7/02	162.3	162.9	161	162.07

Table 3: Cart2 Fluid Consumption Results for Hydrogen at different drive cycles

Test Date	$W_{P, T}$ (g)	W_{Scale} (g)	W_{MF}(g)
9/26/02	135.7	135.5	134.9
9/26/02	137.1	138.1	137.2
9/26/02	150.6	152.1	151.5
9/26/02	149.7	151.1	150.6
9/26/02	65.2	65.5	65.2
9/27/02	192.5	193.8	192.2
9/27/02	64.7	65.4	65
9/27/02	66.2	67.6	65.3

The above data points out that the hydrogen consumption measurement for mass flow measurement and pressure/temperature are quite consistent, especially for the measurement performed after Sept.3, 2002. The hydrogen content change in the flexible tubing of the test setup is considered for each test since that time. Error contributions for each of these three methods are discussed separately below.

Table 4 lists the data with the pressure/temperature and flow measurement methods at different steady state hydrogen flow rates. The H2 consumption data from pressure/temperature and flow measurement methods are very close. Figure 5 illustrates the H2 measurement difference with the two methods.

Table 4

Sequence	Flow rate kg/hr	Initial Pressure P1	Final Pressure P2	Initial Temperature T1	Final Temperature T2	Total H2 (g) (flow)	Total H2 (pressure method)	Difference
1	0.07	961.2	508	68.5	66.6	81.3	80.9	0.49%
2	0.09	1004	502	79.3	70.3	87.1	86.3	0.92%
3	0.09	1063	519	81.8	71.6	92.5	92.7	-0.22%
4	0.24	1193	668	80.8	70.6	86.2	85.4	0.93%
5	0.41	1008	509	76.5	69.7	88.6	87.6	1.13%
6	0.85	1023	517	80.8	71.6	87.2	86.6	0.69%
7	0.14	1051	556	78	69.4	87.3	86.6	0.80%
8	2.1	967	471	80.7	72	86.1	85.6	0.58%
9	2.8	1044	517	81.2	71.4	90	90	0.00%
10	2.8	1137.5	551.8	82.1	70.9	99.2	99.2	0.00%
11	3.65	1021	508	78.7	70	88.9	88.3	0.67%

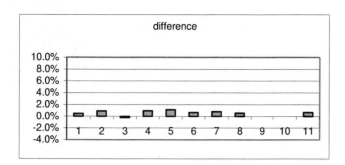

Figure 5. H2 consumption difference with P/T and flow measurement methods

Weight Method Error

A & D's precision scale is quoted to be accurate to 0.1grams (rated up to 30kg max). This is excellent precision, however, its actual precision is 1 gram because its reading drifts with time randomly even though it is quoted to be 0.1 gram. In addition, eliminating experimental error has also proven to be a daunting task. For example, ambient air blowing over the scale can affect its readings by as much as 1 to 2 grams. It was found that to reduce experimental error from the tank weight measurements the environmental conditions around the tank must remain constant. To accomplish this:

The tank weight should not be measured with the flex lines attached to both the output and relief valves of the tank. The stiffness of the flex line changes with the changing pressure of fluid being drained from the tank. This has a significant impact on the weight measurement of the tank, because the attached flex lines provide a moment to the tank.

The electrical connections to the tank (i.e. solenoid power) and pressure sensor should be taped down after the tank has been positioned on the scale. While taping the wires, make sure they are not taut, to avoid affecting the weight measurement. Movement of the electrical wiring does effect the weight measurement. The flex lines create a moments force hanging from the tank.

The tank's position on the scale should not change throughout the test.

H2 content change due to pressure change before and after testing within flexible tubing connecting valve and tank need be considered in the H2 calculation.

The scale should be recalibrated every time the tank is removed from the scale for re-fueling.

The tank should be continuously leak checked to make sure the integrity of the couplings have not been compromised by attaching/detaching the flex lines.

After the pressure change in the flexible tubing had been accounted for (after September 13) the hydrogen consumption measured with scale coincide with the data from the other two methods very well.

b. Pressure/Temperature Method Error.

The sensors used to measure the tank pressures and temperatures are extremely accurate (less than .4% for pressure sensor and 0.5 °C as quoted by the manufacturer). Considering that the full range of the pressure sensor is 4000 psi, the maximum error of the sensor on the pressure measurement will be 1.6psi. The1.6psi error for a small tank used here (33 liters) at 3000 psi will provide about a 0.25g error. The result indicates that the pressure/temperature method should provide very accurate hydrogen consumption measurement. However, they do not record at the same sampling rate. This could contribute to some measurement error, because the instantaneous pressure reading does not always correspond to the temperature reading at that precise moment. In order to reduce error, the tank pressure/temp should be recorded at steady state temp/press conditions. However, it is time consuming to wait for the tank to come to steady state, especially since the tank temperature can reach 90°F after a fill-up to 2300psi.

Measurement of the fluid consumption in the tank can be accomplished by recording the initial tank temperature/pressure state prior to the test and the final tank temperature/pressure after the test has completed. Using NIST (Nat'l Institute of Standard & Technology) experimental data look-up tables for the respective test fluid, the change in pressure/temperature state will give the amount of fluid mass consumed during the test.

c. Mass Flow Meter Experimental Error

Our testing did not reveal any apparent experimental error with the mass flow meter. There is one notable comment. The mass flow measurement shown on the LED screen does not match the true frequency measurement recorded on the meter. There is some filtering in the PLC program, which creates the variation in the values. In addition, the relatively big error of flow meter at low hydrogen flow speed, e.g., during fuel cell system idle etc., also contributes to the mass flow error.

CONCLUSION

The testing results indicated that the pressure/temperature method typically provides more consistent, more precise and traceable results. The flow meter method provides consistent result which is comparable to the pressure/temperature method. The weigh method provides somewhat larger errors as compared with the other two methods because of accuracy limitations and testing errors. Pressure changes in flexible tubing that connected the regulator and gas tank before and after the gas flow testing may contribute about 1-2 grams hydrogen usage error. In addition, the scale itself used in these experiments may contribute more than 1 grams error for each reading.

ACKNOWLEDGMENTS

Authors would like thank Drs. Alex Bogicevic and Ken Hass for the many valuable discussions.

CONTACT

yding@ford.com

Tel: 313-248-4738

2004-01-1013

The Fire Tests with High-Pressure Hydrogen Gas Cylinders for Evaluating the Safety of Fuel-Cell Vehicles

Yohsuke Tamura, Jinji Suzuki and Shogo Watanabe
FC/EV Center, Japan Automobile Research Institute

ABSTRACT

The high-pressure hydrogen gas cylinder of a fuel-cell vehicle is equipped with a pressure relief device (PRD) to prevent the rupture of the cylinder due to heating by fire. Flame exposure tests (bonfire tests) are conducted to evaluate the safety of the cylinder with the PRD, specifically, cylinder resistance to fire and performance of the PRD. In this study, however, fire tests of vehicles equipped with high-pressure cylinders were not required for this test method.

We implemented released-hydrogen flame tests by performing bonfire tests and fire tests on vehicles equipped with hydrogen-filled high-pressure gas cylinders (20,35MPa) to examine safety measures for fuel-cell vehicles. We then investigated the following: the characteristics of the released-hydrogen flame, radiation heat flux from the jet flame, combustion noise, the rate of pressure rise in the cylinder, the venting direction of the PRD, and behavior of fire in conjunction with a gasoline flame. The results demonstrate that it is necessary to specify the venting direction of the PRD, as well as the details of the shielding for PRD to be used to prevent direct flame impingement.

INTRODUCTION

In fuel-cell vehicles, it is necessary to install a pressure relief device (PRD) in the high-pressure hydrogen gas cylinder in order to prevent rupture due to heating by fire. The PRD detects cylinder heating by temperature (through fusible material), pressure (by a rupture disk), or a combination (by a Combination Relief Device), and vents the combustible gas from inside the cylinder to the atmosphere before the cylinder can rupture.

To confirm whether a cylinder is sound, the bonfire test exposes it to fire using draft test methods specified by the International Organization for Standardization (ISO) for high-pressure hydrogen gas cylinders for vehicles (ISO/CD15869). The ISO draft specifies that the bonfire test on the high-pressure gas cylinder be conducted with a PRD installed. The fire source shall be a uniform flame that envelops the entire cylinder, and a metallic shield shall be installed to prevent the flame from directly

contacting the PRD. The combustible gas vented from PRD may be vented at a location away from the fire source via a vent pipe.

Consider a case in which a vehicle equipped with a high-pressure hydrogen gas cylinder encounters a fire. When the PRD is activated, hydrogen gas is vented to the atmosphere via the vent pipe installed on the PRD. If the fire comes into contact with the released combustible gas, a hydrogen jet flame is formed around the vent tube discharge opening. In order to ensure safety during fire fighting and rescue activities, it is necessary to investigate the impact of the hydrogen jet flame when the PRD is activated.

High-pressure cylinders may encounter external fire (e.g., from fuel spilled from a collision, or fire spreading from another vehicle). Additionally, the cylinders may encounter fire from the vehicle itself (e.g., arson or fire from the electrical, exhaust or fuel systems). Therefore, it is necessary to examine the influence of the hydrogen jet flame and the relevance of the bonfire test in case of a vehicle fire caused by different fire sources.

As a first step for assuring the safety of fuel-cell vehicles equipped with high-pressure hydrogen gas cylinders, we investigated three areas: 1) Hydrogen jet flame when the PRD is activated by performing the bonfire tests; 2) Hydrogen jet flame of the vehicle fire tests by simulating a high-pressure (20,35MPa) fuel system for fuel-cell vehicles; and 3) The problems revealed by the bonfire tests.

VENTING FLAME TEST

To examine the best methods for fire fighting and rescue in the event of a fuel-cell vehicle fire when the PRD is activated, we investigated flame size and the influence of the hydrogen jet flame on flame temperature, radiant heat flux, combustion noise, duration of flame, etc. Factors affecting these phenomena include the venting direction and inner diameter of the vent pipe, cylinder internal pressure and capacity, activating method for (and flow characteristics of) the PRD, and location of the fire source. We implemented tests on flame vented from a high-pressure hydrogen cylinder with a pressure of

35MPa when the PRD is activated using three parameters: venting direction of the vent pipe, inner diameter of the vent pipe, and capacity of the cylinder.

EXPERIMENTAL APPARATUS AND PROCEDURE

Definition of Bonfire Test

In order to simulate the state of released-hydrogen flame when the PRD is in operation, we activated the PRD by a Bonfire test applied to high-pressure cylinders for vehicles and observed the hydrogen flame released from the vent pipe. We conducted this test at Powertech Labs at the Provincial Fire Training Center, British Columbia, Canada. The Bonfire test procedure complied with ISO-11439 (Gas Cylinders – High-pressure cylinders for the onboard storage of natural gas as a fuel for automotive vehicles).[1] Cylinder internal pressure and cylinder surface temperature were measured based on this method.

Venting Method

Table 1 indicates the test conditions.

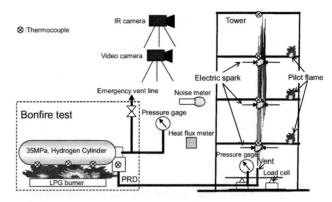

Figure 1. Schematic of testing diagram for hydrogen gas to be released upward (T-1, T-3, T-4, T-5)

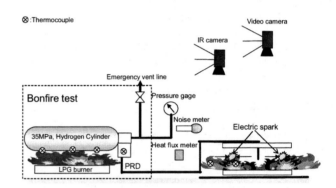

Figure 2. Schematic of testing diagram for hydrogen gas to be released downward (T-2)

Table 1. Test condition

#	Direction of Vented Gas	Internal Diameter of Venting Tube [mm]	Cylinder Type	Cylinder Capacity [Liters]
T-1	Upward	7.2	Type4	45
T-2	Downward	7.2	Type4	45
T-3	Upward	10.7	Type3	34
T-4	Upward	7.2	Type3	34
T-5	Upward	4.2	Type3	34

Top and bottom gas venting directions were examined. Three stainless steel pipes with nominal diameters of 1/2", 3/8", and 1/4" inch were used for the vent pipe, and the actual measured internal diameters were 10.7, 7.2, and 4.2 mm. Two types of cylinders were used: Type 4 (45-liter capacity) and Type 3 (34-liter capacity). A PRD with an activating temperature of 110+/-4 °C (nominal diameter 6 mm) was installed on each of these cylinders. Figure 1 (2) illustrates the test apparatus for venting upward (downward).

To activate the PRD, the cylinder was heated directly with a liquefied petroleum gas flame. When the PRD was activated, hydrogen gas was sent through stainless steel piping (16m length, 1/2" diameter) and released into the atmosphere via the 500mm-long vent pipe. An electric spark or pilot flame ignited the hydrogen gas released from the vent pipe. The test rig for downward venting is shown in Figure 3.

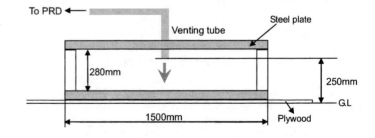

Figure 3. Schematic diagram of venting rig for venting towards the ground. (T-2)

452

The downward venting gas jet is affected by the structure of the vehicle, particularly lower parts of the vehicle. In this study, we assumed a jet flame released from a vehicle with a ground clearance of 280 mm and an overall width of 1.5 m. We therefore installed a vent pipe on the center of a steel plate (1.5m square) and released hydrogen gas from a height of 250 mm from ground level.

Measurement

Flame size, flame temperature, radiant heat flux, combustion noise and duration of flame were measured while the PRD was operating. Since a released-hydrogen flame is difficult to detect visually in direct sunlight, flame size (height and diameter) was measured using images taken by an infrared camera system (Model # TVS8502, AVIO Corp.) with a frame rate 30 Hz.

To measure the released-hydrogen flame, the emissivity of the infrared camera image was compensated until the temperatures indicated by the thermocouple and the infrared camera image were the same. Under those conditions the emissivity was 0.13. At the flame boundary of the released flame during a nighttime test, the temperature of the compensated infrared camera image was 700°C. Observations were made every 1/30 seconds when the temperature boundary by infrared camera image was also 700°C (defined as the flame boundary with continuous flame).

The flame size obtained by this procedure was generalized by the mass flow rate calculated from the volume of pressure decrease in the cylinder or by the static pressure just before spouting from the vent pipe (hereafter referred to as "jet pressure").

Thermocouples (K-type, sheath diameter 3.2 mm) were used to measure flame temperatures. During the upward venting test, measurements were made at four locations above the vent pipe (H=0.4, 2.4, 4.4, and 6.4 m). During the downward venting test, measurements were made at two locations from the vent pipe (1m and 2m) horizontally at a height of 250mm above ground level. We used a thermocouple with a sheath diameter of 3.2 mm (if a thermocouple with a smaller diameter was used, the thermocouple itself would be melted by the hydrogen jet flame, thus invalidating the measurements).

A heat flux gauge (MEDTHERMO Corp., Schmidt-Boetter type, range 0 to 150 kW/m^2) was used to measure radiant heat flux. Figure 4 depicts the measurement points for upward and downward venting.

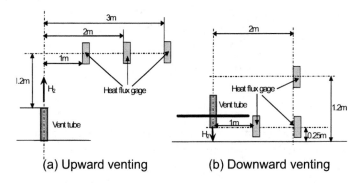

(a) Upward venting   (b) Downward venting

Figure 4. Measurement positions for each heat flux gage

A noise meter (Model #NL-20, RION Corp.) was used to measure combustion noise. Combustion noise was evaluated by the C-Frequency weighting networks, which provide guidelines for the sound volume over the entire frequency band audible to humans, while suppressing the influences of wind noise and other sounds. Horizontal measurement points were located at 4m and 10m from the vent pipe (at a height of 0.5m above the vent pipe discharge opening).

It should be noted that thrust acts on the vent pipe when the PRD is activated and hydrogen gas is released.

Since this data is necessary to establish the conditions for installing the vent pipe, the thrust applied to the vent pipe was measured in the upward venting test. Three load cells (Model #LCFA-500N, KYOWA Corp.) were attached to the fixed portion of the vent pipe, and the thrust was calculated by summing these loads. The resulting data were recorded using a recorder (Model #8842, 12 bit, HIOKI Corp.) at sampling intervals of 1ms.

Cylinder Conditions Just Prior to PRD Activation

Table 2 shows the filling pressure of the cylinder used for each test, the pressure at PRD activation, and the duration for the venting gas obtained from the bonfire test results.

Table 2. Cylinder condition (Filling pressure, Pressure at the PRD activation, Duration)

#	Filling Pressure [MPa]	Pressure at the PRD activation [MPa]	Duration [sec]
T-1	33.55	33.58	61.1
T-2	32.59	32.63	58.6
T-3	29.97	30.77	32.7
T-4	30.42	30.66	32.8
T-5	29.73	30.52	55.3

Using the results obtained under these test conditions, we investigated the influence of different cylinder conditions on the released flame by comparing the venting directions between T-1 and T-2; venting diameters of T-3, T-4 and T-5; and the capacities of the cylinder between T-1 and T-3.

COMPARISON OF THE VENTING DIRECTION

Flame Behavior

Released-hydrogen flames were compared for upward and downward venting. Figure 5 (6) presents consecutive photos of the hydrogen jet flame vented upward (T-1) (downward (T-2)). Figure 7 depicts infrared camera images of T-1. In both cases, the flame rapidly became smaller approximately 10 seconds after PRD activation. Furthermore, intermittent flames were not observed in the hydrogen jet flame under these test conditions, although intermittent flames at intervals of several seconds were observed in a pool fire using hydrocarbon fuels (gasoline, diesel fuel, etc.) when the pool diameter was 20 cm or larger.[2)3)]

Flame Length

Figures 8 and 9 illustrate the flame height and diameter of the hydrogen jet flame obtained from infrared camera images.

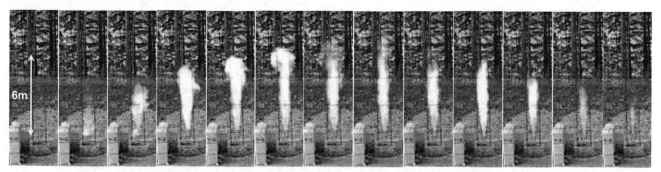

0sec 1/30sec 2/30sec 3/30sec 4/30sec 5/30sec 6/30sec 10/30sec 1sec 2sec 5sec 10sec 20sec

Figure 5.   Visible image of hydrogen jet flame from venting tube during PRD activated in T-1. Hydrogen gas was vented out upward. Filling pressure before bonfire test is 33.55MPa. Maximum pressure just before the PRD activated is 33.58MPa. Venting duration was 61.1 seconds. Vent pipe I.D. is 7.2 mm. Cylinder capacity is 45 Liters.

0sec   2/30sec   4/30sec   10/30sec

1sec   2sec   5sec   10sec

Figure 6.   Visible image of hydrogen jet flame from venting tube during PRD activated in T-2. Hydrogen gas was vented out downward. Filling pressure before bonfire test was 32.59MPa. Maximum pressure just before the PRD activated was 32.63MPa. Venting duration was 58.6 second. Vent pipe I.D. is 7.2 mm. Cylinder capacity is 45 Liters.

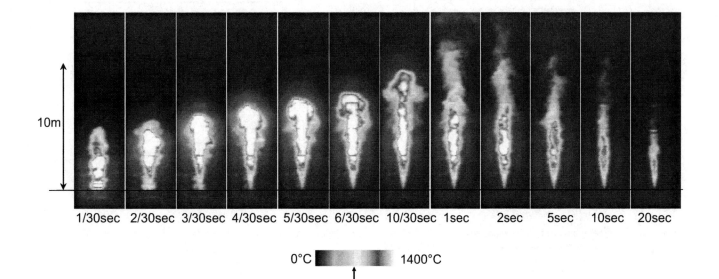

Figure 7.   Infrared image of hydrogen jet flame from venting tube during PRD activated. (Upward, T-1)

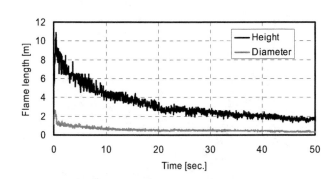

Figure 8. Flame height and diameter for venting upward (T-1). Maximum flame height is 10.9 meters at 0.50 seconds. Maximum flame diameter is 2.6 meters at 0.30 seconds.

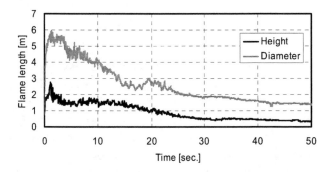

Figure 9. Flame height and diameter for downward venting (T-2). Maximum flame height is 2.7 meters at 1.23 seconds. Maximum flame diameter is 6.0 meters at 1.33 seconds.

In upward venting, the maximum flame height is 10.9m (after 0.5 seconds of PRD activation). In downward venting, the maximum flame diameter is 6.0m (after 1.33 seconds of the PRD activation). In either case, flame height or diameter decreased below 4m after 12 seconds of PRD activation, and the release of burning hydrogen ended within 1 minute.

Flame Temperature

Figure 10 (11) presents the temperatures of jet flames vented upward (downward).

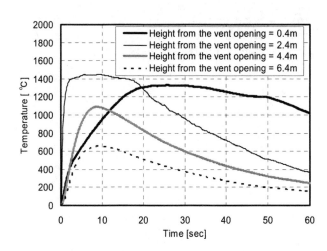

Figure 10. Flame temperature along the center axis of hydrogen jet flame for upward venting (T-1). The thermocouples are fixed on the vertical vent tube discharge opening. Notice that temperature is not corrected for radiation loss.

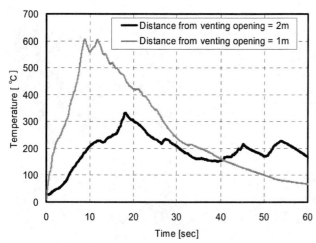

Figure 11. Flame temperature for downward venting (T-2). Heights of the measurement points are the same as the venting tube discharge opening. Notice that temperature is not corrected for radiation loss.

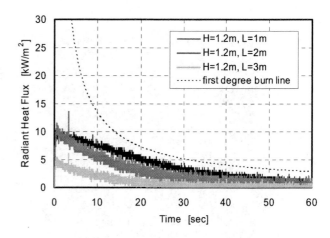

Figure 12. Radiant heat flux for upward venting (T-1). (H = vertical distance from ground level, L = horizontal distance from the venting tube discharge opening). Time axis of the first-degree burn curve represents the continuous exposure time of the human body to the fixed radiant heat. For example, a human body exposed to $10KW/m^2$ for 14 seconds will receive first-degree burns.

In upward venting, the maximum temperature 2.4m above the vent pipe was 1400°C or more (duration 20 seconds), which exceeds the measurement range of the K-type thermocouple and is higher than the temperature in any other position. The temperature at a height of 0.4m gradually rose with time because the area of non-burnt gas at the discharge opening gradually decreased as the flow rate decreased. In downward venting, the maximum temperature at a height of 250 mm from ground level (the same height as the discharge opening), and at a radius of 1m, reached approximately 600°C (duration 5 seconds).

Radiant Heat Flux

Figure 12 (13) illustrates the upward (downward) heat fluxes radiated from the flame toward a plane. These figures also depict the radiant heat flux and permissible time until first-degree burn injuries occur.[4]

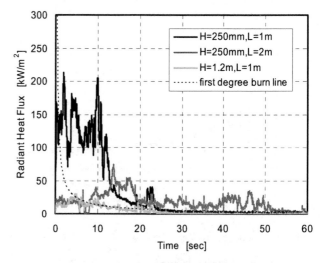

Figure 13. Radiant heat flux for downward venting (T-2). (H = vertical distance from ground level, L = horizontal distance from the venting tube discharge opening).

In upward venting, the maximum heat level was 10 kW/m^2, even at a distance of 1m from the vent pipe, and would not cause first-degree burns. Downward venting increased the risk of flame heat exceeding the first-degree burn level because the pipe was positioned 250 mm from ground level, which would place human legs within 2m of the vent pipe. However, at a height of 1.2m from ground level (which corresponds to the human chest) and within 1m horizontally from the vent pipe, the maximum instant radiant heat flux is just enough to cause first-degree burns (25 kW/m^2) or less. However, the result was influenced by mixed combustion with veneer plates since part of the experiment yard was covered by veneer plates to protect the asphalt. Thus, in actual downward venting, mixed combustion with asphalt road surfaces, grass or exterior materials of the vehicle is relevant and requires further examination.

Combustion Noise Level

Table 3 shows the peak sound pressure level of hydrogen jet flame.

Table 3. Peak sound pressure level (C-Frequency weighting networks) of the venting flame.

#	Vent direction	Peak sound pressure level at 4m from the vent tube [dB]	Peak sound pressure level at 10m from the vent tube [dB]
T-1	Upward	138.5	125.0
T-2	Downward	130.0	127.5

In upward venting, the hydrogen jet flame generated a peak noise level of 138.5 dB at a distance of 4m from the source. The exposure to peak noise level in excess of 135 dB may damage hearing immediately.[5] In both cases, however, noise levels safe for human eardrums will be obtained 10m or more from the vent pipe.

The above results indicate that a hydrogen flame released from a cylinder charged at 35 MPa when the PRD is activated has less effect in upward venting than in downward venting in surroundings where humans may be present.

EFFECT OF THE VENTING DIAMETER

Released-hydrogen flames were compared while varying the internal diameter of the vent pipe.

Flame Size

Figures 14 and 15 illustrate the relationship between flame size and the mass flow rate of a hydrogen jet flame. The flow rate was calculated by the following equation.

The bonfire test causes the temperature of the hydrogen gas inside the cylinder and the internal pressure to rise. The state of the gas in the cylinder when the bonfire test is started can be expressed as

$$P_1 V = m Z_1 R T_1.$$

The state of gas just before the PRD activation during the Bonfire test can be expressed as

$$P_2 V = m Z_2 R T_2.$$

Therefore, gas temperature T_2 just before the PRD activation can be calculated using the following equation:

$$T_2 = \frac{Z_1 P_2}{Z_2 P_1} T_1.$$

Here, P is the cylinder internal pressure, T the gas temperature in the cylinder, Z the coefficient of compression, V the capacity of the cylinder, and m the mass of the gas in the cylinder. Subscript 1 represents the state just before the Bonfire test, and subscript 2, the state just before PRD activation.

When hydrogen is vented by PRD activation, it becomes difficult to calculate the hydrogen gas temperature because the gas in the cylinder is subject to the Joule-Thomson effect as well as heating from outside the cylinder. However, this study assumed that the gas temperature just before PRD activation is maintained after the activation because the temperature of the cylinder during the Bonfire test is constant or decreases slightly. Mass flow rate Δm was calculated from the cylinder capacity, and the change in pressure was calculated using the following equation:

$$\Delta m = \frac{(P_3 - P_2)V}{(Z_3 - Z_2)R T_2}.$$

Here, subscript 3 represents the state after PRD activation.

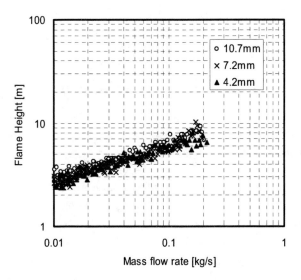

Figure 14. Flame height to mass flow rate for different venting diameters.

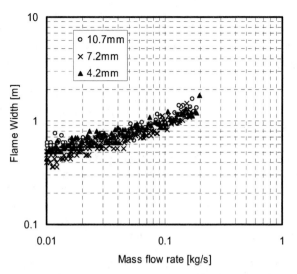

Figure 15. Flame width to mass flow rate for different venting diameters.

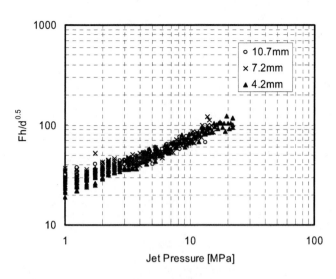

Figure 16. Flame height to jet pressure for different venting diameters.

Flame height and width are proportional to the flow rate to the 0.37th power. However, flame height and width depend only slightly on the venting diameter when the venting flow rate is kept constant. Therefore, the flow field is governed by the Froude number when ambient air is involved via the vent flame through the convection of the hot-air stream. The relationships between the Froude number Fr and the ratio of the flame height Fh [m], or flame width Fw [m], to the venting diameter d [m] can be calculated using the following equation:

$$Fh/d = 18.3Fr^{0.33}$$

$$Fw/d = 6.4Fr^{0.27}.$$

Here, the Froude number is expressed as $Fr = \dfrac{u}{(dg)^{\frac{1}{2}}}$, where u is the venting flow rate [m/s], and g is the gravitational acceleration [m/s^2].

Figure 16 depicts the relationship between the jet pressure, i.e., the static pressure just before the vent pipe, and flame height $Fh/d^{0.5}$.

The flame height is proportional to the jet pressure to the 0.545th power when jet pressure is 1 MPa or higher. Furthermore, flame height is proportional to the internal venting diameter to the 0.5th power when the jet pressure is kept constant. Similar trends are found in flame width.

References 6 and 7 have also reported that flame height and width are proportional to jet pressure when the pressure is 1MPa or higher. In previous research, however, the flame height and width are proportional to the vent's internal diameter to the 1st power when the jet pressure is kept constant, as shown in the following equations.

$$Fh/d = aP^b$$

$$Fw/d = cP^d$$

a, b, c, and d are constants.

The internal diameter of the discharge opening (venting diameter) differs from that in this study because the venting diameter chosen for comparison (0.001 to 0.004 meters[7] and 0.00032 to 0.0117 meters[6]), is smaller than the diameter in this study (0.0042 to 0.0107 meters). Thus, the venting diameter influence was negligible.

From the above results, the following equation can be obtained as an experimental formula for flame height and width.

$$Fh/d^{0.5} = 23P^{0.545}$$

$$Fw/d^{0.5} = 3.5P^{0.545}$$

Here, Fh is the flame length [m], Fw is the flame width [m], d is the internal diameter of the vent tube discharge opening [m], P is the jet pressure [MPa], and the application range of this equation is 1<P<20 MPa, 0.004<d<0.01 mm.

When this equation is applied, doubling the venting diameter leads to a 1.4 times greater flame height as long as jet pressure is maintained.

<u>Flame Temperature, Radiation, Noise, Thrust Force</u>

Other main results are shown in Table 4.

Table 4. Maximum temperature, radiant heat flux, noise level, thrust force for different venting diameters

Vent Internal Diameter	4.2mm (T-5)	7.2mm (T-4)	10.7mm (T-3)
Max. Temperature at 2.4m above vent [°C]	1368	N/A (melted)	1348
Max. Radiant heat flux at 1.2m from vent, height is 1.2m [KW/m^2]	9.83	10.99	10.50
Noise level at 4m from vent [dB]	137.5	137.5	138.5
Max. thrust force [Newtons]	178	280	280

Aside from the thrust decrease at a venting diameter of 4.2 mm, no other remarkable difference was found.

COMPARISON OF THE CYLINDER CAPACITY

Two cases were compared, cylinder capacities of 45 liters (T-1) and 34 liters (T-3). The only noticeable difference was venting time (refer to Table 2). Few differences were found in flame temperature, radiant heat flux, combustion noises or thrust.

VEHICLE FIRE TEST WITH HIGH-PRESSURE HYDROGEN GAS CYLINDERS

When a vehicle fitted with high-pressure cylinders catches fire and the PRD is activated, the hydrogen jet flame is affected by differences in the fire source and/or combustible materials.

It is thus necessary to determine whether the methods of the Bonfire test implemented on standalone high-pressure cylinders are also applicable to vehicle fires. Therefore, we performed vehicle fire tests on vehicles equipped with high-pressure hydrogen cylinders with a pressure of 35 and 20 MPa, and investigated the influences of venting direction of the vent pipe and combustible materials in the surroundings of released flame (gasoline pool fire).

COMPARISON OF VENTING DIRECTIONS

<u>Experiment Apparatus and Procedures</u>

Figure 17 illustrates the test vehicles, and Table 5 outlines the conditions for this test.

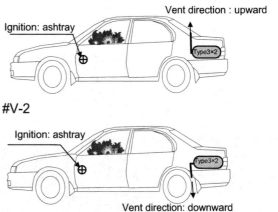

Figure 17. Outline of the test vehicle

Table 5. Test condition for vehicle fire

#	Fire scenario	Fire source	Release point	Cylinder
V-1	Cabin fire	Alcohol solid fuel placed on the ashtray	Top of trunk area	35[MPa], 34[liters]
V-2	Cabin fire	Alcohol solid fuel placed on the ashtray	Bottom of the vehicle	35[MPa], 34[liters]

Gasoline vehicles with a displacement of 1800 cc were used as test vehicles. The gasoline tank of each vehicle was removed and modified for mounting hydrogen gas cylinders. Two Type-3 cylinders were mounted in the trunk. A PRD (the same type as used above venting flame test) was installed.

A vent pipe with a nominal diameter of 3/8" (measured internal diameter 7.2 mm) was installed on the PRD. The vent pipe discharge opening was located on the trunk lid in V-1 and on the underside near the rear wheels in V-2. Generally, although fuel-cell vehicles isolate the passenger compartment from hydrogen fuel system to prevent fuel intrusion in the unlikely event of a hydrogen leak, this isolation was not provided for this test vehicle. Items measured included the temperature at various locations on the vehicle, cylinder internal pressure, radiant heat flux, and combustion noise.

A steel pan (7m long x 2.5m wide x 0.2m high at the edge) was placed on the ground near the center of the test field, and surrounded by banks and concrete blocks to prevent contamination of the soil by residuals. The test vehicle was placed on the pan. The windows on the driver's and passenger's sides were fully opened, and the rear left window was opened approximately 1/3" to allow for the insertion of measurement cables. After activation of the PRD, and confirmation (using a pressure monitor or visual) that the gas in the cylinder had nearly all been released, the test vehicle fire was extinguished by instructors of the fire training center.

Results and Discussion[*]

For V-1 and V-2, where fire in the passenger compartment was assumed, the PRD was activated within 15 minutes and 30 seconds to 17 minutes after the fire began. Figures 18 and 19 shows the side view of the vehicle rear at that time.

| After 10/30 sec. | After 10 sec. |

Figure 18. Vehicle rear side after activation of the PRD. (TEST V-1)

| After 10/30 sec. | After 10sec. |

Figure 19. Vehicle rear side after activation of the PRD (TEST V-2)

In upward venting, the maximum flame height was 10m (at 0.27 seconds after activation of the PRD), and the maximum flame width was 9m (at 0.1 seconds after activation of the PRD). In downward venting in a similar manner, the maximum flame distance from the side face of the vehicle was 5m (at 1.6 seconds after activation of the PRD), and maximum flame height was 2.6 m (1.4 seconds after activation). (The flame width could not be measured due to the presence of a protection wall.) Though both reach their maximum within 0.5 seconds after activation of the PRD, they attenuated instantly. The flame diameter in upward venting and the flame height in downward venting became smaller instantaneously.

Figures 20 and 21 depict the conditions after activation of the PRD in V-1 and V-2 for upward and downward venting, and the radiant heat flux near the vehicle. In test V-2, the heat flux gauge positioned 1.2 m from ground level and 1m from the rear of the vehicle failed to operate due to wire disconnection. Therefore, those results are not shown on the graph.

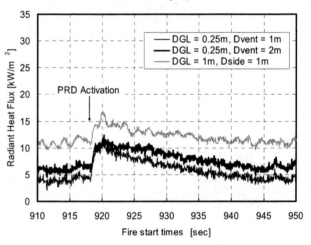

Figure 20. Radiant heat flux (TEST V-1, Upward venting).
DGL: Distance from ground level. Dvent: Distance from the vent end. Dside: Distance from vehicle side.

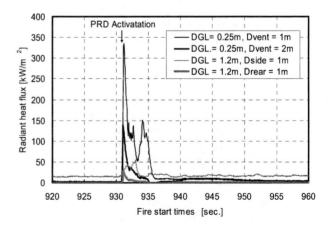

Figure 21. Radiant heat flux (TEST V-2, downward venting).
DGL: Distance from ground level. Dvent: Distance from the vent end. Dside: Distance from vehicle side. Drear: Distance from vehicle rear.

460

In upward venting, the increase in radiant heat flux before and after PRD activation was only 5 to 6 kW/m^2. The maximum radiant heat flux during combustion of materials in the vehicle before PRD activation (600 to 700 seconds after the fire began) was 19 kW/m^2 at 1 m from the side of the vehicle. Therefore, the radiant heat flux from the burning interior and exterior materials was higher than that when the hydrogen jet flame was formed.

In downward venting, at 250mm from ground level and 2m horizontal from the vent pipe (which corresponded to the position of human legs), the radiant heat flux reached 340 kW/m^2 immediately after PRD activation. Eight seconds after PRD activation, however, the influence of radiation from jet flames disappeared. At a height of 1.2m from ground level and a distance of 1m from the side of the vehicle (which corresponds to the human chest location), the maximum value was 40 kW/m^2. Thus, as in the venting test above, in downward venting, the radiant heat flux impinging on materials depends on the height of the vehicle from ground level.

In both cases, the combustion noise when the PRD was activated was 130 dB at a distance of 5m from the vehicle, and 126 dB at a distance of 10m. The noise level at a distance of 10m was roughly the same as that caused by a tire burst, which is 123 dB at a distance of 5m from the vehicle. From the above results, a distance of 10 m from the vehicle is safe under these test conditions, even when the PRD is activated and regardless of venting direction.

COMPARISON OF THE FLAME OF THE VENTING FLAME TEST AND THE VEHICLE FIRE TEST

As shown in Figure 22, a comparison of the flame size of the venting flame test (T-4) to that of the vehicle fire test (V-1) revealed that the internal diameters of the vent tube (=7.2mm) and upward venting were the same.

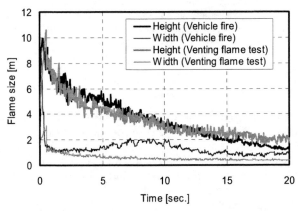

Figure 22. Comparison of the flame size of the vehicle fire test (V-1) and venting flame test (T-4)

Although the flame heights were almost equal, the flame widths differed. While the flame width was 9m immediately after PRD activation in the vehicle fire test, it was 3m in the Bonfire test. Also in downward venting, the flame is wider in the vehicle fire test only immediately after the PRD activation.

Additionally, the total radiant heat flux (Figures 12 and 20) was slightly larger in the vehicle fire test than in the bonfire test because the radiant heat flux from the burning interior and exterior materials of the vehicle fire were added. However, the difference was largest within 1 second following the PRD activation, as explained below.

Before PRD activation, radiant heat flux and a large volume of soot were released from the vehicle body because the interior and exterior materials were burning. However, the volume of soot released from the vehicle was small in the image of released-hydrogen flame formed after the PRD activation (Figure 23).

| Just before PRD activation | 5second after PRD activation |

Figure 23. Soot emitted from vehicle (TEST V-2)

This was due to the soot released from the burning interior and exterior materials of the vehicle that were re-combusted by the released-hydrogen flame. A luminous flame was generated from the combustion of soot, and very high radiant heat flux was generated.7). Therefore, radiant heat flux in the vehicle fire test increased slightly. Immediately after PRD activation, the flame became wider, and the flame surface approached the sensor. This resulted in a larger radiant heat flux because the combustible gas cloud of released hydrogen caught fire after it was diffused and reached the fire source, and the soot in that area was combusted.

These results demonstrate that the hydrogen jet flame released by PRD activation during a vehicle fire affects the combustible materials near the discharge opening of the vent.

461

EFFECT OF COMBUSTIBLE MATERIALS -GASOLINE POOL FIRE -

Influences of combustible materials near the PRD were investigated by implementing tests on vehicle fires caused by gasoline pool fires (fires due to leaking gasoline).

Experiment Apparatus and Procedures

Table 6 outlines conditions for this test, and Figure 24 illustrates the test vehicle.

Table 6. Test conditions for gasoline pool fire

#	Fire scenario	Fire source	Release point	Cylinder
V3	Gasoline pool fire of spills on a road surface	The gasoline pool fire put under the cylinders	Inside the trunk room	21[MPa], 46[liters]

#V-3

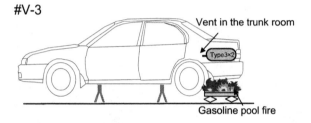

Figure 24. Outline of the gasoline pool fire.

A gasoline vehicle with a displacement of 5000 cc was used. Two Type-3 (46-liter) high-pressure hydrogen cylinders of with a pressure of 21 MPa were mounted in the trunk. A PRD (the same as used in the venting test) was installed. No vent pipe was mounted on the PRD, so hydrogen was released into the trunk. A fire pan (0.78 m long x 1 m wide x 0.1 m deep) filled with 15 liters of gasoline was placed under the vehicle with high-pressure cylinders installed. The level of gasoline was flushed to the edge of the fire pan with water. The heights of the vehicle and fire pan were adjusted by jacking to align the height of the vehicle from the oil level with the height of the vehicle from the road surface. Other test methods were the same as those implemented in T-1 and -2.

Result and Discussion

During the test on gasoline pool fires, the PRD was activated 3 minutes and 47 seconds after ignition. Figure 25 presents a photograph taken immediately after the PRD activation.

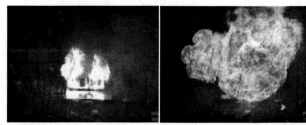

Just before PRD activation   1second after PRD activation

Figure 25. Vehicle after PRD activation (TEST V-3). A fireball formed during the vehicle fire test simulating a gasoline leak.

Within 2 seconds following PRD activation, a fireball more than 10m diameter fireball formed. Radiant heat flux across the PRD activation is shown in Figure 26.

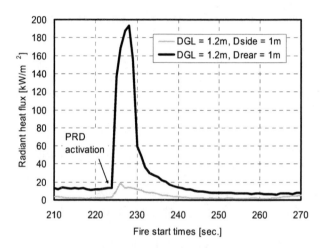

Figure26. Radiant heat flux to the vehicle from a hydrogen jet flame (TEST V-3). A fireball formed from the gasoline pool. The heat flux that appeared grew to an extreme size of 190 kW/m^2 at the rear of the vehicle as the flame surface approached the sensor. DGL: Distance from ground level. Dside: Distance from vehicle side. Drear: Distance from vehicle rear.

The radiant heat flux in the rear of the vehicle was 190 kW/m^2 and the peak lasted 5 seconds. Compared with the vehicle fire test with downward venting in TEST V-2, in gasoline fires the duration of the maximum radiant heat flux was longer, and the damage to the surroundings, larger. The following may have caused the fireball formation.

When the PRD was activated in the trunk where the cylinders were installed, the hydrogen gas vented by the PRD was released through the clearance of the trunk lid.[8] Because the density of hydrogen was lower than that of air, a hydrogen cloud extending from that clearance to the atmosphere above the vehicle was formed and then was ignited by the gasoline pool. For this reason, when a fire is caused by a gasoline vehicle leak after a collision, the hydrogen jet flame causes mixed combustion with other fuels when the PRD is activated. As a result, the fire expands instantaneously. The above results dictate that it is necessary to further study the venting direction and venting location of the vent tube to assure enhanced safety against fire.

COMPARISON OF THE CYLINDER CONDITION OF THE VEHICLE FIRE TEST AND BONFIRE TEST

The location where high-pressure cylinders are mounted may vary depending on the type of vehicle. Therefore, the Bonfire test, which is performed on standalone cylinders, must be applicable to various types of vehicles. In this section, we compare the following: the average temperature of each part of the cylinder, the starting time of PRD activation, and the cylinder internal pressure obtained from the bonfire test (T-3, -4 and -5) and vehicle fire test (V-1 and V-2 on four cylinders total) using the same cylinders and PRDs to extract problems in the bonfire test.

Table 7 shows the surface temperatures of the PRD and cylinder just before PRD activation, starting time of PRD activation, maximum pressure, and the rate of pressure rise, assuming the filling pressure as 100% in the Bonfire test (T-3, -4 and -5) and vehicle fire test (V-1 and V-2).

Table 7. Comparison of the cylinder condition of the vehicle fire test and bonfire test

	Bonfire test (T-3,-4,-5)	Vehicle fire (V-1,-2)
Average cylinder surface temperature before the PRD activation [°C]	590 (408-739)	266 (210-336)
PRD surface temperature before the PRD activation [°C]	281 (221-359)	143 (116-181)
Filling pressure [MPa]	30.0 (29.7-30.4)	34.02 (32.9-35.21)
Time PRD activation [sec]	48.7 (29.6-64.0)	537.2 (451-627)
Max pressure [MPa]	30.5 (30.2-30.7)	39.36 (38.5-41.0)
Rate of pressure rise [MPa/min]	0.512 (0.325-0.687)	0.605 (0.464-0.766)

In the vehicle fire test, the cylinders were not heated immediately because the fire was initiated in the passenger compartment. Therefore, the time from the point where any temperature change occurred near the cylinder until the PRD activation is shown as in the starting time of PRD activation in the vehicle fire test. The rate of pressure rise is defined as follows.

Rate of pressure rise = (Cylinder internal pressure just before the PRD activation – charging pressure)/Starting time of PRD activation

In the vehicle fire test, the cylinder surface temperature just before PRD activation is 266°C, which is lower than the temperature required by ISO11439 (590°C or more). Furthermore, the difference between the cylinder surface temperature and the PRD surface temperature was smaller in the vehicle fire test. In the Bonfire test implemented on standalone cylinders, however, the rate of pressure rise was smaller because the PRD was activated while there was a rise in pressure. The cause of this can be conjectured as follows.

It is difficult for the flame to intrude into a semi-airtight space like a trunk, so there was little chance for the flame to directly impinge on the cylinder or PRD. Therefore, the cylinders mounted in the trunk were heated more slowly than in the Bonfire test. Heat can transfer to the gas in the cylinder over a long time, and thus, the cylinder internal pressure rose until the PRD reached the activation temperature.

There are two sources of vulnerability when cylinders are exposed to fire: the increase in internal pressure and the decrease in cylinder strength due to heating by fire. Therefore, in the Bonfire test, it is necessary to consider both heating of the cylinder and the pressure resistance against the increase in internal pressure. Factors that affect such damage include the fire source and PRD shield; both sway test results. The PRD shield, in particular, delays PRD activation and leads to the increase in cylinder internal pressure, simulating the on-board state.

However, ISO standards contain no detailed specifications for the PRD shield. Therefore, we consider it is necessary to investigate the influence of the PRD shield.

CONCLUSIONS

To examine safety measures for fuel-cell vehicles, we performed tests on vehicles with hydrogen-filled high-pressure cylinders (20,35MPa) installed for which released-hydrogen flames formed during Bonfire tests and fire tests. Our findings are as follows.

1. Released-hydrogen flames vented in an open space from a cylinder filled to 35 MPa when the PRD was activating influenced the surroundings less when vented upward (rather than downward) in areas where humans are present at ground level. In both cases, however, a minimum distance of 10m from the vehicle is required.

2. In the vehicle fire test assuming a gasoline fire, a fireball with a diameter of 10m or more is formed. It is necessary to study the venting direction and venting location of the PRD vent pipe further.

3. The PRD was activated before the high-pressure cylinder ruptured. From this result, we believe that the cylinders that pass the Bonfire test can also maintain their integrity in an actual vehicle fire. However, there are differences in cylinder internal pressure and cylinder surface temperature just before the PRD activation in Bonfire tests on standalone cylinders and vehicle fire tests. In particular, the cylinder internal pressure rises when cylinders are mounted in a semi-airtight space, such as a vehicle's trunk. Therefore, to enhance the reliability of the Bonfire test, it is necessary to specify the details of PRD shield employed in the Bonfire test.

4. When the flow rate increases, the flame height Fh[m] and flame width Fw[m] increase, but do not depend on the venting diameter d[m] .

5. When the jet pressure P[MPa] increases, the flame height Fh[m] and flame width Fw[m] increase. When the jet pressure is kept constant, the jet pressure and flame height are proportional to the venting diameter d[m] to the 0.5th power. The following empirical equation was derived from this result.

$$Fh/d^{0.5} = 23P^{0.545}$$

$$Fw/d^{0.5} = 3.5P^{0.545}$$

This equation applies for 1<P<20MPa, 0.004<d<0.01mm.

6. For 1 second following PRD activation, the radiant heat flux and the flame scale are larger in the vehicle fire test than in the release flame test with hydrogen gas.

In closing, fuel-cell vehicles must coexist with other vehicles using other fuels such as gasoline, and safety issues under these circumstances are important.

ACKNOWLEDGMENTS

This paper summarizes the contents of studies implemented as part of "Establishment of Codes and Standards for Popularization of PEFC System for automobiles (Millennium Project)" of the *New Energy and Industrial Technology Development Organization*.

REFERENCES

1. Gas Cylinders – High pressure cylinders for the on-board storage of natural gas as a fuel for automotive vehicles – ISO 11439, 2000/9/15
2. McCaffrey, B.J., Purely buoyant diffusion flames – Some experimental results, NBSIR79-1910, 1979
3. Frank P. Lees, Loss prevention in the process industries, Butterworth and Company, 1980
4. R,Gambone, et al, Performance Testing of Pressure Relief Devices for NGV Cylinders, U.S.Depertment of Commerce,1997
5. Thomas, P.H., et al, Some experiments on buoyant diffusion flames, Combustion and Flame, vol.5,1961
6. Masaji Iwasaka, et al, Hazard of hydrogen jet flame, The high pressure gas, Vol.16,333,1974.(in Japanese)
7. Keiji Takeno, et al., Experimental study on high-pressure hydrogen diffusion flame, 22th conference proceeding of the hydrogen energy system society of Japan,2003.(in Japanese)
8. Yohsuke Tamura, et al., Evaluation of the high-pressure hydrogen gas cylinders in simulated FCEV fires, JARI Research Journal, Vol.24,No.10,2002.(in Japanese)

Development of Fuel Economy Measurement Method for Fuel Cell Vehicle

Satoshi Aoyagi and Takuya Shirasaka
Honda R&D Co., Ltd. Wako Research Center

Osamu Sukagawa and Naoki Yoshizawa
Honda Motor Co., Ltd.

ABSTRACT

A practical method of fuel economy measurement has been developed for fuel cell vehicles. The weight method was selected for study from among the various fuel economy measurement methods available for pure hydrogen-type fuel cell vehicles as a direct measurement method that was accurate and could be produced at a comparatively low cost. Solutions were provided to the drawbacks of measurement errors caused by pressure variations in the fuel pipes and lowered hydrogen purity due to air entering the pipes. This development project assisted in fuel economy measurements for the Honda FCX, the world's first market-ready fuel cell vehicle; it also provided measurement system for the Environmental Protection Agency (EPA) tests of the same vehicle, which resulted in the world's first confirmatory test of a fuel cell vehicle.

INTRODUCTION

In the accelerated development of fuel cell vehicles of recent years, there has been a desire for similar accelerated development of methods to evaluate performance[1],[2]. In the USA, particularly, in order to respond to official confirmatory tests for pure hydrogen fuel cell vehicles carried out by the EPA (Environmental Protection Agency), it was absolutely essential to develop a fuel economy measurement method that was both highly accurate and practical[3].

On the other hand, one international standard of fuel economy measurements for internal combustion engines such as gasoline or diesel engines is the "carbon balance method" utilizing the fact that the amount of carbon is the same in the intake and exhaust. By measuring the amount of carbon in the exhaust gases, it is possible to determine the amount of fuel consumption. However, with a fuel cell vehicle, that principle cannot be applied, and each manufacturer has adopted their own measurement method. In this development project, a comparative study was made of various fuel economy measurement methods used for pure hydrogen fuel cell vehicles. And an attempt was made to develop a measurement method that can lead to the manufacture of a device that is highly accurate and can be used universally, at a comparatively low cost.

DEVELOPMENT AIMS

There were four development aims.

1. Measurement Accuracy

In the US fuel economy measurement modes of City and Highway, measurement accuracy in the SAE-J2572 standards stipulated by the SAE (Society of Automotive Engineers) must be ±1%.

2. Universal Use

It must be possible to use the method both in Japan and in the USA. It must be possible to measure the amount of fuel consumption for each phase, even in modes composed of multiple phases, such as the US certification fuel consumption mode.

3. Cost of Manufacture

It must be able to be manufactured at a cost lower than the equipment for gasoline vehicles use, the carbon balance method, at no more than ¥2 million.

4. Standardization

Used for the official confirmatory test, the fuel economy value given in the test results must be recognized as an official value.

SELECTION OF MEASUREMENT METHOD

REPRESENTATIVE MEASUREMENT METHODS

1. Weight method

This is a method that directly measures the mass of the pure hydrogen fuel, and this principle is used as a calibrated standards method in carbon balance methods as well. Specifically, the mass of the hydrogen tank, prepared for measurements as shown in Fig. 1, is measured before and after the tests, and the fuel consumption is determined from the difference in the masses. Fuel consumption W(g) is calculated using Equation 1 below. If an accurate measuring device with fewer causes of errors is used, higher measurement accuracy is expected. However, there would be issues such as occurrence of measurement errors due to pressure variations in the fuel pipes or lowered hydrogen purity due to air entering the pipes at the time of connection of the pipe to a hydrogen tank.

$$W = W_i - W_f \qquad (1)$$

W_i : Mass of hydrogen tank before tests (g)
W_f : Mass of hydrogen tank after tests(g)

2. Flow method

This method measures fuel consumption by continuously measuring fuel supply flow and integrating the measurements. Specifically, the value detected by Flow Meter 1 shown in Fig. 1 is integrated, and the fuel consumption W(g) is calculated according to Equation 2. No errors occur due to pressure variations in the pipes but there would be issues such as insufficient accuracy under transient conditions, and at very low flow rates.

$$W = \frac{\sum Q}{22.414} \times m \qquad (2)$$

Q : Flow rate of fuel (NL/sec)
m : Molecular weight of hydrogen 2.016(g/mol)

3. P/T method (pressure method)

Before and after the tests, the hydrogen tank's pressure and temperature are measured, and the fuel consumption is calculated from the ideal gas equation, etc. Specifically, the pressure and temperature of the hydrogen tank are measured before and after the tests using a pressure gauge and thermometer, as shown in Fig. 1. Fuel consumption is calculated using the ideal gas equation, etc., from the changes before and after. Fuel consumption W(g) is calculated using Equation 3 below. In addition to the occurrence of measurement errors due to pressure variations in the pipes, there would be issues due to a large number of incident factors for errors because too many items need to be measured.

$$W = \left(\frac{P_i \times V}{R \times T_i} - \frac{P_f \times V}{R \times T_f} \right) \times m \qquad (3)$$

P_i: Hydrogen tank pressure before tests (Pa)
P_f: Hydrogen tank pressure after tests (Pa)
T_i: Hydrogen tank temperature before tests (K)
T_f: Hydrogen tank temperature after tests (K)
V : Hydrogen tank capacity (m^3)
R : Gas constant 8.314(J/mol•K)
m : Molecular weight of hydrogen 2.016(g/mol)

4. Electrical current method

Utilizing the principle that the electrical current output from the fuel cell stack is proportional to the fuel consumption used in generating electricity, the values detected by the current sensor shown in Fig. 1 are integrated, and fuel consumption calculated. Depending on the test vehicle, it is also necessary to measure, using a flow method or something similar, the hydrogen that is exhausted without being used to generate electricity, called "purged hydrogen". The fuel consumption W(g) is calculated using Equation 4 when the purged hydrogen portion is measured using Flow Meter 2 shown in Fig. 1. No measurement errors due to pressure variations in the pipes occur and highly accurate measurement under transient conditions is expected. However, there would be issues such as removal of water from the purged hydrogen.

$$W = \frac{n \times m \times \sum I}{F} + \frac{\sum Q_p}{22.414} \times m \qquad (4)$$

I : Fuel cell output current (A)
n : Number of cells in the fuel cell
F : Faraday constant 9.64867×10^4 (C/mol)
m : Molecular weight of hydrogen 2.016(g/mol)
Q_p : Purged hydrogen flow rate (NL/sec)

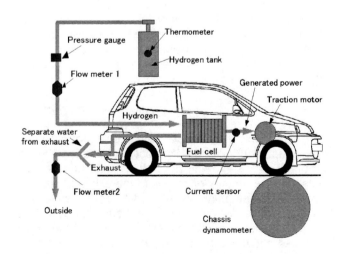

Fig.1. Comparison of various method

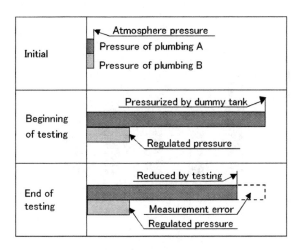

Fig.2. Plumbing pressure and measurement error

ADOPTION OF THE WEIGHT METHOD

For the following reasons, the weight method was adopted in this development project.

1. Because the hydrogen being measured is measured only in a direct manner, there is essentially little cause of error.

2. It has become comparatively easy to obtain an accurate measurement device.

3. The outlook is good for countermeasures concerning the hydrogen remaining inside the pipes.

MEASUREMENT ACCURACY

The major factor affecting weight method measurement accuracy is errors due to pressure variations within the pipes. In addition, there is the problem of a hydrogen purity drop due to air contamination in the pipes during hydrogen tank connection. The following countermeasures were taken, for these issues.

Because errors due to pressure variations within the pipes occur in pressure difference before and after tests, as shown in Fig. 2, errors can be reduced by maintaining a fixed pressure in the pipes before and after testing. Therefore, as shown in Fig. 3, the pressure in the pipes is raised to a specific pressure prior to the test, without using the hydrogen in the measurement tank, by using a dummy tank. By having a pressure regulator in the pipes close to the hydrogen tanks, although pressure variations remain in Plumbing A, as shown in Fig. 3, there is no pressure variation in Plumbing B, and error is reduced. Also, using error e (%) obtained from Equation 5, effects of countermeasures and measurement accuracy were calculated. Fig. 4 and 5 show the effect of countermeasures and Fig. 6 shows measurement accuracy. Finally, it was capable to fully satisfy the SAE standard requiring measurement accuracy of ±1%.

$$e = \frac{\dfrac{(P_{ini} - P_{end}) \times V \times m}{RT} + e_w}{M} \qquad (5)$$

P_{ini} : Pipe pressure before tests (Pa)
P_{end} : Pipe pressure after tests (Pa)
V : Pipe volume (m^3)
M : Molecular weight of hydrogen 2.016(g/mol)
R : Gas constant 8.314(J/mol·K)
T : Hydrogen temperature in pipe (K)
e_w : Minimum measurement weight (g)
M : Total hydrogen consumption (g)

Concerning the hydrogen fuel supplied to fuel cell vehicles, because generally an extremely pure hydrogen – 99.99% class – is required, it is necessary to take appropriate countermeasures to prevent even the smallest amount of air penetrating the pipes when the hydrogen tanks are connected. Therefore, as shown in Fig. 3, by repeatedly discharging the hydrogen fuel in the dummy tank to the outside through the purge valve, any air between Valve A and Valve B is eliminated, without using hydrogen from the hydrogen tank for measurement.

In addition, giving due consideration to safety, a flow limiter valve was added to restrict fuel leaks if a pipe disconnects. A pressure relief valve was added to prevent abnormal pressure rise if the pressure regulator fails. And a breakaway device was added to prevent fuel leaks if the vehicle moves away. These were added along Plumbing A and B, as shown in Fig. 3.

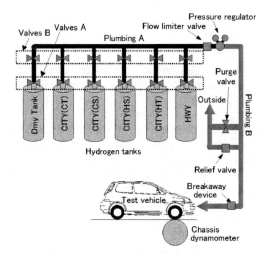

Fig.3. Configuration of weight method system

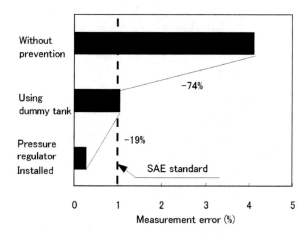

Fig.4. Measurement error (CITY)

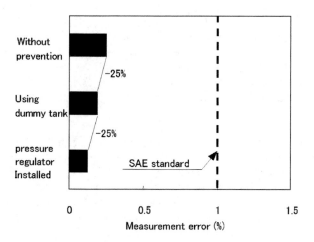

Fig.5. Measurement error (HWY)

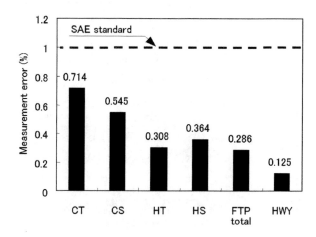

Fig.6. Measurement error of each phase

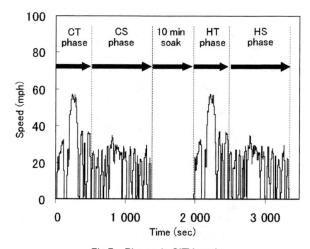

Fig.7. Phases in CITY mode

UNIVERSAL USE

As for the hydrogen tanks, the measurement system was made possible to change the height of Plumbing A connecting the hydrogen tanks, as shown in Fig. 3. Also, it was made easy to replace the connectors on the pipe side that connect the pipe to Valve A, which are the tank valves integrated with the hydrogen tanks themselves. Applying these measures, it is possible to use, for measurement, hydrogen tanks that are readily available in the USA or in Japan, regardless of their size.

 In addition, in order to handle fuel consumption modes that are composed of multiple phases, like the City mode shown in Fig. 7, and so require measurements of fuel consumption for each phase, the number of hydrogen tanks is equal to the number of phases. The hydrogen tanks connected to Plumbing A as shown in Fig. 3 are placed in a rack, and the tanks can be switched by using Valves B. This arrangement makes it possible to measure fuel consumption for each phase.

COST OF MANUFACTURE

Table 1 gives the measurement system specifications. Using readily available parts and avoiding special ones, a simple system was created, that can be manufactured for approximately ¥1.8 million. This has kept the cost to less than that of the carbon balance method used for gasoline vehicles. Figure 8 is a photograph of the measurement system.

Table 1. Measurement system specifications

Type		SUK-1
Dimension (mm)		W 2 000 × D 500 × H 1 940
Fuel supply pressure (MPa)		0.9 ~ 1.0
Max.number of tank		6
Safety device		Pressure relief valve
		Flow limiter valve
		Break away device
Scale	Manufacture	A&D
	Type	GP-30K
	Max.mass (g)	31 000.0
	Min. mass (g)	0.1

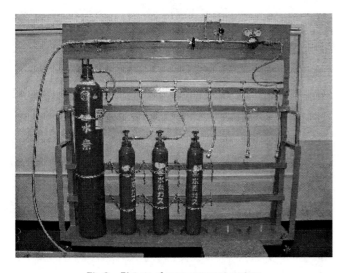

Fig.8. Picture of measurement system

STANDARDIZATION

Keeping in mind automotive emissions regulations, a fuel economy measurement method was required that has a high standard measurement accuracy and reproducibility, particularly for use in the USA. In addition, to be used during official fuel economy certification testing indicates what we think would be one of the test methods applied to other models certified in the USA, and therefore can be one of the standard methods.

EPA and manufactures are now working on the evaluation of measurement methods but according to our evaluation results we believe that weighting method achieved the measurement accuracy and reproducibility required of standard test. As a result, this measurement method and measurement system has been used in the world's first certification fuel economy test, which was implemented for the Honda FCX fuel cell vehicle in 2002. The results of this official test are given in Table 2. Values calculated from the test results are described on the fuel economy labels, which are required to be attached to all vehicles sold in the USA, giving the vehicle's practical fuel economy value. The fuel economy unit used in Table 2 is the distance traveled, in miles, for each 1 kg of hydrogen fuel consumed.

Table 2. Official test result of 2002 model year

Mode	Fuel economy (miles/kg Hydrogen)
CITY	56.8
HWY	61.7

CONCLUSION

By adopting the weight method, reducing measurement errors, and using the US certification fuel consumption modes, the following results were achieved.

1. In both US fuel consumption certification modes of City and Highway, measurement accuracy of total fuel consumed was 0.286% and 0.125%, so this fully satisfies the SAE standard requiring measurement accuracy of ±1%.

2. It is possible to create a fuel economy measurement system that can use hydrogen tanks that are readily available not only in Japan but also in the USA. By adding a hydrogen tank switching mechanism, it is also possible to measure fuel economy for each phase.

3. The cost of manufacturing the fuel consumption measurement system is approximately ¥1.8 million, which meets the target of no more than ¥2 million. This is a lower cost than that of fuel economy measurement systems using the carbon balance method for gasoline vehicles.

4. This measurement method has been used for official fuel economy certification testing, showing that it can be one of the standard measurement methods.

REFERENCES

1. Watanabe, S.: State of the Art of Fuel Cell Vehicle, Journal of Society of Automotive Engineers of Japan, Inc., Vo. 56, No. 1, p. 54-58 (2002)

2. Seko, T.: Present Status and Future Prospects of Alternative Fuel Vehicle, Journal of Society of Automotive Engineers of Japan, Inc., Vol. 53, No. 5, p. 15-20 (1999)

3. Kuroda, E. et al.: Test Methods to Measure Fuel Economy in Direct Hydrogen Fuel Cell Vehicle, JARI Research Journal, Vol. 24, No. 10, p. 49-54 (2002)

VI. MODELLING, CONTROL AND DIAGNOSIS

VI. Modelling, Control and Diagnosis

In conventional engine systems, electronic control is required to maintain the engine at conditions where the efficiency of the engine and after-treatment processes are maximized subject to constraints imposed by exhaust emissions and the requirements for the driveability of the vehicle.

In a fuel cell system, the need for system control remains the same. There is the same requirement for the delivery of torque to road wheels in a timely manner, but the physical and chemical processes through which torque is created are quite different.

Development of the requirement for a control system depends on the propulsion architecture. In a hybrid system the control system must co-ordinate the power supply from the primary unit (the fuel cell stack) and the supply to and from an energy store. This requirement leads to an optimizing strategy in which the relative costs of the energy sources are evaluated and balanced. In the hybrid case, the state of charge of the energy store plays a significant role because the need to capture braking energy is essential to justify the cost and weight of the energy storage device.

The control function must be delivered by an embedded control system, sensors and actuators. Design analyses have shown (2002-01-1930 is an example) that the manufacturing costs of the control system amount to 30% of the total. Some work has already been done on the details– and in particular the sensors that would be needed to provide feedback information. As yet there is little published on the application methodologies that will be used to adjust the control system parameters to meet the particular requirements of efficiency and driveability.

Diagnosis is an essential function in all potential applications of fuel cells. On-board diagnostic tools will be embedded in the powertrain control systems to monitor for changes that will influence the emissions of regulated pollutants. Such monitoring requirements are likely to be stringent for systems that include a fuel processor, but more relaxed for pure hydrogen fuelled systems. The greatest concern with hydrogen systems is the detection of a leak of hydrogen gas and the formulation of control actions to limit its effect.

The design of a fuel cell system and the subsequent design and development of control systems will need to be supported by simulation tools. Simulation allows a range of design options to be investigated in advance of a hardware becoming available. General purpose tools will be extended to cover the dynamic requirements of a system simulation and there are examples of where hardware in the loop (HIL) methods have been employed in fuel cell systems development (2002-01-0409 is an example). Modeling will be needed to support detailed design work that will for example include the control of gas pressures in a fuel cell system. Progressive integration of modeling tools into whole-system modeling frameworks is already taking place and increasingly tools will be available for both system and component selection and optimization.

Control Systems

The authors of paper 2003-01-1137 describe the development of a hydrogen sensor for leak detection in hydrogen fuelled vehicles. The established method of measuring hydrogen concentration is by means of thermal conductivity. Humidity strongly affects the reading, and new developments are needed to compensate for humidity to give an accurate reading.

Paper 2002-01-0100 describes the sub-models of a fuel cell system with particular attention to the air supply and water management functions. The authors use the integrated model to estimate the fuel efficiency of the proposed system design over the New European Drive Cycle (NEDC) as a function of operating pressure. The model, implemented in a commonly-used simulation code is used by the authors to conduct a sensitivity analysis into system design parameters.

The authors of paper 2002-01-0102 describe a powertrain optimization strategy for a hybrid fuel cell propulsion system. Concluding that a long term optimization of powertrain functions is not possible with driving conditions unknown in advance, a short term optimization is more appropriate in which the fuel cost of the power sources available is evaluated.

In paper 2004-01-1298 the authors compare a step by step strategy for a hybrid system similar to the one described in 2002-01-0102. At each control interval the replacement cost of recharging the energy storage device is compared with drawing the same power from the fuel cell stack. In a modification of the scheme, the authors add a scheduling criterion that allows for the variation of efficiency with load in the fuel cell stack.

The authors of 2004-01-1299 describe the extension of a cost based strategy that has already been applied to IC engine hybrid powertrain systems. The control actions are calculated using a cost function that weights the cost of operating the fuel cell stack as well as noise emissions and regulated gas emissions. The authors describe the formulation of the cost function and the selection of components for the simulation of an SUV type vehicle.

Modelling

Paper 2001-01-0543 builds on earlier modeling work at the University of California, Davis and explains the background and motivation to the development of a model of a hybridized fuel cell powered vehicle. The authors describe the sub-models in detail and show results from the integrated model.

In paper 2003-01-1144, the authors describe the sub models that make up a transient model of a fuel cell vehicle that includes a fuel processor for the conversion of liquid and gaseous fuels. The authors employ the model to support the analysis of start-up and shut-down effects, thermal management and the measurement of overall efficiency.

The authors of paper 2004-01-1474 describe an extension to an existing powertrain and vehicle simulation code to include a PEM fuel cell stack as the main source of propulsion power. The authors demonstrate different modes of operation of the model including stand alone operation, assessment of thermal effects and integration with a vehicle model in a hybrid configuration.

VI. MODELLING, CONTROL AND DIAGNOSIS

Control Systems

2003-01-1137

Hydrogen Sensor for Fuel Cell Vehicles

Masaki Tada, Rihito Shoji, Nobuharu Katsuki and Junichi Yukawa
Matsushita Electronic Components Co., Ltd.

ABSTRACT

From the viewpoint of global environment and petroleum energy depletion problems, a hydrogen-based fuel cell is attracting people's attentions as a clean energy source. The object of the present paper is to summarize approaches for achieving high stability and high selectivity in developing hydrogen sensors that detect hydrogen gas, fuel of fuel cells. For gas detection principles, various systems including a semiconductor system exist.

We are developing hydrogen sensors with thermal conductivity principle applied, in which the difference of gas thermal conductivity is utilized with adaptability in applications for fuel cell electric vehicles taken into account. When the environment in automobile applications is assumed, high-selectivity to detect hydrogen alone is required because various gases such as water vapor and exhaust gas (methane, carbon monoxide, etc.) coexist. In particular, the thermal conductivity principle causes big errors at high temperature and high humidity.

To overcome this drawback, we have developed a high-selectivity hydrogen sensor principle by applying the humidity correction principle that we originally developed, in addition to the existing thermal conductivity principle. This development has made it possible to correct the mixed water vapor with high precision and enabled high-accuracy hydrogen gas detection under environment of fuel cell electric vehicles.

INTRODUCTION

In recent years, fuel cells have been attracting attentions of people as a technology effective for protecting global environment. Hydrogen used for fuel of fuel cells basically generates water alone when combusted, and does not discharge air pollutants or carbon monoxide.

Furthermore, since hydrogen is obtained by electrolyzing water by electric power generated by clean energies such as solar energy, hydraulic power, wind power, geothermal power, etc., it will greatly contribute to problems of depletion of fossil fuel resources.

However, since hydrogen provides good diffusibility and is colorless and odorless, it is not noticeable even if it leaks, and provides properties of high inflammability as its explosion limit when mixed with air ranges from 4% to 75%.

Consequently, for fuel cell systems, safety measures against hydrogen leakage are paramount and hydrogen sensors that detect hydrogen leakage are indispensable parts.

With the foregoing for the background, we have continued our development of hydrogen sensors, and we particularly have focused our attention on specifications required as hydrogen sensors for automotive fuel cells as well as detection principles, and narrowed down to the thermal conduction principle.

This paper reports our investigation results on correction techniques of the thermal conductive type hydrogen sensors particularly against interfering gas, humidity.

WORKING ENVIRONMENT OF HYDROGEN SENSORS AND DESIRED SPECIFICATIONS

As the expected location of hydrogen sensors in automobiles would be the cabin inside, hood inside, trunk inside, near hydrogen tank, and near fuel cells.

Based on our investigation results, the main desired specifications of hydrogen sensors are summarized as follows:

Detection concentration range   0%-2%

Detection accuracy   ±0.01%H_2

Temperature range   -30 to 100° C

Humidity range   30-100%R.H. (including condensation)

Gas selectivity   To be free of malfunction due to detection except for hydrogen

Starting time   within 5 seconds

Response   within 5 seconds

For critical points in developing automotive hydrogen sensors, influence of interfering gas should be mentioned. Examples of interfering gas would include exhaust gas (CO, HC, NOx, SPM; Suspended Particulate Matter), humidity, gas released from interior parts and air refresheners, influence by smoking, carbon dioxide discharged from human body, etc.

Though under limited environment, influences of sulfur in hot springs, influences of sodium or chlorine near seashores, and all other various conditions must be assumed.

Sensors with no sensitivity or little susceptibility to these interfering gases are sought for.

Starting capabilities and response are also assumed critical.

With respect to starting capabilities, sensors must be ready for detection before hydrogen gas has reached the fuel cell system after keying to start a car.

As for response, when hydrogen is detected as an automotive system, forcible ventilation and system shutdown are expected to take place, and the detector should response immediately.

Needless to say, in addition to these, there are restrictions such as temperature conditions, power supply conditions, vibration environment, durability requirements, etc., but in this paper, measures against humidity that would exert extremely great influence with respect to influences of the interfering gas discussed above.

SELECTION OF HYDROGEN DETECTION PRINCIPLE

Because there proposed are various detection principles for hydrogen sensors, the details of them are examined and whether the desired specifications stipulated above could be achieved or not were investigated.

For the principle of in-vehicle hydrogen detectors, following discussions will be made on the semiconductor principle, contact combustion principle, and thermal conduction principle which have comparatively simple construction and have already been applied in gas leak detectors, etc.

Semiconductor principle

The semiconductor principle is applied to detect combustible gases by changes of element resistance resulting from gas adsorption and desorption, and is very popularly adopted for home gas leak detectors, etc.

In general, SnO_2 is used for the element. This element is extremely sensitive and can detect several hundreds of ppm of inflammable gas concentration.

However, when the concentration exceeds 2000 ppm, the output becomes saturated and it is difficult to detect highly accurately in the hydrogen concentration region as high as 2%.

In addition, as this element reacts with all of the combustible gases, it has poor gas selectivity.

For example, if traces of combustible interfering gas exist, the element generates big errors because it provides high sensitivity at the low concentration.

Based on this, it is assumed difficult to adopt the semiconductor principle to automobile hydrogen sensors.

Contact combustion principle

The contact combustion principle is a method for detecting calories generated when inflammable gas combusts as temperature changes, and it is a popularly accepted method to detect it as a change in resistance using a platinum temperature measuring resistor.

Since the contact combustion principle combusts inflammable gas, if inflammable gases other than hydrogen are mixed, they cannot be basically distinguished, and poor selectivity results.

In addition, because the performance of this principle is greatly dependent on catalyst combustion performance, in the in-vehicle environment where existence of various gases is assumed, it will become essential to identify the degree of influence on such gases.

In particular, it is important to secure stability of catalyst combustion performance against catalyst poisoning substances such as silicon, hydrogen sulfide, sodium ions, etc. and the difficulty of developing such technology would be assumed extremely high.

Thermal conduction principle

The thermal conduction principle utilizes the thermal conductivity of hydrogen which is considerably higher than other gases, and detects calories depleted by hydrogen from the element heated to a specified temperature as a temperature change.

Unlike the above-mentioned semiconductor or contact combustion principles, the thermal conduction principle detects hydrogen by utilizing the difference of physical properties called thermal conductivity of the gas without causing any hydrogen chemical reactions to take place. It has advantages to estimate the influence on the desk with respect to various gases that can be assumed in the in-vehicle environment.

Furthermore, since the principle does not utilize any chemical reactions, it achieves characteristics of essentially high stability as compared to the two principles mentioned above.

thermal conductivity of gas	
Species of gas	conductivity
Hydrogen	16.82
Air	2.41
Steam	1.58
Carbon monoxide	2.32
Carbon dioxide	1.45
Nitrogen monoxide	1.51
Dinitrogen Dioxide	2.38
Argon	1.63
Chlorine	0.79
Oxigen	2.45
Nitrogen	2.40
Neon	4.65
Ammonia	2.18
Methane	3.02

x10^{-2}W/mK

Figure.1 Thermal conductivity of various gases

Fig. 1 shows thermal conductivity of various gases. The gas thermal conductivity exhibits a tendency to be inversely proportional to molecular weight of gas, and helium gas is focused as interfering specie, but it is assumed not critical with the actual automobile environment taken into account.

In addition, with respect to the humidity, when it exists at high concentration, there is a possibility to cause large errors, scrutiny is required.

After thoroughly investigating the detection principles as above, we chose the thermal conduction principle.

The biggest reason why we chose the thermal conduction principle is that because the principle detects hydrogen utilizing the difference of physical properties of thermal conductivity which the gas possesses without causing chemical reactions of hydrogen, we judged that the principle is suited for in-vehicle environment.

BASIC INVESTIGATION ON THERMAL CONDUCTION PRINCIPLE

For this principle, in general, platinum temperature measuring resistor is used for heating temperature sensing resistor element.

Because the temperature change corresponding to calories depleted by hydrogen is extremely small, the resistance change is also extremely small when platinum temperature measuring resistor with small temperature coefficient is used.

Based on the background described above, on the current development stage, thermistor with large temperature coefficient are used to carry out investigation.

When thermistor are used, in air difficult to transmit heat,

thermistor generate heat in themselves due to heat accumulation and lower resistance, and in hydrogen gas in which heat is easy to transmit, self-heat generation is suppressed and resistance is increased.

However, because this change is only a marginal temperature change and at the same time it is subject to ambient temperature, air is sealed in one of the two cases and measuring gas is introduced into the other case.

In such event, since temperature difference is generated in two elements, these are formed into a bridge and the gas concentration is measured by its differential voltage.

Humidity selectivity of thermal conduction principle

Investigation was made on the selectivity of the thermal conduction principle.

As described above, for interfering gas, exhaust gas, gas released from interior parts and air refresheners, carbon dioxide generated from human body, etc. are assumed, but the concentration is low and is assumed that the influence on sensors would be negligible.

However, with respect to humidity, the concentration up to about 30% is expected, and it is assumed that humidity would exert great influence on the sensor output.

This can be easily assumed because the thermal conduction principle is applied to absolute humidity sensors.

For the thermal conduction principle, characteristics when hydrogen and humidity are introduced around heating elements, respectively, were evaluated with ambient temperature kept constantly to 80° C.

Elements and evaluation

The thermistor elements which were heating elements were in the form of disk 1.4 mm in diameter and 0.2 mm thick, had the surface coated with glass, and were arranged in a case provided with holes for introducing gas to be detected.

To the heating elements, load resistances were connected in series and the heating element temperature is adjusted to about 160° C.

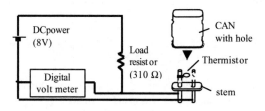

Figure.2 General configuration of element and evaluation circuit

479

Fig. 2 outlines the thermistor element and the evaluation circuit. The output characteristics are measured with air used as a base, mixture gases of 0%, 0.5%, 1.0%, and 2% hydrogen concentration were prepared as detected gas and voltage across both ends of the thermistor element was measured when hydrogen was introduced near the sensor.

By the way, with respect to measurement of magnitude of influence of humidity on output characteristics, the detected gas was humidified by bubbler and evaluated. Fig. 3 shows the results.

In the figure, the abscissa is the hydrogen concentration and the ordinate voltage across both ends of the thermistor element.

Because humidification is carried out by allowing nitrogen and air mixture gas to pass through bubbler, the hydrogen concentration of the detected gas during humidification becomes relatively low.

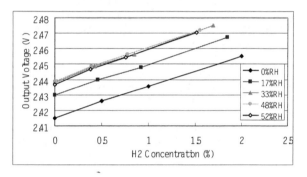

Figure.3 80degC hydrogen concentration output characteristics

Fig.3 indicates that voltage across both ends of thermistor element increases in accord with the hydrogen concentration and varies in accord with the hydrogen concentration, and at the same time, influence by humidity is exhibited as expected by about 1.4% at maximum as hydrogen concentration.

Because the thermal conduction principle is a principle to detect hydrogen concentration by utilizing the difference of physical property called thermal conductivity, influence of humidity is certain to occur, meaning that humidity correction is indispensable with some kinds of means added.

INVESTIGATION OF HUMIDITY CORRECTION MET HOD IN THERMAL CONDUCTION PRINCIPLE

For a means for correcting the humidity, various techniques were investigated and examined including application of general-purpose humidity sensors, etc., but we cannot find any applicable humidity correction technique in existing techniques.

When consideration is given to the concept using a principle based on no chemical phenomenon and the essence to utilize the difference of heat-generating condition of elements, which is a characteristic of thermal conduction principle, we judged best to develop a humidity correction technique based on the thermal conduction principle for the total performance.

We made detailed investigation on the thermal conduction principle and found out that humidity correction would be possible by using two elements with varying heat generating temperature.

Thermal conduction theory of mixture gas

Relationship between element output voltage and thermal conductivity

The thermal conductivity can be defined by Fourier's law by Eq. (1) as follows:

$$Q = -\lambda \qquad A(dT/dy) \qquad \text{Eq. (1)}$$

where, Q denotes the thermal energy transmitted, A the cross sectional area to which heat is conducted at position y on y-axis, dT/dy temperature gradient, and λ thermal conductivity.

To think Eq. (1) with the thermistor element heated assumed, the calorie Q transmitted from thermistor element to surroundings can be defined by the temperature gradient at position y and thermal conductivity λ of atmosphere gas.

Now, think the case in which thermal conductivity λ is the value nearly on the surface of the thermistor element. It is Fig. 4 that schematically expresses the case.

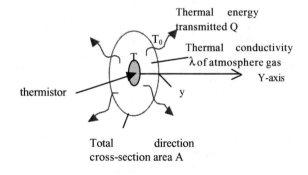

Figure.4 Schematic representation of thermal conduction of thermistor element

From Fig. 4, Eq. (1) becomes the following equation.

$$Q = \lambda \cdot A \cdot (T-T_0)/y \qquad \text{Eq. (2)}$$

Since Q denotes the total calorie of thermistor element, Q is expressed by the following equation:

$$Q = I \cdot V \qquad \text{Eq. (3)}$$

where, I is the current that flows in the thermistor element and V the voltage across both ends of the thermistor element, that is, output voltage, and I can be expressed by Eq. (4) using applied voltage V_c and load resistor Rl as follows:

$$I = (V_c - V)/Rl \qquad \text{Eq. (4)}$$

From Eq. (2) through (4) above, the following equation can be obtained:

$$V \cdot (V_c - V)/Rl = \lambda \cdot A \cdot (T-T_0)/y \qquad \text{Eq. (5)}$$

From Eq. (5), output voltage V can be denoted as a function of thermal conductivity λ of atmosphere gas.

Now, as thermal conductivity λ varies in accord with the kind of gas and humidity and concentration, as a result, output voltage V becomes a function of gas concentration.

Thermal conductivity of mixture gas

Since steam, air, and hydrogen are polyatomic molecules, their thermal conductivity can be considered as a sum of portion (λ_{0tr}) by translational energy and energy (λ_{0int}) by degree of internal freedom.

For the system in which each monoatomic gases are multicomponent mixed, the Sutherland-Wassiljewa type theoretical equation is known. Its general formula is shown in Eq. (6).

$$\lambda_{0m} = \sum(\lambda_{0i}/(1+\sum A_{ij}(x_j/x_i))) \qquad \text{Eq. (6)}$$

where, λ_{0i} denotes thermal conductivity of each monoatomic gas i and x_i and x_j denote mol fractions (concentrations) of components i and j.

In addition, first $\sum$ covers i = 1 to n (n-component system) and the second $\sum$ j = 1 to n where i ≠ j.

A_{ij} is an incidence number and if this and thermal conductivity λ_0 are known, the thermal conductivity in an optional composition can be calculated.

For calculating A_{ij}, a theory with consideration given to the fact in that water vapor has a particle with strong polarity must be used, and investigation results on various approximate expressions have indicated that the approximate expression of Brokaw that can be applied to polar molecules well coincides with measured thermal conductivity results and has been proved best suited.

The Brokaw approximate expression is the expression obtained by dividing the thermal conductivity into translation motion and degree of internal freedom and expressing each by Eq. (6), and can be expressed as follows:

$$\lambda_{0m} = \sum((\lambda_{0tr})_i/(1+\sum(A_{mon})_{ij}(x_j/x_i))) \qquad \text{Eq. (7)}$$
$$+\sum((\lambda_{0int})_i/(1+\sum(A_{int})_{ij}(x_j/x_i)))$$

where, $(\lambda_{0tr})_i$ denotes the thermal conductivity by translation motion of component i, $(\lambda_{0int})_i$ the thermal conductivity by the degree of internal freedom of component i, $(A_{mon})_{ij}$ the incidence number concerning translation motion of monoatomic molecule, and $(A_{int})_{ij}$ the incidence number concerning the degree of internal freedom.

Space did not permit us to insert the details of these parameters and we hope that you would refer to the reference.

Now, when the thermal conductivity of moist air containing hydrogen is found, Eq. (7) should be extended to 3 components.

With respect to i, j, let's define the indices of each component as air 1, water vapor 2, and hydrogen 3, and then, Eq. (7) can be expressed as follows:

$$\lambda_{0m} = (\lambda_{0tr})_1/(1+(A_{mon})_{12}\cdot(x_2/x_1)+(A_{mon})_{13}\cdot(x_3/x_1))$$
$$+(\lambda_{0tr})_2/(1+(A_{mon})_{21}\cdot(x_1/x_2)+(A_{mon})_{23}\cdot(x_3/x_2))$$
$$+(\lambda_{0tr})_3/(1+(A_{mon})_{31}\cdot(x_1/x_3)+(A_{mon})_{32}\cdot(x_2/x_3))$$
$$+(\lambda_{0tr})_1/(1+(A_{int})_{12}\cdot(x_2/x_1)+(A_{int})_{13}\cdot(x_3/x_1))$$
$$+(\lambda_{0tr})_2/(1+(A_{int})_{21}\cdot(x_1/x_2)+(A_{int})_{23}\cdot(x_3/x_2))$$
$$+(\lambda_{0tr})_3/(1+(A_{int})_{31}\cdot(x_1/x_3)+(A_{int})_{32}\cdot(x_2/x_3)) \qquad \text{Eq. (8)}$$

Next discussion is made on the temperature for calculating Eq. (8). Since the thermal conductivity very close to the thermistor surface is found, the temperature at position y is assumed to be equal to the thermistor surface temperature.

Consequently, the thermal conductivity was obtained by calculating the exothermic temperature of actual thermistor element at 80°C and using the temperature obtained.

The thermistor exothermic temperature can be defined by the temperature characteristics of resistance of the following equation2):

$$R = A \cdot T^{-c} \cdot \exp(D/T) \qquad \text{Eq. (9)}$$

where, R denotes resistance (W) at temperature T(K), and A, C, and D constants, which can be defined by measuring resistance at temperature of three points.

The resistance R of heated thermistor element is expressed by the equation below:

$$R = V \cdot RI / (V_c - V) \qquad \text{Eq. (10)}$$

Consequently, from Eq. (9) and (10), the exothermic temperature can be calculated.

The exothermic temperature of the thermistor element used for the recent investigation was 157.88°C at ambient temperature of 80°C or lower.

The thermal conductivity of 3-component-based mixture gas under this temperature condition was calculated. Fig. 5 shows the results.

The thermal conductivity was calculated in such a manner as to find the hydrogen sensitivity at each humidity so that the results can be compared to Fig. 3. The ordinate is the thermal conductivity λ (W/mK).

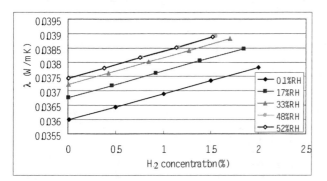

Figure.5 Hydrogen concentration dependency of calculated thermal conductivity at 80degC

By comparing Fig. 3 to Fig. 5, it is possible to confirm that the thermal conductivity obtained by calculation is correlated with the output voltage of thermistor element.

Humidity correction principle in thermal conduction equation

Next discussion will be made on the method for correcting humidity output and finding the hydrogen concentration only.

In Eq. (8), the thermal conductivity λ_{0m1} of moist air containing hydrogen around the thermistor element heated to an exothermic temperature is a function of mol fraction (= concentration) of each component, and the following equation holds for each component.

$$x_1 + x_2 + x_3 = 1 \qquad \text{Eq. (11)}$$

From Eq. (8) and (11), the mol fraction x_1 of air can be

deleted but x_2, x_3 cannot be determined. That is, an equation that expresses the relationship between x_2 and x_3 is required.

Now, if the exothermic temperature of thermistor element is varied, since Eq. (8) has λ_{0m} that consists of parameters λ_{0tr}, λ_{0int}, A_{mon}, A_{int}, the thermal conductivity λ_{0m2} is obtained, which is different from the above-mentioned thermal conductivity λ_{0m1}.

Since another equation that expresses the relation between this thermal conductivity λ_{0m2} and x_2, x_3 can be obtained, the hydrogen concentration x_3 only can be found by mathematically solving the simultaneous equation of two unknowns.

Specific humidity correcting technique

Based on the above-mentioned conception, the hydrogen sensor comprises a total of four elements, with two thermistor detection elements with varying exothermic temperatures and correction reference elements of changes of ambient temperature using a bridge circuit.

In such event, for the element output voltage, the relevant outputs (= differential output = voltage of detection element across both ends - voltage of reference element across both ends) after ambient temperature is corrected by the bride circuit, and calculating by applying the above two equations to these output voltages can find the hydrogen concentration x_3.

However, because the simultaneous equation with two unknowns based on Eq. (8) cannot directly supply strict solutions, calculation becomes essential, and this is not realistic when the application to actual sensors is considered.

Consequently, investigation was made on a method for correcting humidity not through the theoretical equation but by directly from differential output obtained with the hydrogen sensor comprising four elements.

Since this is a method for theoretically finding the correlation between output voltage and concentration as described above, this was investigated with two sets of thermistor elements with different exothermic Sensor configuration and conditions

The sensor configuration is achieved by locating two sets of detection sections comprising a detection element and a temperature correction reference element with no holes in the case and air sealed on high-temperature heating element side and low-temperature heating element side, respectively. Fig. 6 shows the configuration.

The exothermic temperature of each element is adjusted by

changing values of load resistances connected in series and temperature difference is achieved.

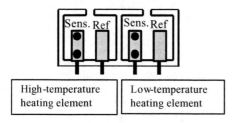

Figure.6 Configuration of detection section

The sensors were evaluated at 80° C, hydrogen concentrations 0, 0.5, 1, 2, 1, 0.5, and 0% which are changed over in that order, and four humidified humidity points of 0 Rh%, 15 Rh%, 55 Rh%, and 80 Rh%. Fig. 7 shows the evaluation results.

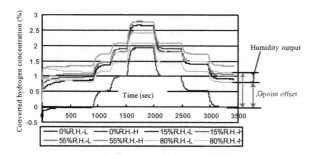

Figure.7 Output of each detection section of humidity correction sensor with exothermic temperature difference provided

In Fig. 7, the abscissa is measuring time and hydrogen gas concentration is changed. The ordinate is converted hydrogen concentration, and is an output standardized by finding output voltage difference of detection elements and reference elements at humidified humidity 0 Rh% at high-temperature heating element side detection section and low-temperature heating element side detection section, respectively, as well as by finding the relevant conversion coefficients so that the output voltage difference of 0% hydrogen concentration (0-900 seconds) becomes 0 and that of 2% (1500-1980 seconds) becomes 2.

Using these conversion coefficients, the output voltage differences under various humidifying conditions were calculated and the results are shown in the figure.

Since as indicated by the figure, the output increases as the humidity increases, it is assumed that each output is the sum of hydrogen concentration output and humidity output.

Consequently, the hydrogen concentration with humidity influences eliminated can be obtained from the difference between the output and humidity errors.

Investigation of humidity correction technique

Because in Fig. 7, the humidity error varies in accordance with the humidity, the value of humidity only (humidity output) must be known in order to subtract the humidity error from the output.

This can be obtained by the difference between the output at the high-temperature heating element side detection section and that of low-temperature heating element side detection section.

The same hydrogen output can be obtained on both sides by standardizing the output voltage of relevant detection sections at 0% and 2% hydrogen sensitivity, but since humidity outputs and hydrogen outputs are of different properties, the outputs cannot become same even if the output voltages containing humidity errors are standardized by the hydrogen concentration conversion coefficient of relevant detection sections.

That is, since these differences correspond to the values subtracting standardized hydrogen outputs of the relevant detection sections, the output due to humidity influences only will be shown.

Fig. 8 shows the output difference corresponding to the humidity output (humidity output = output of high-temperature heating element side detection section- output of low - temperature heating element side detection section).

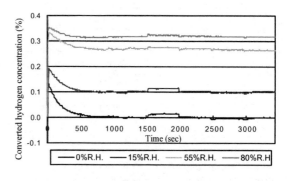

Figure.8 Humidity output of humidity correction sensor by exothermic temperature difference provided

The figure indicates that the humidity output that varied in accord with the humidity can be obtained.

In addition, though hydrogen concentrations vary in the period from 900 to 2940 seconds on the abscissa of Fig. 8, the humidity output is constant without being subject to the influences.

Now, Fig. 9 shows the correlation between humidity output and humidity error. In the figure, the abscissa is the humidity output and the ordinate is the humidity error (unit: converted hydrogen concentration ($\%H_2$)).

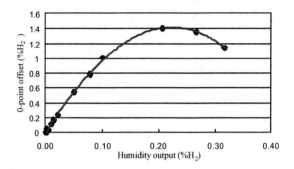

Figure.9 the correlation between humidity output and humidity error

The humidity error can be calculated by the approximation of this relation.

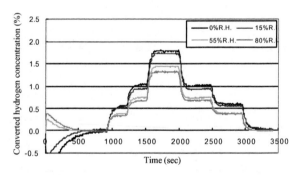

Figure.10 the results of humidity error corrected by the approximation

Fig. 10 shows the results of humidity error corrected by the approximation. As described above, the humidity error can be corrected by the humidity output, which is the difference between the output of high-temperature heating element side detection section and the output of low-temperature heating element side detection section.

It has been clarified that it is currently possible to achieve about ±0.1% correction accuracy to the humidity.

With the foregoing, it has become possible to secure the high selectivity that enables measurement of the hydrogen concentration only even under humidifying environment without any chemical phenomena by adopting the humidity correction technique using the difference of output results of two sets of detection sections having different exothermic temperatures.

CONCLUSION

A basic principle of hydrogen sensor with excellent stability and reproducibility as well as high gas selectivity has been developed by adopting the thermal conduction principle utilizing physical phenomena and developing our unique correction technique for the humidity, which is the biggest interfering gas, for hydrogen sensors for fuel cells, which have been attracting people's attention as a technique effective for protection of global environment.

In particular, for securing high selectivity, a technique using two detection thermistor elements with varying endothermic temperatures is adopted, and since this correction method also utilizes temperature dependency of thermal conductivity, it can be expected that stable output can be secured without using any chemical phenomena.

In the future, further technological development will be energetically carried forward in order to achieve still another critical point of the desired development specifications: high-speed start-up and high response.

REFERENCES

1. K.Makita, "Viscosity and Thermal Conductivity" Baifukan, Tokyo, 1975.

2. E.Habata, et al, "Thermistor-Liquid-Level Sensor" National Technical Report, Vol.26, No.3 June 1980.

3. A National Astronomical Observatory,"Science Chronological Table" Maruzen,Tokyo, 1993.

Dynamic Model of a Load-Following Fuel Cell Vehicle: Impact of the Air System

M. Badami and C.Caldera
Politecnico di Torino

ABSTRACT

Fuel cell vehicles promise to become, in near future, competitive with conventional cars in terms of performance, efficiency and compliance with emission reduction schedules. However, many steps still have to be done, and a series of fundamental choices, such as high vs. low air pressure system options remain unresolved. Modeling can be a powerful instrument to evaluate different components or plant layout, and to predict the dynamic behavior of a fuel cell system.

The first part of this paper illustrates the implementation of a direct engineering dynamic model of a load-following fuel cell vehicle. The modeling techniques, assumptions and basic equations are explained for each subsystem, with special attention to the air supply system, whose dynamic simulation was one of the primary targets of this work.

Some of the simulation results are presented in the second part. The performance of the vehicle was evaluated for the New European Driving Cycle (NEDC), showing the impact of the air system characteristics on the overall energy balance. The dynamic behavior of the components and their interaction through control strategies were also analyzed and other working conditions, such as acceleration runs and maximal velocity were tested.

INTRODUCTION

Over recent years, research into Fuel Cell Vehicles (FCV) has undergone a rapid increment and many automotive companies and research laboratories are now engaged in the study and development of FCV prototypes. The results of these researches have led to the increase of the power density of proton exchange membrane (PEM) fuel cell stacks (as reported in Fig. 1) and to enhance the performance of the subsytems that compose the plant. For this reason the project of building fuel cell vehicles for the market is not so far away as it was before.

Modeling, either to obtain a better understanding of the behavior of the system or to choose the optimal technology that should be adopted, represents a very important instrument for research.

Several research centers are active in this field, such as the National Renewable Energy Laboratory, where the famous ADVISOR (Advanced Vehicle Simulator) program was developed. This reverse engineering program allows the evaluation of performances for a wide range of vehicles (ICE, Electric, Hybrid, Fuel cell vehicles ...). The University of California is also deeply involved in fuel cell vehicle model activity and this is highlighted by the several papers that have been written on the subject ([1], [4], [6], [9]).

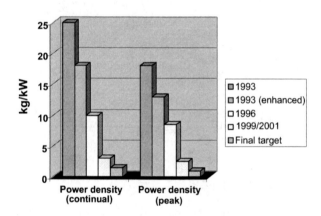

Figure 1. *Increment of power density from 1993 to 2001*

In fuel cell vehicles a very important role for the overall efficiency, is played by the air system. For this reason a careful design of this plant has been attempted, as can be seen from literature, but is still needed. In [4] the authors compared high and low pressure air systems, detecting the different operative conditions of the stack and of the other components and the impact that this choice has on the dimensions, power and efficiency of the system. In [11] different compressor technologies (2-3 bar) for FCS are compared and the impact on the overall efficiency of the system calculated under partial and transitory load. The energy trade-off of the stack, air system and Water and Thermal Management (WTM) has been dealt with in [6], both for high and low pressure systems.

In the present work, a Matlab-Simulink FCV dynamic model has been implemented with a direct-engineering approach. Special care was taken over the air system simulation. The components of the air system are usually simulated, in Matlab-Simulink, using lookup tables of experimental maps; from these tables it is possible to determine the compression power as a function of the operative conditions (pressure and flow rate). This approach, nevertheless, does not allow one to verify whether these components are able to satisfy the needs of the system, above all during transient conditions. The approach used in the present work is to simulate the behavior of the air system components through mathematical equations that take the most important features of the system and the most important causes of loss into account.

The air system model was then implemented in a fuel cell vehicle model, developed on the basis of different works found in literature [9]. The efficiency and consumptions were calculated on a European Cycle NEDC (Figure 2) in order to evaluate the performance of the system; the performance was also detected on 0-50 km/h and 0-100 km/h runs.

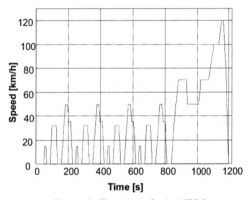

Figure 2. European Cycle NEDC

MODEL OF THE FCV

The scheme of the overall model is presented in Figure 3, where the following subsystems are evident:

- Vehicle
- Electric Motor
- Fuel cell stack
- Fuel feeding system
- Air system
- Water and thermal management system

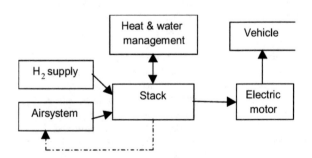

Figure 3. Schematic FCV model

Vehicle

The model of the vehicle has the electric motor torque or the brake torque (during decelerations) in the input, and the vehicle velocity in the output.

The forces acting on the system are synthesized by the gravitation and inertia forces applied to the barycentre of the vehicle, by the aerodynamic drag and by the interaction forces between the tires and the road. The scheme of the vehicle and of the acting forces is presented in Figure 4 (subscripts 1 and 2 refer to the front and rear axle).

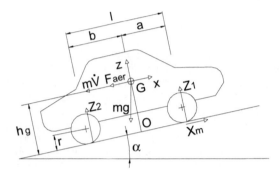

Figure 4. Scheme of the vehicle with acting forces

The vehicle dynamics can be represented by the following differential equation:

$$\dot{V} = \frac{X_m - \sum_i F_i}{m};$$

where V is the vehicle velocity, $X_m = C_m/r$ is the traction force (C_m=wheel torque) and F_is are the following resistent actions (see Table 1 for vehicle parameters):

- The aerodynamic drag: $F_{aer} = \frac{1}{2}\rho V^2 S C_x$;

- The rolling resistence: $F_{rot} = [mg\cos(\alpha) - Z_1]*f$,
  where $f = f_0 + KV^2$ and Z_1 is the vertical load on the front axles

  $$Z_1 = mg\left[\frac{b}{l}\cos(\alpha) - \frac{h_G}{l}sen(\alpha) - \frac{r}{l}f_0 - \left(\frac{\rho S}{2lmg}h_G C_x + \frac{r}{l}K\right)V^2 - \frac{h_G}{lg}\dot{V}\right];$$

  $$Z_2 = mg\left[\frac{a}{l}\cos(\alpha) + \frac{h_G}{l}sen(\alpha) + \frac{r}{l}f_0 - \left(\frac{\rho S}{2lmg}h_G C_x + \frac{r}{l}K\right)V^2 + \frac{h_G}{lg}\dot{V}\right];$$

- The brake action: $F_b = C_b/r$ when needed;

The integration of the previous dynamic equation allows the calculation of the velocity of the vehicle.

The simulation was performed using a *direct engineering* approach: the velocity of the vehicle is adjusted by a control that regulates the electric motor and the brake in a way that is similar to a real driver's behavior.

Table 1. Vehicle parameters

Vehicle mass	m	1570	Kg
Axle base	l	2,54	m
Barycentre-rear axle length	a	1,305	m
Front axle-barycentre length	b	1,235	m
Barycentre height	h_0	0,5	m
Wheel radium	r	0,273	m
Air density	ρ	1,2	g/dm^3
Front area	S	2	m^2
Aerodynamic drag coefficient	C_x	0,3	[-]
Rolling friction	f_0	0,013	[-]
Rolling friction-quadratic term	K	6,5*10^{-6}	s^2/m^2
Road grade	α	0	°
Total gear ratio	τ	6	[-]
Traction type		front	

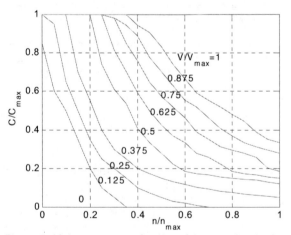

Figure 5 – Scheme of the control strategy of the vehicle

Electric motor

The system that has been considered has two different electric motors, one for the traction and the other for the compressor.

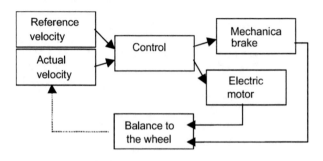

Figure 6. *Motor torque as a function of the speed and voltage*

These components were simulated, including the map of a standard motor, scaled to fit the needs of the system, in an appropriate look-up table.

A nominal voltage of V_{max} = 400V, a maximum torque of C_{max} = 180 Nm and a rotational speed of n_{max}=10000 rpm were considered for the traction motor. The compressor motor, instead, was limited to a voltage of V_{max} = 48V, a torque of C_{max} = 50 Nm and a speed of 4000 rpm. For simplicity, a constant mean efficiency of the motors (η = 0.8) was used instead of the real maps.

An estimate of the mean current required by the motor can be calculated by dividing the mechanical power by the voltage and the efficiency:

$$i = \frac{1}{\eta_m} \cdot \frac{C \cdot \omega}{V}$$

Fuel cell stack

Our reference for the FC stack was a 50 kW prototype. The voltage vs. current density curves are plotted for a single cell for several air pressure levels, in Fig. 7 (T = 80°C).

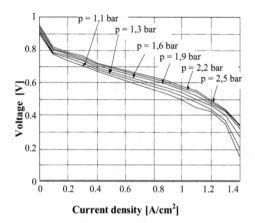

Figure 7. *Voltage vs. current density for a single cell, plotted for different air pressure levels; T = 80°C*

The current i [A] generated by the stack is related to the amount of reactants flowing to the respective electrodes, according to the following expressions:

$$\dot{m}_{air} = 28,97 \cdot \frac{1}{0,21} \cdot \frac{i}{4F} \cdot SR \cdot n_{cells} \text{ [g/s]}$$

$$\dot{m}_{H2} = \frac{i}{F} \cdot n_{cells} \text{ [g/s]}$$

where F is the Faraday constant [As/mol].

The stack performance depends on the availability of reactants at the electrodes: the lack of air or fuel causes a decrement of the cell voltage for a certain current density. This fact can be confirmed by the figure 8, which is based on experimental data, showing the voltage decrement during a constant current test with no air sent to the cathode.

The shortage in one of the reactants makes the voltage go below the nominal curve, moving towards a region of lower efficiency. In order to take into account this fact, an additional efficiency was introduced into the model, to be applied to the nominal voltage curve.

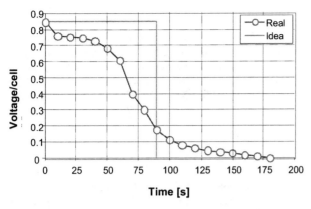

Figure 8. *Voltage decrease during a constant current generation with no airflow, compared to the "ideal" behavior.*

The air-side volume (estimated V_c=20 dm³) of the stack was modeled as a capacity, with an inlet mass flow ($\dot{m}_{IN}$) elaborated by the compressor and an outlet one ($\dot{m}_{OUT}$) leaving the system through the valve or the expander. The mass of air present in the capacity, at a given time step, can be evaluated by the following differential equation[2]:

$$\dot{m}_{IN} - \dot{m}_{OUT} = \frac{d}{dt}m = \frac{d}{dt}\left(\frac{p_2 V_c}{RT_2}\right)$$

$$= \frac{V_c}{RT_2}\cdot\frac{dp_2}{dt} - \frac{p_2}{RT_2^2}\cdot\frac{dT_2}{dt}$$

where p_2 and T_2 are the fluid pressure and temperature in the stack, while R is the universal gas constant over the molecular weight of the air.
A considerable simplification of this expression come from the assumption of constant temperature operation of the stack since PEM fuel cells are usually temperature controlled at about 80°C. In this way one can easily evaluate the pressure inside the capacity (p_2) through the following expression:

$$p_2 = \int\frac{RT_2}{V_c}(\dot{m}_i - \dot{m}_u)dt$$

This value has a very important role on the running conditions of the compressor and the valve/expander, as will be clearly illustrated in the air system section.

Fuel storage

The model considers a direct hydrogen fuel cell system. The hydrogen is fed by a tank of compressed gas, the mass flow being regulated by means of a controllable valve. The subsystem gives the fuel consumption during the simulation in the output.

---

[2] according to the conventions used in the model, subscript 1 indicates the inlet conditions of the compressor, 2 the conditions at the compressor outlet or in the stack and 3 the fluid conditions at the valve or expander exit

Air system

The core of this work was the simulation of the air system. For this reason most of the effort was concentrated on the dynamic modeling of this subsystem, and its interaction with the whole FCS.

Air can be introduced into the stack by means of a simple blower (low pressure system, below 1.4 bar) or a compressor (high pressure system, between 2 and 4 bar). The advantages of adopting a high-pressure configuration can be summarized as follows:

- The higher partial oxygen pressure improves the reactive capability of the cells and facilitates the mass transfer: the voltage of the single cell and the electric power density of the stack increase.
- The amount of water needed to humidify the PEM reduces, simplifying the water on-board balance

High pressures however do represent an additional cost to the entire system, because the compression power increases rapidly with the pressure ratio, as shown in fig. 9.

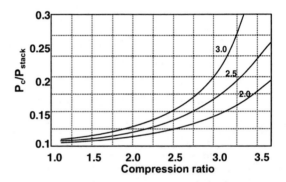

Figure 9. *Power needed for the air compression as a function of pressure ratio for different stoichiometric ratios (SR)*

A detailed comparison between the low and high pressure air management systems for automotive fuel cell systems is reported in [4] where it is shown how complex it can be to find a trade-off for water and thermal management.
Once decided to work at higher pressures than atmospheric pressure, two main layouts can be selected.
In the first solution, the air leaving the stack flows out through a backpressure valve; in the second one, the fluid is elaborated in an expander unit to recover a part of its energy (see Figure 10 and 11).
As previously mentioned, the stack was modeled as a capacity; the inlet air mass flow depends on the pressure and on the rotational speed of the compressor-motor group, while the outlet mass flow is a function of the valve opening state or of the operating point of the expander, which is rotating at the same speed as the compressor. Figure 10 (a and b) explains the reciprocal interactions of the air system components as they are modeled in this work.

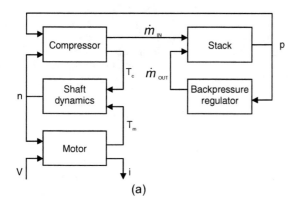

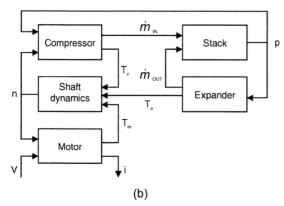

Figure 10. *Model of the air management system in both configurations with a backpressure valve (a) and with an expander (b)*

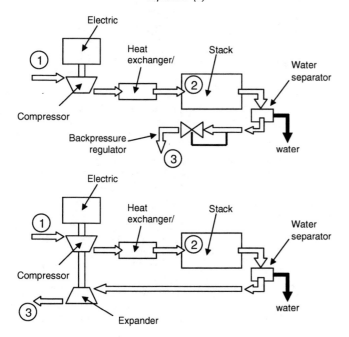

Figure 11. *Air system schemes with a backpressure valve and with an expander unit.*

Compressor

The compressor is, among the fuel cell system auxiliaries, the one with the major weight on the energy balance; for this reason, the choice of the optimal

technology, in terms of efficiency, performance and weight-volume, is one of the most critical factors in an FCV project. The advantages and disadvantages of several technologies, that are suitable for fuel cell mobile applications, are presented in [11] and [13]; in the present work, we have chosen to model a reciprocating compressor that seems to be one of the most efficient technologies even though it has some shortcomings due to its high cost and dimensions.

The analytical procedure that was used in this work, is based on a simplified model of the real compression cycle. The assumptions are as follows:

- The compression and expansion are politropic relations with exponent m and m', respectively
- Pressure losses through the valves can be assumed to be constant along the induction and discharge phases.

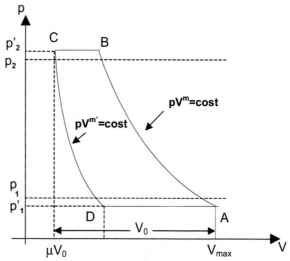

Figure 12. *Pressure-volume diagram for a reciprocating compressor*

Referring to Figure 12, the work done in compressing air during a cycle is given by the following expression:

$$L_c = \left[\frac{m}{m-1} \cdot p'_1 V_A \cdot \left(\beta_i^{\frac{m-1}{m}} - 1 \right) - \frac{m'}{m'-1} \cdot p'_1 V_D \cdot \beta_i^{\frac{1}{m'}} \cdot \left(\beta_i^{\frac{m'-1}{m'}} - 1 \right) \right] \cdot p'_1 V$$

$$= \left[\frac{m}{m-1} \cdot (1+\mu) \cdot \left(\beta_i^{\frac{m-1}{m}} - 1 \right) - \frac{m'}{m'-1} \cdot \mu \cdot \beta_i^{\frac{1}{m'}} \cdot \left(\beta_i^{\frac{m'-1}{m'}} - 1 \right) \right] \cdot p_1 \cdot V$$

where

$\mu = \dfrac{V_c}{V_0}$ is the clearance volume coefficient, and

$\beta_i = \dfrac{p'_2}{p'_1}$ is the internal compression ratio

Pressure levels p'_1 and p'_2 (inner pressures during the filling and the discharge phases) are respectively lower and higher than the corresponding external values (p_1 and p_2), because of the pressure losses through the valves, thus the internal compression ratio can be written as follows:

$$\beta_i = \frac{1+\delta_2}{1-\delta_1} \cdot \beta$$

where β is the external compression ratio and δ_1 and δ_2 depend on the angular speed according to the quadratic law:

$$\delta = k_1 + k_2 \cdot n^2$$

The coefficients k_1 e k_2, functions of the machine geometry and of the thermodynamic operating conditions, were assumed to be constant. In order to simplify the model, δ_1 was considered to be equal to δ_2.

The outlet flow rate of the compressor can be determined through the following equation:

$$\dot{m} = \lambda_v \cdot \rho_1 \cdot iV \cdot n$$

The delivery ratio λ_v can be calculated by the following expression (see Table 2 for nomenclature)

$$\lambda_v = \eta_\Phi (1-\delta_1) \left[1 - \mu \left(\frac{\beta_i^{1/m'}}{\tau} - 1 \right) \right]$$

where $\tau = \dfrac{T_A^{(ad)}}{T_A}$ is the thermal ratio, which takes into account the heat exchange during the induction process ($T_A^{(ad)}$ is the temperature that would be reached at the end of the induction in an adiabatic process) and $\eta_\Phi = \dfrac{\dot{m}}{\dot{m}+\dot{m}_f}$ is the volumetric efficiency. The leakage $\dot{m}_f$ can be approximated with the expression $\dot{m}_f = \alpha \cdot \sqrt{\beta - 1}$, (factor α is a constant that has the dimensions of a mass flow rate).

The power required to compress the air is equal to

$$P_c = \frac{P_i}{\eta_m} = \frac{i \cdot L_c \cdot n}{\eta_m}$$

while the adiabatic efficiency of the compressor is defined by the expression:

$$\eta_{is} = \frac{\dot{m} \cdot L_{c,is}}{P_c}$$

where $L_{c,is} = \dfrac{k}{k-1} RT_1 \left(\beta^{\frac{k-1}{k}} - 1 \right)$ is the specific work under isentropic conditions.

Table 2. Compressor model parameters

Cylinder displacement	V_p	0,385 [dm³]
Number of cylinders	i	0,385 [dm³]
Clearance displacement coefficient	μ	0,1
Leakage coefficient	α	2 [g/s]
Valve loss coefficient	k_1	0,05
Valve loss coefficient (quadratic term)	k_2	8×10⁻⁵ [s²]
Politropic compression exponent	m	1,45
Politropic expansion exponent	m'	1,35
Mechanical efficiency	η_m	0,90
Thermal efficiency	τ	0,95

The previously presented theoretical analysis was applied to a high-performance reciprocating compressor, whose technical characteristics used for the simulation are synthesized in the Table 2. The graphs in Figures 13 and 14 compare the experimental and theoretical performances of the compressor.

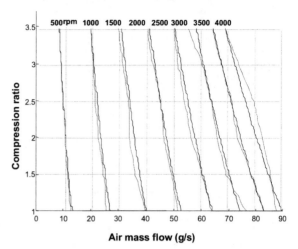

Figure 13. Compression ratio vs. mass flow rate (simulated data - solid line and experimental data - dotted line).

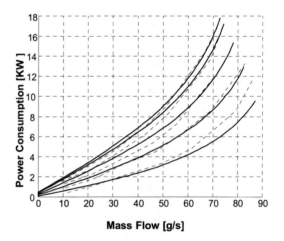

Figure 14. Power vs. mass flow rate, β=1,1.5,2,2.5,3,3.2 (simulated data - dotted line and experimental data - solid line).

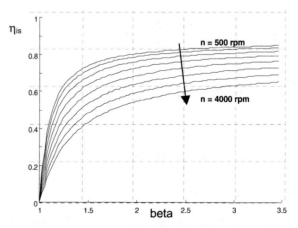

Figure 15. Adiabatic efficiency as a function of the compression ratio, plotted for different angular speeds.

490

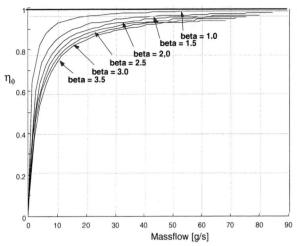

Figure 16. Volumetric efficiency as a function of the mass flow rate, plotted for several compression ratios

Backpressure Regulator

The flow rate through the backpressure regulator can be expressed by the equation

$$\dot{m}_u = A \cdot \frac{p_2}{\sqrt{RT_2}} \cdot f(\Pi, k)$$

where Π is the expansion ratio p_3/p_2. The non-dimensional term f, a function of Π and the fluid type (k), is defined as follows:

$$\begin{cases} f(\Pi, k) = \sqrt{\dfrac{2k}{k-1} \Pi^{\frac{2}{k}} \cdot \left(1 - \Pi^{\frac{k-1}{k}}\right)} & \text{if } \Pi > \Pi_{cr} \\ f(\Pi, k) = \sqrt{k \cdot \left(\dfrac{2}{k+1}\right)^{\frac{k+1}{k-1}}} & \text{if } \Pi \leq \Pi_{cr} \end{cases}$$

where $\Pi_{cr} = 0.528$ for the air. The pressure inside the capacity, once all the other conditions are fixed, is directly proportional to section area A of the nozzle; A PI control of the valve area was used in the model to set a desired pressure level during the simulation.

Expander

The fluid at the stack outlet has a higher pressure and temperature than the external atmosphere; this energy can be partly recovered by an expander. The isentropic work that can be obtained through the expansion between the stack outlet conditions (2) and the external ones (3), considering $p_3 = p_1$, is

$$L_{e,is} = \frac{k}{k-1} RT_2 \cdot \left[1 - \left(\frac{p_1}{p_2}\right)^{\frac{k-1}{k}}\right]$$

The effective work L_i can be estimated by multiplying the isentropic value by the isentropic efficiency $\eta_{e,is}$,

which, in this work is assumed to be constant and equal to 0.8, according to DOE guidelines. The power provided by the expander is given by the following expression:

$$P_e = \dot{m} \cdot \eta_{e,is} \cdot L_{e,is}$$

Shaft dynamics

The angular acceleration of the compressor shaft is expressed by the dynamic equilibrium equation:

$$C_m - C_c = I \cdot \frac{d\omega}{dt}$$

where I is the inertia momentum of the group, C_m is the motor torque (plus the expander when present) and C_c is the compressor torque. The angular speed of the shaft can be calculated by integration:

$$\omega = \int \frac{C_m - C_c}{I} \, dt [rad / s]$$

($I = 0.02$ kgm^2 in the simulations)

Flow rate control

The FC stack is able to generate the required electric power only if the correct amount of reactants - hydrogen and air in this case - is provided. This quantity varies during normal vehicle operations; a control unit, acting on the electric motor to minimize the difference between the required and the actual air mass flow, was set in the model to satisfy this variable demand.

Water management system

Inside a fuel cell stack, water guarantees humidification of the membrane and removal of excess heat. PEM-FC manufacturers suggest relative humidity values close to 100%; the same relative humidity can however be obtained with different water quantities, varying pressure and temperature: figure 17 shows how operating at higher pressure levels, at the same temperature, significantly reduces the amount of water needed to reach a certain humidity.

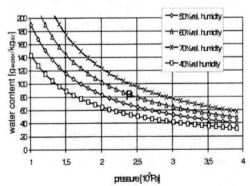

Figure 17. Water content in humid air for different values of relative humidity as a function of pressure. (T= 80°C) [13]

If the pressure and temperature of the fluid are appropriately selected, the water produced during the electrochemical reaction can be sufficient to keep the relative humidity at 100%.

The pressure and temperature couples that allow a neutral water balance, are presented in Fig. 18 for two different stoichiometric ratios.

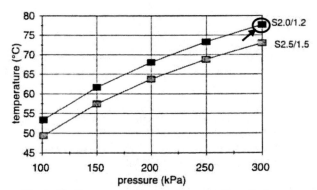

Figure 18. *Pressure and temperature conditions that guarantee a neutral water balance [13]*

In the present work the reference operating point is the one that is highlighted in figure 18 (T≈80°C, SR_{air}=2, p=3 bar): this point assures correct PEM humidification. The other points that will be considered (T≈80°C, SR_{air}=2, p<3) do not allow a neutral water balance, for this reason the introduction of a condenser would be necessary in the actual FCS.

RESULTS

A simulation analysis was performed to evaluate the behavior of the model. The parameters used for the simulation are as follows:

- Air pressure at stack inlet: p = 3 bar
- Stoichiometric ratio: SR = 2
- Constant expander adiabatic efficiency: $\eta_{e,is}$ = 0.8

The times to accelerate the car from 0 to 50 km/h and from 0 to 100 km/h (about 60 mph) were first calculated along with the maximum speed of the car and the fuel consumption over the European Driving Cycle (NEDC). The results are summarized in table 3.

Table 3. *Results of simulation - Performance of the car and fuel consumption*

Performances

Time 0-50 km/h	7.3 s
Time 0-100 km/h	21.0 s
Max. speed	150.6 km/h

Fuel consumption over NEDC driving cycle

System without expander	101.3 g
System with expander	91.8 g

In order to better understand the impact of the air supply system on the overall performance of the vehicle, several simulations were then performed over the NEDC cycle, varying the air pressure level from 1.2 to 3 bar.

Figures 19 and 20 show the overall efficiency and the hydrogen consumption over the NEDC as a function of the air pressure at the outlet of the stack; the efficiency of the system, without taking into account the energy needed for air compression, is presented in Figure 21.

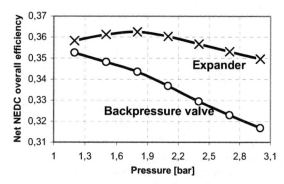

Figure 19. *NEDC Overall efficiency as function of the air pressure; with & without expander configurations*

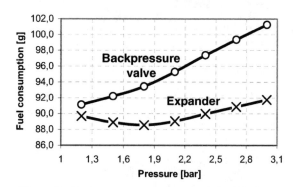

Figure 20. *Hydrogen consumption along the NEDC as a function of the air pressure; with & without expander configurations.*

The expander determines a significant efficiency increment (10% at 3 bar), moreover the efficiency curve with expander shows a maximum (fig. 19) for pressure values of about 1.9 bar. This result cannot however be assumed as a general rule, but as a consequence of the expander and compressor parameters.

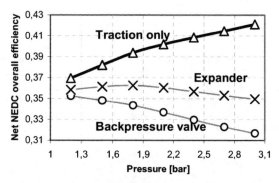

Figure 21. *Comparison between the overall efficiency with and w/o expander and the efficiency obtained without taking into account the energy for air compression*

492

SENSITIVITY TO DESIGN PARAMETERS

As a conclusion of this work, the model was applied to investigate the importance of some compressor parameters on the efficiency of the overall system. It was supposed to enhance the design of the compressor, by reducing some reasons of losses such as the clearance volume, the valve pressure losses and the leakage. The consumption and the overall efficiency were then calculated on the NEDC cycle. All the simulations were conducted for the FCS without expander at the pressure of 3 bar.

Figure 22 shows the efficiency improvement if the causes of loss are reduced by 50%. In particular the following compressor parameters were used for the simulation:

- Clearance displacement coefficient $\mu=0.05$
- Valve loss coefficients: $k_1=0.0025$, $k_2=4\times10^{-5}$ $[s^2]$
- Leakage coefficient: $\alpha=1$ [g/s]

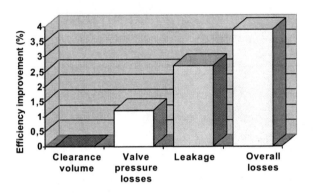

Figure 22. Improvement in the overall efficiency on the NEDC driving cycle obtained halving some causes of loss.

During the driving cycle, the compressor often operates at low speed, with corresponding small mass flow rates; for this reason, as one can easily deduce from the graph, the effect of the leakage is particularly important for the system's efficiency.

CONCLUSIONS

A Matlab-Simulink computational model of a Hydrogen FCV has been implemented. A direct engineering approach to the modeling of the components and their control was used to simulate the behavior of the system under dynamic conditions. Particular attention was paid to the simulation of the air management system, in general, and to the reciprocating compressor, in particular.

A series of simulations were performed along the European Driving Cycle (NEDC), for different values of the pressure level in the stack to try to understand the importance of the compression energy on the overall efficiency of the system. The configurations with and without an expander were taken into consideration and compared. The performance of the system, in terms of time to accelerate the vehicle (0-50 km/h and 0-100 km/h) was analyzed.

Finally, the effect of the reduction of the internal losses of the compressor on the overall efficiency was determined. The analysis shows that a hypothetical reduction of 50% of the compressor losses would lead to a 4% increment of the overall efficiency on the NEDC cycle.

ACKNOWLEDGMENTS

The authors would like to express their gratitude to Dr. Campanile and Dr. Bellerate (Centro Ricerche FIAT) for their help and cooperation during the present research.

REFERENCES

[1] P.Badrinarayanan, S.Ramaswamy, A.Eggert, R.M.Moore, Institute of Transportation Studies, UC Davis, "Fuel cell stack water and thermal management: impact of variable system power operation", SAE Paper 2001-01-0537

[2] F.Barbir, M.Fuchs, A.Husar, J.Neutzier, Energy Partners, "Design and operational characteristics of automotive PEM fuel cell stacks", SAE Paper 2000-01-0011

[3] S.Chalk, P.Davis, J.Milliken, D.Lee, "Fuel cells for transportation – Program Implementation Strategy" – Department Of Energy, Office of Transportation Technologies

[4] J.M.Cunningham, M.A.Hoffman, D.J.Friedman, University of California, Davis, "A comparison of high-pressure and low-pressure operation of PEM fuel cell systems", SAE Paper 2001-01-0538

[5] R.A.J.Dams, P.R.Hayter, S.C.Moore, Wellman CJB Limited, "The development and evaluation of compact, fast response integrated methanol reforming fuel processor systems for PEMFC electric vehicles", SAE Paper 2000-01-0010

[6] D.J.Friedman, A.Eggert, P.Badrinarayanan, J.Cunningham, University of California, Davis, "Balancing stack, air supply and water/thermal management demands for an indirect methanol PEM fuel cell system", SAE Paper 2001-01-0535

[7] M.H.Fronk, D.L.Wetter, D.A.Masten, A.Bosco, General Motors Corporation, "PEM fuel cell system solutions for transportation", SAE Paper 2000-01-0373

[8] R.Füßer, O.Weber, Filterwerk Mann+Hummel GmbH, "Air intake and exhaust systems in fuel cell engines", SAE Paper 2000-01-0381

[9] K.H.Hauer, D.J.Friedmann, R.M.Moore, S.Ramaswamy. A.Eggert, P.Badrinarayanan, University of California, Davis, "Dynamic response

of an in direct-methanol fuel cell vehicle", SAE Paper 2000-01-0370

[10] M.Ogburn, D.J.Nelson, W.Luttrell, B.King, S.Postle, R.Fahrenkrog, Virginia Polytechnic Institute and State University, "System integration and performance issues in a fuel cell hybrid electric vehicle", SAE Paper 2000-01-0376

[11] S.Pishinger, C.Schönfelder, Institute for Combustion Engines, RWTH Aachen, W.Bornscheuer, H.Kindl, A.Wiartalla, FEV Motorentechnik GmbH, Aachen, "Integrated air supply and humidification concepts for fuel cell systems", SAE Paper 2001-01-0233

[12] S. Thomas, M. Zalbovitz, "Fuel cells – Green power", Los Alamos National Laboratory LA-UR-99-3231

[13] A.Wiartalla, S.Pischinger, Institute for Combustion Engines, RWTH Aachen; W.Bornscheuerr, K.Fieweger, J.Ogrzewalla, FEV Motorentechnik GmbH, Aachen, "Compressor expander units for fuel cell systems", SAE Paper 2000-01-0380

CONTACT

Prof. M. Badami
Dipartimento di Energetica
Politecnico di Torino
C.so Duca degli Abruzzi, 24
10129 Torino
ITALY

Tel.: 39-11-5644436
Fax.:39-11-5644599
E-mail: badami@athena.polito.it

2002-01-0102

Optimizing Control Strategy for Hybrid Fuel Cell Vehicle

Gino Paganelli, Yann Guezennec and Giorgio Rizzoni
Center for Automotive Research and Intelligent Transportation, The Ohio State Univ.

ABSTRACT

This paper presents a general formulation of the instantaneous power split between a fuel cell and an electrical accumulator in a charge-sustaining fuel cell hybrid vehicle. The approach proposed in this paper is based on the ECMS (Equivalent Consumption Minimization Strategy) control strategy previously developed for parallel hybrid vehicle applications suitable for real time application and allowing the overall minimization of hydrogen consumption while meeting the driver demand.
This control strategy has been applied to a representative hybrid PEM (Proton Exchange Membrane) fuel cell mid-size vehicle. Using a Hybrid Fuel Cell vehicle simulator, the vehicle performance and energy requirements are estimated. The results provided by the ECMS control strategy approach are also compared to a more basic approach.

INTRODUCTION

Opportunities for reduced energy requirements, environment impact while meeting consumer demands in automotive applications exist although challenges in design optimization are extensive due to the interactions of powertrain components and the influence of changes in component size and operation on overall powertrain performance. These challenges have received attention recently in the context of hybrid electric vehicles powered by traditional internal combustion engines and electric machines. The same challenges apply to fuel cell-powered hybrid electric vehicles.

A hybrid vehicle is powered by at least two energy sources. In the series hybrid case, the node between those two energy sources is electric (power converter), as opposed to the mechanical node of a parallel hybrid configuration. A hybrid fuel cell vehicle falls under the category of a series hybrid.

One problem raised by such architecture and addressed in this paper is how to instantaneously distribute the electrical power requested to the two energy sources to optimize the operation of the vehicle. Obviously the first constraint is that the driver demand or electric motor power requirement must be satisfied at any time, within the power envelopes of the components.

A fuel cell hybrid vehicle is mechanically powered by one or more electric motors while the electrical energy needed is produced by a fuel cell along with an electrical accumulator as shown in Figure 1.

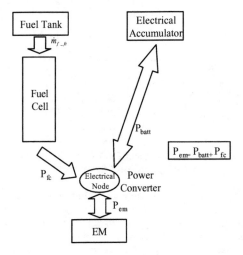

Figure 1: General Architecture of a Hybrid Fuel Cell Powertrain

The pedals position is interpreted as a continuously varying driver's motor power demand $P_{em}(t)$, positive or negative and prorated in function of the maximum power available at the current vehicle speed.

As shown is the Equation 1 below, both the fuel cell and the electrical accumulator contribute to supply the electrical energy needed:

$$P_{em}(t) = P_{fc}(t) + P_{batt}(t) \qquad \text{Eq. (1)}$$

Where $P_{fc}(t)$ is the net power produced by the fuel cell and $P_{batt}(t)$ is the power to/from the electrical accumulator[1,2].

Typically, the electrical accumulator energy storage capability is very limited and it is used as an energy buffer in order to optimize the operation of the powertrain. In other words, the net energy consumption associated to the electrical accumulator over a long drive is negligible as opposed to the energy consumption associated to the hydrogen consumed in the fuel cell and this operation is known as *charge sustaining*.

Practically, the control variable to balance the power distribution is the power produced by the fuel cell $P_{fc}(t)$ through an appropriate power converter, as shown in Figure 2. The power produced or captured by the electrical accumulator $P_{batt}(t)$ being the remaining power needed to satisfy the total power requirement:

$$P_{batt}(t) = P_{em}(t) - P_{fc}(t)$$

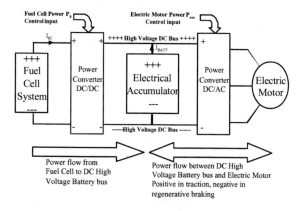

Figure 2: Practical Hybrid Fuel Cell System Hybrid Configuration Schematic

Commonly used hybrid fuel cell control strategy are the SOC-based control, the load following or a combination of both [5-7]. The simplest version of the SOC-based control turns on or off the fuel cell at given accumulator SOC (*State Of Charge*) thresholds. This is generally used with small APU fuel cell. For bigger fuel cell, the fuel cell operation is discretized in several power levels or even continuously (proportional SOC-based control). With the load following approach, the fuel cell power is set to meet the driver request, the electrical accumulator being used as a temporary buffer.

---

[1] The various efficiencies have been neglected in this paper for clarity, but are included in the formulation of the control strategy actually implemented.
[2] Throughout this paper, the power flow at the electrical accumulator is defined "positive" when current is drawn out from the accumulator (discharge); and it is defined "negative" when current flows into the accumulator (charge).

While generally allowing proper operation, those approaches do not embed any energy minimization consideration and consequently do not allow to reach optimal fuel mileage. The following approach proposes to bridge this gap.

CONTROL STRATEGY APPROACH

Ideally, the power distribution has to be optimized to minimize the overall hydrogen consumption over a given driving run, such as:

$$\underset{\{P_{fc}(t)\}}{Min} \sum \dot{m}_{f\_fc}(t) \qquad \text{Eq. (2)}$$

The main problem with this global minimization criterion is that the whole driving schedule has to be known *a priori*, thus real-time control cannot be readily implemented.

To avoid this drawback the ECMS (*Equivalent Consumption Minimization Strategy*) power distribution control strategy, initially developed for parallel hybrid vehicle applications [1-3], proposes to replace the global criterion by a local one, appropriately reducing the problem to a minimization of an equivalent fuel consumption at each time. The criterion becomes at all times:

$$\sum \underset{\{P_{fc}(t)\}}{Min} \dot{m}_{f\_equi}(t) \qquad \forall t \qquad \text{Eq. (3)}$$

where the equivalent fuel flow rate cost function $\dot{m}_{f\_equi}(t)$ is simply defined as the sum of the actual hydrogen consumption rate of the fuel cell $\dot{m}_{f\_fc}(t)$ and the equivalent fuel use rate due to the electrical accumulator $\dot{m}_{f\_batt}$ (positive or negative):

$$\dot{m}_{f\_equi} = \dot{m}_{f\_fc} + \dot{m}_{f\_batt} \qquad \text{Eq. (4)}$$

The global minimization problem and the local one are not strictly equivalent. However, the local one above can be easily used for real-time control as the global one is non causal and hence non realizable.

In order to comprehend the conceptual background of ECMS, one must understand the energy flow in the hybrid powertrain. When the electrical accumulator is currently discharging, the total electrical power supplied to the electric motor is, therefore, the sum of power output from the fuel cell and the electrical accumulator. In this case, at one fixed sample time, some fuel is consumed due to operation of the fuel cell and some current is drawn out from the electrical accumulator. To sustain the electrical accumulator state-of-charge, the energy drawn out from the electrical accumulator (at present) must be recharged (in the future). The necessary power to recharge the electrical accumulator will be provided either from the ICE and/or regenerative braking.

On the other hand, when the electrical accumulator is currently charging, while the fuel cell provides the traction power, a fraction of the power produced is also used to charge the electrical accumulator. This extra energy stored in the electrical accumulator (at present) must later be discharged to maintain the state of charge near target level (to prevent overcharging), which implies less fuel usage to run the vehicle (in the future).

In this conceptual framework of the power flow, the electrical accumulator can be modeled as a virtual auxiliary reversible fuel tank in the powertrain system. This is virtually implying that operating the electrical accumulator in discharge mode consumes extra fuel, and operating the electrical accumulator in charge mode puts back some fuel into the fuel tank for later use. Moreover, with such power flow concept, the electrical accumulator can be treated as a device that is run directly by the chemical energy stored in the fuel tank. Practically, the basis of the ECMS approach is thus to associate the electrical energy stored in the electrical accumulator E_{batt} to an amount of fuel m_{f_batt}. This amount of fuel is estimated by accounting for the average efficiency of the energy path to convert fuel to electricity (see Figure 3).

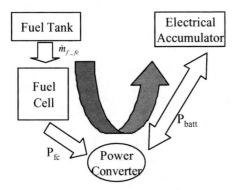

Figure 3: Energy Path to Convert Fuel to Electricity

Consequently, any power flow in or out of the electrical accumulator P_{batt} can be associated to an equivalent fuel mass flow $\dot{m}_{f\_batt}$

$$E_{batt} \Rightarrow m_{f\_batt} \qquad \text{consequently :}$$

$$\frac{dE_{batt}}{dt} = P_{batt} \Rightarrow \dot{m}_{f\_batt}$$

CONTROL STRATEGY IMPLEMENTATION

The above control strategy has been applied to a Hybrid Fuel Cell vehicle simulator developed [4] at the Center for Automotive Research and Intelligent Transportation at the Ohio State University. This quasi-static simulator allows to estimate the vehicle performance and energetic requirements with a very good accuracy. Special care has been devoted in the design of the hybrid fuel cell model used in the simulator in order to

account for practical issues such as temperature effect or fuel cell accessories energy consumption. Those considerations allow very realistic simulation results.

In the chosen configuration a PEM (Proton Exchange Membrane) fuel cell is assumed. The design and operating parameters used are representative of those proposed in automotive application for a mid-size sedan vehicle, they are shown in Table 1.

Table 1: Fuel Cell Stack Input Parameters

Input Parameter	Value
Active Area [cm^2]	400
Number of cells in series	440
Nominal operating temperature [K]	353
Air and hydrogen inlet temperature [K]	333
Air and hydrogen inlet relative humidity [%]	100
Anode pressure [atm]	2
Fuel utilization	0.8
Ambient temperature [K]	273
Stack Current Density for Maximum Power at Nominal Operating Temperature [A/cm^2]	0.98
System Current Density for Maximum Power at Nominal Operating Temperature [A/cm^2]	0.84

The maximum net power available at nominal operating temperature is 86.4 kW. It is associated to a 42 kW maximum power nickel metal hydride battery with typical specific power of 500 W/kg and specific energy of 50 Wh/kg. This configuration is the result of a sensitivity analysis presented in [5] and suitable for the vehicle specifications presented in the simulation section below. The fuel cell system utilizes a variable speed compressor/expander for the air delivery system. Except when stated differently, the accessories needed to operate the fuel cell are assumed to be infinitely variable and ideally controlled to maximize the efficiency of the fuel cell system.

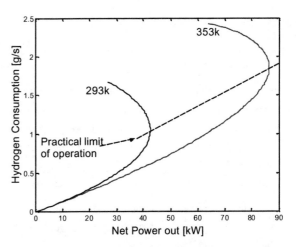

Figure 4: Fuel Cell System Hydrogen Consumption as a Function of Net Power Output for 2 Operating Temperatures

The corresponding characteristic of most interest to our problem is the hydrogen consumption of the fuel cell system versus the net electrical power output. The net fuel cell power output is the power available from the system when accounting for the power needed to power the (required) auxiliaries. This relationship is strongly temperature-dependent, as shown in Figure 4 below for 2 temperatures (cold start at 293 K and nominal operating temperature at 353K) for the fuel cell system considered.

Accordingly to the ECMS approach, a virtual fuel consumption can also be associated with the battery use. First of all, a hydrogen specific consumption is calculated for the battery according to the following equation. Basically, this virtual specific consumption $\overline{SC_{batt}}$ (gr/kW.h) represents the average amount of hydrogen needed to store 1 kW.h of electrochemical energy in the battery using the fuel fell as a charger.

$$\overline{SC_{batt}} = \frac{\overline{SC_{fc}}}{\overline{Eff_{pc}} \cdot \overline{Eff_{ch\_batt}}} \qquad \text{Eq. (5)}$$

where $\overline{SC_{fc}}$ is the average specific consumption of the fuel cell from fuel to electrical energy, $\overline{Eff_{pc}}$ is the average efficiency of the power converter and $\overline{Eff_{ch\_batt}}$ is the average efficiency of the battery during charging.

In order to associate a net electrical power at the battery terminals to an equivalent electrochemical battery power, appropriate efficiencies must be applied. Accordingly, the equivalent hydrogen mass flow $\dot{m}_{f\_batt}$ is calculated as follow:

For positive power flow (battery currently discharging):

$$\dot{m}_{f\_batt} = \frac{\overline{SC_{batt}} \cdot P_{batt}}{3600 \cdot Eff_{dis\_batt}} \qquad \text{Eq. (6)}$$

For negative power flow (battery currently charging):

$$\dot{m}_{f\_batt} = \frac{\overline{SC_{batt}} \cdot P_{batt} \cdot Eff_{ch\_batt}}{3600} \qquad \text{Eq. (7)}$$

where P_{batt} is the instantaneous net power on the battery terminals, Eff_{dis_batt} is the instantaneous discharge efficiency and Eff_{ch_batt} is the instantaneous charge efficiency. This heuristic formulation allows to estimate a total equivalent mass flow $\dot{m}_{f\_equi} = \dot{m}_{f\_fc} + \dot{m}_{f\_batt}$ or fuel cost, for any instantaneous power flow distribution.

For a given electric motor power requirement ($P_{em} = P_{fc} + P_{batt}$), the power distribution with the lowest fuel cost is then selected.

In order to meet the slowly changing charge sustaining constraint, the equivalent mass flow $\dot{m}_{f\_batt}$ is instantaneously biased up or down using a multiplicative penalty factor $pen_{mfbatt}(SOC)$ in order to favor or reduce the use of the electrical accumulator to tend toward the state of charge target. This formulation affects the power distribution between the fuel cell and the battery, but has no effect on the instantaneous dynamic performances of the vehicle, as the driver demand is met.

An on-line algorithm allows to define the best operating point according to the above description at each time. In order to illustrate the algorithm operation, we will assume a 50 kW driver request. Figure 5 represents the optimization algorithm for the fuel cell at nominal temperature (353 K) and the battery state of charge at its target value of 70%. The fuel cell power P_{fc} appears on the X-axis. First of all, provided the component instantaneous power limitations, the valid range of possible fuel cell power is determined so that the driver request P_{em} can be met. The valid range is shown by the two vertical solid lines in the figure. For each fuel cell power candidate, the corresponding battery power P_{batt} needed to meet the driver request is calculated, as shown on Figure 5a. Knowing the fuel cell current operating temperature, the hydrogen used by the fuel cell $\dot{m}_{f\_fc}$ is determined using a pre-computed set of maps and is shown in Figure 5b (analogous to Figure 4). Alternatively the fuel cell system model could be used on line to determine the hydrogen consumption, but the use of pre-computed map is significantly cheaper for real-time computation. As previously described, Figure 5c shows the equivalent battery mass flow $\dot{m}_{f\_batt}$ computed from equations 6 or 7. Note that the plot of $\dot{m}_{f\_batt}$ becomes a straight line because P_{batt} decreases linearly as P_{fc} increases, and because of the constant $\overline{SC_{fc}}$, Eff_{dis_batt} and Eff_{ch_batt} values.

Also, one should note that P_{batt} (and consequently $\dot{m}_{f\_batt}$) become equal to zero when P_{fc} is equal to P_{em}. This is simply because the power requirement is satisfied by the fuel cell alone, and the power requirement at the battery is equal to zero in such case. This is also implying that the operation mode of the battery is toggled between charge mode and discharge mode at $P_{fc} = P_{em}$. Finally, as can be seen in the bottom plot, the optimal fuel cell power is determined by taking the global minimum of $\dot{m}_{f\_equi} = \dot{m}_{f\_fc} + \dot{m}_{f\_batt}$, i.e. 40.5 kW. The battery must provide the remaining fraction of power required.

Power [kW] Needed From the battery P_{batt}

Hydrogen mass flow [gr/s] consumed by the fuel cell $\dot{m}_{f\_fc}$

Equivalent Hydrogen mass flow [gr/s] associated to the battery $\dot{m}_{f\_batt}$

Total equivalent Hydrogen mas flow [gr/s] $\dot{m}_{f\_equi}$

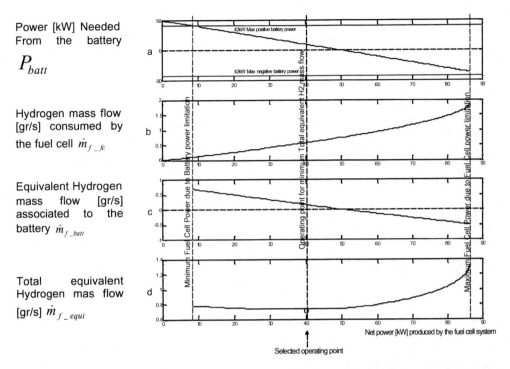

Figure 5: Fuel Cell Operating point Selection at 353 K and 70% SOC for a 50 kW Request

Power [kW] Needed From the battery P_{batt}

Hydrogen mass flow [gr/s] consumed by the fuel cell $\dot{m}_{f\_fc}$

Equivalent Hydrogen mass flow [gr/s] associated to the battery $\dot{m}_{f\_batt}$

Total equivalent Hydrogen mas flow [gr/s] $\dot{m}_{f\_equi}$

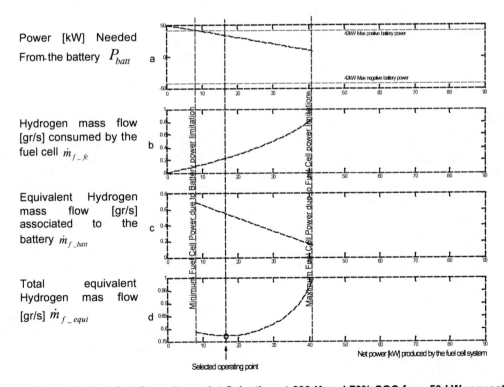

Figure 6 : Fuel Cell Operating point Selection at 293 K and 70% SOC for a 50 kW request

Figure 6 depicts exactly the same scenario as above, but with a fuel cell at ambient temperature (293 K) corresponding to a cold start case. As expected, the valid range of fuel cell power is significantly narrowed because of fuel cell power limitation at low temperatures. The decrease of the fuel cell efficiency at low temperature causes the fuel cell hydrogen consumption curve (Figure 6b) to be steeper than in the previous case. With our assumptions, the shape of the battery equivalent hydrogen consumption is not affected by the temperature (Figure 6c). The resulting total equivalent hydrogen mass flow versus fuel cell power leads to an optimal fuel cell power of 16.5 kW.

Because of the poor efficiency of the fuel cell at low temperature, the algorithm naturally favors the use of the battery during cold start situation, leading to better mileage. This example shows also the necessity of a battery at low temperatures as the fuel cell maximum power may be lower than the driver request (as in the case shown here in Figure 6).

Both examples shown so far were with the battery state-of-charge at the target value, hence demonstrating the ability of the control strategy to choose the operating point of the fuel cell to minimize the instantaneous equivalent fuel consumption unconstrained by the battery state-of-charge. As mentioned earlier, to practically enforce the global state-of-charge constraint, a SOC correction factor is used in order to shift the optimal power split up or down according to the deviation between the actual and target state-of-charge. A non-linear "penalty function" is defined to heuristically embody the desired characteristics. The deviation between the current battery SOC and the target battery SOC is fed into this nonlinear penalty function.

When the current SOC matches the SOC target (0.7 in this case), the penalty factor must take the neutral value 1 enforcing the optimal distribution is kept. Moreover it should be relatively flat around this SOC target allowing the optimal distribution to be (nearly) maintained when the battery state-of-charge is close to the target value. This is required to achieve maximum benefits from the hybridization by letting the powertrain (and hence the state-of-charge of the battery) best utilize the energy sources on board as long as they are (nominally) available. On the other hand, this function becomes significantly larger (or smaller) when the state-of-charge approaches the preset low (or high) limits. This is required to avoid under- or over-charging of the battery regardless of the vehicle and driver demand. The shape and width of this penalty function must be adjusted to reflect the battery charge and discharge characteristics and the desirable range of use of the battery pack selected.

In this paper, as an appropriate example, the following simple cubic-based function has been chosen for this (multiplicative) penalty function:

$$f_{pen} = 1 - (1 - 0.8 \times x_{SOC}) \times x_{SOC}^3 \qquad \text{Eq. (8)}$$

$$x_{SOC} = \frac{SOC - \dfrac{(SOC_L + SOC_H)}{2}}{(SOC_H - SOC_L)}$$

This function satisfies the above constraints and provides the desirable SOC control sensitivity.

The shape of the multiplicative penalty factor f_{pen} is given figure 7 for $SOC_L = 0.6$ and $SOC_H = 0.8$.

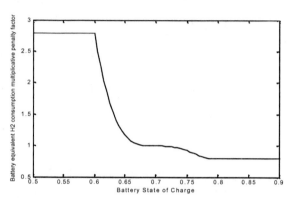

Figure 7 : Penalty Factor *versus* Battery State-of-Charge

The penalty factor is used to modify the basic Equation 4 previously given as follows:

$$\dot{m}_{f\_equi} = \dot{m}_{f\_fc} + f_{pen} \times \dot{m}_{f\_batt} \qquad \text{Eq. (9)}$$

The effect of this penalty factor is to increase or decrease the "cost" of using the battery depending on the error between the targeted SOC and the current SOC. This then results in shifting up or down the optimum fuel cell power in the context of the minimization of the instantaneous equivalent fuel consumption. To illustrate the effect of this penalty factor, reconsider the example previously depicted Figure 5, but assuming a lower battery state of charge at 60% (instead of the 70% target value used in the previous example). Figure 8 depicts this new example. The low state-of-charge causes the battery equivalent hydrogen consumption to be penalized with a 2.8 multiplicative factor as illustrated in Figure 8c). This significantly affects the shape of the total equivalent hydrogen mass flow (Figure 8d), and the selected fuel cell power is then 81.5 kW, causing the excess power from the fuel cell to recharge the battery, hence leading towards a state-of-charge recovery while yet meeting the driver demand.

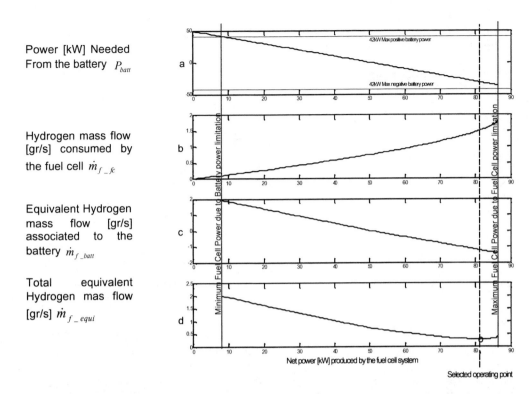

Power [kW] Needed From the battery P_{batt} — a

Hydrogen mass flow [gr/s] consumed by the fuel cell $\dot{m}_{f\_fc}$ — b

Equivalent Hydrogen mass flow [gr/s] associated to the battery $\dot{m}_{f\_batt}$ — c

Total equivalent Hydrogen mas flow [gr/s] $\dot{m}_{f\_equi}$ — d

Figure 8: Fuel Cell Operating Point Selection at 353 K and 60% SOC for a 50 kW Request

The procedure described above (Figures 6, 7 and 8) for discrete situations must be repeated on-line for any variation of the following input variables : battery state-of-charge (SOC), driver power request P_{em} and fuel cell operating temperature. In the following section simulation results using the ECMS approach are presented.

SIMULATION RESULTS

For the simulations a 1450 kg, SUV-type vehicle is represented. Table 2 lists the vehicle specifications used. It is powered by a 54 kW, 162 Nm nominal AC induction electric motor, able to produce twice its nominal power for short periods. It requires up to 126 kW when developing peak power.

Table 2: Vehicle Specifications

Specification	Vehicle	Vehicle with Trailer
Mass [kg]	1452	2360
Frontal area [m²]	2.7	4
Coefficient of drag	0.4	0.75
Coefficient of rolling resistance	0.015	0.015

Table 3 below compares the fuel mileage results provided by the ECMS control strategy approach to those obtained with a more basic control approach. Both the Federal Urban Driving Schedule (FUDS) and Federal Highway Driving Schedule (FHDS) were used for the simulations. For the FUDS simulations, the fuel cell starts at ambient temperature to represent a cold start. For the FHDS simulations, the fuel cell starts at nominal fuel cell operating temperature to represent a fully warmed vehicle.

The initial battery state-of-charge is 0.7 for all the simulations. The fuel usage is corrected for (minor) differences in initial and final battery state-of-charge. The correction is based on the net electricity used (reflected by the difference between battery final state of charge and initial state of charge), using the corresponding average fuel cell system efficiency from that simulation to convert net electricity usage to equivalent fuel consumption. Furthermore, the corrected fuel (hydrogen) usage is then converted to equivalent gasoline usage based on the relative lower heating value of hydrogen and gasoline. This allows to report the results in a more "intuitive" form.

In order to evaluate and contrast the results of the ECMS control strategy, a SOC-based proportional control approach was used. It is reported to lead to very good fuel mileage as explained in [5], as it consistently attempts to operate the fuel cell at a lower current density where the fuel cell system power output is close to the average power demand (which is more efficient than operating the fuel cell with bursts at high current densities). As shown in Figure 9, the SOC-based proportional control is only SOC dependent. While allowing a smooth variation of the fuel cell operating point because of the slow varying nature of the SOC (leading to good fuel efficiency), this control does not allow the fuel cell to temporary deliver the maximum power to satisfy a high driver request. This limits considerably the performance achievable by the vehicle.

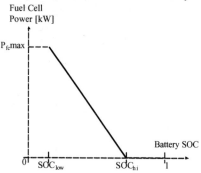

Figure 9: Proportional SOC-Based Control

Three different cases were analyzed with both control strategies. The base case corresponds to the base vehicle and fuel cell system with ideally controlled auxiliaries as specified in the previous section, which is the best-case scenario. In order to test the robustness of the control strategy, two other degraded situations are considered:

- Vehicle with trailer according to the specifications given Table 2, but the same fuel cell system
- Fuel cell system with non-ideally controlled auxiliaries

In the last case, instead of considering infinitely controllable auxiliaries components which ideally track the fuel cell system power demand, we consider the more realistic case of components with a more limited variation range. In those simulations, the minimum power drawn by the compressor is 5 kW corresponding to 20% of the power required by the compressor at maximum fuel cell power (23.03 kW). Similarly, the minimum power of the cooling system is also restricted to 1 kW out of the 3 kW required at maximum load. Moreover the benefit of an expander has been removed in this case. The effect of non-ideal auxiliaries is shown

on the hydrogen consumption vs. net power output curve is shown in the figure 10 below.

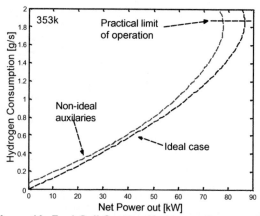

Figure 10: Fuel Cell System Hydrogen Consumption as a Function of Net Power Output with non-ideal auxiliaries

These two cases represent two different types of variations, one a significantly increased load, and the other a fuel cell system with significantly different characteristics. This tests the ability of a control strategy to adapt to either external changes (load) or internal changes (plant).

Let's recall that the SOC-based proportional control is not a realistic approach, since it does not allow to draw the full power potential of the fuel cell at any time if requested. This approach is only given because it provides a good fuel economy and serves as a good basis of comparison.

The results show a constant advantage in fuel economy for the more practically implementable ECMS control strategy. As can be seen on the Table 3 below, the gap between those two approaches becomes larger in the case more realistic of a non-ideal fuel cell system. This is due to the fact that the ECMS formulation implicitly embodies the efficiency characteristics of the underlying plant by accounting for the actual hydrogen consumption of the fuel cell system *vs.* net electrical power output. In the non-ideal auxiliaries case the shape of this relationship is modified and ECMS takes it into account by favoring the most efficient distribution of the operating point of the hybrid system.

Table 3: Simulated Fuel Economy Results

	Base Case: Vehicle only, Ideal Fuel Cell System		Vehicle with Trailer, Ideal Fuel Cell System		Vehicle only, Non-Ideal Fuel Cell Auxiliaries	
	FUDS	FHDS	FUDS	FHDS	FUDS	FHDS
SOC-based Proportional control	62.1 mpg	64.5 mpg	30.38 mpg	23.77 mpg	37.18 mpg	50.2mpg
ECMS Control	63.95 mpg	65.3 mpg	30.79 mpg	24.13 mpg	42.15 mpg	53.74 mpg

CONCLUSIONS

This paper describes a practical formulation for the supervisory control problem in charge-sustaining hybrid fuel cell vehicles. This formulation is based on adjusting the instantaneous power split between the fuel cell and the electrical accumulator to minimize the overall hydrogen consumption using a simple analytical formulation well suited for real-time control. The control strategy is *automatically* adapted to the selection of components. For one specific representative hybrid fuel cell vehicle configuration, the ECMS control strategy demonstrated its robustness under a wide range of driving conditions while enforcing a charge-sustaining operation. Furthermore, the strategy is inherently self-adaptive to the efficiency characteristics of the powertrain components.

The ECMS approach is not limited to the battery/fuel cell configuration proposed in this paper, it can easily be applied to other configurations such as ultracapacitor/fuel cell case where the charge sustaining operation is even more desirable due to the smallest energy capacity of the electrical accumulator.

One must note that the performance of the ECMS is sensitive to following several crucial parameters: most obviously, the average specific consumption of the fuel cell from fuel to electrical energy is one; however, parameters such as the "shape" of the nonlinear penalty function, the target state of charge high and low limits also affect the performance of the ECMS. Further investigation is needed to optimize those parameters for minimizing fuel consumption depending of the current vehicle and environment configurations. Finally, the low transient capability of a fuel cell system compared to an electrical accumulator could also be accounted for while maintaining the fuel efficiency potential of the control strategy described here.

ACKNOWLEDGMENTS

The work presented in this paper was supported by the Ohio State University Center for Automotive Research and Intelligent Transportation Industrial Consortium.

CONTACTS

Dr. Gino Paganelli
Swiss Federal Institute of Technology Zurich (ETH)
ETH Zentrum ML, IMRT
CH-8092 Zurich, Switzerland
Phone (+41) 1-632 5131
Fax   (+41) 1-632 1139
E-mail: paganelli@imrt.mavt.ethz.ch

Pr. Yann Guezennec
Center for automotive Research and Intelligent Transportation
The Ohio State University
930 Kinnear Road
Columbus, OH 43212, USA
Phone 614 292-1910
Fax   614 688-4111
E-mail: guezennec.1@osu.edu

REFERENCES

[1] Paganelli, G., Guerra, T.M., Delprat, S., Santin, J.J., Delhom, M., Combes, E., "Simulation and Assessment of Power Control Strategies for a Parallel Hybrid Car", J. Automobile Engineering, also in *Proc.* of the Institution of Mechanical Engineers IMechE, SAE International, 214, pp. 705-718, 2000.

[2] G. Paganelli, S. Delprat, T.M. Guerra, J. Rimaux, J.J. Santin, "Equivalent Consumption Minimization Strategy For Parallel Hybrid Powertrains", Fall VTC-01 Conference sponsored by VTS (Vehicular Technology Society) and IEEE, Atlantic City, NJ, 2001.

[3] Paganelli, G., Ercole, G., Brahma, A., Guezennec, Y., Rizzoni, G., "General Supervisory Control Policy for the Energy Optimization of Charge-Sustaining Hybrid Electric Vehicles", JSAE Review, Vol. 22, No. 4, pp. 511-518, Oct. 2001.

[4] Boettner, D., Paganelli, G., Guezennec, Y., Rizzoni, G., and Moran, M., 2001, "PEM Fuel Cell System Model for Automotive Vehicle Simulation and Control," accepted in *ASME J. of Energy Resources Technology*, 2001.

[5] D. Boettner, G. Paganelli, Y. Guezennec. M. J. Moran, G. Rizzoni, "Component Power Sizing and Limits of Operation for Proton Exchange Membrane (PEM) Fuel Cell/Battery Hybrid Automotive Applications", Proc. 2001 IMECE Meeting, Symposium on Advanced Automotive Technologies, New York, November 2001.

[6] D. Patton, J. Latore, M. Ogburn, S. Gurski, P. Bryan, D. J. Nelson, "Design and Development of the 2000 Virginia Tech Fuel Cell Hybrid Electric FutureTruck", Society of Automotive Engineers SAE SP-1617, pp.185-202, 2001.

[7] M. Ogburn, D.J. Nelson, W. Luttrell, B. King, S. Postle, R. Fahrenkrog, "Systems Integration and Performance Issues in a Fuel Cell Hybrid Electric Vehicle", Society of Automotive Engineers, Paper 2000-01-0376, 2000.

2004-01-1298

Adaptive Energy Management Strategy for Fuel Cell Hybrid Vehicles

Bruno Jeanneret
French National Institute for Transport and Safety Research

Tony Markel
National Renewable Energy Laboratory

ABSTRACT

Fuel cell hybrid vehicles (FCHVs) use an energy management strategy to partition the power supplied by the fuel cell and energy storage system (ESS). This paper presents an adaptive energy management strategy, created in the ADVISOR^{TM} software, for a series FCHV. The strategy uses a local or "real-time" optimization approach, which aims to reduce total energy consumption at each instantaneous time interval by dynamically adjusting the amount of power supplied by the fuel cell and ESS. Compared with a static control strategy, the adaptive strategy improved the simulated FCHV's fuel economy by 1.4%-8.5%, depending on the drive cycle.

INTRODUCTION

FCHVs combine a fuel cell power system with an ESS. The ESS (e.g., batteries or ultracapacitors) reduces the fuel cell's peak power and transient response requirements. It allows the fuel cell to operate more efficiently and allows recovery of vehicle energy during deceleration.

FCHVs use an energy management strategy to partition the power supplied by the fuel cell and ESS [1]. This paper presents an adaptive energy management strategy for a series FCHV, developed at the National Renewable Energy Laboratory (NREL) using NREL's ADVISOR vehicle modeling software. A similar strategy was developed by NREL for an internal combustion engine parallel hybrid vehicle [2]. The adaptive strategy uses a local or "real-time" optimization approach to reduce total energy consumption at each instantaneous time interval by dynamically determining the ideal power split between the fuel cell and ESS. The strategy does not require knowledge of future driving conditions.

---

* ADVISOR is a trademark of the Midwest Research Institute

FCHV STATIC CONTROL STRATEGY

By default, the ADVISOR software employs a static FCHV control strategy. The default control strategy (CSd) uses several parameters to manage the ESS state of charge (SOC) and power split between the ESS and fuel cell during the drive cycle. The following are the most influential parameters:

cs_min_pwr - The minimum power output set point of the fuel cell system. The requested output power of the fuel cell will be at or above this set point.
cs_max_pwr - The maximum power output set point of the fuel cell system. The fuel cell operates below this level if the ESS SOC becomes too high.
cs_charge_pwr - A factor used in conjunction with the current ESS SOC and the desired range of SOC to modify the fuel cell power output request. It works to bring the ESS SOC back to the center of the desired operating range.

The main objectives are to satisfy the power required by the driver, maintaining the ESS SOC between a low and a high value and allowing maximum ESS power regeneration from vehicle deceleration. When the fuel cell is on, its power command is calculated as the traction power required to propel the vehicle multiplied by a factor depending upon SOC.

Figure 1 provides a graphical representation of the strategy. The minimum and maximum power request set points are shown with dashed lines. If the ESS SOC is greater than the desired SOC level, the fuel cell power request is reduced. As a result, battery energy will be consumed to supplement the traction power needs. Likewise, the fuel cell power request is increased to recharge the battery if the SOC is less than the desired level. The adjustment factor increases proportionally with respect to how far the SOC is from the desired level.

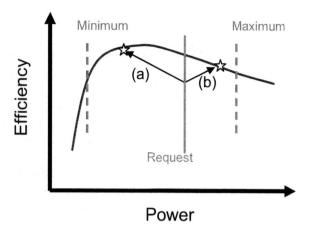

Figure 1: Static vehicle energy management strategy: (a) SOC is high - reduce fuel cell power request, (b) SOC is low - increase power request

These parameters are fixed for a specific vehicle but can be optimized for known driving profiles [3]. As a result, the vehicle may be sub-optimal for other driving profiles. The static strategy does not consider efficiency directly, but, if optimized, it will cluster the operating points in the most efficient operating region. It also does not consider the potential impacts of regenerative braking energy.

FCHV ADAPTIVE CONTROL STRATEGY

An adaptive control strategy optimizes, at each time interval, the distribution between power provided by the fuel cell and the ESS, making two decisions:

1. Turn the fuel cell system on or off
2. If the fuel cell is on, optimally partition power between the fuel cell and ESS to satisfy driving demands

The simulated series FCHV used in this study had a fuel cell and battery, both connected to the power bus of the vehicle. The control strategy developed for this vehicle only sought to optimize fuel economy; however, emissions from reformer exhaust could be integrated easily if appropriate.

ADAPTIVE CONTROL ALGORITHM – The algorithm for the new strategy, Control Strategy 1 (CS1), was implemented in the ADVISOR software. The main idea of this control strategy is to compute at each time interval a global cost function describing the total energy consumption with respect to fuel cell instantaneous power. This function accounts for both fuel cell consumption and an equivalent ESS consumption. In order to be consistent, this equivalent ESS consumption must include the whole efficiency from fuel cell power to ESS pathway (e.g., fuel cell to bus power) and finally to ESS. Once this objective function is obtained, the algorithm finds the global minimum and orders the fuel cell to deliver this optimized power. At each simulation time step, the algorithm follows a four-step process.

Step 1 - Define battery available operating range that satisfies power demand – Ten fuel cell power/battery power couples are chosen to evaluate the objective function. The function depends on the following constraints:

- Driver request
- Minimum and maximum available power from fuel cell system (including maximum transient power)
- Minimum and maximum available power from ESS

Step 2 - Compute energy consumption for each fuel cell available power setting – The fuel cell model is called multiple times to compute the energy consumption for each potential fuel cell net power operating point. The result of this step is a curve that gives fuel cell energy consumption (Jfuel) for each feasible operating point (Figure 2).

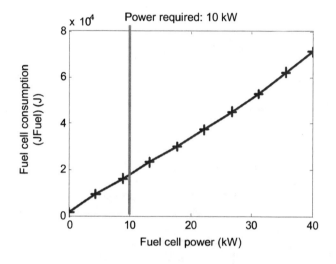

Figure 2: Fuel consumption vs. fuel cell net power

Step 3 - Compute SOC variation and equivalent energy consumption for each ESS available power – The fuel cell is the primary energy source for the vehicle, and the quantity of battery energy is limited. The battery SOC must be maintained either through the recovery of regenerative braking energy or by the fuel cell. A reference curve is created correlating SOC variation and fuel cell consumption at the power requested by the driver assuming no contribution from regenerative braking. Figure 3 illustrates this process for a power request of 10 kW.

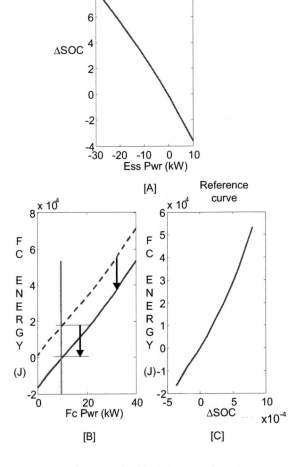

[A]

Reference curve

[B]

[C]

Figure 3: Creating a reference curve

The battery model is run multiple times at each time step to compute variation in SOC for each ESS power requirement (Figure 3A). The fuel cell energy consumption is calculated for each associated fuel cell power (Figure 3B, dashed line). Fuel cell consumption at zero energy storage power is subtracted (Figure 3B, solid line). Finally, the reference curve (Figure 3C) is obtained by relating fuel cell power and ESS power according to the following equation:

ESS Power + Fuel Cell Power = Required Power   Eq. 1

Delta SOC values are corrected by accounting for the free regenerative energy recovered between decelerations. This calculation is made by integrating and averaging the variation in SOC under the following conditions:

- Traction power request is less than zero
- Fuel cell output power is zero

At this time, there are couples of values describing the fuel cell energy consumption and the corresponding SOC variation of the battery. Overall in hybrid vehicle

operations, the battery energy is limited compared with the fuel tank. If the batteries are used during a time step, the SOC decreases (see ΔSOC_{Actual} in Figure 4). In order to have neutral charge behavior, the SOC would eventually need to increase by the same amount. The strategy finds this "replacement" cost by simply flipping the sign of ΔSOC_{Actual}. The cost of the replacement energy (JFuelEq) is obtained via interpolation of the reference curve at $(-1)*\Delta SOC$ values (Figure 4). This provides an estimate of the replacement energy associated with energy storage use.

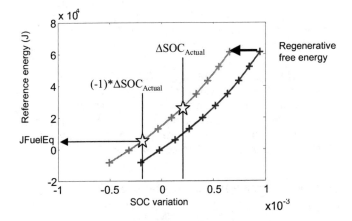

Figure 4: Energy consumption vs. SOC variation

Step 4 - Calculate total energy consumption and find the minimum of the objective function – An example energy function for a 10 kW load is illustrated in Figure 5. JFuel represents the fuel energy consumed by the fuel cell, and JFuelEq is the equivalent fuel consumption associated with the change in SOC of the ESS.

Total Energy = JFuel + JFuelEq   Eq. 2

Figure 5: Energy functions for 10 kW request

Additional Parameters – The strategy as implemented typically will determine that using the ESS is less costly. As a result, it is difficult to balance the SOC over the entire drive cycle. An SOC regulation factor (SOCFactor) has been included to provide additional control over the SOC balancing.

$$\text{Energy} = \text{JFuel} + \text{SOCFactor} \cdot \text{JFuelEq} \qquad \text{Eq. 3}$$

A nominal SOCFactor value of 1 will have no impact. A low factor makes use of the battery less costly, thus the battery is discharged. A high factor favors charging the battery.

Functionality was also included to evaluate the influence of shutting off the fuel cell and its associated auxiliary loads. If this function is activated, the algorithm calculates the equivalent battery energy consumption at the driver requested power without any fuel cell consumption. If this value is lower than the optimum energy value from Step 4, the fuel cell is commanded to be shut down, provide no power, and consume no fuel.

VEHICLE DATA – The set of ADVISOR data files that were used for the simulation of a FCHV represent a standard sedan. The hybrid components included a 20-module lead-acid battery pack and a 75 kW AC induction traction motor. Table 1 provides a summary of the vehicle characteristics that were used in the model.

Table 1: Characteristics of simulated FCHV

Vehicle data file	VEH_SMCAR
Vehicle mass	1240 kg
Aerodynamic drag coefficient	0.335
Frontal area	2.0 m^2
Battery type	Lead-acid, Hawker 26 Ah cell, 20 modules in series
Electric motor	75 kW, 92% peak efficiency
Hydrogen fuel cell	50 kW maximum, 60% peak efficiency

Figure 6 shows the nominal fuel consumption and efficiency maps of the fuel cell system. The efficiency impacts of warm-up from ambient conditions were not considered in this study.

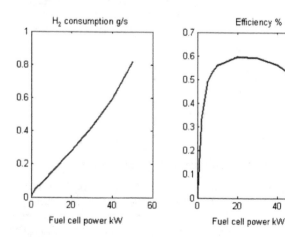

Figure 6: Nominal fuel consumption and efficiency map (fully warm conditions)

RESULTS FOR STEADY STATE REQUIRED POWER – To evaluate the control strategy and the best partitioning of power between the fuel cell and battery, the control strategy model was tested independently. An increasing step profile power request was delivered to the control strategy module to test its response through the range of 0 to 70 kW (Figure 7). The calculations accounted for disconnecting the regenerative energy calculation module and the energy storage SOC.

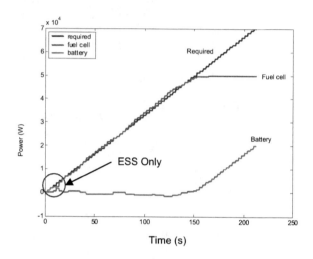

Figure 7: Power split for different power requirements (fuel cell always on) (CS1)

As seen in Figure 7, when the fuel cell is not allowed to shut-off, the strategy does not use fuel cell power when the power request is less than 3 kW. At 3-6 kW, power is split between battery and fuel cell. Above 6 kW, the control strategy reacts in a full load-following manner, and battery power is almost zero. Figure 8 shows the control strategy when fuel cell shut-off is authorized. In this case, the battery provides power until 7 kW of requested power, and then the results are the same as in Figure 7.

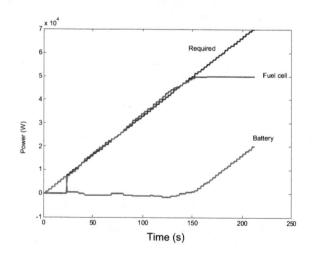

Figure 8: Power split for different power requirements (fuel cell shut-off authorized) (CS1)

SENSITIVITY OF THE CONTROL STRATEGY TO FUEL CELL AND BATTERY EFFICIENCIES

<u>Ideal battery</u> – The control strategy was evaluated with the battery having the following ideal characteristics:

- No internal resistance
- Perfect Coulombic efficiency

Under these conditions, there is no cost for using battery power at each required bus power. As a result, the fuel cell should work at its optimal point at every time step and the battery should work as a load follower to satisfy the power bus. The results were different than what was expected (Figure 9).

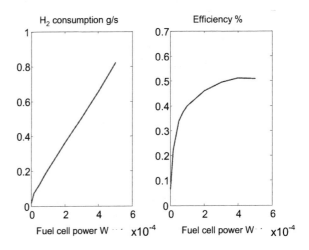

Figure 10: Lower efficiency fuel cell characteristics

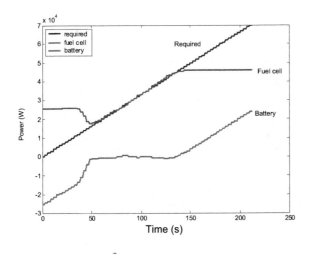

Figure 9: Power split for different power requirements, ideal battery case (fuel cell always on) (CS1)

As Figure 9 shows, when required power is less than 12 kW, the fuel cell works at its optimal operating point and the battery charges. At 12-46 kW the fuel cell is a load follower. Above 46 kW the battery provides complementary power to achieve the requirements. This behavior is explained by the way the equivalent energy storage function is calculated. Even when there are no losses due to charging or discharging the battery (i.e., Figure 3A is linear and the ΔSOC is the battery power divided by the battery open circuit voltage), different fuel cell efficiencies are used to build Figure 3B.

<u>Lower fuel cell efficiency map</u> – The fuel cell consumption map was modified by an arbitrary linear factor depending on fuel cell output power, from 1.5 at 0 kW to 1.0 at maximum output power. Figure 10 shows the resulting consumption and efficiency curves. Figure 11 shows the resulting power split.

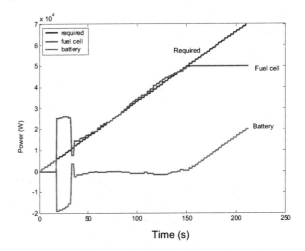

Figure 11: Power split for different power requirements, lower efficiency fuel cell case (fuel cell always on) (CS1)

Because fuel cell efficiency is lower for low output power, the control strategy uses battery power as long as the required power does not exceed 6 kW. Between 6-10 kW loads, the shape of the global function is not monotonic. The global minimum occurs at 25-26 kW fuel cell power (Figure 12). The results are the same as for the nominal fuel cell characteristics once the required power exceeds 10 kW.

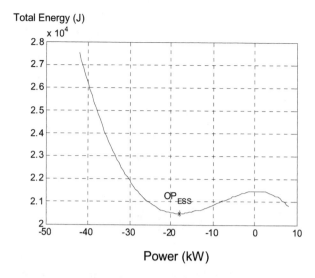

Figure 12: Battery power vs. global energy function for 8 kW required power

MODIFIED ADAPTIVE CONTROL STRATEGY

The initial adaptive strategy (CS1) assumes that similar operating conditions will exist in the future. Different fuel cell efficiencies are accounted for to compute equivalent battery energy, which means energy may be replaced at a non-optimal operating point. This need not be the case. A modified adaptive strategy (CS2) assumes that energy will be replaced at a favorable fuel cell operating point. For example, energy would be replaced when fuel cell efficiency is 60% for a fuel cell with 60% peak efficiency (the plot on Figure 3B would be linear).

RESULTS FOR STEADY STATE REQUIRED POWER – Figure 13 shows the results over a 0 to 70 kW power requirement range using CS2. At less than 25 kW required power, there is little difference between this strategy and CS1 (Figure 7). Above 25 kW the power requirements are shared by the fuel cell and battery, which is different than the results obtained for CS1. The fuel cell system efficiency begins to falloff as the power requirement exceeds 25 kW (Figure 6). Supplementing the fuel cell system with power from the ESS provides a more favorable energy cost solution.

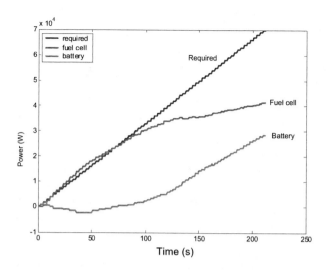

Figure 13: Power split for different power requirements (fuel cell always on) (CS2)

SENSITIVITY OF THE CONTROL STRATEGY TO FUEL CELL AND BATTERY EFFICIENCIES

Ideal battery – Using ideal battery assumptions, the equivalent energy storage function is not affected by different fuel cell efficiencies. The result was as expected (Figure 14). The fuel cell operates at its optimal efficiency point, and the battery provides complementary power (charge or discharge).

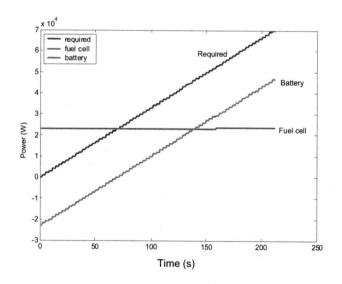

Figure 14: Power split for different power requirements, ideal battery case (fuel cell always on) (CS2)

Lower fuel cell efficiency map – When fuel cell consumption is increased for low output power (Figure 10), the strategy uses battery power to satisfy the demand below 13 kW (Figure 15). The fuel cell satisfies the demand for 14-40 kW required power. Above 40 kW, the strategy with a lower fuel cell efficiency map uses more fuel cell power compared with the strategy with nominal fuel cell characteristics. This result was unexpected. The fuel cell was expected to reach its maximum output power (also the maximum efficiency in this case) at much lower required power.

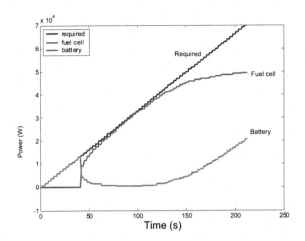

Figure 15: Power split for different power requirements, lower efficiency fuel cell case (fuel cell always on) (CS2)

FUEL CONSUMPTION OVER VARIOUS DRIVING CYCLES USING CS2

EFFECTS OF SOC REGULATION FACTOR ON PARTITIONING OF POWER – The ADVISOR software has two main ways of obtaining SOC balance, both involve modifying the initial SOC value:

- Linear regression between two simulations in charge and discharge relative area
- Iterative modification of the initial SOC to satisfy criteria:
  - Express in relative ΔSOC between beginning and end of cycle. This criterion is insensitive to relative battery energy size and simulation duration.
  - Express in the ratio of equivalent battery energy vs. primary source energy—multiply ΔSOC (Ah) by mean open circuit voltage.

The latter criterion is much more precise and is used in the simulations for this study (convergence criterion is 1% for the ratio of ESS/fuel cell energy).

An SOC correlation factor was introduced to evaluate fuel consumption with charge-sustaining calculations (i.e., SOC of the battery is the same at the beginning and end of the cycle). Figure 16 shows the results of the partition of power for SOC correlation factors of 0.8 and 1.2.

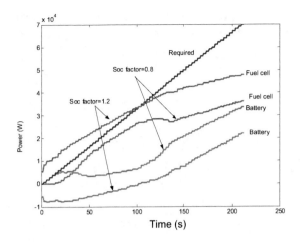

Figure 16: Power split for different power requirements, effect of SOC regulation factor (fuel cell always on) (CS2)

CONSUMPTION RESULTS AT ZERO RELATIVE DEPTH OF DISCHARGE – Table 2 shows fuel consumption results for simulations in which fuel cell shut-off is authorized and not authorized. The default ADVISOR control strategy (CSd) authorizes fuel cell shut-off.

Table 2: Hydrogen consumption for different control strategies over various driving cycles

Drive Cycle	H₂ Consumption L/100 km (g/100 km)			Fuel Economy Gain Using CS2.2
	CS2.1 Shut-off not authorized	CS2.2 Shut-off authorized	CSd Shut-off authorized	100*(CSd-CS2.2)/CSd
IM240	51.0 (918)	48.4 (871)	51.9 (934)	6.7%
10-15 Modes	57.0 (1026)	49.9 (898)	54/55* (972/990)	8.5%
UDDS	53.3 (959)	49.2 (886)	51.9 (934)	5.2%
HWFET	42.8 (770)	42.0 (756)	42.6 (767)	1.4%

*Convergence criteria not satisfied after 15 iterations

Hydrogen consumption can be converted to equate with gasoline consumption (Table 3) using the following relations:

- Hydrogen density = 18 g/L
- Hydrogen low heating value = $12e^3$ J/g
- Gasoline density = 0.749 g/L
- Gasoline low heating value = $42.6e^3$ J/g

511

Table 3: Hydrogen consumption for different control strategies over various driving cycles

Drive Cycle	Gasoline Consumption (L/100 km)			Fuel Economy Gain Using CS2.2
	CS2.1 Shut-off not authorized	CS2.2 Shut-off authorized	CSd Shut-off authorized	100*(CSd-CS2.2)/CSd
IM240	3.45	3.28	3.51	6.7%
10-15 Modes	3.86	3.38	3.66/3.72	8.5%
UDDS	3.61	3.33	3.51	5.2%
HWFET	2.90	2.84	2.88	1.4%

CONCLUSION

The purpose of this study was to create an adaptive control strategy for a FCHV and evaluate its effect on fuel consumption. An initial adaptive control strategy (CS1) was developed that used a range of fuel cell efficiencies for replacement of energy storage energy. This strategy was replaced with a modified strategy (CS2), which used constant fuel cell efficiency.

For very low/very high required power, the feasible battery power leads to charge/discharge of the battery, and values must be extrapolated to compute the replacement energy at $(-1)^*\Delta SOC$. This can be addressed by evaluating the cost of charging or discharging the battery using the battery characteristics, i.e., if battery energy is used, what is the cost of recovering it? To calculate the cost of charging the battery, the following must be summed in the battery model (calculating the cost of discharging the battery is done in the same way):

- Internal loss when discharging
- Net power delivered
- Internal loss when charging
- Coulombic efficiency when charging

This assumes that the battery will be recharged at the same current value. Fuel cell efficiency must also be considered in recharging the battery. This is done in CS2 by assuming that charging will occur (excluding free regenerative energy) at a good fuel cell operating point (i.e., at high efficiency).

The best FCHV fuel economy was achieved using CS2 with fuel cell shut-off authorized. Compared with the static default control strategy, the modified adaptive strategy improved fuel economy by 1.4%-8.5%, depending on the drive cycle.

ACKNOWLEDGMENTS

Bruno Jeanneret would like to thank NREL's staff for welcoming him into the Center for Transportation Technologies and Systems. He would also thank Francois Badin and Keith Wipke for their support to complete these analyses. This work was funded in part by the FreedomCAR and Vehicle Technologies Program of the Office of Energy Efficiency and Renewable Energy in the U.S. Department of Energy and in part by the French National Institute For Transport And Safety Research.

REFERENCES

1. Trigui, R., Badin, F., and Jeanneret, B., "Hybrid light duty vehicles evaluation program," EVS 19 Symposium, October 19-24, 2002.
2. Johnson, V.H., Wipke, K.B., and Rausen, D.J., "HEV Control Strategy for Real-Time Optimization of Fuel Economy and Emissions," FutureCar Congress, April 2000, SAE paper 2000-01-1543.
3. Markel, T. and Wipke, K., "Optimization techniques for hybrid electric vehicle analysis using Advisor," IMECE Proceeding, November 2001.

CONTACT

Bruno Jeanneret is part of the Transport and Environment Laboratory at the French National Institute for Transport and Safety Research. He is involved in hybrid vehicle modeling and evaluation. His Institute provided him the opportunity to spend two months at NREL to work on hybrid vehicle modeling topics. This paper is part of the collaborative work completed during his assignment at NREL.

Tony Markel is part of the Center for Transportation Technologies and Systems at the National Renewable Energy Laboratory. He applies computer modeling and simulation to the evaluation of advanced automotive systems. He has been instrumental in developing the ADVISOR software tool. His technology focus areas include advanced numerical and architectural methods for vehicle systems analysis and fuel cell systems research and development. Tony developed and maintains a distributed computing network that adds to the analysis capabilities of the center. He is currently researching several innovative technologies for fuel cell vehicles and enjoys applying optimization tools to real world analysis problems. He has a B.S.E. in mechanical engineering, with an emphasis in fluid and thermal sciences and is currently attending the University of Colorado to pursue a M.S. in mechanical engineering.

For more information about this paper, contact Tony at tony_markel@nrel.gov. On the Web, see www.ctts.nrel.gov/analysis.

An Application of Cost Based Power Management Control Strategies to Hybrid Fuel Cell Vehicles

Lawrence Buie, Malcolm Fry, Peter Fussey and Chad Mitts

Ricardo plc

ABSTRACT

For fuel cell vehicles to become a commercial reality, a number of challenges must be met including fuel infrastructure, durability, cost, performance, and efficiency. To help address these challenges, hybrid systems that combine fuel cells with an electrical energy storage system such as batteries or ultracapacitors is a potentially important configuration to increase efficiency and extend fuel cell life times by mitigating transient loads. However, the successful implementation of a hybrid fuel cell system requires achieving performance and efficiency benefits that offsets the additional costs, weight, and complexity of the energy storage system. The key to achieving the levels of efficiency required to justify the hybrid system lies in the power management control strategy. To this end, this paper examines the extension of a novel cost based control strategy developed for conventional internal combustion engine hybrid vehicles to a fuel cell platform (Patent WO 02/42110).

INTRODUCTION

The benefits of developing advanced hybrid electric vehicles (HEV) include fuel economy gains and emission improvements. Numerous HEV hardware configurations are being explored in modern vehicles. The parallel hybrid is a common approach where both an internal combustion engine and electric motors can provide propulsion power. The power for the electric motors typically comes from large battery packs onboard the vehicle. The batteries can be recharged via regenerative braking or from the engine through the electrical machine when vehicle tractive power demand is low. Another approach, the series HEV, uses an electric motor as the main propulsion source. The electrical motor draws power from energy storage devices such as batteries or ultracapacitors and from an internal combustion engine coupled with a generator.

One way to further expand the benefits of hybrid electric vehicle technology is to replace IC engines in the series configuration with fuel cell systems. Through careful system design and integration, direct hydrogen fuel cells are capable of efficiently providing power for vehicle propulsion with little or no toxic tailpipe emissions.

Given the complexity and high degree of design flexibility of fuel cell hybrid electric vehicles, advanced supervisory control strategies are required to successfully manage the power distribution and operation of all the various subsystems and components [Refs 1-5]. Commonly, power management control strategies are rule based algorithms that utilize look-up tables to set power demand based on normalized pedal input and battery SOC. While these approaches are reliable, incorporating other vehicle inputs such as emissions, fuel cell operating temperature, membrane water content, and NVH results in complex logic that can be difficult to optimize. Therefore, this paper examines cost based functional control strategies as an alternate approach to rule based systems by implementing the strategy into a model of a direct hydrogen series fuel cell SUV.

VEHICLE CONFIGURATION

A mid-sized SUV was selected as the platform for the cost based controller analysis due to the large public demand for SUVs combined with the potential for significant environmental benefits that can be realized by improving their typically low fuel economy.

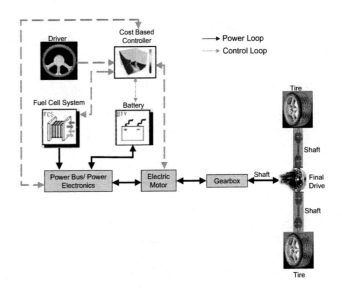

Figure 1. Fuel Cell Series HEV Configuration

The hybrid configuration shown in Figure 1 is a series design with both a battery pack and a fuel cell system that can supply power to the electric motor. The battery pack is recharged by regenerative braking and also by the fuel cell system. The control strategy determines the power distribution between the fuel cell and the battery.

The SUV data used in the simulation are listed in Table 1. This data is estimated from typical 2003 model year 4x2 SUV configurations. The curb weight used in the analysis is adjusted to account for the estimated weight of current fuel cell and electric propulsion technologies in place of the IC engine.

Table 1. Primary SUV Data

Vehicle Mass	2336 kg
CD * A	0.99 m^2
Wheelbase	2.69 m
Final Drive Ratio	3.8
Tire Rolling Radius	0.387 m

COST BASED CONTROL STRATEGY

A key advantage of the cost based control strategy is the ability to continuously control power demands in an adaptive manner. This is accomplished through an overall objective function. The objective function, which may take any functional form, is calculated from vehicle parameters such as:

- Fuel Consumption = FC
- Vehicle Emissions = VE
- Noise Vibration and Harshness Metrics = NVH
- Battery State of Charge = SOC
- Vehicle Performance (Acceleration) = ACC

The objective function can be a simple linear relation where each parameter is adjusted by normalized weighting factors (W) as shown in Equation (1).

$$OBJ(t) = FC(t)*W_{FC} + VE(t)*W_{VE} + NVH(t)*W_{NVH} \quad (1)$$

In the case of the fuel cell objective function used in the SUV analysis, the objective function was a combination of higher order polynomials based on:

- Fuel Cell System Efficiency
- Battery State of Charge

Controlling the power distribution with an objective function involves three primary factors:

- Creation of the objective function
- Establishing the target cost limit

- Implementation of a search routine to change the fuel cell power to achieve the target cost limit thus setting the transient response of the fuel cell system

Figure 2 illustrates the calculated objective function as a function of battery state of charge and fuel cell system power for the analyzed fuel cell SUV.

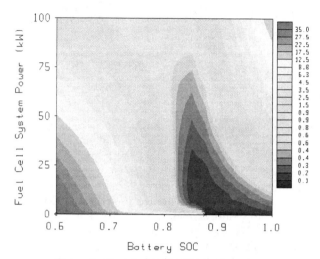

Figure 2. Fuel Cell HEV Objective Function

The objective function is defined in such a way that it can be considered the cost of operating the fuel cell at a given point. For example, the objective function in Figure 2 is primarily related to the fuel cell system efficiency when the battery state of charge is in its target band of 0.825 to 0.875. Therefore, the "cost" is high at low power levels where the fuel cell system efficiency is low and it reaches a minimum where the fuel cell system efficiency peaks. However, as the battery state of charge moves outside of the target band, the objective function becomes dominated by the SOC functions such that when the SOC is low, cost decreases as fuel cell power increases and when the SOC is high, cost increases as fuel cell power increases.

Once the objective function is set, the next key to optimizing the vehicle performance is the selection of the target cost limit. The cost limit can be either adaptive or it can be set to a constant value based on typical driving cycles. For the current SUV analysis, the only two parameters defining the objective function are the battery state of charge and the fuel cell system efficiency. Therefore, a single limit of 0.07 was selected. This represents the value of the objective function at peak fuel cell system efficiency and optimal battery state of charge. However, as more vehicle inputs are added to the objective function, the selection of the cost limit will become more complicated as no single minimum can be selected.

The final aspect of the cost based control strategy is the implementation of the search routine used to determine the target fuel cell power from the objective function.

The important features of the search routine are the response time and rate limits imposed on the fuel cell power output. Generally, mitigating the transient variations on the fuel cell power system will increase its overall efficiency and improve durability. However, this shifts the burden of managing transients to the battery. Therefore, reducing the transients on the fuel cell system must be balanced against maintaining the battery SOC in an acceptable band and evaluating the impact of charge and discharge efficiencies of the batteries on the overall vehicle efficiency. Nonetheless, the ability to incorporate the response characteristics of the fuel cell system into the cost based control strategy is a powerful feature.

DYNAMIC SUV MODEL AND CONTROL STRATEGY IMPLEMENTATION

The control strategy was implemented and tested over numerous vehicle duty cycles using the MSC.EASY5® dynamic simulation software. MSC.EASY5 allows dynamic modeling of powertrain, vehicle, fuel cell, and electrical systems by using built-in library components and modules.

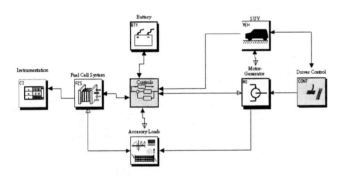

Figure 3. Fuel Cell Series HEV Model

Figure 3 represents the high-level hybrid electric fuel cell SUV model. The battery, fuel cell system, controls, SUV, and accessory load icons in Figure 4 are submodels that contain the base component modules. Each of these subsystems will be described in the following sections. The driver control block is the MSC.EASY5 model that modulates the brake and accelerator positions on the vehicle to meet the target vehicle speed for a given drive cycle.

CONTROLS AND POWER ELECTRONICS

The primary focus of this paper is the operation of the controls submodel that regulates power distribution between the fuel cell stack, battery, and the primary 300 volt bus using the cost based strategy. This model is shown in Figure 4. The main block in this model is the cost based controller. This component contains the FORTRAN code that defines the objective function and search routines. This block also acts as the primary DC/DC converter that ties together the fuel cell and battery to the high voltage bus that in turn interfaces with

the propulsion system motor / generator. A secondary 42-volt bus is also used to power the accessory loads.

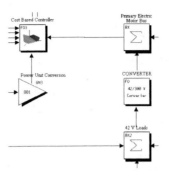

Figure 4. Power Electronics and Controls Submodel

SUV CHASSIS AND DRIVELINE MODEL

The vehicle chassis and driveline components contained in the SUV submodel are shown in Figure 5. This submodel interfaces with the electrical motor / generator through the gearbox.

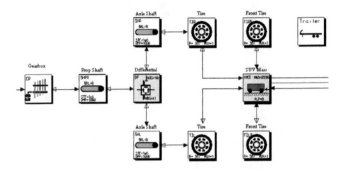

Figure 5. Fuel Cell SUV Driveline and Chassis Submodel

ELECTRIC MOTOR / GENERATOR

The primary drive electric motor / generator provides the propulsion power for the driven wheels and it also replenishes power back to the battery during regenerative braking. The machine is installed on the 300-volt primary bus. The characteristics of the motor / generator are given in Table 2.

Table 2. Propulsion Motor / Generator Summary

Maximum Motor Current	350 Amps
Maximum Generating Current	350 Amps
Rated Power	105 kW
Nominal Efficiency	91%
Motor Supply Voltage	300 Volts

The electric machine duty cycle, EMDC, is used to control the operation of the motor / generator and it varies from -1.0 at full regenerative braking to 1.0 at full motoring power. EMDC is an output from the driver control block and it is used to determine the current from the electric machine via:

$$I_{Electric\ Machine} = EMDC * Maximum\ Current$$

FUEL CELL SYSTEM

The fuel cell system submodel is shown in Figure 6. The cost based controller sets the target fuel cell stack power output.

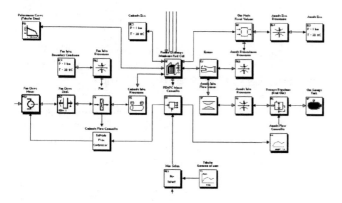

Figure 6. Fuel Cell System Model

The core of the fuel cell system is the low pressure PEMFC stack. The fuel cell polarization curve used in the analysis is shown in Figure 7 and the key stack parameters are given in Table 3.

Table 3. PEMFC Stack Data

Maximum Power	113 kW
Cell Active Area	800 cm^2
Number of Cells in the Stack	400
Anode Stoichiometry Setpoint	1.5
Cathode Stoichiometry Setpoint	2.0

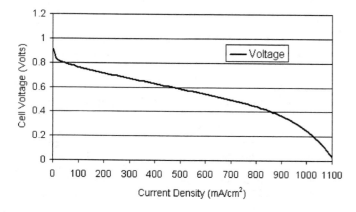

Figure 7. Fuel Cell Polarization Curve

The air on the cathode is supplied via a blower driven by an electric motor on the 42 volt bus. The hydrogen on the anode is supplied from a 10,000 psi hydrogen tank through a pressure regulator. Hydrogen at the exit of the anode is circulated back into inlet via an ejector. The fuel cell power demand is translated into target anode and cathode mass flow rates based on the stoichiometry setpoints. The target mass flows are achieved through local controllers that modulate the pressure regulator on the hydrogen supply and the fan speed on the air supply.

The fuel cell stack was sized based on pulling a 2267 kg trailer up a 6.1% grade at 55 mph. For a sustained and steady hill climb, the batteries will not be utilized and the power will be delivered primarily from the fuel cell system.

BATTERY PACK

The battery pack provides power to the propulsion motor when vehicle demand exceeds the fuel cell system power during a transient. The battery is recharged via regenerative braking and by the fuel cell system when the transient fuel cell power output exceeds the vehicle requirements. The battery specifications are listed in Table 4.

Table 4. Battery Pack Data

Maximum Battery Capacity	8.5 Amp.hr
Efficiency	85.0%
Nominal Voltage	336 Volts
Power – Nominal 50% SOC @ 35°C	70 kW
Battery Management	Integrated

ACCESSORY LOADS

The accessory load submodel, illustrated in Figure 8, contains basic models for the high and low temperature cooling circuits serving the PEM, battery pack, electric propulsion motor, and power electronics as well as a hydrogen purge system.

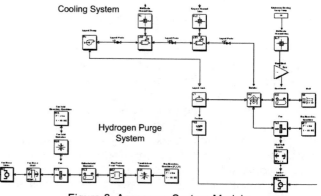

Figure 8. Accessory System Models

The coolant pump, radiator fan, and hydrogen purge fan are all operated off of the 42 volt bus. In addition, a constant load is applied to the 42 volt bus to account for hotel loads such as HVAC, headlights, instrument panel, and radio. The miscellaneous hotel loads remain constant throughout the drive cycle analysis.

ANALYSIS AND RESULTS

The hybrid fuel cell SUV model governed by the cost based control strategy was run over three transient duty cycles:

- USFTP75 drive cycle
- Full power acceleration from 0 km/hr to 96 km/hr
- Trailer tow up a 6.1% grade at 88.5 km/hr

The drive cycle analysis is the primary measure of how the control system performs. The acceleration and hill climb analyses were executed to verify that the control strategy is robust and allows the basic vehicle performance criteria to be successfully achieved during the most demanding vehicle driving conditions.

USFTP75 ANALYSIS

The USFTP75 drive cycle analysis was executed to verify that the propulsion system governed by the cost based control strategy delivers the power required to meet typical driving conditions while maintaining the battery SOC. To this end, Figure 9 illustrates that the vehicle speed matches the corresponding setpoint over the entire cycle.

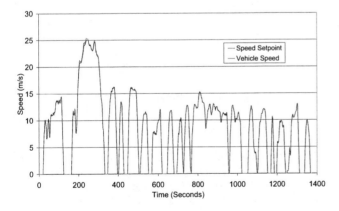

Figure 9. USFTP75 Drive Cycle Vehicle Speed Setpoint and Speed Achieved

To illustrate how the control system responds, Figure 10 shows the output value of the objective function and the corresponding inputs of SOC and fuel cell system power. At the beginning of the drive cycle, the fuel cell system is at a net power of zero. At this condition, all of the power produced in the fuel cell stack drives the accessories and hence the overall system efficiency is zero and the corresponding "cost" value from the objective function is much greater than the limit. Therefore, the controller begins to ramp up the fuel cell

system power to take advantage of the increasing efficiency. However, the vehicle power demand continues to exceed the fuel cell power and hence the battery continues to be depleted. At approximately 110 seconds into the drive cycle, the controller increases the fuel cell power ramp up rate in response to the steadily decreasing battery SOC. This decreases the value of the objective function mainly due to the increasing fuel cell system efficiency.

Simultaneous to the increase in the fuel cell power ramp up rate, the vehicle experiences the first deceleration to a complete stop at approximately 125 seconds into the cycle. This event results in energy generation through regenerative breaking. Therefore, the battery SOC rapidly begins to increase and the "cost" value of the objective function rapidly decreases. Once the battery is recharged, the controller begins to decrease the power of the fuel cell system at a time of 128 seconds.

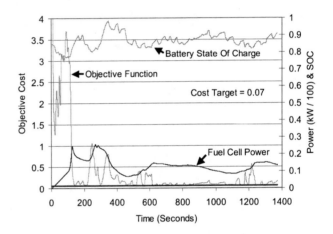

Figure 10. Objective Function Output and Input Over the USFTP75 Drive Cycle

As the vehicle begins the second acceleration cycle at 163 seconds, the fuel cell power is still decreasing due to the battery SOC being above the target value of 0.85. As a result, the series of accelerations between 163 seconds and 280 seconds causes the SOC to decrease again. At approximately 232 seconds, the controller responds to the SOC falling below the target value of 0.85 by increasing the fuel cell power again. This type of control system response is repeated over the rest of the drive cycle.

A key benefit of the cost based control strategy is the mitigation of transients that the fuel cell system experiences. This is illustrated in Figure 11 by showing the battery pack, fuel cell, and propulsion motor / generator power levels during the drive cycle. While the battery and motor power change rapidly in response to vehicle demand, the fuel cell system power output is continuous and relatively steady in comparison. This not only maintains the fuel cell stack in a region of high efficiency as shown in Figure 12, but it also extends the lifetime of the fuel cell system.

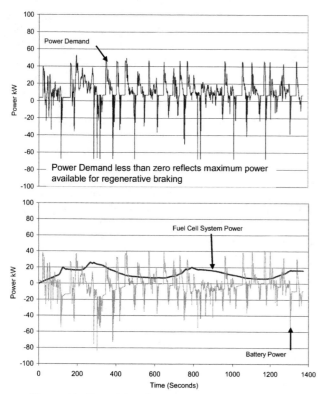

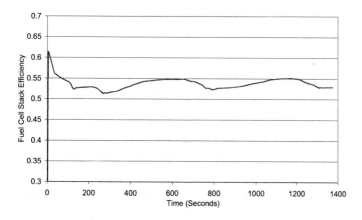

Figure 11. Propulsion System Power Levels Over the USFTP75 Drive Cycle

Figure 12. Fuel Cell Stack Efficiency Over the USFTP75 Drive Cycle

The responsiveness of the fuel cell system is set in the controller through a combination of rate limits and a first order lag applied to the fuel cell power output response. The objective function and cost limit define the target fuel cell power for a given battery SOC. This target fuel cell power is then filtered through a first order lag to determine the actual fuel cell power command signal sent from the controller. This allows the response time of the fuel cell system to be controlled through the time constant of the first order lag. In addition, limits on the rate of change of the fuel cell power determined from the first order lag are also set thus providing another dimension of responsiveness control.

As the responsiveness is increased, the fuel cell power will cycle over a greater range. This will reduce the variation in the battery SOC and the energy losses associated with cycling the fuel cell energy through the battery. However, it will also cause the fuel cell efficiency to vary over a wider band. Balancing the cycling of the battery and the fuel cell requires a detailed study that looks not only at the USFTP75 drive cycle, but also other conditions that may be limiting. While this will be examined in detail in future work, the response of the system to a full power acceleration and a trailer towing transient were evaluated in this paper.

ACCELERATION PERFORMANCE

A full power acceleration was executed to verify that the cost based control strategy applied to the fuel cell propulsion system delivers an acceptable 0 to 96 km/hr (60 mph) time for the vehicle. The acceptance criterion was set at 12 seconds based on typical performance data for SUVs of this size. Figure 13 illustrates the vehicle speed response to a full power input and it demonstrates that the vehicle is capable of meeting the acceleration performance requirement.

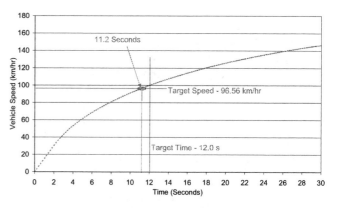

Figure 13. Full Power Acceleration

GRADEABILITY

The second transient condition examined beyond the drive cycle was towing a 2268 kg trailer up a 6.1% grade at a constant speed of 88.5 km/hr (55 mph). For this condition, the vehicle was accelerated from rest to the target speed and then driven for a period of 16 minutes. This is an important test case as the continuous high power demand for a long duration shifts all of the duty to the fuel cell after the initial acceleration period. Furthermore, there is no opportunity to recharge the battery through regenerative braking. Therefore, the control system should ensure that the battery does not deplete too far and that the fuel cell power is increased to meet all of the demand.

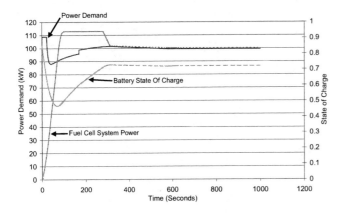

Figure 14. Propulsion System Results for Towing a 2268 kg Trailer up a 6.1% Grade

The results of the hill climb study are shown in Figure 14. During the initial acceleration period, the propulsion system utilizes the battery while the fuel cell is ramping up. Once the fuel cell reaches full power, it begins recharging the battery until it reaches a SOC of just above 0.7. While this is below the optimal SOC of 0.85, the controller does not charge it any more due to the cost of the additional power being very high. At that point, the controller ramps down the fuel cell to match the vehicle demand with no additional power being supplied to or taken from the battery pack. If the vehicle load were decreased or the vehicle were brought to a rest, then the controller would recharge the battery to the optimal point of 0.85 using the fuel cell in a more efficient manner. This system behavior confirms that the control strategy is robust and effective in handling even the most demanding vehicle conditions.

CONCLUSION

Cost based power management control strategies applied to hybrid fuel cell vehicles are an effective way to optimize system efficiency in a robust and continuous manner. This approach uses an objective function combined with target limits in lieu of complex logic routines. In the current study, the objective function was based on the fuel cell system net power and the battery SOC. However, the cost based control strategy is flexible and can incorporate other vehicle inputs such as emissions, fuel cell operating temperatures, fuel cell water balance indicators, and NVH measures.

The cost based control strategy was implemented in an MSC.EASY5 model of a hybrid fuel cell powered SUV. The vehicle was tested over the USFTP75 drive cycle, a full power acceleration run, and a trailer tow up a 6.1% grade. Under all conditions, the power management control strategy met all vehicle performance criteria while effectively managing the battery state of charge. Furthermore, the strategy mitigated the transients on the fuel cell system thus not only increasing efficiency, but also extending the operating lifetimes.

Future work will investigate the incorporation of additional vehicle inputs into the objective function along with optimizing performance through the parametric studies of the applied cost limits, response controls, and objective function characteristics.

ACKNOWLEDGMENTS

MSC.EASY5® is a registered trademark of MSC.Software Corporation.

The authors would like to acknowledge the Ricardo Advanced Engineering Committee for supporting this study.

REFERENCES

1. Horwinski. Hybrid powered automobile. US Patent 4,042,056 August 16, 1977
2. Paice Corporation. Hybrid Vehicles. International Patent WO 00/15455 September 10, 1999
3. Jin-Hwan Jung, Young-Kook Lee, Jung-Hong Joo and Ho-Gi Kim. Power Control Strategy for Fuel Cell Hybrid Electric Vehicles. SAE 2003-01-1136
4. Chu Liang and Wang Qingnian. Energy Management Strategy and Parametric Design for Fuel Cell Family Sedan. SAE 2003-01-1147
5. Gino Pagnelli, Yann Guezennec and Giorgio Rizzoni. Optimizing Control Strategy for Hybrid Fuel Cell Vehicle. SAE 2002-01-0102
6. Ricardo Consulting Engineers Limited. Hybrid Power Sources Distribution Management. International Patent WO 02/42110 May 30 2002
7. M Godoy Simoes and Paulo E. M. Almeida. Neural Optimal Control of PEM-Fuel Cells with Parametric CMAC Networks. Conference Record of the 2003 IEEE Industry Applications Conference. Paper 723-30 vol.2

CONTACT

Chad Mitts
Ricardo, Inc.
40000 Ricardo Drive
Van Buren Twp., MI 48111
(734) 397-6676
chad.mitts@ricardo.com

VI. MODELLING, CONTROL AND DIAGNOSIS

Modelling

2001-01-0543

The Hybridized Fuel Cell Vehicle Model of the University of California, Davis

Karl-Heinz Hauer, R. M. Moore and S. Ramaswamy

University of California, Davis

ABSTRACT

Vehicle manufacturers claim that fuel cell vehicles are significantly more fuel-efficient and emit fewer emissions than conventional internal combustion engine vehicles /1/. A computer model can help to explore and understand the underlying reasons for this potential improvement. In previous published work, the UC Davis Vehicle Model for the case of a load-following Indirect Methanol Fuel Cell Vehicle (IMFCV) has been introduced and discussed in detail /2/.

Because of possible technical barriers with load following vehicles, as well as near term cost issues, hybrid fuel cell vehicle concepts are widely discussed as another fuel cell vehicle option. For load following vehicles, the questions of fast start up and fuel processor dynamics in extreme transient situations, (e.g., during phases of hard acceleration) are not totally resolved at this time. For both of these performance issues, a hybrid design could offer at least an interim solution.

To investigate the potential of the Indirect Methanol Fuel Cell Hybrid Vehicles (IMFCHV) further, the original model for the load following IMFCV has been expanded. First, a battery component model has been added (including a model of a battery controller). Second, a dc-dc converter component model has been included between the fuel cell stack and the battery (to ensure the proper integration of the battery into the overall operation of the electrical system).

The purpose of this work is to explain the underlying modeling philosophy, as well as the algorithms for the hybrid case. The approach followed in this IMFCHV model allows parametric studies to be made over a large range of configurations, and simplifies the effort involved in deciding the optimal design for a given scenario.

Several examples are given to illustrate the capabilities of the model. However the paper is not meant to exhaustively answer specific questions about fuel economy or other vehicle properties. Instead the focus is primarily on the explanation of the algorithms of the simulation, in order to allow a better understanding of the model itself.

INTRODUCTION

Figure 1 shows the propulsion energy flow and the information flow of the Indirect Methanol Fuel Cell Hybrid Vehicle (IMFCHV) model. The figure is for the design configuration of a series hybrid vehicle. The chemical energy stored in the fuel is converted into electrical energy. This electrical energy is then taken and stored in the battery or directly supplied to the electric drive train. The dc-dc converter between the fuel cell stack and battery allows the fuel cell system to operate at a steady state independent from the state of charge of the battery and of the electric load imposed by the drive train and auxiliaries.

All components communicate via a data bus. The data values carried are: battery voltage, acceleration pedal position, motor current, battery current, fuel cell stack current, and other variables necessary for an optimal component interaction.

The work presented in this paper builds up on previous work /2, 3, 4/ and expands the basic model of an indirect methanol fuel cell vehicle to a model for a battery hybrid fuel cell vehicle.

For an IMFCHV, it could be shown, that depending on the fraction of regenerative braking, significant energy savings could be achieved for a given drive cycle. However, in this paper it is shown that, due to the practical need for a combination of mechanical brakes and electrical brakes, only half of the kinetic energy available during braking can be recovered for the case of an US06 cycle. The exact results of such an analysis depend, of course, on the exact driving cycle used for the analysis.

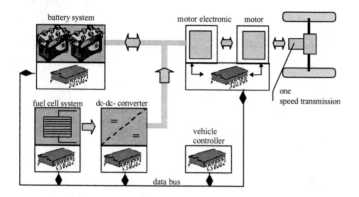

Figure 1: Energy and Information flow in an Indirect Methanol Fuel Cell Hybrid Vehicle (IMFCHV)[1]

Vehicle Model

The principal setup of the simulation model is shown in Figure 2. The model separates the driver block from the rest of the vehicle components. In this way, the driver functions as a controller for the vehicle velocity specified in the drive cycle /5/. This setup is consistent with the general principle of separating control algorithms and component models, as described in /6/. This modeling approach allows a systematic and detailed investigation of hardware limitations separately from limitations due to control issues of the overall system. In addition this separation allows the use of control algorithm for rapid prototyping approaches.

The IMFCHV vehicle model is different from the model of an Indirect Methanol Fuel Cell Vehicle (IMFCV). The differences are the additional battery storage, including the battery controller, and the block "brake assistant". The battery system provides energy storage and decouples the fuel cell system from the drive system. Therefore the drive train now "sees" the battery voltage, and not the stack voltage as in the case of the IMFCV. The brake assistance block determines, based on the brake pedal position, the fractions of mechanical and electrical braking.

The other new block in this model is the "Fuel Cell System and DC-DC" block. This block includes the fuel cell system model plus the model of a dc-dc converter that isolates the fuel cell stack from the battery.

Furthermore the block "electric drive train" is modified such that it allows a bi-directional energy flow. The energy flows from the battery into the drive train during phases of acceleration and vice versa during phases of

---

[1] The accessory power for the fuel cell system e.g. air compressor is drawn from the battery and not from the stack terminals

regenerative braking.

The vehicle shell block stayed unchanged from the model of an Indirect Methanol Fuel Load Following Vehicle (IMFCV) and is described in detail in /2/. This block takes the motor and brake torque as inputs and computes two outputs (vehicle speed, and motor speed based on the vehicle speed) considering the vehicle properties such as mass, aero dynamic drag resistance and tire friction.

The integration of a battery also allows the operation of the fuel cell system in steady state mode, except for the ramp-up and ramp-down phases. In a hybrid configuration it is possible to operate the fuel cell system in only one or in a limited number of operating points. This operational strategy provides better fuel economy for the overall vehicle. It also releases the fuel cell system designer from the duty of designing the system to follow the transient requirements of a drive cycle.

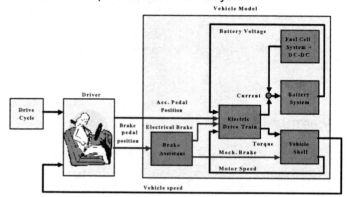

Figure 2: Overview of the hybrid vehicle model

The interaction of the components with each other and the meaning of the two feedback loops for battery voltage and motor speed are described in /2/. The only difference for the IMFCHV from the IMFCV is that on the electric side the fuel cell system is now replaced by the battery system. The fuel cell system therefore can now operate "off the loop" and is released from most dynamic requirements.

The following section explains the properties and specifics of each block in the vehicle model, with the exception of the blocks "Vehicle Shell" and "Fuel Cell System". The setup and functionality of these blocks is explained in detail in /2, 7/.

Brake Assistant

The IMFCHV model allows two different braking modes, namely: electric braking (operating the electric motor as a generator), and mechanical braking with the conventional brake system of the car. The former serves to recharge the battery, while the latter generates heat at the brake pads and discs.

The block "brake assistant" is responsible for dividing the brake pedal signal (as the expression of the drivers desire) into two different brake signals. One determines the electric brake mode and the other determines the mechanical brake mode. In a real hybrid vehicle this function could be integrated into the Anti-Lock Brake System (ABS) or could be done by a totally independent control unit.

From an energy point of view, it seems logical to choose the maximum fraction of electric braking - hence mechanical braking only dissipates the energy into heat. However, due to the fact that electrical braking only works for the wheels connected to the electric motor, electric braking has it limits. These limits are dictated by both safety and convenience. Another limit is the limited torque that the motor can provide, which is normally not enough for hard decelerations or emergencies.

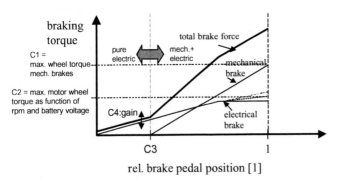

Figure 3: Electrical (regenerative) and mechanical brake forces[2]

Figure 3 explains how the "brake assistant" model allows modifying the total composition of the overall brake torque. Four parameters are available:

- C1 : The maximum mechanical brake torque applied to the wheels
- C2: The maximum electrical brake torque generated by the electric motor (this value is not fixed but depends on the motor characteristics and the supply voltage).
- C3: The brake pedal position up to which only electric braking is possible (The parameter in the model is: FCV.vehicle.regen_exclusive).
- C4: The increase of the electric brake torque with increasing brake pedal position (The parameter in the model is: FCV.vehicle.regen_strength)[3].

---

[2] Note that the braking characteristic shown is only conceptual. It is not a suggestion of an optimal braking characteristic for hybrid vehicles. In this example setup, the varying slope of the characteristic could be disturbing for the driver. More about the issue of convenience and feel for the case of a "blended" brake system can be found in /13/.

[3] The maximum mechanical torque is always applied at the maximum brake pedal position and zero for brake pedal positions below C3. Therefore no separate parameter is necessary to specify the slope of the mechanical brake torque.

The setup of the model allows the variation of the above-mentioned parameters for the "brake assistant". For example, the amount of regenerative braking could be lowered if the battery is not able to receive the maximum possible regenerative energy. Logically such a signal would be generated by the battery controller and than looped to the brake assistant. This function is not incorporated in the current version of the model, but the model structure supports such an implementation.

Drive Train

The electric drive train has been modeled the same way as the electric drive train of an IMFCV with one exception - the motor controller takes not only the acceleration pedal command into account but also considers the brake signal for switching into the electric braking mode. With this signal the motor controller allows for reversing the motor torque in phases of regenerative braking in situations when the relative acceleration pedal position is smaller than the signal that indicates electric braking (normally either one or the other signal is zero). Because the maximum motor torque depends on voltage and motor speed in the same fashion for the generator mode and in the motor mode, the motor characteristic for the acceleration mode can simply be reversed for operating in the generator mode. The result is a motor map with positive motor speed only (no reverse) and symmetric lines of positive and negative maximum motor torque depending on the supply voltage[4] (Figure 4).

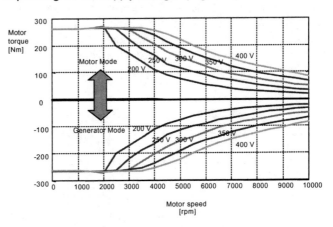

Figure 4: Motor characteristic for the hybrid case

The same argument that applies for the torque-speed-voltage characteristic for the braking mode applies also for the efficiency map of the electric motor. The efficiency in motor and generator mode is symmetric to the line of zero torque (Figure 5).

---

[4] This discussion assumes that the mechanical fraction in the bearings and due to the aero dynamic drag of the rotor could be neglected. If the motor is cooled by forced air, and the cooling fan is sitting on the shaft, this assumption is not automatically justified due to the higher aerodynamic drag losses.

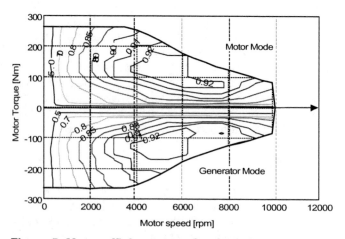

Figure 5: Motor efficiency map for the hybrid case

Battery Model including the Battery controller

The battery system model consists of three blocks: dc-dc converter, battery, and battery controller (Figure 6). Each of these blocks will be explained in detail in the next section. The dc–dc converter is the interface between the fuel cell stack and the battery. The battery block describes the battery properties (such as open circuit voltage and internal resistance as a function of state of charge), as well as the actual battery capacity, as a function of the current applied to the battery.

The battery controller block determines, based on the battery current, the state of charge of the battery and switches the fuel cell system on or off. In addition to this very basic algorithm, other measures for the optimal control of the battery, integrated in a vehicle, could be applied using the model structure.

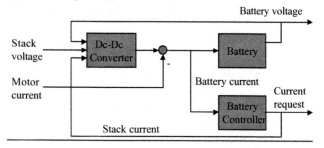

Figure 6: Overview of the battery system model

DC-DC Converter

The fuel cell stack and battery are not directly connected with each other, but are connected through a so-called direct-current to direct-current converter (dc-dc converter). This dc-dc converter allows the operation of the fuel cell stack at one specific predetermined current independent from the battery voltage /8,9/. Without a dc-dc converter between the fuel cell stack and battery such an operation mode would not be possible. This is because

the fuel cell stack current is determined by the voltage difference of the stack and battery divided by the total load resistance. The total load resistance is the sum of stack resistance (R_{fc}) plus cable resistance plus the battery resistance in parallel to the resistance of the electric drive train (Figure 7).

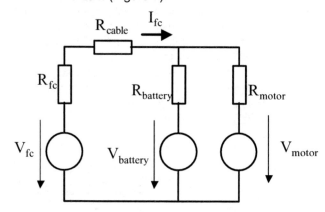

Figure 7: Fuel cell stack current as a function of resistances and voltages in the overall electric system for the configuration without dc-dc converter.

If steady state operation of the fuel cell system is desired only a one-directional dc-dc converter (the energy flow is always from the fuel cell stack to the battery) is necessary to decouple fuel cell stack and battery voltage.

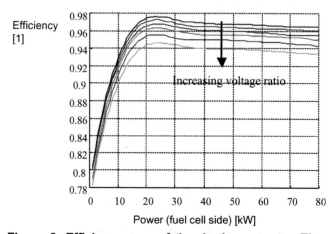

Figure 8: Efficiency map of the dc-dc converter. The ratio of output voltage (battery side) to input voltage is varied from 1.1 to 2.1. The transferred power varies from 1 kW to 80 kW.

The dc-dc converter is modeled with a two-dimensional efficiency map only. Because of the almost immediate response of the dc-dc converter, transient effects are not taken into account. Consequently the model does not consider the effects of current ripple on the fuel cell stack. In an actual design a capacitor between the fuel cell stack and the dc-dc converter would reduce the current ripple in the stack imposed by the dc-dc converter. In this case the stack current ripple is comparable to the stack current ripple in a load following vehicle design caused by the switching of the motor electronics.

The efficiency of the dc-dc converter is defined as the ratio of output power on the battery side versus input power on the fuel cell system side (Equation 1). The efficiency depends on the transferred power and the voltage difference between input (fuel cell stack voltage) and output (battery voltage) /10/. Because the efficiency is not derived from 1st principles in this model, the modeling approach relies on access to experimental data, e.g.: data for the efficiency for different operating points. However because only limited experimental data were available for the power level required in the model, the map in Figure 8 is the result of a scaling process based on data provided by King /10/ for a smaller dc-dc converter. Additional assumptions made are:

- The efficiency of the dc-dc converter depends only on the ratio of output voltage over input voltage, and not on the absolute voltage levels itself
- For a constant voltage ratio, the efficiency increases with increasing power until it reaches it peak and then drops slowly with further increasing power

However, due to the relatively flatness of the curves, and the two point operating strategy of the fuel cell system, it is not expected that a complete dc-dc converter map would significantly alter the results.

$$\eta_{dc-dc} = \frac{P_{dc-dc-bty}}{P_{dc-dc-fc}}$$

$where:$

η_{dc-dc} = efficiency of the dc $-$ dc converter [1]

$P_{dc-dc-bty}$ = output power of the dc - dc

converter on the battery side [W]

$P_{dc-dc-fc}$ = input power of the dc - dc $-$ converter

on the fuel cell stack side [W]

Equation 1

Battery

The battery is modeled by accounting for the internal charge and discharge resistance as a function of the state of charge, and the open circuit voltage as a function of state of charge[5]. The state of charge is calculated using the Peukert relation /11/. Equations 2-4 describe the overall calculation of the battery state of charge.

---

[5] This battery model assumes that the impact of charge and discharge current on the actual battery capacity is the same and independent from the SOC of the battery. Temperature effects are not modeled.

$$\bar{I}_{dchg}(t) = \frac{1}{t} \cdot \int_0^t \left| I_{dchg}(t) \right| \cdot dt$$

$I_{dchg}(t)$ = battery discharge current [A]

$\bar{I}_{dchg}(t)$ = average battery discharge current [A]

t = time [sec]

Equation 2

The actual capacity of the battery is calculated based on the Peukert relation and depends on the average discharge current (Equation 3).

$$C_{actual} = K_1 \cdot \left| \frac{\bar{I}_{dchg}(t)}{I_{noml}} \right|^{K_2}$$

$where:$

K_1 = Peukert constant (1 hour discharge capacity) [Ah];

K_2 = Peukert Exponent [1]

I_{noml} = nominal 1 hour battery discharge current [A]

C_{actual} = actual battery capacity [Ah]

Equation 3

The state of charge of the battery is calculated according to Equation 4. With this calculation the actual state of charge of the battery is related to the actual battery capacity.

$$SOC = \frac{\int_1^t I(t)dt}{C_{actual}} + SOC_{initial}$$

$where:$

SOC = state of charge

$SOC_{initial}$ = initial state of charge [Ah]

$I(t)$ = battery current

Equation 4

Battery Controller

The battery controller has several functions.

First, the battery controller monitors the state of charge of the battery system and, depending on this variable, it activates and deactivates the fuel cell system. Superposed on this mode of activation is the activation of the fuel cell system when the average power drawn by the electric drive train plus auxiliaries exceeds a lower power threshold, or if the acceleration pedal position exceeds a lower threshold.

Second, the battery controller generates a signal that deactivates the ability for regenerative braking in those situations for which continued regenerative braking could lead to overcharging of the battery. This signal is received by the motor controller but is currently deactivated (in the model).

Third, the battery controller sets the current request to the fuel cell system to zero if the system is not ready to supply energy, e.g.; during the warm up phase.

Figure 9 shows the functions of the battery controller in graphical form. The model makes it possible to expand this initial set of functions, if required for specific vehicle designs.

The battery controller would also limit the maximum regenerative power according to the receptivity of the battery system.

The battery control algorithm requests only constant current (single point operating strategy) from the fuel cell system. However other control strategies with varying fuel cell power requests could be incorporated into the battery controller.

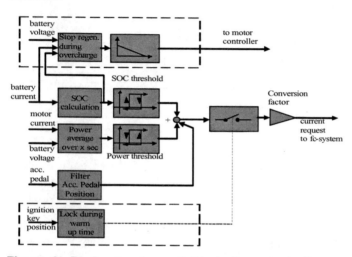

Figure 9: Basic structure of the battery controller (the functions inside the dashed boxes are not activated yet)

Additional functions of the battery controller could include:

- restriction of regenerative braking in cases for which continuing regenerative braking would lead to overcharging and battery damage (limited battery receptivity)
- charging algorithms for the battery charger for charge depleting vehicles
- a test of different algorithms for estimating the state of charge of the battery
- controls for power assist modes in designs in which the battery alone is not able to meet the maximum motor power request (superposition of battery power and fuel cell power)

Simulation results

This section provides examples of simulation results, focusing specifically on the effect of regenerative braking. The vehicle concept discussed in this paper is a Range Extender hybrid vehicle concept. In Range Extender concepts the fuel cell system is operated at the design point only. However the model can also be used to analyze Power Assist concepts. In Power Assist hybrid vehicles the fuel cell system has to be able to follow the dynamics of the drive cycle in some degree. The results provided here are not meant to provide an analysis of hybrid vehicles in general, nor a comparison of hybrid and load following fuel cell vehicles. Instead they are to provide illustrations of the capabilities of the model.

Figures 10 and 11 show the energy flows in the drive train, and in the battery, as a function of the fraction of the braking pedal position that allows for only regenerative braking determined by parameter C3 (see also Figure 3). The underlying drive cycle for both figures is the US06 cycle. The key vehicle parameters are shown in Table 1.

Vehicle:	
Wheel diameter	0.3556 m
total wheel inertia	4 kg*m^2
rolling resistance	0.0085
aerodynamic drag coefficient	0.3
frontal area	2.00 m^2
total vehicle mass (including driver)	1500 kg
Fuel cell system:	
Number of cells	432
maximum stack current (100% fuel and air supply)	400 A @ 0V; 200A @ 223 V
voltage current stack characteristic	internal model (UC-Davis)
Battery System	
Technology	NiMH
Capacity	22 Ah
Number of Modules	25
Number of cells per Module	11
Nominal voltage	330 V
Transmission:	
Number of gears	1
total gear ratio	7.5
transmission efficiency map	assumed /2/
Electric Motor:	
Type	induction motor
motor efficiency map	assumed (Figure 5)
torque-speed-voltage map	assumed (Figure 4)
motor inertia	0.1 kg*m^2 (assumed)

Table 1: Vehicle parameters (excerpt)

Figure 10 shows the net energy consumed by the electric motor, together with the regenerated energy and the mechanical energy dissipated in the brake system (as a function pf parameter C4 in figure 3), assuming that the vehicle is operated over the USO6 driving cycle. The mechanical energy dissipated in the brakes at zero regenerative braking (C4=0) is effectively the total potential of energy that could be recovered by regenerative braking. In this example, the maximum potential energy recovery is 81 Wh/mile.

The electric energy recovered is measured at the dc terminals of the motor power electronics. Therefore the battery losses are not included in Figure 10.

It can be seen that for an exclusive electrical braking fraction of the normalized brake pedal position of 0.3 the point of diminishing returns for recapturing energy by regenerative braking has already been reached at a regenerative braking strength of C4=0.3. A further increase of the parameter C4 (regenerative braking strength) would not significantly increase the amount of captured energy. A further increase of recaptured energy is only possible with a larger region of the brake pedal position reserved for exclusive regenerative braking (increase of parameter C3). However an increase of the pure electric braking regime would leave less of the total pedal displacement for the proper adjustment of the mechanical brakes. The driver would have a harder time adjusting the mechanical brake in an optimal way.

The maximum energy that can be recaptured by regenerative braking is equivalent to the dissipated energy in the mechanical brakes when all the braking is done with the mechanical brakes. In this specific vehicle configuration, and for the USO6 cycle, this upper limit is about 81Wh/mile (or about 23% of the energy the motor requires to follow the USO6 drive cycle). This fraction would be higher in a drive cycle with more accelerations and decelerations per mile.

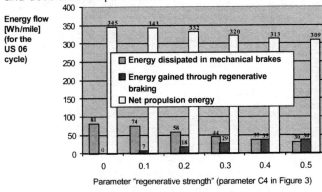

Figure 10: Energy flow in the electric drive train in as a function of the parameter C4 in Figure 3. The parameter C3 (regime of exclusive electric braking) stayed constant as well as parameter C1 (maximum mechanical brake torque). Parameter C2 (maximum generator torque) is implicitly given by the drive train characteristics together with the characteristics of the overall electric system (battery, fuel cell system and dc-dc converter)

However this potential recoverable energy for regenerative braking cannot be used 100%. The reason is that harder decelerations cannot be achieved with the electric brake only. At higher vehicle speed, especially, the maximum motor torque is too small for hard decelerations (Figure 4, /12,13/). Therefore in this regime some of the potential energy has to be dissipated in the mechanical brake. The maximum energy regained from the wheels is 36Wh/mile or about 44 % of the total braking energy. This maximum regeneration has been

achieved with a setting of the parameter "regenerative strength" to C4= 0.5.

With increasing regenerative braking, the energy provided by the fuel cell system and the fuel consumption of the vehicle is decreasing (Figure 11, diamond curve). For the battery, a higher fraction of regenerative energy means higher stress - because it has to accept the regeneration energy (Figure 11, triangles and squares). Quantitatively the battery has to accept and supply ca. 10% more energy for the case of intense use of regenerative braking as compared to no regenerative braking. This value is only valid for the US06 cycle, the applied fuel cell system control strategy, and the assumed vehicle parameters. Other drive cycles and vehicle configurations will give different results. However the principal message is valid for all hybrid vehicles, namely: increasing use of regenerative braking means higher stress for the battery.

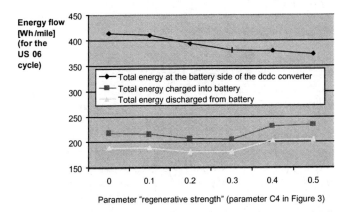

Figure 11: Battery stress [in units of Wh/mile] illustrated in terms of discharge and charge energy as a function of the amount of regenerative braking.

Dynamic aspects of regenerative braking

Figure 12 shows the electrical and mechanical braking power, respectively, for the case of acceleration from 0 to 60 miles/hour followed by a deceleration from 60 to 20 miles/hour. The parameter determining the brake pedal regime of exclusive electrical braking (C3) was held constant at C3=0.3. The parameter determining the strength of the electrical brake (C4) has been varied from C4=0.0 to C4=0.5. It can be seen that with increasing electric braking:

a) the time for deceleration becomes shorter
b) the electric braking power increases
c) the mechanical braking power decreases
d) The mechanical braking power does not go towards zero even in the case of intense use of regenerative braking, because of the hard deceleration requirement.

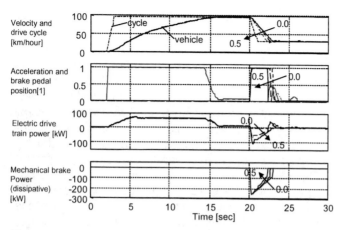

Figure 12: Mechanical and electrical brake for different parameters "regenerative strength". The parameter "regen_exclusive" stayed constant at C3=0.3 in all three simulation runs. Values of the parameter "regenerative strength" illustrated are C4=0.0, 0.3, 0.5.

<u>Next Steps and Conclusion</u>

A model for an indirect methanol fuel cell hybrid vehicle has been developed. The basis of this model is the existing model of a load-following indirect methanol fuel cell vehicle. To obtain the hybrid vehicle model the load-following model was first expanded by including a battery model (with a dc-dc converter and a battery controller). On the drive system side, the ability of regenerative braking (reversion of energy flow) was also added.

Similarly to the existing model for a load following vehicle, the overall setup of the hybrid model allows the modification of each individual component. For example, the fuel cell system can be changed to a system with lower power, use a different fuel, or to an ambient pressure air supply without the modification of other components.

In addition to the strict separation of component models from each other, the component descriptions are also strictly separated from control algorithms. This separation allows the investigation and pretest of control algorithms without changing the complete model. One example for this is the restriction of the electric brake mode in situations when the battery is fully charged.

Further upgrades of the model will include a simplified data bus structure between the different component controllers. This structure will improve the modularity and ease the understanding of the model. It is not expected that simulation results will change because of this upgrade. Other upgrades will include ultra capacitors as an alternative way for energy storage and hybridization.

More specifically it has been shown that only about half of the kinetic energy of the vehicle could be recovered with regenerative braking in the US06 cycle, for the specific setup of the electric brake system and mechanical brake system that is simulated in this paper. The reasons for this result are: the high brake force demand in the US06 cycle, and the limited ability of the electric motor to provide the necessary brake force at higher speeds. To ensure that the vehicle follows the drive cycle as closely as possible, additional mechanical brake torque has to be supplied to provide the required deceleration.

<u>References</u>

/1/	Fuel cells for vehicle applications in cars – bringing the future closer, Ferdinand Panik, Daimler Benz AG, Journal of Power Sources 1997, page 36-38
/2/	A Simulation Model for an Indirect Methanol Fuel Cell Vehicle, Future Technology and Transportation Conference, SAE 2000-01-3083, Karl-Heinz Hauer et. al.
/3/	Technical Requirements of Fuel Cell Powered Electric Vehicles, Global Powertrain Congress, Detroit 1998, Karl-Heinz Hauer, Julius Quissek et. al.
/4/	Dynamic Response of an Indirect Methanol Fuel Cell Vehicle, SAE 2000 conference, SAE 2000-01-0370, Karl-Heinz Hauer et. al.
/5/	A Modular Approach to Powertrain Modeling for the Prediction of Vehicle Performance, Economy and Emissions SAE 960427, R.P.G. Heath and C.Y. Mo Ricardo Consulting Engineers Ltd.
/6/	Control Challenges and Methodologies in Fuel Cell Vehicle Development, Woong-chul Yang, Bradford Bates / Ford Motor Company Nicolas Fletcher, Ric Pow, dbb Fuel Cell Engines
/7/	Simulated Performance of an Indirect Methanol Fuel Cell System, Draft SAE 2001 Anthony Eggert, David Friedman, Sitaram Ramaswamy, Karl Hauer, Joshua Cunningham, Dr. Robert Moore
/8/	Modeling and evaluation of advanced traction systems and new technology with SABER, Department of Electronic and Electrical Engineering, University of Sheffield, J. Li, P.H. Meller et. al.
/9/	Propulsion System Strategies for Fuel Cell Vehicles, SAE 200-01-0369, Kaushik Rajashekara, Delphi Automotive Systems
/10/	Ultra capacitor/Battery Electronic Interface Development, Robert D. King et al. General Electric Company
/11/	Batteries for Electric Vehicles, D.A.J. Rand, Woods, Dell Research Studies Press/ SAE 1998
/12/	Toyota Braking System for Hybrid Vehicle with Regenerative System, Y. Sasaki, et. al., Toyota Motor Corporation, EVS14, 1997
/13/	Electric Vehicle Braking Systems, S.R. Cikanek, K.E. Bailey, Ford Scientific Research Laboratory, EVS 14, 1997

Performances Analysis of PEM Fuel Cell Based Automotive Systems Under Transient Conditions

Luca Andreassi, Stefano Cordiner and Fabio Romanelli
University of Rome "Tor Vergata"

ABSTRACT

The use of Polymeric Electrolyte Membrane Fuel Cells (PEMFC) based power trains as a substitute of Internal Combustion Engines (ICE) in transportation has been demonstrated but still requires both technical development of components and their integration in the system. One of the major technical challenges for automotive application is the system response during rapid variations of operative conditions, which is one of the most significant elements concurring in determining the vehicle drivability (i.e. the power train capability to allow rapid transient phenomena). The power train response under transient conditions depends on the combined action of all the system components performances. This behaviour is still more complicated when hydrogen has to be produced on board and the reformer response has to be taken into account.

In this paper the performances of PEMFC based energy production systems for automotive applications is analysed by means of a simulation model able to describe the performance of all the system components from the on board energy storage to the road load. Different fuel storage/production solutions are analysed and their performances are compared (methanol steam reformer, methane auto thermal reformer and ammonia reformer).

To increase results significance, comparison are made with respect to the European Urban Cycle (ECE+EUDC) and to this aim a simple model of vehicle dynamics has been also developed.

Model predictions have proved to be able to simulate system behaviour and could be used to design system characteristics as well to define its control system.

INTRODUCTION

PEMFC based engines have a great potential for the replacement of internal combustion engine in transportation due to the low pollutant emissions (which drop to zero if the fuel cell is fuelled with pure hydrogen), the high efficiency, the modularity and the capability to perform quick start-up (as the operating temperature is low). Moreover, they use the most innocuous electrolyte (an hydrated proton conducting plastic polymer).

Nevertheless, before PEMFC for automotive applications could be really competitive in terms of overall efficiency, fuel consumption and emission levels, several challenges have to be overcome. This is clearly underlined in Costamagna and Srnivasan [1,2] where the fuel choice and its efficient reforming to obtain pure hydrogen, the improvement of the anodic electro catalysts tolerance to CO, the increase of exchange current density, the improvement of the proton conduction properties of the membrane and the optimisation of thermal, air and water management are all indicated as major issues. Obviously these targets have to be achieved without increasing the cost to reach the DOE goal of 50 $/kg foreseen for the year 2004.

In this scenario, the numerical simulation of PEMFC based system contributes to the better understanding of the influence of design parameters on the physical-chemical phenomena occurring in a fuel cell. Moreover, it is a basic tool to perform the optimisation of structure, geometry, operating conditions and thermal, air and water management. Several research centres and groups proposed numerical approaches to study the PEMFC behaviour. Notable works include those of Bernardi and Verbrugge [3,4] and Springer et al. [5,6] whose models are substantially one-dimensional. Fuller and Newman [7], Garau et al. [8], Yi and Nguyen [9] developed a bidimensional approach. Nevertheless, the field of applicability of these bidimensional models is quite narrow, especially if applied to large-scale fuel cells under high fuel utilization conditions. More recently, Zhao and Liou presented a very interesting three-dimensional approach [10] in order to deeply investigate the cell performances without dramatic penalizations in terms of computational time.

However, it is widely accepted that the deep investigation on the fuel cell behaviour is restrictive if the goal is to find an optimum configuration of a complete PEMFC system for automotive applications. There is, in fact, an enormous need to integrate the major components under typical variable conditions (reformer, stack, electric drive, etc.). Advanced simulation of all the components in the fuel cell vehicle is, then, going to play a greater role in system integration in order to contribute in reducing complexity. In fact, if it is possible to model the stack drive and vehicle constraints at the same time, it is possible to optimise the system (Jost [11], Johnson et al. [12], Pischinger et al. [13], Lee and Lalk [14], Andreassi et al. [15]).

For automotive applications the system development is further complicated by the driveability requirements. On road typical use of vehicle is, in fact, characterised by rapid variations of operating conditions. For this kind of applications, driveability means the power train capability to follow these variations accordingly to some driving styles. Dynamic behaviour has, then, to be implemented also in the model depending on both fuel cell stack and components dynamics. Heat and gas management characteristics could, in fact, cause significant variations in the cell dynamics and voltage under varying load characteristics. Consequently the relationship among the fuel cell, reformer, compressor and heat exchangers has to be evaluated under dynamic conditions.

In this paper a modelling approach to the description of a complete automotive PEMFC system is presented, which can be used as a helpful tool in system design and optimisation. The modelling techniques assumptions and basic equations are illustrated for each subsystem, with special attention to the air system, thermal management and fuel reformer. The model is able to couple the simulation of the electrochemistry processes, which take act within the single fuel cell together with the mass and heat exchanges so allowing a quite detailed simulation of dynamic performances of a stack. The model also allows simulating the transient behaviour of other system components (reformer, heat exchangers, air system components, electric engine).

As an application, the model simulation has been, then, used to evaluate generation system behaviour during transient start-up and shutdown operating conditions. The time required to reach defined conditions, the effect of cooling system control strategies and the influence of design on fuel consumptions have been analysed.

After developed a simple vehicle dynamic model, system performances have then been compared with respect to the ECE+EUDC cycle for a middle size vehicle characteristic of European cities. Different fuel choices have been compared with reference to the fuel economy (Fig. 1).

MODELING A PEMFC BASED POWER GENERATION MODULE

As previously mentioned the system integration is a specific challenge in the development of PEMFC based power generation module. First of all, different system layouts can be potentially used, each of them requiring the optimisation of a large number of configuration variables and their on line control. A complete

generation system is, in fact, constituted of a number of different components (fuel cell stack, compressor, humidification systems, reformer, etc.) linked following different strategies. Moreover the power generation module has to be interfaced with the electric engine on one side and to the fuel delivery system on the other side and its performances have to fulfil the load-speed profile, which comes from the road characteristics and driver requirements.

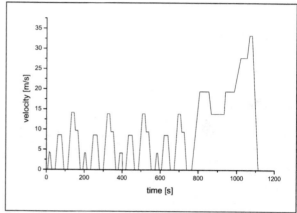

Fig. 1 *European Cycle ECE+EUDC*

In Figure 2 a typical configuration of a complete system is sketched in a schematic way.

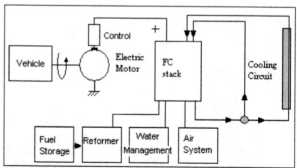

Fig. 2 *Fuel Cell system*

PEMFC for automotive applications can be used both by hybridising the power generation system with other energy source (batteries or Internal Combustion Engines) or by using the fuel cell stacks as the unique on board energy source. Besides the use of the fuel cell as an auxiliary power unit (APU), which is nevertheless of great interest [16], in the hybrid systems batteries provide the additional power required to balance peak power conditions (start-up, acceleration and hill climbing) and to recover the energy lost during braking in start-stop cycles (urban driving). The unique source solution is however preferable as minimises vehicle weight and overall costs. Nevertheless, this choice requires a careful development of all system components dynamics to make them able to follow rapidly varying operating conditions while in the hybrid configuration this task is mainly performed by the electric batteries and their control system.

The numerical simulation of a complete system is then a powerful tool in system integration: it allows to screen among a large number of solutions and to verify the system behaviour and the fulfilment of the project target also limiting the number of experimental tests. Scaling up of laboratory solutions to actual system size and/or downsizing of large stationary plant tested solutions in order to fit automotive requirements are other useful applications of the modelling approach.

The physical and chemical processes involved in system performance have, nevertheless, different characteristic times and dimensions and the modelling activity requires a trade-off between detailed and comprehensive descriptions.

The main components required for a complete model can be identified by following the fuel (or energy vector fluid) from the vehicle tank and describing the main processes that it undergoes.

Following this approach, the first element is represented by the hydrogen source. Hydrogen on-board solution is still not an option for a large diffusion of PEMFC technology both for fuel storage capacity and for distribution issues. [Ye Lwin et al.,[17]; Peppley et al., [18]; Rabou, [19]]. To drive the fuel cell solution to a wide diffusion for transport application an on-board transformation of a more available and traditional fuel is still required. For this reason in this work three different reforming techniques have been tested: a methanol steam reformer, an autothermal reformer and an ammonia cracker. Last solution, even not "traditional", has a great potential in terms of overall fuel efficiency and worldwide availability and is under a number of feasibility studies [20]. Besides the particular fuel utilized, the three reforming methodologies represent different available technologies and the modelling approach followed can be easily extended to other fuels.

The core of the system is the energy production device and is then represented by the fuel cell stack. The basic requirement is to reproduce the polarization curve of the cell so determining, for each operating condition, cell current and voltage as well as the various energy losses [Barbir et al, [21], Ho and Chul Yi, [22]]. Once the experimental characterization of the cell is known, such a task could be performed by using semi-empirical relationships, which give a mathematical description of the different loss sources (activation, ohmic, concentration). However, when large variation of operating conditions is expected or the effect of design and/or ambient parameters has to be analysed, such a simple description is not able to simulate the real behaviour of the fuel cell. A more detailed description of the complex electrochemical and physical processes which take act within the cell is, then, required. [Lee and Lalk, [14]]

Finally, the matching among the different components has to be evaluated with particular reference to transient conditions. In particular, the coupling between the fuel cell stack and electric motor is needed to connect the mechanical power requirement to the electrical power generated. Moreover, as already observed, the single component behaviour under dynamic operating conditions has to be simulated to describe the transient conditions of this energy system for automotive applications.

All the previous mentioned requirements have represented the basis of a complete simulation model, which is here after described. Model components will be described with respect to the requirements of 50 kW FC system supposed to equip a small size car characteristic of European cities. The PEMFC stack main characteristics are reported in Table 1

FUEL CELL STACK		
Net electric Power	50	kW
Voltage	250	V DC
Active area	464	cm^2
Anode H_2 stoichiometric ratio	1.2	
Cathode O_2 stoichiometric ratio	2.0	
Temperature	75	°C

Tab. 1 *50 kW stack characteristics*

ON BOARD H_2 PRODUCTION - Direct storage can be performed by means of high-pressure bottles, metal hydrides or carbon nanotubes. The use of these technologies is straightforward but their development is not completed yet. An on board reformer is then required.

Fuel reforming is a chemical-physical process by which hydrogen is produced by some containing hydrogen. Such chemical process is usually endothermic and requires some additional heat from an external or internal source. As an example, Fig.3 represents the scheme of an autothermal reformer.

Modelling such systems requires a chemical evaluation of the produced hydrogen and a thermodynamic description of the heat required to perform this task. In this section a brief description for each reformer is presented.

Stationary performances of reformers have been simulated by using an equilibrium approach based on the minimization of the Gibbs free energy. Even through equilibrium hypothesis could represent only an ideal target, the results obtained can be helpfully used to proper design the integration system. Moreover, available examples in literature show as the conversion efficiency reaches very high values under various operating conditions. For these reasons the equilibrium assumption could be a useful preliminary investigation tool (Tschauder et al, [23]).

Methanol steam reformer – Methanol is an ideal fuel candidate due to its availability, high energy density and easy storage and transportation and the steam reforming is the most widely technique used. Nevertheless, for automotive applications, this strategy is penalized by its slow start-up.

The reforming process is endothermic and its operating temperature is determined by catalyst formulation and overall system dimensions. Depending on inlet gas composition and system temperature CO appears as one of the final product in a concentration which is not admissible with the 10 ppm limit of the PEM FC. A double stage CO cleaning device has, then, to be included within the system constituted by a shift reactor followed by a preferential oxidation system.

As already mentioned, from a chemical point of view the reformer is modelled as an equilibrium reactor in which the following reaction set takes act [24]:

$$CH_3OH + H_2O \leftrightarrow CO_2 + 3H_2$$
$$CO_2 + H_2 \leftrightarrow CO + H_2O$$

Synthesis gas produced from reforming reactions contains an appreciable amount of CO. Therefore, it is further processed by a CO cleaning-up phase This process is realized in two steps. The first step exploits the properties of the water gas shift reaction.

$$CO + H_2O \leftrightarrow CO_2 + H_2$$

By cooling the reformate gases WGS equilibrium is shifted toward the production of carbon monoxide and a part of CO is converted to CO_2 and a new gas composition is determined. Being the so obtained CO concentration still too high for a PEMFC, a selective oxidation reactor (PROX), which preferentially adsorbs carbon monoxide, rather than hydrogen, has to be inserted in the system.

The heat required or released by the different reformer units is exchanged with the system components or obtained by burning the PEMFC tailpipe gases. Particularly critical is the start up phase when the operating reformer temperature has to be reached shortly. For this reason, this phase is carried out by combustion of methanol until all the components have reached their operation temperatures and full power.

Methane autothermal reformer - Hydrogen can be produced by an autothermal reforming process. Its main characteristic is the very low energy requirements if specifically designed. The reforming process of methane consists of the following chemical reactions [25]:

$CH_4 + H_2O \leftrightarrow CO + 3H_2$	$(\Delta H_{298}° = 206.2 \text{ kJ/mol})$
$CH_4 + 0.5O_2 \leftrightarrow CO + 2H_2$	$(\Delta H_{298}° = -35.6 \text{ kJ/mol})$
$CO + H_2O \leftrightarrow CO_2 + H_2$	$(\Delta H_{298}° = -41.2 \text{ kJ/mol})$

The autothermal reformer integrates a substoichiometric oxidation of methane with a steam reformer. The basic idea is to supply the heat required by the first endothermic steam reforming by a partial oxidation of the fuel (second reaction). For defined oxygen/carbon (O2/C) and steam/carbon (S/C) ratios the reforming process is self sustained and no further heat is required. Other choice for these parameters could, however, lead to higher fuel conversion efficiency and the final choice is, then, a result of trade off analysis. As for the methane steam reformer, CO appears as one of the final products in a non-admissible concentration. Consequently, a WGS and a selective oxidizer have to be used in order to reduce CO concentration to admissible value.

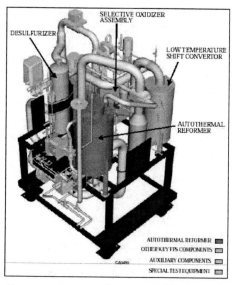

Fig.3 *Auto thermal reformer system assembly by "International Fuel Cell – Series 200 IFPS"*

Ammonia reformer - Ammonia can be conveniently stored in liquid form, has a high energy density (3000 Wh/kg), and the safety issues concerning its storage and handling, even not negligible, are well established. More importantly, it has a high hydrogen density and the product stream (hydrogen/nitrogen) is carbon monoxide free.

The chemical reaction used to reproduce the ammonia cracking is:

$$NH_3 \leftrightarrow \tfrac{1}{2}N_2 + 3/2H_2$$

Which requires 15333 KJ of heat per Kg of H_2 produced.

Reformer stationary behaviour analysis - The different hydrogen production strategies have been compared in terms of temperature, pressure and initial concentrations definition. The model results are summarized in Figs. 4, 5, 6, 7. For the methanol steam reformer, Figures 4 shows the gas composition as a function of temperature for fixed steam/methanol ratio (s/M=1,5), while Fig. 5 the gas composition in the preferred operating temperature (400-500 K) as a function of the steam/methanol ratio at the reformer inlet is presented. In Fig. 6 the gas composition profiles at fixed steam/methane and oxygen/methane (s/C=1.5; O/C=0.5) ratios for the auto thermal reformer are presented. Finally, in Fig. 7 the equilibrium concentrations against temperature are reported for the ammonia cracker. As underlined, a water gas shift is required at the exit of the methanol steam reformer and the auto thermal reformer.

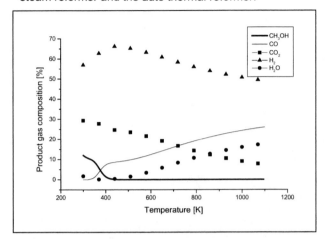

Fig.4 *Equilibrium concentrations as a function of temperature – Methanol steam reformer*

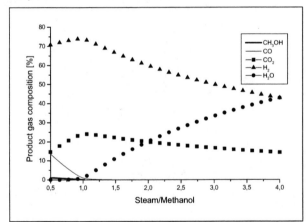

Fig.5 *Equilibrium concentrations as a function of the Steam/Methanol ratio – Methanol Steam reformer*

Kinetics effects (Ammonia Reformer) - The extraction reaction of Hydrogen from ammonia ("cracking") requires high temperature and catalyst. The presence of catalyst is absolutely necessary in order to speed up the cracking reaction. In fact, the membrane of PEM FC is made of perfluorosulphonic acid, whereas ammonia is an alkali. Therefore ammonia must be completely cracked into hydrogen and nitrogen, which requires high temperatures (about 900°C) [26].

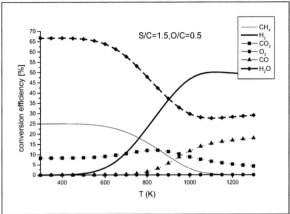

Fig.6 *Equilibrium concentrations as a function of temperature – Auto thermal reformer*

In the present paper, the developed model keeps into account the cracking reaction velocity by assuming a fixed catalyst mass. The catalyst composition is nickel on aluminium oxide catalyst, improved by addition of noble metals like ruthenium.

The reaction velocity has been evaluated as follows:

$$ r = C_1 e^{\frac{-E}{RT}} \left(K p_{NH_3} - p_{H_2}^{1.5} p_{N_2}^{0.5} \right) $$

where the constants could be evaluated by experimental data available in literature [27,28,29] once defined the catalyst mass.

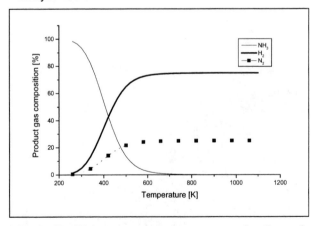

Fig.7 *Equilibrium concentrations as a function of Temperature – Ammonia reformer*

In Fig.8 the ammonia, hydrogen and nitrogen concentrations, numerically evaluated by the model, are plotted versus time for a temperature of 950 K and for a fixed reformer geometry and catalyst mass. It is possible to determine in such a way the minimum time to reach the "optimum" condition which is characterized by the absence of ammonia at the exit of the reformer.

535

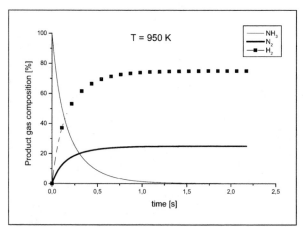

Fig.8 *Ammonia cracking: concentration profiles vs. time*

FUEL-CELL STACK - The PEMFC simulation model used in present simulations is based on the work of Springer et al. [5, 6] and it makes possible the simulation of the current-voltage relationship for steady state conditions with reference to the Membrane Electrode Assembly (MEA) representation reported in Fig. 9. The anode is fuelled with a gas mixture constituted by hydrogen, water vapour and a mix of inert gases (CO_2, N_2, etc.) whose composition will depend on the fuelling options of the whole system (hydrogen stored in a tank or coming out from a reformer). Anode outlet composition depends on fuel utilisation and water back diffusion in the system.

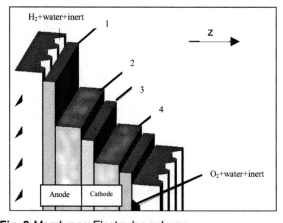

Fig. 9 *Membrane-Electrodes scheme*

The model describes separately the MEA processes (z-direction) and the gas dynamic flow within the anode and cathode side channels.

Accordingly, in the z-direction the mass and energy conservation equations are solved and the diffusion through the electrodes is simulated as follows:

$$\frac{dx_i}{dz} = RT \sum_j \frac{x_i N_j - x_j N_i}{PD_{ij}}$$

$$PD_{ij} = a\left(\frac{T}{\sqrt{T_{ci}T_{cj}}}\right)^b \left((p_{ci}p_{cj})^{\frac{1}{3}}(T_{ci}T_{cj})^{\frac{5}{12}}\left(\frac{1}{M_i}+\frac{1}{Mj}\right)^{\frac{1}{2}}\left(\varepsilon^{\frac{3}{2}}\right)\right)$$

where a = 0.0002745, b = 1.832 per H2 , O2 , N2 e a = 0.000364, b = 2.334 for the steam.

The term $\varepsilon^{3/2}$ is the Bruggemann correction for the diffusion coefficient, which takes into account the electrode porosity.

Using these equations coupled with the knowledge of the molar fluxes N_i it is possible to calculate:

at the anode side

$$\frac{dx_{wA}}{dz} = \frac{RTI}{P_A}\left[\frac{x_{wA}-x_H\alpha}{D_{wH}}+\frac{(1-x_{wA}-x_H)\alpha}{D_{wN}}\right]$$

$$\frac{dx_H}{dz} = \frac{RTI}{P_A}\left[\frac{-(x_{wA}-x_H\alpha)}{D_{Hw}}+\frac{-(1-x_{wA}-x_H)}{D_{HN}}\right]$$

at the cathode side

$$\frac{dx_O}{dz} = \frac{RTI}{P_C}\left[\frac{x_O(1+\alpha)+0.5x_{wC}}{D_{Ow}}+\frac{1-x_{wC}-x_O}{D_{ON}}\right]$$

$$\frac{dx_{wc}}{dz} = -\frac{RTI}{P_C}\left[\frac{(1-x_{wC}-x_O)(1+\alpha)}{D_{wN}}+\frac{0.5x_{wC}+x_O(1+\alpha)}{D_{wO}}\right]$$

This differential system is solved by means of a 4th order Runge-Kutta integration method.

Water molecular flux through the membrane is caused by concentration gradient and by H$^+$ ions dragging effect.

$$N_{WA} = \alpha \cdot I = 2\eta_{dragg}\frac{\lambda}{22}I - \frac{\rho_{dry}}{M_m}D_\lambda\frac{d\lambda}{dz}$$

Using the Zawodzinski [6] correlation is possible to determine the water activity and, therefore the water concentration x_w.

The solution of mass and energy conservation equation gives then the required boundary conditions for the MEA problem. A more detailed description of this model could be found in [15]. The fuel cell stack behaviour characterization is then summarized by the polarization curve and the power profile vs. current by the reference PEMFC stack (Figs. 10, 11). In addition, the electric response of the stack has been calculated in different conditions of external temperature and air relative humidity: as can be observed in Fig. 12 the power generation is strongly dependent on these two variables.

The described model has been validated, with reference to the PEMFC stack data reported in Table 1, against the available literature data [30], demonstrating good agreement.

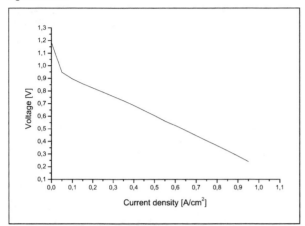

Fig. 10 *Cell Voltage vs. Current*

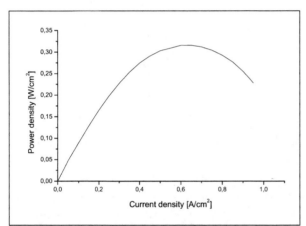

Fig. 11 *Power density vs. Current*

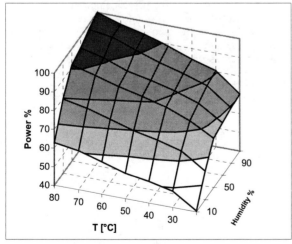

Fig. 12 *Power profile as function of temperature and relative humidity*

AIR SYSTEM - The thermodynamic characteristics of inlet air play an important role in determining both PEM FC and overall system efficiency.

In fact, by increasing working pressure it is possible to have a gain in the voltage value because of the reduction of the cathode activation over voltage (the positive effect on the polarization curve is proportional to the natural logarithm of the pressure increase ratio). Moreover, increasing the working pressure the amount of water required for membrane humidification decreases. Both these effects are particularly interesting in such applications where higher power/volume ratio is required together with self-sustained water operation. Nevertheless, this benefit could be completely counterbalanced by the increase in the power required for the compression.

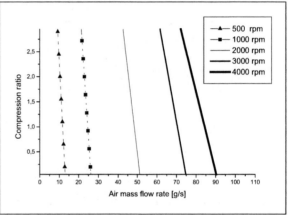

Fig.13 *Compressor map*

In the PEMFC system used for present work a volumetric compressor has been modelled in order to take in proper account the energy required for the compression which is calculated by means of politropic work divided for the politropic efficiency. The performances of different compressors types (e.g., turbo compressors, roots compressor, etc.) can be then analysed by assuming the proper characteristic efficiency and the compression ratio/volumetric flow rates characteristics.

During transient operations, the compressor rotational speed is determined by the instantaneous balance among motor driving torque, the compressor torque and the inertial momentum related to the angular acceleration $d\omega/dt$ as follows:

$$T_M - T_C = I_p \frac{d\omega}{dt}$$

(I_p has been assumed 0.02 kgm^2 [31])

Starting from the knowledge of the angular acceleration which guarantees the needed air mass flow rate, it is, then, possible to evaluate the motor driving torque with respect to the compressor characteristic torque evaluated as follows:

$$P_{comp} = \dot{m}_{air} \cdot l_r = T_C \cdot \omega$$

$$l_r = c_p \cdot T_{amb} \cdot \left(\beta^{\frac{k-1}{k \cdot \eta pol}} - 1 \right)$$

In Fig. 14 the compressor torque evaluated by the compressor power and the compressor characteristic map against time during the ECE+EDC cycle is presented.

Within the air system it is also present a humidifier and at the cell outlet a water separator to recover a part of the water produced.

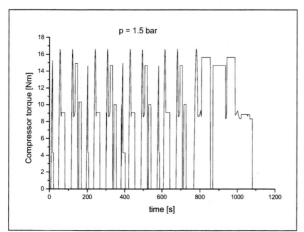

Fig. 14 *Compressor torque vs. time during ECE+EUDC cycle*

THERMAL MANAGEMENT - With reference to the single cell the simulation of transient response requires a model able to evaluate the thermal exchanges between fluid and plates and then to define the stack temperature profile with respect to time (Fig.15).

Under the hypothesis of uniform temperature distribution within the stack, the unsteady energy balance can be written as:

$$M_{plate} \cdot c_{plate} \cdot \frac{dT_{plate}}{dt} = \sum_i h_i S_i \Delta Tml_i + h_{air} S_{air} \left(T_{air} - T_{plate} \right) + \dot{Q}$$

Where M_{plate} is the mass of the fuel cell, c_{plate} is the specific heat of the bipolar plate and dT_{plate}/dt is the temperature change with respect to time and

$$\Delta Tml_i = \frac{\left(T_{iOUT} - T_{plate} \right) - \left(T_{iIN} - T_{plate} \right)}{\ln\left(\dfrac{T_{iOUT} - T_{plate}}{T_{iIN} - T_{plate}} \right)}$$

$$\dot{m}_i \cdot c_{p_i} \cdot \Delta T_i = h_i \cdot S_i \cdot \Delta Tml_i$$

$$\Delta T_i = T_{i\,OUT} - T_{i\,IN}$$

Thermal capacity of bipolar plate can be evaluated as 0.35 kJ/K [32].

Among the terms on the right hand side of the energy balance, which represent the various energy fluxes, the term Q is a source term related to the chemical energy not converted in electric work. This transient model, coupled with the electrochemical model, allows evaluating the stack temperature profile whit respect to time as follows:

$$\frac{dT_{plate}}{dt} = \frac{\sum_i h_i S_i \Delta Tml_i + h_{air} S_{air} \left(T_{air} - T_{plate} \right) + \dot{Q}}{M_{plate} c_{plate}}$$

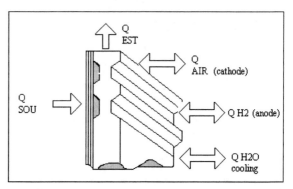

Fig.15 *Schematic representation of thermal bipolar plate fluxes*

ELECTRIC MOTOR - In the generic configuration of Fig.2 two different electric motors have been adopted: a main one for traction and a smaller for compression. Both were simulated according to the motor characteristic map [31] (Fig.16). Electric motors in direct current for transportation are "series excited" because in this way the characteristic curves allow to have the maximum torque during start-up, and consequently maximum acceleration. The engine is interfaced on one side with the vehicle dynamics model and, on the other side, with the fuel cell system by means of a control system, which is briefly described in Appendix 2.

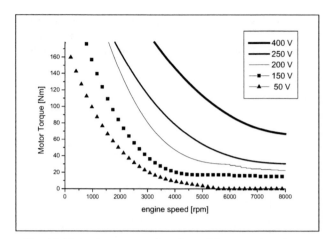

Fig. 16 *Motor torque as a function of the speed and voltage*

VEHICLE DYNAMICS - A vehicle dynamics module is required to transform, for each analysed vehicle, the road load characteristics into mechanic torque required to the power train. The vehicle characterisation used through the paper is reported in Appendix 1.

VEHICLE MASS	M	1570	kg
Axle base	L	2,54	m
Front axle-centre of gravity-length	a	1,235	m
Centre of gravity-rear axle length	b	1,305	m
Centre of gravity height	h_G	0,5	m
Wheel radius	R	0,273	m
Air density	ρ	1,2	kg/m^3
Front area	S	2	m^2
Aerodynamic drag coefficient	C_X	0,33	
Rolling friction	f_0	0,013	
Rolling friction-quadratic term	K	$6,5*10^{-6}$	s^2/m^2
Road grade	α	0	°
Total gear ratio	τ	5,5	
Traction type		Front	

Tab. 2 *Vehicle parameters*

SYSTEM ANALYSIS

The system analysis has been performed with respect to a PEMFC equipped mid-size vehicle typical in European car fleets. The main characteristics of the vehicle are reported in Table 2 and in Table 1 the PEMFC general system characteristics are reported for this application (see also Fig. 22 in the Appendix 1).

As an example of model use, simulation analysis was firstly performed to evaluate the behaviour of the fuel cell system under critical working conditions.

Start-up and shutdown conditions have been investigated to define the stack fundamental parameters, as the characteristic times, and to develop control strategies. Evaluating their influence on the overall system efficiency is, of course, an other major issue.

After that, the analysis of complete vehicle behaviour under transient conditions has been performed. The rapid varying operating conditions of ECE+EUDC have been utilised allowing a meaningful analysis of the FC system performances under real driving conditions.

START-UP AND SHUT-DOWN – In Fig.17 the heating period of the fuel cell stack after a sudden start-up phase during which the system power increases from 0 to 10 kW is presented. In this picture it is possible to notice the stack temperature profile. The equilibrium temperature value is achieved after almost 20 minutes [33]. In the same picture the darker curve represents how temperature profile varies by inserting in the cooling system a thermostatic valve. The time requested to reach the set value is sensibly lower and it is about 6/8 minutes in agreement with [34]. Large room for system integration/optimisation is then expected for this subsystem which is of course required to reach the DOE targets of 1 minute start-up time. The same analysis has been carried out for a shutdown operation and the stack current and temperature profiles for this example are reported in Fig.18. The characteristic data of the cooling network are reported in Table 3

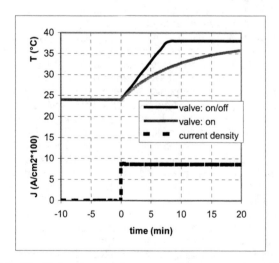

Fig.17 *Plate temperature vs. time assuming the reference temperature as $T_{set} = 38°C$ and $J=0.086$ A/cm^2.*

$T_{i\,IN}$	24 °C
T_{0p}	24 °C
T_{0H2O}	24 °C
Current	0.086 A/cm^2
Pressure	1 bar
L_p	0.126 m

Tab. 3 *Cooling network characteristic data*

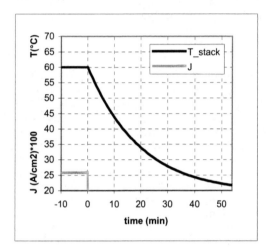

Fig.18 *Transient plot for the shutdown to 0 A/cm^2 $T_0=60°C$.*

ECE+EUDC TEST - The evaluation of the cell behaviour under the dynamic condition defined by the ECE+EUDC test is a preliminary and basilar operation in order to define the system characteristic in terms of overall efficiency and specific fuel consumption. In particular, three different reformers have been tested and for each of them the fuel consumption over the ECE+EUDC test has been evaluated. Starting from the velocity profile of Fig.1, the vehicle dynamics allows to define the torque profile with respect to time as described in appendix and shown in Fig.19. Once known this value, together with the angular speed, the voltage and current profile are determined and reported in Fig.20. In Fig.21 the specific fuel consumption profile is reported too. This value has been obtained by integrating the engine power on the complete driving cycle. As already highlighted, in the present study the analysed configuration is characterized by a direct matching between fuel cell stack and electric motor. Accordingly, road load variations influence directly the electric power requirement so leading the fuel cell working in off-design conditions for large periods.

Researches nowadays are running on the possibility to couple the FC stack with super capacitor in order to recover energy during the vehicle braking phases and provide energy during vehicle acceleration. In this way

the fuel cell operating range would be very close to its normal operating conditions and, as a consequence, in a power range where the efficiency is quite constant at the maximum value.

The total hydrogen mass requested to complete the cycle was 104 g when the cell is operated with pure hydrogen and the overstoichiometric fuel cell could be used to support the endothermic reformer, if needed.

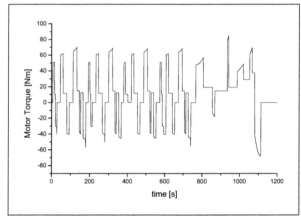

Fig. 19 *Motor torque vs. time Hp: r=0.182.*

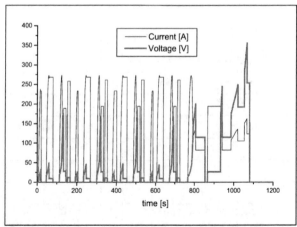

Fig.20 *Voltage, current and angular speed of the electric motor vs. time.*

In Table 4, a comparison among the different studied solutions for the reformer is presented. In this Table the mass flow rate requested for the different fuels (normalised with reference to 1 Watt of electrical power) and the heat consumption measured as W_{term} / W_{elect} is presented.

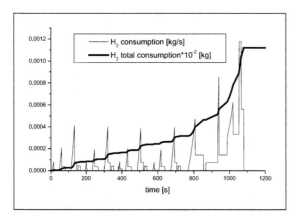

Fig.21 *Hydrogen consumption during the ECE+EUDC test*

	H_l (kJ/kg)	kg/W s	W_{TERM}/W_{ELECT}
Methanol CH_3OH	$2.10 \cdot 10^4$	$1.083 \cdot 10^{-7}$	2.27
Methane CH_4	$4.80 \cdot 10^4$	$5.627 \cdot 10^{-8}$	2.70
Ammonia NH_3	$1.67 \cdot 10^4$	$1.151 \cdot 10^{-7}$	1.92

Tab.4 *Analysis of performances of different reformers*

CONCLUSIONS

The study of the performances of PEMFC based automotive powertrain fuelled with different fuels has been performed by means of a comprehensive simulation model.

In the paper the model has been firstly illustrated in all its components. In particular reformer simulation has been sketched for different reformer techniques (methanol steam, auto thermal and cracking). Both equilibrium and transient reformate composition can be evaluated as a function of reformer temperature, overall dimensions, species concentrations ratios.

System performances have been then tested with reference to different transient conditions. Start-up and shutdown processes have been examined and cooling system performances have been analysed in detail. ECE+EUDC analysis, finally, allowed evaluating with a good detail the energy requirement during on road operating conditions.

On the basis of previous results the model has proved to be able to describe the FC system behaviour and identify system design parameters. It can be, then, a useful design tool in determining an overall system configuration and in driving the development of its control system.

REFERENCES

[1] P.Costamagna, S.Srinivasan, " Quantum Jumps in the PEMFC science and technology from the 1960s to the year 2000 Part I. Fundamental Scientific Aspects" Journal of Power Sources, Volume: 102, Issue: 1-2, December 1, 2001, pp. 242-252

[2] P.Costamagna, S.Srinivasan, " Qanatum Jumps in the PEMFC science and technology from the 1960s to the year 2000 Part II. Engineering, technology development and application aspects" Journal of Power Sources, Volume: 102, Issue: 1-2, December 1, 2001, pp. 253-269

[3] D.M. Bernardi, M.W. Verbrugge, " AIChE J., 37, 1151 (1991)

[4] D.M. Bernardi, M.W. Verbrugge, "Journal of the Electrochemical Society,139, 2447 (1992)

[5] Springer T.E., Wilson M.S.., Gottesfeld S., "Modeling and Experimental Diagnostics in a Polymer Electrolyte Fuel cell" Journal of Electrochemical Society, Vol. 140, n.12, 1993

[6] Springer T.E., Zawodzinski T.A., Gottesfeld S., "Polymer Electrolyte Fuel Cell Model" Journal of Electrochemical Society, Vol. 138, n.8, 1991

[7] T.F.Fuller, J.Newman, Journal of Electrochemical Society, 140, 1218 (1993)

[8] V.Garau, H. Liu, S.Kacac, AIChE J., 44, 2410 (1998)

[9] J.S.Yi, T.V.Nguyen, Journal of Electrochemical Society, 146, 38 (1999)

[10] T.Zhao, H.Liu, "Development and Simplification of a Three-Dimensional PEM Fuel Cell Model", Proceedings of ASME ESDA2002, 6th Biennal Conference on Engineering Systems design and Analysis (2002), n. AES-037

[11] K. Jost, "Advances in fuel cell development", SAE Automotive Engineering, June 2002 p: 67-73.

[12] Johnson R., Morgan C., Witmer D., Johnson T.,"Performance of a proton exchange membrane", International Journal of Hydrogen Energy 26 (2001) 879-887

[13] Pischinger S., Schonfelder, Bornscheuer W., Kindl H.,Wiartalla A.,"Integrated Air Supply and humidificationConcepts for Fuel Cell Systems" SAE paper 2001-01-0233

[14] Lee J.H. and Lalk T.R., "Modeling fuel cell stack systems", Journal of Power Source, 73 pp. 229-241, 1998

[15] L.Andreassi, S.Cordiner, M.Feola, "Analysis of Overall Efficiency of Fuel-Cell Based Power Generations Systems", Proceedings of ASME ESDA2002, 6th Biennal Conference on Engineering Systems design and Analysis (2002), n. AES-004

[16] J.W. Stevenson, P.Singh, "Solid Oxide Fuel Cell Development at Pacific Northwest National Laboratory", Fuel Cell Seminar 2002

[17] Lwin Ye, Wan Ramlii Wan Daud, Abu Bakar Mohamad, Zahira Yaakob,"Hydrogen production from steam-methanol reforming: thermodynamic analysis", International Journal of Hydrogen Energy 25 (2000) 47-53

[18] Peppley Brant A., Amphlett J., Kearns Lyn, Mann R.,"Methanol steam reforming on Cu/ZnO/Al2O3. Part1: the reaction network" Applied Catalysis A: General 179 (1999) 21-29

[19] Rabou L.P.L.M., "Modelling of a variable flow methanol reformer for a polymer electrolyte fuel cell"International Journal of Hydrogen Energy, vol. 20 n. 10 pp 845-848, 1995

[20] EU RTT Project n. ENK6CT2001/00580 ACCEPT – Ammonia Cracking for Clean Energy Power Technology

[21] Barbir F., Fuchs M., Husar A., Neutzler, "Design and Operational Characteristics of Automotive PEM Fuel Cell Stacks" SAE Paper n. 2000-01-0011, 2000

[22] J. Ho, Chul Yi , "A computational simulation of an alkaline fuel cell",Journal of Power Source, 84 pp.87-106, 1999

[23] A.Tschauder, B.Emonts, T.Grube, R. Peters, D.Stolten, "Compact Methanol Reformer Operated in a Fuel Cell Drive System", 2002 Fuel cell Seminar

[24] B. Höhlein, M. Boe, J. Bøgild-Hansen, P. Bröckerhoff, G. Colsman, B. Emonts, R. Menzer, E.Riedel "Hydrogen from methanol for fuel cells in mobile systems: development of a compact reformer", Journal of Power Sources 61 (1996) 143-147

[25] C.Palm, P.Cremer, R.Peters, D.Stolten, "Small-scale testing of a precious metal catalyst in the autothermal reforming of various hydrocarbons feeds", Journal of Power Source 106 (2002) 231-237

[26] U. Bachhiesl, P. Enzinger , Energieinnovation in Europe, Graz 31.01.2002 Drei-Platten-Ammoniak-Cracker zur Wasserstoffherstellung Drei-Platten-Ammoniak-Cracker zur Wasserstoffherstellung

[27] C.A.Vancini, "La sintesi dell'ammoniaca", Ed. Hoepli, 1961, in latin

[28] S.E.Gay-Desharnais, J. Routex, M. Holtzapple, M. Ehsani, " Investigation of Hydrogen Carriers for Fuel Cell Based Transportation", SAE paper n. 2002-01-0097

[29] A.S.Chelappa, C.M.Fischer, W.J.Thomson, "Ammonia decomposition kinetics over Ni-Pt/Al2O3 for PEM fuel cell applications", Applied Catalysis – Elsevier, 2001

[30] www.cfdrc.com/datab/Applications/FuelCells/FC_validate.htm

[31] M.Badami, C.Caldera, "Dynamic model of a Load-Following Cell-Veichle: Impact on the Air System", SAE paper n. 2002-01-0100, 2002

[32] J.C.Amphlett, R.F.Mann, B.A.Peppley, P.R.Roberge, A.Rodrigues, "A model predicting transient responses of proton exchange membrane fuel cells", Journal of Power Sources 61 (1996) 183-188

[33] M. De Francesco, E. Arato, "Start-up analysis for automotive PEM fuel cell systems", Journal of Power Sources 108 (2002) 41-52

[34] K.Weisbrod, J.Hedstrom, J. Tafoya, R.Borup and M.Inbody, "Cold-Start Dynamics of a PEM Fuel Cell Stack", Los Alamos National Laboratory, Los Alamos NM 87545

SIMBOLS

λ: Water content or local ratio H_2O/SO_3^- H^+ in the membrane

α: Ratio of net H_2O flux in membrane to H_2O flux product at cathode

η_{dragg} : Electro-osmotic drag coefficient

v_H: Anode H_2 stoichiometric ratio

v_O: Cathode air excess ratio

η_{pol} : compressor politropic efficiency

c_{pi}: Specific heat i-fluid [kJ/kg K]

$D\lambda$: Diffusion coefficients as a function of λ

h_i: Heat exchange coefficient (i-fluid) [W/m^{2}K]

I: Water molar flux produced at cathode J/2F [mol/(cm^2 s)]

I_p : Compressor shaft inertia momentum [kgm^2]

J: Current density [A/cm^2]

L_p: Bipolar plate side length (the plate is supposed to be squared) [m]

M_{plate}: Bipolar plate mass [kg]

N^i_{HA}: H_2 molar flux at anode inlet [mol/(cm^2 s)]

N^i_{NA}: Inert gas molar flux at anode inlet [mol/(cm^2 s)]

N^i_{NC}: Inert gas molar flux at the cathode inlet [mol/(cm^2 s)]

N^i_{OC}: Oxygen molar flux at cathode [mol/(cm^2 s)]

N^i_{WA}: Water molar flux at anode inlet [mol/(cm^2 s)]

N^i_{WC}: Water molar flux at cathode [mol/(cm^2 s)]

N^L_{NC}: Inert gas molar flux at cathode outlet [mol/(cm^2 s)]

N^L_{OC}: O_2 molar flux at cathode outlet [mol/(cm^2 s)]

N^L_{WC}: water molar flux at cathode outlet [mol/(cm^2 s)]

N_{WA} : Water molar flux through the membrane [mol/(cm^2 s)]

N_{WA}: Molar flux through the membrane [mol/(cm^2 s)]

P_{comp} : Compressor Power [W]

S_{air}: Thermal exchange surface between the plate and air [m^2]

S_i: Thermal exchange surface between the plate and l-fluid [m^2]

T_C : Compressor Torque [Nm]

T_{iIN}: Inlet temperature of the i-fluid [K]

T_{iOUT}: Outlet temperature of the i-fluid [K]

T_M : Engine torque (Compressor driving) [Nm]

T_{plate}: Bipolar plate temperature [K]

x_{HN}: Inert gas molar fraction at anode outlet

x'_{WA}: Water molar fraction (anode inlet)

x'_{WC}: Water molar fraction (cathode inlet)

x_{O4}: O_2 molar fraction (cathode outlet)

x_{ON}: Oxygen molar fraction

x_{W1}: Water molar fraction (anode outlet)

x_{W4}: Water molar fraction (cathode outlet)

APPENDIX 1

The vehicle motion law has been defined considering the following assumptions (see Fig.23).

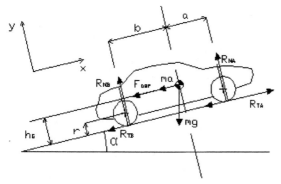

Fig. 23 *Vehicle Scheme*

- The vehicle travels along a rectilinear road and there is not any wind effect

- The tyre masses are negligible

- The tyres don't slide

The forces acting on the system are:

- Gravitation force applied to the centre of gravity of the vehicle $F_G=mg$

- Inertial forces which reduce, considering the simplifying hypothesis, to $F_i= m\dot{v}$ applied to the centre of gravity of the vehicle

- Aerodynamic drag force proportional to the squared value of the velocity as $F_{aer} = 0.5 \rho V^2 SC_X$ assumed to be applied to the centre of gravity of the vehicle.

- The interaction forces between tyres and road R_{tA}, R_{nA}, R_{tB}, R_{nB}. The rolling resistance is kept into account by moving forward of a u distance the normal components of the reaction forces. The rolling friction f_v (=u/r, where u is the distance between the application point of the interaction forces and the tyre centre, evaluated along the x direction, so keeping into account the tyre deformation) is schematised as sum of a constant term f_0 and a quadratic term kv^2.

- The brake action $F_b = C_b/r$

(See Table 2 for the vehicle parameters [30])

The global equilibrium reactions are:

x): R_{tA} - R_{tB} - F_{aer} - $m\dot{v}$ =0

y) : $mgcos\alpha$ - R_{nA} - R_{nB} =0

M):$-R_{nB}*(a+b)+F_{aer}*h_G+mgsen\alpha*h_G+mgcos\alpha*(a+u)+m\dot{v}*h_G=0$

Other two equations derive from the tyres momentum equilibrium.

Driving tyre): $R_{tA}*r + R_{nA}*u-M_{mu}=0$

Drag tyre): $R_{tB}*r - R_{nB}*u =0$

Combining these equations it is possible to obtain the traction force X_m as:

$X_m= (f_0+kV^2)mgcos\alpha+0.5\rho v^2 SC_X+mgsen\alpha+m\dot{v}$

APPENDIX 2

CONTROL SYSTEM

The PWM driving – In present work simulation, the PWM control has been used. This control technique allows to change the velocity without any major penalties in terms of energetic efficiency. The power circuit is driven by a square wave produced by a driver managed, in turn, by a micro controller (Fig.24). If the switching is frequent enough (some kHz), the average current is substantially constant because of the presence of coil and proportional to the duty cycle of the power circuit inlet signal (Fig. 25). It is important to underline how not all the electric motors could be driver in such a described way, because the loss within the magnetic circuit, proportional to the frequency could become unacceptable.

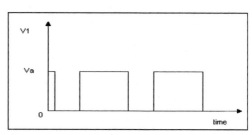

Fig. 24 *Voltage Square wave produced by a driver managed by a micro controller*

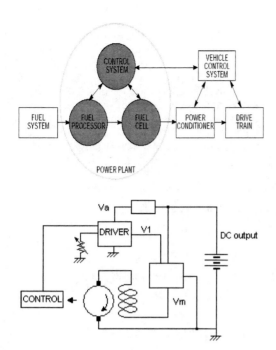

Fig. 25 *Electric motor scheme and control system*

An Integrated Proton Exchange Membrane Fuel Cell Vehicle Model

Syed Wahiduzzaman, Babajide Kolade and Selim Buyuktur
Gamma Technologies Inc.

ABSTRACT

The potential of fuel cells as an automotive power source is well recognized due to their high efficiency and zero tailpipe emissions. However, significant technical and economic hurdles need to be overcome in order to make this technology commercially viable. A proton-exchange membrane (PEM) fuel cell model has been developed to assess some of these technical issues. The fuel cell model can be operated in a standalone mode or it can be integrated with vehicle and fuel supply system models. A detailed thermal model of the fuel cell stack was used to identify significant design parameters that affect the performance of PEM fuel cell vehicles. The integrated vehicle model was used to explore the relative benefits of hybridization options.

INTRODUCTION

Growing environmental concerns, increasing demand and decreasing supply of global crude oil reserves, combined with political factors have increased interest in the search for alternative energy sources. Over the last years vehicle manufacturers, while pushing the limits of the existing technology and introducing new improvements, have started investing in the research and development of new technologies.

Although fuel cells existed for over a century, they have become one of the seriously considered alternatives to today's internal combustion engines because of their reduced emissions and high efficiency characteristics. Although there exist major challenges on the road to making fuel cells viable, steady progress have been made over the last decades (e.g. Matsumoto et al. [1], Ki-Chun et al. [2]). It is now widely accepted that fuel cell power plants will be one of the leading contenders of a viable alternative to IC engines in the future.

Fuel cell powered vehicles offer a different set of challenges and options that need to be examined and assessed in an economic and expedient fashion. Today computer modeling and simulation tools are already serving the conventional transportation industry effectively, helping to reduce costs, save time and increase efficiency. These modeling methodologies now need to be extended to encompass emerging and alternative technologies. In this paper, a fully integrated fuel cell and fuel cell vehicle simulation tool has been presented. The PEM fuel cell model was developed within the scope of an engine system simulation software tool GT-SUITE developed over the last decade by Gamma Technologies. The key feature of this tool is that it is a single integrated code with identical handling of engine, powertrain, vehicle, hydraulics, electrical, thermal and control elements. This tool supports multiple levels of models, so that users can select the appropriate level of complexity and timescale for the problem. With its multi-disciplinary approach and flexibility, the tool is intended to serve the fuel cell and fuel cell vehicle industry by aiding the development of this new technology.

Traditionally, the simulations used in the industry or developed by research organizations (Baschuk and Li [3]) are specialized tools focused on a range of issues dealing with specific subsystems. In the presented integrated modeling tool, all relevant subsystems can be modeled within a single tool with flexible levels of complexity. In the context of a fuel cell powered vehicle, the system may comprise of fuel delivery, water management, hydraulics, cooling, thermal, electrical, mechanical, and control subsystems. An alternative could be the use of specialized tools for modeling subsystems and coupling it through another tool (co-simulation). However, this exacts a heavy price on the productivity and maintainability of the models. Compared to that, an integrated tool offers many advantages:

The subsystem models of an integrated tool are consistent with each other and are naturally coupled in terms of data exchange, while interaction between separate specialized tools may often be limited to a few parameters.

The use of non-integrated third-party tools offers special challenges for maintaining versions that are compatible to each other. Furthermore, the development cycles of these tools are, in general, asynchronous with each other.

In a co-simulation environment, there are steep learning curves – various groups may use different tools and speak different and incompatible technical languages.

In an integrated environment where disparate submodels can coexist (e.g., fluids, thermal, control, mechanical and electrical), various submodels can be developed by specialists in that field and can be shared by all through a common tool (Morel et al. [4], Kolade et al. [5]). Thus, the multi-disciplinary skills available within an enterprise can be harnessed and leveraged in an efficient manner resulting in enhanced productivity.

ELEMENTS OF THE INTEGRATED TOOL

As mentioned above, the integrated tool contains a wide range of elements that are needed to represent an entire engine-powertrain-vehicle system powered by a fuel cell stack. The multi-disciplinary libraries of elements (Morel et al. [4]) needed for the fuel cell powered vehicles are briefly discussed below.

FLUIDS AND THERMODYNAMICS LIBRARY

These elements are needed to model supply (gas storage tanks, supply lines, valves, water injectors, ejectors, compressors, heat exchangers etc.), exhaust and water management systems. In the case of internal combustion engine hybridization, these elements can also be used to model engine performance, EGR, combustion, emissions, and turbocharging, as well as intake and exhaust acoustics.

MECHANICAL LIBRARY

These elements are used to model the kinematics, dynamics, forces and torques of moving parts of the vehicle and powertrain. The elements may be prepackaged at higher (e.g. cranktrain, valvetrain, driveline and vehicle) or more elementary levels (e.g. mass spring, friction objects or damper). Elements of either type can be combined to create an assembly of the desired mechanical system.

HYDRAULICS LIBRARY

This library includes elements such as pumps, injectors, accumulators and various types of control and delivery valves. In general, due to the high operating pressures, these elements require accurate treatment of the density equation of state for compressible fluids, cavitation, and pipe wall compliance. The details of the models are discussed in reference (Kolade et al. [5]).

THERMAL LIBRARY

The thermal library consists of elements that can represent lumped thermal masses, thermal resistances, detailed finite element models and mixed thermal boundary conditions. As with the other libraries, the elements of this library may be coupled with mechanical and flow elements in order to model heat rejection and warm-up issues that are likely to be encountered in thermal management of fuel cell and other on-board systems. Typical applications of thermal elements under transient or steady state operation include:

- Fuel cell and battery warm-up

- Analysis of the entire cooling circuit

- Radiator sizing

- Pump sizing

- Pipe and orifice sizing

- Engine warm-up

- Thermostat specification

- Thermostat cycling

The thermal aspect of the fuel cell is modeled as collection of distributed masses linked to each other through temperature dependent thermal resistances. An ambient condition is imposed on the outer surfaces of the stack. The internal thermal load on the fuel cell is calculated by subtracting the electrical energy from the heat of reaction, taking into account the states of input and output streams. A further 10% parasitic loss is assumed for the auxiliary system components that are not currently implemented into the model.

CONTROL LIBRARY

There are extensive control elements available to design and model active and passive control systems. These elements can be used to model logical, control and math operations, which are typically performed by the control system's central processing unit (CPU). These elements can be used either from within the tool or in combination with Simulink. The control library elements interact with physical systems through sensors and actuators.

ELECTRICAL LIBRARY

These elements are used for modeling various electrical components of electric and hybrid-electric vehicle systems. These include various electro-magnetically driven components such as, solenoids, battery motors, generators and electrical loads. The newest entry into the electrical component library is the PEM fuel cell.

The fuel cell model described in this paper is an electrochemical model of a Proton Exchange Membrane Fuel Cell (PEMFC). The model is based on semi-empirical correlations for cell losses and first-principle approach for open circuit voltage calculation. The governing equations are largely adopted from those proposed by Larminie and Dicks [6]. The model calculates the maximum theoretical voltage (also known as the reversible open circuit voltage) that can be obtained at a given operating condition by calculating the Gibbs free energy of the fuel cell reaction at that condition. The relevant properties, such as enthalpy and entropy, of the working fluids are implemented in the model as a function of temperature and pressure. The actual cell voltage is then calculated by taking into account various irreversibilities that occur during operation. The losses that are being considered are:

- Activation voltage losses that occur as a result of the reaction kinetics taking place at the surface of the electrodes. This voltage loss is defined by the Tafel equation [6].

- Ohmic losses that are generated due to the electrical resistances of the bipolar plates, the electrodes and the resistance of the polymer electrolyte to hydrogen ion flow. For simplicity, the resistances are lumped in a single term that defines an overall resistance for the fuel cell.

- Mass transport losses (or concentration losses) that occur as a result of the change in the concentration of the reactants at the surface of the electrodes.

The details of the activation and the mass transport losses can be found in reference [6]. The actual operating cell voltage is calculated by taking the difference between the reversible open circuit voltage and the total voltage loss caused by the irreversibilities. The input to the fuel cell is a power request, which defines a current density and an operating voltage for the stack. If the power request can not be met the fuel cell supplies power at its maximum capacity. It is theoretically possible, for a given set of fuel cell characteristic parameters, to obtain two solutions of current densities for a specific power request. In such a situation, the lower current density is accepted as the correct solution.

As mentioned earlier, the fuel cell loss models used here are empirical in nature. However, these can be easily tuned with small number of model parameters to correlate with experimentally observed polarization curves. This approach provides a fast execution time even when multiple subsystems are integrated (e.g. vehicle, cooling and electrical systems etc.). The architecture of this tool provides easy means for the incorporation of external models with detailed transport and kinetics mechanisms (e.g. Baschuk and Li [3], Singh and Lu [7]) through user model interface facilities or direct integration. An example of a user model is described in Cantore et al. [8]. Additionally, this tool avoids redundant input and affords easy scalability due to its adoption of hierarchical description of system elements and model building through template libraries (described above), objects and parts akin to modern programming languages. For example, an object can be created once and reused many times within a model to create multiple parts that inherit properties of the object. Thus a wholesale change of the parts properties can be accomplished through corresponding changes in object properties only.

VALIDATION STUDIES

In the validation studies no attempts were made to match a wide range of fuel cell data since corresponding geometric, thermo-physical property and operating condition data were largely unavailable. Consequently, efforts were focused on obtaining correct sensitivities and qualitative behavior of the full cell under steady state and transient operating conditions. Attempts were also made to indirectly deduce functional dependence of model parameters (e.g. internal resistance, exchange current density, charge transfer coefficient etc.) from available polarization data (Barbir et al. [9]). The results are presented in the following section. Thermo-physical properties of the fuel cell materials were obtained from the study conducted by Musser and Wang [10].

STEADY STATE MODEL RESULTS

A steady-state simulation was run on a fuel cell model with an active cell area of 232cm^2 to show the effect of the fuel cell operational temperature on the fuel cell performance. A sweep over 4 different operating temperatures was conducted. Figure 1 shows the polarization curves and the power distribution of the fuel cell obtained at different temperatures. The temperature-dependent parameters are automatically updated as the operating condition changes during simulation. As seen from the figure the model is capable of capturing the trends and simulating the effect of temperature on fuel cell performance. The above results are also consistent with those presented in reference [11].

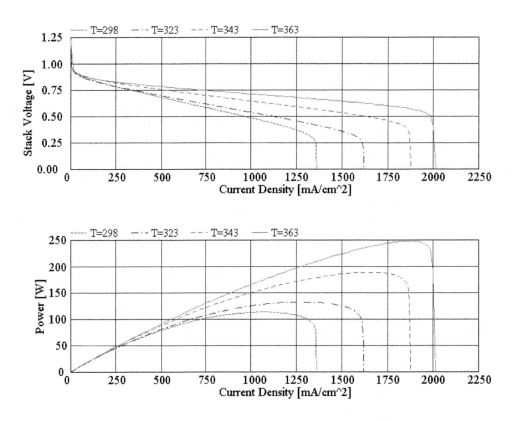

Figure 1 Fuel Cell Polarization Curves and Power Outputs at Different Operational Temperatures

Effects of other model parameters on input or output variables can be implemented into the model in order to capture dependencies that are observed and/or learned from more detailed models. For example, fuel cell characteristic parameters can also be a function of any output variable. This capability enables users to assign dependency of a model parameter (e.g. exchange current density) on one or two system outputs (e.g. stack temperature and current density). This tool allows users to implement parametrical relationships and dependencies in various forms, such as function of time and/or function of one or two variables.

As an example, the pressure effect on some model parameters were implemented using published data (Barbir et al. [9]). With the implementation of these dependencies the tool was able to generate identical polarization curves to the published ones without having to use time intensive and complex loss models. The following data were obtained at 333K from an experimental hardware with an active area of 292cm^2 and consisting of 110 cells. The comparison of predicted and experimental polarization curves is shown in Figure 2. The fuel cell stack used in the subsequent analysis is derived through scaling the previously described fuel cell (by increasing the number of cells but keeping the polarization properties the same).

WARM-UP AND STEP RESPONSE

Transient models are important in predicting behavior of fuel cells under variable load and operating conditions. System heat-up duration and response to sudden load changes are critical factors in the evaluation of fuel cells for transportation applications. In order to assess the transient response behavior, a standalone system model was created. In this model a temperature distribution within the fuel cell is computed by discretizing the stack into ten thermal elements. Coolant is supplied to each element through a fluid circuit represented by 4000 micro-channels. A more detailed thermal management system can be implemented with little additional effort (Morel and Wahiduzzaman [12]). In the current model an overall temperature is used to calculate the stack voltage and current density. A fuel cell stack consisting of 10 cells with an active area of 232cm^2 was simulated using the transient model. A heat up with 0.5kW load and a step load from 0.1kW to 0.5kW was simulated. The stack was divided into 10 regions representing each cell. The contact resistances between individual cells and thermal inertias are modeled by utilizing the thermal primitives that were discussed earlier. Figure 3 shows the average stack temperature profile when the fuel cell generates 0.5kW of power. A proportional-integral-derivative (PID) controller mechanism was incorporated to keep the

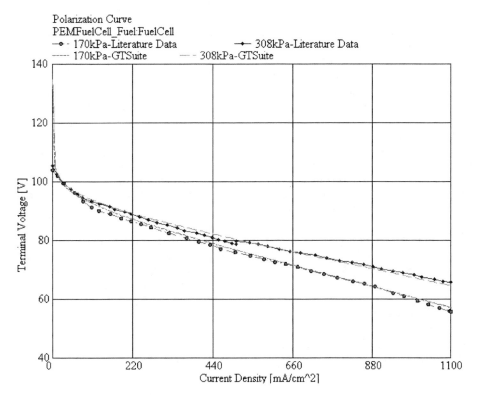

Figure 2 GT-SUITE results vs. Experimental Measurements

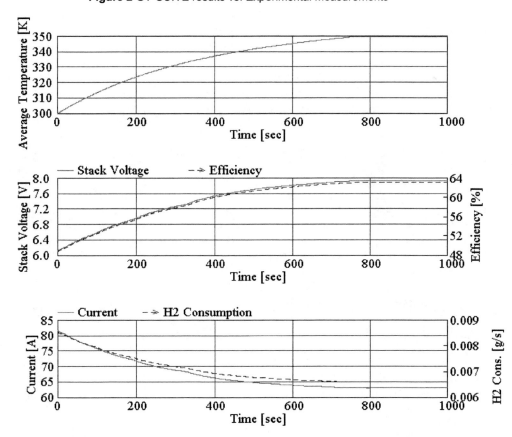

Figure 3 Step Load to 0.5kW

stack temperature around 350K. The inner most cell temperature is sensed and sent to the PID controller and accordingly the PID adjusts the coolant flow in order to keep the temperature at a desired level. In the presented model the heat transfer coefficients are actuated and change depending on the coolant flow rate.

The stack voltage trace, overall efficiency, current density, and reactant gas consumptions are also presented in the same figure. As it can be observed, the fuel cell voltage increases with increasing temperature mainly because of the enhanced reaction kinetics and the decrease in the resistance of the membrane to ion conduction. Gas consumption decreases as a result of the change in the current density, which also decreases in order to keep the fuel cell power output constant. With increased voltage the fuel cell operates at a higher efficiency.

Another case simulated was a step load from 0.1kW to 0.5kW. Figure 4 shows the temperature change during this run. The inner most cell and the outer cell temperatures are plotted separately.

With the chosen inputs only a small temperature difference (~1K) has been observed which is possibly due to the low inter cell contact resistance (.005W/K).

When the voltage curve is observed, it can be seen that the fuel cell voltage drops suddenly when the load is increased. This steep drop is a result of the increase in the current density. All fuel cell losses are a function of the current density and they tend to increase as the current density increases causing the cell voltage to drop. Once the load is changed the cell voltage rises with increasing temperature as is expected.

Although no physical hardware or well characterized physical data were available for a direct comparison of the predicted results, the fuel cell behavior and the trends of the curves are in qualitative agreement with earlier works (Buyuktur [11], Amphlett et al. [13]).

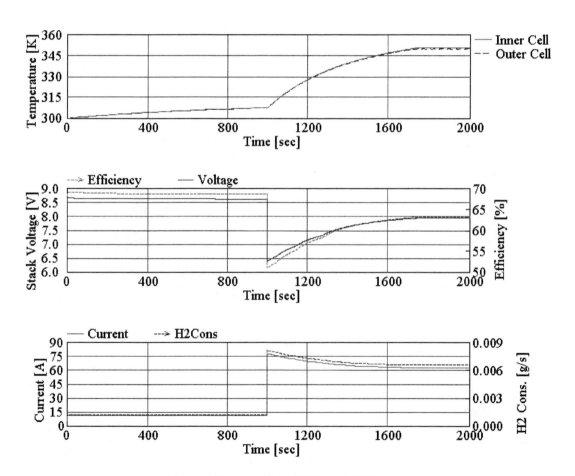

Figure 4 Step Load from 0.1kW to 0.5kW

MOTOR-GENERATOR MODEL

The motor-generator is modeled using two first order coupled ODE. The first ODE represents momentum conservation (motor effect).

$$J\dot{\omega} = K_t I_a - T_{load} - T_{fric} \qquad (1)$$

where:

J : Rotational inertia of the shaft

$\dot{\omega}$: Angular velocity of shaft

K_t : Torque coefficient

I_a : Armature current

T_{load} : Load torque

T_{fric} : Frictional torque

The second equation describes the generator effect.

$$L_{eq}\dot{I}_a = V_{app} - K_b\dot{\omega} - I_a R_{eq} \qquad (2)$$

where:

L_{eq} : Equivalent inductance of motor-generator circuitry

V_{app} : Applied voltage

K_b : Back voltage coefficient

R_{eq} : Equivalent resistance of the motor-generator circuitry

BATTERY MODEL

The performance of the battery is modeled with two sets of equations. The first equation is a quadratic equation that governs the current drawn from the battery.

$$L_{req} = IV_{op} - I^2 R_{int} = IV_t \qquad (3)$$

where:

L_{req} : Requested load

I : Battery current

V_{op} : Battery open-circuit voltage

R_{int} : Battery internal resistance

V_t : Battery terminal voltage

The second equation, equation 5, describes the state of the charge (SOC) that is calculated by integrating the current flowing in or out of the battery.

$$CAP_{used} = \int I\left(V_{op}/V_t\right)\cdot dt \qquad (4)$$

$$SOC = 1 - CAP_{used}/CAP_{total} \qquad (5)$$

where:

CAP_{used} : Used battery capacity

CAP_{total} : Total battery capacity

The open circuit voltage and internal ohmic resistance of the battery are functions of the battery's SOC and average temperature and may be represented by characteristic equations and tables. These variables are allowed to depend on the direction of flow of current; that is, they depend on the charging or discharging of the battery.

The heat generation rate in a battery cell is derived from the Gibbs free energy relationship. The development of this equation, equation 6, is similar to that employed in [14].

$$q = I\left(V_{op} - V_t + T\frac{dV_{op}}{dT}\right) \qquad (6)$$

where:

q :   Heat generation rate

T :   Cell-averaged battery temperature

Detailed calculations of the battery warm-up, taking into account the geometry of the battery and fluid flow conditions, may performed by using the thermal library.

VEHICLE MODEL

GT-SUITE allows two levels of representation of the vehicle. In the detail level, a vehicle model is constructed and configured by the user by combining/connecting various elements from the driveline dynamics library. This library consists of various high level components such as engine, clutch, tire, shaft, transmission, driver, road etc.

The elements allow modeling of elastic driveshafts and axles, the stiffness and damping characteristics of which may be specified or calculated from material properties. There are also low-level components (e.g. spring, mass and damper) that can be assembled to create models of higher level components.

A pre-programmed vehicle model is also available in this tool. In this model a typical driveline and vehicle has been pre-connected using the components available in the detailed level. In the current study this type of model is used where the driveline, vehicle and road are kinematically connected, allowing the system to be represented by a single degree of freedom and effective inertia.

This level thus differs from the detailed modeling in the sense that it is restricted to rigid shafts. There is also no model of tire slip and the vehicle is kinematically "connected" to the road, with rolling resistance applied to the tires.

VEHICLE MODEL CONFIGURATIONS

Two configurations are presented in this paper:

1. Purely fuel cell powered vehicle

2. Vehicle powered by fuel cell and battery in series

   configuration

The results are presented in the following sections.

Fuel cell powered vehicle

In the fuel cell only configuration the fuel cell is directly connected to the electric motor, which drives the vehicle. A driver module that defines a vehicle speed is built in the model. Schematic of the model is given in Figure 5.

The vehicle simulated has a mass of 1500kg. Fuel cell is sized such that, with the input parameters, a rated power of approximately 80kW is obtained at 80°C, 1.85A/cm^2 and 0.46V/cell. The stack consists of 400 cells with an active area of 232cm^2 per cell. The fuel cell subsystem model is presented in Figure 6. The thermal model was turned on during the simulation and the vehicle was driven over the UDDS cycle starting from cold start (27°C). A PID controller has been built into this fuel cell model to control the operating temperature of the stack.

The average temperature profile of the fuel cell during the UDDS cycle together with the average stack efficiency and the vehicle speed trace are presented in Figure 7.

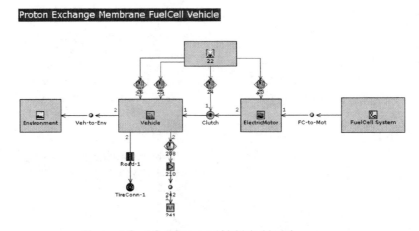

Figure 5 Fuel Cell Powered Vehicle Model

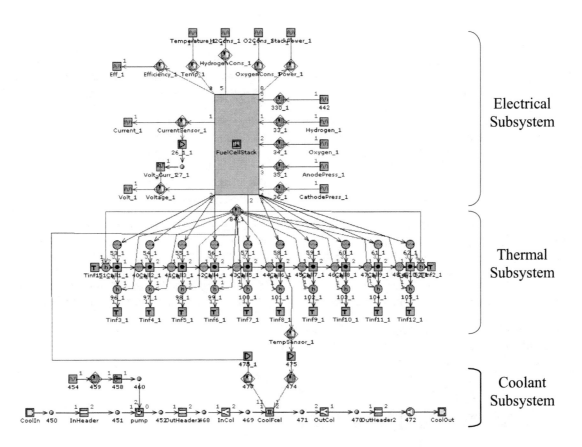

Figure 6 Fuel Cell System Module

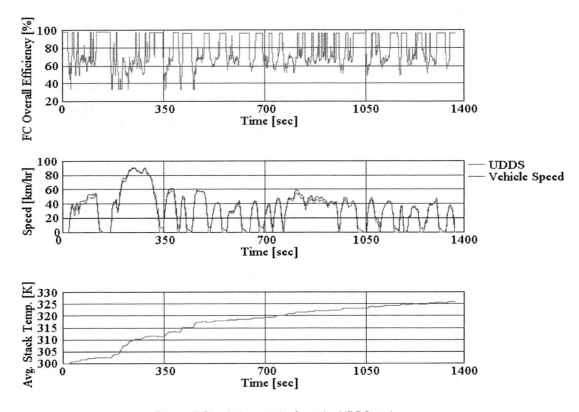

Figure 7 Simulation results from the UDDS cycle

As observed from the figure, the fuel cell operates at high efficiency values, mostly between 55 to 70%. The fuel cell efficiency is based on the lower heating value of the fuel. The equivalent gasoline consumption of the vehicle during the UDDS cycle is calculated to be 69.93mpg. With heating/cooling turned off, the stack temperature reached approximately 53°C at the end of the cycle. The maximum power generated by the fuel cell was around 29kW during the entire cycle.

Fuel cell and Battery in Series Configuration

Another simulated case is a fuel cell and battery hybrid vehicle. The fuel cell stack and the battery were connected serially in this model. The battery drives the vehicle and the fuel cell is used to control the battery state of charge (SOC). A schematic of the vehicle system is given in Figure 8. The battery is capable of producing approximately 85kW and the same fuel cell was used as in the fuel cell-only vehicle example. The vehicle weight was adjusted to take into account for the battery. The Coulombic efficiency of the battery was assumed to be 0.9.

Two different cases, one with regenerative braking and one without regenerative braking, were run. In the regenerative braking case the battery was allowed to charge back during braking. Figure 9 shows the battery SOC, fuel cell power and fuel cell efficiency outputs from the UDDS cycle for this vehicle. The fuel cell power increases as the battery SOC drops. Both parameters reach balanced state where the fuel cell power reaches an average value that is enough to compensate for the power drawn from the battery.

In the regenerative braking mode, the SOC of the battery was maintained within a desired range making also use of the power recovered during braking. An additional advantage of regenerative braking was that the fuel cell operated at higher efficiency regions (lesser current density). As Figure 9 shows, the fuel cell efficiency in the regenerative braking case was higher throughout the UDDS cycle.

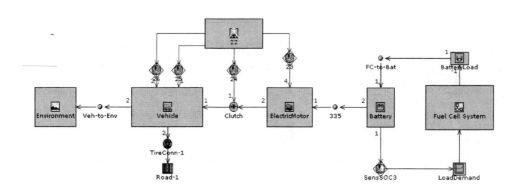

Figure 8 Fuel Cell - Battery Hybrid Vehicle

A simple logic was implemented in order to control the SOC of the battery by supplying power directly form the stack. The controller was designed such that the fuel cell keeps the battery SOC within a given range (0.6-0.8). If the SOC falls below the minimum value the fuel cell increases its power to bring the battery SOC within the desired range. If the upper limit of this range is reached or exceeded, the fuel cell produces minimal or no power. More rigorous control strategies could be easily implemented into the model if needed.

The thermal model was also turned on during the simulation and the cycle was run from cold-start. Without any external heating, only with the heat generated by the fuel cell, the stack temperature reached approximately 38°C in the no-regenerative braking case and 35°C in the case with regenerative braking. The combination of operating the fuel cell at a higher efficiency point and making use of the braking power by charging the battery results in a lower fuel consumption. The equivalent gasoline mileages obtained from the two cases are presented in Table 1.

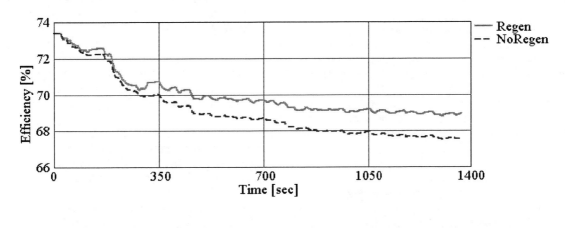

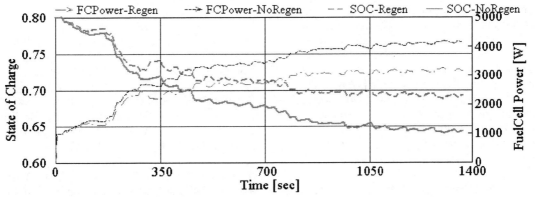

Figure 9 Regenerative Braking Effects

TABLE 1
Fuel Consumption Comparison

Equivalent Gasoline Consumption	No Regenerative Braking	Regenerative Braking
UDDS (gallons)	0.0597	0.0474
SOC Recovery (gallons)	0.0591	0.0401
Total Consumption (gallons)	0.1188	0.0875
MPG	68	78

The difference in the SOC of the battery between the beginning and the end of the UDDS cycle was taken into account while calculating the fuel consumption of the vehicle. This was accomplished by allowing the fuel cell to recharge the battery to its original SOC at end of the UDDS cycle.

The fuel economy improvements of the presented configurations (fuel cell only, fuel cell-battery hybrid) over the traditionally powered vehicles are as a result of the onboard efficiency improvement. The tank to wheel efficiency of a fuel cell powered vehicle is significantly higher due to the high efficiency characteristic of the fuel cell itself. To make a true comparison between the two technologies (IC engine and fuel cell) the well-to-wheel efficiencies should be taken into account. Such studies are being conducted and a recent example could be found in the publication made by Kreith and West [15].

CONCLUSION

A fuel cell model was developed and integrated with comprehensive multi-disciplinary simulation tool, GT-SUITE. It is shown that the model is capable of reproducing experimentally observed polarization characteristics using empirical loss models and adaptive implementations of dependencies on fuel cell operating state. The multi-disciplinary nature of the tool allowed integration of thermal management and control subsystems.

The fuel cell model was incorporated into a vehicle model as a sole power plant and also in a hybrid scheme where battery carried the vehicle load and load transients. The fuel cell was operated only to maintain the SOC of the battery under a relatively constant power demand. In the second scheme both regenerative and non-generative braking modes were assessed. It is shown that with proper control strategy the fuel cell can be operated at the high efficiency regions of the polarization curve by letting the conventional battery take up the load transients. Additionally, regenerative braking can recoup a significant fraction of braking power loss. Thus, it was demonstrated that a fuel cell hybrid with regenerative braking has a high potential for high tank-to-wheel efficiency.

Finally, it was demonstrated that a complex multidisciplinary model could be built with the described integrated environment. This environment provides modeling of vehicle systems that include fluids, thermal, mechanical, control and electrical subsystems. The adoption of hierarchical description of system elements through template libraries, objects and parts gives the tool a high level of modularity and flexibility. This allows efficient building and editing of models, and importing and combining models from disparate disciplines. Furthermore, the code is not limited to the specific fuel cell modeling techniques used in this work. User written models or external codes can be linked to this tool through user model facility if more sophisticated treatments are required and made available.

REFERENCES

1. Ki-Chun, L., Seo-Ho, C., Soo-Whan, et al., "Hyundai Santa Fe FCV Powered by Hydrogen Fuel Cell Power Plant Operating Near Ambient Pressure", SAE Paper 2002-01-0093, 2002.

2. Matsumoto, T., Watanabe, N., Sugiura, H., Ishikawa, T., "Development of Fuel-Cell Hybrid Vehicle", SAE Paper 2002-01-0096.

3. Baschuk, J.J., Li, X., "Modelling of polymer electrolyte membrane fuel cells with variable degrees of water flooding", Journal of Power Sources 86 (2000) 181-196.

4. Morel, T., Keribar, R., Leonard, A., "Virtual Engine/Powertrain/Vehicle Simulation Tool Solves Complex Interacting System Issues", SAE Paper 2003-01-0372, 2003.

5. Kolade, B., Boghosian, M.E., Reddy, P.S., Gallagher, S., "Development of a General Purpose Thermal-Hydraulic Software and its Application to Fuel Injection Systems", SAE Paper 2003-01-0702, 2003.

6. Larminie, J., and Dicks, A., " Fuel Cell System Explained", John Wiley & Sons, 2000.

7. Singh, D., Lu, D.M., Djilali, N., "A two-dimensional analysis of mass transport in proton exchange membrane fuel cells", International Journal of Engineering Science 37 (1999) 431-452.

8. Cantore, G., Monotorsi, L., Mauss, F., et al., "Analysis of a 6-Cylinder Turbocharged HCCI Engine Using Detailed Kinetic Mechanisms", ICE-Vol. 38, Spring Technical Conference, ASME, 2002.

9. Barbir, F., Fuchs, M., Husar, A., Neutzler, J., "Design and Operational Characteristics of Automotive PEM Fuel Cell Stacks", SAE Paper 2000-01-0011, 2000.

10. Musser, J., Wang, C.Y., "Heat Transfer in a Fuel Cell Engine", 34th National Heat Transfer Conference, August 20-22, 2000.

11. Buyuktur, S., "Development of a Proton Exchange Membrane Fuel Cell System Model for a Vehicle Simulator and for Studying Transient Behaviors of a Fuel Cell System", MS Thesis, Univ. of Michigan, Ann Arbor, 2001. For a copy-e-mail: buyuktur@superonline.com

12. Morel, T., Wahiduzzaman, S., "System Model of Engine Thermal Management", VTMS London UK, 1999.

13. Amphlett, J.C., Mann, R.F., Peppley B.A., Roberge, P.R., Rodrigues, A., "A model predicting transient responses of proton exchange membrane fuel cells", Journal of Power Sources 61 (1996) 183-188.

14. Al Hallaj, S., Maleki, H., Hong, J.S., Selman, J.R., "Thermal modeling and design considerations of lithium-ion batteries", Journal of Power Sources 83 (1999) 1-8.

15. Kreith, F., West, R.E., "Gauging Efficiency, Well to Wheel", Mechanical Engineering Power 2003.

Bibliography

1 Policy, Fleet Trials, Public Reactions

SAE Papers
- Fuel cell vehicles: Technology development status and popularization issues. SAE Technical Paper No. 2002-21-0036.
- Fuel cells at General Motors. SAE Technical Paper No. 2002-21-0072.
- A hybrid and fuel cell vehicle future? SAE Technical Paper No. 2002-01-1908.
- What FutureCar MPG levels and technology will be necessary? SAE Technical Paper No. 2002-01-1898.
- Fuel cell development program in support of the Army transformation. SAE Technical Paper No. 2002-01-2143.

Other Publications
- The Hydrogen Economy: Opportunities, Costs, Barriers and R&D Needs. National Academy of Engineering. National Academies Press. Washington DC. 2004.
- Hughes, Wm.L. Comments on the hydrogen fuel cell as a competitive energy source. Power Engineering Society Summer Meeting. 2001. IEEE. Volume 1. 15-19 July 2001. Pages 726-730.
  Williams, M.C. Fuel cells and the world energy future. Power Engineering Society Summer Meeting, 2001. IEEE. Volume 1. 15-19 July 2001. Page 725.
- Ordubadi, F. PEM fuel cells and future opportunities. Power Engineering Society Summer Meeting, 2001. IEEE. Volume 1. 15-19 July 2001. Pages 710-716.
- Ramakumar, R. Fuel cells-an introduction. Power Engineering Society Summer Meeting, 2001. EEE. Volume 1. 15-19 July 2001. Pages 702-709.
- Smith, J.A.; Nehrir, M.H.; Gerez, V.; Shaw, S.R. A broad look at the workings, types, and applications of fuel cells. Power Engineering Society Summer Meeting, 2002. IEEE. Volume 1. 21-25 July 2002. Pages 70-75.
- Cook, B. Introduction to fuel cells and hydrogen technology. Engineering Science and Education Journal. Volume 11, Issue 6. Dec. 2002. Pages 205-216.
- Laughton, M.A. Fuel cells. Power Engineering Journal (see also Power Engineer). Volume 16, Issue 1. Feb. 2002. Pages 37-47.
- Ortmeyer, T.H.; Pillay, P. Trends in transportation sector technology energy use and greenhouse gas emissions. Proceedings of the IEEE. Volume 89, Issue 12. Dec. 2001. Pages 1837–1847.
- Farooque, M.; Maru, H.C. Fuel cells-the clean and efficient power generators. Proceedings of the IEEE. Volume 89, Issue 12. Dec. 2001. Pages 1819–1829.
- Ellis, M.W.; Von Spakovsky, M.R.; Nelson, D.J. Fuel cell systems: efficient, flexible energy conversion for the 21st century. Proceedings of the IEEE. Volume 89, Issue 12. Dec. 2001. Pages 1808-1818.
- Oman, H. Performance, life-cycle cost, and emissions of fuel cells. Aerospace and Electronic Systems Magazine, IEEE. Volume 17, Issue 9. Sept. 2002.

2 Fuel Issues

Fuel Technology

SAE Papers
- Investigation of Hydrogen Carriers for Fuel Cell Based Transportation. SAE Technical Paper No. 2002-01-0097.
- Synthetic hydrocarbon fuel for APU application: The fuel processor system. SAE Technical Paper No. 2003-01-0267.

Fuel/Vehicle Analysis

SAE Papers
- Well to wheels analysis of SUV fuel cell vehicles. SAE Technical Paper No. 2003-01-0415.
- Well to wheel energy use and greenhouse gas emissions for various vehicle technologies. SAE Technical Paper No. 2001-01-1343.
- Comparing Apples and Apples: Well to Wheels Analysis of Current ICE and Fuel Cell Vehicle Technologies. SAE Technical Paper No. 2004-01-1015.
- Comparing estimates of fuel economy improvement via fuel cell powertrains. SAE Technical Paper No. 2002-01-1947.
- System comparison of hybrid and fuel cell systems to internal combustion engines. SAE Technical Paper No. 2002-21-0070.

Other Publications
- Züttel, P. Wenger, S. Rentsch, P. Sudan, Ph. Mauron and Ch. Emmenegger., LiBH4 a new hydrogen storage material. Journal of Power Sources. Volume 118. Issues 1-2. 25 May 2003. Pages 1-7. http://www.sciencedirect.com/science/article/B6TH1-486G7WM-2/2/1fbcb42a48a88041056f7e41eb97040c.
- U. Casellato, N. Comisso, G. Davolio and G. Mengoli. Electrolytic hydrogen storage in reluctant intermetallic systems. Journal of Power Sources. Volume 118. Issues 1-2. 25 May 2003. Pages 237-247. http://www.sciencedirect.com/science/article/B6TH1-4817CWK-5/2/5323f97266267feefd3a8a72702792ef.
- Pier Paolo Prosini, Alfonso Pozio, Sabina Botti and Roberto Ciardi. Electrochemical studies of hydrogen evolution, storage and oxidation on carbon nanotube electrodes. Journal of Power Sources. Volume 118. Issues 1-2. 25 May 2003. Pages 265-269. http://www.sciencedirect.com/science/article/B6TH1-481232P-8/2/78d829cac5acdaf192b08db80c50cc56.
- Dragica Lj. Stojic, Milica P. Marceta, Sofija P. Sovilj and Scepan S. Miljanic. Hydrogen generation from water electrolysis--possibilities of energy saving. Journal of Power Sources. Volume 118. Issues 1-2. 25 May 2003. Pages 315-319. http://www.sciencedirect.com/science/article/B6TH1-4830431-2/2/60a461441dbc4a22d65e0a0213bd3165.
- Adam Wojcik, Hugh Middleton, Ioannis Damopoulos and Jan Van herle. Ammonia as a fuel in solid oxide fuel cells. Journal of Power Sources. Volume 118. Issues 1-2. 25 May 2003. Pages 342-348. http://www.sciencedirect.com/science/article/B6TH1-483BVBJ-1/2/a338867cf774b439998a8ddf245b3ecd.

3   Systems Design and Evaluation

Architectures

SAE Papers
- Progress in the Development of PEM Fuel cell engines for transportation. SAE Technical Paper No. 2001-01-0540.
- Automotive ambient-pressure PEM fuel cell development at UTC fuel cells. SAE Technical Paper No. 2002-01-1897.
- Hybrid storage system: an optimization case. SAE Technical Paper No. 2002-01-1914.
- Fuel Cell Stack and Fuel Cell Powertrain for Automotive Application. SAE Technical Paper No. 2004-01-1005.

Other Publications
- Bo Yuwen; Xinjian Jiang; Dongqi Zhu. Structure optimization of the fuel cell powered electric drive system. Power Electronics Specialist, PESC. 2003 IEEE 34th Annual Conference. Volume 1. June 15-19, 2003. Pages 391-394.
  Gao, Y.; Ehsani, M. Systematic design of fuel cell powered hybrid vehicle drive train. Electric Machines and Drives Conference. IEMDC 2001. IEEE International. Pages 604–611.
- Koyanagi, F.; Uriu, Y.; Yokoyama, R. Possibility of fuel cell fast charger and its arrangement problem for the infrastructure of electric vehicles. Power Tech Proceedings. IEEE Porto. Volume 4. 10-13 Sept. 2001. 6 pp.
- Pera, M.C.; Hissel, D.; Kauffmann, J.M. Fuel cell systems for electrical vehicles. Vehicular Technology Conference, 2002. IEEE 55th. Volume 4. 6-9 May 2002. Pages 2097-2102.
  Gay, S.E.; Hongwei Gao; Ehsani, M. Fuel cell hybrid drive train configurations and motor drive selection. Vehicular Technology Conference, 2002. IEEE 56th. Volume 2. 24-28 Sept. 2002. Pages 1007-1010.

Auxiliary Power Units

SAE Papers
- Solid oxide fuel cell auxiliary power unit~A paradigm shift in electric supply for transportation. SAE Technical Paper No. 2000-01-C070.
- Analysis of a planar solid oxide fuel cell based automotive auxiliary power unit. SAE Technical Paper No. 2002-01-0413.
- Fuel Cell Auxiliary Power Systems: Design and Cost Implications. SAE Technical Paper No. 2001-01-0536.
- Liquid-fueled APU fuel cell system for truck application. SAE Technical Paper No. 2001.01-2716.
- Army strategy for utilizing fuel cells as auxiliary power unit. SAE Technical Paper No. 2001-01-2792.
- Vibration and shock considerations in the design of a truck-mounted fuel cell APU system. SAE Technical Paper No. 2002-01-3050.
- Demonstration of a proton exchange membrane fuel cell as an auxiliary power source for heavy trucks. SAE Technical Paper No. 2000-01-3488.

Other Publications
- Bertoni, L.; Gualous, H.; Bouquain, D.; Hissel, D.; Pera, M.-C.; Kauffmann, J.-M. Hybrid auxiliary power unit (APU) for automotive applications. Vehicular Technology Conference, 2002. IEEE 56th. Volume 3. 24-28 Sept. 2002. Pages 1840-1845.
- Wolfgang Winkler and Hagen Lorenz. Design studies of mobile applications with SOFC-heat engine modules. Journal of Power Sources. Volume 106. Issues 1-2. 1 April 2002. Pages 338-343. http://www.sciencedirect.com/science/article/B6TH1-44VX1D8-2/2/9e0ba3706724d987d6fe6ecb9e4ecf2b.

Vehicle Development and Test

SAE Papers
- Hyundai Santa Fe FCV powered by Hydrogen Fuel Cell Power Plant Operating Near Ambient Pressure. SAE Technical Paper No. 2002-01-0093.
- Design and Integration Challenges for a Fuel Cell Hybrid Electric Sport Utility Vehicle. SAE Technical Paper No. 2002-01-0095.
- Ford's Zero Emission P2000 fuel cell vehicle. SAE Technical Paper No. 2000-01-C046.
- The development of Ford's P2000 fuel cell vehicle. SAE Technical Paper No. 2000-01-1061.
- New hybrid bus prototype for clean urban transportation. SAE Technical Paper No. 2003-01-0419.
- A comparison of energy use for a direct hydrogen hybrid versus a direct hydrogen load-following fuel cell vehicles. SAE Technical Paper No. 2003-01-0416.
- Degree of Hybridisation Modelling of a Fuel Cell Hybrid Sports Utility Vehicle. SAE Technical Paper No. 2001-01-0236.
- Development of Fuel Cell Hybrid Vehicle. SAE Technical Paper No. 2002-01-0096.
- Degree of hybridization modeling of a hydrogen fuel cell PNGV class vehicle. SAE Technical Paper No. 2002-01-1945.
- Design of a Fuel Cell Hybrid City Bus for the Italian Historical Towns. SAE Technical Paper No. 2004-01-1007.

Other Publications
- Parten, M.; Maxwell, T. Development of a PEM fuel cell vehicle. Vehicular Technology Conference, 2001. IEEE VTS 54th. Volume 4. 7-11 Oct. 2001. Pages 2225-2228.

4   Component Development

Air Supply

SAE Papers
- Performance of amorphous carbon coating in turbocompressor air bearings. SAE Technical Paper No. 2002-01-1922.
- Integrated air supply and humidification systems for fuel cell systems. SAE Technical Paper No. 2001-01-0233.
- PEM fuel cell air management at part load. SAE Technical Paper No. 2002-01-1912.

Electrical Systems

SAE Papers

- Comparison of thermal performance of different power electronic stack constructions. SAE Technical Paper No. 2002-01-1917.
- Analysis of fuel cell/battery~Capacitor hybrid sources used for pulsed load applications. SAE Technical Paper No. 2002-01-3219.
- A soft-switched dc/dc converter for fuel cell vehicle applications. SAE Technical Paper No. 2002-01-1903.

Other Publications

- Bo Yuwen; Xinjian Jiang; Dongqi Zhu. Structure optimization of the fuel cell powered electric drive system. Power Electronics Specialist, PESC. 2003 IEEE 34th Annual Conference. Volume 1. June 15-19, 2003. Pages 391–394.
- Rocco, T.; Duke, R.M.; Round, S.D. Development and control of an alkaline fuel cell power conditioner. Power Electronics Specialist, PESC. 2003 IEEE 34th Annual Conference. Volume 1. June 15-19, 2003. Pages 379-384.
  Cheng, K.W.E.; Sutanto, D.; Ho, Y.L.; Law, K.K. Exploring the power conditioning system for fuel cell. Power Electronics Specialists Conference. PESC. 2001. IEEE 32nd Annual. Volume 4. 17-21 June 2001. Pages 2197-2202.
- Mekhiche, M.; Nichols, S.; Kirtley, J.L.; Young, J.; Boudreau, D.; Jodoin, R. High-speed, high-power density PMSM drive for fuel cell powered HEV application. Electric Machines and Drives Conference. IEMDC 2001. IEEE International. Pages 658–663.
- Fahimi, B.; Gao, Y.; Ehsani, M. On the suitability of switched reluctance motor drives for 42 volts super high speed operation: application to automotive fuel cells. Industrial Electronics Society, 2001. IECON '01. The 27th Annual Conference of the IEEE. Volume 3. 29 Nov.-2 Dec. 2001. Pages 1947-1952.
- Azli, N.A.; Yatim, A.H.M. DSP-based online optimal PWM multilevel control for fuel cell power conditioning systems. Industrial Electronics Society. IECON '01. The 27th Annual Conference of the IEEE. Volume 2. 29 Nov.-2 Dec. 2001. Pages 921-926.
- Jalili, K.; Farhangi, S.; Saievar-Iranizad, E. Sensorless vector control of induction motors in fuel cell vehicle using a neuro-fuzzy speed controller and an online artificial neural network speed estimator. Proceedings of the 2001 IEEE International Conference on Control Applications. 5-7 Sept. 2001. Pages 259–264.
- Yeary, M.; Sangsun Kim; Enjeti, P.; King, G. Design of an embedded DSP system for a fuel cell inverter. IEEE International Conference on Acoustics, Speech, and Signal Processing. Volume 4. 13-17 May 2002. Pages IV-3824-IV-3827.
  Di Napoli, A.; Crescimbini, F.; Guilii Capponi, F.; Solero, L. Control strategy for multiple input DC-DC power converters devoted to hybrid vehicle propulsion systems. Proceedings of the 2002 IEEE International Symposium on Industrial Electronics. Volume 3 , 26-29 May 2002. Pages 1036-1041.
- Di Napoli, A.; Crescimbini, F.; Solero, L.; Caricchi, F.; Capponi, F.G. Multiple-input DC-DC power converter for power-flow management in hybrid vehicles. Conference Record of the 37th Industry Applications Conference. Volume 3. 13-18 Oct. 2002. Pages 1578-1585.
  Li Chen; Davis, R.; Stella, S.; Tesch, T.; Fischer-Antze, A. Improved control techniques for IPM motor drives on vehicle application. Conference Record of the 37th Industry Applications Conference. Volume 3. 13-18 Oct. 2002. Pages 2051-2056.
- Spiazzi, G.; Buso, S.; Martins, G.M.; Pomilio, J.A. Single phase line frequency commutated voltage source inverter suitable for fuel cell interfacing. 33rd Annual Power Electronics Specialists Conference. IEEE. Volume 2. 23-27 June 2002. Pages 734-739.

- Andersen, G.K.; Klumpner, C.; Kjaer, S.B.; Blaabjerg, F. A new green power inverter for fuel cells. IEEE 33rd Annual Power Electronics Specialists Conference. Volume 2. 23-27 June 2002. Pages 727-733.
- Gopinath, R.; Sangsun Kim; Jae-Hong Hahn; Webster, M.; Burghardt, J.; Campbell, S.; Becker, D.; Enjeti, P.; Yeary, M.; Howze, J. Development of a low cost fuel cell inverter system with DSP control. IEEE 33rd Annual Power Electronics Specialists Conference. Volume 1. 23-27 June 2002. Pages 309-314.
- Krein, P.T.; Balog, R. Low cost inverter suitable for medium-power fuel cell sources. IEEE 33rd Annual Power Electronics Specialists Conference. Volume 1. 23-27 June 2002. Pages 321-326.

Fuel Cell Stacks

SAE Papers
- New powerful catalysts for auto-thermal reforming of hydrocarbons and Water-Gas shift reaction for on-board hydrogen generation in automotive PEMFC applications. SAE Technical Paper No. 2001-01-0234.
- Balancing stack, air supply, and water/thermal management demands for an indirect methanol PEM fuel cell system. SAE Technical Paper No. 2001-01-0535.

Other Publications
- B. Zhu, X. T. Yang, J. Xu, Z. G. Zhu, S. J. Ji, M. T. Sun and J. C. Sun. Innovative low temperature SOFCs and advanced materials. Journal of Power Sources. Volume 118, Issues 1-2. 25 May 2003. Pages 47-53. http://www.sciencedirect.com/science/article/B6TH1-48FK67F-1/2/a592952fbd70fd7a7b89a79c34638cea.
- Nobuyoshi Nakagawa and Yikun Xiu. Performance of a direct methanol fuel cell operated at atmospheric pressure. Journal of Power Sources. Volume 118, Issues 1-2. 25 May 2003. Pages 248-255. http://www.sciencedirect.com/science/article/B6TH1-48B023J-3/2/f5fe0ddd6acfb043a363c940abd5147c.
- R. J. Gorte, H. Kim and J. M. Vohs. Novel SOFC anodes for the direct electrochemical oxidation of hydrocarbon. Journal of Power Sources. Volume 106, Issues 1-2. 1 April 2002. Pages 10-15. http://www.sciencedirect.com/science/article/B6TH1-44X03CH-2/2/7d3e1fd448830fe2888b25ba68f5f3b5.
- Stanley, K.G.; Wu, Q.M.J.; Vanderhoek, T.; Nikumb, S.; Walker, Z.M.; Parameswaran, M. Fabrication of a micromachined direct methanol fuel cell. IEEE Canadian Conference on Electrical and Computer Engineering. Volume 1. 12-15 May 2002. Pages 450-454.
  Riezenman, M.J. Metal fuel cells (Zn-air fuel cells). Spectrum. IEEE. Volume 38, Issue 6. June 2001. Pages 55-59.
  J. Doyon, M. Farooque and H. Maru, The Direct FuelCell(TM) stack engineering. Journal of Power Sources. Volume 118, Issues 1-2. 25 May 2003. Pages 8-13. http://www.sciencedirect.com/science/article/B6TH1-4830431-5/2/df3dd75ef5ac954d7283d4488b2d461d.
- E. Middelman, W. Kout, B. Vogelaar, J. Lenssen and E. de Waal, Bipolar plates for PEM fuel cells, Journal of Power Sources, Volume 118, Issues 1-2, 25 May 2003, Pages 44-46. http://www.sciencedirect.com/science/article/B6TH1-48F5KX2-1/2/4dc16d45af174e6ae9d0082e70ff6f5e.
- Saisset, R.; Turpin, C.; Astier, S.; Lafage, B. Study of thermal imbalances in arrangements of solid oxide fuel cells by mean of bond graph modeling. IEEE 33rd Annual Power Electronics Specialists Conference. Volume 1. 23-27 June 2002. Pages 327-332.

- Sang-Hee Kwak, Tae-Hyun Yang, Chang-Soo Kim and Ki Hyun Yoon. The effect of platinum loading in the self-humidifying polymer electrolyte membrane on water uptake. Journal of Power Sources. Volume 118, Issues 1-2. 25 May 2003. Pages 200-204. http://www.sciencedirect.com/science/article/B6TH1-4851CHS-3/2/0bfe55797ece801e3c453c343aa75f7a.
- Michael A. Priestnall, Vega P. Kotzeva, Deborah J. Fish and Eva M. Nilsson. Compact mixed-reactant fuel cells. Journal of Power Sources. Volume 106, Issues 1-2. 1 April 2002. Pages 21-30. http://www.sciencedirect.com/science/article/B6TH1-44TV9WR-G/2/4fc0e68c42afac8c5a5debf09ce092d7.
- M. R. Haines, W. K. Heidug, K. J. Li and J. B. Moore. Progress with the development of a CO2 capturing solid oxide fuel cell. Journal of Power Sources. Volume 106, Issues 1-2. 1 April 2002. Pages 377-380. http://www.sciencedirect.com/science/article/B6TH1-44TV9WR-6/2/3141b4f7468b0fb1f83cb9e71fca94d0.
- S.-Y. Ahn, S.-J. Shin, H. Y. Ha, S.-A. Hong, Y.-C. Lee, T. W. Lim and I.-H. Oh. Performance and lifetime analysis of the kW-class PEMFC stack. Journal of Power Sources. Volume 106, Issues 1-2. 1 April 2002. Pages 295-303. http://www.sciencedirect.com/science/article/B6TH1-44VG3HH-7/2/482a9cd0dfdfb021a43125c03cbbd6b2.
- Michael W. Fowler, Ronald F. Mann, John C. Amphlett, Brant A. Peppley and Pierre R. Roberge. Incorporation of voltage degradation into a generalised steady state electrochemical model for a PEM fuel cell. Journal of Power Sources. Volume 106, Issues 1-2. 1 April 2002. Pages 274-283. http://www.sciencedirect.com/science/article/B6TH1-44V1X6D-2/2/ed57b43674a20f8f97be0c636f803d87.
- R. Peters, R. Dahl, U. Klüttgen, C. Palm and D. Stolten. Internal reforming of methane in solid oxide fuel cell systems. Journal of Power Sources. Volume 106, Issues 1-2. 1 April 2002. Pages 238-244. http://www.sciencedirect.com/science/article/B6TH1-44TV9WR-J/2/8667d2f88cf2dab8f03c16764c1ed779.

Water and Thermal Management

SAE Papers
- Adaptation of traditional cooling airflow management to advanced technology vehicles. SAE Technical Paper No. 2002-01-1968.
- Fuel cell thermal management with micro-coolers. SAE Technical Paper No. 2002-01-1913.

Other Publications
- Stinivasan, P.; Sneckenberger, J.E.; Feliachi, A. Dynamic heat transfer model analysis of the power generation characteristics for a proton exchange membrane fuel cell stack. Proceedings of the 35th Southeastern Symposium on System Theory. 16-18 March 2003. Pages 252–258.

Fuel Processors

SAE Papers
- Plate type methanol steam reformer using new catalytic combustion for a fuel cell. SAE Technical Paper No. 2002-01-0406.
- Fuel flexible, fuel processors (F3P), Reforming Infrastructure, Fuels for Fuel Cell Vehicles. SAE Technical Paper No. 2001-01-1341.
- Integrated fuel processor development. SAE Technical Paper No. 2002-01-1886.
- Steam reformer/burner integration and analysis for an indirect methanol fuel cell vehicle fuel processor. SAE Technical Paper No. 2001-01-0539.
- How to minimize the clean-up system of a fuel processor by using a power electronic. SAE Technical Paper No. 2002-01-1874.

- Auto-thermal reforming catalyst development for fuel cell applications. SAE Technical Paper No. 2002-01-1884.
- Reforming petroleum-based fuels for fuel cell vehicles: Composition-performance relationships. SAE Technical Paper No. 2002-01-1885.

Other Publications

- P. Marty and D. Grouset. High temperature hybrid steam-reforming for hydrogen generation without catalyst. Journal of Power Sources. Volume 118, Issues 1-2. 25 May 2003. Pages 66-70. http://www.sciencedirect.com/science/article/B6TH1-47XWMPC-2/2/eb01ec106e87d9b0809bac36569e94d2.
- L. Hartmann, K. Lucka and H. Köhne. Mixture preparation by cool flames for diesel-reforming technologies. Journal of Power Sources. Volume 118, Issues 1-2. 25 May 2003. Pages 286-297. http://www.sciencedirect.com/science/article/B6TH1-4834JRR-2/2/aaa762e3063be1d8a15de65f1f01c787.
- Ersoz, H. Olgun, S. Ozdogan, C. Gungor, F. Akgun and M. Tiris. Autothermal reforming as a hydrocarbon fuel processing option for PEM fuel cell. Journal of Power Sources. Volume 118, Issues 1-2. 25 May 2003. Pages 384-392. http://www.sciencedirect.com/science/article/B6TH1-48GP6XF-3/2/39086776c57c4a66040387c9b9ff1769.
- Wolfgang Ruettinger, Oleg Ilinich and Robert J. Farrauto. A new generation of water gas shift catalysts for fuel cell applications. Journal of Power Sources. Volume 118, Issues 1-2. 25 May 2003. Pages 61-65. http://www.sciencedirect.com/science/article/B6TH1-481FT6H-2/2/1fcbf4fae776a6cbd0e1301f934fe2c9.
- Bård Lindström and Lars J. Pettersson. Development of a methanol fuelled reformer for fuel cell applications. Journal of Power Sources. Volume 118, Issues 1-2. 25 May 2003. Pages 71-78. http://www.sciencedirect.com/science/article/B6TH1-4817CWK-4/2/deadd4869f4e432a9303fac86f63e295.
- Arana, L.R.; Baertsch, C.D.; Schmidt, R.C.; Schmidt, M.A.; Jensen, K.F. Combustion-assisted hydrogen production in a high-temperature chemical reactor/heat exchanger for portable fuel cell applications. 12th International Conference on Transducers, Solid-State Sensors, Actuators and Microsystems. Volume 2. June 9-12, 2003. Pages 1734-1737.
  Arana, L.R.; Schaevitz, S.B.; Franz, A.J.; Jensen, K.F.; Schmidt, M.A. A microfabricated suspended-tube chemical reactor for fuel processing. The Fifteenth IEEE International Conference on Micro Electro Mechanical Systems. 20-24 Jan. 2002. Pages 232-235.
  B. Emonts, J. Bøgild Hansen, T. Grube, B. Höhlein, R. Peters, H. Schmidt, D. Stolten and A. Tschauder. Operational experience with the fuel processing system for fuel cell drives. Journal of Power Sources. Volume 106, Issues 1-2. 1 April 2002. Pages 333-337. http://www.sciencedirect.com/science/article/B6TH1-44V1X6D-7/2/788f8c78e9e553081a3e63ce777d25e4.
- N. Muradov. Emission-free fuel reformers for mobile and portable fuel cell applications. Journal of Power Sources. Volume 118, Issues 1-2. 25 May 2003. Pages 320-324. http://www.sciencedirect.com/science/article/B6TH1-481232P-4/2/60d7a7985f7172487891fc28e2cd2d75.
- Johan Agrell, Henrik Birgersson and Magali Boutonnet. Steam reforming of methanol over a Cu/ZnO/Al2O3 catalyst: a kinetic analysis and strategies for suppression of CO formation. Journal of Power Sources. Volume 106, Issues 1-2. 1 April 2002. Pages 249-257. http://www.sciencedirect.com/science/article/B6TH1-44TV9WR-F/2/26fa28ee3a17ac03be8dba44e4417679.
- Palm, P. Cremer, R. Peters and D. Stolten. Small-scale testing of a precious metal catalyst in the autothermal reforming of various hydrocarbon feeds. Journal of Power Sources. Volume 106, Issues 1-2. 1 April 2002. Pages 31-237. http://www.sciencedirect.com/science/article/B6TH1-45J82CF-3/2/d1bf4ebc3cc7cac7340ca1c04544204c.

Energy Storage

SAE Paper
- Battery for a fuel cell HEV application. SAE Technical Paper No. 2002-01-1976.

Other Publication
- Gagliardi, F.; Pagano, M. Experimental results of on-board battery-ultracapacitor system for electric vehicle applications. Proceedings of the 2002 IEEE International Symposium on Industrial Electronics. Volume 1. 8-11 July 2002. Pages 93-98.

Materials

SAE Papers
- Platinum: Too precious for fuel cell vehicles? SAE Technical Paper No. 2002-01-1896.
- Gold--A future role in automotive pollution control? SAE Technical Paper No. 2002-01-2148.

Other Publications
- Karnik, S.V.; Hatalis, M.K.; Kothare, M.V. Towards a palladium micro-membrane for the water gas shift reaction: microfabrication approach and hydrogen purification results. Journal of Microelectromechanical Systems. Volume 12, Issue 1. Feb. 2003. Pages 93-100.
- Don Cameron, Richard Holliday and David Thompson. Gold's future role in fuel cell systems. Journal of Power Sources. Volume 118, Issues 1-2. 25 May 2003. Pages 298-303. http://www.sciencedirect.com/science/article/B6TH1-48GP6XF-2/2/1932e4f4722d40c427141209ee43d0a6.

5   Development, Testing and Lifecycle Issues

Lifecycle

SAE Paper
- Environmental Evaluation of Direct Hydrogen and Reformer-Based Fuel Cell Vehicles. SAE Technical Paper No. 2002-01-0094.

Testing Methods

SAE Papers
- Electric and hybrid vehicle testing. SAE Technical Paper No. 2002-01-1916.
- Efficiency analysis in a direct methanol fuel cell with a measurement of methanol concentration. SAE Technical Paper No. 2001-01-0237.
- Pressure and gas matrix independent dilution system for fuel cell gas and exhaust distribution monitoring. SAE Technical Paper No. 2002-01-1680.
- Transient Measurement in a gasoline fuel cell fuel processor. SAE Technical Paper No. 2001-01-0232.

Other Publications

- T. Romero-Castañón, L. G. Arriaga and U. Cano-Castillo. Impedance spectroscopy as a tool in the evaluation of MEA's. Journal of Power Sources. Volume 118, Issues 1-2. 25 May 2003. Pages 179-182. http://www.sciencedirect.com/science/article/B6TH1-48BTYGP-1/2/ea0f809c16c51b94c3a2675408f417ca.
- F. Boccuzzi, A. Chiorino and M. Manzoli. FTIR study of methanol decomposition on gold catalyst for fuel cells. Journal of Power Sources. Volume 118, Issues 1-2. 25 May 2003. Pages 304-310. http://www.sciencedirect.com/science/article/B6TH1-481232P-5/2/06c6fea30f7a6fda6da905f44d90c083.
- Lukas, M.D.; Lee, K.Y.; Ghezel-Ayagh, H.; Abens, S.G.; Cervi, M.C. Experimental transient validation of a Direct FuelCell(R) stack model. Power Engineering Society Summer Meeting. IEEE. Volume 3. 15-19 July 2001. Pages 1363-1368.
- Schupbach, R.M.; Balda, J.C. A versatile laboratory test bench for developing powertrains of electric vehicles. Proceedings of the IEEE 56th Vehicular Technology Conference. Volume 3. 24-28 Sept. 2002. Pages 1666-1670.
- Gui-Jia Su; Peng, F.Z.; Adams, D.J. Experimental evaluation of a soft-switching DC/DC converter for fuel cell vehicle applications. Power Electronics in Transportation. Oct. 24-25, 2002. Pages 39–44.
- Claycomb, J.R.; Brazdeikis, A.; Le, M.; Yarbrough, R.A.; Gogoshin, G.; Miller, J.H. Nondestructive testing of PEM fuel cells. IEEE Transactions on Applied Superconductivity. Volume 13, Issue 2. June 2003. Pages 211-214.

Design Models and Methods

Other Publications

- Rubin, S.H.; Ceruti, M.G. Application of extended plausible-reasoning theory to fuel-cell design. IEEE International Conference on Systems, Man, and Cybernetics. Volume 1. 7-10 Oct. 2001. Pages 304-309.
- Julien Godat and Francois Marechal. Optimization of a fuel cell system using process integration techniques. Journal of Power Sources. Volume 118, Issues 1-2. 25 May 2003. Pages 411-423. http://www.sciencedirect.com/science/article/B6TH1-48765TY-2/2/afb1aebcc50e119ba33814c0ea79a2a5.

6   Modeling, Control and Diagnosis

Modelling and Simulation

SAE Papers

- An analysis of start-up for an operational fuel cell transit bus. SAE Technical Paper No. 2000-01-3471.
- Identification of response limiting processes in an indirect methanol fuel cell bus powertrain. SAE Technical Paper No. 2002-01-2855.
- Concept study of a methanol fuel cell vehicle. SAE Technical Paper No. 2002-21-0069.
- Simulated performance of an indirect methanol fuel cell system. SAE Technical Paper No. 2001-01-0544.
- Application of Modeling Techniques to the Design and Development of Fuel Cell Vehicle Systems. SAE Technical Paper No. 2001-01-0542.
- Comparison of start up and transient response with ASPEN model predictions for a 50kWe autothermal fuel processor. SAE Technical Paper No. 2003-01-0807.

Other Publications

- Correa, J.M.; Farret, F.A.; Gomes, J.R.; Simoes, M.G. Simulation of fuel-cell stacks using a computer-controlled power rectifier with the purposes of actual high-power injection applications. IEEE Transactions on Industry Applications. Volume 39, Issue 4. July-Aug. 2003. Pages 1136-1142.

- Akella, S.; Sivashankar, N.; Gopalswamy, S. Model-based systems analysis of a hybrid fuel cell vehicle configuration. Proceedings of the 2001 American Control Conference. Volume 3. 25-27 June 2001. Pages 1777-1782.

- Jemei, S.; Hissel, D.; Pera, M.C.; Kauffmann, J.M. Black-box modeling of proton exchange membrane fuel cell generators. IEEE 2002 28th Annual Conference of the IECON (Industrial Electronics Society). Volume 2. Nov 5-8, 2002. Pages 1474-1478.

- M. Roos, E. Batawi, U. Harnisch and Th. Hocker. Efficient simulation of fuel cell stacks with the volume averaging method. Journal of Power Sources. Volume 118, Issues 1-2. 25 May 2003. Pages 86-95. http://www.sciencedirect.com/science/article/B6TH1-4808MSS-9/2/b3838feadfd94c7e18653b04cd405f5d.

- Orehek, M.; Robl, C. Model based design in the development of thermodynamic systems and their electronic control units. Proceedings of the 2002 International Conference on Control Applications. Volume 2. 18-20 Sept. 2002. Pages 707-712.

- Correa, J.M.; Farret, F.A.; Canha, L.N. An analysis of the dynamic performance of proton exchange membrane fuel cells using an electrochemical model. The 27th Annual Conference of the IEEE Industrial Electronics Society. Volume 1. 29 Nov.-2 Dec. 2001. Pages 141-146.

- Lasseter, R. Dynamic models for micro-turbines and fuel cells. IEEE Power Engineering Society Summer Meeting. Volume 2. 15-19 July 2001. Pages 761-766.

- Yerramalla, S.; Davari, A.; Feliachi, A. Dynamic modeling and analysis of polymer electrolyte fuel cell. IEEE Power Engineering Society Summer Meeting. Volume 1. 21-25 July 2002. Pages 82-86.

- Lukas, M.D.; Lee, K.Y.; Ghezel-Ayagh, H. An explicit dynamic model for direct reforming carbonate fuel cell stack. IEEE Transactions on Energy Conversion. Volume 16, Issue 3. Sept. 2001. Pages 289-295.

- Iwan, L.C.; Stengel, R.F. The application of neural networks to fuel processors for fuel-cell vehicles. IEEE Transactions on Vehicular Technology. Volume 50, Issue 1. Jan. 2001. Pages 125-143.

- L. Petruzzi, S. Cocchi and F. Fineschi. A global thermo-electrochemical model for SOFC systems design and engineering. Journal of Power Sources. Volume 118, Issues 1-2. 25 May 2003. Pages 96-107. http://www.sciencedirect.com/science/article/B6TH1-488W029-1/2/51f3ff7cb834b4cb5ff2923f1bbc560c.

- S. B. Beale, Y. Lin, S. V. Zhubrin and W. Dong. Computer methods for performance prediction in fuel cells. Journal of Power Sources. Volume 118, Issues 1-2. 25 May 2003. Pages 79-85. http://www.sciencedirect.com/science/article/B6TH1-4834JRR-1/2/26142f8bb4e7f407e4a898c8bef6d31d.

- Tomoyuki Ota, Michihisa Koyama, Ching-ju Wen, Koichi Yamada and Hiroshi Takahashi. Object-based modeling of SOFC system: dynamic behavior of micro-tube SOFC. Journal of Power Sources. Volume 118, Issues 1-2. 25 May 2003. Pages 430-439. http://www.sciencedirect.com/science/article/B6TH1-485PCC2-1/2/1fe9f0392602cebc2595c58020b41a7d.

- Onoda, S.; Lukic, S.M.; NAsiri, A.; Emadi, A. A PSIM-based modeling tool for conventional, electric, and hybrid electric vehicles studies. Proceedings of the 2002 IEEE 56th Vehicular Technology Conference. Volume 3. 24-28 Sept. 2002. Pages 1676-1680.

Control and Diagnosis

SAE Papers
- Power control strategy for fuel cell hybrid electric vehicles. SAE Technical Paper No. 2003-01-1136.
- An analysis of shutdown for an operational fuel cell transit bus. SAE Technical Paper No. 2001-01-2778.
- The Effects of Start-up and Shutdown of a Fuel Cell Transit Bus on the Drive Cycle. SAE Technical Paper No. 2002-01-0101.

Other Papers
- Pukrushpan, J.T.; Stefanopoulou, A.G.; Huei Peng. Modeling and control for PEM fuel cell stack system. Proceedings of the 2002 American Control Conference. Volume 4. 8-10 May 2002. Pages 3117-3122.
- Jurado, F.; Jose Ramon Saenz. Adaptive control of a fuel cell-microturbine hybrid power plant. IEEE Transactions on Energy Conversion. Volume 18, Issue 2. June 2003. Pages 342-347.
- Sakhare, A.R.; Davari, A.; Feliachi, A. Control of stand alone solid oxide fuel cell using fuzzy logic. Proceedings of the 35th Southeastern Symposium on System Theory. 16-18 March 2003. Pages 473-476.
- Sedghisigarchi, K.; Feliachi, A. H-infinity controller for solid oxide fuel cells. Proceedings of the 35th Southeastern Symposium on System Theory. 16-18 March 2003. Pages 464-467.
- Kozlowski, J.D.; Byington, C.S.; Garga, A.K.; Watson, M.J.; Hay, T.A. Model-based predictive diagnostics for electrochemical energy sources. IEEE Proceedings from the Aerospace Conference. Volume 6. 10-17 March 2001. Pages 3149-3164.
- Candusso, D.; Valero, I.; Walter, A.; Bacha, S.; Rulliere, E.; Raison, B. Modelling, control and simulation of a fuel cell based power supply system with energy management. IEEE 2002 28th Annual Conference of the IECON (Industrial Electronics Society). Volume 2. Nov 5-8, 2002. Pages 1294–1299.
- Tolbert, L.A.; Fang Zheng Peng; Cunnyngham, T.; Chiasson, J.N. Charge balance control schemes for cascade multilevel converter in hybrid electric vehicles. IEEE Transactions on Industrial Electronics. Volume 49, Issue 5. Oct. 2002. Pages 1058-1064.

Other Applications of Fuel Cell Systems

Telecoms

Other Publications
- Lei Xia; Bentley, J. Fuel cell power systems for telecommunications. Twenty-Third International Telecommunications Energy Conference. 14-18 Oct. 2001. Pages 677–682.
- LeSage, B.C. Solid oxide fuel cell products for telecom applications. Twenty-Third International Telecommunications Energy Conference. 14-18 Oct. 2001. Pages 667-670.
- Smith, W.F.; Giancaterino, J. Telecom back-up power systems based upon PEM regenerative fuel cell technology. Twenty-Third International Telecommunications Energy Conference. 14-18 Oct. 2001. Pages 657–661.
- Green, K.; Wilson, J.C. Future power sources for mobile communications. Electronics & Communication Engineering Journal. Volume 13, Issue 1. Feb 2001. Pages 43-47.
- E. Varkaraki, N. Lymberopoulos and A. Zachariou. Hydrogen based emergency back-up system for telecommunication applications. Journal of Power Sources. Volume 118, Issues 1-2. 25 May 2003. Pages 14-22. http://www.sciencedirect.com/science/article/B6TH1-487DN0G-1/2/6c68b20f175a62739df0f1bec6d20851.

Distributed Generation

Other Publications

- Kato, N.; Kurozumi, K.; Susuki, N.; Muroyama, S. Hybrid power-supply system composed of photovoltaic and fuel-cell systems. Twenty-Third International Telecommunications Energy Conference (INTELEC 2001). 14-18 Oct. 2001. Pages 631–635.
- Choi, W.; Enjeti, P.; Howze, J.W. Fuel cell powered UPS systems: Design considerations. IEEE 34th Annual Conference on Power Electronics Specialist. Volume 1. June 15-19, 2003. Pages 385-390.
- McDowall, J.A. Opportunities for electricity storage in distributed generation and renewables. IEEE/PES Transmission and Distribution Conference and Exposition. Volume 2. 28 Oct.-2 Nov. 2001. Pages 1165-1168.
- Rahman, S. Fuel cell as a distributed generation technology. IEEE Power Engineering Society Summer Meeting. Volume 1. 15-19 July 2001. Pages 551-552.
- Swift, W.; Fiskum, R. Fuel cells for buildings program. IEEE Power Engineering Society Summer Meeting. Volume 1. 15-19 July 2001. Pages 717-718.
- Eskander, M.N.; El-Shatter, T.F.; El-Hagry, M.T. Energy flow and management of a hybrid wind/PV/fuel cell generation system. IEEE 33rd Annual Power Electronics Specialists Conference. Volume 1. 23-27 June 2002. Pages 347-353.
- Tuckey, A.M.; Krase, J.N. A low-cost inverter for domestic fuel cell applications. IEEE 33rd Annual Power Electronics Specialists Conference. Volume 1. 23-27 June 2002. Pages 339-346.
- Santi, E.; Franzoni, D.; Monti, A.; Patterson, D.; Ponci, F.; Barry, N. A fuel cell based domestic uninterruptible power supply. Seventeenth Annual IEEE Applied Power Electronics Conference and Exposition. Volume 1. 10-14 March 2002. Pages 605-613.
- Zhenhua Jiang; Dougal, R.A. Control design and testing of a novel fuel-cell-powered battery-charging station. Eighteenth Annual IEEE Applied Power Electronics Conference and Exposition. Volume 2. 9-13 Feb. 2003. Pages 1127-1133.
- Whitney Colella. Design options for achieving a rapidly variable heat-to-power ratio in a combined heat and power (CHP) fuel cell system (FCS). Journal of Power Sources. Volume 106, Issues 1-2. 1 April 2002. Pages 388-396. http://www.sciencedirect.com/science/article/B6TH1-44VX1D8-C/2/69ff4284f7f252726c29068b8c80fd44.
- Evgueniy Entchev. Residential fuel cell energy systems performance optimization using "soft computing" techniques. Journal of Power Sources. Volume 118, Issues 1-2. 25 May 2003. Pages 212-217. http://www.sciencedirect.com/science/article/B6TH1-482GBMY-8/2/0e8501128dbe839b1259bd6b6126ede1.
- Jan Van herle, F. Maréchal, S. Leuenberger and D. Favrat. Energy balance model of a SOFC cogenerator operated with biogas. Journal of Power Sources. Volume 118, Issues 1-2. 25 May 2003. Pages 375-383. http://www.sciencedirect.com/science/article/B6TH1-48717DJ-5/2/4ea7a8b05646d6fd720e7cbfc43a4e83.
- Azra Selimovic and Jens Palsson. Networked solid oxide fuel cell stacks combined with a gas turbine cycle. Journal of Power Sources. Volume 106, Issues 1-2. 1 April 2002. Pages 76-82. http://www.sciencedirect.com/science/article/B6TH1-44TV9WR-7/2/ab7a3177e187746ec861c050a79eb5db.

Aerospace Applications

Other Publications
- Oman, H. Fuel cells power aerospace vehicles. Aerospace and Electronic Systems Magazine. IEEE. Volume 17, Issue 2. Feb. 2002. Pages 35-41.
- Oman, H. Hydrogen-fueled pollution-free transportation. Aerospace and Electronic Systems Magazine. IEEE. Volume 17, Issue 1. Jan. 2002. Pages 34-40.

Small Scale and Premium Power

Other Publications
- van der Merwe, J.B.; Turpin, C.; Meynard, T.; Lafage, B. The installation, modelling and utilisation of a 200 W PEM fuel cell source for converter based applications. IEEE 33rd Annual Power Electronics Specialists Conference. Volume 1. 23-27 June 2002. Pages 333-338.
- Jarvis, L.P.; Cygan, P.J.; Roberts, M.P. Fuel cell/lithium-ion battery hybrid for manportable applications. The Seventeenth Annual Battery Conference on Applications and Advances. 15-18 Jan. 2002. Pages 69–72.
- Hahn, R.; Krumm, M.; Reichl, H. Thermal management of portable micro fuel cell stacks. Ninteenth Annual IEEE Semiconductor Thermal Measurement and Management Symposium. March 11-13, 2003. Pages 202-209.
- Mu Chiao; Lam, K.B.; Liwei Lin. Micromachined microbial fuel cells. The IEEE Sixteenth Annual International Conference on Micro Electro Mechanical Systems. Kyoto. Jan. 19-23, 2003. Pages 383-386.
- Kyong-Bok Min; Tanaka, S.; Esashi, M. Silicon-based micro-polymer electrolyte fuel cells. The IEEE Sixteenth Annual International Conference on Micro Electro Mechanical Systems. Kyoto. Jan. 19-23, 2003. Pages 379-382.
- Young Ho Seo; Young-Ho Cho. A miniature direct methanol fuel cell using platinum sputtered microcolumn electrodes with limited amount of fuel. The IEEE Sixteenth Annual International Conference on Micro Electro Mechanical Systems. Kyoto. Jan. 19-23, 2003. Pages 375-378.
- Jarvis, L.P.; Cygan, P.J.; Roberts, M.P. Hybrid power source for manportable applications. Aerospace and Electronic Systems Magazine. IEEE. Volume 18 Issue 1. Jan. 2003. Pages 13–16.
- Luciano Cardinali, Saverio Santomassimo and Marco Stefanoni. Design and realization of a 300 W fuel cell generator on an electric bicycle. Journal of Power Sources. Volume 106, Issues 1-2. 1 April 2002. Pages 384-387.
- http://www.sciencedirect.com/science/article/B6TH1-44TV9WR-D/2/3d86d0d30194edb8a06339f81a13fc78.
- C. K. Dyer. Fuel cells for portable applications. Journal of Power Sources. Volume 106, Issues 1-2. 1 April 2002. Pages 31-34. http://www.sciencedirect.com/science/article/B6TH1-44V1X6D-C/2/f35a0268cb8f61ae93f9126f65dff04c.

Marine Applications

SAE Papers
- Primary and regenerative fuel cells for UAV applications. SAE Technical Paper No. 2000-01-3660.
- Analysis and simulation of a UAV power system. SAE Technical Paper No. 2002-01-3175.

Other Publications

- Lawton, R.; Bash, J.F.; Barnett, S.M. Marine applications of fuel cells. Oceans '02 MTS/IEEE. Volume 3. Oct. 29-31, 2002. Pages 1785-1791.
- Adams, M.; Halliop, W. Aluminum energy semi-fuel cell systems for underwater applications: the state of the art and the way ahead. Oceans '02 MTS/IEEE. Volume 1. Oct. 29-31, 2002. Pages 199–202.
- Dow, E.G. Future (past?) of metal/aqueous semi fuel cell hybrid energy source. Proceedings of the 2002 Workshop on Autonomous Underwater Vehicles. 20-21 June 2002. Page 177.
- Hasvold, O. Alkaline Al/hydrogen peroxide semi-fuel cell. Proceedings of the 2002 Workshop on Autonomous Underwater Vehicles. 20-21 June 2002. Page 167.
- Barbir, F. Proton exchange membrane fuel cell technology status and applicability for propulsion of autonomous underwater vehicles. Proceedings of the 2002 Workshop on Autonomous Underwater Vehicles. 20-21 June 2002. Pages 65-69.
- Winchester, C.; Govar, J.; Banner, J.; Squires, T.; Smith, P. A survey of available underwater electric propulsion technologies and implications for platform system safety . Proceedings of the 2002 Workshop on Autonomous Underwater Vehicles. 20-21 June 2002. Pages 129-135.
- Haberbusch, M.S.; Stochl, R.J.; Nguyen, C.T.; Culler, A.J.; Wainright, J.S.; Moran, M.E. Rechargeable cryogenic reactant storage and delivery system for fuel cell powered underwater vehicles. Proceedings of the 2002 Workshop on Autonomous Underwater Vehicles. 20-21 June 2002. Pages 103-109.
- Hasvold, O.; Johansen, K.H. The alkaline aluminium hydrogen peroxide semi-fuel cell for the HUGIN 3000 autonomous underwater vehicle. Proceedings of the 2002 Workshop on Autonomous Underwater Vehicles. 20-21 June 2002. Pages 89-94.
- Adams, M.; Halliop, W.Aluminum energy semi-fuel cell systems for underwater applications: the state of the art and the way ahead. Proceedings of the 2002 Workshop on Autonomous Underwater Vehicles. 20-21 June 2002. Pages 85-88.
- Swider-Lyons, K.E.; Carlin, R.T.; Rosenfeld, R.L.; Nowak, R.J. Technical issues and opportunities for fuel cell development for autonomous underwater vehicles. Proceedings of the 2002 Workshop on Autonomous Underwater Vehicles. 20-21 June 2002. Pages 61–64.
- Medeiros, M.G.; Patrissi, C.J.; Tucker, S.P.; Carreiro, L.G.; Dow, E.G.; Bessette, R.R. The development of a magnesium-hydrogen peroxide semi-fuel cell. Proceedings of the 2002 Workshop on Autonomous Underwater Vehicles. 20-21 June 2002. Pages 51-56.
- Hagerman, G. Wave energy systems for recharging AUV energy supplies. Proceedings of the 2002 Workshop on Autonomous Underwater Vehicles. 20-21 June 2002. Pages 75-84.

Rail Transport

Other Publications

- Gay, S.E.; Ehsani, M. Design of a fuel cell hybrid tramway. IEEE 56th Vehicular Technology Conference. Volume 2. 24-28 Sept. 2002. Pages 995-997.
- Moghbelli, H.; Gao, Y.; Langari, R.; Ehsani, M. Investigation of hybrid fuel cell (HFC) technology applications on the future passenger railroad transportation. Proceedings of the 2003 IEEE/ASME Joint Railroad Conference. April 2003. Pages 39-53.
- Swanson, J.D. Light rail systems without wires? Proceedings of the 2003 IEEE/ASME Joint Railroad Conference. 22-24 April 2003. Pages 11-22.

Commercialization of Fuel Cell Technology

- Pilkington, A. Technology commercialisation: patent portfolio alignment and the fuel cell. Technology Management for Reshaping the World. PICMET '03: Portland International Conference on Management of Engineering and Technology. 20-24 July 2003. Pages 400-407.
- Dyerson, R.; Pilkington, A. All steering in the same direction? Patterns of patent activity and the development of fuel cell technology. 2002 IEEE International Engineering Management Conference. Volume 2. 18-20 Aug. 2002. Pages 713-718.

About the Editor

Richard Stobart received his first degree in Mechanical Engineering from the University of Cambridge. He worked for Ricardo Consulting Engineers in test bed automation and gasoline engine development before moving to the consulting firm Arthur D Little as an automotive technology specialist. In 2001 he was offered the Chair in Automotive Engineering at the University of Sussex in the UK, and now heads up Automotive Systems research, an area of interest that includes both emerging powertrain systems and improved control methods for conventional engines. He also serves as a co-organizer of the Fuel Cell Technology Session at the SAE World Congress.